Lecture Notes in Network

Volume 121

The series "Lecture Notes in Networks and Systems" publishes the latest developments in Networks and Systems—quickly, informally and with high quality. Original research reported in proceedings and post-proceedings represents the core of LNNS.

Volumes published in LNNS embrace all aspects and subfields of, as well as new challenges in, Networks and Systems.

The series contains proceedings and edited volumes in systems and networks, spanning the areas of Cyber-Physical Systems, Autonomous Systems, Sensor Networks, Control Systems, Energy Systems, Automotive Systems, Biological Systems, Vehicular Networking and Connected Vehicles, Aerospace Systems, Automation, Manufacturing, Smart Grids, Nonlinear Systems, Power Systems, Robotics, Social Systems, Economic Systems and other. Of particular value to both the contributors and the readership are the short publication timeframe and the world-wide distribution and exposure which enable both a wide and rapid dissemination of research output.

The series covers the theory, applications, and perspectives on the state of the art and future developments relevant to systems and networks, decision making, control, complex processes and related areas, as embedded in the fields of interdisciplinary and applied sciences, engineering, computer science, physics, economics, social, and life sciences, as well as the paradigms and methodologies behind them.

** Indexing: The books of this series are submitted to ISI Proceedings, SCOPUS, Google Scholar and Springerlink **

More information about this series at http://www.springer.com/series/15179

Pradeep Kumar Singh · Wiesław Pawłowski ·
Sudeep Tanwar · Neeraj Kumar ·
Joel J. P. C. Rodrigues ·
Mohammad Salameh Obaidat

Editors

Proceedings of First International Conference on Computing, Communications, and Cyber-Security (IC4S 2019)

 Springer

Editors
Pradeep Kumar Singh🆔
Department of Computer Science
and Engineering
Jaypee University of Information
Technology
Waknaghat, Solan, Himachal Pradesh, India

Sudeep Tanwar
Department of Computer Engineering
Institute of Technology
Nirma University
Ahmedabad, Gujarat, India

Joel J. P. C. Rodrigues
The National Institute of
Telecommunications (Inatel)
Santa Rita do Sapucaí, Brazil

Wiesław Pawłowski
Faculty of Mathematics, Physics,
and Informatics
University of Gdańsk
Gdańsk, Poland

Neeraj Kumar
Department of Computer Science
and Engineering
Thapar Institute of Engineering
and Technology
Patiala, Punjab, India

Mohammad Salameh Obaidat
King Abdullah II School of Information
Technology
University of Jordan
Amman, Jordan

ISSN 2367-3370　　　　　　ISSN 2367-3389　(electronic)
Lecture Notes in Networks and Systems
ISBN 978-981-15-3368-6　　　ISBN 978-981-15-3369-3　(eBook)
https://doi.org/10.1007/978-981-15-3369-3

This Springer imprint is published by the registered company Springer Nature Singapore Pte Ltd.
The registered company address is: 152 Beach Road, #21-01/04 Gateway East, Singapore 189721,
Singapore

Preface

The International Conference on Computing, Communications, and Cyber-Security (IC4S-2019) targeted the researchers from different domains of communication and network technologies at a single platform to show their research ideas. There are mainly four technical tracks of the conference: communication and network technologies, advanced computing technologies, data analytics and intelligent learning, and latest electrical and electronics trends. The conference is proposed with the idea of an annual ongoing event to invite researchers on a single platform to exchange their ideas and thoughts. It will definitely evolve as an intellectual asset in the long term year after year. The International Conference on Computing, Communications, and Cyber-Security (IC4S-2019) was held at Chandigarh on 12–13 October 2019 in association with Southern Federal University, Russia, Institute of Technology and Sciences, Mohan Nagar, Krishna Engineering College, Ghaziabad, as academic partners and technically sponsored by the CSI Chandigarh chapter, India. We are highly thankful to our valuable authors for their contribution and our Technical Program Committee for their immense support and motivation towards making the first IC4S-2019 a grand success. We are also grateful to our keynote speakers— Dr. Zdzislaw Polkowski, Jan Wyzykowski University, Polkowice, Poland, Dr. Sanjay Sood, Associate Director, CDAC Mohali, India, Dr. Pljonkin Anton, Southern Federal University, Russia, and Prof. (Dr.) Madhuri Bhavsar, Nirma University, Ahmedabad—for sharing their technical talks and enlightening the delegates of the conference. We express our sincere gratitude to our publication partner, Springer, LNNS Series, for believing in us.

Waknaghat, Solan, India	Dr. Pradeep Kumar Singh
Gdańsk, Poland	Dr. Wiesław Pawłowski
Ahmedabad, India	Dr. Sudeep Tanwar
Patiala, India	Dr. Neeraj Kumar
Santa Rita do Sapucaí, Brazil	Dr. Joel J. P. C. Rodrigues
Amman, Jordan	Dr. Mohammad Salameh Obaidat
October 2019	

Contents

Data Analytics and Intelligent Learning

Security and Privacy Issues

About the Editors

Dr. Pradeep Kumar Singh is an Associate Professor at the Department of CSE at Jaypee University of Information Technology (JUIT), Waknaghat, Himachal Pradesh. He is a life member of CSI, IEI and was promoted to senior member grade of the CSI and ACM. He is an Associate Editor of IJISC Romania and IJAEC, IGI Global, USA. He has published 75 research papers and received three sponsored research project grants worth Rs 25 Lakhs. He has a total of 412 Google Scholar citations, an H-index of 12 and an i-10 Index of 15.

Dr. Wiesław Pawłowski is an Assistant Professor at the University of Gdańsk, Poland. His scientific interests include applications of logic in computer science, formal modeling and specification of software systems, and more recently, semantics-based approaches to interoperability, especially in the area of IoT. He has a total of 485 Google Scholar citations, an H-index of 11 and an i-10 Index of 12.

Dr. Sudeep Tanwar is an Associate Professor at the Nirma University, Ahmedabad, India. His current interests include routing issues in WSN, integration of sensors in the cloud, computational aspects of smart grids and blockchain technology. He has authored or co-authored more than 70 technical research papers and has authored three books. He is an associate editor of the Security and Privacy Journal and is a member of the IAENG, ISTE and CSTA.

Dr. Neeraj Kumar is an Associate Professor at Thapar Institute of Engineering and Technology, Deemed University, India. He was a Postdoctoral Research Fellow at Coventry University, UK. His research focuses on distributed systems, security and cryptography, and body area networks. He has published more than 150 research papers in leading journals and conferences in the areas of communications, security and cryptography.

Dr. Joel J. P. C. Rodrigues is a Professor at the National Institute of Telecommunications (Inatel), Brazil; senior researcher at the Instituto de Telecomunicações, Portugal. He is the editor-in-chief of the International Journal on E-Health and Medical Communications and editorial board member of several respected journals. He has authored or co-authored over 700 papers in refereed international journals and conferences, 3 books, and 2 patents.

Prof. (Dr.) Mohammad S. Obaidat is an internationally known academic/ researcher/scientist/ scholar. He has published over 850 refereed technical articles—including over 400 journal articles, more than 65 books, and numerous book chapters. He is editor-in-chief of 3 scholarly journals and editor of several international journals. He is a Life Fellow of IEEE and a Fellow of SCS. He has a total of 8787 Google Scholar citations, an H-index of 44 and an i-10 Index 222.

Communication and Network Technologies

State of the Art: A Review on Vehicular Communications, Impact of 5G, Fractal Antennas for Future Communication

Abdul Rahim, Praveen Kumar Malik and V. A. Sankar Ponnapalli

Abstract In recent years, there is an advancement towards the applications of vehicular communications and leading research area as there is a lot of scope regarding enhancing safety measures, mobility, security and comfort. As the technology is moving ahead towards the 5G, there is a direct impact on the future of vehicular communications. In this paper, we will discuss the various applications of vehicular communication, the antennas used for the communication using 5G technology and the impact of Internet of vehicles. We will also present the impact of fractal antennas towards 5G and vehicular communication applications and various emerging areas in the domain of autonomous vehicles.

Keywords 5G · Vehicular communication · Mobility · Autonomous vehicles

1 Introduction

History goes backwards to the first prototype of the automated highway system which was introduced for the improvement in safety, comfort, speed and efficiency [1]. Back days these were the four basic objectives on which the entire automation system was developed. As the era changes, many more objectives were introduced such as road safety, time management, accident avoiding systems and smart traffic lights with ambulance priority [2].

To accommodate these many smart features for the vehicles, there is an immense need of the bandwidth and the latest technology for multi-band features. The vehicular system should accommodate in GSM-1800/1900, along with DCS-1800 (digital communication system), PCS-1900 (personal communication service), UMTS

A. Rahim (✉) · P. K. Malik
Lovely Professional University, Phagwara, Punjab, India
e-mail: rahim.mrecw@gmail.com

P. K. Malik
e-mail: pkmalikmeerut@gmail.com

V. A. Sankar Ponnapalli
Sreyas Institute of Engineering & Technology, Hyderabad, India
e-mail: vadityasankar3@gmail.com

© Springer Nature Singapore Pte Ltd. 2020
P. K. Singh et al. (eds.), *Proceedings of First International Conference on Computing, Communications, and Cyber-Security (IC4S 2019)*, Lecture Notes in Networks and Systems 121, https://doi.org/10.1007/978-981-15-3369-3_1

(Universal Mobile Telecommunications System), LTE2600 (Long-Term Evolution), radio band ISM 2.4G (Industrial, Scientific, and Medical), WLAN (wireless local area network), Bluetooth, WiMAX (Worldwide Interoperability for Microwave Access), IEEE802.11p protocol based Vehicle-to-everything, DSRC (dedicated short-range communications) and WAVE (wireless access in vehicular environments) communication bands [3].

As the communication can be handled between vehicles, vehicle to network/infrastructure, vehicle to devices and vehicle to everything, there is a need of direct communication as well as multi-hop communication. With the emerging concepts of millimetre wave, where the radio frequency spectrum ranges from 30 to 300 GHz, which is much higher than the present spectrum range of below 6 GHz [4]. We will discuss more about the impact of millimetre spectrum over vehicular communication and the benefits of wider spectrum in this paper.

As more number of features are to be accessed using the vehicular communication, a variety of compact and powerful antennas are required. This article provides a complete overview of the various antennas that can be used to accommodate all the needs and features of vehicular communication. The rest of the paper is organized as follows. Vehicular network architecture (VNA) is illustrated in 1.1 and a brief description of vehicular communication applications (VCA) in 1.2. Importance of communication and how it is evolved are described in Chap. 2. Existing proposals are illustrated in Chap. 3. Fractal geometry and types of fractal antennas with various designs are discussed in Chap. 4, and finally concluding remarks are provided in Chap. 5.

1.1   Vehicular Network Architecture (VNA)

The type of communication referred to vehicular networks is entitled as vehicle-to-everything communication which is sub-categorized as (i) vehicle-to-vehicle (V2V) communication which is ad hoc and the communication can be established in two ways depending on the range of communication between the two vehicles as shown in Fig. 1. Multi-hop communication comes to existence when the vehicles are not in the range [5]. Most of the time this is useful at parking lots where the empty space can be detected way before the vehicle enter the parking area. (ii) Vehicle-to-network (V2N) communication is used for transferring the data towards the cloud, fog and grid networks as shown in Fig. 2. This type of communication basically needs infrastructure where the nearest roadside units can be connected using hotspots and the data can be transferred to a cloud [6]. Few more communications are possible such as vehicle-to-pedestrian (V2P), vehicle-to-infrastructure (V2I) and vehicle-to-device (V2D) in vehicle-to-everything communication.

Establishing these many varieties of communication, a powerful system architecture of the vehicular network is required which must provide in-vehicle domain, ad hoc domain and infrastructure domain components [7].

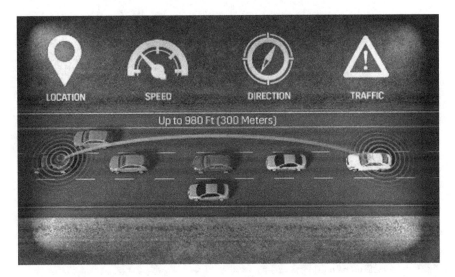

Fig. 1 Vehicle-to-vehicle communication [3]

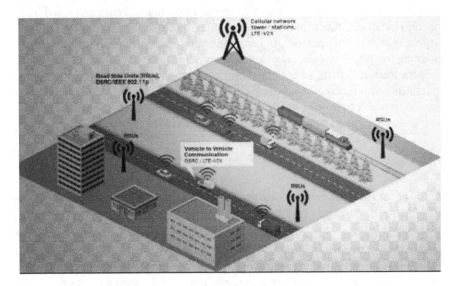

Fig. 2 Vehicle-to-network communication [5]

In-vehicle domain [8] is used for information of the vehicle like the fuel consumption, temperature in the vehicle, opening and closing of the sunroof using voice control, etc., and these can be achieved using human–machine interface along with microcontrollers in a controller area network (CAN) [9]. Different wireless technologies such as Bluetooth, Wi-Fi and GPS [10] are used to achieve the in-vehicle domain as shown in Fig. 3.

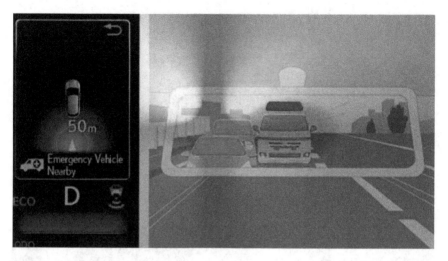

Fig. 3 In-vehicle domain [3]

Ad hoc domain is used to connect V2V as the connection must be established spontaneously and sometimes multi-hop is needed as discussed earlier. A controller can be helpful to extend the range of the communication usually called as communication controller unit (CCU) [11].

Infrastructure domain is a combination of wireless and wire communication where the roadside wireless devices can connect with the infrastructure which is a wired connection basically to provide a wireless hotspot using switches and routers as shown in Fig. 4. The wired connection is helpful to connect the authorities for collecting the information and controlling the access.

1.2 Vehicular Communication Applications (VCA)

There are numerous applications using the vehicular technology in order to provide safety, comfort, security and free flow moment. These applications can be accessed using either a sequential programming or event-driven programming. They can be categorized using the concepts like driver driven, vehicle driven, passenger driven and infrastructure driven [7]. According to the above features, these applications can be broadly classified into safety and non-safety applications. Safety applications are those applications which are directly related to the vehicles on the road [1]. It can be an information provided to the driver about the road surface ahead, or may be weather status, or may be the traffic ahead or may be road maintenance ahead or accident prone areas, pedestrian crossing ahead or any mishap occurs ahead, etc. As an example in the month of December, in Delhi a thick fog blocks the view of the driver due to which every year many vehicles collide and accidents

Fig. 4 Vehicle-to-infrastructure communication [3]

occur too frequently as shown in Fig. 5. Using the vehicular application, a warning message can be broadcasted which can be received by all the vehicles which are in that area such that the drivers can be more cautious and can avoid the collision. Non-safety applications are those applications which are used to provide services

Fig. 5 Accidents due to thick fog [6]

related to repairs, maintenance information [1], fuel refill stations, motels nearby [12], route guidance, navigation, etc. Now we will quickly access the history of communication and how 5G can make an impact towards the VCAs. Communication between vehicles can be ad hoc as specified by the author [11].

2 Communications

From 1980s, a rapid growth is observed in radio technologies because of multi-directional evolution with the launch of analogue cellular systems, which is considered as the first-generation cellular systems. As the second generation evolved, wireless technology came to existence. The wireless technology advanced to digital from the analogue world. This era of digital wireless communication systems consistently fulfills the growing needs of human beings for 2G to now 5G, where each generation is defined as a set of telephone network standards, which detailed the technological implementation of particular mobile system [13].

The major aim of wireless communication is to provide high quality and reliable communication [14], and each generation represented a big step in that direction. Each generation has requirements that specify things like throughput, delay, etc. that need to be considered [15].

Up to 4G, the improvement observed in the means of data capacity standard, multiplexing techniques, switching types, services, handoff types and frequency ranges.

As 5G technology evolved, which is driven by OFDM [16] as shown in Fig. 6, whose frequency band is in between 3 and 300 GHz with a mobility of 500 kbps and a latency of less than 1 ms is used. Many 5G networks are operated on high frequencies

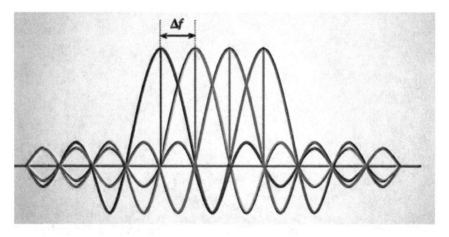

Fig. 6 OFDM [17]

called as mm waves which have the benefit to carry lots of data but are limited in the range [17]. The main drawback is that they are easily blocked by common objects.

Millimetre-wave (mm-wave) frequency ranges from 30 to 300 GHz which is much higher than the spectrum used presently in vehicular communication. If this spectrum is used wisely with better modulation and using various coding schemes with the help of beam forming, multiple-input–multiple-output techniques and using fractal geometry lead to high data rate, reduce interference and spectral reusability. IEEE 802.11ad [8] is used for the mm-wave architecture at 60 GHz frequency band over short distances ranging of 10 m. Millimetre-wave technology is also used for the automotive radar systems which are used at a high frequency of around 70 GHz and above [4]. Radar systems are used for increasing the range of communication with vehicle-to-everything communication. In order to accommodate these many various communication systems, the various authors proposed a dedicated spectrum for millimetre wave [15], which can include roadside communication, infrastructure communication, signal processing and trans-receivers in vehicles [18].

International Telecommunications Union provides the radio spectrum usage [19] for the worldwide use in which some portion of spectrum is provided for the mobile communication. At present, 3G and 4G communications are used around the world wide which is concentrating now on the next generation which is expected to standardize by the year 2020. According to the international treaty, every four years the spectrum is revised and resolutions will be carried out for the frequency band for the future use [7]. 5G network technically must have capabilities of supporting multitasking, multi-application and service provider for different networks. The bit rate must be around 20 Gbps with the mobility support up to 500 km/h with a frequency band of 700 MHz to 100 GHz [4]. The major challenges according to the author, frequent beam formation needed to track vehicles and must maintain beam alignment with low latency.

As Internet of things has an impact towards the latest technology, even Internet of vehicles as shown in Fig. 7 will establish an impact with the mm-wave communication and the frequency standards used. The communication data and the vehicles information will be sent to the cloud in order to share and access the information, which as vehicles are increasing the information, data volume will be huge and continuously increasing which must be preserved in a separate cloud for vehicle communication. The information from vehicles will be transferred to the nearest transceiver which is referred to as EDGE. The data will be transferred to a fog [9] which in turn transfers to the cloud. Using EDGE, the vehicles can offload the data, through which the latency can be minimized, in turn improves faster access and helps to enhance safety and convenience.

Fog can be used as a middle level between the cloud and edge, to facilitate faster data transfer from the end user and to minimize the latency, to improve response time and can be used to route data to the cloud.

Fig. 7 Internet of vehicles
[10]

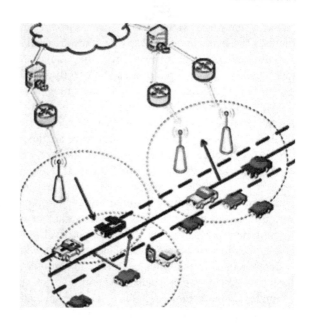

3 Existing Proposals

Many authors proposed various antenna designs for the 5G communications [4, 15, 18], and hence, we can discuss a little about these antennas that can be used for vehicular communications. The first antenna in this category is a conformal antenna which can be designed on the surface of a carrier, which is helpful for saving the space and there is no damage on the mechanical structure [12]. The conformal antenna can be a micro-strip, strip line or crack antenna. The shape of this antenna is cylindrical, which is helpful to maintain a proper angle between the main lobe and the plane of array. This antenna can be used for frequency of 35 GHz, and the results are suited for vehicular communication as shown in Fig. 8. With the addition of conformal antenna, bandwidth achieved was greater than 10%, and the gain was more than 10 dB, and the first side lobe level was reduced to -16 dB. At very high frequencies, the major problem that can be seen is the uncontrolled radiation pattern. For vehicular communication, the pattern must be as directional as possible. In order to achieve as well-controlled radiation patter, the author [2] proposed switching of polarization among linear and orthogonal circularly polarized states using magneto-electric dipole antenna. This method can be used for the future antenna designs for the mm-wave communication. When this type of dipole antenna with a micro-strip line is used, there is a possibility of high mutual coupling. To achieve a low mutual coupling, the author [9] proposed insertion of rectangular slot along with the folded micro-strip line, by this way low mutual coupling can be achieved.

As for the vehicular applications are concerned, ultra-wideband antennas are majorly used which are achieved using a 2×2 MIMO pattern as proposed [12] and

Fig. 8 Conformal antenna [20]

as shown in Fig. 9. As the frequency is increased, the bandwidth must be improved, in order to improve it a recessed ground plane is used [7]. For antenna arrays with different impedance characteristics, the author suggested an epsilon-near-zero impedance matching circuit [4], for impedances of 50, 100 and 150 Ω.

When a number of radiators are increased for arrays, there is a possibility of cross-polarization current on the radiators and low port isolation, which can be eliminated using dual-polarized antenna using vector synthetic mechanism, which has inherent advantages of high port isolation and low cross-polarization. When accessing many papers, we came across an interesting antenna which is operated in 5G bands. The

Fig. 9 MIMO antenna [12]

structure consists of hexagonal split-ring resonator (SRR) and a closed ring resonator (CRR). The proposed antenna [4] was modelled upon 0.254 mm thick Rogers substrate and has a low profile of 6×8 mm^2. It has a peak gain of 4.63 dbi with more than 80% radiation efficiency throughout its operating range. The proposed antenna has 2.83 GHz bandwidth and 7.92 GHz at 38 GHz, respectively.

For digital beam forming, a 64-channel massive multiple-input, multiple-output trans-receiver antenna was proposed [8] which consist of 16 columns and 4 rows with half-wavelength element spacing that is provided.

The next important aspect is the size of the antenna, as a number of radiators are increased the size automatically increased. One of the remedies is to use an antenna with a staircase type to reduce the overall size, and the author named it as zeroth-order resonance antenna [21]. To increase more directivity, an antenna was proposed with eight elements with very less envelope correlation coefficient [15] which is helpful for isolation. Wherever the antenna design is discussed, for better bandwidth and multi-band application, fractal antenna comes to existence. Let us discuss now the fractal antenna and how it can be significant in vehicular communications.

4 Fractal Antennas

Researchers found interesting about the fractal geometry implementing on antenna design which majorly concentrated on two areas in which the first area deals with design and analysis of the fractal antenna elements [22], and the second area is about the application of fractal concepts on designing of the array antennas. The fractal antenna possesses recursive nature which leads to develop rapid beam forming algorithms [23].

One of the major usages of fractal antenna is for ultra-wideband technology applications like imaging system, vehicular radar applications and measurement systems. New designs emerged for fractal geometry for various applications [20] and their designs can have an impact on 5G in near future.

The basic fractal models [24] were analysed for various applications at different frequencies which have no much effect on improving neither the bandwidth nor the gain, so the basic models need to be modified in order to achieve greater results.

For good impedance matching and constant gain, a new design was proposed [25] which is used for the frequency range of 3.1–10.6 GHz and the design was studied for current distribution, radiation pattern and group delay and the results were satisfactory, but still the enhancement can be done.

The enhancement can be carried out in two ways in which one of the idea is to add more antennas as the concept of arrays [18], or use of fractal layers [25] are used. In the second method is to use one fractal antenna with two separate layers [17] such that the size of the antenna need not be compromised. The increase in antenna elements can also provide significant amount of improvement as artistic array antennas. These antennas can be significant while using for wearable on-body applications [22] and vehicular communication applications. In the present scenario, fractal antennas can be most effectively used for vehicular-to-everything communication. Researchers can progress using the fractal antennas for various applications in 5G communication.

5 Conclusion

This paper provides a complete review of vehicular communication systems. We highlighted various types of communications available in the vehicular architecture. We presented the details such as in-vehicle domain, ad hoc domain and infrastructure domain. We discussed the safety and non-safety applications and the impact on the environment. We also highlighted the antennas used for 5G and the remedies provided by various authors on the problems faced in 5G communications. Finally, we discussed the novelty of fractal antennas and the impact on the future applications. We hope that this review article provides meaningful insight and leads towards the research on fractal antennas for the vehicular communication applications.

References

1. Malik P.K., Parthasarthy, H., Tripathi, M.P.: Alternative mathematical design of vector potential and radiated fields for parabolic reflector surface. In: Unnikrishnan, S., Surve, S., Bhoir, D. (eds.) Advances in Computing, Communication, and Control. ICAC3 2013. Communications in Computer and Information Science, vol. 361. Springer, Berlin, Heidelberg (2013)
2. Mondal, T., Maity, S., Ghatak, R., Chaudhuri, S.R.B.: Compact circularly polarized wide-beamwidth fern-fractal-shaped microstrip antenna for vehicular communication. IEEE Trans. Veh. Technol. **67**(6), 5126–5134 (2018)
3. Madhav, B.T.P., Anilkumar, T., Kotamraju, S.K.: Transparent and conformal wheel-shaped fractal antenna for vehicular communication applications. AEU—Int. J. Electron. Commun. **91**, 1–10 (2018)
4. Ullah, H., Tahir, F.A., Ahmad, Z.: A dual-band hexagon monopole antenna for 28 and 38 GHz millimeter-wave communications. In: 2018 IEEE International Symposium on Antennas and Propagation & USNC/URSI National Radio Science Meeting, Boston, MA, pp. 1215–1216 (2018)
5. Jameel, F., Wyne, S., Nawaz, S.J., Chang, Z.: Propagation channels for mmWave vehicular communications: state-of-the-art and future research directions. IEEE Wirel. Commun. **26**(1), 144–150 (2019)
6. Peng, H., Liang, L., Shen, X., Li, G.Y.: Vehicular communications: a network layer perspective. IEEE Trans. Veh. Technol. **68**(2), 1064–1078 (2019)
7. Malik, P., Parthasarthy, H.: Synthesis of randomness in the radiated fields of antenna array. Int. J. Microwave Wirel. Technol. **3**(6), 701–705 (2011)
8. Yang, B., Yu, Z., Lan, J., Zhang, R., Zhou, J., Hong, W.: Digital beamforming-based massive MIMO transceiver for 5G millimeter-wave communications. IEEE Trans. Microw. Theory Tech. **66**(7), 3403–3418 (2018)
9. Trivedi, H., Tanwar, S., Thakkar, P.: Software defined network-based vehicular adhoc networks for intelligent transportation system: recent advances and future challenges. In: Singh, P., Paprzycki, M., Bhargava, B., Chhabra, J., Kaushal, N., Kumar, Y. (eds.) Futuristic Trends in Network and Communication Technologies. FTNCT 2018. Communications in Computer and Information Science, vol. 958. Springer, Singapore (2019)

10. Singh, P.K., Nandi, S.K., Nandi, S.: A tutorial survey on vehicular communication state of the art, and future research directions. Veh. Commun. **18**, 100164 (2019). ISSN 2214-2096
11. Malik, P.K., Singh, M.: Multiple bandwidth design of micro strip antenna for future wireless communication. Int. J. Recent Technol. Eng. **8**(2) (2019). ISSN: 2277-3878
12. AL-Saif, H., Usman, M., Chughtai, M.T., Nasir, J.: Compact ultra-wide band MIMO antenna system for lower 5G bands. Wirel. Commun. Mobile Comput. **2018**, Article ID 2396873, 6p (2018)
13. Seker, C., Güneser, M.T., Ozturk, T.: A review of millimeter wave communication for 5G. In: 2018 2nd International Symposium on Multidisciplinary Studies and Innovative Technologies (ISMSIT), Ankara, pp. 1–5 (2018)
14. Kumar, A., Gupta, M.: A review on activities of fifth generation mobile communication system. Alexandria Eng. J. **57**(2), 1125–1135 (2018)
15. Kumaran, V., Rajkumar, S., Thiruvengadam, S.: Performance analysis of orthogonal frequency division multiplexing based bidirectional relay network in the presence of phase noise. Am. J. Appl. Sci. **10**, 1335–1344 (2013). https://doi.org/10.3844/ajassp.2013.1335.1344
16. Gholibeigi, M., Sarrionandia, N., Karimzadeh Motallebi Azar, M., Baratchi, M., van den Berg, H.L., Heijenk, G.: Reliable vehicular broadcast using 5G device-to-device communication. In: WMNC 2017: 10th IFIP Wireless and Mobile Networking Conference, Sept 2017, pp. 25–27. IEEE, Valencia, Spain (2017)
17. Rajkumar, S., Thiruvengadam, J.S.: Outage analysis of OFDM based cognitive radio network with full duplex relay selection. IET Signal Process. **10**(8), 865–872 (2016)
18. Yan, K., Yang, P., Yang, F., Zeng, L.Y., Huang, S.: Eight-antenna array in the 5G smartphone for the dual-band MIMO system. In: 2018 IEEE International Symposium on Antennas and Propagation & USNC/URSI National Radio Science Meeting, Boston, MA, pp. 41–42 (2018). Design and Implementation of Multi Array Fractal Antenna for 5G Vehicle to Vehicle Communication 12
19. Kubacki, R., Czyżewski, M., Laskowski, D.: Microstrip antennas based on fractal geometries for UWB application. In: 2018 22nd International Microwave and Radar Conference (MIKON), Poznan, pp. 352–356 (2018)
20. Werner, D.H., Ganguly, S.: An overview of fractal antenna engineering research. IEEE Antennas Propag. Mag. **45**(1), 38–57 (2003)
21. Malik, P.K., Tripathi M.P.: OFDM: A Mathematical Review. J. Today's Ideas–Tomorrow's Technol. **5**(2), 97–111 (2017)
22. Arif, A., Zubair, M., Ali, M., Khan, M.U., Mehmood, M.Q.: A compact, low-profile fractal antenna for wearable on-body WBAN applications. IEEE Antennas Wirel. Propag. Lett. **18**(5), 981–985 (2019)
23. Rahim, A., Malik, P.K., Sankar Ponnapalli, V.A.: Fractal antenna design for overtaking on highways in 5g vehicular communication ad-hoc networks environment. Int. J. Eng. Adv. Technol. (IJEAT). ISSN: 2249–8958, **9**(1S6), 157–160 (2019)
24. Mishra, G.P., Maharana, M.S., Modak, S., Mangaraj, B.B.: Study of Sierpinski fractal antenna and its array with different patch geometries for short wave Ka band wireless applications. Procedia Comput. Sci. **115**, 123–134 (2017)
25. Sankar Ponnapalli, V.A., Jayasree, P.V.Y.: Thinning of Sierpinski fractal array antennas using bounded binary fractal-tapering techniques for space and advanced wireless applications. ICT Express **5**(1), 8–11 (2019)

Abdul Rahim is a Ph.D. student in department of Electronics and Communication at Lovely Professional University, Punjab, India. He received his B.E. in Electronics and Communication from Osmania University and M.Tech. in VLSI system design for JNTUH. His current research interest include Vehicular Communication Applications, Fractal Antenna Design at milli meter frequencies.

Dr. Praveen Kumar Malik A professional qualified and experienced person with extensive knowledge and skills in Antenna and Embedded System. He is working as Professor in the department of Electronics and Communication, Lovely Professional University, Punjab, India. He is B.Tech., M.Tech. and Ph.D. from Electronics and Communication system. His major area of interest is signal processing, Antenna, and Embedded systems. He is having more than 10 papers in international refereed journals. Having more than 10 papers in the international and national conference also.

V. A. Sankar Ponnapalli received his B.Tech. in Electronics & Communication Engineering from JNT University Kakinada in 2011, M.Tech. in RF & Microwave Engineering and the Ph.D. degrees from GITAM (Deemed to be University) Visakhapatnam in 2013 and 2018 respectively. He is currently working as an Associate Professor in the Department of Electronics and Communication Engineering, Sreyas Institute of Engineering and Technology Hyderabad. He has authored or co-authored books, book chapters, and more than 25 research articles. He is acting as a reviewer and editorial board member for various international journals and conferences. His research interests include microwaves, antennas, and intelligent transportation systems etc.

Energy Enhancement of TORA and DYMO by Optimization of Hello Messaging Using BFO for MANETs

Navjot Kaur, Deepinder Singh Wadhwa and Praveen Kumar Malik

Abstract In specially appointed systems, the errand of steering is very perplexing and loaded with difficulties because of their dynamic conduct. The absence of foundation in remote correspondence is additionally one of the issues which are related to the current affixed framework. In such a condition of undertaking, generally client gets crack outcomes. The steering conventions as often as possible utilize changes of topology, while simultaneously constraining the effect of interest these progressions on remote assets. Trade of hi message movement is favored for finding the neighborhood connect. Because of neighbor disclosure messages, high overhead is made in the DYMO convention.

Keywords Dynamic MANET on demand · Temporally ordered routing algorithm · Bacterial foraging optimization algorithm

1 Introduction

With the expanding request of information traffic, cell phones become a significant zone of future. The figuring surroundings which contain infra-organized and foundationless versatile systems are then required to be less clogged. Remote local region organizes and bolsters IEEE 802.11 convention as a base. The innovation wherein portable hub constantly speaks with the distinctive base stations inside milliseconds, subsequently a quick remote connection is required for bouncing between the hub and its neighbor base stations. A MANET is that the one which contains a lot of portable hosts which may speak with one another and move around at their own will. MANET is related to unstructured system which can be actualized for various

N. Kaur · D. S. Wadhwa (✉)
BGIET Sangrur Punjab, Sangrur, India
e-mail: wadhwadeepinder@gmail.com

N. Kaur
e-mail: knavjot911@yahoo.com

P. K. Malik
Lovely Professional University, Phagwara, Punjab, India
e-mail: pkmalikmeerut@gmail.com

© Springer Nature Singapore Pte Ltd. 2020
P. K. Singh et al. (eds.), *Proceedings of First International Conference on Computing, Communications, and Cyber-Security (IC4S 2019)*, Lecture Notes in Networks and Systems 121, https://doi.org/10.1007/978-981-15-3369-3_2

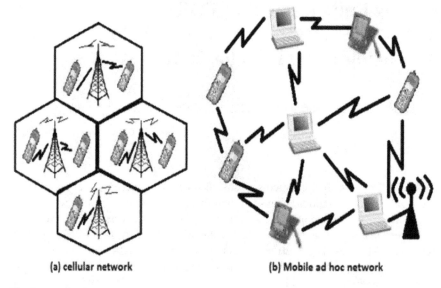

(a) cellular network   (b) Mobile ad hoc network

Fig. 1 Mobile ad hoc network [1]

applications at normal interims. Nodes can go about as hosts further as switches (Fig. 1).

1.1 MANET's Route Selection Process

One of the extraordinary worries that obliterate the exhibition of a specially appointed system is the technique for steering that is executed in a network. Routing happens for the most part in three stages, i.e., finding, picking and safeguarding courses from source to goal hub conveying the information bundles. Along these lines, course choice technique is ought to be more précised with the target to decrease vitality utilization. Hubs move in partner degree and at dynamical speed, generally coming about to connecting issues. The high caliber and moreover each development of hubs in specially appointed system cause individual connections between hosts and goal that intrudes on regularly.

1.2 Routing Protocols

The traffic is routinely circulated with the assistance of directing hubs and is performed by shrewd steering convention. In table-driven/proactive directing conventions, hubs sporadically trade steering information and resolve to stay up with the

Fig. 2 MANET routing
protocols [2]

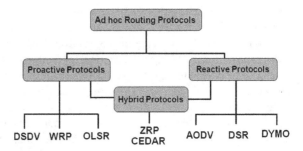

latest directing information. Inside the accompanying segments, we have a tendency to first portray the two classes of steering conventions in further detail. We tend to then run down the difficulties looked by directing conventions (Fig. 2).

2 Literature Review

One of the ad hoc Routing Protocols are used in MANET and VANET are Temporally-ordered routing algorithm (TORA) protocol [3]. These Routing Protocols are separated into two sections, i.e., table-driven and on demand. In on request, directing conventions by producing directing packets and neighbor disclosure message turn out high overhead like TORA Protocol. In order to beat such issues, a totally extraordinary topic in specially appointed systems that upheld Intelligent TORA (I-TORA) is arranged. Simulation results of the anticipated neighbor revelation system upheld quality in fluctuated eventualities and contrasted and the underlying TORA exhibit that I-TORA performs much better. In [4], the authors audited the working of DYMO routing protocol contrasted and the predominant TORA convention. Since its partner degree raised adaptation of TORA convention, it is conjointly called TORAv2 that goes for improving TORA by expelling pointless decisions and receiving succeeding choices from DSR like way aggregation. Besides, its exhibition beats the conventional TORA convention once it includes gigantic systems with sizable measure of nodes and dynamical topology. These units still having a few difficulties confronting remote appointed systems concerning their exhibitions in various outcomes anyway adjacent to this each convention which having their very own merits and demerits. Remote impromptu systems are turning out to be progressively prevalent in view of their influence employments. In [5], the authors discussed passes that MANET is an unorganized arrangement inside which the nodes are commonly versatile and independent. Nodesgo about as hosts further as routers. This quality may prompt weakness in MANET. Bacterial foraging optimization algorithm (BFOA) likely could is a bio-motivated algorithmic program. This algorithmic program mimics the conduct of bacterium which can be successfully applied in fluctuated fields. All through this, BFO is used in MANET for improving security exploitation parameters PDR (Packet Delivery Ratio) overhead and yield. There are three parameters on which the capacity of MANET steering conventions is based that are assessed and contrasted with which BFTORA is found. Here, the topic is anticipated by applying BFO on-request MANET steering convention TORA and it becomes BFTORA. The exhibition of the following conventions, for example TORA, BFTORA and DSDV, is analyzed and bolstered three parameters PDR, overhead and yield. In [6], research shows that most limited way directing is unrealistically prudent because it spares time and monetarily valuable regarding import. Lately, the directing disadvantage has been well self-tended to exploitation smart enhancement procedures. Properties of several protocols were studied, simulated and compared [7–9]. All through this paper, will examine these calculations on fluctuated remote systems. We have ton of calculations to determine directing disadvantage in mounted systems besides dynamic systems like MANETs, and it is appallingly troublesome.

3 Problem Formulation and Major Issues

In mobile ad hoc networks exploitation PDAs, power consumed by no sleep energy bugs may be a difficult issue and for the course foundation and support local connection property data is amazingly vital. It demonstrated that our anticipated theme supress the undesirable messages with no extra delay. For native link property data employed in neighbor disclosure sporadically trading hello messages is utilized, though in such antiquated hello electronic correspondence conspires no beginning/end condition is represented. In view of the no rest vitality bugs, hello messages persistently channel batteries when cell phones do not appear to be being used. A versatile hello electronic correspondence subject for neighbor revelation is anticipated by stifling pointless hello messages.

4 Proposed Protocol Optimization Using BFOA

4.1 Introduction to BFOA

Bacteria search optimization algorithmic program (BFOA), anticipated by Passino [10], might be a substitution comer to the group of nature-roused enhancement calculations. For over the past five decades, improved calculations like Genetic Algorithms [1, 11], Evolutionary Programming (EP) [5], Evolutionary Strategies (ES) [12]. As of late characteristic swarm intrigued calculations like Particle Swarm optimisation (PSO) [10], Ant Colony optimisation (ACO) [13] have discovered their methodology into this area and confirmed their effectiveness. Following indistinguishable pattern of swarm-based calculations, Passino anticipated the BFOA in [10]. Use of group search system of a swarm of E. coli bacterium in multi-ideal work advancement is the key arrangement of the new algorithmic program. Analysts attempt to breed BFOA with appropriate algorithms in order to explore its local and world hunt properties separately. It is as of now been applied to several universe issues and confirmed its adequacy over numerous variations of GA and PSO.

4.2 Bacteria Foraging Optimization Algorithm (BFOA)

During the search activity of the essential bacterium, bundle of tensile flagella accomplishes velocity. It likewise encourages the partner degree *E. coli* smaller-scale living being to tumble or swim that square measure of two fundamental tasks is performed by a miniaturized scale creature at the hour of search. After they turn the flagella inside the dextrorotatory course, every flagellum pulls on the cell. That prompts the moving of flagella severally, and in the long run the small-scale creature tumbles with lesser assortment of tumbling while in an exceptionally unsafe spot it tumbles frequently to look out a supplement inclination. Moving the flagella at interims counterclockwise heading causes the small-scale creature to swim at an extremely snappy rate. Inside the above-named algorithmic program, the bacterium experiences taxis, and any place they want to move toward a supplement slope and stay away from unwholesome environment. Generally, the bacterium moves for an all-inclusive separation in an inviting environment. Figure 3 depicts anyway dextrorotatory, and counter dextrorotatory development of a small-scale living being shows it in supplement goals.

To search the minimum of J (′) where′ (for example could be a *p*-dimensional vector of genuine numbers), and that we don't have estimations or partner degree

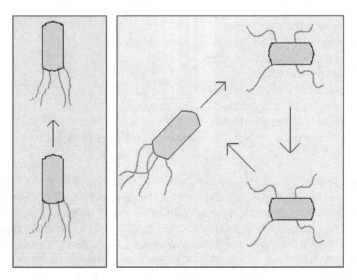

Fig. 3 Bacterium activities [5]

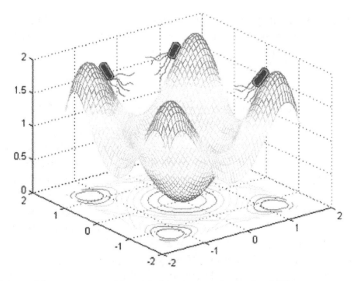

Fig. 4 A bacterial swarm on a multimodal objective function surface [13]

systematic description of the slope J ( ). BFOA genuinely has four instruments found in an undeniable smaller-scale living being framework, for example taxis, swarming, reproduction and disposal dispersal to determine this non-inclination advancement drawback. A virtual micro-organism is extremely one preliminary goal (might be known as an inquiry operator) that proceeds onward the handy surface to locate the overall ideal specialist (Fig. 4).

Let

K: Dimension of the search house,
M: Maximum variety of bacterium within the population,
Kc: The amount of chemotactic steps,
Ks: Length of movement,
Kre: the amount of replica steps,
Ked: The amount of elimination–dispersal events,
Ced: Elimination–dispersal chance,
B (i): The dimensions of the step taken within the random direction fixed by the tumble.

(i) **Chemotaxis**: This technique simulates the development of partner degree *E. coli* cell through swimming and tumbling by means of flagella. Naturally, partner degree *E. coli* micro-organism will move in two different manners. It will swim for a measure of your time inside a similar heading or it will tumble and switch back and forth between these two methods of activity for the entire time frame. Assume (j, k, l) speaks to i-th small-scale creature at jth chemotactic, kth productive and lth disposal dispersal step. $C(i)$ is the size of the progression taken inside the irregular heading fixed by the tumble (run-length unit). At that point in processing chemotaxis, the development of the bacterium might be represented by

$$\theta^i(j+1,k,l) = \theta^i(j,k,l) + C(i)\frac{\Delta(i)}{\sqrt{\Delta^T(i)\Delta(i)}} \qquad (1)$$

(ii) **Swarming**: A surprising bunch of conduct has been found for some motile types of bacterium such as *E. coli* and *S. typhimurium*, any place byzantine and stable spatio-temporal patterns (swarms) fashioned in strong supplement medium. The cell-to-cell correspondence in *E. coli* swarm is additionally depicted by the ensuing work.

$$J_{cc}(\theta, P(j,k,l)) = \sum_{i=1}^{S} J_{cc}(\theta, \theta^i(j,k,l))$$

$$= \sum_{i=1}^{S}\left[-d_{\text{attractant}}\exp\left(-w_{\text{attractant}}\sum_{m=1}^{P}(\theta_m - \theta_m^i)^2\right)\right]$$

$$+ \sum_{i=1}^{S}\left[\text{repellent}\exp\left(-w_{\text{reppellent}}\sum_{m=1}^{P}(\theta_m - \theta_m^i)\frac{1}{2}\right)\right] \qquad (2)$$

S is that the total variety of bacterium, and P is that the variety of variables to be optimized, that square measure gift in every micro-organism purpose within the p-dimensional search domain.

$\theta = [\theta_1, \theta_2, \ldots . \theta_p]\frac{1}{T}$ is a point in the p-dimensional search domain [1, 3–5, 10, 12–14].

(iii) **Reproduction**: The littlest measure of sound bacterium in the end kicks the bucket, though every one of the more advantageous bacteria (those yielding lower estimation of the objective capacity) abiogenetically splits into two bacteria and they are set in the equivalent location. As a consequence of this, the swarm size stays steady.

(iv) **Elimination and Dispersal**: Gradual or sharp changes inside the local surroundings, for example a significant local ascent of temperature, could execute a gaggle of bacterium which are by and by in an area with a high grouping of supplement slopes. Occasions will manifest itself in such a way the vast majority of the bacterium in a field could be killed or a species is appropriated into a fresh out of the plastic new area. To simulate this improvement in BFOA, some bacterium goes to exchanged state in a self-assertive way with a less expectation. Then again, the novel substitutions are haphazardly started at the rummage space.

5   Results and Comparison

5.1   *Implementation Results*

Primary target of our examination is decrease of vitality utilization and system overhead. We can do that just if hello interim is framed relative to the occasion interim
of a node. In the event that the welcome interim neglected to consider the occasion
interim of a node, at that point the peril of causation bundle through partner degree
distant connection will expand (Pfdd). In our investigation, we tend to square quantify considering the situation of partner degree urban town and exponential traffic
conveyance for our analysis. We tend to expect the CDF of occasion interim in Fig. 5
and furthermore the supposition that is considered for the one 00 occasion, and any
place it is examined for one hour and is given by:

$$X(\alpha, \beta) = 1 - e^{-\alpha/\beta} \tag{3}$$

We contemplate $X(\alpha, \beta)$ as chance of the entire no. of outcomes delimited by
exponential distribution.

Figure 6 shows the comparison of energy consumption between DYMO and
DYMO-AH. Each node is assigned with one joule of energy from the start. Over the
long haul, expanding DYMO devours energy rapidly on account of the occasional
hello electronic communication subject and along these lines additional energy utilization and battery channel occur. Once DYMO-AH is applied, undesirable hello

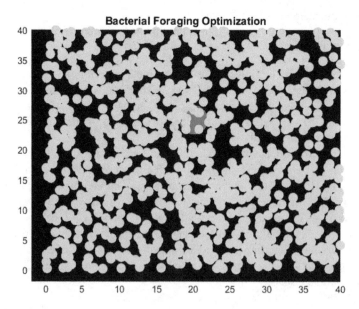

Fig. 5 Bacterial foraging

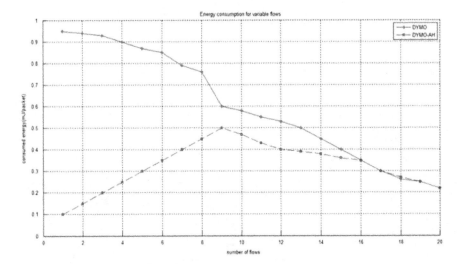

Fig. 6 Energy consumption for variable flows

messages are being stifled and the energy is expended in littler sum due to the less spreading and accepting of hello messages (Fig. 9).

Also, Fig. 7 illustrates the consumed energy per received packet over varying numbers of flows at $t = 60$ min. When the number of flows is less than 5, the more energy is saved. The amount of saved energy is decreased as the number of flows

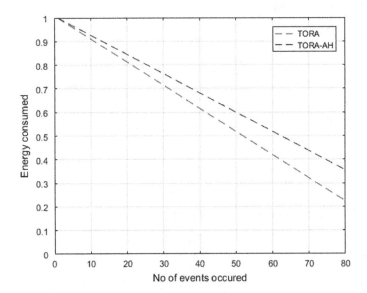

Fig. 7 Energy consumption with the number of events

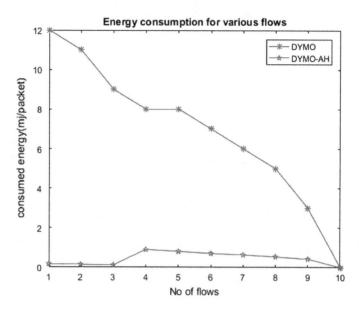

Fig. 8 Energy consumption with the number of events

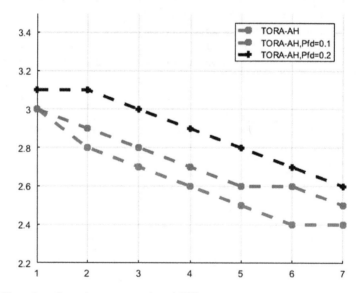

Fig. 9 Throughput for various max speeds and PFD

increased because of the nodes that are participating in forwarding the data packets more.

Figure 8 shows the comparison of the consumed energy DYMO and DYMO-AH with the more number of events, respectively.

Figure 6 shows the comparison of utilization among DYMO and DYMO-AH. Each node is assigned with one joule of energy from the start. Over the long haul, expanding DYMO devours vitality rapidly on account of the occasional hi electronic communication subject and along these lines additional vitality use and battery channel occur. Once DYMO-AH is applied, undesirable hi messages are being stifled and the vitality is expended in littler sum due to the less spreading and accepting of hi messages. Figure 9 shows the impact of PFD on the output and hello proportion as the varying maximum speed. Packets that are successive are going to be forwarded through a sound link. That's why a high PFD doesn't considerably decrease the output.

5.2 Comparisons

When we compare the output of TORA and TORA-AH, it is shown that energy consumed in TORA is more than TORA-AH. Table 1 shows the comparison of remaining energy. Energy consumption in TORA is increasing at a very fast rate as the number of nodes is increasing. On the other hand in TORA-AH, it remains almost constant as the number of nodes increases. Here, energy given to nodes is 160 J.

Similarly, comparison between DYMO and DYMO-AH is shown in Table 2. Here in the results, we conclude that performance of DYMO-AH is much better than DYMO.

After that, we have a tendency to do comparison between TORA-AH and DYMO-AH. We are able to see in Table 3 that DYMO-AH consumes less energy than TORA-AH.

Now, comparison is finished on the premise of event prevalence. Every node has at first one joule of energy. Comparing TORA and TORA-AH, we terminated that with increasing no. of events energy consumption is reduced more in TORA-AH whereas it goes down in TORA (Table 4).

Table 1 Comparison of remaining energy in TORA and TORA-AH	Remaining energy		
	No. of nodes	TORA	TORA-AH
	5	159.992	159.995
	10	159.985	159.989
	15	159.976	159.983
	20	159.970	159.980
	25	–	159.971

Table 2 Comparison of remaining energy in DYMO and DYMO-AH

Remaining energy		
No. of nodes	DYMO	DYMO-AH
5	159.994	159.997
10	159.987	159.993
15	159.978	159.986
20	159.972	159.984
25	–	159.979

Table 3 Comparison between TORA-AH and DYMO-AH

Remaining energy		
No. of nodes	TORA-AH	DYMO-AH
5	159.995	159.997
10	159.989	159.993
15	159.983	159.986
20	159.980	159.984
25	159.971	159.979

Table 4 Energy consumed by TORA and TORA-AH w.r.t no. of events occurred

Energy consumed		
No. of events occurred	TORA	TORA-AH
10	0.92	0.95
20	0.85	0.89
30	0.75	0.83
40	0.67	0.78
50	0.58	0.71
60	0.49	0.66
70	0.40	0.60
80	0.31	0.54

Similarly, comparison between DYMO and DYMO-AH is shown in Table 5. Due to the periodic hello messaging scheme, more energy is consumed in DYMO and battery drain is more. When the adaptive scheme of DYMO is applied, i.e., DYMO-AH, due to the suppressed hello messages the energy consumed is less (Table 5).

Results of DYMO-AH are far better than TORA-AH. DYMO-AH maintains energy stable and reduces energy consumption significantly.

Table 5 Energy consumed by DYMO and DYMO-AH w.r.t. no. of events occurred

Energy consumed		
No. of events occurred	DYMO	DYMO-AH
10	0.95	1.00
20	0.92	0.99
30	0.88	0.99
40	0.84	0.98
50	0.80	0.97
60	0.76	0.96
70	0.71	0.96
80	0.66	0.95

Table 6 Comparison of remaining energy between TORA, TORA-AH and TORA-AHBFO

Remaining energy			
No. of nodes	TORA	TORA-AH	TORA-AHBFO
5	159.992	159.995	159.999
10	159.985	159.989	159.998
15	159.976	159.983	159.997
20	159.970	159.980	159.996
25	–	159.971	159.994

5.3 Optimization Results

By using the BFOA technique during optimization, the results are much better than adaptive scheme. As seen in Table 6, we concluded that energy consumed by the TORA-AHBFO is almost same by increasing the number of nodes as compared to TORA and TORA-AH (Fig. 10 and Table 7).

Here, we have concluded that when using the optimization technique bacterial foraging optimization algorithm (BFOA) we have reduced energy consumption approximately up to 85%.

6 Result and Conclusion

In this investigation, a fresh out of the plastic new approach, known as adaptive "hello" messaging scheme is anticipated to determine the issues related to battery utilization and system overhead. The welcome electronic correspondence topic means to downsize undesirable hello messages while neighbor revelation and conjointly to decide a dependable association between the source node and the goal node. This can be one among the fundamental issues that impressively affect the presentation of

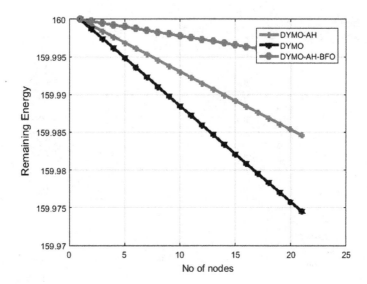

Fig. 10 Energy consumption with variable nodes of DYMO

Table 7 Comparison of remaining energy between TORA, TORA-AH and TORA-AHBFO

Remaining energy			
No. of nodes	DYMO	DYMO-AH	DYMO-AHBFO
5	159.994	159.997	159.999
10	159.987	159.993	159.998
15	159.978	159.986	159.996
20	159.972	159.984	159.995
25	–	159.979	159.994

the MANETs. We have demonstrated that battery utilization can be decreased significantly more by utilizing the advancement method BFOA when contrasted with global optimization algorithm.

References

1. Han, SY., Lee, D.: An adaptive hello messaging scheme for neighbor discovery in on-demand manet routing protocols. IEEE Commun. Lett. **17**(5), 1040–1043 (2013)
2. Sivakumar, P., Srinivasan, R., Saranya, K.: An efficient neighbor discovery through hello messaging scheme in MANET routing protocol. Int. J. Emerg. Technol. Adv. Eng. **4**(1), 169–173 (2014)
3. Mostajeran, E., Noor, RM., Keshavarz, H.: A novel improved neighbor discovery method for an intelligent-TORA in mobile ad hoc networks. In: International Conference of Information and Communication Technology (ICoICT), pp. 395–399. https://doi.org/10.1109/icoict.2013. 6574608 (2013)

4. Gupta, A.K., Sadawarti, H., Verma, A.K.: Implementation of DYMO routing protocol. Int. J. Inf. Technol. Model. Comput. **1**(2), 49–57 (2013)
5. Gulia, P., Siha, S.: Enhance security in MANET using bacterial foraging optimization algorithm. Int. J. Comput. Appl. **84**(1), 32–35 (2013)
6. Kulvir, K., Sonia, G.: Optimized routing in wireless ad hoc networks. IJECT **2**(1) (2011)
7. Shivahare, BD., Wahi, C., Shivhare, S.: Comparison of proactive and reactive routing protocols in mobile ad-hoc network using routing protocol property. Int. J. Emerg. Technol. Adv. Eng. **2**(3), 356–359 (2012)
8. Giruka, V.C., Singhal, M.: Hello protocols for ad-hoc networks: overhead and accuracy trade-offs. In: Sixth IEEE International Symposium on a World of Wireless Mobile and Multimedia Networks. https://doi.org/10.1109/WOWMOM.2005.50 (2005)
9. Dubey, S., Shrivastava, R.: Energy consumption using traffic models for MANET routing protocols. Int. J. Smart Sens. Ad Hoc Netw. **1**(1), 84–89 (2011)
10. Kaur, I., Rao, A.L.N.: A framework to improve the network security with less mobility in MANET. Int. J. Comput. Appl. **167**(10), 21–24 (2017)
11. Singh, SB., Ambhaikar, A.: Optimization of routing protocol in MANET using GA. Int. J. Sci. Res. (IJSR) **1**(2), 23–26 (2012)
12. Kumar, N., Anitha, A.: A study of differing techniques for reduction of power dissipation over the MANET. Int. J. Adv. Res. Comput. Sci. Softw. Eng. **3**(11) (2013)
13. Kokila, G., Karnan, M.M., Sivakumar, M.R.: Immigrants and memory schemes for dynamic shortest path routing problems in mobile ad-hoc networks using PSO, BFO. Int. J. Comput. Sci. Manag. Res. **2**(5) (2013)
14. Shanmugha Priya, B.: Mitigating superfluous flooding of control packets MANET. Int. J. Innov. Res. Comput. Commun. Eng. **2**(1), 29–34 (2014)

Horseshoe-Shaped Multiband Antenna for Wireless Application

Vineet Vishnoi, Praveen Kumar Malik and Manoj Kumar Pal

Abstract This paper presents the usefulness of horseshoe-shaped antenna: mathematically based on baker's transformation. The proposed antenna has the property of filling a plane using higher-order iterations and exploited in realization of a multiband resonant antenna. The effect of additional iterations resulting in the reduction of resonant frequency is near-logarithmic pattern. The designed antenna shows multiple frequency bands ranging from 1.01 to 7.60 GHz. It has been also observed that the proposed prototype antenna has 75% efficiency, directivity up to 11.5 dBi and gain of about 10 dB. The antenna characteristics have been studied using IE3D v.14 simulation software based on method of moment (MoM) and also experimentally verified using VNA network analyzer. Simulation and experimental results are in good agreement and demonstrate the performance of the design methodology and the proposed antenna structures.

Keywords Baker's transformation · Horseshoe-shaped antenna · IE3D · Multiband antennas

1 Introduction

State-of-the-art wireless telecommunication system needs antenna with wider bandwidth, multifunctional and compact in size than conventional antennas. As antenna length may be order of two or large for efficient radiation, this limits the performance of other parameters like bandwidth, gain and efficiency [1, 2]. This generates a new family of antenna termed as fractal coined by Mandelbrott [3], and a lot of work

V. Vishnoi (✉)
Inderprastha Engineering College, Ghaziabad, Uttar Pradesh, India
e-mail: vishnoivineet@gmail.com

P. K. Malik
Lovely Professional University, Jalandhar, Punjab, India
e-mail: pkmalikmeerut@gmail.com

M. K. Pal
Bharat Sanchar Nigam Limited, Kanpur, Uttar Pradesh, India
e-mail: manoj1976bsnl@gmail.com

© Springer Nature Singapore Pte Ltd. 2020   33
P. K. Singh et al. (eds.), *Proceedings of First International Conference on Computing, Communications, and Cyber-Security (IC4S 2019)*, Lecture Notes in Networks and Systems 121, https://doi.org/10.1007/978-981-15-3369-3_3

has been compiled through various researchers on such class of antenna subfamily [4–10]. The fractal mathematics is too old, but some of fractal geometries find its wide application in antenna geometry, and from the last two decades, few of such fractal geometries were investigated in antenna design and become popular such as Koch curves, Minkowski curves and Sierpinski carpets which are investigated for antenna design [11–15] as fractal geometry-based antennas have diverse application in several fields of science and engineering. The multiband characteristics of fractal antenna are due to their self-similarity of fractal geometry qualitatively associated with their multiband characteristics of antenna, along with space-filling properties resulting to the miniaturization of antenna without degradation in antenna parameters such as gain, bandwidth and efficiency. These antennas resonate at frequencies in a near-logarithmic interval. The individual bands at these resonant frequencies are generally small, and allocation of these frequency bands founds arbitrarily in its frequency spectrum [16–21, 22–24].

Apart from this, other important features of this fractal antenna include low profile, low cost, conformability and ease of integration with active devices. There is limit on bandwidth and gain of such antennas. Fortunately, a lot of communication systems do not require large bandwidths; therefore, it is not an important problem. Many techniques have been used to reduce the size of antenna, such as using dielectric substrates with high permittivity [17–21, 22–25], applying resistive or reactive loading [18], increasing the electrical length of antenna by optimizing its shape [19] and utilizing strategically positioned notches on the patch antenna [20, 21, 22–26]. As in this regard, the earlier article [21] published on horseshoe-shaped antenna using multi-L-slots does not represent the true horseshoe shape, but in the present article, we present a truly horseshoe-shaped pattern which is used here in antenna design. This consists of baker's transformation curve patterns (22), which have several important characteristics hitherto unexplored in antenna engineering.

This paper is organized as follows. Section 2 describes the mathematics of horseshoe geometry. Sect. 3 presents the simulated and measured results of proposed antenna. Results and discussion is provided in Sect. 4.

2 Horseshoe Fractal Design Methodology

One of the simplest planar dynamical systems with a fractal attractor is the so-called baker's transformation [9] because it resembles the process of repeatedly stretching a piece of dough and folding it in two. Let $E = [0, 1] \times [0, 1]$ be the unit square. For fixed $0 < \lambda < \frac{1}{2}$, we define the baker's transformation $f: E \to E$ by

$$f(x, y) = \begin{cases} (2x, \lambda y), & (0 \leq x \leq 1/2) \\ (2x - 1, \lambda y + 1/2), & (1/2 \leq x \leq 1) \end{cases} \tag{1}$$

(a)

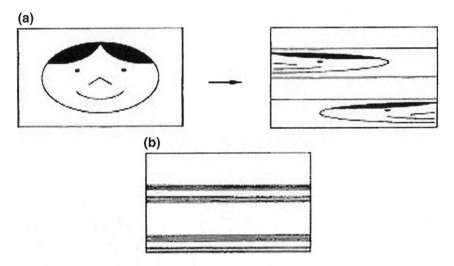

(b)

Fig. 1 Baker's transformation. **a** Its effect on the unit square. **b** Its attractor

This transformation may be thought of as stretching E into a $2 \times \lambda$ rectangle, cutting it into two $1 \times \lambda$ rectangles and placing these above each other with a gap of $\left[\frac{1}{2} - \lambda\right]$ in between; see Fig. 1. Then, $E_k = f^k(E)$ is a decreasing sequence of sets, with E_k comprising 2^k horizontal strips of height λ^k separated by gaps of at least $(\frac{1}{2} - \lambda)\lambda^{(k-1)}$. Since $f(E_k) = E_k + 1$, the compact limit set $F = E_k$ satisfies $f(F) = F$. (Strictly speaking, $f(F)$ does not include part of F in the left edge of the square E, a consequence of f being discontinuous. However, this has little effect on our study.) If $(x, y) \in E$, then $f^k(x, y) \in E_k$, so $f^k(x, y)$ lies within distance λ^k of F. Thus, all points of E are attracted to F under iteration by f. If the initial point (x, y) has $x = 0, a_1, a_2, \ldots$ in base 2 and $x \neq \frac{1}{2}$, then it is easily checked that

$$f^k(x, y) = (0, a_{k+1}, a_{k+2}, \ldots, y_k) \tag{2}$$

where y_k is some point in the strip of E_k numbered $a_k, a_{k-1}, \ldots, a_1$ (base 2) counting from the bottom with the bottom strip numbered 0. Thus, when k is large, the position of $f^k(x, y)$ depends largely on the base 2 digits a_i of x with i close to k. By choosing an x with base 2 expansion containing all finite sequences, we can arrange for $f^k(x, y)$ to be dense in F for certain initial (x, y), just as in the case of the tent map. Further analysis along these lines shows that f has sensitive dependence on initial conditions and the periodic points of f are dense in F, so that F is a chaotic attractor for f. Certainly, F is a fractal—it is essentially the product $[0, 1]F_1$, where F_1 is a Cantor set that is the IFS attractor of $S_1(x) = x, S_2(x) = 1/2 + \lambda x$. Since $\dim_H F_1 = \dim_B F_1 = \log 2/\log \lambda$, $\dim H_F = 1 + \log 2/\log \lambda$, using Corollary 2. The baker's transformation is rather artificial, being piecewise linear and discontinuous. However, it does serve to illustrate how the 'stretching and cutting' procedure results in a fractal attractor. The closely related process of 'stretching and folding' can occur for continuous functions

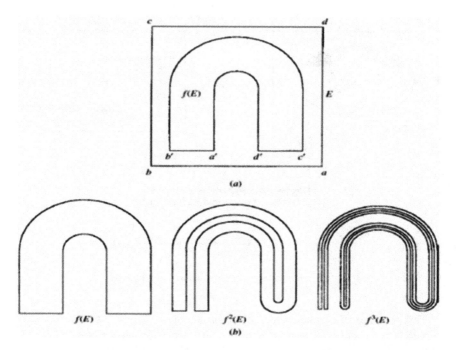

Fig. 2 A horseshoe map. **a** The square E is transformed, by stretching and bending, to the horseshoe $f(E)$, with a, b, c, d mapped to a', b', c', d', respectively. **b** The iterates of E under f form a set that is locally a product of a line segment and a cantor set

on plane regions. Let $E = [0, 1] \times [0, 1]$ and suppose that f maps E in a one-to-one manner onto a horseshoe-shaped region $f(E)$ contained in E. Then, f may be thought of as stretching E into a long thin rectangle which is then bent in the middle. This figure is repeatedly stretched and bent by f, so that $f^k(E)$ consists of an increasing number of side-by-side strips; see Fig. 2.

We have $E \supset f(E) \supset f^2(E), \ldots, \supset f^k(E)$, and the compact set $F = \bigcap_{k=1}^{\infty}$ attracts all points of E. Locally, F looks like the product of a Cantor set and an interval. A variation on this construction gives a transformation with rather different characteristics; see Fig. 2. If D is a plane domain containing the unit square E and $f: D \to D$ is such that $f(E)$ is a horseshoe with ends and arch lying in a part of D outside E that is never iterated back into E, then almost all points of the square E (in the sense of plane measure) are eventually iterated outside E by f. If $f^k(x, y) \in E$ for all positive k, then $(x, y) \in \bigcap_{k=1}^{\infty} f^{-1}(E)$. With f suitably defined, $f^{-1}(E)$ consists of two horizontal bars across E, so $\bigcap_{k=1}^{\infty} f^{-1}(E)$ is the product of $[0, 1]$ and a Cantor set. The set $F = \bigcap_{k=-\infty}^{\infty} f^k(E) = \bigcap_{k=0}^{\infty} f^k(E) = \bigcap_{k=1}^{\infty} f^{-k}(E)$ is compact and invariant for f and is the product of two Cantor sets. However, F is not an attractor, since points arbitrarily close to F are iterated outside E.

The sets $\bigcap_{k=1}^{\infty} f^{-k}(E)$ and $\bigcap_{k=0}^{\infty} f^k(E)$ are both products of a Cantor set and a unit interval. Their intersection F is an unstable invariant set for f. A specific example

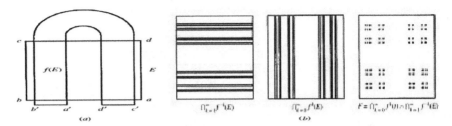

Fig. 3 A horseshoe map. **a** An alternative horseshoe map. **b** The square E is transformed, so that the arch and ends of $f(E)$ lie outside E

of a 'stretching and folding' transformation is the Henon map $f: R2 \rightarrow R2$ (Fig. 3).

$$f(x, y) = (y + 1 - ax^2, bx) \tag{3}$$

where a and b are constants. (The values $a = 1.4$ and $b = 0.3$ are usually chosen for study. For these values, there is a quadrilateral D for which $f(D) \subset D$ to which we can restrict attention.) This mapping has Jacobian b for all (x, y), so it contracts area at a constant rate throughout $R2$; to within a linear change of coordinates, Fig. 2 is the most general quadratic mapping with this property. The transformation in Fig. 2 may be decomposed into an area-preserving bend, a contraction and a reflection, the net effect being horseshoe-like; see Fig. 4. This leads us to expect f to have a fractal attractor, and this is borne out by computer pictures, Fig. 4. Detailed pictures show banding indicative of a set that is locally the product of a line segment and a Cantor-like set. The diagrams show the effect of this successive transformation on a rectangle. Estimates suggest that the attractor has box dimension of about 1.26 when $a = 1.4$ and $b = 0.3$. Detailed analysis of the dynamics of the Henon map is complicated. In particular, the qualitative changes in behavior (bifurcations) that occur as a and b vary that are highly intricate. Many other types of stretching and folding are possible. Transformations can fold several times or even be many-to-one; for example the ends of a horseshoe might cross. Such transformations often have fractal attractors, but their analysis tends to be difficult (Fig. 5).

3 Simulation Setup

The resonant properties of the proposed antenna have been predicted and optimized using a frequency domain three-dimensional full-wave electromagnetic field solver (IE3D-Zeland), as the proposed horseshoe antenna is designed on the patch size of 3.8×5.5 cm, using scale factor one-third on glass–epoxy material having height 1.6 mm and dielectric constant 4.5 as shown in Fig. 6.

As explained above, the base shape for horseshoe geometry is designed on rectangular patch of specific dimension as provided above, and for further iterations,

Fig. 4 Henon map may be decomposed into an area-preserving bend, followed by a contraction and followed by a reflection in the line $y = x$

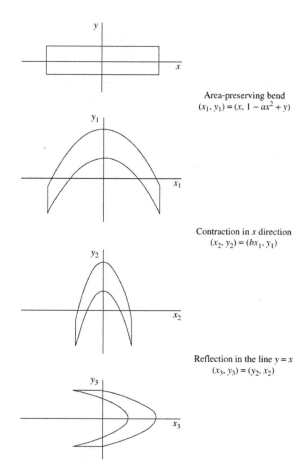

Area-preserving bend
$(x_1, y_1) = (x, 1 - ax^2 + y)$

Contraction in x direction
$(x_2, y_2) = (bx_1, y_1)$

Reflection in the line $y = x$
$(x_3, y_3) = (y_2, x_2)$

the geometry is scaled by a factor of one-third keeping outer dimension of the patch invariant as 3.87×5.56 cm for first iteration and second iteration which are generated in the same manner as its originator base shape as shown in Fig. 7.

After completion of simulation setup, IE3D provides various antenna parameters through its easily accessible user graphics format for analysis point of view, and here, all such typical antenna parameters are analyzed viz return loss ($S11$), directivity, antenna gain and its radiation/antenna efficiency.

Fig. 8 illustrates the comparison in between return loss for base shape of horseshoe and its iterated geometries, and it is observable that lowering of resonance frequency form 1.229 to 1.034 Ghz along with incorporation of multiple frequency bands from 1.034 to 7.6 GHz

Similarly, other important antenna parameters such as directivity, gain and efficiency for horseshoe's base shape and its iterated geometries are compared in Figs. 9, 10 and 11, respectively, from Fig. 5, directivity approaches up to 13 dBi for second iteration, and similarly, gain approaches up to 21 and 10 dB on an average (Fig. 10).

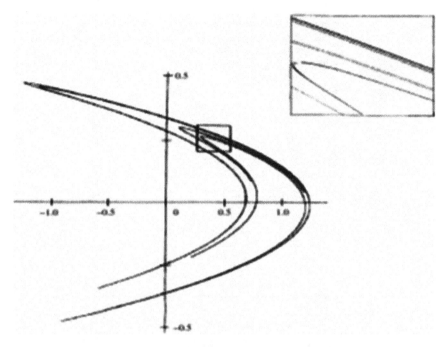

Fig. 5 Iterates of a point under the Henon map showing the form of the attractor. Banding is apparent in the enlarged portion in the inset

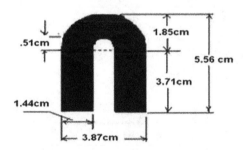

Fig. 6 Proposed prototype antenna with design dimensions

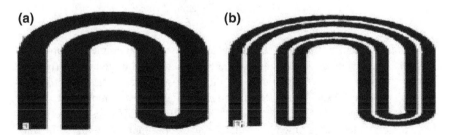

Fig. 7 Design antenna. **a** First iteration. **b** Second iteration

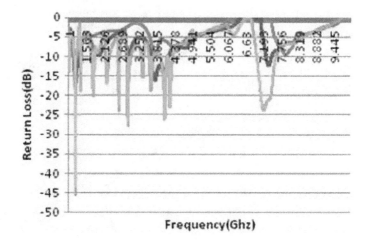

Fig. 8 Return loss ($S11$) curve for second iteration of horseshoe antenna

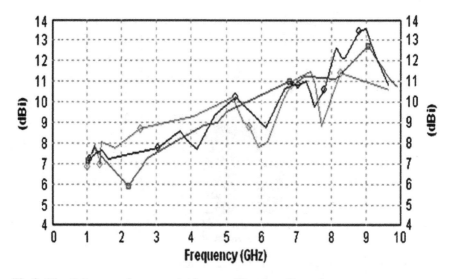

Fig. 9 Directivity versus frequency plot for second iteration of horseshoe antenna

In continuation to this, from Fig. 11, it is also observable that efficiency approaches up to 75%, so all antenna parameters provide satisfactory optimized values (Fig. 12).

Apart from above-shown parameters, the proposed antenna has the promising radiation patterns for elevation ($\phi = 0°$ and $\phi = 90°$) and azimuth plane ($\theta = 900°$) in Fig. 13 at frequencies 1.22 GHz, 1.821 GHz and 6.935 GHz, respectively. After the radiation pattern, experimental results are also superimposed (with red in color) on simulated radiation pattern for comparison point of view in the respective figures.

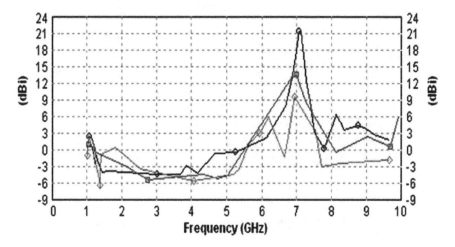

Fig. 10 Total gain versus frequency plot for second iteration of horseshoe antenna

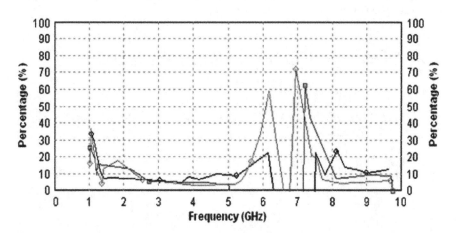

Fig. 11 Efficiency (antenna and radiation) versus frequency plot for second Iteration of horseshoe antenna

On comparing in between simulated results and measured results for return loss as shown in Fig. 12, it is quite promising that for proposed antenna both the curves, simulated and measured results follow each other with high degree of accuracy, and the variation in between these two curves can also be anticipated on the basis of design accuracy.

Since all the three geometries of proposed horseshoe-shaped antenna are analyzed on IE3D simulation software, in this regard, a combined comparative graph is shown in Figs. 6 and 7, its details are provided in Table 1 for mutual comparison of their return loss ($S11$) plots for observing the effect of higher-order iteration, and from the curve and table, it is quite noticeable that nos. of resonating frequency samples

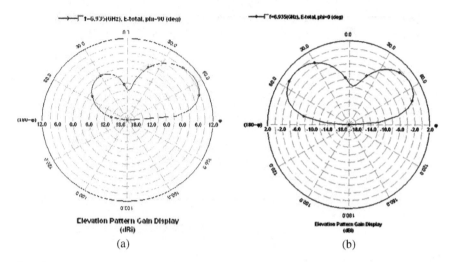

Fig. 12 Radiation curve for simulated antenna for second iteration of horseshoe antenna at frequency 6.935 GHz for elevation **a** $\phi = 0°$, **b** $\phi = 90°$

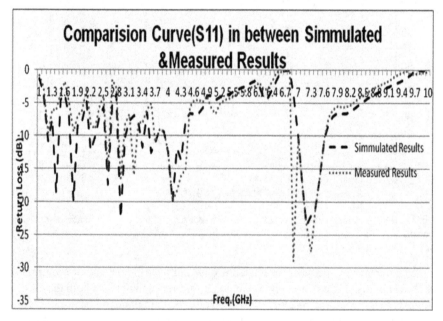

Fig. 13 Comparison curve in between simulated and measured (in black) return loss ($S11$) for second iteration of horseshoe antenna

Table 1 Comparison table between the base shape of first and second iterations

S. No.	Frequency (GHz)	Second iteration	First iteration	Base shape
1	1.034	−11.68	−2.73	−4.08
2	1.229	−30.08	−20.12	−14.06
3	1.48	−18.18	−5.29	−5.56
4	1.8	−20.13	−8.36	−4.5
5	2.19	−10	−6.73	−3.4
6	2.22	−16	−6.21	−3.46
7	2.47	−10	−5.06	−3.41
8	2.6	−20.81	−3.9	−3.3
9	2.86	−10	−6.17	−2.42
10	2.91	−27.38	−11.54	−2.1
11	2.99	−10	−8.34	−1.48
12	3.36	−14	−5.01	−1.6
13	3.58	−10	−7.8	−4.52
14	3.64	−18.37	−10.02	−6.18
15	3.7	−10	−13.7	−8.07
16	3.95	−10	−12.58	−5.54
17	4	−25	−13.85	−5.4
18	4.25	−22	−8.71	−6.9
19	4.34	−10	−7.1	−7.9
20	6.98	−10	0	0
21	7.2	−25	−5.8	0
22	7.6	−10	−8.39	−5.54

go on incrementing while approaching higher-order iteration (as shown in Table 1). It is also remarkable that about 1.22.

Moreover, from above discussion, it is also noticeable that the proposed antenna peruses significant directivity from 7 to 11.5 dBi (Fig. 5) for wide range of frequency, and similarly, Fig. 6 depicts the curve of total gain versus frequency plot, as from the curve, it is noteworthy that total gain approaches up to 10 dB value. The other important antenna parameters are efficiency of proposed antenna geometry, from Fig. 7, the curve illustrates antenna and radiation efficiency versus frequency, from the curve, it is notable that radiation efficiency approaches 100% for a significant band of freq, also, the antenna efficiency approaches 70% in this frequency range, the above values are quite promising for a good antenna design, and the proposed antenna is well suited for wireless applications.

4 Conclusion

A novel horseshoe-shaped fractal antenna for wireless application has been proposed, constructed and tested. The proposed antenna has simple iterative geometry. As the proposed antenna design provides adoptability of various frequencies ranging from 1.03 up to 7.21 GHz and provides characteristics such as 100% radiation efficiency, highly permissible gain up to 10 dB and directivity up to 11.5 dBi. These promising characteristics are found higher than earlier design traditional wireless antennas. The proposed antenna has good directional-radiation characteristics. Furthermore, this antenna has many advantages such as easy fabrication, low cost and compact in size. Therefore, it is found quite suitable for wireless/WLAN applications.

References

1. Bahl, J., Bhartia, P.: Microstrip Antennas. Artech House, Dedham, Ma (1981)
2. Lizzi, L., Viani, F., Zeni, E., Massa, A.: A DVBH/GSM/UMTS planar antenna for multimode wireless devices. IEEE Antennas Wirel. Propag. Lett. **8**, 568–571 (2009)
3. Madelbrot, B.B.: The Fractal Geometry of Nature. Freeman, New York (1983)
4. Puente, C., Romeu, J., Cardama, A.: Fractal-shaped antennas. In: Werner, D.H., Mittra, R. (eds.) Frontiers in Electromagnetic, pp. 94203 (1999)
5. Cohen, N.: Fractal antenna applications in wireless telecommunications. In: Proceedings of Electronics Industries Forum of New England, pp. 4349 (1997)
6. Puente, C., Pous, R., Romeu, J., Garcia, X.: Antennas fractals of multi fractals. Eur. Pat. ES **2**,112–163 (1998)
7. Peitgen, H.-O., Henriques, J.M., Penedo, L.F. (eds.) Fractals in the Fundamental and Applied Sciences. Amsterdam, North Holland (1991)
8. Cherepanov, G.P., Balankin, A.S., Ivanova, V.S.: Fractal fracture mechanics. Eng. Fract. Mechan. **51**, 9971033 (1995)
9. Werner, D.H., Haupt, R.L., Werner, P.L.: Fractal antenna engineering: the theory and design of fractal antenna arrays. IEEE Antennas Propagat. Mag. **41**, 3759 (1999)
10. Werner, D.H., Werner, P.L., Jaggard, D.L., Jaggard, A.D., Puente, C., Haupt, R.L.: The theory and design of fractal antenna arrays, In: Werner, D.H., Mittra,R. (eds.) Frontiers in Electromagnetic, pp. 94203 (1999)
11. Puente, C., Romeu, J., Pous, R., Cardama, A.: On the behavior of the sierpinski multiband fractal antenna. IEEE Trans. Antennas Propagat. **46**, 517524 (1998)
12. Puente, C., Romeu, J., Pous, R., Garcia, X., Benitez, F.: Fractal multiband antenna based on the sierpinski gasket. Electron. Lett. **32**(1), 12 (1996)
13. Puente, C., Romeu, J., Bartoleme, R., Pous, R.: Fractal multiband antenna based on sierpinski gasket. Electron. Lett. **32**, 12 (1996)
14. Baliarda, C.P., Romeu, J., Cardama, A.: The koch monopole: a small fractal antenna. IEEE Antennas Propag. Mag. **48**(11), 1773–1781 (2000)
15. Best, S.R.: On the resonant properties of the koch fractal and other wire monopole antennas. IEEE Antennas Wirel. Propag. Lett. **1**, 74–76 (2002)
16. Puente, C., Romeu, J., Bartolome, R., Pous, R.: Perturbation of the sierpinski antenna to allocate operating bands. Electron. Lett. **32**, 2186–2188 (1996)
17. Lo, T.K., Hwang, Y.: Microstrip antennas of very high permittivity for personal communications. In: Proceedings of 1997 Asia Pacific Microwave Conference, pp. 253256 (1997)

18. Sinati, R.A.: CAD of Micro Strip Antennas for Wireless Applications. Artech House, Norwood, MA (1996)
19. Wang, H.Y., Lancaster, M.J.: Aperture-coupled thin-film superconducting meander antennas. IEEE Trans. Antennas Propag. **47**, 829836 (1999)
20. Waterhouse, R.: Printed Antennas for Wireless Communications. Wiley, Hoboken, NJ (2007)
21. Kwak, K.-S., Choi, S.-H.: Horseshoe shaped antenna for dual-band WLAN communications with multi L slot. Microw. Technol. Lett. **51**(2), 463–465 (2009)
22. Malik, P., Parthasarthy, H.: Synthesis of randomness in the radiated fields of antenna array. Int. J. Microwave Wirel. Technol. **3**(6), 701–705 (2011). https://doi.org/10.1017/S1759078711000791
23. Malik, P.K., Parthasarthy, H., Tripathi, M.P.: Axisymmetric excited integral equation using moment method for plane circular disk. Int. J. Sci. Eng. Res. **3**(3), 1–3 (2012, March). ISSN 2229-5518
24. Malik, P.K., Singh, M.: Multiple bandwidth design of microstrip antenna for future wireless communication. Int. J. Recent Technol. Eng. **8**(2), 5135–5138 (2019, July). ISSN: 2277-3878. https://doi.org/10.35940/ijrte.B2871.078219
25. Malik, P.K., Parthasarthy, H., Tripathi, M.P.: Analysis and design of Pocklingotn's equation for any arbitrary surface for radiation. Int. J. Sci. Eng. Res. **7**(9), 208–213 (2016, September)
26. Budhiraja, I., Aftab Alam Khan, Mohd., Farooqi, M., Pal, M.K.: Multiband stacked microstrip patch antenna for wireless applications. J. Telecommun. **16**(2), 925–931 (2012, October). (I.F.-1.7)

A Review Paper on Performance Analysis of IEEE 802.11e

Harpreet Singh Bedi, Kamal Kumar Sharma and Raghav Gupta

Abstract Quality of service is an important parameter in wireless communication systems. To provide the good data delivery and end-to-end delivery to its users, a system or a network must work on its services. Providing quality of services like excellence throughout and less delay are the main tasks and responsibility of IEEE 802.11e. In this paper, various techniques were discussed to improve the network performance of IEEE 802.11e, and results were compared with existing techniques for better experiments and results.

Keywords MAC · Delay · Throughput · PCF · DCF · EDCF

1 Introduction

A communication network is the main and essential part and block of any organization. In this paper, various techniques were discussed to overcome the network quality of IEEE 802.11e, and the results were compared with existing techniques for better experiments and results. The network must do a lot of support from data collection to data receiving for different types multimedia applications. Networks should provide secure, predictable, measurable, and from time to time assured features. To maintain the QoS by means of the delay, extend variant (jitter), range of frequency, and data loss quantities on a system turns into the secret to a profitable net business solution. Thus, QoS is the set of strategies to manipulate community services. To provide QoS to such sort of utilization, administration support is must. The IEEE 802.11e has basically two components, i.e., DCF and the incorporated arrangement known as PCF. As both the medium access delay and PHY layer of IEEE 802.11e

H. S. Bedi (✉) · K. K. Sharma · R. Gupta
Department of Electronics and Communication, Lovely Professional University, G. T. Road, Phagwara, Punjab 144411, India
e-mail: harpreet.17377@lpu.co.in

K. K. Sharma
e-mail: Kamal.23342@lpu.co.in

© Springer Nature Singapore Pte Ltd. 2020   47
P. K. Singh et al. (eds.), *Proceedings of First International Conference on Computing, Communications, and Cyber-Security (IC4S 2019)*, Lecture Notes in Networks and Systems 121, https://doi.org/10.1007/978-981-15-3369-3_4

Fig. 1 Partitioned network

are intended for quality transmission, the first 802.11 preferred does not consider. Thus, to supply QoS guide IEEE 802.11, trendy gathering has indicated any other IEEE 802.11e standard (Fig. 1).

1.1 Medium Access Control Layer

Medium access control layer is the second layer of data layer in OSI model which deals with data transmission. It gives control to channel devices that make it profitable for a few devices to convey inside a numerous entrance that combines a mutual medium. The device that makes the MAC is alluded to be controller mechanism [1]. This layer gives a full-duplex intelligent control in multipoint arrangement. This channel may give single-cast, dual-cast, or communicate with each other and devices.

1.2 Distributed Coordination Function (DCF)

It is one of the best methods and protocol of IEEE 802.11e. It works on the ideology of CSMA/CA which is carrier-sense multiple access with collision avoidance. This function works with a single transmission queue with works in agenda of FIFO input. It contains a combined MAC based on checking and features of the channel availability, whether busy or free. If the medium is not idle, then MAC must wait for the time until the channel is free, and it must send an extra time interval which is (DIFS). If the medium stays idle during DIFS deference, this layer then converts the back off process by collecting any back off counter. For every interval, during which the channel will remain idle, the back off counter range is decreased [2]. If a device does not get right to use the channel in first cycle, it stops its back off counter, looks for the channel to be free again for DIFS frame, and starts the process again. When the link expires, the device accesses the medium in Fig. 3. Each device must identify CW which is used to predict the counter [3].

An Ack frame will be transmitted by the receiver for receiving and collecting the data [4]. This frame is sent over a SIFS who is smaller than DIFS. As soon as SIFS is smaller than DIFS, the signal and its information are protected from other

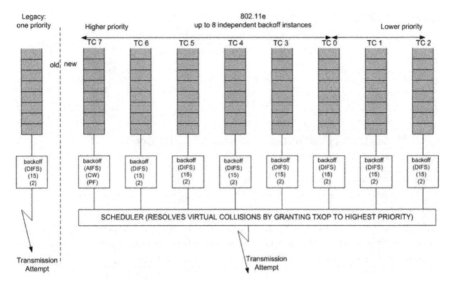

Fig. 2 Prioritization of data with back off window

station in Fig. 2. The size of contention window is less, and data will be lost [5]. If the size is doubled, then back off algorithm will be performed. All the quantities of MAC depend upon the bit frame and size of the contention window. The value of frame (DIFS) will be equal to frame and twice of slot time against the physical layer (Fig. 3).

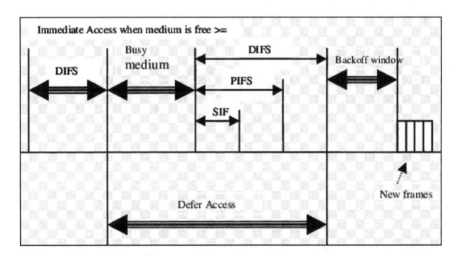

Fig. 3 Timing relationship for DCF

Table 1 Access categories

Priority	Access category (AC)	Designation
1	0	Videoconferencing
2	1	Online
3	2	Back off
4	3	VoIP

1.3 Enhanced Distributed Coordination Function (EDCF)

It is used to supply organized QoS with the aid of upgrading the dispute-based DCF. It offers separated, conveyed access to the far-flung mode for QoS stations (QSTAs) utilizing 8 one of a kind purchasers wants (UPs). Before coming into the MAC layer, each datum bundle bought from the higher layer is allotted a precise consumer need esteem. Instructions to label a want an incentive for every bundle of execution issue [6]. This system characterizes four various first-in FIFO lines, get to classifications (ACs) that offer help for the conveyance of site visitors with stations. This relative prioritization is mounted from IEEE 802.1d scaffold particular [7]. Various kinds of utilizations (e.g., videoconferencing traffic, online site visitors, and back off traffic) can be coordinated into quite a number of ACs. For every AC, an improved variant of the DCF is called as an upgraded appropriated coordination work (EDCF) [8].

Different priority of data is given in Table 1 with respect to different access categories. Videoconferencing is given the highest, and least is given to voice-over-Internet applications.

The figure shows the design model with different queues, where each AC behaves like a virtual station in Fig. 4.

The timing diagram of EDCA is shown in Fig. 5. The access category with the smallest frame has the highest priority [9]. The range of frames and contention window size, which are referred to as the EDCA parameters, are referred to as frames (beacon).

In this paper, the research was done with respect to medium where one can send one data frame with transmission opportunity round.

2 Literature Review

Barry, M. et al. (2001): This paper investigates service differentiation depends on the IEEE 802.11 appropriated coordination work (DCF), initially intended to help best-exertion information administrations. The virtual MAC assessments key MAC-level measurements identified with administration quality, for example, delay, defer variety, bundle crash, and parcel misfortune. We demonstrate the productivity of the

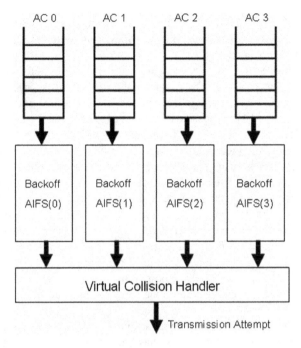

Fig. 4 Access categorization with back off window

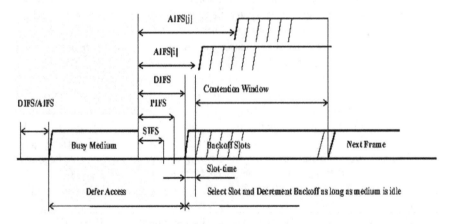

Fig. 5 Timing relationship for EDCF

virtual MAC calculation and think about fundamentally covering cells and exceedingly bursty traffic blends [10]. A virtual source (VS) calculation uses the virtual MAC to assess application-level administration quality.

Shklyaeva, A. et al. (2007): The subsequent parameters of the two systems, for example, media access delay, start to finish delay, bundle line size, lining deferral, or bit error rate, were compared [11].

Vardakas, John S. et al. (2009): In this paper, the MAC defer investigation is performed depending on basic likelihood hypothesis (restrictive probabilities) while staying away from the mind-boggling Markov chain strategy [6]. A far-reaching investigation of the MAC postponement is introduced by giving higher snapshots of the MAC defer conveyance [12]. To the extent which the lining postponement is concerned, we give the mean lining delay by considering a lining framework with one line for each access category (AC) per versatile station, with a solitary server (the remote medium), basic to every portable station, and a Markov adjusted Poisson process as information that communicates the burst idea of Internet traffic.

Liu, X. et al. (2009): This paper displays an all-inclusive investigative model dependent on Bianchi's model in IEEE 802.11, by considering the gadget suspending occasions, unsaturated traffic conditions, just as the impacts of blunder inclined channels [5]. In view of this model, we reinfer a shut structure articulation of the normal administration time. The exactness of the model is approved through broad reproductions. The examination is additionally instructional for IEEE 802.11 systems under restricted load.

Jansang, A. et al. (2009): A queueing scientific model for assessing the normal parcel postponement of bit rate (VBR) has been proposed in this article [7]. The outcomes demonstrate that standard acknowledgment strategy cannot ensure the QoS. The article likewise demonstrates that tolerating a VBR stream with a modest quantity over mean information rate ought to have the option to deal with the issue without corruption of acknowledged streams.

Matin, M. A. et al. (2010): This research has focused on single hop, access point-based environments, and network-related key performance indicator (KPI) for performance analysis of ad hoc networks. Thus, the authors have extended the analysis to a further level based on real-time actual frames of video transmission with subjective [13] and objective video quality testing metrics. In this paper, the authors analyzed the real-time video streaming quality for QoS with 802.11e EDCA (enhanced distributed channel access) specification over 802.11 DCF for MPEG-4 (ffmpeg) in MANET/VANET environment with random mobility and multihop scenario. The results show that 802.11e EDCA is better performing for frame delay, frame delivery, frame jitter, PSNR, and MOS over 802.11 DCF with very low and medium mobility.

Jiménez, A. et al. (2018): Local area networks dependent on the IEEE 802.11 standard speak to the biggest arrangement of remote system innovation around the world, and such systems have quickened the development of mixed media applications. This kind of substance requires unique settings to guarantee nature of administration, particularly on account of constant video spilling. As of now, the cutting edge for the most part centers around tackling the QoS issue over conventional IEEE 802.11 systems [14], yet few works attempt to assess the presentation of QoS instruments over programming characterized WLAN arrange situations. The software-defined networking (SDN) engineering improves, rearranges, and programs the system and the board as indicated by explicit needs, making the structure and sending of new control arrangements simpler and quicker to actualize. This paper exhibits an examination of throughput and defers measurements conduct of QoS systems from IEEE 802.11 standard, for example, confirmation control, actualized over a reenactment situation dependent on NS-3 coordinated with SDN.

J. Y. Lee et al. (2012): In this paper, an exact prototype of EDCA model was developed which covers almost all the important aspects of QoS [15]. From the verification of the prototype, we can find the stable throughput and relate the identification of the proposed model with different models, and their results were also compared.

Sunghyun et al. (2003): In this article, the author has analyzed a contention-based access mechanism, i.e., EDCF alongside with original MAC protocol [14]. This layer provides different access to frames and priorities to them. In this paper, contention-free burst allows MAC transmission at a single opportunity, and during this process, overall system performance is increased without much loss of frames and data.

Crow, Brian P. et al. (1997): The main points of this paper were to provide a thorough study of both the DCF and the PCF over a CFP repetition time period. The detailed model is providing a streamlined data which can be sent over DCF without any delay and prioritize the data as per access categories. It also includes free error channel which is transmission data over a wireless channel [16]. Data prioritization is also done where idle voice stations can send data with packets for sending the data.

Navpreet Kaur et al. (2012): In this paper, the result shows the improvement for the QoS [17], and this came at a decline in the quality of data as shown in Fig. 6.

The delivery of data to access categories for different priorities of data, i.e., higher priority, will transmit the data first as compared to other, and overall system performance was improved as compared to other existing techniques.

Karthik Chowdhary et al. (2016): This paper proposes a new D-cast method to reduce the peak-to-average power ratio (PAPR) by minimizing the overall bit error rate as well as the out-of-band interference (OBI) in the orthogonal frequency division multiplexed signals [18]. One of the main drawbacks of OFDM signals is that they suffer from high PAPR. Although there are multiple PAPR reduction techniques, comparisons have shown that the companding technique and D-cast method are more effective in the reduction of PAPR to improve the signal quality in addition to being less complex. This paper includes the comparison between various PAPR reduction methods and shows how the proposed method turns out to be the best one.

Navdeep Kumar et al. (2012): In this paper, different techniques based on convolution coding (CC) and techniques based on Wi-Max system [19]. In this article, the number of symbols in source encoded is more in designed manner to overcome the objectives of the paper, i.e., error detection and correction, which depends upon signal-to-noise ratio, and hence, the paper quality and high bit error rate will be achieved with less SNR.

Shekhar Verma, et al. (2016): The paper proposed a new algorithm which depends on selection criteria of performance of a network by possibility of winning of a station and possibility of collisions in the station which is calculated by taking path account the contention window and AIFS. By this algorithm, maximum data can easily be transmitted between the nodes to the destination without any data loss [20]. We have

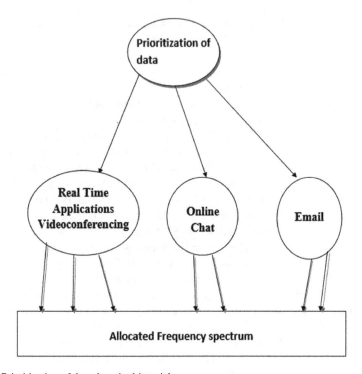

Fig. 6 Prioritization of data done in this article

taken four cases from the station where we are getting the maximum possibility, and there we will send the data first and will be having the better QoS. We have analyzed that by using this algorithm, i.e., 90% data has been transferred successfully to the next station.

Marko Hännikäinen et al. (2016): The design objective has been to develop a simple, multimedia service capable protocol that provides sufficient medium utilization efficiency and guarantees quality of service (QoS) parameters.

Eko F. Cahyadi et al. (2019): The EDCA gives a lot better value in total end-to-end delay and jitter, especially in real-time services compared with the DCF method. This result is majorly affected by the classification process that is deployed in EDCA, whereas for traffic, such voice and video gain more advantages to utilize the network resources than the best effort. On the other hand, since the DCF scenario does not employ a traffic prioritization, it has a better performance in throughput and packet loss parameters compared with EDCA, primarily in the number of nodes 16 upwards. This case is ultimately caused by the saturation in the link that causes an overload in the channel, so that the DCF method can perform a lot better.

Prabhjot Kaur et al. (2016): Video data packet must be sent earlier to continue its streaming as compared to a data packet of text mail. This packet priority is calculated using fuzzy terms—high, medium, and low.

Mohammed A. Al-Maqria et al. (2016): The proposed scheme is a feedback-based scheme in which the station sends information with each packet sent about the next arrival time of the next frame. Based on this, in each SI period, the QAP will selectively poll stations that are ready to transmit in order to reduce the poll overhead.

3   Conclusions

The outcomes got from simulation from various techniques demonstrate that EDCF gives productive instrument to support separation alongside gives nature of administration to the WLAN [21]. In any case, the outcomes include some significant pitfalls of abatement in nature of the lower traffic data. The securing of the radio channel by the higher-need traffic is considerably more forceful than for the lower need. Higher-need traffic profited, while lower-need traffic endured [22]. As far as general execution (under the utilized recreation conditions in this specific investigation of QoS of wireless LAN), DCF performs imperceptibly well than EDCF. This occurs because of reason that in EDCF instrument, every AC capacity acts like a virtual station for medium access, so more crash will be normal for EDCF situation. In any case, as far as quality of administration for postpone touches applications (like videoconferencing), EDCF beats DCF.

References

1. Syed, I., Roh, B.-h.: Delay analysis of IEEE 802.11e EDCA with enhanced QoS for delay sensitive applications. In: 2016 IEEE 35th International Performance Computing and Communications Conference (IPCCC). IEEE (2016). https://doi.org/10.1109/PCCC.2016.7820668
2. Jansang, A., Phonphoem, A., Paillassa, B.: Analytical model for expected packet delay evaluation in IEEE 802.11e. In: International Conference on Communications and Mobile Computing (2009)
3. Khaki, Z.G., Bedi, H.S.: Transient correction using EDFA: in-line optical fiber with feedback. In: International Conference on Computing Sciences (2012)
4. Lakrami, F., El Kamili, M.: An enforced QoS scheme for high mobile adhoc networks. In: Proceedings of the International Conference on Wireless Networks and Mobile Communications (WINCOM15), Marrakech, Morocco, 20–23 Oct 2015. IEEE Explorer (2015)
5. Liu, X., Zeng, W.: Throughput and delay analysis of the IEEE 802.15.3 CSMA/CA mechanism considering the suspending events in unsaturated traffic conditions. In: 2009 IEEE 6th International Conference on Mobile Adhoc and Sensor Systems. https://ieeexplore.ieee.org/document/5337030
6. Chauhan, N.S., Kaur, L.L.: Implementation of QoS of different multimedia applications in WLAN. Int. J. Comput. Appl. **62**(8) (2013)
7. Jansang, A., Phonphoem, A., Paillassa, B.: Analytical Model for expected packet delay evaluation in IEEE 802.11e. In: 2009 WRI International Conference on Communications and Mobile Computing, Yunnan (2009)

8. Al-Maqria, M.A., Othmana, M., Mohd Alib, B., Mohd Hanapia, Z.: Packet-based polling scheme for video transmission in IEEE 802.11e WLANs. In: The 7th International Conference on Ambient Systems, Networks and Technologies (ANT 2016)
9. Vijay, B.T., Malarkodi, B.: Improved QoS in WLAN using IEEE 802.11e. In: Procedia Computer in Twelfth International Multi-conference on Information Processing-2016 (IMCIP-2016)
10. Barry, M., Campbell, A.T., Veres, A.: Distributed control algorithms for service differentiation in wireless packet networks. In: Proceedings IEEE INFOCOM 2001. Conference on Computer Communications. Twentieth Annual Joint Conference of the IEEE Computer and Communications Society (Cat. No. 01CH37213), Anchorage, AK, USA (2001)
11. Shklyaeva, A., Kubanek, D.: Analysis of IEEE 802.11e for delay sensitive traffic in wireless LANs. In: Proceedings of the Sixth International Conference on Networking (ICN'07)
12. Vardakas, J.S., Logothetis, M.D.: End-to-end delay analysis of the IEEE 802.11e with MMPP input-traffic. In: 2009 International Symposium on Autonomous Decentralized Systems, Athens, pp. 1–6 (2009). https://doi.org/10.1109/ISADS.2009.5207380
13. Jiménez, A., Botero, J.F., Urrea, J.P.: Admission control implementation for QoS performance evaluation over SDWN. In: 2018 IEEE Colombian Conference on Communications and Computing (COLCOM), Medellin (2018)
14. Choi, S., Del Prado, J., Sai Shankar, N., Mangold, S.: IEEE 802.11e contention-based channel access (EDCF) performance evaluation. In: IEEE International Conference (2003)
15. Lee, J.Y., Lee, H.S.: A performance analysis model for IEEE 802.11e EDCA under saturation condition. IEEE Trans. Commun. 57(1), 56–63 (2009)
16. Crow, B.P., Widjaja, I., Kim, J.G., Sakai, P.T.: IEEE 802.11 wireless local area networks. IEEE Commun. Mag. 35, 116–126 (1997)
17. Singh, H., Kaur, N., Singh, B.: Real time applications and delay analysis of IEEE 802.11e. In: National Conference on Innovative & Emerging Technologies in Computing Methodology (IECM-2012)
18. Chowdhary, K., Singh, H.: Performance comparison of companding techniques and new D cast method for reduction of PAPR in OFDM. Int. J. Control Theory Appl. 9, 217–222 (2016)
19. Kumar, N., Singh, T.: Overview of performance of coding techniques in mobile WiMAX based system. In: International Conference on Technical and Executive Innovation in Computing and Communication (TEICC 2012)
20. Verma, S., Singh, B., Singh, H.: A survey on enhancing the quality of service in wireless sensor networks. Int. J. Control Theory Appl. (2016)
21. Cahyadi, E.F.: The QoS trade-off in IEEE 802.11n EDCA and DCF for voice, video, and data traffic. In: EEE International Conference on Consumer Electronics—Taiwan (ICCE-TW), At Yilan, Taiwan (2019)
22. Sharon, O., Alpert, Y.: Coupled IEEE 802.11ac and TCP goodput improvement using aggregation and reverse direction. Wirel. Sensor Netw. 8, 107–136 (2016)

Voice-Controlled IoT Devices Framework for Smart Home

Ravi Ranjan and Aditi Sharma

Abstract The escalating growth in automation industries and technologies has revealed astonishing opportunities for digitally controllable devices that can change the houses that we live into a smart home automation having wirelessly controlled smart appliance with the help of radio transmissions, Bluetooth, or over Internet, where the owner can access any device of his home from any location through mobile and can control the basic operations. Internet of Things (IoT)-based home electronic appliances control system which has been designed with the help of ESP8266-based NodeMCU IoT development board has been discussed in this article. The voice control system can command the basic operations like switching on and off any home electronic device from anywhere using Blynk application installed on the mobile phone, or it can be commanded over Google Assistant also. The model provides an affordable system for home automation by using the smart phones they already have. The voice-controlled automated homes will be able to make the life of differently abled persons. Voice-automated system can further be improved by using machine learning to detect the patterns of users to provide them better service by understanding their habits and provide them services without even asking daily.

Keywords Blynk · Google Assistant · IFTTT · IoT · NodeMCU · Voice control · Home automation

1 Introduction

Internet of Things (IoT) is a network of interconnected and uniquely identifiable computing devices embedded in everyday artifacts [1]. It is an idea by which devices can be monitored and analyzed from a remote location. The devices that IoT can make communicate can belong to a same house, making a simple house into a digitally connected, intelligent and smart home, and it can enhance the security of the house by

R. Ranjan · A. Sharma (✉)
DIT University, Dehradun, India
e-mail: aditi.sharma@dituniversity.edu.in

R. Ranjan
e-mail: ravi.ranjan@dituniversity.edu.in

© Springer Nature Singapore Pte Ltd. 2020
P. K. Singh et al. (eds.), *Proceedings of First International Conference on Computing, Communications, and Cyber-Security (IC4S 2019)*, Lecture Notes in Networks and Systems 121, https://doi.org/10.1007/978-981-15-3369-3_5

Fig. 1 Generalized IoT
architecture

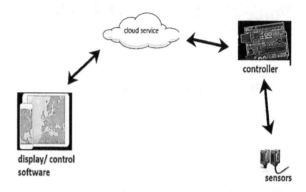

detecting any intruder and alerting the owner on mobile device through message or notification. IoT-based smart homes can serve many purposes and can save electricity, and you can switch off devices like fan, AC, geyser, or lights from your office, if you forget to do it while leaving the home [2]. IoT architecture may vary from system to system depending on its application. A generalized architecture of IoT is shown in Fig. 1. IOT devices need not to always maintain a two-way communication between all the components of the IoT system, and it may also be one way or may vary from system to system. IoT has progressed toward becoming way of life of an individual with extraordinary potential [3]. Within the past few years, the "home network" has begun to evolve, as the personal computer(s) are connected to various other home appliances, such as a security system, messaging system, and heating/lighting.

IoT has opened ways to conceal all the necessities of human dealings using such a "home network", and it is now possible to allow a home device to control and communicate with various other home devices. A few voice-controlled home automation devices exist in the market, but they are not as affordable for the common people, like "Amazon Echo" second generation cost Rs. 14,999 in India [4]. The voice-controllable home automation devices can become a support system for physically disabled persons, and their lives can be made a little simpler [5, 6]. Voice-controlled home automation devices are developed as a project by us which cost only Rs. 3350.

This device can be customized too as Google assistance works on machine learning, so it updates itself on the basis of the input that is being provided to it. Since all the commands will be communicated through Google assistance only, it can store and analyze the pattern of commands being passed into the Blynk app to communicate to the devices to perform the basic operation. We can also optimize this by applying data mining techniques to find the pattern and that can help to make the home automation more user-friendly [7]. Pattern analyzing will help to find the daily time of switching on or off some of the daily consumable devices, like geyser, as it is usually switched on by the user in the morning and usually everyone switches it off while going to the office, so after analyzing the regular pattern, the application can switch the geyser on and off on regular intervals on weekdays like switch on at 5 am and switch off at 9 am, and on weekends either on following pattern of weekend or if there is no regular timing then wait for users' command on weekend. Similarly, for other devices too like AC in summers or tube lights, etc.

IoT has been introduced in our day-to-day life through smart refrigerators, smart washing machines, and many other appliances [8]. IoT also covers the personal physical devices like fitness watch or even the mobile phone to the Internet and analyzing them using a Web interface at a remote location. As per the survey conducted in 2018, people are trusting IoT-based devices, even though they have concern for security, but the idea of smart home feels comfortable to the users [1]. In this article, a hardware design has been developed that can control home electronic appliances such as fans, lights, AC, door, alarms, electronic door locks, and television over the Internet with the help of two controls for a single system. For a demonstration purpose, an eight-channel relay system has been designed which can switch analog or digital appliances and is controlled by an ESP8266 System on Chip (SoC)-based microcontroller named as NodeMCU [9]. User interfacing devices such as mobile or laptop sends the data to the cloud/server, and then, this data is analyzed on server and is transferred to controller which further performs an appropriate action in response to the command received from the cloud/server. Lastly, the voice control system is built over an IFTTT platform using Google Assistant and Blynk.

In the next section, the literature of home automation devices has been discussed, followed by the proposed model containing software design and hardware design. Section 4 contains the analysis of the working model. And in the last section, the model has been concluded with the future scope for the project.

2 Related Work

Even though the idea of smart homes is relatively new for common people, extensive research work has been conducted in the field in past few years. The concept commenced when the idea of Internet of Things and Internet of Systems took an upsurge in the researchers. With the commencement of fourth revolution of industry, even smart factories have been set up in Germany. In [10], researchers' procurement and investigated different sensor information can be utilized as crosswise over smart homes. They proposed a design for separating relevant data by investigating the information procured from different sensors and provide context-aware services. In [11], they reduced power dissipation and proposed methods to conserve energy in smart homes utilizing IoT. It utilizes cameras for perceiving human exercises through picture handling methods. The need for common standards and protocols with the aim to achieve sustainable IoT-based applications has been discussed by authors in [12] focusing particularly on smart homes. The authors in [13] have discussed about difficulties and issues that arise in IoT-based smart home systems by conducting a survey and have also provided possible solutions. In 2002, Sriskanthan used the concept of Bluetooth for connecting the devices for home automation, the concept was good but since the range of Bluetooth is a few meters, so it never got popular even being a novel technique. Khusvinder et al. in 2009 have used Bluetooth and Wi-Fi for home automation using sensors [14]. Mittal et al. in 2015 have proposed an Arduino-based voice-controlled home automation, in which they proposed to customize the application for users' voice only [15]. Hamdan in 2019 proposed a dual model for IoT-based home automation, in which appliances can be both monitored and controlled through even voice signals and gestures too. The model has two phases, in first, smart phones are used with virtual switches and sliders to monitor and control the appliances, and in second phase, chat-based system is used to communicate commands either by text or by audio message [16]. Mahamud also in 2019 has created a Web portal for passing the commands to the Wi-Fi enabled Web server at the home to control the devices [17]. Amrita and Ansari in [18] have developed a home automation system that is voice controllable, where they used raspberry pie as controller, they also used Blink and IFTTT, but they have used the mic to record the voice commands that follow specific operations. Since they do not have large dataset to understand every command, it has limited functionality. Though similar works are carried out in [16–18], we have integrated Google Assistant with the existing systems to make it voice controllable. The model and working have been discussed next.

3 Proposed Model

We have proposed a home automation system with voice control, but to reduce the cost, as we need to make the system more affordable for common people, we have used the Google Assistant, rather than developing our own natural language processing system. This will make the interface more user-friendly and accurate, since the Google Assistant has been trained with large dataset, and it improves every day with the input of different styles it is receiving every day. The complete system may be divided into two segments: system hardware and system software. The hardware and software components of the proposed system are discussed next and shown in the form of block diagram in Fig. 2.

Fig. 2 Block diagram of proposed system

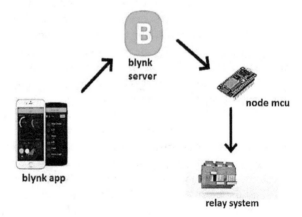

3.1 Hardware Design

The system hardware comprises of 8 relays, 12 V constant power supply source, NodeMCU board, 7805 voltage regulator, heat sink, ULN2803 IC, PC817 Opto-coupler IC, and IN4007 diodes (for reverse current protection). The pin diagram of NodeMCU is shown in Fig. 3.

NodeMCU: It is an IoT development board consisting of ESP8266, SoC integrating a 32-bit microcontroller. It has 16 GPIOs and a processing unit which is operating at a clock frequency of 80–160 kHz [19]. NodeMCU is Lua script-based programmable device, and it can also be programmed by using Arduino IDE [7].

PC817: It is a 4 pin Opto-coupler IC. The purpose of this IC is to isolate the development board from the external circuitry in order to avoid loading effect on it.

ULN2803: It is an 18 pin IC, and it has eight Darlington pairs in it having eight inputs and eight outputs. The purpose of this IC in the system is to meet the current requirements of relays. Its each output provides a maximum current of 500 mA [20].

Relays: Relays serve the purpose of switching element in the system. Appliances are connected to Common (COM) and Normally Open (NO) terminals of relay.

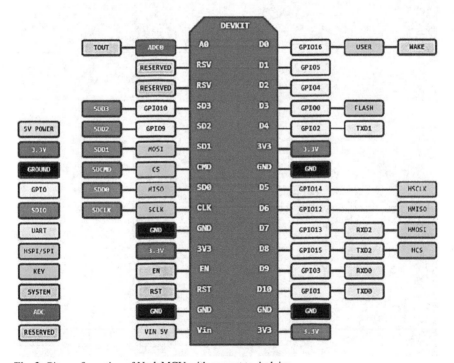

Fig. 3 Pin configuration of NodeMCU with respect to Arduino

3.2 System Software

The software's used for the system are: Blynk interface and Google Assistant, where Google Assistant provides a voice controlling ability to the system.

Blynk: Blynk is an IoT platform which has applications for both Android and iOS which facilitates building of a graphical interface as shown in Fig. 4, for controlling the system simply by dragging and dropping some widgets such as buttons, sliders, and maps. Blynk server is on different IP addresses for different countries, and for India, IP address of Blynk server is 188.166.177.186 [21].

Fig. 4 Blynk graphical user interface of our system

Fig. 5 Google Assistant interface of system

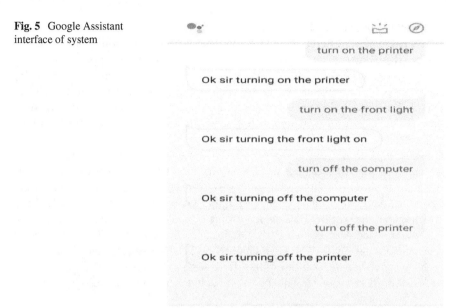

IFTTT: IFTTT basically provides a platform to synchronize or interlink two or more software in order to perform a specific task. It works as a triggering agent, i.e., whenever it receives an input through Google Assistant, it checks, if there exists any command for this text by using the conditional statements. If the message or command passed through Google Assistant has any pre-defined function set for it in the IFTTT database, then that particular action will be performed, so it sets the trigger for execution of that particular command. We have synchronized Blynk and Google Assistant for building a voice control system [22].

Google Assistant: Google Assistant voice control system is a Google account secured system. It is only operable on specific Google account on which it is configured [5]. This voice control system is built using IFTTT platform, and its interface is depicted in Fig. 5. We have used IFTTT to synchronize Blynk server and Google Assistant. In IFTTT, one must decide that how and by which means we want to trigger the system and what action has to be performed associated with that trigger. This combination of trigger and action is called recipe.

4 Analysis

A voice-activated command interface module for interacting with a variety of home-based electronic devices to allow a remotely-located homeowner to communicate with, command and control, various electronic devices over voice using Google

Assistant has been developed [7]. Devices contain ports, each port can communicate with an associated different type of communication interface with other electronic devices [6]. Model also includes a voice network communication port for receiving the voice commands from the homeowner and a data network communication port for transmitting, monitoring, and controlling information between the electronic devices and the homeowner. In operation, the command interface module is responsive to voice commands received from a remote user. Figure 6 depicts the simple working of the designed system:

NodeMCU is an active low device, and it is programmed in such a manner that whenever button sends ON signal, GPIO switches to ground which in turn activates the transistor inside the Opto-coupler IC [23]. As a result, current starts flowing from collector to emitter terminal of transistor, and via emitter, it reaches to ULN2803 which ignites the ULN2803 IC, which further allows the current to flow through the coil of relay and activates a switch, i.e., it connects Common (COM) terminal of relay to Normally Open (NO) terminal of relay [24]. Data or command from Google Assistant or Blynk user interface is transferred to Blynk server which further sends it to NodeMCU, specifying it with the help of authentication key as depicted in Fig. 7.

Fig. 6 Functional block diagram of the system

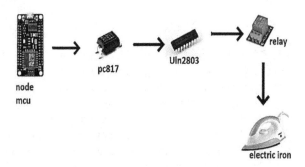

Fig. 7 Directional flow of data between the software's and the designed hardware's IoT system

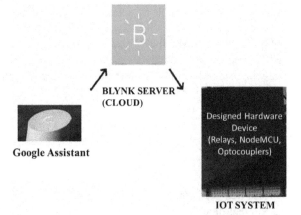

Blynk provides an easy manner to control our system from anywhere in the world by the help of an authentication key which is specific for specific project. We have authentication key by creating our project in the Arduino IDE code, which acts as a kind of address of development board for Blynk helping it to specify where the command has to be sent.

For our system, specific phrase is set for specific relay allowing Google Assistant to control it. When a specific phrase is sent to Google Assistant by either typing or sending a voice command, Google Assistant responds and gets triggered in accordance with the sent phrase. It further sends a command to Blynk server by sending a Web request through the Webhooks in IFFT using URL.[1]

5 Conclusion

Computers have been deployed in a variety of arenas to gain efficiencies, reduce cost, and increase productivity. Miniaturization and portability have made personal computers more accessible and a more valued tool in many business environments. Many existing IoT systems provide only single type of interface for home automation applications, but in this paper, we have integrated Google Assistant with the existing systems in order to make it voice controllable. Moreover, the presented system eliminates the need to install any extra software in user's mobile phone to make it voice controllable, as Google Assistant is one of the inbuilt/pre-installed applications in the Android smart phones. From the literature, it can be found that that most of the applications developed for the purpose of home automation are Android based, thereby restricting the iOS mobile users for controlling the systems. The Blynk application used in this paper allows even an iOS user to control switching of home appliances with the Blynk interface. Even an iOS user can control the appliances using voice commands by simply installing Google Assistant application in his/her phone. This application still has the restriction that through Blink app we are just able to perform the basic operations, the device should have intelligence to perform the operations by its own, like while a user gives the command for starting the geyser in the morning, the application has the intelligence to switch on the coffee machine, and similarly as a user commands to switch on the lights after getting back home, the TV or AC should start by itself by detecting the position of the user. It should be able to learn from the daily routine of the user to provide better service.

References

1. Aldossari, M.Q., Sidorova, A.: Consumer acceptance of Internet of Things (IoT): smart home context. J. Comput. Inf. Syst. 1–11 (2018)

---

[1]http://ip(blynk)/authtoken/update/pin.

2. Asadullah, M., Raza, A.: An overview of home automation systems. In: 2016 2nd International Conference on Robotics and Artificial Intelligence (ICRAI), Pakistan pp. 27–31. IEEE (2016)
3. Durani, M., Sheth, M., Vaghasia, M., Kotech, S.: Smart automated home application using IoT with Blynk app. In: 2nd International Conference on Inventive Communication and Computational Technology (ICICCT), Coimbatore, pp. 393–397 (2018)
4. Piyare, Rajeev: Internet of things: ubiquitous home control and monitoring system using android based smart phone. Int. J. Internet of Things 2(1), 5–11 (2013)
5. Gill, K., Yang, S.-H., Yao, F., Xin, L.: A zigbee-based home automation system. IEEE Trans. Consum. Electron. 55(2), 422–430 (2009)
6. Ruman, M.R., Das, M., Mahmud, S.M.I.: IoT based smart security and home automaton system. Asian J. Convergence Technol. (AJCT) (2019)
7. Muhammad, G., Rahman, S.M.M., Alelaiwi, A., Alamri, A.: Smart health solution integrating IoT and cloud: A case study of voice pathology monitoring. IEEE Commun. Mag. 55(1), 69–73 (2017)
8. Pavithra, D., Balakrishnan, R.: IoT based monitoring and control system for home automation. In: Global Conference on Communication Technologies (GCCT), Thuckalay, pp 169–173 (2015)
9. Kang, B., Park, S., Lee, T., Park, S.: IOT based monitoring system using tri-level context making model for smart home services. In: IEEE International Conference on Consumer Electronics (ICCE), Las Vegas, pp. 198–199 (2015)
10. Gupta, P., Chhabra, J.: IoT based smart home design using power and security management. In: 2016 International Conference on Innovation and Challenges in Cyber Security (ICICCS-INBUSH), Noida, pp. 6–10. IEEE
11. JeyaPadmini, J., Kashwan, K.R.: Effective power utilization and conservation in smart homes using IoT. In: International Conference on Computing of Power, Energy, Information and Communication (ICCPEIC), Chennai, pp. 0195–0199 (2015)
12. Eslambolchi, H., Hushyar, K., Tofighbakhsh, M.: U.S. Patent No. 8,654,936. U.S. Patent and Trademark Office, Washington, DC (2014)
13. Kamilaris, A., Pitsillides, A.: Towards interoperable and sustainable smart homes. In: IST-Africa Conference & Exhibition, Nairobi, pp. 1–11 (2013)
14. Pande, K., Pradhan, A., Nayak, S.K., Patnaik, P.K., Champaty, B., Anis, A., Pal, K.: Development of a voice-controlled home automation system for the differently-abled. In: Bioelectronics and Medical Devices, pp. 31–45. Woodhead Publishing (2019)
15. Mittal, Y., Toshniwal, P., Sharma, S., Singhal, D., Gupta, R., Mittal, V.K.: A voice-controlled multi-functional smart home automation system. In: 2015 Annual IEEE India Conference (INDICON), New Delhi, pp. 1–6. IEEE (2015)
16. Hamdan, Omar, Hassan Shanableh, Inas Zaki, A. R. Al-Ali, and Tamer Shanableh. "IoT-based interactive dual mode smart home automation." In *2019 IEEE International Conference on Consumer Electronics (ICCE)*, Las Vegas, pp. 1–2. IEEE, 2019
17. Mahamud, M.S., Zishan, M.S.R., Ahmad, S.I., Rahman, A.R., Hasan, M., Rahman, M.L.: Domicile-an IoT based smart home automation system. In: 2019 International Conference on Robotics, Electrical and Signal Processing Techniques (ICREST), Bangladesh, pp. 493–497. IEEE (2019)
18. Maharaja, A., Ansari, N.: Voice controlled automation using raspberry pi. Available at SSRN 3366895 (2019)
19. ElKamchouchi, H., ElShafee, A.: Design and prototype implementation of SMS based home automation system. In: 2012 IEEE International Conference on Electronics Design, Systems and Applications (ICEDSA), Kuala Lumpur, pp. 162–167. IEEE (2012)
20. Vishwakarma, S.K., Prashant, U., Kumari, B., Mishra, A.K.: Smart energy efficient home automation system using IoT. In: 2019 4th International Conference on Internet of Things: Smart Innovation and Usages (IoT-SIU), Ghaziabad, India, pp. 1–4. IEEE (2019)
21. Kodali, R.K., Jain, V., Bose, S., Boppana, L.: IoT based smart security and home automation system. In: 2016 International Conference on Computing, Communication and Automation (ICCCA), pp. 1286–1289. IEEE (2016)

22. Gaikwad, P.P., Gabhane, J.P., Golait, S.S.: A survey based on smart homes system using Internet-of-Things. In: International Conference on Computation of Power, Energy, Information and Communication (ICCPEIC), Chennai, pp. 0330–0335 (2015)
23. Sangeetha, S.B.: Intelligent interface based speech recognition for home automation using android application. In: 2015 International Conference on Innovations in Information, Embedded and Communication Systems (ICIIECS), Coimbatore, pp. 1–11. IEEE (2015)
24. Jat, D.S., Limbo, A.S., Singh, C.: Voice activity detection-based home automation system for people with special needs. In: Intelligent Speech Signal Processing, pp. 101–111. Academic Press (2019)

Comprehensive Analysis of Social-Based Opportunistic Routing Protocol: A Study

Shubham Singh, Pawan Singh and Anil Kumar Tiwari

Abstract Today's network becomes more complex due to which performances of network decrease like message delivery ratio, overhead ratio, average latency, cost, etc. A new technique comes to forward the message on the basis of social infrastructure. In this paper, we analyze and compare the result of social-based opportunistic routing protocol with different mobility traces, and we compare the BubbleRap, DLife (daily life routine routing algorithm) and dLifeComm (daily life routine community-based routing algorithm) social routing protocol with different movement models and real traces with respect to delivery ratio, overhead ratio, and average latency. Comparison is based on the real human traces and synthetic mobility model. We use Opportunistic Network Environment (ONE) simulator for result analysis. As from the simulation result, delivery ratio of BubbleRap is better than DLife and dLifeComm when node follows the community-based movement.

Keywords Mobile social network · Opportunistic routing · Daily routine · Delay-tolerant network · Network dynamics

1 Introduction

Nowadays, everyone is socially connected through the Internet like Facebook, Twitter, Instagram, and WhatsApp. Due to that, network becomes more complex, and its performance decreases in terms of message delivery ratio, latency, and cost. An investigation estimates that the quantity of advanced mobile phone clients is more than 2.5 billion of every 2019, with more than 36% of the total population to utilize a cell

S. Singh (✉) · P. Singh · A. K. Tiwari
Amity University, Lucknow 226010, India
e-mail: singhshubham070@gmail.com

P. Singh
e-mail: pawansingh51279@gmail.com

A. K. Tiwari
e-mail: aniltiwari19640@gmail.com

© Springer Nature Singapore Pte Ltd. 2020
P. K. Singh et al. (eds.), *Proceedings of First International Conference on Computing, Communications, and Cyber-Security (IC4S 2019)*, Lecture Notes in Networks and Systems 121, https://doi.org/10.1007/978-981-15-3369-3_6

phone by 2018. The report additionally expresses that the most non-gaming versatile application (named as Apps) in 2018 are those that encourage interpersonal interaction (like Facebook, WhatsApp, Google) [2]. The current delay-tolerant network directing conventions primarily center around the plans to improve the probability of finding sharp ways [18].

Mobile social network (MSN) is informal communication where people with comparative interests chat and interface with each other through their cell phone and additionally tablet [3]. Much like electronic person-to-person communication, portable mobile social network happens in virtual networks. Many electronic long-range interpersonal communication destinations, for example, Facebook and Twitter, have made portable applications to give their clients' movement and constant access from anyplace they approach the Internet. Moreover, local portable informal organizations have been made to enable networks to be worked around versatile usefulness. In Fig. 1, it shows that how mobile social network is working. All devices like mobile, laptop, and others gave request for data through Internet, and Internet provides the information from service provider.

Opportunistic Networks (OppNets) are a subset of mobile ad hoc networks (MANETs) in which versatile hubs arrange availability by means of short-run remote correspondence [11]. Routing and sending of messages is a difficult task in Opportunistic systems because of no fixed topology and the irregular network of the nodes [13]. It is unimaginable to expect to know the total system topology in opportunistic network [7].

In Fig. 2, an assortment of correspondence innovations is utilized, for example, cellular mobile communication, wired PC system, and microwave hand-off, showing that Opportunistic network consists of different heterogeneous systems. Some vehicular hubs are associated with Opportunistic network, and their portability regularly cause visit correspondence interface disturbance and enormous delay, which make topology of Opportunistic network exceptionally unique and adaptable.

At the point when the association with the Internet is not accessible during system administration exercises, an opportunistic method shows the encounter between portable human-conveyed gadgets for trading data [14]. In opportunistic networks, nodes communicate intermittently based on store-carry-forward paradigm while exploiting node mobility [1]. Sending the message from source to destination in opportunistic network requires multihop communication [10]. A social-based opportunistic routing algorithm is introduced to increase the message delivery ratio [5].

In social correspondence, cell phones can be viewed as socialization hubs in informal communities [15]. In BubbleRap, Hui et al mainly focused on defining community and centrality. As network topology changes rapidly and the node is mobile, they build a routing algorithm that uses contact to allow people to interact without network infrastructure [5].

In Fig. 3, in the event that a hub has a message foreordained for another hub, this hub first rises the message to various leveled positioning trees utilizing the worldwide positioning until it comes to a the hub which is in a similar network as the goal hub. At that point, the neighborhood positioning framework is utilized rather

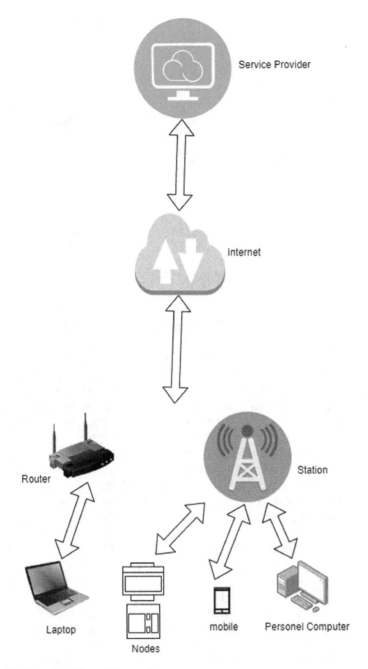

Fig. 1 Working of mobile social network

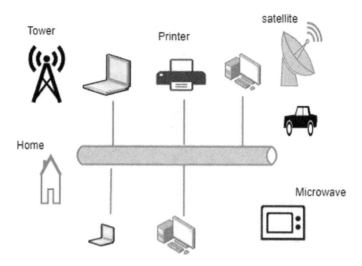

Fig. 2 Architecture of opportunistic network [16]

than the worldwide positioning, and the message keeps on rising through the nearby positioning tree until the goal has come to or the message lapses [5].

Social relationship between the nodes is very low with respect to the topology. Another important task for them is to detect the social movement in a decentralized way. Therefore, they detected different communities, out of which some people may interact with more people within the community known as hub (high centrality), and some people may interact more people with people of different communities known as gateway [17].

They defined two terms, hub and gateway, for writing their algorithm. Suppose if someone receives a packet to deliver a message to different community, then it can be accomplished through the gateway node. DLife algorithm is based on the daily routine work in which it divides daily routine work in the time interval, and it finds that how strongly they are connected with time interval; on this basis, they proposed the algorithm [8].

The main advantage of DLife over BubbleRap is that DLife topology varies less than the topology of BubbleRap [8]. DLife divides the social relationship into the time interval on the daily routine basis in which human shows that how long they are connected over a time period [8].

DLife contains two utility functions: First, Time-Evolving Contact Duration contains the social interaction among pair of users in the same daily interval of time, over consecutive days; second, utility function is Time-Evolving Contact Duration importance (TECDi) that contains the importance of user, based on its node degree and social strength toward its neighbors, in different intervals of time [8].

They also write a dLifeComm (daily life routine community-based routing algorithm) which is community-based in which they divide the community in a different intervals of time and observed that in which interval they are highly connected [8].

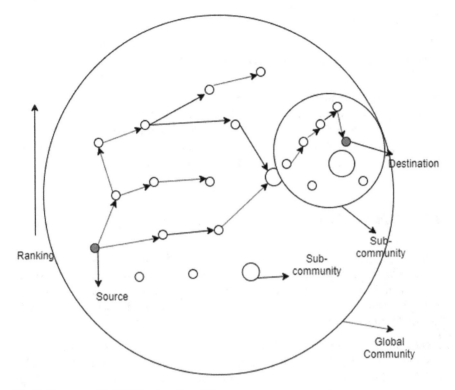

Fig. 3 Structure of BubbleRap algorithm [5]

The rest of the paper is explained as follows: Section 2 explains research contribution, and Sect. 3 describes the simulation setup, different mobility model, and real-world traces. The result is analyzed in Sect. 4. The conclusion of this paper and future work is described in Sect. 5.

2 Research Contributions

We take a social routing algorithm and movement model and real traces for our research. Our research contribution is as follows:

- We show a systematic and comprehensive comparison of social-based routing protocol, namely BubbleRap, DLife, and dLifeComm with different movement models (CAHM, map-based) and real traces (Infocom05, Infocom06, Real traces, and Cambridge) with respect to delivery probability, overhead ratio, and average latency.
- We compare the different routing algorithms through graph.

3   Simulation Setup And Description

3.1   ONE (Opportunistic Network Environment) Simulator

We use the ONE simulator for experimental result. DTN simulators only focus on forwarding message from source to destination that is on routing, but ONE simulator is a combination of routing, visualization, and mobility model. It is capable of generating node movement with different trace models [6]. It can import mobility data from real-world traces or any other. It is written in Java [6].

3.2   Mobility Model and Real-World Traces

In this paper, we use five scenarios referred to as CAHM, Map-Based Movement, Infocom05, Infocom06 and Reality dataset, in which CAHM and Map-Based Movement is a mobility model, whereas the other three are real-world datasets. Infocom05 and Infocom06 are collected by Haggle project, and Reality dataset is from the MIT Reality Mining Project.

- **Infocom05**: Dataset is generated from the 50 students who attended the Infocom student workshop. Every student is holding Bluetooth device for four days [12].
- **Infocom06**: As Infocom05 generated on a small scale, it is generated on a large scale. A total of 80 people carry Bluetooth device for five days. The area spread on three floors, and the participant is classified into four-part based on their affiliation [12].
- **CAHM**: It is the community-based movement model. Community Aware Heterogeneous Human Mobility Model (CAHM) is based on human movement and finds different community and overlap community, and based on that information, it sends the data [9].
- **Reality**: This data is generated by MIT in which students and faculty carrying 100 devices for nine months [3].
- **Map-Based Movement**: In this movement model, node movement follows the path on the previously defined map. It includes three map-based movement model: 1) Shortest Path Map-Based Movement (SPMBM), 2) Random Map-Based Movement (MBM), and 3) Routed Map-Based Movement (RMBM) [4].
- **Cambridge**: In this movement model, we use 36 nodes. Data was collected in a different location for two months while Cambridge University students doing their daily routine work [8].

3.3 Simulation Parameter for Different Movement Model

4 Simulation Result

In this part, the simulation results of social routing protocols like BubbleRap, DLife, and dLifeComm with different movement models like CAHM, Infocom05, Infocom06, and Reality are represented. Average latency, delivery ratio, and overhead ratio are used to measure performance. We use ONE simulator for our result, and parameters used in implementation are shown in Table 1.

In Fig. 4, we have shown the comparison among different social routing protocols with different movement models with respect to delivery probability. In this implementation, we run the simulation in a different scenario with changing rngSeed. This simulation result shows that BubbleRap routing protocol is better than the DLife and dLifeComm routing protocols, and also BubbleRap gives better delivery probability than CAHM movement model.

In Fig. 5, we have provided the comparison of different social routing protocols with different movement models on the basis of overhead ratio. We perform the simulation in different scenario with changing the value of rngSeed. This simulation result shows that when we use Community Aware Heterogeneous Human Mobility Model with BubbleRap routing protocol, it gives less overhead ratio than other social routing protocols.

In Fig. 6, we compare the different social routing protocols with different movement models with respect to average latency. Average latency is defined as average time taken by the message to send from source to destination. In delay-tolerant network, the social routing algorithm can bear delay, but more delay reduced the performance of the network. In this simulation result, we found that for large scenario,

Table 1 Simulation parameter settings

Parameter	Value
Simulation time	86,400 s
Interface	Bluetooth, Wi-Fi
Number of nodes	Real traces- 41, 98, 97, 36 mobility model 100
Buffer size	100 M
Transmit speed	250 k(2 mbps)
Movement model	CAHM, Infocom05, Infocom06, Reality, MapBased, Cambridge
TTL of message	3000 s
Routing protocol	BubbleRap, DLife, dLifeComm
Message size	50 K,1 M
Message interval	75, 90 s
Warm-up time	5000 s

Fig. 4 Comparison of
delivery probability

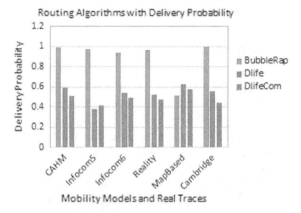

Fig. 5 Comparison of
routing protocol

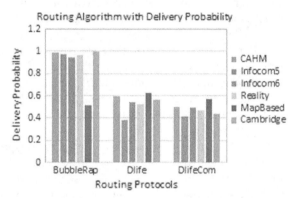

Fig. 6 Comparison of
overhead ratio

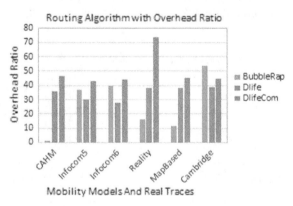

it becomes more connected through which it reduces the average latency. Here, selection of movement model has a important role on the social routing protocol.

In Fig. 7, it compares the performance of social routing algorithm with different movement models with respect to the delivery probability. Simulation result shows that BubbleRap routing protocol is better than the DLife and dLifeComm routing protocols.

In Fig. 8, it shows the comparison of Real Traces with CAHM. Before comparing any movement model with real traces, we need to set the simulation parameter as in real traces scenario. In this simulation, we match the number of contact per second, number of nodes, and world size with reality and Infocom06.

We set cell sizes 160 and 80 to compare CAHM with reality and Infocom06, respectively. So from simulation result, it shows that social routing algorithm behaves similarly in both the implementations. Hence, we can say that CAHM can represent real-world mobility scenario.

Fig. 7 Comparison of average latency

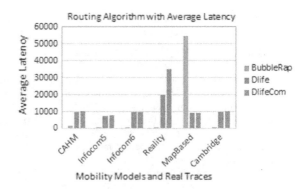

Fig. 8 Comparison of CAHM and real traces

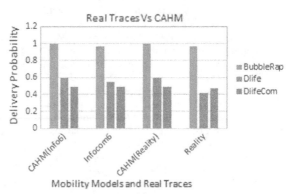

5 Conclusion and Future Direction

In this paper, we analyze the performance of social routing algorithms like Bub-bleRap, DLife, and dLifeComm, and their performance is tested with different real-world traces and mobility model. The study concludes as:

Movement model, number of contact per second, and message interval have an important factor for the performance of social routing algorithm. Opportunistic network has lower overhead ratio and average latency when node follows CAHM movement model. BubbleRap routing protocol has better performance regarding delivery ratio and overhead ratio. Delivery ratio of DLife is better than dLifeComm when we use MapBased movement model. In future, we can take other social-based routing protocols which cover the selfish problem, limited resource for node to improve the delivery ration.

References

1. Arastouie, N., Sabaei, M., Bakhshi, B.: Near-optimal online routing in opportunistic networks. Int. J. Commun. Syst. **32**(3), e3863 (2018)
2. Costantino, G., Maiti, R.R., Martinelli, F., Santi, P.: Losero: a locality sensitive routing protocol in opportunistic networks. In: Proceedings of the 31st Annual ACM Symposium on Applied Computing, vol. 14, no. 7, pp. 644–650 (2016)
3. Eagle, N., Pentland, A.S.: Reality mining: Sensing complex social systems. Personal Ubiquitous Comput. **10**(4), 255–268 (2006)
4. Hossen, S., Rahim, M.S.: Impact of mobile nodes for few mobility models on delay-tolerant network routing protocols, pp. 1–6 (2016)
5. Hui, P., Crowcroft, J., Yoneki, E.: Bubble rap: social-based forwarding in delay-tolerant networks. IEEE Trans. Mob. Comput. **10**(11), 1576–1589 (2011)
6. Keränen, A., Ott, J., Kärkkäinen, T.: The one simulator for dtn protocol evaluation, vol. 55, pp. 1–10 (2009)
7. Mendes, P., Sofia, R.C., Soares, J., Tsaoussidis, V., Diamantopoulos, S., Sarros, C.A.: Information-centric routing for opportunistic wireless networks,vol. 2, pp. 194–195 (2018)
8. Moreira, W., Mendes, P., Sargento, S.: Opportunistic routing based on daily routines. In: 2012 IEEE International Symposium on a World of Wireless, Mobile and Multimedia Networks (WoWMoM), vol. 6, pp. 1–6 (June 2012)
9. Narmawala, Z., Srivastava, S.: Community aware heterogeneous human mobility (cahm): model and analysis. Pervasive Mob. Comput. **21**, 119–132 (2015)
10. Pal, S., Saha, B.K., Misra, S.: Game theoretic analysis of cooperative message forwarding in opportunistic mobile networks. IEEE Trans. Cybern. **47**(12), 4463–4474 (2017)
11. Rashidibajgan, S., Doss, R.: Privacy-preserving history-based routing in opportunistic networks. Comput. Secur. **84**, 244–255 (2019)
12. Scott, J., Gass, R., Crowcroft, J., Hui, P., Diot, C., Chaintreau, A.: CRAWDAD dataset cambridge/haggle (v. 2009-05-29) (May 2009)
13. Sharma, D.K., Dhurandher, S.K., Agarwal, D., Arora, K.: krop: k-means clustering based routing protocol for opportunistic networks. J. Ambient Intell. Humanized Comput. **10**(4), 1289–1306 (2019)
14. Socievole, A., Caputo, A., De Rango, F., Fazio, P.: Routing in mobile opportunistic social networks with selfish nodes. Wirel. Commun. Mob. Comput. pp. 1–15 (2019)

15. Wu, J., Chen, Z., Zhao, M.: Weight distribution and community reconstitution based on communities communications in social opportunistic networks. In: Peer-to-Peer Networking and Applications, vol. 12, no. 1, pp. 158–166 (2019)
16. Wu, Y., Zhao, Y., Riguidel, M., Wang, G., Yi, P.: Security and trust management in opportunistic networks: a survey. Secur. Commun. Netw. **8**(9), 1812–1827 (2015)
17. Xiao, M., Wu, J., Huang, L.: Community-aware opportunistic routing in mobile social networks. IEEE Trans. Comput. **63**(7), 1682–1695 (2014)
18. Zeng, Y., Xiang, K., Li, D., Vasilakos, A.V.: Directional routing and scheduling for green vehicular delay tolerant networks. Wirel. Netw. **19**(2), 161–173 (2013)

An Efficient Delay-Based Load Balancing using AOMDV in MANET

Roshani S. Patel and Pariza Kamboj

Abstract Mobile ad hoc network is composed of mobile nodes without any fixed infrastructure and centralized control. Due to mobility, these nodes are free to roam anywhere resulting in temporary a topology of the networks. Multipath routing always performs better than single-path routing in MANET. We are preferring the multipath routing because it shares the load among the nodes by distributing the traffic among different alternate paths and thus reduces the possibility of congestion occurrence in the network. Balancing the traffic among various routes becomes a challenging issue in the routing because of the dynamic topology in MANET. This paper describes an efficient delay-based load balancing routing protocol using AOMDV to improve the performance of the network in terms of end-to-end delay, load balancing, and throughput. A new data_packet_counter variable is used in the routing table of every node. For every data packet forward, the counter variable of the node is incremented by one. The total load on a particular path is calculated by adding the counter for all nodes on that path. New packets are then forwarded through the less loaded path. The proposed system proved to be less complex and simple; it decreases the end-to-end delay and the packet loss ratio. Also proposed system improves the packet delivery ratio, throughput, and normalized routing load in case of increased scalability also.

Keywords Mobile ad hoc network · Multipath routing · Load balancing · Congestion adaptation · AOMDV

1 Introduction

Mobile ad hoc network (MANET) is an infrastructure-less decentralized network that is a collection of wireless mobile nodes. The nodes in the network work as hosts as well as routers and they communicate with each other using multi-hop wireless links

R. S. Patel · P. Kamboj (✉)
Sarvajanik College of Engineering and Technology, Surat, India
e-mail: pariza.kamboj@scet.ac.in

R. S. Patel
e-mail: roshani.16101993@gmail.com

© Springer Nature Singapore Pte Ltd. 2020
P. K. Singh et al. (eds.), *Proceedings of First International Conference on Computing, Communications, and Cyber-Security (IC4S 2019)*, Lecture Notes in Networks and Systems 121, https://doi.org/10.1007/978-981-15-3369-3_7

[1]. In the last few years, the usage of mobile devices in most of the applications has been increased multifold due to which significant amount of data is transmitted from these devices leading to heavy traffic load [2]. The structure of the MANET changes frequently and unpredictably due to the mobility of nodes [3]. In addition, links in the network have limited bandwidth, and nodes have limited battery power. Moreover, dense network areas in MANET face more frequent link failures and longer delays [4]. Due to these constraints, one of the key challenges in MANETs is to develop a dynamic routing protocol that can discover multiple routes between a source node and a destination node to effectively transfer the packets from the source node to the destination node.

There is an enormous need to structure an efficient routing protocol that should be completely versatile to dynamic topology and mobility, able to calculate quick and simple routes, and perform maintenance process. Single-path routing protocols mostly utilize shortest-path routing methodology, so the center of the system attracts more congestion in contrast to the border of the system. Due to this reason, it causes more overheads and devours data transfer capacity. In single-path routing, congestion becomes a routine when a network is heavily loaded [1]. To overcome this problem, multipath routing protocols are presented. Multipath routing protocol discovers all possible paths between source and destination. The data packets are forwarded using one path, and the rest of the paths are kept as backup paths to be used in case the current one is broken. In our paper, we are using ad hoc on-demand multipath distance vector routing protocol (AOMDV) [5] because it discovers loop-free and link disjointed paths with less route discovery and less overhead. AOMDV also ensures the paths select based on the shortest route [6].

Neighbor nodes in MANET share the wireless channel so finding congestion-aware routes using IEEE 802.11 MAC still remains a challenging problem [7]. In MANET, the main objective of load balancing is to distribute the traffic over multiple routes and prevent early expiration of overloaded nodes due to excessive power consumption in transferring more packets. Load balancing is a challenging issue in MANETs due to its special characteristics like dynamic topology, lack of centralized administrator, and presence of node mobility which causes more link breaks that leads to congestion and delay for the whole network.

Over several years, many load balanced ad hoc routing multipath routing protocols have been proposed and most of them are based on on-demand routing protocols [8]. They combine load balancing strategies with route discovery and prefer a less loaded path over other paths available from source to destination. The routing protocols are generally categorized into three types based on their load balancing techniques [9]:

1. When the load balancing is achieved by attempting to avoid nodes with high link delay is called as a delay-based load balancing.
2. When the load balancing is achieved by eventually distributing traffic load among various nodes are called as traffic-based load balancing.
3. By combining the features of both delay-based and traffic-based load balancing which is called as hybrid load balancing.

Different load balanced ad hoc routing protocols utilize different load metrics like active path, traffic size, packets in interface queue, channel access probability, and node delay to mention a few. Here, the term load metrics is defined as how actively a node is engaged in receiving and forwarding packets over wireless media.

In this paper, we are going to propose an efficient delay-based load balancing approach by utilizing the AOMDV routing protocol. This approach gives better performance in terms of packet delivery ratio, delay, throughput, packet loss ratio, and normalized routing load because a lightly loaded and congestion free path is selected for packet transferring by this approach.

The rest of this paper is organized as follows. Section 2 describes the related work that has been done in the field of congestion-aware adaptability and load balancing. Section 3 gives a detailed explanation of the proposed system. The results achieved from the proposed system are presented in Sects. 4 and 5 conclude this paper.

2 Related Work

A wireless MANET is a self-organized and rapidly deployable network. Due to the flexible nature of MANET, it attracts various real-world applications having dynamic network topology. At the same time, MANET also has various weaknesses such as limited battery power, bandwidth, and computational power. Finding an efficient routing protocol in MANET becomes a very crucial task because of the dynamic topology and constraint battery resource of the nodes. This problem becomes more crucial in dense network areas due to recurrent link failures in MANET [4]. This front of MANET has attracted the attention of the research community, and hence the research is going on multipath routing, load balancing, congestion control, and congestion adaptive techniques.

The single-path routing protocols use a shortest-path routing approach, so the center of the network gets more congestion compared to the perimeter of the network. Many congestion awareness and adaptive mechanisms based on multipath have been proposed in the literature. This section briefs the existing work done in congestion-aware adaptability in multipath routing for MANET.

Authors proposed a path efficient ad hoc on-demand multipath distance vector (PE_AOMDV) [9] routing protocol which computes the load on a node based on the number of routes passing through that node. They employed two variables for each node, one is active path (AP) threshold representing the maximum number of paths passing through a node, and second is AP counter representing the active number of paths passing through a node currently. AP counter and AP threshold are added to the routing table and route request packets of the PE-AOMDV, respectively. A node becomes eligible to broadcast route request (RREQ) packet if and only if its AP counter value is below the RREQ packets' AP threshold value. For every new path establishment communication, the AP counter of the node is incremented by one and this node then rebroadcasts the RREQ packet, if the number of hops in the RREQ packet is less than or equal to the last hop count recorded in the routing table. This way, the overloaded nodes are avoided from becoming the parts of the paths, and utilizing this scheme an on-demand routing protocol is able to distribute the traffic load evenly on multiple nodes in the network. Therefore, this technique exhibits a better performance in terms of packet delivery ratio, throughput, and normalized routing overhead.

Authors in [10] introduced a routing strategy for MANET named as multipath load balancing technique for congestion control (MLBCC) to efficiently distribute load among different paths. In this strategy, two major functions are performed during the transmission of information over the multipath. At the first stage, congestion is identified by utilizing the arrival rate and the outgoing rate of packets at a specific interval time T. And at the second stage, gateway nodes are chosen by utilizing the link cost and the path cost to productively distribute the load by choosing the most optimum path. The gateway nodes are the relay nodes of the established reliable paths. At any point of time, when the load on a particular path crosses a maximum threshold value (indicates over-burdened path), the future traffic is adjusted by the gateway nodes. The packets are divided, equally distributed, and sent through various arbitrarily selected paths in a round-robin manner. This way, the accessibility of paths and nodes over these paths are determined for transmission with an assurance of equal load distribution among all the available paths.

Tashtoush et al. [11] have presented a Fibonacci multipath load balancing (FMLB) protocol that balances packets' forwarding over multipath passing through these mobile nodes by utilizing Fibonacci succession. FMLB protocol is responsible for adjusting the packet transmission over these chosen paths and selecting the paths as per their hop counts. This protocol finds numerous routes between the source and the destination. Paths with less number of hops are considered first. A Fibonacci weight is given to each of these paths. The source node forwards its packets over these chosen paths based on their Fibonacci weight. So, it mitigates congestion in an ideal way. This appropriate distribution improves the delivery ratio and also reduces the clog.

The downside of the protocol FMLB in [11] is improved upon by the authors in [12] using Fibonacci sequence-based load balancing routing protocol which is named as congestion-aware Fibonacci sequence-based multipath load balancing (CA-FMLB). The authors portray this protocol in three stages: route discovery phase, load balancing phase, and route maintenance phase. In the route discovery phase, when a source node broadcasts RREQ toward the destination, it saves the sending time in the routing table and when the route request (RREP) packet arrives at the source node, its receiving time is also saved in the routing table. Based on these RREQ's sending time and RREP's receiving time, a round-trip time (RTT) is determined. In the load balancing phase, the round-trip time proves more suitable in deciding the congestion status in the network. The majority of the routing protocols choose the best route according to the minimum number of hops. If a route desires less number of hops than other routes between the same pair of source and destination, then this route is prescribed to other future transmissions. It is assumed that the route with a few numbers of hops takes less time. But this does not hold true in every case. In case the intermediate nodes on such shortest paths are already participating in prior communications, then this assumption might not withstand. Therefore, an RREQ packet transmitting through such a busy shortest path might face delay in arriving at the destination node. To overcome such issues, round-trip time proves more appropriate for the selection of a route. Utilizing round-trip time, traffic jam on any route can be determined in a fair manner. Multiple node-disjoint paths between a source and destination pair are sorted in increasing order of round-trip time. Data packets of incoming traffic are then distributed over these paths using the Fibonacci sequence numbers.

A congestion awareness and adaptive routing for MANET have been proposed in [13, 14]. The congestion adaptive routing protocol (CRP) handles the congestion proactively at the very first place before it occurs and becomes adaptive if it still happens. Each node warns the previous node on the route if there is a chance of its becoming congested. This previous node then uses a "bypass" route to the first non-congested node on the primary route, thus bypass the potential congestion area. All incoming traffic onwards is split over these two primary and bypass routes, thereby lessening the chance of congestion occurrence effectively.

In [15], load balanced congestion adaptive routing (LBCAR) has been proposed as a solution to both congestion adaptive routing and load balancing simultaneously. Two metrics, link cost and traffic load density associated with a route are employed by LBCAR, to find out the congestion status. Based on these metrics, this protocol selects a route with low traffic load density and maximum lifetime for transmission. Besides, incoming traffic is split and forwarded through the non-congested routes, making less traffic coming to the congested node. By diverting the traffic load from a congested node to less busy nodes congestion can stay away, and in turn, the traffic load is balanced. At the same time, the probability of packet loss is also reduced.

Authors in [16] presented a novel approach of load balancing. This technique is conceptualized on estimation routing based on queue length and data rate adaptation which was implemented on a multipath routing protocol, i.e., AOMDV. A sender sends data over multiple disjoint paths provided by AOMDV. With every transmission, it measures end-to-end delay as well as acknowledgment delay difference during normal time (without congestion). Congestion was introduced in the network scenario, and again acknowledgment delay difference was computed by the sender. In case of the increased value of acknowledgment delay difference, the sender decreases its data transmission rate. This bandwidth estimation strategy through acknowledgment delay difference is employed in every TCP sender node and as indicated by it every sender node changes its essential data sending rate and thus minimizes congestion in the system. The second methodology for congestion control is a dynamic queue management strategy which is implemented at every node of a network to limit the congestion due to queue flood. These two strategies help in reducing congestion, and at the same time, the level of receiving data in the network is increased. Additionally, the average end-to-end delay of the network is also decreased.

A new protocol, load balancing AOMDV (LB-AOMDV) has been devised by authors in [17] for achieving a better load balancing mechanism. They extended the format of the RREP packet by including another field called buffer_size for each route which indicates the traffic load on the route. The traffic load is referred to as the sum of buffer_size of intermediate nodes falling on the route between the source and the destination. At a point, when an adjacent node gets an RREP packet, buffer_size field of RREP is incremented with the size of the buffer of the node. When the RREP packet is received by the source node, it divides the value of the buffer-size field by the hop count of the route through which it has been received in order to find out the congestion level. Every node sorts the list of the routes in ascending order as per the value of the buffer size of each route. Every node sends data packets through a route with a minimal buffer size. The LB-AOMDV protocol selects three paths between the source and the destination node. The packets sent by source node are scheduled among these three paths according to the round-robin (RR) algorithm which works better than AOMDV in terms of average delay and packet delivery ratio.

The LB-AOMDV multipath routing in [17] has been stretched out in [18] by utilizing the quality of service (QoS) parameter to make it suitable for multimedia traffic. QoS routing requires not only discovering a path from a source to a destination but a path that fulfills the end-to-end QoS requirement, often given in terms of bandwidth, delay or loss probability. The objective of QoS routing in MANET is to optimize network resource utilization while satisfying specific application requirements. Quality of service is more difficult to accomplish, especially in ad hoc networks in comparison with their wired counterparts, because the wireless bandwidth is shared among adjacent nodes and the network topology changes unpredictably as the nodes move. Trouble in supporting QoS in MANET environments is due to node mobility, routing overhead, and limited battery life. Achieving an application-specific QoS constrained from source to destination during transmission is the objective of this new protocol called QLB-AOMDV. To exchange the crucial information for achieving QoS specification, RREQ messages were used. Every node in the system estimates the quality of links with its one-hop neighbors. A node can evaluate the link delay by utilizing the information in the RREQ message. For that reason, the authors restructured the RREQ packet by adding two new fields which indicate the received time of the packet (Tr) and the transmission delay of the packet (delay). Every node finds the route_list fields that fulfill the end-to-end QoS requirements by the ascending value of delay. Every node forwards the data packets by utilizing the path with the minimum delay.

3   Delay-Based Load Balancing in MANET

Many routing protocols in the literature concerned with balancing traffic load are yielding many performance benefits but very few have commented in-depth about the adverse effects of improper distribution of load. Improper distribution of load can cause congestion in the network which impacts network parameters like lessen throughput, low packet delivery ratio, and less average end-to-end delay.

Our proposed system delay-based load balancing in MANET works on two principles, firstly multipath finding, and secondly selection of less congested paths out of the identified paths. Distribution of traffic load is accomplished in a fair manner to achieve an efficient load balancing. The proposed system is illustrated with a flowchart in Fig. 1 followed by a detailed explanation of its functioning.

Step 1: Route discovery process is done as per AOMDV routing protocol, and all possible paths between source and destination are found out.
Step 2: After finding all possible paths between source and destination, total_load on each path is computed which was set to 0 initially.
Step 3: One additional field, i.e., the data_packet_counter was added in the routing table of each node. After that one path out of the identified path list, starting from the source node is selected.
Step 4: Source node then sends a data packet to its next one-hop neighboring nodes. When a one-hop neighboring node receives a data packet, the data_packet_counter

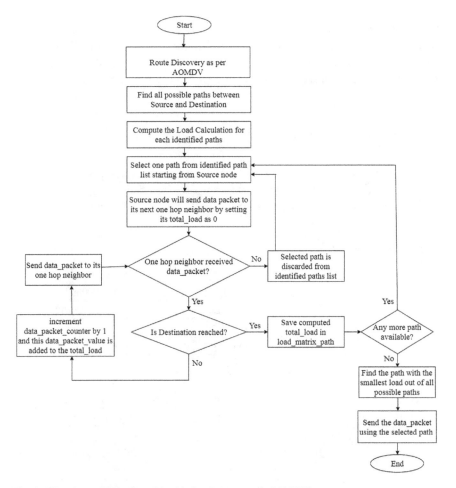

Fig. 1 Flowchart of delay-based load balancing system in MANET

of the receiving node is incremented by one and this value of data_packet_counter is added to the total_load. And this process is continued for each intermediate node on the path until the data packet reaches the destination.

Step 5: Once the data packet is reached to the destination node, the value of total_load is transferred to the load_matrix_path which indicates the total load on the path. Repeat steps 4 and 5 for each identified path.

Step 6: After finding the load on each identified path, find the path with the smallest load out of all possible paths.

Step 7: Send data_packet to the destination using this selected path.

4 Experimental Results

The proposed system is implemented in simulator NS-2.35 [19], and the implementation results are compared with that of the existing protocol PE-AOMDV. Evaluation parameters such as packet delivery ratio, throughput, delay, and packet loss ratio are taken into account. Simulation parameters are given in Table 1.

4.1 Packet Delivery Ratio

Packet delivery ratio can be represented as the ratio of successively received packets by a destination to the transmitted packets by a source node during the simulation time. Table 2 and Fig. 2 show that when the node scalability is increased, the packet

Table 1 Simulation parameters

Parameter	Value
Simulator	NS-2.35
Simulation area	1000 * 1000 m
Number of nodes	25, 50, 75, 100, 125
Source type	CBR
Transmission range	250 m
Simulation time	150 s
Packet size	512 bytes
Propagation radio model	Two-way ground
Simulation model	Random waypoint
Antenna	Omni
MAC layer	IEEE 802.11b
Maximum speed	1 m/s
Routing protocol	AOMDV

Table 2 Packet delivery ratio (in %)

No. of nodes	PE-AOMDV	Proposed system
25	95.67	99.28
50	96.32	99.12
75	95.81	99.17
100	93.09	99.03
125	93.83	99.02

Fig. 2 Number of nodes versus packet delivery ratio (in %)

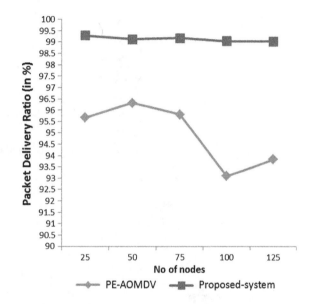

delivery ratio of the proposed system decreases by a small fraction but remains more in comparison with PE-AOMDV. Increased packet delivery ratio is the consequence of selecting a lightly loaded path and also less packets are dropped due to multipath.

4.2 Delay

Delay is defined as the time taken to forward every data packet from source to the destination. More end-to-end delay is an indication of congested path due to heavy traffic load. Table 3 and Fig. 3 show that the delay of the proposed system is maintained low compared to PE-AOMDV when the node scalability is increased. This faster transferring of data packets happened due to the fact that the selected path is less congested because of light traffic load and also load balancing is done by distributing packets over all possible paths between a source and destination pair.

Table 3 Delay (in seconds)

No. of nodes	PE-AOMDV	Proposed system
25	99.75	99.75
50	128.42	99.70
75	192.69	99.71
100	148.33	99.68
125	196.32	99.69

Fig. 3 Number of nodes versus delay (in seconds)

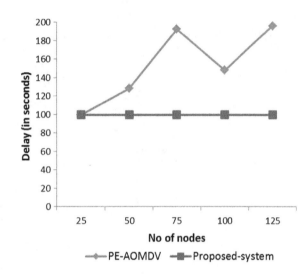

4.3 *Throughput*

Throughput measures the rate of successful message delivery over a channel. It increases when connectivity is better. Throughput (in kbps) is calculated by how many packets from the source are received at the destination during total simulation time. Table 4 and Fig. 4 show that the proposed protocol's throughput is maintained higher in case of increased node scalability also. On the contrary, the throughput of PE-AOMDV protocol becomes lesser when node scalability is increased.

Table 4 Throughput (in kbps)

No. of nodes	PE-AOMDV	Proposed system
25	213.13	361.31
50	227.09	337.62
75	230.86	334.56
100	208.08	355.94
125	192.41	361.43

Fig. 4 Number of nodes versus throughput (in kbps)

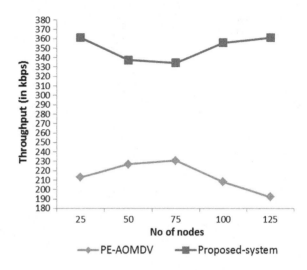

---PE-AOMDV ---Proposed-system

Table 5 Packet loss ratio (in %)

No. of nodes	PE-AOMDV	Proposed system
25	4.32	0.71
50	3.67	0.88
75	4.18	0.82
100	6.90	0.97
125	6.16	0.93

4.4 Packet Loss Ratio

Incorrect routing information, mobility, congestion, and link failure are the main reasons behind the packets' drop.

From Table 5 and Fig. 5, we conclude that there are fewer packet drops compared to existing protocol because data packets are routed through multiple paths and also these selected paths are less congested.

4.5 Normalized Routing Load

Normalized routing load is the total number of routing packets transmitted per data packet delivered to the destination. It is calculated by dividing the total number of routing packets sent (includes forwarded routing packets as well) by the total number of data packets received. Table 6 and Fig. 6 show that there is less overhead in the proposed system compared to PE-AOMDV as less number of RREQ and RREP

Fig. 5 Number of nodes versus packet loss ratio (in %)

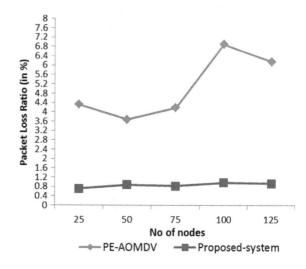

Table 6 Normalized routing overhead

No. of nodes	PE-AOMDV	Proposed system
25	1.411	0.711
50	2.433	1.537
75	3.682	2.367
100	5.255	3.234
125	6.367	4.120

Fig. 6 Number of nodes versus normalized routing overhead

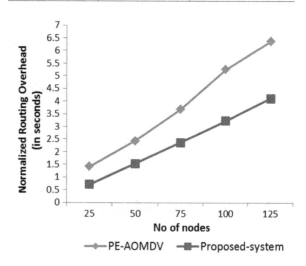

packets are forwarded during communication because of equal load distribution among multiple creates less chance of link breaks, hence routes re-establishing happens rarely.

5 Conclusion

An appropriate load balancing which ensures the distribution of traffic load considering the shortest paths as well as less congested paths is very crucial in MANET. Improper distribution of the load results in the occurrence of congestion which may lead to long delay in packet forwarding through the route. This sometimes even result in packet loss, therefore, degrades the network performance too.

This Delay-Based Load Balancing algorithm is based on AOMDV protocol which provides loop-free, and link disjoint paths multiple paths based on shortest route with less route discovery and less overhead. This algorithm proves to be very simple, less complex, and robust as far as balancing of the load is concerned because the algorithm considers only shortest paths and selects the path which is less congested and lightly loaded among all the possible paths between source and destination. This, in turn, exhibits good performance in congestion management and results in overall better network performance in terms of packet delivery ratio, throughput, packet loss ratio, and normalized routing load.

References

1. Sharma, R.: Survey of load balancing and load sharing through AOMDV protocol in MANET. Int. J. Sci. Eng. Res. **4**(5), 2269–2275 (2013)
2. Sahu, R., Sharma, S., Rizvi, M.: ZBLE: energy efficient zone-based leader election multipath routing protocol for MANETs. Int. J. Innov. Technol. Explor. Eng. **8**(9), 2231–2237 (2019)
3. Sharma, B., Chugh, S., Jain, V.: Investigation of adaptive multipath routing for load balancing in MANET. Int. J. Eng. Adv. Technol. **2**(5), 65–71 (2013)
4. Aouiz, A.A., Boukli Hacene, S., Lorenz, P.: Channel busyness based multipath load balancing routing protocol for ad hoc networks. IEEE Netw. **33**(5), 118–125 (2019)
5. Marina, M., Das, S.: Ad hoc on-demand multipath distance vector routing. Wirel. Commun. Mob. Comput. **6**(7), 969–988 (2006)
6. Wannawilai, P., Sathitwiriyawong, C.: AOMDV with sufficient bandwidth aware. In: Proceedings of 10th IEEE International Conference on Computer and Information Technology, pp. 305–312 (2010)
7. Gawas, M., Gudino, L., Anupama, K.: Congestion-adaptive and delay-sensitive multirate routing protocol in MANETs: a cross-layer approach. J. Comput. Netw. Commun. **2019**(Article ID 6826984), 1–13 (2019)
8. Alghamdi, S.: Load balancing maximal minimal nodal residual energy ad hoc on-demand multipath distance vector routing protocol (LBMMRE-AOMDV). J. Wirel. Netw. **22**(4), 1355–1363 (2016)
9. Mounagurusamy, K., Eswaramurthy, K.: A novel load balancing scheme for multipath routing protocol in MANET. J. Comput. Inf. Technol. **24**(3), 209–220 (2016)

10. Mallapur, S., Patil, S., Agarkhed, J.: Load balancing technique for congestion control multipath routing protocol in MANETs. Wirel. Pers. Commun. **92**(2), 749–770 (2016)
11. Tashtoush, Y., Darwish, O., Hayajneh, M.: Fibonacci sequence based multipath load balancing approach for mobile ad hoc networks. Ad Hoc Netw. **16**, 237–246 (2014)
12. Naseem, M., Kumar, C.: Congestion-aware fibonacci sequence based multipath load balancing routing protocol for MANETs. Wirel. Pers. Commun. **84**(4), 2955–2974 (2015)
13. Tran, D.A., Raghavendra, H.: Congestion adaptive routing in mobile adhoc networks. IEEE Trans. Parallel Distrib. Syst. **17**(11), 1294–1305 (2006)
14. Tran, D.A., Raghavendra, H.: Routing with congestion awareness and adaptivity in mobile ad hoc networks. Int. J. Distrib. Sensor Netw. In: Proceedings of 10th IEEE Wireless Communications and Networking Conference (WCNC), NewOrleans, LA, USA (Mar 2005)
15. Kim, J., Tomar, G., Shrivastava, L., Bhadauria, S., Lee, W.: Load balanced congestion adaptive routing for mobile ad hoc networks. Int. J. Distrib. Sens. Netw. **10**(7), 1–10 (2014)
16. Shukla, A., Sharma, S.: Queue length based load balancing technique using with AOMDV protocol in MANET. Int. J. Sci. Eng. Res. **4**(10), 506–511 (2013)
17. Tekaya, M., Tabbane, N., Tabbane, S.: Multipath routing mechanism with load balancing in ad hoc network. In: International Conference on Computer Engineering & Systems, Cairo, pp. 67–72 (2010)
18. Tekaya, M., Tabbane, N., Tabbane, S.: Multipath routing with load balancing and QoS in ad hoc network. Int. J. Comput. Sci. Netw. Secur. **10**(8), 280–286 (2010)
19. "Marc Greis' Tutorial for the UCB/LBNL/VINT Network Simulator "ns"", Isi.edu, http://www.isi.edu/nsnam/ns/tutorial/nsbasic.html

Metaheuristic-Based Intelligent Solutions Searching Algorithms of Ant Colony Optimization and Backpropagation in Neural Networks

Abraham Ayegba Alfa, Sanjay Misra, Kharimah Bimbola Ahmed, Oluwasefunmi Arogundade and Ravin Ahuja

Abstract Computation is concerned with the validation of algorithm, estimation of complexity and optimization. This requires large dataset analysis for the purpose of finding the unknown optimal solution. Aside the intricacies in completing tasks, this process is expensive and time inefficient; attaining solutions with conventional mathematical approaches are unrealizable. Search algorithms were advanced to improve solution approaches for optimization problems by finding the possible sets of solution to a particular problem as contained in search space. However, metaheuristic algorithms suggest three solutions to optimization problems on the basis of the application areas in real-life situations including: near optima, the optimal or the best solution. This paper analyses the decision-making processes of two nature-inspired search algorithms namely: Backpropagation search algorithm and ant colony optimization (ACO). The results revealed that, backpropagation search algorithm without ACO training trailed those trained with ACO for MSE, RMSE, RAE and MAPE. Again, forecasts errors estimated in the neural network set-up were smaller due to directional search mechanism of the ACO as against the approach provided in neuro-fuzzy rules set tuning by Rajab and Sharma (Soft Comput 23:921–936, 2017) [1]. There is need to consider metaheuristic algorithms approaches to obtain better solutions or nearest optimal values to the optimization problems in neural networks.

A. A. Alfa · K. B. Ahmed
Kogi State College of Education, Ankpa, Nigeria
e-mail: abrahamsalfa@gmail.com

K. B. Ahmed
e-mail: bimbola.k.ahmed@gmail.com

S. Misra (✉)
Covenant University, Otta, Nigeria
e-mail: sanjay.misra@covenantuniversity.edu.ng

O. Arogundade
Federal University of Agriculture, Abeokuta, Nigeria
e-mail: arogundadeot@funaab.edu.ng

R. Ahuja
Vishvakarma Skill University, Gurgaon, Haryana, India
e-mail: ravinahujadce@gmail.com

© Springer Nature Singapore Pte Ltd. 2020
P. K. Singh et al. (eds.), *Proceedings of First International Conference on Computing, Communications, and Cyber-Security (IC4S 2019)*, Lecture Notes in Networks and Systems 121, https://doi.org/10.1007/978-981-15-3369-3_8

Keywords Neural networks · Search algorithms · Best solution · Nearest optima · Decision making · Metaheuristic · Swarm intelligence · Optimization

1 Introduction

The art of choosing the best alternative among obtainable collection of options is known as optimization. In other words, it is one key quantitative mechanism in network of decision making whereby decisions are carried out to optimize one or many objectives in already prescribed collection of circumstances [2]. At present, there are on-going fields of study in order develop diverse schemes for achieving solutions to tasks. These include: technical analysis, statistical analysis, time series analysis, fuzzy logic systems and artificial neural networks. Neural network is one of the most patronized estimators for future trends of entities such as stocks [3].

In fact, majority of research make efforts to model swarm intelligence by considering insect intelligent behaviours to contribute to the development of certain metaheuristic, which simulate insects' skills in the course of solving common problems. The combined intelligence of social insect colonies is attained through interactions among these insects. In many cases, these have been adapted in solving scientific and everyday life optimization problems such as stock price forecast [4]. Neural networks were built largely to cover all kinds of mathematical models of natural nervous systems. Mostly, neural networks are used for stochastic problems with discrete and non-linear characteristics represented by neurons as transfer functions [5].

Several learning rules/algorithms have found relevance for the tuning of neural networks weights in non-deterministic problems domain including particle swarm optimization (PSO), genetic algorithm (GA), backpropagation, Hebbian, perception and ant colony [5]. ACO is used to simulate natural foraging behaviour of ants such as adaptation and cooperation for attaining the shortest route from the nest to a food source and back. ACO is swarm-based intelligence for search solution space architecture for paths leading to defined solution targets. ACO produces decisions based on heuristics for large datasets to get best solution. The paper compares the neural networks and ACO searching algorithms in attaining best solution in search space.

The remaining parts of the paper are structured as follows: Sect. 2 is swarm intelligence and problems solving. Section 3 is the methodology. Section 4 is the discussion of results. Section 5 is the conclusion.

2 Swarm Intelligence and Problems Solving

Swarm intelligence is a decentralizing, self-organizing, emergent and cooperative behaviour of small animal species such as bee, ant, fish and bird. These animals coexist and carry out complex tasks and functions as a group. However, this idea has been applied to several fields of computer science such as particle swarm and ant colony optimizations to solve complex computational problems. Some swarm intelligence exhibited by animals as include [6]:

Stigmergy: In the course of moving food to nest or back from nests, social animals discharge special chemical referred to as pheromones, which is used some sorts of indirect communication between the animals and environment. Pheromones support ants in many ways in decision making whenever the choice appropriate path is needed for further movement.

Foraging: Whenever, these social animals (such as ants) move from nests to food source and back; pheromones are continuously released to build a pheromone trail. The highest pheromone concentration level perceived by smell senses assists in the choice of the maximum probabilistic path because of strong level of pheromone.

Autocatalysis or Positive Feedback: A huge volume of pheromones available on a specific path increase the choice of ants for such as a path. Whenever, there are a large number of social animals to release bigger volume of pheromone on particular path, the chances are that the path with the biggest concentration becomes the common to ants.

2.1 Metaheuristic Algorithms

Optimization is a procedure that seeks a best, or optimal, solution for a problem. In general optimization problems, more than one local solution exists [7]. This results in a situation in which a good optimization scheme is required to search the best solution beyond the neighbourhood that does not mislead the search process and restrict it to a local solution. In practice, the optimization algorithms are expected to effectively attain solution from local and global searches. Mostly, methods such as combinatorial and mathematical were deployed in resolving several kinds of optimization problems. Only recently, the existence of large search space solutions for optimization problem makes conventional mathematics schemes less applicable.

Consequently, numerous metaheuristic [8–11] optimization solutions (such as combinatorial) evolved to deal complex optimization problems [12]. This type of optimization is desirable when there is discrete collection of feasible solutions in the search space. But, finding all the possible solutions is unattainable in large search space. Metaheuristic algorithms are probable ways of solving these problems, which offer to produce near optimal in minimum time and cost [2]. Broadly, [2] categorized metaheuristic forms into three including trajectory-, population- and hybrid-based methods.

Trajectory: These methods consider a single solution in search space. The process offers knowledge concerning effectiveness of the algorithm and its strength in the problem being solved such as tabu search, simulated annealing, greedy randomized adaptive search and variable neighbourhood search.

Population: These methods perform processes using a collection of solution whose effectual outcomes are different solution or evolve into fresh solutions such as genetic algorithms, ant colony and particle swarm optimizations.

Hybrid: The concept for hybridization of diverse algorithms is to harness the interconnected behaviour of several optimization approaches. This can be achieved by combining population-based schemes (such as ant colony optimization) and evolutionary computation that makes use of local search schemes to offer high quality solutions through further solution processing such as constraint programming, problem relaxation, dynamic programming, stochastic optimization and tree search based metaheuristic methods.

2.2 Ant Colony Optimization Search Algorithm

Fuzzy logic controls ACO is a population-based scheme used to resolve combinatorial optimization problems inspired by the foraging behaviour of ants in their capability to seek the shortest route from a food source to the nest. ACO was conceived by Macro Dorgio, Alberto Colorni and Vittorio Maniezzo after a study on combinatorial optimization that relies on iterative process for constructing solution with simple agents' population repeatedly. This is probabilistically regulated through heuristic information of the defined sample of problem as well as a common memory holding experiences of the ants in preceding iterations [13].

ACO is a mathematical application of real ants, nest and food represented as node, path between nodes depicted with arc. Then probabilistic rule matching each arc is specified to assist the movement between decision making and vertices [6]. In essence, ACO algorithm mimics the ants' natural behaviour of for adaptation and collaboration techniques. According to [14], the ACO algorithms are based on certain assumptions as follows:

Firstly, each path followed by an ant matches to a candidate solution for a given problem.

Secondly, when an ant follows a path, the amount of pheromone released on the path is relative to the quality of the matching candidate solution for the specific problem.

Third, when an ant wishes to choose between two or more paths, the path(s) having highest pheromone volume are more attractive to the ant.

Finally, upon completion of certain iterations, the ants converge to the path, which is most probable to a near-optimum, the optimum or solution for the given problem.

One notable procedure for adopting ACO algorithm in modelling real-life problems was provided by [13]. These include:

1. Determination of input variables ranges with corresponding names. Construction graph is identical to the problem graph in the modelling stages. The set of components (X) is identical to the set of nodes (Y) (that is, $X = Y$). The connections match to the collection of arcs, and each connection has a weight

that matches the distance between nodes. The states of the problem are the set of all possible partial tours for a travelling salesman problem (TSP).

2. The TSP constraints are linked to all the cities to be visited at least once. This constraint is prescribed if an ant at each construction step selects the next city only among unvisited.

3. Pheromone trails and heuristic information in the TSP are the desirability to visit a city directly after the last city. Typically, the heuristic information is inversely proportional to the distance between the cities for a straightforward choice.

4. Solution construction: An ant initially positioned in a randomly selected start city and at each step, iteratively adds one unvisited city to its partial tour. The solution construction stops whenever there is unvisited city.

2.3 Neural Networks

Neural network (NN) is a mathematical process built to learn, visualize and validate neural networks approaches [15]. NN is considered to be a model-free estimator because it does not rely on a predefined form for the source data. Again, the data structure of NN can be easily modified in order to map from an input dataset supplied to features or associations available inside the data. The amount of inputs to the network matches to external depiction of the problem to be resolved [15]. Whenever NNs are successfully trained, other actions such as prediction, estimation, classification or simulation with fresh dataset in the similar source can be achieved. In effect, these operations of NNs make use of numerous training (or learning) algorithms as depicted in Fig. 1 [16].

Majority of NNs operate on numerical dataset to identify and evaluate different problems modelled as given by Eq. (1) [16].

Given a dataset, $\{a_j.b_j\}j^M = 1$, from an indefinite function:

$$y = f(x) \tag{1}$$

From Eq. (1), reasonable estimates of the function can be derived by means of numerical algorithms. These processes begin with the choice of appropriate data

Fig. 1 A neural network layout [16]

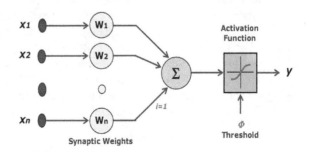

type structure most useful for prevailing problem being modelled. The networks are trained and evaluated with another dataset to ensure satisfactory mapping is attained [16]. NNs have been built from general approaches in solving problems similar those of humans or biological ideas such patterns identification [15].

There are several types NNs as follows: *Multilayer perceptron (MLP)* data flows forward to the output steadily without feedback connections. *Recurrent* network is different MLP because of support for feedback connections. NN is a computational system inspired by the structure, processing and learning capability of the biological brains, which possess enormous simple neuron-like processing elements. More so, there are many weighted connections between the elements with distributed depiction knowledge throughout the connection, which is acquired through a process of learning [17]. NNs are mostly applicable for solving diverse problems due to large parallelism, learning ability, generalization capability, distributed representation and fault tolerance [17].

NNs make use of the *leaning* process on the training dataset by means of connection weights. There are two main ways of learning:

1. Supervised learning utilizes the correct output for every input pattern given to the network. These weights are chosen to allow the network to create output which is closest to the prescribed correct output (such as backpropagation algorithm).
2. Unsupervised learning uses input patterns of the training dataset supplied without a correct output associated with each training dataset. The unique structure or relationships in the training dataset or data patterns are explored organized and categorized (such as Kohonen algorithm). The *hybrid learning* integrates the supervised and unsupervised learning by appraising the weights from both algorithms [16].

2.4   Related Works

Particle swarm and ACO were applied to resolve the problems of Quality-of-Service (QoS) requests such as jitter, bandwidth, delay and packets loss rate in multicast applications by [14]. Mainstream Internet multicast applications, namely video conferencing, distance learning and online simulations require to send information from source to desired destinations. These are categorized into QoS constrained multicast routing problem, which make use of non-deterministic algorithm. Consequently, an intelligent algorithm based on swarming agent (that is, a hybrid ACO/PSO) to enhances the multicast tree.

Again, [18] proposed hybridization solution of PSO and ACO algorithms in optimal designing of truss structures problem. In this approach, PSO explored the design space. And ACO carried out a local search for the best solution determined by PSO. In addition, ACO served as the global optimizer; and the local optimizer was PSO.

An ACO was applied in solving job shop scheduling problem by [19]. In this, fuzzy logic is used for scheduling the sequence of job and it can be optimized using ACO. In a separate study, [20] proposed an ACO for solving travelling salesman problem in which effective distribution strategy and information entropy are piloted. Also, it makes use of local optimization heuristic in solving travelling salesman problem with the basic ACO. The proposed algorithm offered improved outcome over ACO algorithm.

More so, [21] advanced the use of ACO to solve the vehicle routing problem. In this approach, a new method was created through increases in pheromone concentration known as ant weight strategy. This mutation operator is used resolve the vehicle routing problem.

In order to improve on the existing neuro-fuzzy system rules for predicting stock prices, [1] proposed a tuning technique on the rules lists with multi-technical factors. The essence was to build efficient and interpretable and accurate predictive model through fuzzy set parameter updates.

3 Methodology

This paper proposed tuning the outcomes search algorithms for neural networks in order to create nearest optima or solution or forecast of Nigeria stock market. The proposed structure of the neural network is composed of input, process and output as shown in Fig. 2.

The input component taken was raw unprocessed data for the neural network search solution model. The supervised learning approach used two sets of neural networks, which provides actual output for the input supplied. This component specifies and enforces entry formats for the different datasets fed into the model.

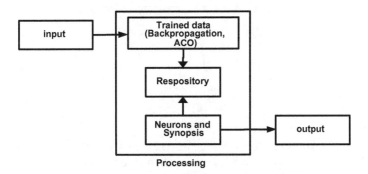

Fig. 2 Structure of NN search solutions tuning model

The Unilever PLC share price for 1 year period served as input for the various neurons and multilayers of the neural networks.

The processing component comprises three hidden operation namely:

1. The trained data converts input data into meaningful information with search algorithms to assist the neural networks to learn all the obtainable characteristics needed for arriving at nearest optimal or best solution.
2. The repository stores, retrieves or updates information about previous search operations. It warehouses diverse information, features and characteristics of search problems.
3. The neuron and synopsis are used to establish connections and formulate patterns for diverse neuron attributes and related links. These are located in the first layer of the two multilayered neural networks discussed in this paper. The search algorithms utilize these features and characteristics and stored in the repository to match problem in solution space vectors. These neurons locate search solution in a problem space vector through formation rules for input and matching outcomes.

The *output component* is concerned with the display of neural networks outcomes after processing operation. The meaningful and measurable data or information is evaluated. It supports best decision making about future trends in data at a later time. The input provided and the hidden multilayers of neural network produce valuable results for analysis such as charts, graphs and tables.

3.1   Mathematical Model of NN Tuning

The proposed tuning approach for the NN forecasting model utilizes the swarming agent nature-inspired attributes using the initial N multicast tree patterns built in random manner. Then, the key quantity w is predetermined for target space w for every pattern of multicast tree. It can be represented as represented by Eq. (2):

$$Y_i = \int_{i=1}^{N} p_i \odot \int_{q=1}^{w} K_q \tag{2}$$

where,

Y_i is the multicast tree pattern i
p_i is the ith attribute of pattern i in each pattern at global space
K_q is the qth attribute of pattern i in each pattern at local space
w is the predetermined key of target work space
N is the quantity of the patterns generated.

In addition, each pattern generated has uniquely matching agent pattern. Afterwards, the mobile agents, n, are produced for the purpose of discovering the fit patterns in arbitrarily large patterns. Again, a new set of patterns are chosen

Table 1 Tuning algorithm for neural networks

INPUT: Pre-processed dataset
Output: Backpropagation and ACO forecasts

1	**INPUT** pre-processed dataset of stock prices of Unilever PLC
2	**SELECT** training dataset in quantity of 60%
3	**CHOOSE** test dataset in quantity of 30%
4	**CONSTRUCT** the neural network (NN) with layers
5	**SETUP** the built NN with backpropagation algorithm
6	**RUN** the new model with training dataset
7	**VALIDATE** trained NN with designated test dataset
8	**FORECAST** share stock prices with new dataset
9	**RETRAIN** the NN set-up with ACO algorithm and **GOTO** step 4
10	**OUTPUT** the backpropagation and ACO algorithms based forecasts
11	**TERMINATE** the forecasts modelling processes and RETURN results

on the basis of fitness degree or value. In the search space, the mobile agents go through these patterns by interacting with adjourning particle agents in a dynamic manner. Analogous to natural ants, at the starting point, the particle agents are disseminated across the search uniformly; each pattern local attributes are identified through pheromones deposits. At the ending point, the particle agents come together at fittest pattern by means of numerous iterations. The pheromone concentration of a pattern or sets of patterns is used to establish the likelihood of attaining best fit pattern; otherwise, its attributes offer clues to the fittest pattern. ACO algorithm uses patterns and its attributes on the basis of fitness values.

3.2 Algorithm of Tuning Model

The neural network is created to perform the training, testing and evaluation using the dataset obtained from pre-processing activities as indicated in the algorithm presented in Table 1.

The experimental parameters for search algorithms deployed with neural network are contained in Table 2.

4 Discussion of Results

In this paper, the outcomes of deploying NN trained with backpropagation and ACO algorithms are intended at solving the search space/optimization problem (that is, stock price forecasts for UniLever PLC). The contributions of deploying

Table 2 Parameter for simulations

Item	Values
Input variable	3
Output variable	1
Hidden layer	1
Neuron in hidden layer	50
Training algorithm	Backpropagation, ACO
Initial learning rate	0.001
Initial momentum coefficient	0.1
Transfer function	Sigmoid
Maximum training cycles	10,000
Selection function	Tournament selection
Population	80

Table 3 Search solutions of the learning algorithms

Algorithms	Backpropagation	ACO
RAE	0.03113291566	0.01046835364
MSE	20.80515809	2.352275584
RMSE	4.561267159	1.533713006
MAPE (%)	0.163857451	0.055096598
Std. error of the estimate	0.9187087494	0.6163915701

these distinct search/learning algorithms for finding the best solutions using NN for selected evaluation parameters are presented in Table 3.

In Table 3, the degree of nearness of NNs best solution forecast for the share price (Unilever PLC) problem which is determined after the learning processes are complete for backpropagation and ACO. In the case of RAE, backpropagation algorithm gave the biggest deviations from the desired solution as against those of ACO. Also, the MAPE values showed that NN with ACO offered the nearest optimal or best solution for the optimization problem. The same outcomes are reported for MSE, RMSE and MAPE with ACO having best solution to forecasting problem.

More so, backpropagation search algorithm used for forecasts performed poorly against ACO search algorithm leading to the need for further optimization of backpropagagtion with ACO. The ACO search algorithm is more beneficial because of errors minimizing capability during the searching processes for neighbourhood variables in solution space. This implies that, the search algorithms such as ACO provide easy route/direction in order to obtain nearest optimal with minimal errors.

The results comparisons of the proposed heuristic-based NN model and fuzzy logic based heuristic in [1] are presented in Table in Table 4.

Table 4 Benchmarking of metaheuristic search algorithms

Metric	Proposed model	Neuro-fuzzy system [1]
RMSE	1.534	30.995
MAPE (%)	0.06	0.35

From Table 4, the proposed system is based on the metaheuristic tuning approach offered by NN as against the heuristic tuning approach of neuro-fuzzy. The proposed model outperformed the existing model in [1] on the basis of RMSE and MAPE evaluation metrics. This is attributed to the ACO capability to arrive at the nearest optima-solution considering the local and global solution spaces.

5 Conclusion

Pure metaheuristic is analogous to pattern designs in which appropriate knowledge is acquired and processed for the purpose of constructing extensive and far reaching optimization algorithms about a problem. In general, a single algorithm can be applied to numerous metaheuristic as well as the incorporation of contemporary methods of finding or searching for optima within solution spaces. Nowadays, several fresh natured-inspired optimization algorithms exist in attempts to solve optimal for a target problem. On the other hand, search algorithms make use of numeral neighbourhood, specified directions, population size and intermediate variables.

This paper discussed two prominent search algorithms for finding best solution, optima or near optima to forecasting problems. The first one is the backpropagation, which was used to perform best solution search for the obtainable dataset. The second one is the ACO algorithm, which was applied to the same dataset in order to arrive at the best solution. The outcomes of the two approaches revealed that ACO algorithm produced the best solution than backpropagation algorithm and the benchmark model in [1] on the bases of RAE, MAPE, MSE and RMSE parameters measured. However, best solution can be attained with metaheuristics [8–11] or hybridized learning algorithms in case of combinatorial optimization problems other than forecasting in NN environment.

Acknowledgements We would like to acknowledge the sponsorship and support provided by Covenant University through the Centre for Research, Innovation, and Discovery (CUCRID).

References

1. Rajab, S., Sharma, V.: An interpretable neuro-fuzzy approach to stock price forecasting. Soft. Comput. **23**(3), 921–936 (2017)
2. Baghel, M., Agrawal, S., Silakari, S.: Survey of metaheuristic algorithms. For combinatorial optimization. Int. J. Comput. Appl. **58**(19), 21–31 (2012)
3. Khan, Z.H., Alin, T.S., Hussain, A.: Price prediction of share market using ANN. Comput. Expert Syst. Appl. **38**(1), 9196–9206 (2011)
4. Turanoglu, E., Ozceylan, E., Kiran, M.S.: Particle swarm optimization and artificial bee colony approaches to optimize of single input-output fuzzy membership functions. In: Proceedings of the 41st International Conference on Computers & Industrial Engineering, pp. 542–548 (2012)
5. Pal, A., Chakraborty, D.: Prediction of stock exchange share price using ANN and PSO. Int. J. Eng. Sci. **80**(1), 62–70 (2014)
6. Vyas, S., Sanadhya, S.: A survey of ant colony optimization with social network. Int. J. Comput. Appl. **107**(9), 17–21 (2014)
7. Said, G.A., Mahmoud, A.M., El-Horbaty, E.M.: A comparative study of meta-heuristic algorithms for solving quadratuc assignment problem. Int. J. Adv. Comput. Sci. Appl. **5**(1), 1–6 (2014)
8. Okewu, E., Misra, S.: Applying metaheuristic algorithm to the admission problem as a combinatorial optimisation problem. Front. Artif. Intell. Appl. Adv. Digit. Technol. **282**, 53–64 (2016)
9. Crawford, B., Soto, R., Johnson, F., Vargas, M., Misra, S., Paredes, F.: A scheduling problem for software project solved with ABC metaheuristic. ICCSA **4**, 628–639 (2015)
10. Crawford, B., Soto, R., Peña, C., Riquelme-Leiva, M., Torres-Rojas, C., Misra, S., Johnson, F., Paredes, F.: A comparison of three recent nature-inspired metaheuristics for the set covering problem. ICCSA **4**, 431–443 (2015)
11. Crawford, B., Soto, R., Johnson, F., Vargas, M., Misra, S., Paredes, F.: The use of metaheuristics to software project scheduling problem. ICCSA, Part V, LNCS **8583**, 215–226 (2014)
12. Alazzam, A., Lewis, H.W.: A new optimization algorithm for combinatorial problems. Int. J. Adv. Res. Artif. Intell. **2**(5), 63–68 (2013)
13. Darquennes, D.: Implementation and applications of ant colony algorithms. A master thesis, Department of Information Technology, University of Namur, Belgium, pp. 1–101 (2005)
14. Patel, M.K., Kabat, M.R., Tripathy, C.R.: A hybrid ACO/PSO based algorithm for QoS multicast routing problem. Ain Shams Eng. J. **2**(1), 113–120 (2014)
15. Moustafa, A.A.: Performance evaluation of artificial neural networks for spatial data analysis. Contemp. Eng. Sci. **4**(4), 149–163 (2011)
16. Chakraborty, R.: Fundamentals of Neural Networks: Soft Computing Course Lecture Notes. Department of Computer Science, Indian Institute of Technology, Madras, pp. 7–14 (2010)
17. Su, Y.: An investigation of continuous learning in incomplete environments. A Ph.D. thesis. University of Nottingham, UK, pp. 1–180 (2005)
18. Gholizadeh, S., Fattahi, F.: Serial integration of particle swarm and ant colony algorithms for structural optimization. Asian J. Civ. Eng. (Build. Hous.) **13**(1), 127–146 (2012)
19. Surekha, P.: Solving fuzzy based job shop scheduling problems using GA and ACO. J. Emerg. Trends Comput. Inf. Sci. **1**, 95–102 (2010)
20. Hlaing, S.Z.S., Khine, M.A.: An ant colony optimization algorithm for solving traveling salesman problem. In: International Conference on Information Communication and Management, vol. 16, pp. 54–59 (2011)
21. Bin, A.Y., Zhong-Zhen, Y., Baozhen, Y.: An improved ant colony optimization for vehicle routing problem. Eur. J. Oper. Res. **196**, 171–176 (2009)

Evaluating Cohesion Score with Email Clustering

Abhishek Kathuria, Devarshi Mukhopadhyay and Narina Thakur

Abstract An email has become one of the prime ways of communication for individuals or organizations and has emerged as an important research field to categorize emails and enable users for easy data segregation, topic modeling, spam detection, network analysis for investigative and analytical purposes. The paper aims to cluster the emails comprising of 500,000 emails taken from the Enron email dataset which was obtained by the Federal Energy Regulatory Commission during its investigation of Enron's collapse, based on the relevance of the words to the whole corpus. The proposed algorithm calculates the cohesion score of each cluster using intra-cluster similarity. This paper implements two unsupervised clustering algorithms for the email clustering process, namely k-means and hierarchical clustering and evaluates the cosine similarity of all the words from each cluster to evaluate the semantic similarity pervading through each cluster. The emails were clustered into three groups and the cohesion score was obtained for each cluster which measured the intra-cluster similarity. The proposed method helped in the computation of the score distribution among the clusters, as well as the intra-cluster similarity. Cluster 1 obtained the highest cohesion score among all the three clusters by attaining the cohesion score 0.1655 while using the k-means algorithm and the cohesion score of 0.2513 while using the hierarchical clustering algorithm.

Keywords Email clustering · TF-IDF · Clustering techniques · K-means · Hierarchical

A. Kathuria (✉) · D. Mukhopadhyay · N. Thakur
Bharati Vidyapeeth's College of Engineering, New Delhi, India
e-mail: abhishekkathuria40@gmail.com

D. Mukhopadhyay
e-mail: debom97@gmail.com

N. Thakur
e-mail: narinat@gmail.com

© Springer Nature Singapore Pte Ltd. 2020
P. K. Singh et al. (eds.), *Proceedings of First International Conference on Computing, Communications, and Cyber-Security (IC4S 2019)*, Lecture Notes in Networks and Systems 121, https://doi.org/10.1007/978-981-15-3369-3_9

1 Introduction

Emails are a part of quotidian life. Emails are used in personal as well as professional life as an economical, reliable and instant way of communication. They serve as an archival tool for many people, as many users never discard some messages because of the significance of the information attached in the proximate future, for example, as a reminder of upcoming events and outstanding issues. However, systems dealing with data mining like information retrieval systems, search engines might be prone to certain unresolved ambiguities, leading to their performance degradation. Name discrimination and email clustering are two different aspects of textual classification as emails are clustered based on their contextual and semantic similarity. The growth of the Internet has led to an exponential increase in the number of digital documents being generated; hence, the analysis of the textual information within emails has become a notable field of study under the category of data mining. The two principal algorithms that are used in this paper for clustering are k-means clustering and hierarchical clustering. Hierarchical clustering is used for obtaining an in-depth analysis of the cluster as well as determining the basis of clustering for each data point, while k-means are used for an efficient and fast information retrieval clustering system. Hierarchical clustering creates clusters that contain the pretrained ordering from top to bottom. The challenges in the email clustering domains have been addressed by many researchers. Alsmadi (2015) [1] discusses various issues which arise in email clustering. Most frequently used words have been collected from the dataset using NLP techniques in these emails and inverse document frequency has been used to find the importance of these words for the document they appear in. The clustering technique used here is k-means. Generally, clustering algorithms are divided into two broad categories—hard and soft clustering methods [2]. Hard clustering algorithms differentiate between data points by specifying whether a point belongs to a cluster or not, that is, absolute assignment, whereas in soft clustering, each data point has a varying degree of membership in each cluster. Dimensionality reduction methods can be considered as a subtype of soft clustering; for documents, these include latent semantic indexing (truncated singular value decomposition on term histograms) and topic models. Other algorithms involve graph-based clustering, ontology supported clustering and order sensitive clustering.

Given a clustering technique, it can be beneficial to automatically derive human-readable labels. An email comprises numerous attributes such as sender's address, receiver's address, subject, message body, etc. Email mining is a process which is a subpart of text mining. It refers to a method of discovering insightful patterns and information from large email data. There are various applications of email mining such as email summarization, categorization, spam filtering, etc.

The main idea on which the clustering is based is the distance of elements. As a result of clustering, nearby elements are grouped into common sets. Through this paper, we intend to gain insights and determine important and valuable terms that occur in the message body of our dataset. This involves an in-depth data analysis and gaining valuable deductions from the message body text. The clustering will eventually help us in determining various things such as about the company from where the dataset has been taken, the important and powerful people involved, the users who communicated frequently through email about a specific topic, the main topics discussed in the company and many others.

Our approach includes clustering the emails based on the features obtained through the message body. The two most simple and less expensive clustering algorithms, namely k-means clustering and hierarchical clustering, are used, which are the most viable options for pre-clustering, as they reduce the space into disjoint smaller subspaces. The elbow method has been used in the k-means clustering algorithm for the determination of clusters. The determination of the optimal number of clusters has been a key problem for many researchers such as Mark Ming-Tso Chiang (2010) [3].

In our paper, for finding the optimal number of clusters in k-means clustering, a more precise method called the 'silhouette method' has been used. In the hierarchical clustering algorithm, the bottom-up approach for clustering called the agglomerative clustering method has been used. A dendrogram has been plotted for the determination of the number of clusters. The clusters have been determined based on the number of points intersected by the threshold line. For the determination regarding the selection of a cluster for gaining concentrated insights, a cohesion score has been calculated which determines the intra-cluster similarities.

The rest of the paper is organized as follows: Related work is discussed in short in Sect. 2, proposed cohesion-based system is described in Sect. 3, experimental setup is included in Sect. 4, the results and analysis are presented in Sect. 5, conclusion is given in Sect. 6 and the references are mentioned in the last section.

2 Related Works

A lot of research has been done on clustering [3–7] in the past decades. Collecting an archive of emails for analysis can be done for several purposes. While some of them focus on presentation reforming, others focus on investigating the efficiency of various clustering algorithms. The performance of experimented k-means clustering algorithms has been evaluated by Alsmadi (2015) [1]. Mark Ming-Tso Chiang (2010) discussed the most contentious problem in k-means clustering, that is, selecting the correct value of the number of clusters [3]. Chiang concentrated on analyzing the performance of an intelligent version of k-means, known as ik-means, which uses anomalous pattern (AP) clusters for initializing k-means clustering. The results showed that this adjusted k-means or ik-means outperformed several other methods, concerning centroid and data recovery, but overestimated the numbers of clusters in the case of small in-between cluster spreads. Andrew Lensen (2017) introduced the coherent approach consisting of three stages for selection of the optimal number of clusters, k, and performing concurrent feature selection along with clustering [8]. In the first stage, Kest, which is an estimate of k, was determined using the silhouette method. The second stage consisted of using the Kestvalue to perform a guided search by particle swarm optimization (PSO) for calculating the number of clusters. The final stage used the centroid representation to perform a localized pseudo search for fine-tuning the solution obtained in stage two. Email categorization has been addressed and worked upon by Azizpour (2018) [4]. Here, similarity measure for text processing (SMTP) has been used for clustering emails. The efficiency of similarity algorithms like Euclidean distance, cosine similarity, extended Jaccard coefficient and dice coefficient and SMTP has been compared by implementing them with a k-means clustering algorithm. For that, four clusters were created and similar emails were put into them. It was observed that the results obtained by using SMTP with the k-means clustering algorithm were better than others; M. Basavaraju (2010) used a text-based clustering approach for spam detection [9]. The datasets used here were ling spam corpus (more description). The spam detection techniques used here were based on the vector-based model. The clustering algorithm incorporated features of both k-means and the BIRCH algorithm. It was observed (result) that the k-means clustering algorithm worked best with smaller datasets while the combination of BIRCH with KNNC worked better with large datasets.

Huang (2008) used a mixed-initiative approach to hierarchically cluster emails [10]. In this approach, the hierarchies are first decided by the computer and then revised through repetitive user feedbacks to make them more meaningful and useful. An edge modification ratio is defined to compare the resulting hierarchies against a reference one. The paper concluded by stating hierarchical clustering helps a user understand the results better than flat clustering and any mixed-initiative system should acknowledge subsequent variation in user feedbacks to create effective strategies for retraining models.

Ercan G (2008) addressed the problem of text summarization, that is, forming extracts using lexical cohesion [11]. Summarization of text includes selecting the most representative sentences. Ercan employed the lexical cohesion of the text structure as a means of evaluating significant sentences. For the linear text segmentation by topic, Pérez (2010) employed clustering cohesion as a criterion for evaluation [12]. The results were then fed to their proposed incremental overlapping clustering algorithm to assign the input stream to a topic cluster.

Klebanov (2008) proposed an automatic method of text analysis aimed at discovering patterns of lexical cohesion, that is, groups of words with similar meaning [13]. Political speeches were analyzed and groups of words with related meaning were extracted and compared with manual analysis of political rhetoric. Hence, the authors concluded that lexical cohesion is a viable method for quantitative and qualitative analysis of the text.

2.1 Comparison Table

The following Table 1 describes the objectives, approach, pros and cons of various state of the art methods.

Table 1 Comparison between various state-of-the-art methods

Author	Year	Objectives	Approach	Pros	Cons
Alsmadi [1]	2015	Clustering and classification of email contents	Using NGrams to develop a feature matrix	High true positives rates in almost all of the cases	Larger feature matrix had to be developed
Chiang [3]	2010	Determining the right no. of clusters in k-means	Using Intelligent k-means clustering	Outperforms other evaluation measures in centroid and data recovery	Overestimates the number of clusters
Basavaraju [9]	2010	Efficient spam mail classification using clustering techniques	Using j-means and BIRCH	High precision rates for both the clustering models	K-means works well with smaller datasets
Xie [5]	2016	Deep embedded clustering (DEC)	Propose to iteratively refine clusters with auxiliary target distribution	DEC is significantly less sensitive to the choice of hyperparameters	No method to find optimal clusters

3 Proposed Cohesion Evaluation-Based Cluster System

3.1 Problem Formulation

After preprocessing, the processed data with the message body is obtained in the form of a data frame. For each message body, the text is lemmatized and converted into lowercase. The stop words are removed by using the Natural Language Toolkit (NLTK) corpora. The term frequency-inverse document frequency matrix was generated using the term frequency-inverse document frequency (TF-IDF) vectorizer. The TF-IDF is a mathematical measure that evaluates the importance of a word to a document in a document corpus based on the frequency of the word occurring in a document and its relevance to the corpus. We used the TF-IDF vectorizer instead of TF-IDF transformer as it counts the words, finds the IDF values and calculates the TF-IDF scores simultaneously. The cleaned data was fed to the TF-IDF matrix to obtain the TF-IDF score for each word in each document. The TF-IDF score is calculated by using the formula as shown in Eq. (1).

$$w_{t,d} = \left(1 + \log_{10}\left(f_{t,d}\right)\right).\log_{10}\frac{N}{df_t} \tag{1}$$

Here, $\log_{10} \frac{N}{df_t}$ is the inverse document frequency score calculated by taking the log of N, which is the total number of documents in the corpus and f_t is the long-term frequency. The long-term frequency of a term t in d is given by $f_{t,d}$ and $w_{t,d}$ gives the TF-IDF score. Using the TF-IDF features, k-means clustering algorithm and hierarchical clustering algorithm are applied. The k-means clustering algorithm is an unsupervised clustering algorithm which determines the optimal number of clusters using the elbow method. It works iteratively by selecting a random coordinate of the cluster center and assigns the data points to a cluster. It then calculates the Euclidean distance [14] of each data point from its centroid, and based on this, it updates the data point positions as well as the cluster centers. For minimizing the within cluster sum of squares (WCSS), the formula in Eq. (2) is used:

$$\arg\min_S (x + a)^n = \sum_{i=1}^{k} \frac{1}{2|S_i|} \sum_{x,y \in S_i} \|x - y\|^2 \tag{2}$$

where S_i is the mean of the points, x contains the observations in a d-dimensional vector and k is the number of cluster centers. As the elbow method can sometimes obfuscate the deduction for the optimal number of clusters, silhouette analysis can prove to be a more precise method. A silhouette value always lies in the range $[-1, 1]$ where $+1$ depicts that the data point under consideration is in close proximity with the assigned cluster and faraway from its neighboring cluster, whereas -1 depicts that the data point under consideration is in close proximity with its neighboring cluster and faraway from the assigned cluster. Let us consider a data point, 'x.' Let 'A' be the assigned cluster to this data point and 'B' be the neighboring cluster. Hence, the silhouette score, $S(x)$, can be given by Eq. (3).

$$S(x) = \frac{M(x) - N(x)}{\max(M(x) - N(x))} \tag{3}$$

where $M(x)$ is the mean distance of the point 'x' with respect to all the data points in the assigned cluster. And $N(x)$ is the mean distance of the point 'x' with respect to all the data points in the neighboring cluster. Hierarchical clustering creates clusters which consist of a predetermined ordering from top to bottom. In this paper, agglomerative method is used for hierarchical clustering. It is a bottom-up approach where each observation is assigned to its own cluster and each data point is considered as a separate cluster. The distance between each cluster is calculated using Ward's method and two similar clusters are combined together. This process is continued until there is only one cluster left. The Ward's method for calculating distance is given by the following formula:

$$\Delta(A, B) = \sum_{i \in A \cup B} \left\| \vec{x_i} - \vec{m}_{A \cup B} \right\|^2 - \sum_{i \in A} \left\| \vec{x_i} - \vec{m}_A \right\|^2 - \sum_{i \in B} \left\| \vec{x_i} - \vec{m}_B \right\|^2 \tag{4}$$

$$= \frac{n_a n_b}{n_a + n_b} \left\| \vec{m_A} - \vec{m}_B \right\|^2 \tag{5}$$

In Eqs. (4) and (5), m_j is the cluster center of j and n_j is the number of points inside the cluster. Δ refers to the cost of combination of A and B. Basically, the sum of squares is zero in the beginning and grows gradually as the clusters are merged. This method is responsible for keeping this growth as minimal as possible. The cohesion score [15] is calculated by using the cosine similarity formula [16] as discussed in Eq. (6):

$$\cos \theta = \frac{d_2 \cdot q}{\|d_2\| \|q\|} \tag{6}$$

In Eq. (6), d_2 and q are the TF-IDF vectors and θ is the angle between the vectors. The obtained cohesion score identifies the intra-cluster similarity between the clusters.

3.2 Architecture

The message body of each email (obtained from the 'body' column of the data frame) was processed before constructing a TF-IDF matrix, that is, stop words were removed, and each word was lemmatized. These processed messages were used to construct a TF-IDF matrix [17]. Figure 1 clearly depicts that this matrix was fed to the k-means and hierarchical clustering functions to obtain three clusters. The words from each cluster were extracted and their pairwise cosine similarity was used as a measure of cohesion within the clusters.

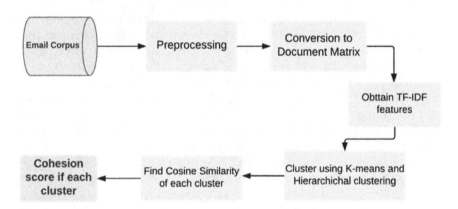

Fig. 1 Proposed cohesion evaluation-based cluster system

4   Experimental Setup

4.1   Dataset

The Enron email dataset was collected and prepared by the Cognitive Assistant that Learns and Organizes (CALO) project. It contains data from about 150 users, mostly senior management of Enron, organized into folders [18]. The corpus contains a total of about 0.5 M messages. The original data included approximately 500,000 emails generated by employees of the Enron Corporation. These emails were read as a.csv file, where the data was split into three columns, namely index, message id and raw message. This raw message contained all the fields present in an email format. From this raw message, the message body was obtained; hence, a data frame of 500,000 fields was constructed with the columns as 'body,' 'to' and 'from.' It is on the column of 'body,' that is, the message body of each email that the cluster analysis was implemented and evaluated.

4.2   Evaluation Measure

When TF-IDF transformation is applied to a text corpus, real-valued vectors of the features (words) for each data point (document) are obtained. Cosine similarity is a common measure to measure cohesion within clusters in text mining [19]. Although the extent to which the quality of clustering is judged is usually determined by comparing how tightly packed a cluster is, and the distance of that cluster from other clusters, it proves to be a trivial and naïve method for analyzing the clustering. A more sophisticated and precise method, for analyzing the quality of clustering, called the cohesion analysis, is discussed in this paper. This cohesion analysis shows the intra-cluster similarity by using cosine similarity. The cosine similarity can be thought of as a similarity measure which calculates the dot product between two nonzero vectors. This cosine score varies between 1 and zero, with zero implying that the vectors are perpendicular, that is, highly unrelated and one implying perfect correlation or practically speaking, they are the same vectors. Hence, all the words from each cluster were extracted, and for each cluster, a matrix was constructed and the cosine similarity score for each pair of the word was calculated. Cosine similarity is a widely used metric where the vector magnitude is not important; as this usually happens while we are working with the textual data which is represented by the word counts. Hence, an average score across this matrix was calculated. This is the cohesion score for the cluster.

5   Result Analysis and Discussion

This section deals with the implementation and evaluation of the clustering algorithms. In this paper, the Enron email dataset containing 500,000 emails was collected. From every email, the message body was extracted and a data frame of these message bodies was created using Pandas [20]. Removal of stop words is done along with lemmatizing. Lemmatizing is the process of reducing each word to its lemma [21]. Next, the email bodies are converted into a document-term matrix using TF-IDF. The term frequency-inverse document frequency (TF-IDF) is the mathematical statistic that is intended to reflect how important a word is to a document in a collection or corpus. The next step is writing a function to get the TF-IDF features out of all the emails. Using document vectors, clustering is done using k-means and hierarchical clustering algorithm. K-means assigns each data point to a different cluster, ensuring that the distance between that cluster center and the data point is minimum compared to the distance with other cluster centers [22]. For finding the number of centroids, the 'elbow method' is used. In this, the sum of squared error (SSE) value is calculated for different values of k (that is, number of clusters) by clustering the dataset following each value of k [23]. The point on the graph where a 'hinge' occurs is considered to be the optimal value of k. Figure 2 shows the elbow method for k-means algorithm. Thus, by looking at the graph, the total number of clusters can be either 2 or 3.

In order to find the optimal number of clusters, silhouette score is used. As the number of cluster centers given by the elbow method lies in the range of [2, 9], we will use n_clusters for determining the optimal number of cluster centers by using the silhouette scores, where n_clusters is a number in the range [2, 9]. The following silhouette scores were obtained for each cluster which is shown in Table 2.

It can be inferred from Table 2 that the value of n_clusters as 4 and 5 are a bad pick as they have the least average silhouette score for the given set of data. The average

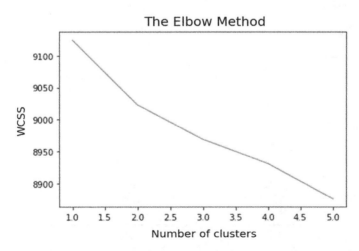

Fig. 2 Elbow method

Table 2 Average silhouette scores

Number of clusters	Average silhouette score
n_clusters = 2	0.65
n_clusters = 3	0.74
n_clusters = 4	0.58
n_clusters = 5	0.41

sihouette score of n-clusters = 2 is 0.65 which is greater than that of n_cluster value of 4 and 5 but is still less than the average silhouette score of n_clusters = 3. Hence, Table 2 clearly depicts that the average silhouette score for n_clusters value of 3, is the highest. Hence, 3 is the optimal number of clusters. After training, the following three clusters are obtained, which are shown in Fig. 3.

After performing these steps, the cosine similarity is calculated for all the terms using the TF-IDF scores of each term for every cluster. For hierarchical clustering [24, 25], the maximum difference between the Euclidian distance of the level of each pair of group is chosen. Then, a horizontal line is passed in the middle of the maximum difference of that Euclidean distance. The number of points the line cuts gives the number of cluster centers. From the dendrogram [26] plotted in Fig. 4, the number of cluster centers is given as three.

The cosine similarity was calculated for all the terms for each cluster using the TF-IDF scores of each term. Cosine similarity is a common technique used to measure the cohesion within the clusters in the field of data mining [18]. Cosine similarity is a measure of similarity between two nonzero vectors of an inner product space that measures the cosine of the angle between them as in (7). The cosine of 0° is 1, and it is less than 1 for any other angle.

To find the cosine distance of one email and all the others, one just needs to compute the dot products of the first vector with all of the others as TF-IDF vectors

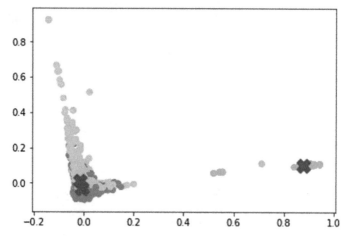

Fig. 3 Clusters

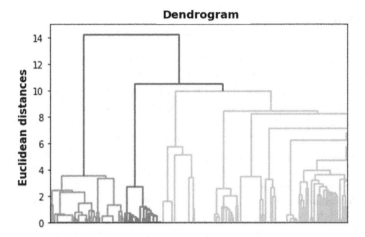

Fig. 4 Dendrogram

are already row normalized. Scikit-learn already provide pairwise metrics (also known as kernels in machine learning parlance) that work both for dense and sparse representations of vector collections [27]. In this case, the dot product is also known as the linear kernel [18]. The cohesion score of the three clusters is given in Table 3.

The cohesion scores given in Table 3 depict that the majority of the information and relevant insights can be extracted through cluster 1 as it has the highest cohesion score. By using the k-means algorithm, the cohesion score of cluster 1 is 0.1655, which is the highest as compared to the cohesion scores of 0.1011 and 0.1252 for the cluster 0 and cluster 2, respectively. For the hierarchical clustering algorithm, the least cohesion score of 0.1421 is obtained by cluster 2 while the highest score is obtained by cluster 1 which corresponds to 0.2513. Moreover, it can also be observed from Table 3 that the cohesion score obtained by cluster 0 is 0.1900. All these observations further strengthen the relevancy of the textual content and help in obtaining dependable and precise insights by choosing a specific cluster. Hence, it can be inferred that all the three clusters can be utilized for getting the holistic insights and specific chosen clusters can be used to gain pertinent and relevant insights.

Table 3 Cohesion score for each cluster

Clustering algorithms	Cohesion score		
	Cluster 0	Cluster 1	Cluster 2
K-means	0.1011	0.1655	0.1252
Hierarchical	0.1900	0.2513	0.1421

6   Conclusion

Email clustering utilizes several natural language processing and data mining activities such as: text parsing, stemming, classification, clustering, etc. There are many reasons for carrying out clustering whether in real time or historical. This may include reasons such as: spam detection, subject or folder classification, information extraction, etc. The clustering of this email dataset revealed quite a few interesting aspects. One of them was the score distribution among the clusters as well as the similarity of the clusters obtained by the two different clustering algorithms. The cohesion score helped in concentration of analysis to a specific cluster for better insights. Further investigation could explain as to the reasons for this skewed score distribution as well as the resulting conclusions from this similarity of clusters. Soft clustering methods like fuzzy clustering could be utilized to determine the degree of membership each data point has to each cluster, opening up further avenues of investigation.

References

1. Alsmadi, I., Alhami, I.: Clustering and classification of email contents. J. King Saud Univ.-Comput. Inf. Sci. 27(1), 46–57 (2015)
2. Chen, M.: Soft clustering for very large data sets. Comput. Sci Netw Secur. J 17(11), 102–108 (2017)
3. Chiang, M.M.-T., Mirkin, B.: Intelligent choice of the number of clusters in k-means clustering: an experimental study with different cluster spreads. J. Classif. 27, 3–40 (2010)
4. Azizpour, S., Giesecke, K., Schwenkler, G.: Exploring the sources of default clustering. J. Financ. Econ. 129(1), 154–183 (2018)
5. Xie, J., Girshick, R., Farhadi, A.: Unsupervised deep embedding for clustering analysis. In: International Conference on Machine Learning, pp 478–487 (2016)
6. Nayak, P., Devulapalli, A.: A fuzzy logic-based clustering algorithm for WSN to extend the network lifetime. IEEE Sens J 16(1), 137–144 (2015)
7. Ferrari, D.G., De Castro, L.N.: Clustering algorithm selection by meta-learning systems: a new distance-based problem characterization and ranking combination methods. Inf. Sci. 301, 181–194 (2015)
8. Lensen, A., Xue, B., Zhang, M.: Using particle swarm optimisation and the silhouette metric to estimate the number of clusters, select features, and perform clustering. In: European Conference on the Applications of Evolutionary Computation, pp. 538–554. Springer, Cham (2017)
9. Basavaraju, M., Prabhakar, D.R.: A novel method of spam mail detection using text based clustering approach. Int. J. Comput. Appl. 5(4), 15–25 (2010)
10. Huang, Y., Mitchell, T.M.: Exploring hierarchical user feedback in email clustering. Email 8, 36–41 (2008)
11. Ercan, G., Cicekli, I.: Lexical cohesion based topic modeling for summarization. In: International Conference on Intelligent Text Processing and Computational Linguistics, pp. 582–592. Springer, Berlin (2018)
12. Klebanov, B.B., Diermeier, D., Beigman, E.: Lexical cohesion analysis of political speech. Polit. Anal. 16(4), 447–463 (2008)
13. Pérez, R.A., Pagola, J.E.M.: Text segmentation by clustering cohesion. In: Iberoamerican Congress on Pattern Recognition, pp. 261–268. Springer, Berlin (2010)

14. Behrens, T., Schmidt, K., Viscarra Rossel, R.A., Gries, P., Scholten, T., MacMillan, R.A.: Spatial modelling with euclidean distance fields and machine learning. Eur. J. Soil Sci. **69**(5), 757–770 (2018)
15. Rathee, A., Chhabra, J.K.: Improving cohesion of a software system by performing usage pattern based clustering. Procedia Comput. Sci. **125**, 740–746 (2018)
16. Kulkarni, A., Pedersen, T.: Name discrimination and e-mail clustering using unsupervised clustering of similar contexts. J. Intell. Syst. **17**(1–3), 37–50 (2008)
17. Reed, J.W., Jiao, Y., Potok, T.E., Klump, B.A., Elmore, M.T., Hurson, A. R.: TF-ICF: a new term weighting scheme for clustering dynamic data streams. In: 2006 5th International Conference on Machine Learning and Applications (ICMLA'06), pp. 258–263. IEEE (2006)
18. Hermans, F., Murphy-Hill, E.: Enron's spreadsheets and related emails: a dataset and analysis. In: 2015 IEEE/ACM 37th IEEE International Conference on Software Engineering vol. 2, pp. 7–16. IEEE (2015)
19. Al-Anzi, F.S., AbuZeina, D.: Toward an enhanced Arabic text classification using cosine similarity and latent semantic indexing. J King Saud Univ-Comput. Inf. Sci **29**(2), 189–195 (2017)
20. Bernard, J.: Python data analysis with pandas. In: Python Recipes Handbook, pp 37–48. Apress, Berkeley, CA (2016)
21. Gupta, R., Jivani, A.G.: Analyzing the stemming paradigm. In: International Conference on Information and Communication Technology for Intelligent Systems, pp 333–342. Springer, Cham (2017)
22. Capó, M., Pérez, A., Lozano, J.A.: An efficient approximation to the K-means clustering for massive data. Knowl. Based Syst. **117**, 56–69 (2017)
23. Syakur, M.A., Khotimah, B.K., Rochman, E.M.S., Satoto, B.D.: Integration k-means clustering method and elbow method for identification of the best customer profile cluster. In: IOP Conference Series: Materials Science and Engineering, vol. 336, no. 1, p. 012017. IOP Publishing (2018)
24. Zhou, S., Xu, Z., Liu, F.: Method for determining the optimal number of clusters based on agglomerative hierarchical clustering. IEEE Trans. Neural Netw. Learn. Syst. **28**(12), 3007–3017 (2016)
25. Day, W.H., Edelsbrunner, H.: Efficient algorithms for agglomerative hierarchical clustering methods. J. Classif. **1**(1), 7–24 (1984)
26. Ferreira, L., Hitchcock, D.B.: A comparison of hierarchical methods for clustering functional data. Commun Stat. Simul. Comput **38**(9), 1925–1949 (2009)
27. Kent, D., Toris, R.: Adaptive autonomous grasp selection via pairwise ranking. In: 2018 IEEE/RSJ International Conference on Intelligent Robots and Systems (IROS), pp. 2971–2976. IEEE (2018)

Congestion Control for Named Data Networking-Based Wireless Ad Hoc Network

Farkhana Muchtar, Mosleh Hamoud Al-Adhaileh, Raaid Alubady, Pradeep Kumar Singh, Radzi Ambar and Deris Stiawan

Abstract There is lack of complete congestion control solution that is optimized or practical with the Named Data Networking (NDN)-based MANET environment. All the existing suggested solutions are either for general NDN which is not optimized for MANET environment or congestion control solution for incomplete NDN-based MANET. Therefore, we recommend a complete congestion control solution specifically for NDN-based MANET which we call Standbyme Congestion Control or simply called Standbyme. Standbyme design optimized for NDN-based MANET needs in reducing network congestion's bad effect such as goodput reduction, increment of number of packet loss or increment of delay in NDN-based MANET. Through the testbed experiment, we did by comparing Standbyme

F. Muchtar (✉)
School of Computing, Faculty Engineering, Universiti Teknologi Malaysia, 81310 Skudai, Johor, Malaysia
e-mail: farkhana@gmail.com

M. H. Al-Adhaileh
Deanship of E-Learning and Distance Education, King Faisal University, Al Ahsa, Kingdom of Saudi Arabia
e-mail: madaileh@kfu.edu.sa

R. Alubady
Information Technology Department, College of Information Technology, University of Babylon, Hillah, Iraq
e-mail: alubadyraaid@itnet.uobabylon.edu.iq

P. K. Singh (✉)
Department of CSE & IT, Jaypee University of Information Technology, Waknaghat, Solan, Himachal Pradesh 17334, India
e-mail: pradeep_84cs@yahoo.com

R. Ambar
Faculty of Electrical & Electronic Engineering, Universiti Tun Hussein Onn Malaysia, 86400 Batu Pahat, Malaysia
e-mail: aradzi@uthm.my

D. Stiawan
Faculty of Computer Science, Universitas Sriwijaya, Jalan Srijaya Negara, 30139 Palembang, Indonesia
e-mail: deris@unsri.ac.id

© Springer Nature Singapore Pte Ltd. 2020 121
P. K. Singh et al. (eds.), *Proceedings of First International Conference on Computing, Communications, and Cyber-Security (IC4S 2019)*, Lecture Notes in Networks and Systems 121, https://doi.org/10.1007/978-981-15-3369-3_10

with other congestion control methods we selected for comparison, i.e., a practical congestion control for NDN (PCON) congestion control and best effort link reliability protocol (BELRP) congestion control, indicating Standbyme was able to drastically reduce network congestion in NDN-based MANET. Without sacrificing the performance of NDN-based MANET, Standbyme has also reduced bad effect of network congestion through better approach of congestion prevention and reduction in MANET environment.

Keywords MANET · Mobile ad hoc network · NDN · Named Data Networking

1 Introduction

Network congestion refers to congestion of network traffic when the network traffic rate exceeds the capacity that can be handled by network resources being used. Congestion network problem in MANET is a critical issue because of two major factors, namely (1) the rate of network congestion that occurs in MANET is far much higher than infrastructured network, and at the same time (2) MANET has limited resources especially network bandwidth and energy supply at each mobile node [1].

Based on the literature by Kanellopoulos [1], Lochert et al. [2], it can be concluded that MANET has higher congestion network than infrastructured network because of the three major factors which are as follows: (1) Available network bandwidth in MANET is actually limited shared resources and each network traffic flow will compete with each other on usage of the limited bandwidth resource. (2) Wireless network has higher network congestion rate compared to wired network due to wireless signal interference and noise, network packet contention and collision, bit error and hidden terminal phenomenon. (3) Network of MANET is dynamic due to movement of mobile node in MANET, and it will trigger network congestion.

According to Kang et al. [3], Seddik-Ghaleb et al. [4], Vyas and Deshpande [5], Sharma et al. [6], Vadivel and Bhaskaran [7], network congestion not only reduces the throughput of network traffic in MANET but it also causes energy wastage due to excessive retransmission process of network packets that failed to be sent due to network congestion. Energy wastage occurs due to high-frequency rate of dropped packet retransmission due to high level of congestion rate [2].

Although there are some researchers that mentioned issues related to congestion control solution for Named Data Networking (NDN)-based MANET such as Amadeo et al. [8, 9] and Li et al. [10–12], the discussions made are not elaborated in depth and at the same time the proposed congestion control solutions are often incomplete to be stated as a congestion control solution for NDN-based MANET. Past research also tends to observe the effect of improving the performance of suggested congestion control solution but pay less attention toward increasing energy efficiency of NDN-based MANET when utilizing the congestion control they have developed.

The objective of this experiment is to identify, design and develop an optimum and practical congestion control solution for NDN-based MANET. In Sect. 2, we will discuss congestion control solution for NDN as suggested by previous study. Then in Sect. 3, we discuss the design and architecture of this research called Standbyme Congestion Control. We then describe how the experiments were conducted in Sect. 4, and the analysis of the experiment results is discussed in Sect. 5. Finally, we summarize the findings we have gained from this research and what is our next planning for the continuation of this research.

2 Related Work

Real mobility mechanism in MANET testbed facility is very important in order to obtain results that are accurate and realistic. However, real mobility implementation in MANET testbed can be very challenging. Nonetheless when done correctly, real mobility mechanisms provide accurate result for experiments conducted in MANET testbed [13]. Node mobility in network simulator is represented by mobility model that determines the location, velocity and acceleration of the mobile node in MANET [4]. However, in real testbed experiment, there are various ways that have previously been used by researchers to create mobility mechanisms in MANET testbed facilities either using virtual mobility or real mobility [5].

In this section, we will be observing existing congestion control solution for NDN in general and congestion control solution for NDN-based MANET in order to gain better insights into what previous studies have already accomplished, and the achievements that they have yet to attain thus tries to tap into opportunities for us to establish the issues of choice in our investigation.

2.1 Congestion Control for NDN in General

To facilitate our discussion on existing congestion control for NDN in general, we have grouped previous work based on the categories we have combined from [13, 14].

2.1.1 Receiver-Driven Versus Hop-by-Hop Congestion Control for NDN

Receiver-based congestion control for NDN refers to congestion control that is only performed by receiver nodes (consumer nodes) particularly with congestion avoidance approach such as interest shaping [14]. Examples of receiver-driven congestion control for NDN are ConTug [15], CCTCP [16], RAAQM [17] and PCON [18].

Hop-by-hop congestion control for NDN on the other hand refers to congestion control that is not only performed by receiver nodes (consumer nodes) but also includes the ones performed by intermediate nodes such as network routers in infrastructured network and intermediate nodes in wireless ad hoc network. Hop-by-hop congestion control method allows congestion avoidance action to be performed directly by intermediate nodes without the need to wait for congestion avoidance actions from consumer nodes [14, 19–21].

Examples of previous work on congestion control for NDN using hop-by-hop congestion control are hop-by-hop interest shaping mechanism (HoBHIS) [22, 23], an improved hop-by-hop interest shaper [24], popularity-based congestion control [25], RCP-based congestion control protocol [26], multipath flow control (MFC) [27], rate-based congestion control [28, 29] and hop-by-hop window-based approach (HWCC) [13].

2.1.2 Window-Based Traffic Shaping Versus Rate-Based Traffic Shaping

To provide clear consistent information that eliminates confusion, we take the initiative to use the term traffic shaping to represent flow control, interest sending rate, interest control protocol and interest shaping. Traffic shaping in congestion control for NDN refers to the control on the amount of interest packets to be forwarded at one time to determine how much data packet will be received based on the current network condition.

There are two types of traffic shaping methods used in congestion control, namely window-based traffic shaping and rate-based traffic shaping. Window-based traffic shaping refers to setting on the amount of interest packets sent at any one time based on current window size. Window size is only increased if all previous interest packets receive all the data packets matching and the rate of increase follows the algorithm that is used as slow start and AIMD. Examples of previous work on congestion control for NDN using window-based traffic shaping methods are ICP [30], ICTP [31], CCS [32], CCTCP [16], RAAQM [17], CHoPCoP/pCHoPCoP [33], PCON [18] and HWCC [13].

Unlike window-based traffic shaping method, rate-based traffic shaping *is* the number of interest packets that is going to be forwarded based current available local link bandwidth estimations. There are several methods that can be used to perform bandwidth estimations such as leaky bucket algorithm [34], token bucket algorithm [35], characterizing interest aggregation [36] and average occupancy of the PIT [37].

Among the existing congestion control solutions for NDN that uses rate-based traffic shaping method are congestion control in stateful forwarding [34], ECN for NDN [38], HoBHIS [22, 23], SECN [38], HIS [24] dan HR-ICP [19].

2.1.3 Local Congestion Detection Versus Congestion Notification

There are two main options to detect network congestion in NDN which are whether it is detected locally or notified by other nodes using congestion notification. Local congestion detection means the mechanism by which any node in the MANET is located to detect congestion at that local link node. While congestion notification refers to the congestion signal sent by the nodes that detect local link congestion to downstream neighbor nodes to notify neighbor nodes that there are network congestion occurs on the node which is sending the congestion notification.

There are several approaches have been proposed to detect local link congestion within the NDN. The most popular approach is to use RTO (timeout) to detect network congestion on local links. Examples of previous work using this method are ICP [30], ICTP [31], CCTCP [16] and HR-ICP [19]. RTO-based local congestion detection is the simplest method. However, at the same time, it is not practical to use it in NDN as the content can be obtained from different paths and sources and has different RTO values [14].

There are also congestion control solutions for NDN that uses queue length (buffer size) monitoring method to detect network congestion such as CHoPCoP/pCHoPCoP [33], ECN for NDN [38] and PCON [18]. This method is inherited from packet loss-based congestion detection methods from TCP congestion control such as TCP Reno. PCON uses more advanced approach to perform queue length monitoring called CoDel AQM that detects packet loss without the need to wait for buffer over flow to occur on the interest packet queue.

There are various forms of congestion notification used in previous studies such as congestion mark and congestion NACK. Congestion mark is some extra network packet header that piggybacked data packet. Congestion mark provides additional information related to congestion control such as network congestion status and window size or rate size suggested based on current available local link bandwidth. On the other hand, congestion NACK is the negative acknowledgment sent to downstream neighbors including consumer node when network congestion is detected.

2.2 Congestion Control for NDN-Based MANET

Not much can be found on researches that discuss network congestion issues in wireless ad hoc network environment including MANET. Existing congestion control solution for NDN mainly sets the assumption that link capacity is fixed and already known, hence this assumption certainly cannot be true for wireless networks [39]. In fact, all researches previously conducted never have provided a complete congestion control solution for NDN-based MANET but rather merely focus on specific issues pertaining to network congestion only.

E-CHANET [8] and self-regulating interest rate control (SIRC) [9] are among the earliest studies that propose congestion control solution for NDN-based MANET.

Both solutions focus on the interest sending rate and data packet transmission rate with the help of data gap (E-CHANET) or RTO (SIRC) in determining the interest sending rate to be set. We think that both solutions are still not optimal for the NDN-based MANET as they are receiver-driven congestion control and congestion control strategy which is proposed using interest rate only.

While chunk-switched hop pull control protocol or CHoPCoP [33, 40] is claimed by the author as a congestion control solution for wired NDN and wireless NDN including NDN-based MANET. But, like E-CHANET and SIRC, CHoPCoP is also a receiver-driven congestion control and congestion avoidance that relies solely on control over interest sending rates.

Finally, congestion control solution proposed in the literature of Li et al. [10–12] has only interest sending rate mechanism based on estimated bandwidth alone. This is because the focus of the literature is on how the energy efficiency of NDN-based MANET can be achieved through integration of solutions using cross-layer methods to several components such as congestion control, forwarding strategy, link scheduler and several other components.

3 Contribution of This Study

Most of the previous congestion control solution for NDN is for wired networks and is less practical for use in MANET environments as it still maintains receiver-driven congestion control methods, relying only on interest sending rate for congestion avoidance purposes, less congestion detection methods for wireless network environment, suitable only for single source scenario and ignores multiple source scenario.

This research focuses on the design and development of congestion control solution for NDN-based MANET. Therefore, the congestion control solution proposed in this research, which is called Standbyme Congestion Control, is the first fully optimized, optimal and practical congestion control solution for MANET environment.

4 Standbyme Congestion Control as Suggested Solution

Based on what we learned from previous related work regardless of whether it is related to NDN infrastructured, host centric MANET or NDN-based MANET, we have designed and developed a congestion control solution suggestion that is named as Standbyme Congestion Control. Standbyme Congestion Control consists of three main components, namely (i) accurate local congestion detection for NDN-based MANET (ii) hop-by-hop congestion notification and (iii) congestion avoidance mechanisms for NDN-based MANET as shown in Fig. 1.

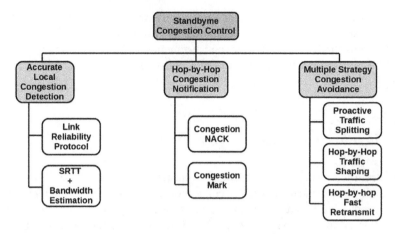

Fig. 1 Standbyme architecture

4.1 Local Congestion Detection

Local congestion detection approach used in Standbyme Congestion Control needs to be suitable for use in NDN-based MANET environments. Accurate and efficient local congestion detection method is very important for the implementation of congestion control in wireless network [41]. The efficiency of local congestion detection for MANET environment is not limited only from the perspective of network bandwidth alone but it also takes into consideration the limitations of energy resources on each mobile nodes in MANET.

To achieve accurate and efficient local congestion detection in NDN-based MANET environment, we propose to use two levels of detection using two different approaches. The first method of network congestion detection is done through the link reliability protocol approach adopted in the link protocol, namely NDNLPv2 in NFD. The second method is to use local link bandwidth estimation information through the SRTT probing adapting method to validate local link congestion detected through the first stage to prevent false positives.

First stage local link congestion detection was performed using the energy efficient link reliability protocol (EELRP), a link reliability protocol adapted from the best effort link reliability protocol (BELRP) Vusirikala et al. [20] specific to the MANET environment. The modifications performed on BELRP to EELRP are (1) sequence checking only implemented on link layer compared to BELRP that implement sequence checking at link layer and network layer. (2) Additional callback function for forwarding strategy when link loss caused by network congestion is detected to allow further action in forwarding strategy if network congestion is detected locally. (3) Combining sequence checking and RTO to detect link loss because in some condition, the number of network packet is not enough to confirm link status through sequence checking.

Second stage local link congestion detection is more toward confirmation of local link congestion event detected in the first stage through the local link bandwidth estimation method. Local link bandwidth estimation is measured using the adaptive SSRT probing method borrowed from the adaptive SRTT-based forwarding strategy [42].

4.2 Hop-by-hop Congestion Notification

Apart from local congestion detection, congestion detection in Standbyme is also conducted through the use of explicit congestion notification (ECN) method to detect network congestion that occurs in other remote nodes. The main reason to allow congestion notification on remote neighbour nodes is to allow proactive action by downstream neighbour nodes when one or more nodes in content path facing network congestion by implementing traffic shaping to reduce traffic load and if alternate path exist, downstream neighbour nodes can use alternate path instead of congested path.

As Standbyme Congestion Control is designed specifically for NDN-based MANET needs, hop-by-hop congestion notification method is the right choice because it is impractical to assume that only consumer node needs to receive congestion notification in MANET environment. By taking into consideration the dynamic environment and network bandwidth differences between one hop to another, hop-by-hop congestion notification is found to be the most suitable method of choice [43].

4.3 Multiple Strategy Congestion Avoidance

As for congestion avoidance mechanism, we choose to use three different methods and combine it to be used as a congestion avoidance solution in Standbyme Congestion Control, namely (1) proactive traffic splitting for multipath content routing, (2) hop-by-hop traffic shaping and (3) fast retransmit and fast recovery. Although all three methods are different, each of these methods is interconnected with each other to create a complete congestion control solution for NDN-based MANET needs.

4.3.1 Proactive Traffic Splitting for Multipath Content Routing

Existing multipath content routing approach used in most forwarding strategies including best route forwarding strategy in NDN Forwarding Daemon (NFD) is done for the purpose of fault-tolerant content routing using adaptive forwarding strategy. Network traffic will only be forwarded using other alternate paths when current best route encounters problems such as packet drop, high congestion rate and link failure [44].

To optimize the use of multipath content routing in NDN-based MANET, it is more practical to use multipath content routing not only to function as fault tolerance but also can be used for network load and energy balancing since network bandwidth and energy source are very limited in MANET. The use of load distribution method is a cheap and practical way to increase the availability of network bandwidth in MANET, and this at the same time increases energy conservation for each mobile node. Load distribution method in Standbyme Congestion Control can become as an early preventive measure to avoid network congestion from occurring using proactive traffic splitting approach that avoids network traffic from being concentrated at any one single route if multipath is available.

Roulette wheel selection algorithm which is a popular selection method used in genetic algorithm, evolutionary algorithms and complex network modeling is selected as a selection method for proactive traffic splitting within Standbyme Congestion Control. The Roulette wheel selection algorithm uses the probabilistic selection method that has the highest fitness or score or highest probability of choice. For content path optimization in proactive traffic splitting, the next hop content path with the highest available bandwidth has the highest probability of being selected as the next hop content path during interest forwarding.

4.3.2 Hop-by-Hop Traffic Shaping for Congestion Avoidance

Traffic shaping is congestion avoidance method that reduces the demand so that it matches the available supplied network bandwidth by reducing interest sending rate that pass through the selected content path [38]. We next proceed to elaborate details on how traffic shaping is done in Standbyme Congestion Control and how the hop-by-hop traffic shaping mechanism can be used to reduce network congestion and how it avoids other worse negative effects if the network congestion still occurs. We introduce three states of traffic shaping in Standbyme Congestion Control, each state has its own distinct traffic shaping rate behaviors, and we present those differences accordingly. Figure 2 shows the summary of all the traffic shaping states.

i. **Normal State**
    Normal state indicates normal network traffic conditions without the occurrence of network congestion. In normal state, interest sending rate is changed based on the modified slow-start traffic shaping method. The purpose of using slow-start method is to achieve network bandwidth efficiency until maximum limit of interest sending rate [M AXi(t)].

ii. **Fast Restransmit**
    In fast retransmit state, interest sending rate does not change constantly. This is because in the fast retransmit state, all the resources especially network band width are used only for the purpose of retransmitting all the interest packets that fail to send or who fail to receive packet data as a result of network congestion

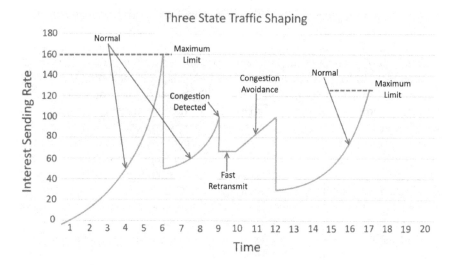

Fig. 2 Three states of traffic shaping in Standbyme Congestion Control

without waiting for timeout. Further discussion on the mechanisms pertaining to fast retransmission in Standbyme Congestion Control can be seen in Sect. 4.3.3.

iii. **Congestion Avoidance State**

Congestion avoidance state on the other hand refers to the traffic shaping used when network congestion is detected at current local link or when it receives congestion notification from upstream neighbor nodes. Contrary to the normal state, interest rate is linearly increased until it reaches a new maximum limit of interest rate [M AXi(t)] and drops exponentially similar to additive-increase/multiplicative-decrease (AIMD) traffic shaping method.

4.3.3 Fast Retransmit and Fast Recovery

Some of the interest packets failed to send to the next hop neighbor node or failed to receive matched packet data due to network congestion. Consumer node will perform fast retransmission without the need to wait for timeout period. Fast recovery refers to the mechanism that ensures fast retransmission is conducted and given priority to do so. When fast retransmission is performed, other congestion avoidance such as traffic shaping will not take place. This mechanism is to ensure that the dropped packet to be delivered is given higher priority instead of the new network traffic.

5　Testbed Design and Facility

Experiments carried out in this research use real-world testbed method to obtain more accurate and realistic experimental result. To meet this need, a private testbed facility was developed with a testbed management system called testbed on MANET or abbreviated ToM. At the same time, some hardware and tools are used to meet the needs of testbed facilities such as the use of Banana Pro and Banana Pi M1 + Single Board Computer (SBC) (see Fig. 3) to represent mobile devices and ESP8266 microcontroller as remote control for each mobile device remotely.

5.1　*Experiment Design and Analysis of Result*

This article covers only the first stage of the experiment, which is an experiment performed under static network topology. No node mobility occurred when the experiment was run because the focus was first on the effectiveness of Standbyme Congestion Control solution in addressing network congestion caused by wireless ad hoc network environment compared to two existing congestion control solutions, namely PCON and BELRP.

Two network topologies were selected in the first phase of this experiment, the baseline network topology and the dumbbell network topology as shown in Fig. 3.

For performance metric, we choose goodput, packet loss rate and RTT average (delay) to measure the efficiency of congestion control solution. All three measurements were obtained using the statistic result generated by the application ndncatchunks. Ndncatchunks and ndnputchunks are applications that we use to implement file transfer transaction based on NDN protocol to generate network traffic when experiments are performed.

To generate network traffic in the experiments that we conducted, we utilized file transfer method in NDN by combining ndnputchunks application in the provider node and ndncatchunks application in the consumer node. A PDF file of NFD Developer

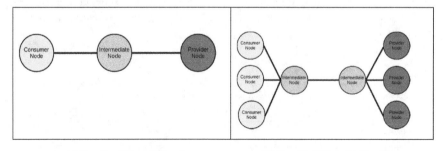

　　　(a) Baseline Network Topology　　　　　　(b) Dumbbell Network Topology

Fig. 3　Static network topology scenarios for experiment

Guide sized at 2063.6 kilobytes is used to create 469 data segments that will be made as data packets for content retrieval during the experiment.

5.1.1 Baseline Network Topology

Baseline network topology is the most basic multihop network topology scenario involving only one consumer node, one provider node and one intermediate node between consumer node and provider node. The purpose of this baseline network topology used in this experiment is to benchmark the results obtained when implemented in the most basic wireless ad hoc network conditions.

The experiment results obtained from the baseline network topology scenario (see Table 1) indicated that there are significant differences when utilizing Standbyme and PCON as well as BELRP from the perspective of efficiency of congestion control usage in the Kruskal–Wallis H test.

The result from the experiment in Table 1 also indicated that network congestion only occurred at links between consumer nodes and intermediate nodes but does not occur at links between intermediate nodes and provider nodes because there is no congestion mark received by the consumer node during the experiment. Based on Kruskal–Wallis H test, all of the parameters that we have selected for as comparison, namely goodput, packet loss rate and RTT average proved significantly different. Hence, comparisons can be performed between Standbyme with PCON and BELRP.

Based on the Kruskal–Wallis H significant test for the results of all three parameters from three different congestion control is significantly different. Goodput produced while using Standbyme is slightly better than goodput when using PCON but much better than goodput when using BELRP.

At the same time, packet loss rate when using Standbyme is slightly lower than packet loss when using PCON but much lower than packet loss rate when using BELRP. Similarly, the delay between packet data where RTT average when using Standbyme is slightly lower than when using PCON but is significantly lower than RTT average when using BELRP.

What can be deduced in the baseline network topology scenario is that the efficiency of Standbyme is slightly better than PCON but much better than BELRP. A summary of the experiment result from the baseline topology scenario can be seen in Fig. 4.

Table 1 Baseline network topology experiment result

Parameters	Congestion control solution		
	Standbyme	PCON	BELRP
Time elapsed (millisecond)	12872.29	12760.12	13190.99
Goodput (kilobit/second)	282.52	2.82	273.27
Packet loss rate (%)	1.72	2.82	5.80
RTT avg (millisecond)	241.65	393.59	379.00

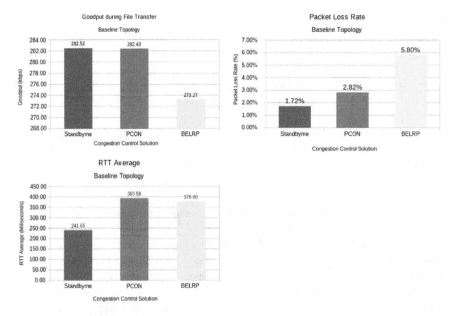

Fig. 4 Experiment results from baseline network topology scenario

5.1.2 Dumbbell Network Topology

Dumbbell network topology scenario is performed for observing the efficiency of the congestion control solution to control network congestion in situation where more than one content retrieval occurs at any one time in NDN-based MANET. Network traffic such as this occurs in actual wireless ad hoc network where more than one transaction occurring at any one time exists. This scenario is created to enable us to prove that Standbyme approach is able to reduce the bad effect such as the decrease in goodput, the increase of packet loss rate and delay caused by network congestion even though it is triggered by more than one transaction occurring at the same time.

Experimental results obtained from the experiments carried out in the dumbbell network topology scenario (see Table 2) is also calculated in the significance test using Kruskal–Wallis H test before comparisons were made on the three congestion

Table 2 Dumbbell network topology experiment result

Parameters	Congestion control solution		
	Standbyme	PCON	BELRP
Time elapsed (millisecond)	127852.04	196219.99	195919.55
Goodput (kilobit/second)	282.52	22.99	21.92
Packet loss rate (%)	24.73	37.18	36.07
RTT avg (millisecond)	97093.69	12587.21	552043.69

control solutions. Comparison of Standbyme between two existing congestion control solutions for NDN, namely PCON and BELRP, is conducted to evaluate whether Standbyme does improve congestion control efficiency compared to PCON and BELRP.

Results from the Kruskal–Wallis H test prove that goodput, packet loss rate and RTT average (delay) results for all three congestion control solutions are significantly different and valid to be compared for analysis.

Based on the Kruskal–Wallis H significant test,the results of all three parameters from the different congestion control was significant for comparison. In contrast to the baseline network topology scenario, experiment results using Standbyme showed significantly better results than using PCON and BELR as a congestion control solution.

The average value of goodput when using Standbyme in dumbbell topology is much better than goodput when using PCON or BELRP. In addition, packet loss rate when using Standbyme in the dumbbell network topology scenario is significantly lower than packet loss rate when using PCON or BELRP. Finally, the RTT average when using Standbyme is significantly lower than when using PCON and BELRP.

What can be deduced for this dumbbell network topology scenario is that Standbyme Congestion Control is more efficient than PCON and BELRP in a wireless ad hoc network environment. A summary of the experiment result with the dumbbell network topology scenario can be seen in Fig. 5.

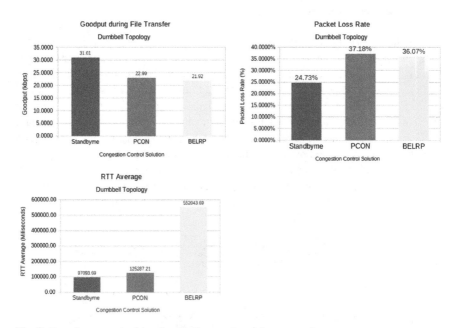

Fig. 5 Experiment results from dumbbell network topology scenario

6 Conclusion and Future Work

6.1 Conclusion

In the first phase of this research, the focus was on the effectiveness of Standbyme Congestion Control in addressing network congestion triggered by network traffic and other factors related to wireless ad hoc network only such as wireless communication and ad hoc network routing without involving node mobility. The purpose is to isolate network congestion caused by wireless communication and also ad hoc network environment with congestion control caused by node mobility.

Based on the analysis of the experiment result, it can be concluded that the efficiency of Standbyme as congestion control for NDN-based wireless ad hoc network is better than PCON and far better than BELRP. Standbyme Congestion Control improves goodput in the NDN-based wireless ad hoc network while reducing packet loss rate and delay compared to PCON and BELRP.

In the next phase of our research, our experiments shall involve real node mobility to see the effectiveness of the Standbyme Congestion Control solution in managing network congestion in MANET as a result of the combination of network traffic, wireless ad hoc network and node mobility compared to existing congestion control solution for NDN, namely PCON and BELRP.

References

1. Kanellopoulos, D.: Congestion control for manets: an overview. ICT Express (2018). https://doi.org/10.1016/j.icte.2018.06.001. URL https://www.sciencedirect.com/science/article/pii/S2405959518302285
2. Lochert, C., Scheuermann, B., Mauve, M.: A survey on congestion control for mobile ad hoc networks. Wirel. Commun. Mob. Comput. 7(5), 655–676 (2007). https://doi.org/10.1002/wcm.524. URL http://onlinelibrary.wiley.com.ezproxy.psz.utm.my/doi/10.1002/wcm.524/abstract.00149
3. Kang, J., Zhang, Y., Nath, B.: Accurate and energy-efficient congestion level measurement in ad hoc networks. In: IEEE Wireless Communications and Networking Conference, 2005, vol. 4, pp. 2258–2263 (2005). https://doi.org/10.1109/wcnc.2005.1424867
4. Seddik-Ghaleb, A., Ghamri-Doudane, Y., Senouci, S.M.: TCP computational energy cost within wireless mobile ad hoc network. In: 2008 33rd IEEE Conference on Local Computer Networks (LCN), pp. 522–524 (2008). https://doi.org/10.1109/lcn.2008.4664220
5. Vyas, G.S., Deshpande, V.S.: Performance analysis of congestion in wireless sensor networks. In: 2013 3rd IEEE International Advance Computing Conference (IACC), pp. 254–257 (2013). https://doi.org/10.1109/iadcc.2013.6514230
6. Sharma, N., Gupta, A., Rajput, S.S., Yadav, and V.K.: Congestion Control Techniques in MANET: a survey. In: 2016 Second International Conference on Computational Intelligence Communication Technology (CICT), pp. 280–282 (2016). https://doi.org/10.1109/CICT.2016.62
7. Vadivel, R., Bhaskaran, V.M.: Adaptive reliable and congestion control routing protocol for MANET. Wirel. Netw. 23(3), 819–829 (2017). https://doi.org/10.1007/s11276-015-1137-3. Bibtex: Vadivel2017 bibtex ISBN: 1127601511373

8. Amadeo, M., Molinaro, A., Ruggeri, G.: E-CHANET: routing, forwarding and transport in information-centric multihop wireless networks. Comput. Commun. **36**(7), 792–803 (2013). https://doi.org/10.1016/j.comcom.2013.01.006. URL http://www.sciencedirect.com/science/article/pii/S0140366413000248

9. Amadeo, M., Molinaro, A., Campolo, C., Sifalakis, M., Tschudin, C.: Transport layer design for named data wireless networking. In: 2014 IEEE Conference on Computer Communications Workshops (INFOCOM WKSHPS), pp. 464–469 (2014). https://doi.org/10.1109/infcomw.2014.6849276

10. Li, C.C., Xie, R.C., Huang, T., Liu, Y.j.: Cross-layer congestion control in named-data multihop wireless networks. Adhoc Sens. Wirel. Netw. **39**(1–4), 61–95 (2017)

11. Li, C., Xie, R., Huang, T., Liu, Y.: Jointly optimal congestion control, forwarding strategy and power control for named-data multihop wireless Network. IEEE Access **5**, 1013–1026 (2017). https://doi.org/10.1109/access.2016.2634525

12. Li, C.C., Xie, R.C., Huang, T., Liu, Y.j.: Jointly optimized congestion control, forwarding strategy, and link scheduling in a named-data multihop wireless network. Front. Inf. Technol. Electron. Eng. **18**(10), 1573–1590 (2017). https://doi.org/10.1631/fitee.16001585. URL https://link.springer.com/article/10.1631/fitee.16001585

13. Kato, T., Bandai, M., Yamamoto, M.: A congestion control method for named data networking with hop-by-hop window-based approach. IEICE Trans. Commun. 2018EBP3045 (2018). https://doi.org/10.1587/transcom.2018EBP3045. URL ˋhttps://www.jstage.jst.go.jp/article/transcom/advpub/0/advpub_2018EBP3045/_article/-char/ja/

14. Ren, Y., Li, J., Shi, S., Li, L., Wang, G., Zhang, B.: Congestion control in named data networking—a survey. Comput. Commun. **86**, 1–11 (2016). https://doi.org/10.1016/j.comcom.2016.04.017. URL http://www.sciencedirect.com/science/article/pii/S0140366416301566

15. Arianfar, S., Nikander, P., Eggert, L., Ott, J.: ConTug: A receiver-driven transport protocol for content-centric networks. In: IEEE ICNP, vol. 2010 (2010)

16. Saino, L., Cocora, C., Pavlou, G.: CCTCP: A scalable receiver-driven congestion control protocol for content centric networking. In: 2013 IEEE International Conference on Communications (ICC), pp. 3775–3780 (2013). https://doi.org/10.1109/icc.2013.6655143

17. Carofiglio, G., Gallo, M., Muscariello, L., Papali, M.: Multipath congestion control in content-centric networks. In: 2013 IEEE Conference on Computer Communications Workshops (INFOCOM WKSHPS), pp. 363–368 (2013). https://doi.org/10.1109/infcomw.2013.6970718

18. Schneider, K., Yi, C., Zhang, B., Zhang, L.: A practical congestion control scheme for named data networking. In: Proceedings of the 3rd ACM Conference on Information-Centric Networking, ACM-ICN'16, pp. 21–30. ACM, New York, NY (2016). https://doi.org/10.1145/2984356.2984369. URL http://doi.acm.org/10.1145/2984356.2984369

19. Carofiglio, G., Gallo, M., Muscariello, L.: Joint hop-by-hop and receiver-driven interest control protocol for content-centric networks. In: Proceedings of the Second Edition of the ICN Workshop on Information-Centric Networking, ICN'12, pp. 37–42. ACM, New York, NY (2012). https://doi.org/10.1145/2342488.2342497. URL http://doi.acm.org/10.1145/2342488.2342497

20. Vusirikala, S., Mastorakis, S., Afanasyev, A., Zhang, L.: Hop-by-hop best effort link layer reliability in named data networking. Technical Report Technical Report NDN-0041, NDN (2015)

21. Mejri, S., Touati, H., Malouch, N., Kamoun, F.: Hop-by-hop congestion control for named data networks. In: 2017 IEEE/ACS 14th International Conference on Computer Systems and Applications (AICCSA), pp. 114–119 (2017). https://doi.org/10.1109/aiccsa.2017.36

22. Rozhnova, N., Fdida, S.: An effective hop-by-hop interest shaping mechanism for CCN communications. In: 2012 IEEE Conference on Computer Communications Workshops (INFOCOM WKSHPS), pp. 322–327 (2012). https://doi.org/10.1109/INFCOMW.2012.6193514

23. Rozhnova, N., Fdida, S.: An extended hop-by-hop interest shaping mechanism for content-centric networking. In: 2014 IEEE Global Communications Conference, pp. 1–7 (2014). https://doi.org/10.1109/glocom.2014.7389766

24. Wang, Y., Rozhnova, N., Narayanan, A., Oran, D., Rhee, I.: An improved hop-by-hop interest shaper for congestion control in named data networking. SIGCOMM Comput. Commun. Rev. **43**(4), 55–60 (2013). https://doi.org/10.1145/2534169.2491233. http://doi.acm.org/10.1145/2534169.2491233

25. Park, H., Jang, H., Kwon, T.: Popularity-based congestion control in named data networking. In: 2014 Sixth International Conference on Ubiquitous and Future Networks (ICUFN), pp. 166–171 (2014). https://doi.org/10.1109/icufn.2014.6876774

26. Lei, K., Hou, C., Li, L., Xu, K.: A rcp-based congestion control protocol in named data networking. In: 2015 International Conference on Cyber-Enabled Distributed Computing and Knowledge Discovery, pp. 538–541 (2015). https://doi.org/10.1109/cyberc.2015.67

27. Li, C., Huang, T., Xie, R., Zhang, H., Liu, J., Liu, Y.: A novel multi-path traffic control mechanism in named data networking. In: 2015 22nd International Conference on Telecommunications (ICT), pp. 60–66 (2015). https://doi.org/10.1109/ict.2015.7124658

28. Kato, T., Bandai, M.: Congestion control avoiding excessive rate reduction in named data network. In: 2017 14th IEEE Annual Consumer Communications and Networking Conference, CCNC 2017, pp. 108–113 (2017). https://doi.org/10.1109/ccnc.2017.7983090

29. Kato, T., Bandai, M.: Avoiding excessive rate reduction in rate based congestion control for named data networking. J. Inf. Process. 26, 29–37 (2018). https://doi.org/10.2197/ipsjjip.26.29. URL https://www.jstage.jst.go.jp/article/ipsjjip/26/0/26_29/_article/-char/ja

30. Carofiglio, G., Gallo, M., Muscariello, L.: ICP: Design and evaluation of an Interest control protocol for content-centric networking. In: 2012 Proceedings IEEE INFO-COM Workshops, pp. 304–309 (2012). https://doi.org/10.1109/infcomw.2012.6193510

31. Salsano, S., Detti, A., Cancellieri, M., Pomposini, M., Blefari-Melazzi, N.: Transport-layer issues in information centric networks. In: Proceedings of the second edition of the ICN workshop on Information-centric networking, pp. 19–24. ACM, Helsinki, Finland (2012). https://doi.org/10.1145/2342488.2342493

32. Fu, T., Li, Y., Lin, T., Tan, H., Tang, H., Ci, S.: An effective congestion control scheme in content-centric networking. In: 2012 13th International Conference on Parallel and Distributed Computing, Applications and Technologies, pp. 245–248 (2012). https://doi.org/10.1109/pdcat.2012.43

33. Zhang, F., Zhang, Y., Reznik, A., Liu, H., Qian, C., Xu, C.: A transport protocol for content-centric networking with explicit congestion control. In: 2014 23rd International Conference on Computer Communication and Networks (ICCCN), pp. 1–8 (2014). https://doi.org/10.1109/icccn.2014.6911765

34. Yi, C., Afanasyev, A., Moiseenko, I., Wang, L., Zhang, B., Zhang, L.: A case for stateful forwarding plane. Comput. Commun. **36**(7), 779–791 (2013). https://doi.org/10.1016/j.comcom.2013.01.005. URL www.sciencedirect.com/science/article/pii/S0140366413000236

35. Ndikumana, A., Ullah, S., Thar, K., Tran, N.H., Park, B.J., Hong, C.S.: Novel co-operative and fully-distributed congestion control mechanism for content centric networking. IEEE Access **5**, 27691–27706 (2017). https://doi.org/10.1109/access.2017.2778339

36. Dabirmoghaddam, A., Dehghan, M., Garcia-Luna-Aceves, J.J.: Characterizing interest aggregation in content-centric networks. CoRR abs/1603.07995 (2016). URL http://arxiv.org/abs/1603.07995

37. Abu, A.J., Bensaou, B., Abdelmoniem, A.M.: Leveraging the pending interest table occupancy for congestion control in CCN. Dubai, Arab United Emirates (2016)

38. Zhou, J., Wu, Q., Li, Z., Kaafar, M.A., Xie, G.: A proactive transport mechanism with explicit congestion notification for NDN. In: 2015 IEEE International Conference on Communications (ICC), pp. 5242–5247 (2015). https://doi.org/10.1109/icc.2015.7249156

39. Ahlgren, B., Hurtig, P., Abrahamsson, H., Grinnemo, K.J., Brunstrom, A.: ICN congestion control for wireless links. In: 2018 IEEE Wireless Communications and Networking Conference (WCNC), pp. 1–6 (2018). https://doi.org/10.1109/wcnc.2018.8377396

40. Zhang, F., Zhang, Y., Reznik, A., Liu, H., Qian, C., Xu, C.: Providing explicit congestion control and multi-homing support for content-centric networking transport. Comput. Commun. **69**, 69–78 (2015). https://doi.org/10.1016/j.comcom.2015.06.019. URL http://www.sciencedirect.com/science/article/pii/S0140366415002352

41. Wan, C.Y., Eisenman, S.B., Campbell, A.T.: CODA: congestion detection and avoidance in sensor networks. In: Proceedings of the 1st International Conference on Embedded Networked Sensor Systems, SenSys'03, pp. 266–279. ACM, New York, NY (2003). https://doi.org/10.1145/958491.958523. URL http://doi.acm.org/10.1145/958491.958523

42. Lehman, V., Gawande, A., Zhang, B., Zhang, L., Aldecoa, R., Krioukov, D., Wang, L.: An experimental investigation of hyperbolic routing with a smart forwarding plane in NDN. In: 2016 IEEE/ACM 24th International Symposium on Quality of Service (IWQoS), pp. 1–10 (2016). https://doi.org/10.1109/iwqos.2016.7590394

43. Ren, Y., Li, J., Shi, S., Li, L., Wang, G.: An explicit congestion control algorithm for named data networking. In: 2016 IEEE Conference on Computer Communications Workshops (INFOCOM WKSHPS), pp. 294–299 (2016). https://doi.org/10.1109/infcomw.2016.756208

44. Bouacherine, A., Senouci, M.R., Merabti, B.: Multipath forwarding in named data networking: flow, fairness, and context-awareness. In: Obaidat, M.S. (ed.) E-Business and Telecommunications, pp. 23–47. Springer, Berlin

A Comparative Review of Various Techniques for Image Splicing Detection and Localization

Amandeep Kaur, Navdeep Kanwal and Lakhwinder Kaur

Abstract Today, society is completely dependent on the utilization of Internet. With the increase in use of social media, millions of pictures are daily uploaded to Internet, providing opportunities for hackers to forge images. Various image editing softwares have opened the ways to image forgery, making forged images to look authentic. The manipulations of content have dissolved image trustworthiness and validation. Advancement in image forensics has introduced a number of image forgery detection techniques, to reestablish the realness in digital media. This paper endeavors to reveal various kinds of image forgery and its recognition techniques. The paper has also presented the performance of the existing splicing techniques by using quality metrics MCC and F-measure. Further, the paper evaluates different state of the art in splicing techniques, and it has been observed that the CFA artifact-based splicing localization achieves an accuracy of 99.75%. This paper features the significance of splicing localization and possible future research work in it.

Keywords Image forensics · Splicing · Image forgery detection · Localization

1 Introduction

In an era of digitization, digital images are believed to be the intrinsic conveyor of information. An image is a portrayal of the external form of a person or thing in art and represents the truth of what has happened, notwithstanding that it does no longer believe. The wide availability of inexpensive image manipulation softwares such as Microsoft Paint, corelDRAW, Sumo Paint and Adobe Photoshop has proliferated

A. Kaur (✉) · N. Kanwal · L. Kaur
Punjabi University, Patiala, Punjab, India
e-mail: baman939@gmail.com

N. Kanwal
e-mail: navdeepkanwal@gmail.com

L. Kaur
e-mail: mahal2k8@gmail.com

© Springer Nature Singapore Pte Ltd. 2020   139
P. K. Singh et al. (eds.), *Proceedings of First International Conference on Computing, Communications, and Cyber-Security (IC4S 2019)*, Lecture Notes in Networks and Systems 121, https://doi.org/10.1007/978-981-15-3369-3_11

the tampering of images or to create new ones [31]. Image forgery is an emerging branch in image forensics, where contents of image are tampered without leaving any visual traces of tampering, to generate fake images or to mislead people [23]. Further, image forgery detection demonstrates whether an image is authentic or forged, and forgery localization is the process of identifying the tampered region in a forged image. Today, people are not concerned about an image to be forged or not, until it causes any distress. They are substantiated as a token of trust by everyone and everywhere. So, to ensure the authentication of images, this paper attempts to review the various passive forgery detection techniques, mainly focusing upon the localization of splicing forgery.

1.1 Need for Image Forgery Detection

Digital imaging is emerging in various fields like sports, legal services, news, intelligence, medical technology and others to convey relevant information. At the same time, tampering is involved in all these fields to distort the actual information [15, 23], where the truth cannot be recognized by the human eye. The proliferation in manipulated information in images has diminished the trust in digital media. It is very common that a face of a person is replaced by the other person to hide something or to convey some interpreted information. In this context, an example of sports video manipulation can be cited, depicting the risk of tampering in favor of one of the competitors for monetary gains [30]. Therefore, it is a significant task to detect image forgery before the information leads to malicious consequences [32]. There is a need to detect forgery to authenticate information available in images—captured from charge-coupled device (CCD) cameras [34] and complementary metal–oxide semiconductor (CMOS).

1.2 Types of Image Forgery

Image forgery can be performed in a number of ways by performing different operations such as adding, deleting or copying some parts of image within an image or to another image, with an aim of leaving no traces for human visual detection. Images are modified using various techniques such as copy-move forgery, image splicing forgery and image resampling.

Copy-Move forgery/cloning: Copy-move is popular, easy to perform and most common image forgery technique [4]. Both operations copy and move are performed on the same image. Some segments of the image are copied and pasted to some other segments of the same image in order to conceal details or duplicate some information [3]. Since the copied segments are taken from the same image, so the color, texture and noise component will remain consistent with the rest of

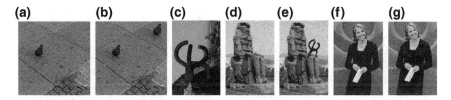

Fig. 1 Copy-move: **a** is an original image of the bird and **b** is the copy-move forged image. Splicing forgery: **c** and **d** are spliced together to create a new image (**e**). Resampling forgery: Image **g** of Katie Couric is resampled to give her a trimmer waistline and a thinner face as given in image (**f**)

the image, making forgery visually undetectable [35]. Figure 1a and b depicts an example of copy-move image forgery.

Image splicing forgery: Image forgery using splicing necessitates replacing fragments of images from one or more dissimilar images to the host image [7]. In order to make forged regions imperceptible, geometric transformations such as rotation, scaling, skewing and stretching are performed [31]. It can also be done without performing post-processing operations such as matting, blending and smoothing of boundaries among spliced regions [32]. If a new image is created using two images of dissimilar backgrounds, then it is difficult to make the boundaries indiscernible [30]. Figure 1c, d and e is an example of splicing forgery.

Image resampling: Image resampling is the process of resizing, stretching, skewing, flipping and rotating portions of images to get a resampled image [19, 31]. Suppose to get composite image of two persons, height of one person may need to be resized or stretched with respect to other person [2]. Figure 1f and g is an example of resampling forgery.

1.3 Image Forgery Detection Techniques

Image forgery is likely to introduce inconsistencies by disturbing the underlying statistics properties in natural images [7]. Any manipulations in images lead to correlations in forged and original image segments. Figure 2 gives the classification of images forgery techniques.

Active Approach: In this approach, certain authentication information must be implanted into the image during its creation. This approach is also limited to notably equip digital cameras. Active approach is further categorized into: digital watermark and digital signature.

Digital Signature: Digital signature is a cryptographic strategy where hash function is used to generate the digital signature which is implanted in an image to demonstrate the authenticity of digital messages [27]. If any forgery attack is ex-

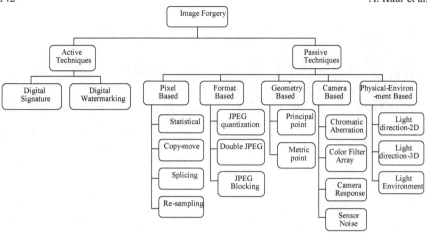

Fig. 2 Classification of image forgery detection techniques

ecuted, the digital signature is destroyed and the image will be detected as forged [16].

Digital Watermarking: Watermark is implanted into an image at the time of its creation and extracted to authenticate the host image whether it is forged or not [27]. Any attempt of forgery in the watermarked image will change the watermark, which will remain a proof for an image of being forged.

Passive Approach: This approach does not require any historical information about the image. There is no hidden information in the image during its creation. This approach is based on the statistical and visual methods. Statistical methods determine the pixel values of an image to detect forgery, whereas visual methods require visual clues like inconsistencies and light deformation in the image, without the necessity of any special hardware or software to detect forgery [30]. Passive techniques are classified as follows:

Pixel-based techniques accentuate on the pixels, which are chief building blocks in an image. They are categorized into copy-move, splicing and resampling to detect statistical peculiarities at pixel levels in forged images [2, 11]. Copy-move forgery has strong interrelationships across forged and authentic parts, so it becomes a strong evidence to detect the forgery [32]. Splicing muddles the high-order Fourier statistics of the image that insight can be used to detect splicing [19]. Resampling/interpolation detectors and forged images containing resampled regions can be detected [23].

Format-based techniques are preferable primarily in JPEG formats. JPEG compression introduces blocking artifacts that are helpful in detecting forgery. This technique is categorized as follows:

(a) JPEG Quantization: Digital image camera manufacturers and image processing softwares utilize JPEG quantization tables to make balance between compression ratio and image quality. So, this differentiation causes different blocking artifacts in the captured image, and these inconsistencies could be used to detect forgery [38].

(b) Double JPEG: Some portion of the JPEG image is forged using image editing tools and is rewritten to the memory in JPEG format. In the complete process, forged portions get doubly compressed and degree of compression of this portion will vary from the rest of the image. Eventually, this serves as an evidence of image forgery [24].

(c) JPEG blocking: Forgery operations such as splicing, resampling and local object operations cause artifact inconsistencies to be blocked in images affected blocks, which will help to determine the manipulations an image has undergone [38].

Camera-based techniques: An image acquired using digital cameras undergo various processing steps such as white adjusting, color correlation, gamma correction, quantization and filtering and JPEG compression. The process from image acquisition to saving on computer depends upon the models of the digital camera being used and its artifacts [19]. Various camera-based techniques are as follows:

(a) Chromatic aberration: Chromatic aberration is the result of failure of optical system to perfectly focus light of all wavelengths on the sensor, which occur in two forms longitudinal and lateral. Both forms cause various color imperfections in the image [18].

(b) Camera response function: Most of the camera responses are linear, which means the intermediate pixels along edges of image will be linear combination of neighboring colors. Blurring artifacts such as introduced by photoshop causes nonlinear CRF. At the same time, this property is exploited to detect forgery [8].

(c) Color filter array (CFA): In the process of CFA-based forgery detection, almost all digital cameras make use of a sensor in collaboration with a CFA. The evaluation of missing color samples is called as CFA interpolation. These missing color samples are interpolated from the recorded samples, to acquire the three-channel color image. Such interpolations add some correlations to the image that could be destroyed during forgery [29].

(d) Sensor noise: The pattern noise is present in almost every image captured by camera sensor, as the image undergoes several processing operations such as demosaicing, white balancing, color correction, quantization, gamma correction, filtering and JPEG compression. Photo-response non-uniformity (PRNU) and fixed-pattern noise (FPN) are the two principal constituents of the pattern noise. Regions that lack pattern noise are determined as forged regions [21].

Geometry-based techniques are used to measure objects in the real world and their positions relative to the camera using principal point method and metric point method [2].

(a) Principal Point: Principal point lies near the center of the image, in the authentic images. Attempt of forgery moves the principal point proportionally as compared to authentic image. The inconsistency in position of principal point in image serves as an evidence of forgery [13].

(b) Metric point: Metric measurement requires only one image to rectify planar surfaces imaged under perspective projection. It has the capability to make real-world measurements from planar surfaces [13].

Physical Environment-based techniques are based on the exact match among the object, light and the camera. Creation of forgery causes inconsistencies in lighting, shadowing, etc., which reveals an evidence of forgery. Consider an example of spliced picture of two movie stars, created by using their individual pictures captured in different environmental conditions. So, it is hard to match the exact lighting conditions under which the each picture was originally captured (e.g., the sunlight on a clear day and the other in a room). Such lighting inconsistencies across an image are used as an evidence of forgery [2]. These techniques are categorized as:

(a) Light direction-2D, 3D: Direction of light source can be estimated in images different objects, variability in the direction of lighting can be used as an evidence of image forgery. Because exact lighting match requires shadow creation or deletion in images that leave traces of forgery [17].
(b) Light Environment: There could be an enormous amount of lights positioned in any number of places, to make complex lighting scene. They detect lighting variations from different parts of the image. The lighting environment can effectively detect splicing forgeries in low-resolution and high-compression images [26].

2 Methodology

Input Image: An image taken into consideration for performing further operations. Image Preprocessing: Various operations such as DCT or DWT, cropping, transforming RGB image into suitable color space may be performed over an input image for better classification performance.

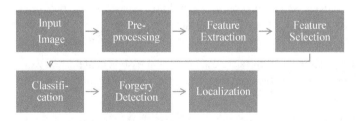

Fig. 3 Generalized structure of image forgery detection and localization

Feature Extraction: The purpose of extracting feature is to extract the specific representation of data that highlights its relevant information. This technique involved the redundant information, thus reducing the efficiency of classifier.

Feature Selection: Based on the selected features, feature selection aims at selecting the smallest number of features, thereby removing redundancy and improving the efficiency of classifier.

Classification: Some of the classifiers used for classification are SVM, LDA, Naive Bayes, logistic regression, FLD, FLDA and others. The goal of classifier is to identify forged and authentic images.

Localization: Localization is a post-processing operation that focuses on locating the forged part in the host image.

As stated in the above steps, the generalized structure of image forgery detection and localization is outlined in Fig. 3.

3   Literature Survey

This section presents a detailed analysis of the state-of-the-art splicing localization techniques. A strategy is proposed in [37], using Bayer CFA and gradient-based demosaicking to obtain the natural counterpart of the image. Then, that image and the test image were compared to classify each pixel to determine authentic and forged region. A binary image was obtained after classification, which was localized for the spliced region after post-processing. Digital Video and Multimedia Lab's (DVMM) uncompressed spliced image dataset was used to test the proposed method. But, there is no evaluation of the compressed images. Authors introduced a method based on the derivation of a unified statistical model in [6], which was applicable where aligned or nonaligned JPEG double compression (i.e., A-DJPEG or NA-DJPEG) was applied. The effectiveness of this method was computed by testing on realistic tampered images. This method was applied to two cases, when part of image was singly compressed as well as when both parts were doubly compressed—one with the A-DJPG and other part with NA-DJPG and produced FI measure results on uncompressed images = 0.306, on compressed = 0.067 and rescaled = 0.068. This method does not work when both parts were double compressed with same grid shift and when post-processing operations such as resizing were applied between both compressions. It is stated in [14] that each $c * c$ block of forged image probably contains CFA artifacts. Green channel was extracted from the forged image, and the error of prediction was calculated using a fixed predictor. Then, the weighted local variance was estimated, and features were obtained for each block. Expectation–maximization (EM) algorithm was applied to estimate the Gaussian mixture model (GMM) parameters. Eventually, a map was generated to indicate the likelihood of forgery. Mean or median filter was used to highlight the localized regions. This method was tested on original dataset and had claimed accuracy for bicubic predictor to be 99.75%, bilinear predictor was 98.45%, gradient-based predictor was 99.75%

and median predictor was 99.54%.This method was not applicable for cameras using super CCD sensors, and less effective in flat regions and sharp edges, and localization of forged regions decreased with JPEG compression below 95%. The authors in [22] introduced a method for locating the spliced region considering that images from different origins were inconsistent with local noise levels, introduced by the sensors or post-processing operations. Average F-measure result produced on Columbia uncompressed image splicing detection evaluation (CUISDE) dataset was 0.373 by this paper; on compressed images, it was 0.119, and on rescaled images, it was 0.137. Another paper [1] presented a technique based on the first digit features and SVM classifier to locate the forged regions in single and double JPEG compressed images. The developed technique was tested on uncompressed color image dataset (UCID) and was found inefficient in dealing with flat and heavily compressed image regions. This method failed to localize small region forgeries. Another limitation of this method was fixed localization window size of 64 * 64. Further, [36] has proposed a technique for natural images for tampering localization. The proposed technique was tested on host images of Canon EOS 450D and the methodology achieved accordingly from target 97.85% on Canon, 98.18% on Digimax and 95.87% on Nikon. In case of the host images from Nikon D200, the claimed performance on target was 93.12% on Canon, 96.46% on Digimax and 92.84% on Nikon. In continuation to its testing with different cameras of host and target image, the author had tested the technique for the same image source of target and source image. The average accuracy results evaluated for camera mode Canon EOS 450D were 93.26, 95.86, 95.64 and 97.58% for the post-forgery operations such as JPEG compression, resizing and rotation, blurring, respectively. For another camera D200, the average accuracy results were 97.8% on blurring, 91.67% on JPEG compression, 94.09% on resizing and 93.51% on rotation. The performance comparison for the proposed method was 95.54%, and 93.86% for Canon EOS 450D and Nikon D200, respectively. This method can be extended to various design choices such as features, model and decision strategy to obtain better performance. The author also suggested that better decision strategy would be used in the future to recover the shape of localized forgery. Authors of [5] made use of partial blur-type inconsistency to determine splicing localization. Basically, this method was used for spliced blurred images. First, partition the image into blocks to extract local blur-type characteristics. These local characteristics were then classified as either out of focus or motion based. Eventually, human was involved to conclude whether the image was tampered or authentic. This method performed well on the original dataset, taking into account different spliced region sizes as 100 * 100, where TPR was 95.1%, TNR was 94.4% and claimed accuracy was 94.5%. When the spliced region size was 200 * 200, the method resulted in TPR as 96.8%, TNR was 95.3% and accuracy was 95.4%. When the spliced region size was further increased to 512 * 384, the TPR was 95.1%, TNR was 96.6% and the accuracy was 96.3%. An expectation–maximization algorithm is employed in [9] and segmentation into authentic and forged parts of the image to detect the splicing forgery detection and localization. This method was tested on Columbia uncompressed dataset and had better accuracy than the method presented in [5]. The authors claimed average F-measure 0.283 on images without compression, 0.228 on compressed images and 0.266 on rescaled images. Later, [39] has proposed to segment these images into nonoverlapping blocks to localize splicing forgery. Blockwise noise-level estimation was performed on a test image using the principle component analysis (PCA). K-means clustering was used to segment the forged region from the authentic region. The author had claimed average precision to be 0.975 on CUISDE dataset. This method was not vigorous to JPEG compression, and it did not work well when splicing is performed on forged and original images with similar noise levels. Another method [10] was further proposed as an improvement for splicing localization which started with feature extraction from image; then autoencoder network will label different blocks to detect the forgery. The author used a loss function which may incorporate weightage to redefine discriminative labeling. This method confirmed better performance than [6, 22], and [9] showing ostu threshold results as 0.415, 0.372 and 0.382 on basic, compressed and rescaled images, respectively, whereas optimal threshold claimed 0.418, 0.378 and 0.389 on basic, compressed and rescaled images, respectively. In 2017, [28] introduced a method with single or double compressed input images. Features were extracted using the first DCT coefficients, by using Benford's law and Markov chain. These features were used by the SVM model to detect and locate the forged region. The results were evaluated on UCID dataset, and average result claimed on single compression patch (SCP) was 71.84% and on double compression patch (DCP) was 61.82%. This method showed poor performance in case of double compression attack. This method had computational overhead. Authors of [33] used convolutional network for localization of forgery and proposed to use single-task fully convolutional network (SFCN), MFCN and edge-enhanced MFCN. The proposed method was tested on CASIA v1.0, Columbia uncompressed carvalho, DARPA/NIST dataset and had claimed average $F1$ score ranging from 0.42 to 0.58 in SFCN, from 0.42 to 0.60 in MFCN and from 0.48 to 0.61 in case of edge-enhanced MFCN. The average MCC score claimed by SFCN was 0.37–0.45, by MFCN was 0.39–0.49 and by edge-enhanced MFCN was 0.41–0.57. The limitation of this method was performance which was degraded in case of compression as well as blur attack. A method [20] evolved conditional random field and deep neural networks to determine splicing localization. This method used three FCNs and a conditional random field (CRF). FCN-CRF method achieved TPR to be 82.6%, whereas FPR is only 7.2% in genuine scenarios. The other TPRs were lower than 70% when the FPR was 7.2%, and the FPRs were above 20% when the TPR was 82.6%. Improvement in detection of very small forged objects was needed in the proposed method. A method is proposed in [25] for stabilized as well as non-stabilized videos exploiting the PRNU-related noise traces. The method was validated on own dataset captured by smart phones. The author had claimed an accuracy of 90%. Limitations of this method were different frames which correlate only if the frames were captured by the same device. Accuracy of this method dropped in the case of stabilized videos. Further, [12] computed the dissimilarity between the original image and its median filtered version to acquire the RGB noise image. Sliding window had produced noise feature for detection of forged region, and eigen-related feature to localize the forged region. The results were evaluated on the CUISDE dataset and claimed $F1$ to be 51% and MCC to be 39%. Future works could overcome the limitation of only focus on feature extraction, and machine learning could be used in the future. Random forest, SVM and convolutional neural networks could be applied to future work.

3.1 Comparative Analysis of the Existing Splicing Localization Techniques

This paper reviews various splicing localization techniques, along with their demerits. Benchmarked image forgery datasets are used to assess the performance and validate the results of various splicing localization techniques. The authors have utilized miscellaneous datasets such as DVMM CUISDE, UCID, CASIA, DARPA/NIST nimble challenge 2016 SCI dataset and original datasets. The analysis also provided various types of cameras such as Canon, Nikon, Digimax, Sony, Kodak and smartphone cameras to capture images. Table 1 gives the comparative analysis of various splicing localization methods. This paper presented the performance of splicing localization methods by comparing F-measure, accuracy and precision.

Low F-measure results on compressed images is 0.067, and high F-measure result is 0.570. In case of uncompressed images, poor F-measure result is 0.306 and higher F-measure result is 0.611, whereas F-measure in case of rescaled images varies in the range of 0.068–0.389. Average MCC results on compressed images vary from 0.420 to 0.570, and on uncompressed images, it varies from 0.390 to 0.479. Precision of splicing localization on different cameras varies from 92.82%-98.18%. Precision for JPEG compressed images ranges from 67.0 to 91.9%. The poor accuracy result presented is 48.83% and the best accuracy result is 99.75%.

4 Conclusion

In this paper, we presented that digital technology has accelerated the act of forgery in images as it is the crucial need of an hour to reinstate the trust of people in images. This paper assessed various passive blind forgery detection techniques and a brief survey on splicing localization methodologies. Splicing localization has been performed by both block-based and feature-based images. This paper also presented a comparative analysis of various splicing localization methods. In splicing localization, the prime drawback is its performance on compressed images and inability to localize forgery in small regions. We have concluded that the F-measure results produced by [33] show the best performance, whereas F-measure results produced by [6] are the worst considered in our analysis. Best MCC results are given by [33]. Based on accuracy, the best results of splicing localization are given by [14]. The worst precision results are given by [39], and the best precision is given by [36]. There is a scope to expand the passive blind forgery to video and audio forgery. A considerable amount of future work can be accomplished by introducing machine learning to splicing localization.

Table 1 The comparative analysis of various splicing localization methods

S. No.	Title/publication year	Method used	Dataset used	Accuracy/precision/F-measure/MCC	Camera	Gaps
1	Image splicing localization based on re-demosaicing (2012)	Combination of CFA and gradient-based interpolation algorithm	DVMM's CUISDE	–	–	No evaluation for compressed images
2	Image forgery localization via fine-grained analysis of CFA artifacts (2012)	CFA artifacts analysis	Original dataset	Accuracy: Bicubic predictor: 99.75% Bilinear: 98.45% Gradient based: 99.75% Median: 99.54%	Canon EOS 450D, Nikon D50, Nikon D90, Nikon D7000	Not applicable with cameras using super CCD sensors. Less potent in case of flat area or sharp edges. Poor localization when JPEG compression is below 95%
3	Image forgery localization via block-grained analysis of JPEG artefacts (2012)	Unified statistical model and Bayesian approach	–	F-measure: uncompressed: 0.306, compressed: 0.067, rescaled: 0.068	Nikon D90, Canon EOS 450D, Canon EOS 5D	Not applicable when post-processing operations like resizing is applied. Does not work in the presence of double JPEG compression with same grid shift

(continued)

Table 1 (continued)

S. No.	Title/publication year	Method used	Dataset used	Accuracy/precision/F-measure/MCC	Camera	Gaps
4	Exposing region splicing forgeries with blind local noise estimation (2013)	Blind noise estimation algorithm and Kurtosis phenomenon	Columbia uncompressed image splicing detection evaluation dataset	F-measure: uncompressed: 0.373, compressed: 0.119, rescaled: 0.137	Canon 400D, Canon digital rebel 100, canon EOS-60D digital camera, SONY DSC–H20 digital camera	Localization fails when noise variance between forged and original image is small. Method fails in case of large regions with distinct texture
5	Spicing forgeries localization through the use of the first digit features (2014)	Benford's law using DCT and to evaluate compression attacks	uncompressed color image dataset (UCID)	–	Nikon D70, Nikon D200	Impotent to handle flat and heavily compressed image regions. Inability to localize small region forgeries. Fixed localization window of 64 ∗ 64

(continued)

Table 1 (continued)

S. No.	Title/ publication year	Method used	Dataset used	Accuracy/ precision/F-measure/ MCC	Camera	Gaps
6	A feature-based approach for image tampering detection and localization (2014)	Camera-based technique	–	Precision: Camera-based localization: (a) Canon EOS 450D Digimax: 98.18%, Canon: 97.85%, Nikon: 95.87% (b)Nikon D200 Digimax: 96.46%, Canon: 93.12%, Nikon: 92.84% Performance comparison: (a) Canon EOS 450D: 95.54% (b) Nikon D200: 93.86% Processing-based localization: (a) Canon EOS 450D: Blurring: 97.28–97.90%, JPEG compression: 79.61–99.34% Resizing: 90.20–98.87% Rotation: 89.57–99% (b)Nikon D200: Blurring: 96.99–98.33%, JPEG compression: 78.43–98.46%, Resizing: 85.14–99.07%, Rotation: 84.81–98.79%	Canon EOS 450D, Nikon Coolpix S5100, Digimax 301, Sony DSC S780	Lack of design choices and techniques such as features, model, decision strategy. Better decision strategy required, to recover the shape of forgery

(continued)

Table 1 (continued)

S. No.	Title/ publication year	Method used	Dataset used	Accuracy/ precision/F-measure/ MCC	Camera	Gaps
7	Image splicing localization based on blur-type inconsistency (2015)	Partial blur-type inconsistency	Original dataset	Accuracy: 96.3%	–	Human decision is involved
8	Splicebuster: A new blind image splicing detector (2015)	Expectation–maximization algorithm	Columbia dataset	Accuracy: 90%	Canon EOS 450D, Canon IXUS 95IS, Nikon D200, Nikon Coolpix S5100, Digimax 301, Sony DSC S780	Robustness to JPEG compression and other post-processing operations are required
9	Single image splicing localization through autoencoder-based anomaly detection(2016)	Autoencoder anomaly detection	Synthetic dataset	F-measure: uncompressed: 0.418 compressed: 0.378 rescaled: 0.389	Smartphones: Huawei P7 mini, Nokia Lumia 925, LG D855, Samsung GT-S7580, Samsung GT-S7580, Apple iPhone 5s, camera samsung ES15	Deep investigation to extract features directly from data by using neural network is required

(continued)

Table 1 (continued)

S. No.	Title/ publication year	Method used	Dataset used	Accuracy/ precision/F-measure/ MCC	Camera	Gaps
10	Image splicing localization using PCA-based noise-level estimation (2016)	Principal component analysis (PCA) algorithm and K-means algorithm	Columbia uncompressed image splicing detection evaluation dataset	Precision: no post-processing: 0.721, JPEG QF = 95: 0.670, down sampling 20% : 0.670	Canon G3, Nikon D70, Canon 350D rebel XT, Kodak DCS330	Not robust to JPEG compression. Fails when forged image is spliced with host image of same noise level
11	Improved image splicing forgery localization with the first digits and Markov model features (2017)	Markov DCT for evaluating single compression attack and double compression attack	uncompressed color image dataset (UCID)	Accuracy: 48.83–92.91%	–	Poor localization in case of double compression attack
12	Image splicing localization using a multitask fully convolutional network (2017)	single-task fully convolutional network (SFCN) and multitask fully convolutional network (MFCN) and edge-enhanced MFCN	Trained network using: CASIA v2.0, Tested the trained models on: CASIA v1.0, Columbia uncompressed Carvalho, DARPA/NIST nimble challenge 2016 SCI datasets	Average F-measure for different datasets: compressed: SFCN: 0.42–0.48, MFCN: 0.42–0.52, Edge-enhanced MFCN: 0.54–0.57, on uncompressed: SFCN: 0.44–0.58, MFCN: 0.47–0.60, Edge-enhanced MFCN: 0.48–0.61 Average MCC score: compressed: SFCN: 0.42–0.45, MFCN: 0.42–0.49, Edge-enhanced MFCN: 0.52–0.57, on uncompressed: SFCN: 0.37–0.42, MFCN: 0.39–0.46, edge-enhanced MFCN: 0.40–0.48	–	Performance degradation in case of compression, additive noise as well as blur attack

(continued)

Table 1 (continued)

S. No.	Title/ publication year	Method used	Dataset used	Accuracy/ precision/F-measure/ MCC	Camera	Gaps
13	Locating splicing forgery by fully and convolutional networks and conditional random field (2018)	deep neural networks and conditional random field	CUISDE and CASIA v2	Precision: 91.9%	–	Detection of small objects is difficult. Over-fitting issue–lack of enough forged images for training
14	Color noise-based feature for splicing detection and localization (2018)	Study on sliding window effects (size and sliding step) and local evaluation algorithm	Columbia uncompressed image splicing detection evaluative dataset	F-measure: 0.51, MCC: 0.39	–	Only focused on feature extraction. No focus on machine learning

References

1. Amerini, I., Becarelli, R., Caldelli, R., Del Mastio, A.: Splicing Forgeries Localization Through the Use of First Digit Features, pp. 143–148 (2014)
2. Ansari, M.D., Ghrera, S.P., Tyagi, V.: Pixel-based image forgery detection: a review. IETE J. Educ. **55**(1), 40–46 (2014)
3. Ardizzone, E., Bruno, A., Mazzola, G.: Copy-move forgery detection by matching triangles of keypoints. IEEE Trans. Inf. Forensics Secur. **10**(10), 2084–2094 (2015)
4. Asghar, K., Habib, Z., Hussain, M.: Copy-move and splicing image forgery detection and localization techniques: a review. Aust. J. Forensic Sci. **49**(3), 281–307 (2017)
5. Bahrami, K., Kot, A.C., Li, L., Li, H.: Blurred image splicing localization by exposing blur type inconsistency. IEEE Trans. Inf. Forensics Secur. **10**(5), 999–1009 (2015)
6. Bianchi, T., Piva, A.: Image forgery localization via block-grained analysis of jpeg artifacts. IEEE Trans. Inf. Forensics Secur. **7**(3), 1003–1017 (2012)
7. Birajdar, G.K., Mankar, V.H.: Digital image forgery detection using passive techniques: a survey. Digit. Investig. **10**(3), 226–245 (2013)
8. Chen, C., McCloskey, S., Yu, J.: Image splicing detection via camera response function analysis. In: Proceedings of the IEEE Conference on Computer Vision and Pattern Recognition, pp. 5087–5096 (2017)
9. Cozzolino, D., Poggi, G., Verdoliva, L.: Splicebuster: a new blind image splicing detector. In: 2015 IEEE International Workshop on Information Forensics and Security (WIFS), pp. 1–6. IEEE, New York (2015)
10. Cozzolino, D., Verdoliva, L.: Single-Image Splicing Localization Through Autoencoder-Based Anomaly Detection, pp. 1–6 (2016)
11. Deshpande, P., Kanikar, P.: Pixel based digital image forgery detection techniques. Int. J. Eng. Res. Appl. (IJERA) **2**(3), 539–543 (2012)
12. Destruel, C., Itier, V., Strauss, O., Puech, W.: Color noise-based feature for splicing detection and localization. In: 2018 IEEE 20th International Workshop on Multimedia Signal Processing (MMSP), pp. 1–6. IEEE, New York (2018)
13. Farid, H.: Image forgery detection. IEEE Signal Process. Mag. **26**(2), 16–25 (2009)
14. Ferrara, P., Bianchi, T., De Rosa, A., Piva, A.: Image forgery localization via fine-grained analysis of CFA artifacts. IEEE Trans. Inf. Forensics Secur. **7**(5), 1566–1577 (2012)
15. Huynh, T.K., Huynh, K.V., Le-Tien, T., Nguyen, S.C.: A survey on image forgery detection techniques. In: The 2015 IEEE RIVF International Conference on Computing & Communication Technologies-Research, Innovation, and Vision for Future (RIVF), pp. 71–76. IEEE, New York (2015)
16. Ilcheva, Z., Lazarov, N.: A digital watermarking scheme for image tamper detection. In: Proceedings of the 15th International Conference on Computer Systems and Technologies, pp. 100–107. ACM (2014)
17. Johnson, M.K., Farid, H.: Exposing digital forgeries by detecting inconsistencies in lighting. In: Proceedings of the 7th workshop on Multimedia and Security, pp. 1–10. ACM (2005)
18. Johnson, M.K., Farid, H.: Exposing digital forgeries through chromatic aberration. In: Proceedings of the 8th Workshop on Multimedia and Security, pp. 48–55. ACM (2006)
19. Kashyap, A., Parmar, R.S., Agrawal, M., Gupta, H.: An evaluation of digital image forgery detection approaches. arXiv preprint arXiv:1703.09968 (2017)
20. Liu, B., Pun, C.M.: Locating splicing forgery by fully convolutional networks and conditional random field. Signal Process. Image Commun. **66**, 103–112 (2018)
21. Lukáš, J., Fridrich, J., Goljan, M.: Detecting digital image forgeries using sensor pattern noise. In: Security, Steganography, and Watermarking of Multimedia Contents VIII, vol. 6072, p. 60720Y. International Society for Optics and Photonics (2006)
22. Lyu, S., Pan, X., Zhang, X.: Exposing region splicing forgeries with blind local noise estimation. Int. J. Comput. Vis. **110**(2), 202–221 (2014)
23. Mahdian, B., Saic, S.: Blind methods for detecting image fakery. IEEE Aerosp. Electron. Syst. Mag. **25**(4), 18–24 (2010)

24. Malviya, P., Naskar, R.: Digital forensic technique for double compression based jpeg image forgery detection. In: International Conference on Information Systems Security, pp. 437–447. Springer, Berlin (2014)

25. Mandelli, S., Bestagini, P., Tubaro, S., Cozzolino, D., Verdoliva, L.: Blind detection and localization of video temporal splicing exploiting sensor-based footprints. In: 2018 26th European Signal Processing Conference (EUSIPCO), pp. 1362–1366. IEEE, New York (2018)

26. Mazumdar, A., Jacob, J., Bora, P.K.: Forgery detection in digital images through lighting environment inconsistencies. In: 2018 Twenty Fourth National Conference on Communications (NCC), pp. 1–6. IEEE, New York (2018)

27. Mousavi, S.M.: Image authentication scheme using digital signature and digital watermarking. **16** (2013)

28. Patil, B., Chapaneri, S., Jayaswal, D.: Improved image splicing forgery localization with first digits and Markov model features. In: 2017 IEEE International Conference on Intelligent Techniques in Control, Optimization and Signal Processing (INCOS), pp. 1–5. IEEE, New York (2017)

29. Popescu, A.C., Farid, H.: Exposing digital forgeries in color filter array interpolated images. IEEE Trans. Signal Process. **53**(10), 3948–3959 (2005)

30. Qazi, T., Hayat, K., Khan, S.U., Madani, S.A., Khan, I.A., Kołodziej, J., Li, H., Lin, W., Yow, K.C., Xu, C.Z.: Survey on blind image forgery detection. IET Image Process. **7**(7), 660–670 (2013)

31. Qureshi, M.A., Deriche, M.: A bibliography of pixel-based blind image forgery detection techniques. Signal Process. Image Commun. **39**, 46–74 (2015)

32. Redi, J.A., Taktak, W., Dugelay, J.L.: Digital image forensics: a booklet for beginners. Multimed. Tools Appl. **51**(1), 133–162 (2011)

33. Salloum, R., Ren, Y., Kuo, C.C.J.: Image splicing localization using a multi-task fully convolutional network (MFCN). J. Vis. Commun. Image Represent. **51**, 201–209 (2018)

34. Sharma, V., Jha, S., Bharti, D.R.K.: Image forgery and it's detection technique: a review. Int. Res. J. Eng. Technol. (IRJET) (2016)

35. Shivakumar, B., Baboo, L.D.S.S.: Detecting copy-move forgery in digital images: a survey and analysis of current methods. Glob. J. Comput. Sci. Technol. (2010)

36. Verdoliva, L., Cozzolino, D., Poggi, G.: A Feature-Based Approach for Image Tampering Detection and Localization, pp. 149–154 (2014)

37. Wang, B., Kong, X.: Image Splicing Localization Based on Re-demosaicing, pp. 725–732 (2012)

38. Ye, S., Sun, Q., Chang, E.C.: Detecting digital image forgeries by measuring inconsistencies of blocking artifact. In: 2007 IEEE International Conference on Multimedia and Expo, pp. 12–15. IEEE, New York (2007)

39. Zeng, H., Zhan, Y., Kang, X., Lin, X.: Image splicing localization using PCA-based noise level estimation. Multimed. Tools Appl. **76**(4), 4783–4799 (2017)

Analysis and Synthesis of Performance Parameter of Rectangular Patch Antenna

Vivek Arora and Praveen Kumar Malik

Abstract This paper demonstrates the detailed analysis and synthesis of the different performance parameters of microstrip rectangular patch antenna. The effect of the length of antenna, width of antenna and dielectric constant with respect to applied frequency is demonstrated. The performance of the different parameters is explained mathematically and simulated with the help of MATLAB software. Some critical performance parameters like width of the antenna feed and effective wavelength are also taken into consideration.

Keywords Dimension · Length · Performance · Size · Width

1 Introduction

Antenna is an essential component for any interface of wired and/or wireless communication. In satellite, mobile phones and radar communication, microstrip antenna plays a vital role. The invention of the microstrip antenna concept has been attributed to many sources and the earliest includes the USA [1]. Lewin investigated radiation from strip-line discontinuities; at that time, the emission of unwanted radiation from thin strip-line circuits was well appreciated and subsequently the dimensions of the substrate [2]. Additional studies were undertaken in the late 1960s by Kaloi who studied basic rectangular and square configuration. During the last decade, microstrip antenna engrossed considerable attention of researchers and industry persons due to the demand of its versatile use of in different engineering fields in wireless communication. Microstrip antenna is having many advantages like small size, lightweight and ease of manufacturing, but it suffers many serious disadvantages also like small gain, lower bandwidth and low efficiency as well [3]. All the performance parameters of the microstrip antenna (gain, S11, VSWR, efficiency and bandwidth) are

V. Arora (✉)
Ganga Institute of Technology and Management, Kablana, India
e-mail: vivekgrover72@gmail.com

P. K. Malik
Lovely Professional University, Phagwara, India
e-mail: pkmalikmeerut@gmail.com

© Springer Nature Singapore Pte Ltd. 2020
P. K. Singh et al. (eds.), *Proceedings of First International Conference on Computing, Communications, and Cyber-Security (IC4S 2019)*, Lecture Notes in Networks and Systems 121, https://doi.org/10.1007/978-981-15-3369-3_12

dependent on some dimensional entities. These dimensional entities are width of the antenna, length of the antenna, feed line length and dielectric constant value used in the antenna [4]. Almost all the research articles which are using the rectangular microstrip antenna are bound to use some predefined calculation and equation for their design and research. Here, we are going to explain that what would be the effect of these dimensional entities on the performance of the antenna [5]. Majorly, we have taken our research toward the effect of the length of the patch, width of the patch and dielectrics constant used between the patch and ground with respect to the frequency applied to the antenna [6].

2 Micro Strip Antenna Design

Architecture of any rectangular microstrip antenna is as shown in Fig. 1. Mainly, it consists of a ground plane, dielectric substrate and a radiating patch. Shape and size of the patch can be circular, rectangular, hexagonal, etc. [7]. There are different feed methods also for antenna like: wave port feed, coaxial feed and lumped feed.

In case of rectangular microstrip antenna, there are some thumb rules for taking the decision of length and width of the patch. Height of the dielectric substrate and dielectric constant also plays an important role during the transmission of radio waves from patch of the antenna [8]. Basic fundamental for the design of antenna is as follows. Width of rectangular microstrip patch antenna is given as:

$$W = \frac{C}{2F_r} \times \sqrt{\frac{2}{E_r + 1}}$$

where "C" is the speed of light (3×10^8 m/s), E_r is the dielectric constant of the substrate and F_r is the resonant or solution frequency of the proposed antenna.

Method to determine the effective dielectric constant of the substrate is given as follows:

$$E_{reff} = \frac{E_r + 1}{2} + \frac{E_r - 1}{2}\left(1 + 12\frac{h}{W}\right)^{-\frac{1}{2}}$$

"h" is defined as thickness or height of the substrate, and "W" is the width of the antenna [9].

Effective lengths of the antenna at resonate frequency is given as:

$$L_{eff(eff.length)} = \frac{C}{2F_r\sqrt{E_{reff}}}$$

Fig. 1 a Basic design of a rectangular microstrip antenna. **b** Basic design of a rectangular microstrip antenna

(a)

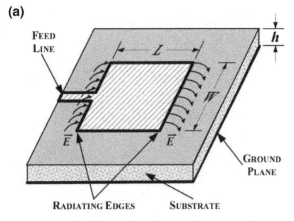

Basic design of a rectangular microstrip antenna

(b)

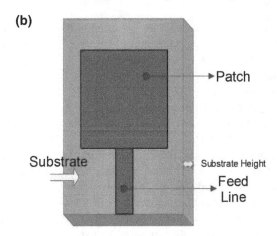

Basic design of a rectangular microstrip antenna

In the below equation, possible length extension of the antenna is given [10].

$$\Delta L = 0.412.h \left(\frac{(E_{\text{reff}} + 0.3)}{(E_{\text{reff}} - 0.258)} \times \frac{\left(\frac{W}{h} + 0.264\right)}{\left(\frac{W}{h} + 0.8\right)} \right)$$

In nutshell, total length of the antenna can be defined as:

$$L = L_{\text{eff(eff.length)}} - 2\Delta L$$

Effective wavelength of the rectangular microstrip patch antenna can be defined as:

$$\lambda_g = \frac{\lambda}{\sqrt{E_{\text{reff}}}}$$

Distance between the feed and lower side of antenna can be computed as for given input impedance of 50 Ω [11–13].

$$x \cong \frac{L}{2\pi} \cos^{-1}\left(\frac{Z_{\text{in}}}{300}\right)$$

3 Experimental Result

A MATLAB-based program is written for the calculation of length and width of the antenna. While designing the antenna, the following constants are assumed. Height of the substrate is assumed constant as 1.6 mm. Input and output impedance of the antenna is assumed to be 50 Ω. Speed of light is taken as 3×10^8 m/s. Operating frequency is taken from 1 GHz to 12 GHz, i.e., up to X band.

First, we would investigate the effect of the frequency on the width of the antenna. From graph, it is obvious that if we increase the operating frequency, then width of the antenna decreases. At 1 GHz, it is 7.693 mm, and at 21 GHz, height of the antenna is 92.32 mm. We can state from Fig. 2a and b that up to 12 GHz, the width of the antenna cannot be more than 92.32 mm. 20.86 mm is the mean value of the dimension of the rectangular antenna.

Figure 3 shows the relation of width of antenna versus frequency at different values of dielectric constant also. In the following figure, range of the dielectric constant is taken as 1–12 with a space of 1. It is clear from Fig. 3 that for lower dielectric constant, width of the antenna is more, while for higher dielectric constant, width of the antenna required is less. It is also seen from Fig. 3 that if we increase the value of frequency, then requirement of the width of the antenna is also decreased. Typical value of width at 1 GHz under these conditions is more than 120 mm, and at frequency 12 GHz, it is about 20 mm.

Figure 4 shows the relation of length of antenna versus frequency at different values of dielectric constant also. In Fig. 4, range of the dielectric constant is taken as 1–12 with a space of 1. It is clear from Fig. 4 that for lower dielectric constant, length of the antenna is more, while for higher dielectric constant, length of the antenna required is less. It is also seen from Fig. 4 that if we increase the value of frequency, then requirement of the length of the antenna is also decreased. Typical value of length at 1 GHz under these conditions is about 120 mm, and at frequency 12 GHz, it is less than 15 mm.

From Fig. 5, we can indicate that selective dielectric constant and effective dielectric constant are almost accord in nature. There is no change in the value of selected

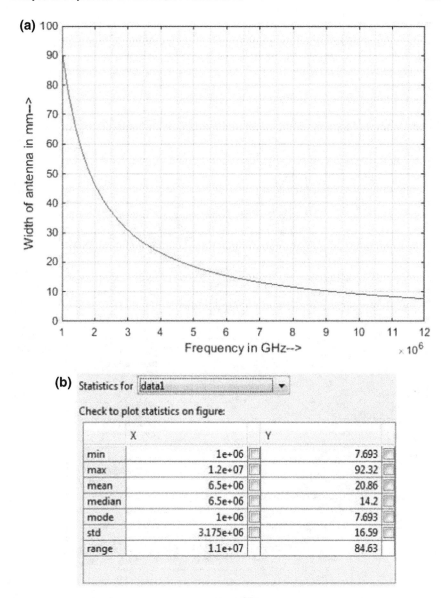

Fig. 2 **a** Width of the antenna versus frequency curve. **b** Data statics for the calculation of width of antenna

dielectric constant and effective dielectric constant calculated from the simulation for different frequencies from 1 to 12 GHz.

Figure 6 shows the relation between the length of the antenna and frequency in GHz. It shows that if we increase the frequency of the antenna, length will decrease.

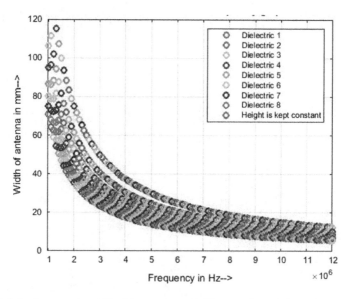

Fig. 3 Relation between the width of the antenna and different values of dielectric and frequencies

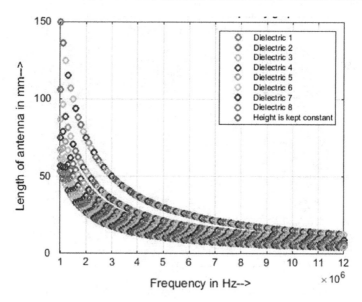

Fig. 4 Length of the antenna versus frequency curve

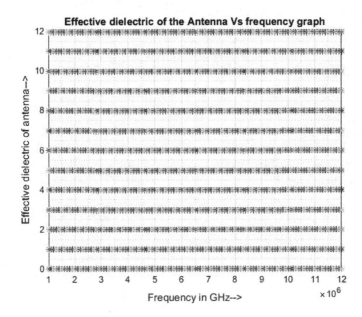

Fig. 5 Relation between the dielectric constant versus frequency

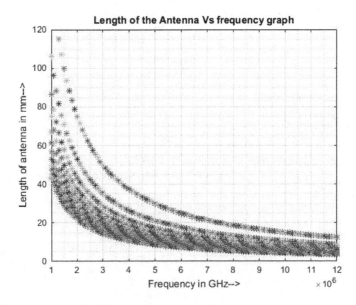

Fig. 6 Relation between the dielectric constant versus frequency

4 Code for Simulation

The following MATLAB program is used for calculating different parameters of the proposed antenna.

```
clc;
clear all;
warning off;
Er=4.28;
Er = input('Enter the value of Dielectric constant: ');
%One can put the variable value of dielectric constant

Fr = 2.4e6;
Fr1 = input('Enter the operating frequency in GHz: ');
%One can put the variable value of frequency

Fr = Fr1*1,000,000;
C = 3e8;
Z0 = 50;
Zin = 50;
h = 1.6e-3;
h = input('Enter the height of dielectric in mm: ');
%One can put the variable value of height in mm

lemda = C/Fr;
W = 19.3;

for Fr = 1e6:100,000:12e6
    for Er = 1:1:8
W = C./((2*Fr)*(sqrt((Er + 1)/2)));
Ereff = (Er + 1)./2 + ((Er-1)./2).*(1./(1 + 12.*(h./W)));
Leff = C/(2*Fr*sqrt(Ereff));
delL = 0.412*h*(((Ereff + 0.3)/(Ereff-0.258)))*((W/h + 0.264)/(W/h + 0.8));
L = Leff-delL;

plot(Fr,Ereff,'o','LineWidth',2)
plot(Fr,Ereff)
hold on;
    end
end

grid on;
grid minor;
axis([1e6 12e6 1 8])
title('Width of the Antenna Vs frequency graph')
xlabel('Frequency in Hz- > ')
ylabel('Length of antenna in mm- > ')
```

legend('Dielectric 1','Dielectric 2','Dielectric 3','Dielectric 4','Dielectric 5','Dielectric 6','Dielectric 7','Dielectric 8','Height is kept constant')

fprintf('\nWidth of the antenna is %3.3f mm\n', W);
fprintf('\nEffective dielectric constant of the antenna is %3.3f\n', Ereff);
fprintf('\nEffective length of the antenna is %3.3f mm\n', L);

A = (Z0/60)*(sqrt((Er + 1)/2)) + ((Er-1)/(Er + 1))*(0.23 + 0.11/Er);
B = (377*pi)/(2*Z0*sqrt(Er));
w = (2*h/pi)*(B-1-reallog(2*B-1) + ((Er-1)/(2*Er))*(reallog(B-1) + 0.39-0.61/Er));

Z2 = sqrt(2*Z0);
Z1 = sqrt(Z0*Z2);
lemdag = (lemda/Ereff);
x = (L/((2*pi))*acos(Zin/300));

5 Conclusion

It is evident from the study that there are substantial effects on the length and width of the rectangular microstrip antenna by changing the operating frequency and dielectric constant. We can conclude that at lower frequency, length and width required for the designing of antenna are more as compared to the width and length required to operate at higher frequency. We can state that requirement of the width of the antenna is also more as compared to the requirement of the length of the antenna. There is no effect on the effective dielectric constant while designing the antenna at these frequencies.

Acknowledgements Author would like to acknowledge his thanks for the support given by the Lovely Professional University, Phagwara. Sincere thanks are also invocation to Dr. Rajesh Singh for their valuable support and motivation.

References

1. Balanis, C.A.: Antenna Theory-Analysis and Design. Wiley Publication (2018)
2. John, K., Ronald, M., Khan, A.: Antenna and Wave Propagation. Mcgraw Higher Ed. ISBN: 9780070671553 (2010)
3. Werfelli, H., Tayari, K., Chaoui, M., Lahiani, M., Hamadi, G.: Design of Rectangular Microstrip Patch Antenna. https://doi.org/10.1109/atsip.2016.7523197

4. Li, W., Liu, B., Zhao, H.: Parallel rectangular open slots structure in multiband printed antenna design. IEEE Antennas Wirel. Propag. Lett. **14**, 1161–1164 (2015). https://doi.org/10.1109/LAWP.2015.2393632
5. Malik, P., Parthasarthy, H.: Synthesis of randomness in the radiated fields of antenna array. Int. J. Microw. Wirel. Technol. **3**(6), 701–705 (2011). https://doi.org/10.1017/S1759078711000791
6. Malik, P.K.: Hardware Design of Equilateral Triangular Microstrip Antenna Using Artificial Neural Network. TECHNIA—I.J.C.S.C.T **3**(2) (Jan 2011)
7. Chen, T., Y. Chen, Jian, R.: A wideband differential-fed microstrip patch antenna based on radiation of three resonant modes. Int. J. Antennas Propagat. **2019**(Article ID 4656141), 7 pages (2019)
8. Cao, Y.F., Cheung, S.W., Yuk, T.I.: A multiband slot antenna for GPS/WiMAX/WLAN systems. IEEE Trans. Antennas Propag. **63**(3), 952–958 (2015)
9. Tekkouk, K., Ettorre, M., Le Coq, L., Sauleau, R.: Multibeam SIW slotted waveguide antenna system fed by a compact dual-LAYER Rotman lens. IEEE Trans. Antennas Propag. **64**(2), 504–514 (2016)
10. Yoon, I., Oh, J.: Millimeter-wave thin lens using multi-patch incorporated unit cells for polarization-dependent beam shaping. IEEE Access **7**, 45504–45511 (2019)
11. Zheng, B., Wong, S., Feng, S., Zhu, L., Yang, Y.: Multi-mode bandpass cavity filters and duplexer with slot mixed-coupling structure. IEEE Access **6**, 16353–16362 (2018)
12. Malik, P.K., Singh, M.: Multiple bandwidth design of micro strip antenna for future wireless communication. Int. J. Recent Tech. Eng. **8**(2), 5135–5138, ISSN: 2277-3878 (2019). https://doi.org/10.35940/ijrte.B2871.078219
13. Malik, P.K., Parthasarthy, H., Tripathi, M.P.: Alternative mathematical design of vector potential and radiated fields for parabolic reflector surface. In: Unnikrishnan, S., Surve, S., Bhoir, D. (eds.) Advances in Computing, Communication, and Control. Communications in Computer and Information Science, ICAC3 2013, vol. 361. Springer, Heidelberg (2013)

Vivek Arora Professionally qualified with more than fifteen years of experience with B.E. and M.Tech. in ECE, and he worked as Principal in Shri Balaji Institute of Engineering and Technology, Sampla, and in Six Sigma Institute of Technology and Science, Rudrapur. He is currently working in Ganga Institute of Technology and Management, Kablana (Jhajjar).

Praveen Kumar Malik is a professionally qualified and experienced person with extensive knowledge and skills in Embedded System and Antenna. He is working as Professor in the Department of Electronics and Communication, Lovely Professional University, Punjab, INDIA. He is B.Tech., M.Tech. and Ph.D. from Electronics and Communication system. His major area of interest is Antenna and Embedded systems. He is having more than 10 papers in international refereed journals, and he is having more than 10 papers in the international and national conference also.

Advanced Computing Technologies and Latest Electrical and Electronics Trends

Fog Computing Research Opportunities and Challenges: A Comprehensive Survey

Shaheen Parveen, Pawan Singh and Deepak Arora

Abstract Fog computing is the new buzzing word in the world of technology which complements the cloud computing and adds the functionality to the Internet of Things (IoT). The major functionality includes lowering the latency rate, improving the security system and creating a smart world in terms of networking. This paper is the survey which represents the idea behind fog computing and its need in the future. We have systematically derived the concept of fog computing, its working, study on some recent surveys and research, its application areas, emerging challenges and advantages in today's technological world. The main focus of the paper is on the importance of fog computing, its opportunities and emerging challenges. Study on this emerging topic leads to explore on some uncovered areas which require the focus of researchers and practitioner. It may also facilitate to discover the solution for the emerging challenges and making the pathways for others. This is the comprehensive study and survey which will help the researchers to explore the new ideas and solution to the new and existing problems. Although it is in its early stage of growth, there are lots of opportunities for the researchers.

Keywords Cloud computing · Fog computing · Edge computing and network · Internet of Things (IoT)

1 Introduction

Fog computing sometimes also considered as edge computing is the extensive version of cloud computing [1]. Fog computing model proposed by cisco is the replication of the development model of cloud computing application where software as a service

S. Parveen (✉) · P. Singh · D. Arora
Department of Computer Science and Engineering, Amity School of Engineering and Technology, Amity University Lucknow Campus, Lucknow, Uttar Pradesh, India
e-mail: shaheenparveen1407@gmail.com

P. Singh
e-mail: pawansingh51279@gmail.com

D. Arora
e-mail: darora@lko.amity.edu

© Springer Nature Singapore Pte Ltd. 2020
P. K. Singh et al. (eds.), *Proceedings of First International Conference on Computing, Communications, and Cyber-Security (IC4S 2019)*, Lecture Notes in Networks and Systems 121, https://doi.org/10.1007/978-981-15-3369-3_13

(SaaS) laid on infrastructure as a service (IaaS) and platform as a service (PaaS). It is not the replacement or substitution of cloud computing rather acting as an intermediately providing the functionality to the Internet of Things (IoT) [2]. There are many issues in cloud computing which evolves the idea of fog computing and came into public arena for the improvement of the same. Initially, it was considered hypothetical but after its implementation, it is found beneficial for end-users and the organization as well. The main purpose of fog computing is to reduce the latency rate so that real-time and time-sensitive data can be fetched and processed easily [3]. The systems which require real-time data such as flights, trains, oil agencies may be get benefitted if the latency of data is minimized. The requirement of this solution leads the research on fog computing and its security concern as well. Fog computing also known as fogging is used to remove the traffic of network by implementing it close to end-users. While the cloud is the process of centralizing the users' data onto the network, fogging uses the concept of distribution of the same. The communication between the users and end nodes can be minimized by the implementation of this decentralization scheme of fog computing [4].

The organization of the paper is as follows: Sect. 2 of this paper explains the working of the system of fog computing. In Sect. 3, the recent survey on fog computing is discussed. Section 4 provides detail information regarding the application of fog computing. Section 5 illustrates the current emerging challenges in the arena of fog computing. In Sect. 6, the advantages/benefits of fog computing are presented. The architectures of some of the fog computing simulators are described in Sect. 7. The conclusion and future scope of this study are discussed in Sect. 8.

2 Working of the System

2.1 Fog Nodes

Fog nodes act as an intermediary or middleware for the servers of cloud and end-user devices. It receives data from the edges of network using any protocol in real time and transiently stores it often for 1–2 h. It gives milliseconds response time by running different IoT applications for analysis and controlling in real time. After processing, it sends data summaries back to the cloud server periodically. The working of the fog is reflected by the architecture of the fog in Fig. 1.

2.2 Cloud Platform

It receives summaries from many fog nodes and aggregates it. After receiving, it starts performing analysis to get the insight of business. Once the analysis has done, it can send the new rule to application of fog nodes for these insights. Fog nodes

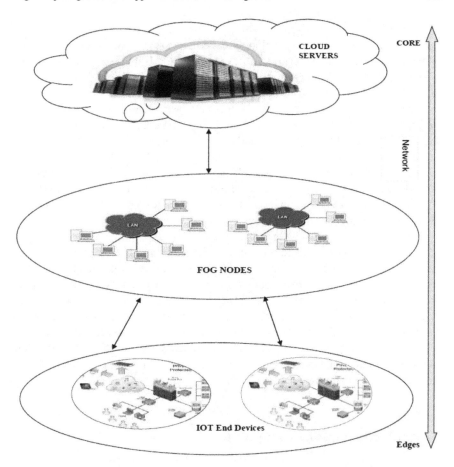

Fig. 1 Fog architecture

can be implemented through IoT application on the edges of network. The nearest fog node receives the device request from the edges. The most time-sensitive data processed on fog nodes and the data which can wait for minutes sent to one of the nearest aggregate fog nodes, and the least time-sensitive data may be sent to the cloud for the analysis.

3 Recent Surveys on Fog Computing

There is the quantity of surveys and instructional exercises which formally characterize fog computing and the related research challenge. Flavio Bonomi, Rodolfo Milito, Jiang Zhu and Sateesh Addepalli proposed the concept of fog platform to support IoT devices for resource-constrained [1]. L. M. Vaquero and

L. Rodero-Merino have given an outline of the idea of fog computing as far as empowering innovations and developing patterns in usage trend. They additionally examine the future challenges [2]. M. Yannuzzi, R. Milito, R. Serral-Gracia, D. Montero and M. Nemirovsky examined a portion of the difficulties in the application of IoT and showed that fog computing platform is a capable and promising empowered agent for the same [3]. A few surveys are available at a large scale on fog ubiquitously by Stojmenovic et al. [4–6]. S. Yi, C. Li and Q. Li discussed about the meaning of fog computing and firmly interconnected ideas. They presented the application of situation and talk about the future challenges too [7].

There are some good research publications on various application spaces, i.e. vehicular ad hoc networks (VANETs), radio access networks (RANs) and the Internet of Things [8–10]. Dasterji suggested the concept of fog computing alongside its attributes. They present different applications that get benefits of fog computing and discussed its future challenges also [11]. M. Chiang, S. Ha, C. L. I, F. Risso and T. Zhang illustrated an instructional exercise on fog computing. They differentiate between cloud computing, fog computing and edge computing. They additionally presented the benefits of fog computing and discussed the upcoming research challenges [12]. The latest related paper is exhibited by Mahmud et al. They examined a scientific classification of fog computing as indicated by its difficulties and highlights. They revealed the contrasts between versatile edge computing, portable cloud computing and its extension called fog computing. They additionally contemplated fog environment setup, networking devices and different measurements of the same [13].

4   Application of Fog Computing

Fog computing can contribute to the wide range of applications for those systems which need reliability of the Internet connection such as connection car, smart home, smart traffic lights and health care and activity tracking and augmented reality. We can drive an automatic car with the help of fog nodes which would update the traffic in real time, and we can also add automatic parking features to the vehicles with the advancement and implementation of fog computing system.

Traffic system can also be handled in real-time sensor by opening the way upon receiving the signals from smart sensing flashlight. It will control the movement of vehicles through sensing the pedestrian and vehicles nearby it by sending the warning to fog nodes.

There are many applications for smart home but each one has the different platform, so it is difficult to run these systems in unified manner. Fog nodes may be used to integrate these applications and operate the same in real time by proper coordination. It will also provide the storage to these independent devices and act as a single interface for communicating to various applications.

Healthcare system has major issues which could be resolved through fogging. We can provide the storage to critical report for real-time processing. As fog nodes has low latency rate for accessing the information, it will also help to response in critical events such as medical report and activity tracking for the patient in real time.

Fog computing will help in augmented reality by accumulating virtual information into the real world. There is a brain–computer game which fills the gap between fog and cloud computing by adding the functionality of real-time processing game with augmented system.

5 Emerging Challenges

There are many challenges (as shown in Fig. 2) which become obstacle in the pathway of development and deployment of fog computing; some of them are scalability, security, complexity, heterogeneity, latency, dynamicity, energy consumption and resource management.

Scalability in fog computing decides the capability of the fog server to respond as many devices as it can. It is highly needed to that fog system which could increase its capacity proportionally with the growth of IoT and end-user.

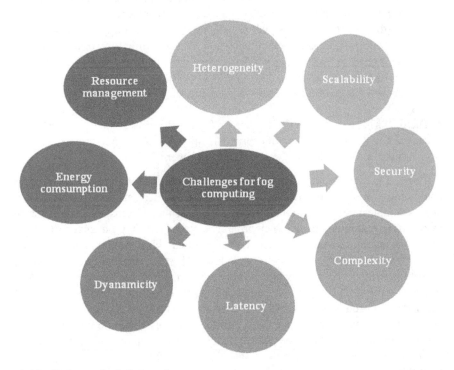

Fig. 2 Challenges for fog computing

- **Scalability** in fog computing decides the capability of the fog server to respond as many devices as it can. It is highly needed to that fog system which could increase its capacity proportionally with the growth of IoT and end-user.
- **Security** measures should be applied in such a way that it could be protected and controlled more than cloud computing as fog nodes are susceptible and vulnerable to the network. Security and privacy issues may include man-in-the-middle attack, intrusion detection, malicious nodes detection in fogging and data protection and management issues.
- **Complexity** of the fog nodes increases due to the different manufacturers and its own product design. Choosing the optimal product for the better performance may be the hectic work for the developer.
- There are numerous IoT gadgets and sensors which are structured by various manufacturers which increase the **heterogeneity** of fog computing. These gadgets have different abilities such as sensors and computing powers. The administration and coordination of such heterogeneous networks of IoT gadgets and the determination of the proper assets will turn into a major test.
- Although the need of fog computing has come into existence due to high **latency** of cloud computing which has the motive to provide the low latency rate to the edge devices, there are many cases in which latency of fog computing may also increase which is biggest issue and challenge for researcher.
- One of the major highlights of IoT gadgets is the capability to advance and change working process of their organization. This **dynamicity** test will change the internal properties and execution of IoT gadgets. The handheld gadgets experience the ill effects of programming and equipment maturing due to this. Therefore, fog systems will require programmed and smart reconfiguration of the whole structure of system.
- There are a huge number of devices used in fog which are distributed in network that can lead to high consumption of energy. So, minimizing the **energy consumption** is also the challenge for researcher and practitioner.

Fog end gadgets are regularly network gadgets outfitted with extra capacity and computing power. Nonetheless, it is troublesome for these gadgets to synchronize and coordinate with the traditional fog servers. Consequently, sensible administration of **resource management** in fog environment is required for productive task of the same. The major challenges also include the functional split and architectural between cloud computing level and fog computing level. There are other perspectives which need the attention for the development of fog computing such as fault tolerance, interoperability and allocation of resources and its scheduling. Implementation of fog nodes in different locations with different ownerships and managing its security and privacy is the big concern for us. Similarly, the platform may be differed for the implementation of the nodes so we would have to keep our concentration also on the portability of the same. The system is distributed in different regions, so fault tolerance should be higher, and if there is system failure, the management of replacement should be provided without any delay. Some systems must be connected with wireless, so reliability of system may be also the big issue. Because of the diversity of the systems, there is also the need of some standard architecture for fog computing systems so that it could not create any problem in the performance of the system. There must be ensured connectivity with fog nodes in the case of cloud networking connection failure which will increase the reliability of this system.

Modelling of the fog-based network might be also a big challenge as it needed the huge amount of the data from IoT and cloud-based systems because of its dynamic nature, and there is requirement to research and investigate the workload efficiency and energy consumptions for this system.

6 Advantages/Benefits of Fog Computing

There are many advantages of the fog computing if we encounter all the challenges. First of all, we are in the way to minimize the latency rate for accessing the network as fog nodes are located in the neighbourhood of the end-user, and it will be more beneficial if we make the system reliable by keeping the uninterrupted connectivity and in case of failure of the cloud connectivity. It will also improve the efficiency of the system and reduce traffic of data of network. It will also help cloud computing by reducing the amount of data transferred amongst the network for processing and storing. The goal of implementing the fog system could be the security measurement and compliance of the network also. Applications related to smart city need to process sensor information on an ongoing premise, where fog devices can play a noteworthy job. The fog processing approach additionally empowers continuous sensor-based healthcare administrations. Smart traffic lights and augmented reality could be achieved through the improvement of fog computing. It can play a vital role in real-time processing devices which need uninterrupted and reliable connection. It is also helping the machine to expand the business and for its agility.

Fog computing will definitely increase the serviceability of the network because of its proximity to users' range. It will decentralize the network which will also minimize the direct dependency onto the cloud network. The restriction in the growth of IoT world can be resolved by this which would be the biggest achievement of the systems. It must be the milestone for the networking if it is implemented by taking all precautions.

7 Fog Computing Simulators

There are many simulators present for the implementation and simulation of fog environment in which **iFogsim, MyiFogsim, Yet Another Fog Simulator (YAFS)** and **FogNetSim++** are the names of few. Choosing the best according to one's skill may be critical in some extent if proper analysis and overview on these tools are not provided. The following is the quick analysis of architecture and comparison of tools which will give the better understanding to beginners.

iFogSim developed by Gupta et al. [14] is the fog simulator based on CloudSim which provides the resource management techniques such as latency, power

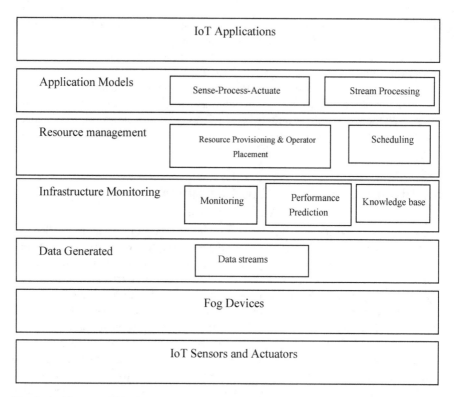

Fig. 3 Architecture of iFogSim

consumption, network utilization and cost estimation. iFogSim comprises the layered architecture (as shown in Fig. 3) in which each layer is responsible for some functionality and transmitting the data to higher layer for facilitating the operations on upper layers. It depends upon CloudSim simulator for implementing the functionality of fog devices and includes the classes such as FogDevices, sensor, actuators and tuples, application (AppModule, AppEdge, AppLoop) entities. **YAFS** is adding the properties of dynamic and customized strategies such as allocation and failure of the new application modules. It has six main classes named topology, core, application, selection, placement and population. The relationship amongst these classes may be seen in Fig. 4. **Core** class acts like a controller which collects messages of events, stores it in raw format and integrates all others fog devices, process life cycle, policies such as selection, population, placement and custom controls. **Topology** is called through controller (core) class by other classes. Other processes such as **placement**, **selection** and **population** are used to allocation of software, selection of processes and allocation of workload and characterization. Stats compute simulation results such as response time, latency and resource utilization.

FogNetSim++ is the toolkit which is based on graphical user interface (GUI) which also adds on the scheduling techniques for the fog nodes configuration. It

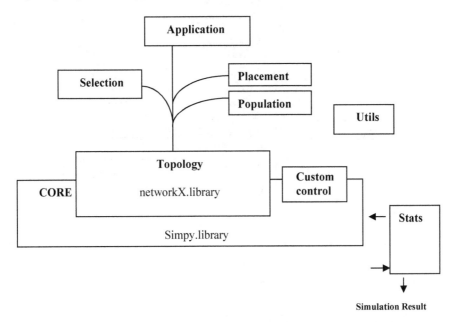

Fig. 4 Architecture of YAFS

uses INET framework on OMNET++ simulator. NED is used for the definition of topology which has the ability to create mesh, ring, tree and graph structure easily [15]. The architecture of the FogNetSim++ is provided in Fig. 5.

There is a common example in iFogSim and YAFS named VR game in which virtual reality is augmented through fog devices. It is the game where four players residing on four different and distant locations are connected through EEG wireless headset with their android smartphone. For winning this game, the brain concentration power of the player should pass a certain threshold value after which a beam produces. This beam is used by player to pull the target object towards himself through his brain's continuous concentration or focus. One who has higher concentration power succeeds in grabbing that target object by competing with opponent's concentration power of brain. Whereas iFogSim uses java programming language, YAFS uses Python and FogNetSim++ uses C++ language for its implementation. Table 1 gives the glimpse of the comparison of the characteristics of fog simulators: iFogsim, YAFS and FogNetSim++.

8   Conclusion and Future Scope

Fog computing is although in its early stages of growth but this literature is showing the complete paradigm of the same. We have presented the basic idea of fog computing, and later, we have stated the various surveys by different researchers

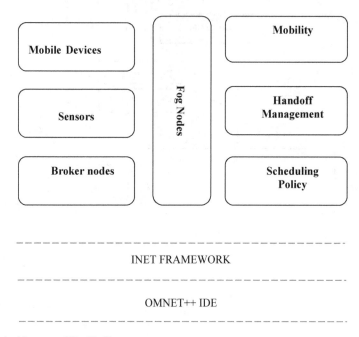

Fig. 5 Architecture of FogNetSim++

Table 1 Comparison of the characteristics of fog simulators

Tools name	Language	Topological structure	Traceability of result	Topology definition
iFogSim	Java	Tree	No	API
YAFS	Python	Graph	Yes	API, JSON, Brite, CAIDA…
FogNetSim++	C++	Versatile	Yes	NED

in this paper. We have also discussed about the challenges, simulation tools for fog computing and its importance. Portability, security, privacy, interoperability and serviceability are also discussed here for the sake of improvement of the system.

The motive of this survey is to give an overview of fog computing, working of fog computing in coordination with cloud computing, application areas, the research-based challenges, its advantages and present tools for simulation. There are many challenges addressed for its implementation which could be faced in the future. We can get the solution for developing the fog nodes for the betterment and advancement of networking. Using the prototype for creating the fog environment and comparing the result with different constraints would not be feasible. We need simulators for getting the feel of fog computing which could be also used for further analysis. There are also in built example which could be further improved and analysed.

The future scope of this paper would lead the reader in quick overview of aforesaid concepts. This paper would also help the research beginners in the field of fog computing to select the best tool according to their skills and requirements. Although fog computing is at its infant stage, it is having the future scope which could be developed for the flexibility and speeding of the networks. We have many opportunities despite many challenges which encourage the research work on this current topic.

References

1. Bonomi, F., Milito, R., Zhu, J., et al: Fog computing and its role in the internet of things. In: The Proceedings of the First Edition of the MCC workshop on Mobile Cloud Computing, pp. 13–16. ACM, Helsinki, Finland (2012)
2. Vaquero, L.M., Rodero-Merino, L.: Finding your way in the fog: towards a comprehensive definition of fog computing. ACM SIGCOMM Comput. Commun. Rev. 44(5), 27–32 (2014)
3. Yannuzzi, M., Milito, R., Serral-Gracia, R., Montero, D., Nemirovsky, M.: Key ingredients in an IoT recipe: fog computing, cloud computing, and more fog computing. In: The Proceedings of 19th IEEE International Workshop on Computer Aided Modeling and Design of Communication Links and Networks (CAMAD), pp. 325–329. IEEE, Athens, Greece (2014)
4. Stojmenovic, I.: Fog computing: a cloud to the ground support for smart things and machine-to-machine networks. In: The Proceedings of the Australasian Telecommunication Networks and Applications, pp. 117–122. IEEE, Melbourne, Australia (2014)
5. Stojmenovic, I., Wen, S.: The fog computing paradigm: scenarios and security issues. In: The Proceedings of the Federated Conference on Computer Science and Information Systems, pp. 1–8. IEEE, Warsaw, Poland (2014)
6. Stojmenovic, I., Wen, S., Huang, X., Luan, H.: An overview of fog computing and its security issues. Concurr. Comput. Pract. Exp. 28(10), 2991–3005 (2015)
7. Yi, S., Li, C., Li, Q.: A survey of fog computing. In: The Proceedings of the 2015 Workshop on Mobile Big Data, pp. 37–42. ACM, Hangzhou, China (2015)
8. Kai, K., Cong, W., Tao, L.: Fog computing for vehicular ad-hoc networks: paradigms, scenarios, and issues. J. China Univ. Posts Telecommun. 23(2), 56–96 (2016)
9. Yan, S., Peng, M., Wang, W.: User access mode selection in fog computing based radio access networks. In: The Proceedings of the IEEE International Conference on Communications (ICC), pp. 1–6. IEEE, Kuala Lumpur, Malaysia (2016)
10. Chiang, M., Zhang, T.: Fog and IoT: an overview of research opportunities. IEEE Internet Things 3(6), 854–864 (2016)
11. Dastjerdi, A., Gupta, H., Calheiros, R., Ghosh, S., Buyya, R.: Internet of Things: Principles and Paradigms, 1st edn. Morgan Kaufmann, Burlington, MA (2016)
12. Chiang, M., Ha, S., Chih-Lin, I., Risso, F., Zhang, T.: Clarifying fog computing and networking: 10 questions and answers. IEEE Commun. Mag. 55(4), 18–20 (2017)
13. Mahmud, R., Kotagiri, R., Buyya, R.: Internet of Things—Technologies, Communications and Computing, 16th edn. Springer, Singapore (2018)
14. Gupta, H., Dastjerdi, A.V., Ghosh, S.K., Buyya, R.: iFogSim: a toolkit for modeling and simulation of resource management techniques in the Internet of Things, Edge and Fog computing environments. Softw. Pract. Exp. 47(9), 1275–1296 (2016)
15. Qayyum, T., Malik, A.W., Khan, M.A., Khalid, O., Khan, S.U.: FogNetSim++: a toolkit for modeling and simulation of distributed fog environment. IEEE Access 6, 63570–63583 (2018)

IIGPTS: IoT-Based Framework for Intelligent Green Public Transportation System

Akhilesh Ladha, Pronaya Bhattacharya, Nirbhay Chaubey
and Umesh Bodkhe

Abstract Today, growing urbanization coupled with the high demands of daily commute for working professionals has increased the popularity of public transport systems (PTS). The traditional PTS is in high demand in urban cities and heavily contributes to air pollution, traffic accidents, road congestion, increase of green house gases like oxides of carbon (OC), methane (CH_4), and oxides of nitrogen (NO_x) emissions. PTS also suffers from the limitations of preset routes, privacy, crowd, and less space for passengers. Some PTS are densely crowded in few routes, whereas in other routes they are not crowded at all. Thus, the aforementioned limitations of toxic emissions coupled with load management and balancing in PTS are a critical issue. The paper proposes *IIGPTS*: IoT-based framework for Intelligent Green Public Transportation System that addresses the mentioned issues by measuring emission sensor readings with respect to varying parameters like passenger density (total carrying load), fuel consumption, and routing paths. The performance of *IIGPTS* is analyzed at an indicated time slot to measure emissions and load on the system. Then, the simulation is performed based on dynamic routes by increasing passenger load and measuring the service time for user traffic as requests, also considering the request drops. The results obtained by *IIGPTS* framework indicate a delay of 104 s for 10,000 requests spread across an entire day, which is negligible considering the load. Thus, *IIGPTS* can intelligently handle varying capacity loads at varying routes, with fewer emissions, making the realization of green eco-friendly PTS systems a reality.

A. Ladha · P. Bhattacharya (✉) · U. Bodkhe
Institute of technology, Nirma University, Ahmadabad, Gujarat, India
e-mail: pronoya.bhattacharya@nirmauni.ac.in

A. Ladha
e-mail: akhilesh.ladha@nirmauni.ac.in

U. Bodkhe
e-mail: umesh.bodkhe@nirmauni.ac.in

A. Ladha · N. Chaubey
Gujarat Technological University, Ahmadabad, Gujarat, India
e-mail: nirbhay@ieee.org

N. Chaubey
Ganpat University, Mehsana, Gujarat, India

© Springer Nature Singapore Pte Ltd. 2020
P. K. Singh et al. (eds.), *Proceedings of First International Conference on Computing, Communications, and Cyber-Security (IC4S 2019)*, Lecture Notes in Networks and Systems 121, https://doi.org/10.1007/978-981-15-3369-3_14

Keywords IoT · Emission sensors · Green transport systems · Dynamic routing

1 Introduction

In today's modern era, growth of transport systems is necessary for sustainable economic growth of a nation. With the rise of independent vehicles, there has been a humongous growth in traffic volumes that fuel many issues like higher delays, fuel consumption, accidents, green house emissions, and much more. These issues lead to air pollution problems, and quality of life of modern citizens is at high risk. According to the report by Government of India, approximately 234,000 Indian rupees are being spent in the road growth [16]. According to the report of the World Bank depicted in Fig. 1a, investment in transport sector will be close to $500 billion (3.7% of GDP) over the course of next 10 years, with US $1 trillion to be spent on various upcoming research projects under infrastructure development, as presented in five year plan (2013–18). According to Forbes report, shown in Fig. 1, daily passenger commutes through various models of travel which is close to 200.4 million. Thus, there is a growing need to manage the deterioration of PTS throughout the country, while at the same time, sustaining maximum passengers over a decided route.

The situation is even worse at the international front. According to the report of Texas transportation institute, people are almost stuck about 42 h a year in traffic, with PTS almost wasting close to 3 billion gallons of fuel, the price of which is estimated close to $160 billion [3]. There is a constant need to build proper public transport ecosystems so that it could fit the needs of the people and also satisfy the budget constraints of the government. However, owing to the dynamic nature of traffic in urban areas, the problem of optimizing resources is an uphill task. At global front, depicted in Fig. 2a, as per word energy outlook, transport sector contributes to about 23% of global CO_2 emissions, and by 2030, developing countries in Asia are going to face more toxic and health problems. With the advances in information and communications technologies (ICT), communications infrastructures and hardware have matured enough to address this growing issue. ICT tools integrated with low-powered IoT devices can lead to the development of smart and intelligent transport systems (ITS). ITS will change the dynamics of travelers experience by ensuring safety of passengers, as well as reducing the growing emission rate and high fuel consumption. This is depicted in Fig. 2, which shows a high compounded annual growth rate (CAGR) of India close to 11% as compared to USA and EU by 2022. Thus, it could lead to the development of greener transport vehicles emitting less toxic gases and ensuring high mobility, sustainability, and economic development.

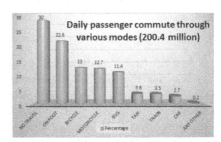

(a) Projects currently active for transportation in India [1].

(b) Percentage of daily passenger commute through various modes of travel in India [2].

Fig. 1 Statistics of public transport in India

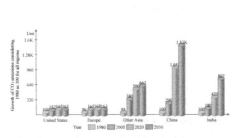

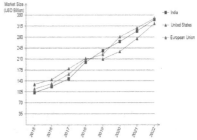

(a) CO_2 emissions for international market till 2030 [5].

(b) CAGR growth comparison of India, US and EU till 2022 [12].

Fig. 2 Growth rate and emissions levels in international transport sector

1.1 Motivation

The study of intelligent public transport systems that are eco-friendly is the need of the global society. Studies proposed till now have mainly focused on emissions, mileage, and battery life of AVs through the use of sensors. Some studies also suggested the use of software-defined networking (SDN) to manage routing paths and data flows through a sequence of pre-defined routes, so that emissions are minimized. None of the studies have focused on measuring a mathematical relation between total load of the vehicles operating on various routes and measured gas emissions. Hence, motivated by the same, *IIGPTS* proposes a novel idea to derive a mathematical relation between the mileage, total load on the vehicle, and gas emissions levels.

1.2   Research Contributions

The following are the major contributions of the paper:

- A mathematical formulation is presented between the mileage of the vehicle, the total load on the vehicle, and gas emissions measured by the vehicle through IoT sensors.
- Analysis is performed with the measure of the possible effects of important parameters like road condition, traffic condition, type of fuel used and fuel quality, tire pressure, the carburetor maintenance on the emissions level of the vehicle.
- An algorithm is presented for the intelligent PTS based on emission readings which decide emission levels for static and dynamic routes.

The precise and rapid alerting of systems which are monitored by nodes for toxic gases requires sensors to be installed at the node. This will minimize emissions and also accidents due to hazardous compounds, many of which are explosive in nature.

1.3   Organization of the paper

The paper is organized into five sections. Section 2 discusses the related work. Section 3 discusses the system components and the proposed architecture. Section 4 discusses the performance evaluation. Finally, Sect. 5 discusses the possible conclusion and future work.

2   Related Work

Design of intelligent PTS is a relatively new emerging field, with only few work of noteworthy importance. Hu et al. [13] monitored the lane assignment strategies for autonomous vehicles (AVs) which are connected through sensor networks and established a trade-off balance between load efficiency and toxic emissions safety. The study derives useful results to minimize the emissions level to a certain threshold in the network. The limitation is that the study does not provide particular bifurcation of toxic components like carbon monoxide, hydrogen sulfide, chlorine, bromine, hydrogen chloride, hydrogen fluoride, nitric oxide, nitrogen dioxide, sulfur dioxide, ammonia, hydrogen cyanide, phosgene, benzene, formaldehyde, methyl bromide, arsine, phosphine, boranes, silane, germane, and many more with the balanced load in the vehicle . These compounds are mainly fund in variety of situations in heavy industries and can effect vehicles in enclosed parking areas. Brown et al. [6] researched on possible impacts of *Kaya identity method* (KIM) to discuss the possible impacts of toxic gases. The results presented few toxic compounds as parameters, while other parameters are not discussed. Wadud et al. [18] compared the effects of travel duration, time and market prices of petrol and concluded that carbon emissions greatly depend on trip distances, and time of travel during the day. Authors in [9] proposed the opportunities and barriers in deploying green autonomous vehicles (GAV). Studies are carried out to predict the required amount of CO_2 emissions and CO emissions by a vehicle. Gawron et al. [11] studied the imapct of green house gas (GHG) emissions of AVs and highlighting the direct and indirect influences on vehicle battery life. The studies show a 9% reduction in energy and GHG emissions of AVs as compared to normal vehicles. Similar studies are also carried out by [14, 19], and [8] which proposed efficient strategies to reduce GHGs emissions and increase vehicles life. Ang et al. [4] derived a mathematical model to predict fuel consumption based on characteristics like vehicle speed, acceleration/deceleration, and number of stops taken by the public transport vehicle. Christopher Prey et al. [10] in 2007 made study on emission of vehicle of different fuel type, where he showed that hydrogen-based vehicle emits less hazardous gases than diesel one in same running condition. Lynette et al. [7] found the relation between load on vehicle (vehicle weight + passengers weight) and fuel consumption, and the result clearly shows that as load on vehicle increases the fuel consumption increases. Another dimension called driving characteristic was introduced in [15] by Ozkan et al., mentioning that a rash driving (i.e., driving with more acceleration, frequent breaks, bad road conditions) will result in more fuel consumption. Xianghao Shen et al.[17] defined a new term "comfort perception," which represents the overall perception of passenger on travel comfort in context to passenger load factor and in-vehicle time.

3 *IIGPTS*: System components and architecture

In this section, the architecture of the proposed *IIGPTS* is presented which serves passengers on daily basis while vehicle emissions are kept in check. The system components are listed, and then the network interactions are presented.

3.1 *System Components*

The various system components presented in the proposed architecture are:

- **Network Assumption**: The design of IoT-based PTS is modeled as a graph $G(V, E)$ where V represents the set of transport systems (buses, heavy vehicles, trucks), where the type of vehicle is indicated as PTS_{type}, and E denotes the possible routes that the buses traverse throughout the day. For every edge $(i, j) \in E$, there is a cost c_{ij} which includes the operational and maintenance cost of running the PTS over the edge (i, j). The proposed model aims to find a mathematical relation to minimize c_{ij}.
- **User Requests**: Any user $U = \{U_1, U_2, \ldots, U_n\}$ can generate a travel ticket $T = \{T_1, T_2, \ldots, T_n\}$ from PTS. A user request is modeled as a quadruple $\{L_i^S, L_i^D, T_i, PTS_{id}\}$ where L_i^S is the source location for any ith ticket, L_i^D is the destination

location for the ith ticket, T_i denotes the time at which the ticket is generated, and PTS$_{id}$ denotes the vehicle registration ID of the PTS. $|L_i^S, L_i^D|$ denotes the total number of randomly generated over a period of time (a day), and $|$PTS$_{id}|$ denotes the count of PTS that are servicing the request $|L_i^S, L_i^D|$.

- **PTS terminals**: A PTS terminal is modeled as an quintuple $\{A_i, D_i, C_i, S_i, \text{MAX}_t, \tau(R)\}$ where A_i denotes the arrival time of an ith PTS to its designated stop, D_i denotes the departure time of the PTS from its designated stop, C_i depicts the cost of the route in terms of stops, counted as hops, S_i denotes the customer satisfaction in the route, and MAX$_t$ denotes the maximum time the PTS can take to travel on the route, and $\tau(R)$ denotes the total response time of the system to formulate the PTS$_{id}$ to service the request.

- **IoT sensors**: The types of sensors which are included in the model are:

  - **Electro-chemical sensors**: Galvanic cell and constant potential electrolysis. On applied voltage, it can detect carbon monoxide, nitric oxide, phosgene, hydrogen cyanide, arsine, phosphine, boranes, silane, germane, etc.
  - **Micro-emission sensors**: Improvements in the palladium grid make the detection of hydrogen-based compounds as soluble in carbon compounds; hence, micro-emission sensors are required to find the emissions effects and readings for the solubles in our detection system.

- **Route decision**: The PTS route decision is denoted as $R = \{R_i^{\text{curr}}, t_i^0, F_i, \eta_i, Q_i, R_i^{\text{prev}}, \delta\}$, where R_i^{curr} denotes the geo-location of PTS at current time, t_i^0 denotes the initial start time of the PTS from designated stop, F_i denotes the fuel indicator to service the request, η_i denotes the emission indicator of the vehicle, Q_i denotes the passenger capacity, R_i^{prev} denotes the previous requests stored a historical transactions, and δ denotes the percentage of dropped requests on a given route R.

3.2 Architecture of IIGPTS

The architecture of *IIGPTS* consists of three modules, as depicted in Fig. 3 namely

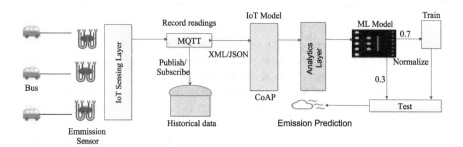

Fig. 3 *IIGPTS* framework

- **Bus Module**: The bus module considers a given set of buses $B = \{B_1, B_2, B_3, \ldots, B_n\}$ and measured load on each vehicle $L = \{l_1, l_2, l_3, \ldots, l_m\}$. The carbon emission data is measured using the emission sensors for effect of hazardous gases like carbon dioxide (CO_2) and oxides of nitrogen (NO_x) based on a dynamic or a pre-defined run by following a bottom up approach to compute the GHG emissions of an ith vehicle on a given load L_j. The relation can be given as follows:

$$\text{GHG}_i = \sum_f \sum_{j=i-1}^{i} (\text{Sales}_{fj}^V \times (1 - \alpha) \times \text{LI}_e) \tag{1}$$

where $Sales_{fj}^V$ denotes the number of sales of vehicles with a fuel type f in a year j, and α is the fraction of total distance powered by fuel. The data is sent using Message Queuing Telemetry Transport (MQTT) standard in publish–subscribe fashion on a periodic basis. The published record is demonstrated as follows:

$$R = \{\text{TS}_i, R_i^{\text{curr}}, \gamma\} \tag{2}$$

where TS_i denotes the current timestamp, and γ denotes the location accuracy. The location of the bus is updated on the server module in suitable data exchange formats as requested by the user application. Any ith user can access the data using his *User_ID* and *Trip_ID*, denoted as a ticket request T, as described earlier. Assuming location information sent consists of a byte information, the published readings are 32 bytes, with outer IP header of 20 bytes and frame MAC header of 18 bytes; thus, the communication overhead for each published message is 70 bytes, which is relatively low. The time complexity to store n records as set of discrete data points is $O(n)$.

- **Server Module**: The server module handles all incoming readings and runs scripts to handle the serviceability of the message, in case of poor wireless communication signals. Data records with poor accuracy are denoted by L_i^{poor} and are interpolated based on standard deviation σ and mean μ of location points P.

$$L_i^{\text{poor}} = \mu + 2.5 * \sigma \tag{3}$$

The intelligent PTS considers two points L and L', designated as obtained location coordinates, and let α_L denotes the current location reading. A path $P(P_1, P_2, \ldots, P_{k-1}, P_k, P_{k+1}, \ldots, P_l)$ consists of a user source L_S^i and L_i^D; the result set R considers a dynamic route pattern as a subset of the available paths P^* and selects a path which minimize emissions. To accelerate the same, historical transactions H considered. The annual survival rate $SR_{i \leftarrow j}$ denotes the serviceability of the vehicle against the green house emission GHG_i that is given in the equation:

$$\text{SR}_{i \leftarrow j} = \sum_t (\alpha \times \text{VMT}_i^{V,t} \times \text{LHV}_f \times \rho_f) \tag{4}$$

where $\text{VMT}_i^{V,t}$ is the travel distance of vehicle i at time t, LHV_f denotes the lower heating value of fuel type f, and ρ_f denotes the density of the fuel type. Considering the reading of the IoT sensors $S = \{S_1, S_2, \ldots, S_k\}$ deployed over the vehicles, the readings are sent to a server S and then load $L = \{l_1, l_2, l_3, \ldots, l_m\}$ is computed as follows:

$$L = \text{LHV}_f \times \text{PC}_j^{V,t} \tag{5}$$

where $\text{PC}_j^{V,t}$ denotes the power consumption of the vehicle at a given time t. After computing the power analysis, once the load is set, we measure the readings of the sensors for trip completion and analyze the emissions. The optimizing condition is to $\sum_{\max} L_i$ while $\sum_{\min} \text{GHG}_i$. Considering n to be number of trips for a particular PTS, with m historical similar trips, and h be the number of hops, a scan through historical transactions could be performed as bitmap vector in $O(m)$ time. Thus, the total computational time is $O(nh + m)$.

- **Machine Learning Model**: The readings from the server module are arranged as an ordered sequence of records S^R. A proposed Algorithm 1 is now presented which computes the intelligent route decision in real time via learning paths through historical transactions H. The time complexity of the algorithm requires real-time additions or deletion of routes, supposing there are p stops from L_i^S to L_i^D, then the overall time taken by the algorithm is $O(p(k - (i + 1)))$. The messages exchanges during the transfer are set of possible paths, i.e., $O(p^2)$.

An IoT sensor-based module reads data from different deployed sensors and sends the data to server. The PTS installed with the sensors records the actual trips conducted at various time instants in day, noted as slots. The real-time traffic condition is imposed in the simulation while increasing the passenger load slowly in the system. System checks in duration of 60 s for starting of trip and later reads data at rate of 20 records per seconds.

4 Performance Evaluation of *IIGPTS*

4.1 *Experimental Setup*

As experimental setup, we have installed *MH-Z16* for reading the CO_2 emissions on university bus working on two shifts, picking staff and students in two slots. The first slot starts from 7:00 AM to 8:50 AM in morning, and second slot is from 9:00 to 10:50 AM, following the same route. Passenger count is noted at every stop. The readings collected by the emission sensor are sent to server which captures data 20 times every second. The server is built on Raspberry Pi processor, and GSM is employed for network connectivity and bridging.The setup simulation is carried out on a virtual machine running Ubuntu Linux *v16.04* LTS with 1 virtual CPU core. The simulation parameters are depicted in Table 1.

Algorithm 1 *IIGPTS*: Route Decision Prediction

Input: Sequence set $T = \{S_1, S_2, \ldots, S_n\}$ of ordered records S^R, published record R, real-time location of PTS. L_i^{curr}.

Output: Route Decision R_{des} optimally selected to minimize c_{ij} by minimizing GHG_i.

Initialization: $i = 0$, $j = 1$, and $k = 0$

1: **procedure** ROUTE_DECISION(T, R, L_i^{curr})
2:     $R_i^{curr} \leftarrow Request\_Source(L_i^{curr}, t_i^0, PTS_{type})$
3:     $T_i \leftarrow Request\_Dest(L_i^D, PTS_{id}, C_i)$
4:     ***callprocedure*** $Check\_Route(R_i^{curr}, T_i, E_i, L_i^S L_i^D)$
5: **end procedure**

6: **procedure** CHECK_ROUTE(R_i^{curr}, T_i, E_i, $L_i^S L_i^D$)
7:     **for** $i \leftarrow L_i^S$ to L_i^D **do**
8:         **for** $j \leftarrow i + 1$ to k **do**
9:             $R_i^{prev} \leftarrow Historical\_Record(R_{j-1}, Rj, w(j-1:j))$
10:            **if** ($R_j^{prev} == H$) **then**
11:                $\Psi \leftarrow (A_j, MAX_j, \eta_i)$
12:            **else**
13:                $R_{des} \leftarrow Update\_Path(R_j^{curr}, Q_j, L_j)$
14:                $\eta \leftarrow |L_i^S, L_i^D|$
15:                $\tau(R) \leftarrow (R_{des}, \Psi)$
16:            **end if**
17:            **for** $k \leftarrow 1$ to L_i^D **do**
18:                $\zeta \leftarrow Train\_Model(S^R, P, L_i^S, \delta)$
19:                $R \leftarrow (P^*, SR_{i \leftarrow j})$ as defined in Eq. 4
20:                $L \leftarrow LHV_f \times PC_j^{V,t}$
21:                $Fetch\_Route(\mu, \rho_f, S, L)$
22:                $\delta \leftarrow Compute\_Drop\_Probability(\zeta, R, \eta, PTS_{type})$
23:                **if** ($L_k^{poor} == TRUE$) **then**
24:                    $L_k^{poor} \leftarrow \mu + 2.5 * \sigma$
25:                    $R_{des} \leftarrow Update\_Path(R_j^{curr}, Q_j, L_j, \tau(R))$
26:                **end if**
27:                $GHG_i = \sum_f \sum_{j=i-1}^i (Sales_{fj}^V \times (1 - \alpha) \times LI_e)$
28:                $c_{ij} \leftarrow min(GHG_i)$
29:            **end for**
30:        **end for**
31:    **end for**
32: **end procedure**

4.2   Simulation Results

To perform the experimental setup, we have considered a route map depicted in Fig. 6. The map consists of 10 nodes (PTS) with set of possible routes, i.e., 20 edges (Roads) between them. The PTS running IoT sensors are run on this route, and the emission readings are measured for the round trip journey. The emission readings are recorded for two slots, for 2 hours, as mentioned in Section 4.1. The following is depicted in Figure 4. The load L_i for the ith PTS at the beginning of journey from 7:00 to 7:50 am, which is 0. The average emission η_i is recorded as 322 g/km. At At peak load of 57 persons during 8:30–8:50 am, the measured emission value is 498 g/km. Overall, the average value of emission after two rounds is 438 g/km. This

Table 1 Simulation parameters

Symbols	Meaning	Description
$|(L_i^S, L_i^D)|$	# of request	Number of request generated randomly using poison distribution spread across the entire day of PTS working
$|(PTS_{id})|$	# of buses	Number of buses running throughout day handling the requests
$\tau(R)$	Response time	Total time taken by system to decide on which bus will satisfy what request
δ	% of request dropped	Request which are not satisfied in percent by algorithm
L_i	Load	Total load on bus {passenger count}
η_i	Emission	Average emission of carbon dioxide (CO_2) in g/km of vehicle traveling
PTS_{type}	Vehicle type	The vehicle observed is of type bus with seating capacity of 50

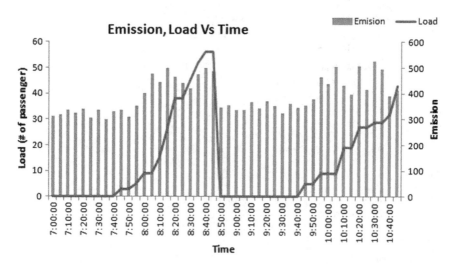

Fig. 4 Variation of emission and load measured at fixed time slot (4 hours)

shows that there is release of GHG gases at no loads as well. Also, there is a direct proportional relationship between increasing load L_i and emissions η_i in the setup.

As depicted in Algorithm 1, requests are randomly generated in the route map, denoted by $|L_i^S, L_i^D|$ for the ith PTS during an entire day of operation. The route decision is dynamic for a given PTS. The response time $\tau(R)$ is to satisfy the increasing requests and is computed, as depicted in Table 2(a) . The results of the same are shown in Figure 5a for given PTS_{type}. For 3 buses and 1000 requests, system takes 13 s to find optimal route. As buses are increased, for 6 buses, and 10,000 requests, the proposed approach takes 104 seconds. Requests drops, denoted by δ, are also considered with varying loads, as shown in Table 2(b). At high loads, small buses are not sufficient to handle the requests, hence drop increases. Around 35%

Table 2 Dynamic Routing statistics table

(a) Response time versus # of request

(b) Request drop probability versus # of request

$\|L_i^S, L_i^D\|$	$\tau(R)$				$\|L_i^S, L_i^D\|$	δ			
	3-buses	4-buses	5-buses	6-buses		3-buses	4-buses	5-buses	6-buses
100	1.29	1.51	1.04	1	100	29	25	15	11
200	2.58	2.84	2.02	1.9	200	29.5	16.5	22.5	14.5
300	4.35	4.72	3.3	3.23	300	35.33	17.33	22	17
500	6.82	8.07	5.95	5.57	500	27.6	23.4	21	16.6
1000	13.66	14	11.06	11.71	1000	35.5	23	19.2	15.1
2000	48.95	21.5	20.61	24.58	2000	34.9	22.65	22.75	14.9
5000	86	60	68	54.6	5000	32.3	24.94	21.9	14.3
10,000	367	104	113	104	10,000	33.9	24.13	21.06	14.81

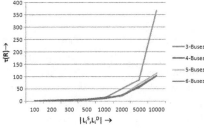

(a) Effect on response time with increased request loads.

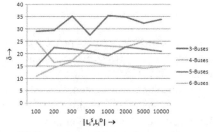

(b) Percentage of dropped requests on increasing request loads

Fig. 5 Relation between response time and request drop probability with number of request to be scheduled

of the requests are dropped for 3 buses, but for 6 buses at 10,000 requests, the drop is significantly reduced to only 15% of the requests. The following is depicted in Figure 5. Thus, $\tau(R)$ increases with increased requests loads but δ can be minimized by increasing the number of buses, indicated by $\|PTS_{id}\|$.

Figure 6a shows the comparison between the load L_i carried by PTS_{id} and emission η_i for the ith PTS on the route map, as in Fig. 6. The total generated requests are 1000, and number of buses running are 3. The run was simulated for 50 hops with an average number of passengers computed is 33.5 for dynamic route in comparison with 35.2 for static route, and results in approximate 6% less emission of (CO_2) over dynamic route patterns.

5   Conclusions and Future Work

India is the third largest emitter of hazardous gases like CO_2 and NO_x and with the current population growth trends and usage of public transport; the numbers

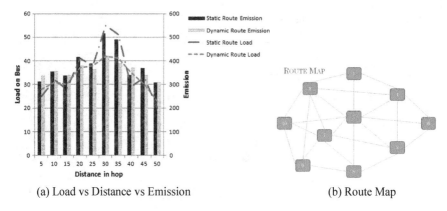

(a) Load vs Distance vs Emission   (b) Route Map

Fig. 6 Simulation results for route map

will be on the rise. The proposed framework *IIGPTS* measures the relation between passenger load carried by the bus and emission released. A direction to minimize the emission is presented in the work; thus, the amount of hazardous gases released into the atmosphere is mitigated. By adding dynamicity into the nature of route selection, *IIGPTS* lowers the emissions by dynamically selecting optimal routes for PTS and has an efficient load management strategy. Thus, the proposed framework is useful in designing smart PTS that is eco-friendly, as well as carry many passengers on a given pre-specified PTS route. As part of the future work, authors would like to explore more about the actual gain in reduction of hazardous gases by calculating dynamic route selection of PTS based on others parameters like load balanced routes, comfort levels of passenger during long journeys, and efficient ticket scheduling mechanism. Indian traffic is non-lane-based; hence, efficient deep learning (DL) strategies needs to be explored for real-time congestion notifications and predicted delays on a much larger-scale map. DL models will analyze traffic conditions at various points and events during the day and thus will be useful in updation of bias of ML models for emission predictions in more better way, and thus optimizing the performance even better.

References

1. India: Safe, clean, affordable, and smart transport. https://www.worldbank.org/en/country/india/brief/india-safe-clean-affordable-smart-transport (2014), accessed: 26-08-2019
2. Smart transportation: a key building block for a smart city. http://www.forbesindia.com/blog/who/smart-transportation-a-key-building-block-for-a-smart-city (2018), accessed: 26-08-2019
3. Texas a and m transportation urban mobility report. https://mobility.tamu.edu/umr/ (2018), accessed: 26-08-2019
4. Ang, B.W., Fwa, T.F.: A study on the fuel-consumption characteristics of public buses (1989)

5. Bank, A.D.: Transport and carbon dioxide emissions: forecasts, options analysis, and evaluation. http://www.indiaenvironmentportal.org.in/files/Transport-CO2-Emissions.pdf (2009), accessed: 26-08-2019

6. Brown, A., Gonder, J., Repac, B.: An analysis of possible energy impacts of automated vehicles. In: Road Vehicle Automation, pp. 137–153. Springer (2014)

7. Cheah, L.W., Bandivadekar, A.P., Bodek, K.M., Kasseris, E.P., Heywood, J.B.: The trade-off between automobile acceleration performance, weight, and fuel consumption. SAE Int. J. Fuels Lubr. **1**, 771–777 (06 2008)

8. Din, S., Paul, A., Rehman, A.: 5g-enabled hierarchical architecture for software-defined intelligent transportation system. Comput. Netw. **150**, 81–89 (2019)

9. Fagnant, D.J., Kockelman, K.: Preparing a nation for autonomous vehicles: opportunities, barriers and policy recommendations. Trans. Res. Part A: Policy Pract. **77**, 167–181 (2015)

10. Frey, H., Rouphail, N., Zhai, H., Farias, T., Gonçalves, G.: Comparing real-world fuel consumption for diesel- and hydrogen-fueled transit buses and implication for emissions. Transp. Res. Part D: Transp. Environ. **12**, 281–291 (2007)

11. Gawron, J.H., Keoleian, G.A., De Kleine, R.D., Wallington, T.J., Kim, H.C.: Life cycle assessment of connected and automated vehicles: sensing and computing subsystem and vehicle level effects. Environ. Sci. Technol. **52**(5), 3249–3256 (2018)

12. Glover, R.: On-demand transportation market size, share & trends analysis report by service type (e-hailing, car rental, car sharing), by vehicle type (four wheeler, micro mobility), and segment forecasts, 2018 - 2025. https://www.grandviewresearch.com/industry-analysis/on-demand-transportation-market (2018), accessed: 26-08-2019

13. Hu, J., Kong, L., Shu, W., Wu, M.Y.: Scheduling of connected autonomous vehicles on highway lanes. In: 2012 IEEE Global Communications Conference (GLOBECOM), pp. 5556–5561. IEEE, New York (2012)

14. Jochem, P., Babrowski, S., Fichtner, W.: Assessing CO_2 emissions of electric vehicles in Germany in 2030. Transp. Res. Part A Policy Pract. **78**, 68–83 (2015)

15. Özkan, M.S., Özener, O., Yavasliol, I.: Optimization of fuel consumption of a bus used in city line with regulation of driving characteristics (2012)

16. Sen, R., Siriah, P., Raman, B.: Roadsoundsense: acoustic sensing based road congestion monitoring in developing regions. In: 2011 8th Annual IEEE Communications Society Conference on Sensor, Mesh and Ad Hoc Communications and Networks, pp. 125–133. IEEE, New York (2011)

17. Shen, X., Feng, S., Li, Z., Hu, B.: Analysis of bus passenger comfort perception based on passenger load factor and in-vehicle time. SpringerPlus **5**(1), 62 (2016)

18. Wadud, Z., MacKenzie, D., Leiby, P.: Help or hindrance? The travel, energy and carbon impacts of highly automated vehicles. Transp. Res. Part A Policy Pract. **86**, 1–18 (2016)

19. Zhang, J., Wang, F.Y., Wang, K., Lin, W.H., Xu, X., Chen, C.: Data-driven intelligent transportation systems: a survey. IEEE Trans. Intell. Transp. Syst. **12**(4), 1624–1639 (2011)

Integrating the AAL CasAware Platform Within an IoT Ecosystem, Leveraging the INTER-IoT Approach

Gianfranco E. Modoni, Enrico G. Caldarola, Marco Sacco,
Katarzyna Wasielewska, Maria Ganzha, Marcin Paprzycki, Paweł Szmeja,
Wiesław Pawłowski, Carlos E. Palau and Bartłomiej Solarz-Niesłuchowski

Abstract *CasAware* is an Ambient-Assisted Living platform, which aims at improving level of comfort and well-being of inhabitants, while optimizing energy consumption. A key feature for a successful realization of such a platform is its integration with other available/deployed IoT solutions. Indeed, this integration has to facilitate smooth communication between the CasAware platform and devices, and other IoT devices, in particular to enable the exchange of data sets among them. In this paper, we introduce an approach, followed in CasAware, to realize such integration. Specifically, the proposed solution exploits the guidelines of the INTER-IoT project, which proposed a framework for inter-platform communication. So far, various existing IoT platforms have been plugged into this framework, originating from multiple application fields, thus demonstrating the advantages of such integration, capable of disregarding the specific application context. The idea behind the herein presented study is that an INTER-IoT-based approach can guarantee enhancement

G. E. Modoni (✉) · E. G. Caldarola · M. Sacco
Institute of Industrial Technologies and Automation, National Research Council, Bari, Italy
e-mail: gianfranco.modoni@itia.cnr.it

E. G. Caldarola
e-mail: enrico.caldarola@itia.cnr.it

M. Sacco
e-mail: marco.sacco@itia.cnr.it

K. Wasielewska · M. Ganzha · M. Paprzycki · P. Szmeja · B. Solarz-Niesłuchowski
Systems Research Institute Polish Academy of Sciences, Warsaw, Poland
e-mail: katarzyna.wasielewska@ibspan.waw.pl

M. Ganzha
e-mail: maria.ganzha@ibspan.waw.pl

M. Paprzycki
e-mail: marcin.paprzycki@ibspan.waw.pl

P. Szmeja
e-mail: pawel.szmeja@ibspan.waw.pl

B. Solarz-Niesłuchowski
e-mail: Bartlomiej.Solarz-Niesluchowski@ibspan.waw.pl

© Springer Nature Singapore Pte Ltd. 2020
P. K. Singh et al. (eds.), *Proceedings of First International Conference on Computing, Communications, and Cyber-Security (IC4S 2019)*, Lecture Notes in Networks and Systems 121, https://doi.org/10.1007/978-981-15-3369-3_15

of interoperability between CasAware and other platforms, thus promoting a unified view of the data, from client's perspective, within the complete IoT ecosystem.

Keywords INTER-IoT · Internet-of-Things (IoT) · CasAware · Smart home · IoT platforms · Semantic interoperability

1   Introduction

With the rapidly growing number of devices that produce and consume data, Internet-of-Things (IoT) is not just a matter of connecting things to the Internet. The real challenge, in exploiting IoT to its full potential, is to link *things/IoT artifacts* into synergistic networks, endowed with virtual and physical components, which work together with a high degree of interoperability [1, 2]. In order to enable such interoperability, it is crucial to identify valid methods of making things more cooperative and collaborative, providing them with capabilities to exchange information in a meaningful way (i.e., making it semantically understood by communicating parties) [3].

Here, we focus on integration, within an IoT ecosystem, of the CasAware platform [4], which is the main outcome of an Italian research project funded by Lombardy region. CasAware is an Ambient-Assisted Living solution, which aims at optimizing energy consumption within a household, exploiting cooperation and collaboration among domestic devices, enabled by the IoT paradigm.

Success of the CasAware platform depends on its capability to interoperate with other existing IoT solutions. Since these, typically, come from various vendors, lack of interoperability can be caused by adoption of different (not directly compatible) communication interfaces, software stacks, operating systems, and hardware [5]. Moreover, lack of widely accepted IoT interoperability standards further worsens the problem. One example of a possible CasAware integration involves meteorological observation platform, allowing to adjust the living environment parameters (e.g., indoor temperature, brightness, etc.) in response to the available meteorological

M. Ganzha
Department of Mathematics and Information Sciences, Warsaw University of Technology, Warsaw, Poland

M. Paprzycki
Department of Management and Technical Sciences, Warsaw Management Academy, Wasaw, Poland

W. Pawłowski
Faculty of Mathematics, Physics, and Informatics, University of Gdańsk, Gdańsk, Poland
e-mail: w.pawlowski@inf.ug.edu.pl

C. E. Palau
DCOM—Universitat Politecnica de Valencia, Valencia, Spain
e-mail: cpalau@dcom.upv.es

data. Obviously, also many other usage scenarios of the CasAware solution require interoperability [6].

The paper describes our experience in integrating CasAware with other IoT platforms leveraging methodology, tools, and framework of the European research project INTER-IoT [7, 8]. INTER-IoT aims at designing, implementing, and testing a set of tools, along with a methodology that helps to achieve interoperability among IoT artifacts (platforms/systems/applications) on different layers (device, network, middleware, applications and services, data, and semantics). So far, various IoT platforms, from multiple application domains, have been plugged into the INTER-IoT ecosystem, demonstrating the advantages of the proposed approach, which, in particular, allows disregarding the specific application context. In what follows, we discuss how the CasAware platform can utilize advantages brought by the INTER-IoT project, i.e., (1) exchange of meaningful information between plugged platforms in a seamless and smart way; (2) simplified communication between client and CasAware back-end services, by exploiting tested, robust communication mechanisms provided within INTER-IoT implementation.

The remainder of the paper is structured as follows. Section 2 outlines state of the art of interoperability between IoT platforms. Next, in Sect. 3 the CasAware system, with its architectural levels paired with a usage scenario that will guide presentation of the application of the INTER-IoT methodology, is described. Section 4 presents the INTER-IoT approach. We follow, in Sect. 5 with the description of integration of the CasAware system with INTER-IoT components/products. Finally, Sect. 6 draws conclusions and discusses future research directions.

2 Interoperability of IoT Artifacts

The concept and approaches to interoperability of IoT artifacts evolved over time. According to [9], the first wave of IoT application emphasized on connecting sensors interfacing with the physical world using lightweight communication protocols such as Constrained Application Protocol (CoAP) [10] and Extensible Messaging and Presence Protocol (XMPP) [11], mainly within *smart city* domain. Subsequently, traditional Internet representational state transfer protocol (REST) idea has been used for similar applications, where event-centric approaches had been implemented to reduce the number of transmitted messages [12].

Later on, the second wave has come, together with the idea of "smart objects", i.e., devices that incorporate a certain degree of intelligence, making them able to "understand" the *environment* and react to external stimuli it provides. With growing real-world awareness, smart objects provide support for increasingly complex solutions [13]. In industrial applications, for example, existing physical objects like containers and tools, as well as procedures, like quality control, have been converted into smart objects equipped with embedded sensors, wireless connectivity, and computational capabilities, so they could communicate, interpret sensor data, make decisions, and cooperate with each other. With the introduction of smart objects, the Internet-of-Things vision became a reality [14, 15].

Finally, the third wave has involved the use of *semantically enriched* data, acquired from *heterogeneous* sources in a, possibly, *multi-domain/cross-domain* environment. The need for semantic techniques/technologies became obvious once the IoT domain started getting congested with multiple platforms/applications using different (essentially incompatible) communication protocols and data models [16]. Because of the proliferation of vendors-specific solutions, many organizations are attempting to promote standardization in order to enable/guarantee interoperability between applications. For example, the OpenIoT and AllSeen alliances have developed the OpenIoT platform and an IoT framework. The Internet Engineering Task Force (IETF) and XMPP Standards Foundation are trying to align their messaging protocols, CoAP and XMPP, with other protocols [9]. Numerous approaches have also been used to align and integrate different communication protocols or data models instead of focusing on creating a *single* standardized one [17, 18].

By leveraging theoretical and technological approaches developed within the *Semantic Web*, data produced and processed by IoT artifacts progressively becomes semantically enriched or even semantics-based. In the case of raw sensor data, for instance, the enrichment usually refers to contextual information, compliant with corresponding data models. To facilitate the "semantic enrichment," a number of standards/proposals for data formats and IoT-related *ontologies* emerged [19]. For instance, for sensor data we can mention (i) OGC Sensor Web Enablement (SWE), established by the Open Geospatial Consortium, and including following important specifications: Observation Measurement (OM), Sensor Model Language (SensorML), and Sensor Observation Service (SOS); (ii) Semantic Sensor Network (SSN) ontology, developed by W3C provides a standard for modeling sensor devices, sensor platforms, knowledge of the environment and observations; and (iii) Semantic Sensor Observation Service (SemSOS).

Although the utilization of these standards supports the integration of Semantic Web with sensor-based applications, the IoT interoperability challenge is far from being solved. A real, semantic-based, IoT-specific architecture is required to provide *semantic interoperability* between IoT artifacts [9, 20, 21]. One of the first initiatives, in this direction was the already mentioned OpenIoT project, funded by Europe Union's framework program. OpenIoT focuses on developing open-source middleware for IoT interoperability using linked sensor data. At the heart of OpenIoT lies the old revision of W3C Semantic Sensor Networks (SSNX) ontology, which provides a common standards-based model for representing physical and virtual sensors [22]. It also uses several well-known vocabularies and relations (e.g., PROVO provenance ontology, LinkedGeoData and basic geo vocabulary, LSM live sensor data management vocabulary, etc.), as well as custom pilot-specific ontologies to model the necessary concepts [23]. FIWARE [24] is another EU sponsored platform for IoT, enabling a market-ready open-source solution, which combines components that enable the connection of IoT with Context Information Management and Big Data services in the cloud.

Another IoT solution developed by the AllSeen Alliance, in conjunction with the Open Connectivity Foundation (OCF), is AllJoyn [25], an open-source software framework that makes it easy for devices and applications to discover and communicate with each other, freeing developers from the details of the transport layer, the manufacturer-specific differences, and so forth, when they develop IoT applications.

3 CasAware Project Overview

The CasAware project aims at improving inhabitants level of comfort and well-being, while optimizing energy consumption and enforcing security. Specifically, it provides a context-aware system, which monitors the dwellers' behavior in order to provide customized services related to their safety in the house, appliance management, and energy consumption management [4]. Customized services are also enabled by the integration of CasAware with other IoT platforms, external to CasAware, which provide further information to the CasAware ecosystem.

3.1 CasAware Architecture

The proposed system leverages combined exploitation of various technologies, ranging from Big Data to the Semantic Web. In particular, it takes advantage of ontological modeling techniques to formalize and represent knowledge of different domestic concepts (house, appliances, and the information they share and exchange). Furthermore, the ontology-based approach enables the use of reasoning to infer new insights into already known facts. The system is currently being developed at the STIIMA CNR's IoT Living Lab in Lecco and is to be tested with real inhabitants (both able-bodied and impaired). Specifically, the CasAware system is structured into five layers, depicted in Fig. 1 (see, also [4]).

The first layer is a physical network of interconnected and interacting devices. Connection between these devices and people can be realized by leveraging active transponders that are suitably equipped with sensors connected to various receivers to transmit acquired home telemetry. Second layer is a semantic data model that formally represents knowledge of the home environment and behaviors of its inhabitants. The semantic model can be expressed as a set of ontologies, which are "formal, explicit specifications of a shared conceptualization" [26]. For their formalization, Resource Description Framework (RDF) [27], Web Ontology Language (OWL) [28], and Semantic Web Rule Language (SWRL) [29] (standard languages, endorsed by the W3C) are used. Since CasAware's main purpose is to enhance the dwellers' comfort, semantic models represent (i) interactions among household devices and (ii) dwellers' behaviors. A good starting point for the development of needed ontologies consists of data models developed within research projects PEGASO [30] and Apps4aME [31]. The third layer contains a reasoner and a set of inference rules

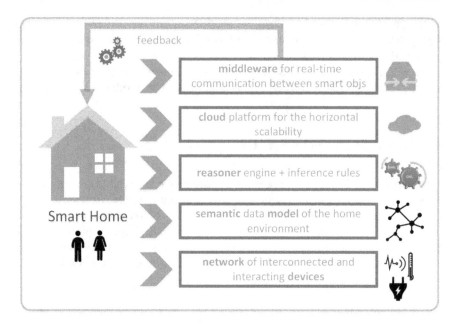

Fig. 1 A layered-based architecture for CasAware project

that, together, allow to entail and extract new knowledge, by exploiting data already acquired by the sensors. The fourth layer is a cloud platform that guarantees horizontal scalability of the whole infrastructure. The cloud hosts the semantic database, where the semantic model is persisted (both instances and corresponding meta-model), leveraging the exposed SPARQL engine [3, 5]. Finally, the fifth layer is a Semantic Middleware (SM) that allows near real-time communication among devices. It is based on the semantic model of the third Layer. In addition, integration with INTER-IoT components takes place in this (SM) layer. Specifically, thanks to the SM, the CasAware platform exposes two types of requests: *publish* and *subscribe*, through a publisher/subscriber mechanism, implemented by the SM component [5] (see Fig. 2), leveraging the MQTT protocol [32].

3.2 CasAware Integration—Motivating Scenario

In [6], authors describe a typical scenario involving CasAware IoT platform. The first described use case highlights the benefits of connecting the CasAware platform to other IoT solutions. Let us summarize it here.

Let us consider Antonio, who is 41 years old and suffering from hypotension. CasAware system is aware that Antonio gets up every morning at 7:00 and leaves home to work at 7:45. It also knows that in regular conditions, it takes Antonio about

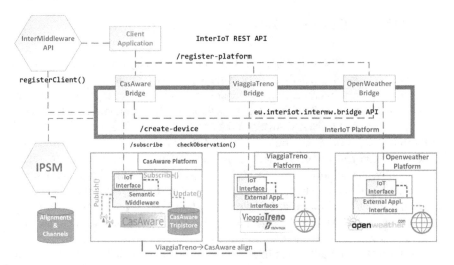

Fig. 2 Integration architecture of CasAware within Inter-IoT

45 min to reach work, by walking 800 m to the nearest train station, and then taking a train that stops at a station that is right next to the building where he works. CasAware processes real-time information regarding train and weather conditions acquired from two other IoT platforms (*Viaggiatreno*[1] for train information, and *Openweather*[2] for weather conditions). Thus, CasAware can estimate potential slowdowns on Antonio's usual path to work (e.g., due to heavy rain, delay of the train, etc.), which in turn can cause a delay of N minutes in his arrival at work. In this case, according to acquired information, CasAware sends information to the alarm clock and wakes Antonio M minutes earlier than usual (to catch train that leaves M minutes earlier). Similarly, the coffee machine starts M minutes earlier, and the coffee is ready as soon as Antonio enters the kitchen. In addition, earlier ignition of the heating system is needed, to ensure a warm environment when Antonio rises. The system will then be switched off, as soon as Antonio leaves the house, thus optimizing gas consumption. When Antonio tries to open the door to leave the house, the door does not open if he has not taken the medicine for the hypotension. Finally, in case of rain or snow, the umbrella stand near the door blinks, reminding Antonio to take the umbrella with him.

It should be highlighted that, compared to CasAware, Viaggiatreno and Meteo.com use completely different data model representation. Thus, the need for translation to a data model is used within the CasAware platform.

---

[1] http://www.viaggiatreno.it.
[2] http://openweather.com.

4 INTER-IoT Project Overview

The EU H2020 INTER-IoT project [7] aims at design and implementation of a set of tools and a methodology, dedicated to achieving interoperability between heterogeneous IoT artifacts. Interoperability mechanisms work at various layers of the hardware/software stack, i.e., devices (D2D), network (N2N), middleware (MW2MW), application and services (AS2AS), and data and semantics (DS2DS). The INTER-IoT solution enables the creation of IoT ecosystems, in which uni or bidirectional communication can be established at any layer of the hardware–software stack. Additionally, the solution offers INTER-API that allows to configure and interact with each layer's components. In the context of CasAware, (semantic) integration is performed at Middleware to Middleware (MW2MW) layer, where INTER-IoT component INTER-MW enables the data exchange among CasAware and the other platforms—*Viaggiatreno* and *Openweather*.

INTER-MW, internally, uses Inter-Platform Semantic Mediator (IPSM) component [21, 33, 34] that performs (streaming) semantic translation. It should be stressed that the INTER-IoT interoperability approach does not require changes to the connected platforms/artifacts (e.g., the CaseAware platform). To achieve interoperability, a connector (Bridge) needs to be constructed from a provided template that specifies interfaces to be implemented. Part of the Bridge functionality is to perform syntactic translation—a two-way change of data formats. As an internal message format, INTER-IoT uses RDF in JSON-LD serialization. Every message consists of two graphs *metadata* and *payload*, where payload represents the "core content" of the message. Therefore, for instance, if the source platform uses XML based on XSD, then Bridge should translate the message to JSON-LD, where the *payload* is expressed in the source platform's ontology/vocabulary (if the platform natively uses RDF, then no syntactic translation is necessary). The semantic translation to/from a *central ontology* (INTER-IoT model, extendable for specific deployments) and source/target artifact's ontology is performed by the Inter-Platform Semantic Mediator (IPSM) component. The translation is based on *alignments*, i.e., mappings between structures in source/target ontology and the central ontology. The central ontology is, in most part, deployment specific, but always extends the common "core"—the Generic Ontology for IoT platforms(GOIoTP) [35–37].

5 Integrating CasAware into INTER-IoT

Let us now describe, in some detail, how the CasAware architecture has been integrated with INTER-IoT components and products. The main idea consists of interfacing CasAware platform through the creation of a corresponding *Bridge* and installing it into the INTER-IoT infrastructure. The Bridge component acts as a mediator that is able to handle the connection of CasAware platform with all other platforms, connected to the INTER-IoT ecosystem at the middleware layer. Its realization is based on a generic interface, which provides a structured template that enables to easily develop a new platform-specific instances. Once the CasAware Bridge is created, the CasAware platform is able to register in the INTER-IoT ecosystem by calling the *registerPlatform* method, and subsequently operates with the other interface methods such as *registerDevice*, *observe*, *actuate*, and so forth.

The methods implemented within the Bridge include logic that is platform-specific and requires bidirectional communication with INTER-IoT middleware components. In this way, on the one hand, a client application or devices deployed on a platform can request information originating in CasAware via the REST services provided by the INTER-MW API. On the other hand, any device connected to CasAware can request information updates coming from any other of the INTER-MW connected IoT platforms/artifacts. For example, when a device of any IoT platform (e.g., a temperature sensor) intends to write a new value in the repository, it makes an *update* call to the Semantic Middleware (SM), which will result in a *publish* call to the SM and then in turn in a *a SPARQL update* operation in the underlying Triple Store (Fig. 4). Conversely, when a client intends to be acknowledged about specific CasAware device values update, it expresses it by calling the *observe* directive via INTER-MW bridge API, which gets translated into a *subscribe* request to the SM (Fig. 5). As soon as the CasAware receives the answer, it elaborates the received information.

Figure 3 shows the projects tree related to the CasAware Dashboard Demo Client and the underlying platforms such as CasAware Emulator, ViaggiaTreno and Open-Weather. In the order of the proposed demo to demonstrate the integration of the mentioned platforms through the INTER-IoT layers, diverse Java Maven modules have been created, the main one being the CasAware Dashboard. This latter is in charge of registering itself and all the platforms to the INTER-IoT middleware and creating the devices from which data will be observed, by exploiting the interfaces exposed by the platform's Bridge (each platform is equipped with a specific bridge as a separated jar module). Thus, the client is able to smoothly collect and integrate

> casaware-bridge [casaware-bridge develop]
> casaware-interiot [casaware-interiot master]
> > casaware-platform-emulator [casaware-platform-emulator develop]
> > CasAwareDashboard [CasAwareDashboard develop]
> viaggiatreno-bridge [intermw_viaggiatreno develop]
> > viaggiatreno-interiot [intermw_viaggiatreno develop]
> > viaggiatreno-platform-emulator [intermw_viaggiatreno develop]
> openweather-bridge [intermw_openweather develop]
> openweather-interiot [intermw_openweather develop]
> > openweather-platform [intermw_openweather develop]

Fig. 3 Eclipse projects tree of CasAware Dashboard Demo Client

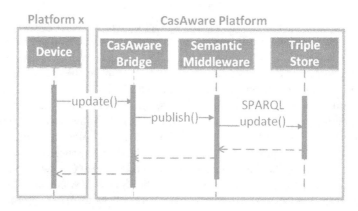

Fig. 4 Sequence diagram to write a value into CasAware

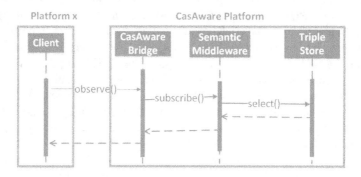

Fig. 5 Sequence diagram representing an observation request

data coming from various platforms at application level disregarding the different levels of heterogeneity characterizing them.

In addition to the Bridge, another INTER-IoT product that CasAware uses is the IPSM. Indeed, since CasAware and other IoT artifacts have to exchange data, an interoperability at the semantic layer is required. For this reason, it is essential to reconcile and mediate the ontology handled by applications and devices connected to CasAware and the data models behind other IoT platforms integrated within the INTER-IoT deployment. For this purpose, GOIoTPex, a more specific version of GOIoTP offered by INTER-IoT, is used as the central ontology. The alignments (or pairs of alignments) between GOIoTPex and the involved platforms data models are part of the configuration required to establish communication with INTER-MW. Specifically, they are internally used by the IPSM, which is used to translate and align the commonalities (overlapping concepts and expressions) between the CasAware application ontology with GOIoTPex and vice-versa (Fig. 6). In order to define the alignments, it was necessary to formally represent the corresponding mappings. A

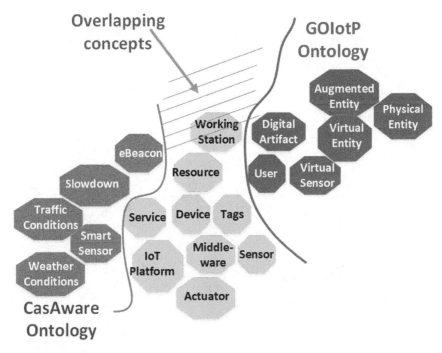

Fig. 6 Overlapping between GOIoTP and CasAware ontology

simple example of such a representation, using the IPSM Alignment Format (IPSM-AF) is given in the listing below:

```
<Alignment
            xmlns="http://www.inter-iot.eu/sripas#"
      xmlns:rdf="http://www.w3.org/1999/02/22-rdf-syntax-ns#"
      xmlns:sripas="http://www.inter-iot.eu/sripas#"
      xmlns:sosa="http://www.w3.org/ns/sosa/"
      xmlns:iiot="http://inter-iot.eu/GOIoTP#"
      xmlns:iiotex="http://inter-iot.eu/GOIoTPex#"
      xmlns:cas="http://itia.cnr.it/CasAwareOntology#"
      name="CasAware_CO_align" version="1.0.11" creator="ITIA-
         ↪ CNR" description="Alignment between CasAware
         ↪ application ontology and INTER-IoT central ontology
         ↪ .">
  <onto1>
    <Ontology about="http://itia.cnr.it/CasAwareOntology#">
      <formalism>
        <Formalism name="OWL2.0" uri="http://www.w3.org/2002/07/
           ↪ owl#"/>
      </formalism>
    </Ontology>
  </onto1>
  <onto2>
    <Ontology about="http://inter-iot.eu/GOIoTPex#">
```

```
<formalism>
  <Formalism name="OWL2.0" uri="http://www.w3.org/2002/07/
      ↪ owl#"/>
</formalism>
  </Ontology>
</onto2>
<steps>
  <step order="1" cell="1_device"/>
</steps>
<map>
  <Cell id="1_device">
    <entity1>
      <sripas:node_CTX>
        <rdf:type rdf:resource="&cas;Sensor" />
      </sripas:node_CTX>
    </entity1>
    <entity2>
      <sripas:node_CTX>
        <rdf:type rdf:resource="&iiot;IoTDevice" />
      </sripas:node_CTX>
    </entity2>
    <relation>=</relation>
  </Cell>
</map>
</Alignment>
```

In the example, two concepts from relevant ontologies (CasAwareOntology and GOIoTP) are aligned. The *cells* of the alignment represent (possible) single *steps* in the corresponding translation process. In the listing, a single example step is defined through which entity1: http://itia.cnr.it/CasAwareOntology#Sensor is postulated to be equivalent to entity2: http://inter-iot.eu/GOIoTP#IoTDevice The actual alignment is comprised of much more cells that describe mappings with varying degree of complexity.

In Fig. 7, an excerpt of the CasAware Dashboard Panel, i.e., a GUI-based tool able to implement the motivating scenario previously described, thanks to the integration of three different IoT platforms is shown. In the figure, the graphical widgets whose activation follows what described in the scenario are shown: (a) the external temperature, humidity percentage, and precipitation forecast (information received from *Openweather* platform); (b) the next available train (information received from *Viaggiatreno* platform; (c) the alarm clock with the set time, the coffee machine with a time indicator set to the activation time, the indoor temperature indicator, two alarms alerting the inhabitant in case of rain and if he/she forgets to take the daily prescribed pill (information handled by the CasAware platform).

The proposed approach has been initially implemented and is currently being tested. It has been established that the translation works, as expected, for all connected artifacts.

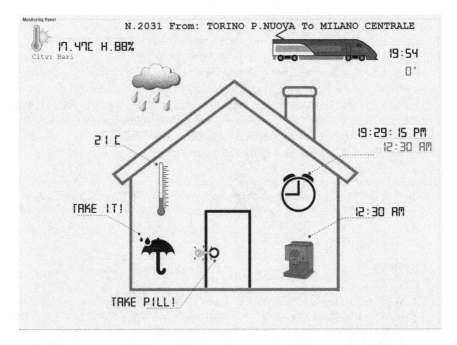

Fig. 7 Screenshot from the CasAware Dashboard Panel

6 Concluding Remarks

The paper presents and discusses the use of INTER-IoT approach and software to integrate the CasAware platform with other IoT artifacts (platforms and devices). The integration was explained within a domain-specific (Ambient-Assisted Living) application. To achieve interoperability, appropriate Bridges have been implemented and alignments between data representations created. Prototype implementation has been completed and is being tested across a number of scenarios outlined in [6]. While the tests are successful for a single deployment, one of the interesting open questions is that of scalability. In other words, how will the proposed approach behave in the case of multiple CasAWARE households being connected to various artifacts (including the two external platforms) via INTER-IoT middleware/services. Future tests will evaluate how the current approach scale, and if it does not, how will it have to be adapted to obtain sufficient scalability. We plan to investigate this question in the near future.

Separately, upon reflections based on the initial implementation, we came to the following conclusions. (1) Instantiation of a Bridge, which connects an artifact to the INTER-IoT environment, is not very difficult and can be achieved by following templates provided by the INTER-IoT project. (2) Creation of alignments is more complex and requires, at least basic, knowledge about ontologies and semantic technologies. Without such knowledge/understanding what is an ontology, how it is represented and how it is used in practical applications, and creation of correct alignments is a relatively complex and time-consuming process, even if appropriate documentation and examples are provided. Here, let us observe that it is a well-known fact that semantic technologies are not very popular, even though they hold a lot of promise. This is, at least in part, because they require knowledge that is not easily available (e.g., few universities include semantic technologies in their CS/CE curriculum). Nevertheless, there must be a way to help alignment developers, for instance, by development of appropriate tools. Here, the AFE tool proposed within the scope of INTER-IoT is a step in the right direction.

Acknowledgements The research reported in this paper has been partially funded by the European Union EU-H2020-ICT under the grant agreement No: 687283, INTER-IoT and by Regione Lombardia Bando Linea RS per Aggregazioni under the grant agreement No: 147152, CasAware.

References

1. Ali, M.I., Ono, N., Kaysar, M., Griffin, K., Mileo, A.: A semantic processing framework for IOT-enabled communication systems. In: International Semantic Web Conference, vol. 2, pp. 241–258 (2015)
2. Blackstock, M., Lea, R.: IOT interoperability: a hub-based approach. In: 2014 International Conference on the Internet of Things (IOT), pp. 79–84. IEEE, New York (2014)
3. Modoni, G.E., Veniero, M., Trombetta, A., Sacco, M., Clemente, S.: Semantic based events signaling for AAL systems. J. Ambient Intell. Humaniz. Comput., pp. 1–15 (2017)
4. Spoladore, D., Caldarola, E.G., Mahroo, A., Modoni, G.E.: Casaware: a semantic-based context-aware system enabling ambient assisted living solutions. ERCIM NEWS **113**, 50–51 (2018)
5. Modoni, G.E., Veniero, M., Sacco, M.: Semantic knowledge management and integration services for aal. In: Italian Forum of Ambient Assisted Living, pp. 287–299. Springer, Berlin (2016)
6. Mahroo, A., Spoladore, D., Caldarola, E.G., Modoni, G., Sacco, M.: Enabling the smart home through a semantic-based context-aware system. In: Second International Workshop on Pervasive Smart Living Spaces, IEEE International Conference on Pervasive Computing and Communication. IEEE, New York (2018)
7. INTER-IoT project. http://www.inter-iot-project.eu
8. Fortino, G., Savaglio, C., Palau, C.E., de Puga, J.S., Ganzha, M., Paprzycki, M., Montesinos, M., Liotta, A., Llop, M.: Towards Multi-layer Interoperability of Heterogeneous IoT Platforms: The INTER-IoT Approach, pp. 199–232. Springer International Publishing, Cham (2018). https://doi.org/10.1007/978-3-319-61300-0_10
9. Desai, P., Sheth, A., Anantharam, P.: Semantic gateway as a service architecture for IOT interoperability. In: 2015 IEEE International Conference on Mobile Services (MS), pp. 313–319. IEEE, New York (2015)
10. Bergmann, O., Hillmann, K.T., Gerdes, S.: A coAP-gateway for smart homes. In: 2012 International Conference on Computing, Networking and Communications (ICNC), pp. 446–450. IEEE, New York (2012)
11. Saint-Andre, P.: Extensible messaging and presence protocol (XMPP): Core. Tech. rep. (2011)
12. Elias, S., Shivashankar, S., Manoj, P., et al.: A rest based design for web of things in smart environments. In: 2012 2nd IEEE International Conference on Parallel Distributed and Grid Computing (PDGC), pp. 337–342. IEEE, New York (2012)

13. Kortuem, G., Kawsar, F., Sundramoorthy, V., Fitton, D.: Smart objects as building blocks for the internet of things. IEEE Internet Comput. **14**(1), 44–51 (2010)
14. Beigl, M., Gellersen, H.W., Schmidt, A.: Mediacups: experience with design and use of computer-augmented everyday artefacts. Comput. Netw. **35**(4), 401–409 (2001)
15. Mattern, F.: From smart devices to smart everyday objects. In: Proceedings of Smart Objects Conference, pp. 15–16 (2003)
16. Bandyopadhyay, S., Bhattacharyya, A.: Lightweight internet protocols for web enablement of sensors using constrained gateway devices. In: 2013 International Conference on Computing, Networking and Communications (ICNC), pp. 334–340. IEEE, New York (2013)
17. Caldarola, E.G., Rinaldi, A.M.: A multi-strategy approach for ontology reuse through matching and integration techniques. In: Quality Software Through Reuse and Integration, pp. 63–90. Springer, Berlin (2016)
18. Lenzerini, M.: Data integration: a theoretical perspective. In: Proceedings of the Twenty-First ACM SIGMOD-SIGACT-SIGART Symposium on Principles of Database Systems, pp. 233–246. ACM (2002)
19. Ganzha, M., Paprzycki, M., Pawłowski, W., Szmeja, P., Wasielewska, K.: Towards common vocabulary for IoT ecosystems—preliminary considerations. In: Intelligent Information and Database Systems, 9th Asian Conference, ACIIDS 2017, Kanazawa, Japan, Apr 3–5, 2017, Proceedings, Part I. LNCS, vol. 10191, pp. 35–45. Springer, New York (2017)
20. Ganzha, M., Paprzycki, M., Pawłowski, W., Szmeja, P., Wasielewska, K.: Semantic interoperability in the Internet of Things: an overview from the INTER-IoT perspective. J. Netw. Comput. Appl. **81**, 111–124 (2017). March
21. Ganzha, M., Paprzycki, M., Pawłowski, W., Szmeja, P., Wasielewska, K.: Towards semantic interoperability between Internet of Things platforms. In: Gravina, R., Palau, C.E., Manso, M., Liotta, A., Fortino, G. (eds.) Integration, Interconnection, and Interoperability of IoT Systems, pp. 103–127. Springer, Berlin (2017)
22. Soldatos, J., Kefalakis, N., Hauswirth, M., Serrano, M., Calbimonte, J.P., Riahi, M., Aberer, K., Jayaraman, P.P., Zaslavsky, A., Žarko, I.P., et al.: Openiot: open source internet-of-things in the cloud. In: Interoperability and Open-Source Solutions for the Internet of Things, pp. 13–25. Springer, Berlin (2015)
23. Ganzha, M., Paprzycki, M., Pawlowski, W., Szmeja, P., Wasielewska, K.: Semantic technologies for the IOT-an inter-IOT perspective. In: 2016 IEEE First International Conference on Internet-of-Things Design and Implementation (IoTDI), pp. 271–276. IEEE, New York (2016)
24. Ramparany, F., Marquez, F.G., Soriano, J., Elsaleh, T.: Handling smart environment devices, data and services at the semantic level with the fi-ware core platform. In: 2014 IEEE International Conference on Big Data (Big Data), pp. 14–20. IEEE, Berlin (2014)
25. Alliance, A.: Alljoyn framework. Linux Foundation Collaborative Projects. https://allseenalliance.org/framework (visited on 09/14/2016)
26. Guarino, N., Oberle, D., Staab, S.: What is an ontology? In: Handbook on Ontologies, pp. 1–17. Springer, Berlin (2009)
27. Rdf. online, https://www.w3.org/RDF/
28. Owl. online, http://www.w3.org/TR/owl2-manchester-syntax/
29. Horrocks, I., Patel-Schneider, P.F., Boley, H., Tabet, S., Grosof, B., Dean, M., et al.: SWRL: a semantic web rule language combining OWL and RuleML. W3C Member submission 21, 79 (2004)
30. Modoni, G., Sacco, M., Candea, G., Orte, S., Velickovski, F.: A semantic approach to recognize behaviours in teenagers. In: SEMANTICS Posters&Demos (2017)
31. Modoni, G., Doukas, M., Terkaj, W., Sacco, M., Mourtzis, D.: Enhancing factory data integration through the development of an ontology: from the reference models reuse to the semantic conversion of the legacy models. Int. J. Comput. Integrat. Manuf. **30**(10), 1043–1059 (2017)
32. Banks, A., Gupta, R.: Mqtt version 3.1. 1. OASIS standard 29 (2014)
33. Ganzha, M., Paprzycki, M., Pawłowski, W., Szmeja, P., Wasielewska, K.: Semantic technologies for the IoT—an Inter-IoT perspective. In: 2016 IEEE First International Conference on Internet-of-Things Design and Implementation (IoTDI), pp. 271–276. IEEE, Berlin, Germany (2016 Apr)

34. Ganzha, M., Paprzycki, M., Pawłowski, W., Szmeja, P., Wasielewska, K.: Streaming semantic translations. In: 21st International Conference on System Theory, Control and Computing (ICSTCC), 19-21 Oct. 2017, Sinaia, Romania, Proceedings, pp. 1–8. IEEE, New York (2017)
35. Generic ontology for IoT platforms. https://docs.inter-iot.eu/ontology/
36. Ganzha, M., Paprzycki, M., Pawłowski, W., Szmeja, P., Wasielewska, K.: Alignment-based semantic translation of geospatial data. In: 2017 3rd International Conference on Advances in Computing, Communication & Automation (ICACCA)(Fall). pp. 1–8. IEEE, New York (2017)
37. Szmeja, P., Ganzha, M., Paprzycki, M., Pawłowski, W., Wasielewska, K.: Declarative ontology alignment format for semantic translation. In: 2018 3rd International Conference On Internet of Things: Smart Innovation and Usages (IoT-SIU), pp. 1–6. IEEE, Berlin (2018)

An IoT-Based Solution for Smart Parking

Mauro A. A. da Cruz, Joel J. P. C. Rodrigues, Gustavo F. A. Gomes,
Pedro Almeida, Ricardo A. L. Rabelo, Neeraj Kumar and Shahid Mumtaz

Abstract Inefficient parking solutions in metropolitan areas contribute to the creation of traffic jams that lead to stress for the drivers and contribute toward greenhouse emissions. Since parking issues affect most of the citizens, it is hard to imagine future smart cities with efficient parking solutions. With the introduction of the Internet of things (IoT), objects are able to connect to the Internet and may collect data in real time. The potential of these data is enormous and may allow cities to move from estimates to concrete data, which enable precise predictive models to be created and improves urban area planning and scaling. For this reason, parking solutions would benefit from the introduction of the IoT concept of connecting everything. The paper overviews the most common parking solutions, analyzing their pros and cons. Then, it introduces an IoT-oriented smart parking system that can be deployed indoor and/or outdoor. With the proposed solution, drivers can remotely monitor the vacancy status of parking spots in real time through a mobile app (or even Web). Furthermore, the solution acts as a repeater, extending the wireless signal to parking slots that are

M. A. A. da Cruz · G. F. A. Gomes · P. Almeida
National Institute of Telecommunications (INATEL), Santa Rita do Sapucaí, MG, Brazil
e-mail: gustavofelipe@gea.inatel.br

P. Almeida
e-mail: pedro_almeida@gec.inatel.br

M. A. A. da Cruz
University of Haute Alsace, Colmar, France
e-mail: maurocruzter@gmail.com

J. J. P. C. Rodrigues (✉) · R. A. L. Rabelo
Federal University of Piauí (UFPI), Teresina, PI, Brazil
e-mail: joeljr@ieee.org

R. A. L. Rabelo
e-mail: ricardoalr@ufpi.edu.br

J. J. P. C. Rodrigues · S. Mumtaz
Instituto de Telecomunicações, Lisbon, Portugal
e-mail: smumtaz@av.it.pt

N. Kumar
Thapar Institute of Engineering and Technology, Patiala, Punjab, India
e-mail: neeraj.kumar@thapar.edu

© Springer Nature Singapore Pte Ltd. 2020
P. K. Singh et al. (eds.), *Proceedings of First International Conference on Computing, Communications, and Cyber-Security (IC4S 2019)*, Lecture Notes in Networks and Systems 121, https://doi.org/10.1007/978-981-15-3369-3_16

distant from the primary access point. This solution is evaluated, demonstrated, and validated through a prototype, and it is ready for use.

Keywords Internet of things · Smart parking · Vehicle parking · Outdoor · Indoor · Wi-fi

1 Introduction

The growing number of vehicles on the streets coupled with poor infrastructure management creates issues like traffic jams that lead to other complications. These complications cause delays and stress on drivers and also contribute toward greenhouse emissions. In metropolitan areas, these issues affect the majority of drivers and are aggravated when searching for parking spaces [1]. Various drivers park blocks away from their destination after searching for many minutes or resort to improper parking, which causes more traffic and sometimes results in parking tickets. Surprisingly, these parking issues also occur on private parking lots, such as shopping malls, and it happens since cars became popular.

With the increasing population in urban areas, the lack of parking vacancies has been subject of various research, and intelligent parking solutions seek to aid the drivers to locate empty spaces efficiently. One of the earliest parking solutions consisted of parking meters that are used up to today and generate revenue for municipalities. However, parking meters can be expensive depending on the zone and result in parking tickets when the driver miscalculates the parking time. In the literature, there are alternative solutions for monitoring the status of parking spaces through various type of sensors, such as ultrasonic, magnetometers, and so on. However, most solutions only display the parking status through monitors or lights located at the parking lot.

The existing parking solutions that rely on sensors would benefit from the disruptive shift introduced by the Internet of things (IoT), where every object can be connected to the Internet and may exchange data among them [2]. The introduction of the IoT concept in parking solutions would allow drivers to remotely verify the parking status of parking spots through mobile apps (or through the Web) and reduce the amount of search time. Recently, some low-cost collaborative solutions resort to crowdsensing with several users reporting on the current parking status. Unfortunately, crowdsensing solutions are only reliable with many active users and are vulnerable to malicious users [3, 4]. In this sense, the paper proposes an IoT-based smart parking solution that uses magnetic sensors and can be deployed either indoor or outdoor, to meet the parking demands of smart cities.

Unlike other smart parking solutions available in the literature, this paper follows an IoT-based approach in which the parking spots data is published on an IoT middleware and consulted in real time through a mobile app or through the Web. The solution is possible because a magnetic field sensor is placed in each parking spot which guarantees precision of over 99%. The data is published through Wi-fi using the Message Queuing Telemetry Transport (MQTT) protocol. The solution is powered directly from the power grid instead of a battery because batteries need periodic replacements and replacing thousands of units requires a dedicated workforce. Furthermore, for the batteries to be replaced, they would have to be accessible, and this would make it more difficult to protect against hazards, such as dust and water. The proposed solution is evaluated, demonstrated, validated, and it is ready for use.

The remainder of this paper is organized as follows. Section 2 discusses parking system approaches available in the literature. Section 3 introduces the proposal and its main features, and Sect. 4 shows the experiments in a real environment that allows the evaluation, demonstration, and validation of the parking solution and discusses the obtained results. Section 5 concludes the paper and suggests further research works.

2 Background and Related Work

Since cars became popular in the twentieth century, drivers started having difficulties in parking their automobiles, which has been caused by various reasons that range from the drivers' disciple to the city infrastructure. One of the first attempts to reduce the amount of time spent searching for vacant spots is based on parking meters, which also generated revenue for the municipalities. A parking meter is a device that is placed at the roadside and allows users to park for a period of time paid by them. The issue with parking meters comes from the fact that they rely on parking inspectors that are responsible for monitoring the status of the parking meter. In modern implementations, the solution is connected to a central server where the user registers its license plate and pays through a mobile app, and the inspectors scan the plates of the parked cars. Despite various efforts, parking is still a big issue in metropolitan areas.

A modern low-cost solution that is easy to maintain considers a large group of individuals collectively sharing parking spots data. This phenomenon follows a crowdsensing approach, and data is generally analyzed and used to predict any aspect of common interest [5]. A prominent example of crowdsensing is the Waze app that estimates the fastest route based on the amount of traffic [6]. The issue with crowdsensing in smart parking is that registered users must continuously share data in order to turn the solution effective, and it is very vulnerable to malicious users that send false data [3, 4]. Solutions such as Park here! [7] try to infer the parking status of a user through various smartphone sensors like accelerometers and gyroscope. Other solutions such as TruCentive [8] try to tackle the reliability issue through a framework based on game theory that adjusts the user reputation bonus according to the number of honest users.

In terms of deployment, some solutions use infrared and ultrasonic sensors in smart parking because they work on the principle of reflected waves to measure the distance between the sensor and an object. The difference is that infrared reflects light waves and ultrasonic reflects sound waves. One advantage from ultrasound over infrared is that ultrasound does not receive interference from sunlight. The disadvantage of both sensors is that they detect the presence of all types of surfaces, which make them challenging to deploy in outdoor scenarios because it is difficult to protect them against hazards such as dust and water. Published studies like the one presented in [9] implement an indoor parking system with ultrasonic sensors, and three colored lights deployed on the ceiling of each parking slot. Each light serves as a visual hint to the users, and they can easily identify when a spot is available, reserved, or occupied. However, this solution is not IoT-oriented, and users can only detect the status of a spot by observing the lights locally.

A solution that is similar to ultrasound and infrared uses optical sensors. They perceive variations in the light and should be deployed in places where the vehicle obstructs the regular light. The disadvantage of optical sensors comes from the fact they would require a constant light source, and this could be an issue especially in outdoor parking. Laser sensors operate in a similar principle like infrared and present a disadvantage that, unlike previous solutions, the laser beam emitted by them is visible. Both laser and optical sensors suffer from a similar issue as the infrared and ultrasonic.

Magnetometers are also used on parking solutions. They are sensors used to measure the earth magnetic field. Its adoption in smart parking applications is useful because enclosing it in any plastic material does not interfere with its measurements and makes it easy to deploy. Most magnetometers cannot detect modifications if the vehicle is distant and is generally placed on the soil. Those that can detect changes at long range are considerably more expensive than previously mentioned solutions. Studies such as the presented in [10] use magnetometers in the middle of vacant spots to determine its occupancy, but the data collected by them is not available over the Internet. An alternative to magnetometers is pressure sensors. They can be used to detect vehicles, but they deteriorate with frequent usage.

Video cameras provide a visual method for monitoring and detect parking slot status and can also be used for safety reasons. The issue with this visual method is that camera position must maximize the number of monitored spots and guarantee that vehicles do not block the view of another spot. This type of solution is vulnerable to environmental modifications such as weather or tree branches that obstruct line of sight. Furthermore, replicating it in different locations demands additional effort because the algorithm is often not generic and must be re-trained for each new camera location. In [11], it is proposed and implemented a parking system based on video cameras, but the data is not available for users through the Internet.

Regarding safety and easy payment methods for users, some studies suggest deploying QR codes on parking spaces, so drivers can pay for usage time [12]. This solution, although exciting, can be abused by scanning a different parking spot because there is no way of certifying that a given user was in a particular spot. A more efficient and automatic solution could consider the use of an RFID tag on a vehicle and calculating the amount of time to scan the tag and stop reading it to determine the amount to pay. In public spaces, the RFID reader could be located on a parking meter and private parking lots, and the reader could be located at the entrance. Regarding parking meters, solutions such as [13] could invalidate the need for an inspector but would require the meters to be equipped with a camera that scans the license plate. However, this type of solution requires the vehicle license plate to be in the camera line of sight.

The solution proposed in this paper is different from the approaches available in the literature because it follows an IoT approach, which means that its data is remotely accessible through the Internet and each parking place sensor is considered an object in the IoT network. The system uses magnetic sensors, which can be easily protected against hazards like water, dust and even hits without compromising their functionalities. Shielding this type of sensor against the mentioned hazards costs less than shielding cheaper sensors like ultrasonic or infrared sensors. Moreover, magnetometers are easy to deploy, and unlike other methods, they will work properly regardless of the weather conditions.

3   Proposal and Construction of the Proposed Parking Solution

The work presented in this paper proposes a smart parking solution for monitoring the parking spots, and it can be deployed in both indoor and outdoor environments. The proposed solution follows an IoT approach, and any user can consult its data through a mobile app or through the Web, which reduces the amount of time that drivers spend searching for vacant spots. The system indirectly reduces greenhouse emissions and improves people comfort.

3.1   System Design

One of the main objectives of the presented solution is the facility to deploy either indoor or outdoor. This is achieved through a magnetometer (ref. MAG3110) and a Wemos D1 mini microcontroller. Both the Wemos and the MAG3110 are small and can be easily shielded with a level of protection IP 67 ingress protection, which means the solution is protected against dust and water submersion after being placed on a proper enclosure. In addition, the system is equipped with an accurate calibration algorithm that requires minimal human intervention.

The system uses Wi-fi communication, and regarding Wi-fi coverage, it can act as a Wi-fi repeater, extending the coverage in areas where the signal is weak or inexistent. The Wemos is a low-cost microcontroller with a built-in Wi-fi module (the ESP8266) and can operate in a temperature ranging from -40 to $+125$ °C (Celsius degrees). Shielding the solution is the most expensive part of the system since the MAG3110 is also inexpensive. In terms of energy supply, the solution receives direct input from the power grid instead of a battery because parking solutions generally involve several

Fig. 1 Proposed system architecture displaying one of the prototypes extending the signal coverage to a sensor that is too far from the main Wi-fi access point (AP)

spots. Therefore, if batteries are used, they would need periodic replacements, which mean that they should be easily accessible and it could compromise the level of ingress protection (IP). Furthermore, continually replacing multiple batteries is not practical and counter-intuitive to the minimal human intervention concept of IoT.

The MAG3110 is a digital three-axial magnetic sensor that communicates with microcontrollers through the I2C (Integrated Circuit) protocol. It detects the presence of automobiles through the variation they generate over the earth' magnetic field, thus the importance of the calibration process. The sensor can operate in a temperature ranging from −40 to +85 °C (Celsius degrees) and is well suited for the application. Then, the microcontroller processes the data collected by the sensor and transmits it to the In.IoT [14], a middleware platform for the Internet of things. The data is sent through the Message Queuing Telemetry Transport (MQTT) protocol, which is a popular application-layer protocol for IoT [15]. Figure 1 shows the system architecture of the proposed solution.

3.2 System Calibration

The calibration process allows the solution to be deployed anywhere and lasts 40 s. During this process, the maximum and minimum values of each sensors axis are recorded. During calibration, the parking spot should be free of any metal objects that are not usually present, and the adjacent spots do not interfere with the process (vehicles can be placed in the adjacent spots). After the calibration process is complete, the system is ready to use, and the user can visualize the spot status via a mobile app or Web. If an object interferes with the calibration process, the administrators can repeat the process, as presented in the flowchart of Fig. 2.

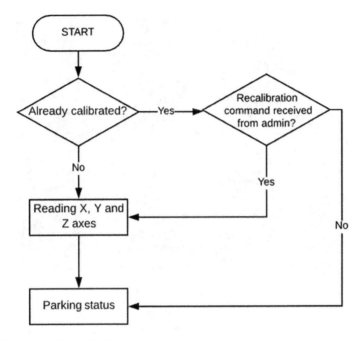

Fig. 2 Flowchart of the calibration algorithm

4 System Evaluation, Demonstration, and Validation

The proposed smart parking solution described in the previous section was evaluated, demonstrated, and validated using the research methodology of a case study through a real prototype. It was conceived according to the comprehensive description offered in the previous sections. Figure 3 shows the real prototype of the smart parking system and its components that were deployed in a real environment (at the Inatel campus). The system was calibrated and evaluated in two separate use cases. The first use case considered that the calibration process was not well-performed, and the detection accuracy was not good, so the administrator sent a command to recalibrate the device when the spot was vacant.

In the second use case, it is verified if a prototype located further away from the primary access point that could not reach Wi-fi coverage can benefit from the repeater functionality of another device. The experiments performed for each use case were successful, and the system is reacting accordingly. Therefore, the solution is qualified and ready for monitoring the occupancy status of parking spaces.

From the experiments, it is possible to observe the solution can adapt to any spot after a successful calibration process, which can be repeated, if necessary (only system administrators can perform this task). Using the real prototype, the sensor data is validated through the console of the Wemos D1 mini. The data gathered by these IoT devices can be useful to other devices that belong to other applications. For

Fig. 3 Real prototype of the proposed parking system installed in a real environment

this reason, the data is transmitted to In.IoT [14], a middleware platform for IoT that allows the solution sharing its information with other IoT applications and gadgets [16]. Figure 4 shows the parking status visualization on the developed mobile app. As may be seen, it is easy to identify the free spots (in green color) as well as the occupied spots (in red color). Furthermore, the application is integrated with Google maps, and the users can use the "directions" feature and drive toward the chosen parking space. Figure 5 shows a visualization in street maps, note that the streetmaps are not a real-time image and only the colored dots reflect the vacancy status.

5 Conclusion and Future Works

Parking issues affect drivers in metropolitan areas, and it is certainly a topic that, allied with IoT technology, can revolutionize and facilitate the life of the population. With the introduction of IoT, several solutions are able to provide real-time data to all users. The potential of these data is gigantic and could allow cities to move from estimates to precise/real data, which enable predictive models to be created and improves urban areas planning and scaling. For this reason, parking solutions would undoubtedly benefit from the IoT concept of connecting objects to the Internet.

This paper proposed a smart parking system that can be deployed both indoor and outdoor following an IoT approach and publishing data in the In.IoT middleware platform. The solution includes a mobile app (accessible also via Web) that allows users to verify the parking spot status in real time. The calibration algorithm allows its deployment with minimal human intervention. A system administrator can remotely

Fig. 4 Smart parking mobile app screenshot showing the vacant spots (in green color) and occupied (in red color)

send a recalibration command if the calibration process was not well-performed. The solution uses a MAG3110 for sensing the spot status and a Wemos D1 Mini for data transmission. Both the Wemos and the Mag can be easily shielded with an IP 67 level of protection, enabling an outdoor deploy. Regarding energy supply, the solution receives direct input from the power grid instead of a battery because parking solutions generally involve several spots. Therefore, if batteries were used, they would need periodic replacements, which mean that they would have to be easily

Fig. 5 Smart parking
mobile app screenshot (the
streetmaps version), the
colored dots reflect the
vacancy status

accessible and it could compromise the level of ingress protection (IP). Moreover, continually replacing batteries is not practical and counter-intuitive to the minimal human intervention concept of IoT. The system was evaluated, demonstrated, and validated in a real environment through a prototype, and it is ready to use.

Future works may consider replacing the three-axial magnetometer by a single axis alternative because it would be cheaper, and the Z-axis is the one that determines the spot occupancy status. A limitation presented the proposed solution is the fact that it cannot accurately detect motorcycles unless they park directly on top of the sensor, which depends on the driver. Regarding the mobile app, a notification feature that alerts users when new parking spots are available could be included. Finally, real-time images would enrich the application because the streetmaps image does not provide an accurate visual representation of the parking lot.

Acknowledgements This work has been supported by FCT/MCTES through national funds and when applicable co-funded EU funds under the Project UIDB/EEA/50008/2020; by FADCOM— *Fundo de Apoio ao Desenvolvimento das Comunicações*, presidential decree nº 264/10, November 26, 2010, Republic of Angola; and by Brazilian National Council for Scientific and Technological Development (CNPq) via Grants No. 431726/2018-3 and 309335/2017-5.

References

1. Lin, T., Rivano, H., Le Mouel, F.: A survey of smart parking solutions. IEEE Trans. Intell. Transp. Syst. **18**(12), 3229–3253 (2017)
2. Ray, P.P.: A survey on Internet of Things architectures. J. King Saud Univ. Comput. Inf. Sci. **30**(3), 291–319 (2018)
3. Owoh, N.P., Mahinderjit Singh, M.: Security analysis of mobile crowd sensing applications. Appl. Comput. Inf. (Oct 2018)
4. Pouryazdan, M., Fiandrino, C., Kantarci, B., Soyata, T., Kliazovich, D., Bouvry, P.: Intelligent gaming for mobile crowd-sensing participants to acquire trustworthy big data in the Internet of Things. IEEE Access **5**, 22209–22223 (2017)
5. Zhang, X., Yang, Z., Sun, W., Liu, Y., Tang, S., Xing, K., Mao, X.: Incentives for mobile crowd sensing: a survey. IEEE Commun. Surv. Tutor. **18**(1), 54–67 (1st quart. 2016)
6. Wang, G., Wang, B., Wang, T., Nika, A., Zheng, H., Zhao, B.Y.: Defending against sybil devices in crowdsourced mapping services. In: Proceedings of the 14th Annual International Conference on Mobile Systems, Applications, and Services—MobiSys '16, Singapore, Singapore, pp. 179–191 (June 26–30)
7. Salpietro, R., Bedogni, L., Di Felice, M., Bononi, L.: Park here! A smart parking system based on smartphones' embedded sensors and short range communication technologies. In: 2015 IEEE 2nd World Forum on Internet of Things (WF-IoT), Milan, Italy, pp. 18–23 (Dec 14–16)
8. Hoh, B., Yan, T., Ganesan, D., Tracton, K., Iwuchukwu, T., Lee, J.-S.: TruCentive: A game-theoretic incentive platform for trustworthy mobile crowdsourcing parking services. In: 2012 15th International IEEE Conference on Intelligent Transportation Systems, Anchorage, AK, USA, pp. 160–166 (Sept 16–19, 2012)
9. Kianpisheh, A., Mustaffa, N., Limtrairut, P., Keikhosrokiani, P.: Smart Parking System (SPS) architecture using ultrasonic detector. Int. J. Softw. Eng. Its Appl. **6**(3), 51–58 (2012)
10. Yuan, C., Fei, L., Jianxin, C., Wei, J.: A smart parking system using WiFi and wireless sensor network. In: 2016 IEEE International Conference on Consumer Electronics-Taiwan (ICCE-TW), Natou, Taiwan, pp. 1–2 (May 27–29, 2016)
11. Hammoudi, K., Melkemi, M., Benhabiles, H., Dornaika, F., Hamrioui, S., Rodrigues, J.J.P.C.: Analyzing and managing the slot occupancy of car parking by exploiting vision-based urban surveillance networks. In: 2017 International Conference on Selected Topics in Mobile and Wireless Networking (MoWNet 2017), Avignon, France, pp. 1–6 (May 17–19, 2017)
12. Bechini, A., Marcelloni, F., Segatori, A., Low-effort support to efficient urban parking in a smart city perspective. In: Advances onto the Internet of Things: How Ontologies Make the Internet of Things Meaningful, pp. 233–252. Springer International Publishing (2014)

13. Ramaswamy, P.: IoT smart parking system for reducing green house gas emission. In: 2016 International Conference on Recent Trends in Information Technology (ICRTIT), Chennai, India, pp. 1–6 (Apr 8–9, 2016)
14. Rodrigues, J.J.P.C., da Cruz, M.A.A.: In: IoT, Registry request of Computer Program in Brazil— RPC No. BR 512018051862-1 (Oct 2018)
15. Alghamdi, K., Alqazzaz, A., Liu, A., Ming, H.: IoTVerif: an automated tool to verify SSL/TLS certificate validation in android MQTT client applications. In: Proceedings of the Eighth ACM Conference on Data and Application Security and Privacy (CODASPY 2018), Tempe, AZ, USA, pp. 25–102 (Mar 19–21, 2018)
16. da Cruz, M.A.A., Rodrigues, J.J.P.C., Al-Muhtadi, J., Korotaev, V.V., De Albuquerque, V.H.C.: A reference model for Internet of Things Middleware. IEEE Internet Things J. 5(2), 871–883 (2018)

Online Monitoring of Solar Panel Using *I–V* Curve and Internet of Things

Praveen Kumar Malik, Rajesh Singh and Anita Gehlot

Abstract For better efficiency and performance of solar panels, they need to be monitored regularly. Power generated by panels depends on several factors such as sun availability, the time instant of the day, light intensity, fill factor and light-generated current. This paper tells about monitoring solar panels online at different time intervals by studying the *I–V* curve of a solar panel at various intervals of time. And by using the curve, power generated, efficiency and fill factor are calibrated at the required intervals, and after that, the average of all the above three is calculated. All this data is collected online using Internet of things and finally is compared with the standard data available to judge the performance of the solar panel.

Keywords Internet of Things · Number of suns · Fill factor · Cloud server

1 Introduction

Due to the everlasting, replenishing and omnipresence of solar energy, a photovoltaic power system for the production of electricity is well known and employed and highly popular these days [1]. Electricity is generated when the photon in the incident light falls on the photovoltaic cell and due to the phenomenon of photoelectric effect excites the electrons in the semiconductors resulting in the flow of electrons to produce current. Greater the number of suns higher is the current generation and higher the amount of voltage [2]. Which could measures the current and voltage, at circuit current *I*sc and fill factor are attained which vary accordingly with the varying intensity throughout the day performance and the efficiency of the solar cell can be assessed using the presented project varying intensities, temperature. Using

P. K. Malik (✉) · R. Singh · A. Gehlot
Lovely Professional University, Phagwara, India
e-mail: pkmalikmeerut@gmail.com

R. Singh
e-mail: srajsssece@gmail.com

A. Gehlot
e-mail: eranita5@gmail.com

© Springer Nature Singapore Pte Ltd. 2020 225
P. K. Singh et al. (eds.), *Proceedings of First International Conference on Computing, Communications, and Cyber-Security (IC4S 2019)*, Lecture Notes in Networks and Systems 121, https://doi.org/10.1007/978-981-15-3369-3_17

the current and voltage values, open-circuit voltage and short-circuit voltage could be observed [3]. The above parameters combined with intensity and temperature and support in examining the performance of the solar cell [4]. With the help of open-circuit voltage, power generated can also be calculated at the particular maximum intensity of light obtained during the day, wherein the $I–V$ characteristics have been obtained graphically using the MATLAB and Simulink environment [5]. With the variation in the maximum power rating and the power obtained during the process of current generation, fault in the cell can directly be assessed by the manufacturers using the IoT [6]. The presented idea is cost-effective as no climate sensors, simulations or kind are required for error detection plus the maintenance in remote places becomes much user-friendly and handy [7]. For the maintenance of the used equipment, the information from the sensors is gathered as fault diagnosis [8]. In this efficacy of the photovoltaic (PV) structure is proportionate to the solar energy. To achieve maximum efficiency which can be achieved through perpendicular positioning of PV cell with sun radiations, tracking of sun is important. Further, Saini et al. have proposed FPGA-based surveillance system [9]. Singh et al. described that two-axis solar tracking system's power gain is more when compared to the fixed systems [10, 11]. An efficient technique is discussed to monitor the large area lightning system, without actually visiting the location [12, 13]. Solar panels are the most reliable renewable energy source nowadays, and hence, we need to make them more efficient and long lasting for which their performance monitoring is important [14–16]. Although efficiency depends on many factors, most influencing factor is sunlight intensity falling on the panel. Since the solar panel is a series collection of solar cells which is again the combination of diodes which has nonlinear $I–V$ curve, hence the performance of solar panel depends on current and the voltage generated by the panel.

2   Contribution

This paper proposes an idea that how current and voltages generated by solar panel will be sampled at different time intervals of day and send to an Android app through IoT. This app plots $I–V$ curve at different time intervals and finds the area under that curve, fill factor and efficiency, and hence, by comparing these parameters obtained for different days, the performance of panel can be judged. When light falls on diode, "light-generated current" is produced which transforms the diode equation and $I–V$ curve of the diode?

$$(I - V \text{ curve})_{\text{solar cell}} = (I - V \text{ curve})_{\text{solar cell diode in dank}} + (I - V \text{ curve})_{\text{solar cell diode in light}}$$

Power can be extracted from diode when its $I–V$ curve is shifted to fourth quadrant. And hence, the diode current equation becomes

$$I = I_0 \left[\exp\left(q V / nkT\right) - 1 \right]$$

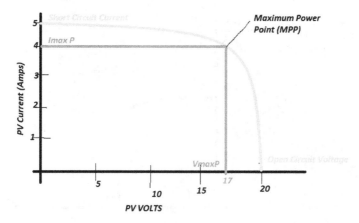

Fig. 1 Current versus voltage graph with varying intensities

where I = Net diode current; I_0 = Dark saturation current, V = Applied terminal voltage; q = Electron charge (absolute value); k = Boltzmann's constant; T = absolute temperature in Kelvin, and n is identity factor ranging 1–2. Figure 1 shows the variation in current and voltages with respect to change in solar radiations.

Light-generated current totally depends on the intensity of sunlight. Different time intervals are taken into consideration and observation because sunlight intensity is different at different times hence performance at every point, that is, when intensity is low and when it is high has to be assessed.

3 Methodologies

Considering the scenario of India, solar radiation intensity is at its peak between 12 and 3 pm. Radiation intensity is minimum between 4 and 5 pm and is moderate between 9 and 10 am.

1. Light intensity on solar panels is called the number of suns.
2. 1 sun = 1 kW/m^2.
3. Effect of shunt resistance becomes important at low level.

At light intensities:

1. **Low**—light-generated current value is low and also low is the voltage generated, and hence, the area under the curve (*I–V*) is less, shown in Fig. 2a.
2. **High**—light-generated current is high (near to short-circuit current) and high is the voltage generated, and hence, the area under the curve (*I–V*) is high, shown in Fig. 2b.
3. **Moderate**—light-generated current is moderate and moderate is the voltage generated, and hence, the area under the curve is moderate.

Fig. 2 **a** *I–V* graph at low to moderate intensity, **b** *I–V* graph at maximum intensity

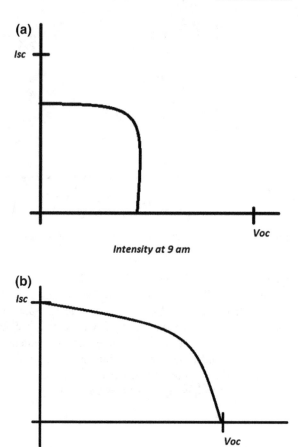

Using IoT in this project is highly beneficial and time efficient since it uses onsite storage which is much faster than online storage cutting down the time on downloading of data. Using this commendable tool, the data of analysis of short-circuit current and open-circuit voltage and the power can directly be accessed by the manufacturer as well as the users. This enables the easy handling and survey of the errors in the solar cells, their detection and quick rectification.

All the data collected will be given as an input to current and voltage sensor, respectively, which will be interfaced with Arduino which will be programmed and interfaced with ESP (IoT module). Finally, IoT module will send all data to the Android app.

4 Hardware Descriptions

The following components described below are used to design a setup for the system.

Arduino—It is a microcontroller board based on AT mega 328 running at 8 MHz with external resonator (0.5% tolerance), has 14 digital input/output pins and eight analog inputs operating at 3.3 V. We are using in our project as a controlling device which takes current and voltage input from the solar panel and sends data to the ESP8266 module which works on UART protocol. The current and voltage sensor input are given to analog pin of microcontroller, and data will be displayed on 16 * 4 LCD display with four bits used data at a time.
Voltage sensor—It is able to determine and measure the voltage supply by converting voltage measured between two points of an electrical into signal proportion to the voltage. Current sensor—It gives current measurement for both AC as well as DC signals, operates from 5 V and analog output voltage proportional to current measured on the sensing terminal and can simply be used as a microcontroller ADC to read the value. Wi-fi module—It is a Wi-fi module that can: Send *AT* commands from a microcontroller via a USB to serial adapter, which mostly uses for testing and setup.

GPRS modem GPRS is a device that operates wirelessly using a SIM and communicates using a set of AT commands. The block diagram representation of the analogies used to collect data and display it on the cloud is as shown in Figs. 3a and 4a. Figure 3a shows the data analysis and sending the procured data to the server using the Node MCU controller or the Wi-fi shield, ESP8266. Figure 3b that clearly displays the circuit diagram of the aforementioned procedure.

Figure 4a shows the data manipulation, monitoring and sending the obtained data to the server using the GPRS module. Following it is Fig. 4b that clearly displays the circuit diagram of the aforementioned procedure incorporating a GPRS/GSM module.

5 Software Description and IoT Implementation

This description is for the Android application, which shows the curve formed at different time intervals and calculates power generated for the day, daywise efficiency and fill factor of the solar panel. Figure 5 shows flowchart of data flow using the mentioned analogies of ESP8266/Node MCU and GPRS modem for solar panel online monitoring. Data is stored and analyzed in the cloud, and application is based on the cloud architecture. Proposed system is important where big data is involved, and computationally, intensive data analysis is required. This can be explained by tracking package handling.

(a)

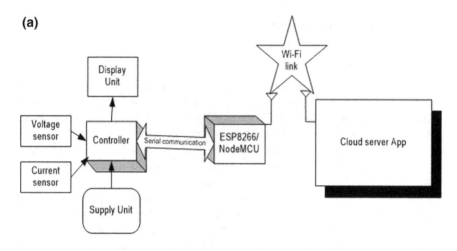

(b)

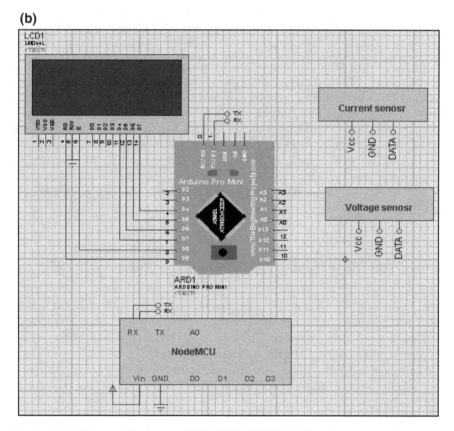

Fig. 3 **a** Block diagram of the system with ESP8266/NodeMCU, **b** Circuit schematics of the system with ESP8266/NodeMCU

(a)

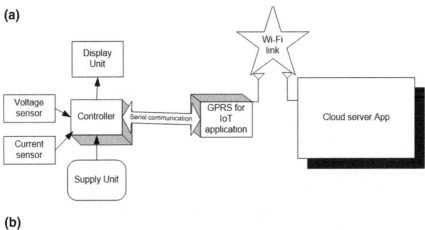

(b)

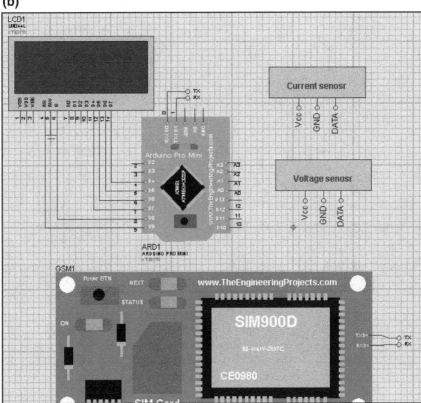

Fig. 4 **a** Block diagram of the system with GPRS modem, **b** Circuit schematics of the system with GPRS modem

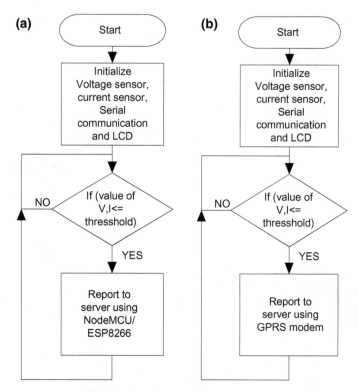

Fig. 5 **a** Flowchart of system data using ESP8266/NodeMCU and GPRS modem, **b** Flowchart of system data using ESP8266/NodeMCU and GPRS modem

6 Conclusion and Future Scope

This paper concludes that solar panels are monitored on the basis of current and voltage generation by the virtue of sunlight intensity. Current and voltage values are recorded at different time intervals, i.e., 9–10 am, 12–1 pm and 3–4 pm. Then, these generated current and voltage values are sensed by current and voltage sensor, respectively, and sent online by using ESP modem which is interfaced with Arduino board having +5 V supply. These values are received by an Android application designed for our hardware, and its software is synchronized with hardware. The paper can be of importance for the solar cell manufacturing companies in monitoring their product and surveying the performance without many heavy efforts. On the same hand, it would be highly beneficial for the users, who have got the panels installed, to keep a check on the product and the output and service it gives.

This not only aids in studying the variation in outputs of the panel but also reduces the efforts of rectifying the error in the panels. This allows the users to keep themselves updated with the day-to-day performance of the panels from different companies which is readily available through the Internet, thus helping the customer in buying the best product with verified results regarding the performance. Additionally, the paper also readily provides the option for testing and monitoring the panel in remote places, where the physical survey is not feasible giving timely conditions of the panels. In the present scenario, the unconventional sources such as solar energy are extensively used for various purposes and operations, and due to immense use, there are various fluctuations and discrepancies in the output results of the solar panels used for the purpose. During the day, due to the difference in radiation intensity, different values of current and voltage are obtained at a maximum value at a particular intensity. Thus, the variation in the maximum power obtained sensitizes to perceive an error. In future, this work can be extended as a very useful tool for the online market survey. Any buyer can see the results obtained by panels of different manufacturing brands and also at different places having different weather conditions. Different uses of the antenna could also be implemented in the future work [17–20].

References

1. Chang, C.-H., Zhu, J.-J., Tsai, H.-L.: Model-based performance diagnosis for PV systems. In: SICE Annual Conference 2010, Proceedings of IEEE (2010)
2. Emery, K., Osterwald, C.: Measurement of photovoltaic device current as a function of voltage, temperature, intensity, and spectrum. Solar Cells **21**(1-4), 313–327 (1987)
3. Ahmed, H.A., et al.: Enhancement in solar cell efficiency by luminescent down-shifting layers. Adv. Energy Res. **1**(2), 117–126 (2013)
4. Warfield, D.B., Paul, G.: Performance monitor for a photovoltaic supply. U.S. Patent No. 7,333,916 (19 Feb 2008)
5. Chouder, A., Silvestre, S.: Automatic supervision and fault detection of PV systems based on power losses analysis. Energy Convers. Manag. **51**(10), 1929–1937 (2010)
6. Xu, X., Zuo, Y., Wu, G.: Design of intelligent Internet of things for equipment maintenance. In: 2011 International Conference on Intelligent Computation Technology and Automation (ICICTA), vol. 2. IEEE, New York (2011)
7. Whitmore, A., Agarwal, A., Da Xu, L.: The Internet of Things—a survey of topics and trends. Inf. Syst. Front. **17**(2), 261–274 (2015). Internet of Things technology (2015)
8. Muselli, M., Notton, G., Canaletti, J.L., Louche, A.: Utilization of meteosat satellite derived radiation data for integration of autonomous photovoltaic solar energy systems in remote areas. Energy Convers. Manage. **39**(1/2), 1–19 (1998)
9. Saini, A., Mittal, M., Singh, S.: An FPGA based efficient surveillance system: a split processing approach. Int. J. Imaging Robot. **18**(3) (2018)
10. Bajpai, R., Singh, R., Gehlot, A., Singh, K., Patel, P.: Water Management, Reminding Individual and Analysis of Water Quality Using IOT and Big Data Analysis (2019). https://doi.org/10.2139/ssrn.3394697
11. Singh, R., Kumar, S., Gehlot, A., Pachauri, R.: An imperative role of sun trackers in photovoltaic technology: a review. Renew. Sustain. Energy Rev. **82** (2017). https://doi.org/10.1016/j.rser.2017.10.018
12. Gehlot, A., Singh, R.K., Mishra, R.G., Kumar, A., Choudhury, S.: IoT and Zigbee based Street Light Monitoring System with LabVIEW (2016)
13. Das, P.K., Malik, P.K., Singh, R., Gehlot, A., Gupta, K.V., Singh, A.: Industrial hazard prevention using raspberry Pi. In: Singh Tomar, G., Chaudhari, N., Barbosa, J., Aghwariya, M. (eds.) International Conference on Intelligent Computing and Smart Communication 2019. Algorithms for Intelligent Systems. Springer, Singapore (2020)

14. Mittal, M., Tanwar, S., Agarwal, B., Goyal, L.M. (eds.): Energy Conservation for IoT Devices: Concepts, Paradigms and Solutions. Studies in Systems, Decision and Control, pp. 1–356. Springer Nature Singapore Pvt Ltd., Singapore (2019). ISBN: 978-981-13-7398-5

15. Patel, D., Narmawala, Z., Tanwar, S., Singh, P.K.: A systematic review on scheduling public transport using IoT as tool. In: Panigrahi, B., Trivedi, M., Mishra, K., Tiwari, S., Singh, P. (eds.) Smart Innovations in Communication and Computational Sciences. Advances in Intelligent Systems and Computing, vol. 670. Springer, Singapore, pp. 39–48 (2018)

16. Tanwar, S., Kumar, N. (eds.): Multimedia Big Data Computing for IoT Applications: Concepts, Paradigms and Solutions. Intelligent Systems Reference Library, pp. 1–425. Springer Nature Singapore Pvt Ltd., Singapore (2019). ISBN: 978-981-13-8759-3

17. Malik, P., Parthasarthy, H.: Synthesis of randomness in the radiated fields of antenna array. Int. J. Microwave Wirel. Technol. **3**(6), 701–705 (2011). https://doi.org/10.1017/S1759078711000791

18. Malik, P.K., Parthasarthy, H., Tripathi, M.P.: Alternative mathematical design of vector potential and radiated fields for parabolic reflector surface. In: Unnikrishnan, S., Surve, S., Bhoir, D. (eds.) Advances in Computing, Communication, and Control. ICAC3 2013. Communications in Computer and Information Science, vol. 361. Springer, Berlin, Heidelberg (2013)

19. Malik, P.K., Singh, Madam.: Multiple bandwidth design of micro strip antenna for future wireless communication. Int. J. Recent Technol. Eng. **8**(2), 5135–5138 (2019, July). ISSN: 2277-3878. https://doi.org/10.35940/ijrte.B2871.078219

20. Malik, P.K., Tripathi, M.P.: OFDM: a mathematical review. J. Today's Ideas–Tomorrow's Technol. **5**(2), 97–111 (2017, December). https://doi.org/10.15415/jotitt.2017.52006

Classification of Chest Diseases Using Convolutional Neural Network

Rakesh Ranjan, Anupam Singh, Aliea Rizvi and Tejasvi Srivastava

Abstract Chest radiography (Chest X-ray) is the most common image taken in medical field for diagnosis of any condition that might be affecting chest or nearby area. Due to a shortage of proficient radiologists, application of this technique has been limited. To overcome this problem, we have come with a solution where we are designing a computer-aided diagnosis for chest X-ray using convolutional neural networks (CNN). CNN is a supervised deep learning that has achieved a great recognition in medical field for automatic and adaptive learning through its various layers. It has proven to be much faster and proficient than human radiologists in diagnosing medical conditions. To make a precise resultant classifying neural network, a large amount of labeled dataset is required alternatively a pretrained CNN using a large labeled dataset can also be used with some sufficient fine-tuning. In this study, by using 108,948 frontal view X-ray images of 32,717 unique patients each diagnosed with any one of the 14 lung diseases, we will build deep learning neural network using CNN from scratch which will analyze the chest X-ray and will diagnose the disease and classify within the 15 classes, i.e., 14 pre-known lung diseases and a healthy pair of lungs.

Keywords CNN · Convolutional layer · Pooling layer

R. Ranjan (✉) · A. Rizvi · T. Srivastava
Pranveer Singh Institute of Technology, Kanpur, Uttar Pradesh, India
e-mail: iiirakeshranjan@gmail.com

A. Rizvi
e-mail: aliearizvi@gmail.com

T. Srivastava
e-mail: tejasvisrivastava1998@gmail.com

A. Singh
University of Petroleum and Energy Studies, Dehradun, Uttarakhand, India
e-mail: anupam.singh@ddn.upes.ac.in

© Springer Nature Singapore Pte Ltd. 2020
P. K. Singh et al. (eds.), *Proceedings of First International Conference on Computing, Communications, and Cyber-Security (IC4S 2019)*, Lecture Notes in Networks and Systems 121, https://doi.org/10.1007/978-981-15-3369-3_18

1   Introduction

Chest radiography (Chest X-ray) is one of the commonly conducted radiological examinations of X-ray. It is the most reasonably priced and simple medical imaging technology.

A chest X-ray can produce images of chests, lungs, heart, airway, and blood vessel. Various diseases like pneumonia, pneumothorax, interstitial lung diseases, cardiac arrest, broken bone, hiatal can be diagnosed with the help of chest X-ray. Radiograph is an encroaching diagnosing test that helps physicians to diagnose and treat medical conditions of the patient.

Approximately 40,000 chest X-rays are produced per hospital every year. Major problem is that radiologists are not enough qualified to analyze the chest X-rays. Fallacy and incompatibility in radiology are very recurrent these days, with an estimated daily rate of 3–5% of the studies reported. Analysis of chest radiography entirely be contingent on the skills and expertise of radiotherapists as the X-ray has no structural details and overlapping with various parts of human body may perhaps cover up the septic tissues. And, several X-rays are tough to analyze when the injury has low variation or overlap with large pulmonary vessels. Radiologists take time to analyze and prepare report moreover, sometimes they work for late hours, and extreme tiredness may lead to misdiagnosis.

Therefore, the misidentification of diseases through X-ray image is very high. It is outlined that approximately 20–50% of lung tumors are left unnoticed or miss-examined on lung X-rays, while most of chest X-rays can be analyzed, respectively, by the next expert. Amateur radiotherapists are at times not sure about the diagnosis, and at times, they would not get the chance to get it verified from the second expert; hence, it can result in misdiagnosis. Radiologists analyzing performance cannot be always accurate, and some errors are unavoidable.

For the diagnosis of these chest X-rays automatically an artificial intelligence-based system known as computer-aided diagnosis (CAD) was developed. CAD can analyze several medical issues such as lymph nodule (LN) detection [1], automatic vertebrae detection [2], interstitial lung disease (ILD) classification [3, 4], automatic coronary calcium scoring [5]. Although newly developed CAD systems are based on the high-resolution computed tomography (CT) images or magnetic resonance imaging (MRI), while discarding the older X-rays that are commonly put to use. The layering images in CTs and MRIs show better specifics and also yield images of more advanced degree of signal-to-noise ratio that makes it simpler to function. Practically, so many X-ray images are there but very less X-ray datasets are there that are as elaborated as MRI and CT datasets.

Conventional CAD system is established on the idea of handcrafted image characteristics, and characteristics are utilized for studying a discrete or binary classifier. The implementation of this procedure is entirely based on the extricated characteristics, and researchers take a long duration to present a good set of characteristics, primarily for the X-rays. Generally, the conventional CAD systems are combination of lesion detection and false-positive reduction.

Convolutional neural network (CNN) has become trendy these days because of its matchless execution in semantic segmentation [6–9] and image classification [10–13]. Radiologists can even be substituted by CNN as it is end-to-end network and training a CNN is very simple. Computer engineers without having any medical knowledge can build these CNN models. But to utilize all methodologies of CNN, there comes a challenge that is the dataset which is required to get CNN trained. A CNN-based method needs proper labeled data for supervised learning which may take much time and effort to gather clinically. To address this problem, there are two widely used techniques to come through the discussed problem. Data augmentation, the first technique which uses the affine transformations such as translation, rotation, and scaling to generate more data from the existed data and the second technique is transfer learning, it is the method where we repurposed one task to another task, i.e., a model which is pretrained on a large dataset is been used as a starting point and a fine-tuning is done at end levels of model with the target dataset to achieve the model goal.

In our project, we used 108,948 front view X-ray images of 32,717 different patients each diagnosed with any one of the 14 lung diseases to train our CNN model and use image classification for diagnosing of disease. Here, every disease is defined as class and we have to categorize x-ray images in one or many classes. For doing this, we permit several disease labels for every X-ray image. We train our model to check three things. Firstly, Is the image of the X-ray normal? Secondly, Do the image of X-ray has disease labeled as x? Thirdly, X-ray image has what all disease labels?

2 Related Works

2.1 Visual Cortex's Receptive Fields

Experiments performed by Wiesel and Hubel in the 1950–1960s conveyed that monkey and cat's cortical visual areas have neurons that separately react to minor areas of the field of vision. If the eyes are not in motion, the area of vision of space inside which vision of stimuli has an impact on the ejection of single neuron is called the receptive field. Adjacent cells possess analogous and coinciding receptive field. The location and size of the receptive fields differ methodically through cortex to create the completed map of vision of space. Every hemisphere's cortex symbolizes the contralateral view of the visual field.

Wiesel's and Hubel's research paper in 1968 recognized two types of primary visual cell in brain [14]:

- Simple cells: their output is augmented by linear edges possessing specific orientations in its receptive fields.
- Complex cells: they possess greater receptive field whose output is unresponsive to identical position of edges in field.

Wiesel and Hubel also conveyed a cascading model of simple cell and complex cell for the purpose of recognition of patterns [15, 16].

2.2 CNN Architecture Origin

'Neocognitron' [17], the origin of convolutional neural network architecture was proposed in 1980 by Fukushima [18–20]. Fukushima was motivated by the work done by Wiesel and Hubel. The neocognitron presented the two primary layers of CNN: down sampling layer and convolutional layer. The down sampling layers have units that have receptive fields covering patch of former convolution layer. These types of units generally evaluate the mean of activations of every unit in the patch. The down sampling provides guidance to accurately classify the object in visual scene however while the object is relocated. The convolutional layers have unit who has receptive field covering a patch of former layers. The weight vectors of these units are known as filter. Filters can be shared by units.

In an alternative of neocognitron known as cresceptron, rather than utilizing Kunihiko's spatial average, J. Weng proposed a procedure known as max-pooling where the down sampling units compute maximum of the activation of units in the patch [21]. Generally, modern CNNs use max-pooling [22].

Many unsupervised and supervised learning algorithms had been presented over the time for training the weight of neocognitron [17]. Although in today's scenario backpropagation is used to train the architecture of CNN.

The first neural network that needs units to be placed at many network positions to possess shared weight is neocognitron. In 1988, neocognitron was adapted for analyzing time-varying signal [23].

2.3 Recognition of Image by CNNs Trained Using Gradient Descent

The system for recognizing manually written ZIP Codes [24] required convolution within which kernel coefficient was designed by hand arduously [25]. Backpropagation was used by LeCoun [25] in 1989 to understand convolutional kernel coefficient straightaway from the image of the handwritten number. Thus, the learning was completely automatic and hence was more effective than the handwritten coefficient design. Moreover, it was suitable for broad range of image recognition problem. Hence, this methodology laid the foundation of present-day computer vision.

2.4   CNN for Lung Nodule Detection

With the emerging functionalities of CNN, it has become a foundation for nodule detection and it is an efficient way to improve survival rate as lung cancer is one of the leading causes to death.

Some CNN-based models for detection of lung nodules are listed in Table 1 [26].

Table 1   CNN-based models for lung CAD systems

Reference	Application	Method	Dataset	Result
Shi et al. (2019) [27]	Lung nodules detection	VGG-16	CT scans	Sensitivity: 87.2% FP/scan: 0.39
Savitha et al. (2019) [28]	Lung cancers classification	Optimal deep neural network (ODNN)	Clinical	Accuracy: 94.56% Sensitivity: 96.20% Specificity: 94.20%
Zhao et al. (2018) [29]	Lung nodules classification	LeNet, AlexNe	LIDC	Accuracy: 82.20% AUC: 0.877
Dey et al. (2018) [30]	Lung nodules classification	3D-DCNN	LUNA16	Competition performance metric (CPM): 0.910
Nishio et al. (2018) [31]	Lung cancer CAD system	DCNN	Clinical	Accuracy: 68%
Anton et al. (2017) [32]	Lung CT CAD system	ResNet	LIDC-IDRI	Sensitivity: 91.07% Accuracy: 89.90%
Ding et al. (2017) [33]	Lung CAD system	DCNNs	LUNA16	Sensitivity: 94.60% FROC: 0.893
Dou et al. (2017) [34]	Lung nodules detection	3D-CNN	LUNA16	Sensitivity: 90.50% FP/scan: 1.0
Cheng et al. (2016) [35]	Lung lesions classification	OverFeat	LIDC	Sensitivity: 90.80% ± 5.30
Li et al. (2016) [36]	Lung nodules classification	DCNN	LIDC	Sensitivity: 87.10% FP/scan: 4.62

3   Convolutional Neural Network

Convolutional neural network is a deep learning algorithm which takes an image as an input and assigns some weights and biases to various objects in the image and makes it different from one another. And, it is done by the neurons out of which it is made up of. CNN is very similar to regular neural network but attains a great advantage, i.e., in regular neural network, we have an input image and then we transfer it through various hidden layers and at last each hidden layer make a set of neurons to form fully connected layer also known as output layer; this output layer will result to overfitting for large image as it holds connection with each feature unlike the fully connected layer in CNN which is accountable for only important once.

This whole process of creating CNN is done by three layers

- Convolutional Layer
- Pooling Layer
- Fully Connected Layer.

I.   **Convolutional Layer**:

Convolutional layer consists of set independent filters (aka Kernels) which are convolved with the image and we get convolved feature matrix by performing matrix multiplication between K and portion I of the image as shown in Fig. 1 [37]. Each shift of kernel/filter is dependent on its **stride** value. It shifts till it parses the complete width and again starts from extreme left by moving down as per the stride value until the whole image is covered. To maintain the convolved feature dimensionality as per

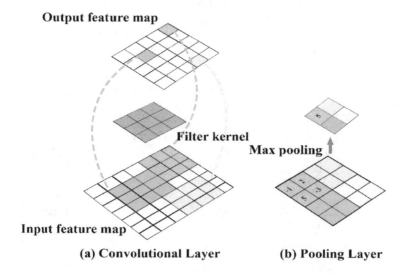

(a) Convolutional Layer   (b) Pooling Layer

Fig. 1 Illustration of **a** convolutional layer and **b** pooling layer. The output feature maps are calculated by convolving filter over the input image. In pooling layer, every 2×2 patch is sample down by max-pooling operation

image, **Padding** is used. In padding, extra pixels are added outside the image. After the convolution layer, usually an activation layer is added, generally, ReLu (rectified linear unit) function $f(x) = \max(0, x)$ is used.

II. **Pooling Layer**:

Pooling layer is used to reduce computational power which is required to process the data through dimensionality reduction. The pooling layer operates at each individual layer or depth slice independently and spatially resizes it. The most common and used pooling is max-pooling in which max value from the portion of the image covered by kernel is return hence also performs de-noising. Other than max-pooling, pooling can also perform other functions such as average pooling or $L2$-norm pooling. In average pooling, we calculate average of each patch as same as we used to find max in max-pooling.

III. **Fully Connected Layer**:

In fully connected layer, neurons have full connections to all activations in the previous layers. Since the previous convolutional layers hold the information of local features that have been computed. These local features are further sent to fully connected layer, only the relevant features are forwarded. Therefore, fully connected layer holds composite and aggregated information from all the layers of the model. Flattening of image is performed into a column vector. The flattened output further goes though forward and backpropagation to finally classify the images.

4 Methods

4.1 Dataset and Preprocessing

The dataset images of X-ray are obtained from the Kaggle provided by the NIH Clinical Center. Totally, 108,948 frontal view X-ray images of 32,717 unique patients who have taken multiple scans which are of size 1024×1024 pixels. Additional structured data is given which consist following data for each chest X-ray-'Image_Index', 'Finding_Labels', 'Follow_Up_#', 'Patient_ID', 'Patient_Age', 'Patient_Gender', 'View_Position', 'Original_Image_Width', 'Original_Image_Height' 'Original_Image_Pixel_Spacing_X', 'Original_Image_Pixel_Spacing_Y', 'Unnamed' in which irrelevant columns which has no variance or provided minimal information about the diagnosed patient is removed. In Patient_Age, column patient whose age is in months or days is removed to get homogeneous data. After which images are resized to 256×256 pixels because classification of CNN must be done in fix-sized inputs because of the fully connected layers.

Before developing CNN model, all the resized images must be converted into a single array which will then be appended to a NumPy array. CNN model was trained using Keras with the Tensorflow backend. The model is inspired by the

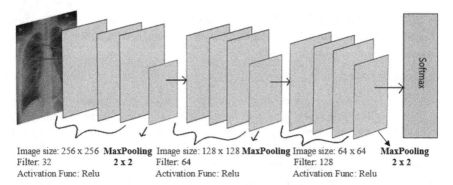

Image size: 256 x 256 **MaxPooling** Image size: 128 x 128 **MaxPooling** Image size: 64 x 64   **MaxPooling**
Filter: 32   **2 x 2**   Filter: 64   Filter: 128   **2 x 2**
Activation Func: Relu   Activation Func: Relu   Activation Func: Relu

Fig. 2 Illustration of CNN model of four layers in which the first three layers each consist of convolutional layer followed by pooling layer and the fourth layer is the fully connected layer, i.e., SoftMax

VGG architectures [11]. In which the created NumPy array was first divided into training and testing models, and training data is used to create the model and train. From 112,120 chest X-ray images, 44,800 images are separated for testing which is again divided into X_test and Y_test. The remaining 67,320 X-rays are divided into 33,600 X_train and 33,600 Y_train images. Next, reshaping of input dataset is done to the expected shape that the CNN model required that is reshaping from (number of elements, width, height) to (number of elements, depth, width, height) where feature is the depth of the image, i.e., if the input image is colorful, the value of feature will be 3 else for grayscale image like X-ray image feature is 1. Therefore, we have to explicitly declare depth of the image for further process. The last step of preprocessing is normalizing data value to range [0, 1] (Fig. 2).

4.2 Preprocessing Class Labels

After conversion of image to *np*-array and then division of testing and training, data is still in one-dimensional array rather being 15 distinct classes in which disease will be classified. To overcome this, data is split into 15 distinct class labels using method np_utils.to_categorical() in Keras which takes two arguments-name of training, testing dataset, and number of classes.

4.3 CNN Classification Model

To fir the data in CNN model, first we have to declare the hyperparameters that will control the size of output volume, i.e., depth, stride, and padding. Depth corresponds to the number of filters that would be used in the convolutional layer, each learns to extract unique feature. Stride is the number of shifts per pixel which is by default 1. Padding is used to allow to use spatial size of the output volume.

There will be total three sets of convolutional and pool layers.

First set of pool layer consists, $K = 32$ number of filters each of size (2×2) size of input image in this set is 256×256 pixels.

Output volume of size will be $H \times W \times D$

Where H and W are calculated by the formula

$$\frac{(N + 2P - F)}{S} + 1 \tag{1}$$

where N is either H (height) or W (width), P (padding), S (stride), and F (spatial extent).

D(depth) will be the number of filters that were used during processing.

After getting the feature map, activation function is applied to it where most commonly used activation Relu is used. Two more layers of convolutional and activation function are used to get well-extracted feature following to which pooling layer, i.e., MaxPool, is used. It is mostly used to reduce the number of parameters and computation of network making it faster and also used to control overfitting. It operates independently on each and every depth slice which was obtained by convolutional layer and resizes it. MaxPool layer accepts a volume of size $H \times W \times D$ and gives output of volume $H_1 \times W_1 \times D_1$ where W_1 and H_1 are calculated by formula

$$\frac{(N - F)}{S} + 1 \tag{2}$$

where N is either H (height) or W (width) and D_1 will be as same as the value of depth in input.

The above process is repeated twice at last of which we get a full activation network which has to be wind up to get the weights/bias of the input image which will be done by the last step of CNN, i.e., fully connected layer in which dropout is also used to reduce overfitting in model which slightly improved the result. Some of the input chest X-ray images can be seen in Fig. 3 which is taken from Chest X-ray14 [38].

5 Conclusion

We performed our experiment on Jupyter Notebook with 8 Titan X GPUs. Experiment is done with the split data as discussed above. The result is of accuracy 54% which proves Luke Oakden-Rayner [39] argument as despite regularization, and rectifying the class imbalances, the model has learned to give less accurate results as required.

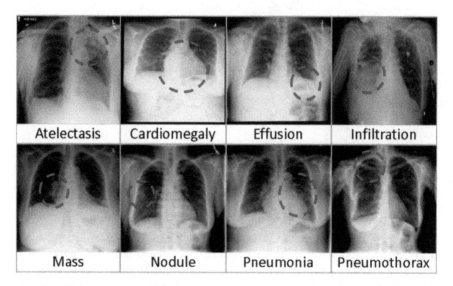

Fig. 3 Eight chest diseases diagnosed in chest X-rays from Chest X-ray14

Since CNN model is based on X-ray labels being correct and accurate the result differs.

References

1. Roth, H., Lu, L., Liu, J., Yao, J., Seff, A., Cherry, K., Kim, L., Summers, R.: Improving computer-aided detection using convolutional neural networks and random view aggregation. IEEE Trans. Med. Imaging (2015)
2. Chen, H., Shen, C., Qin, J., Ni, D., Shi, L., Cheng, J.C.Y., Heng, P.A.: Automatic localization and identification of vertebrae in spine CT via a joint learning model with deep neural networks. Springer, New York (2015)
3. Shin, H.C., Roth, H.R., Gao, M., Lu, L., Xu, Z., Nogues, I., Yao, J., Mollura, D., Summers, R.M.: Deep convolutional neural networks for computer-aided detection: CNN architectures, dataset characteristics and transfer learning. IEEE Trans. Med. Imaging **35**(5), 1285 (2016)
4. Yang, S., Cai, W., Huang, H., Yun, Z., Yue, W., Feng, D.D.: Locality-constrained subcluster representation ensemble for lung image classification. Med. Image Anal. **22**(1), 102–113 (2015)
5. Wolterink, J.M., Leiner, T., Viergever, M.A., Isgum, I.: Automatic coronary calcium scoring in cardiac CT angiography using convolutional neural networks. Med. Image Anal. **9349**, 589–596 (2016)
6. Shelhamer, E., Long, J., Darrell, T.: Fully convolutional networks for semantic segmentation. IEEE Trans. Pattern Anal. Mach. Intell. **39**(4), 640 (2017)
7. Mostajabi, M., Yadollahpour, P., Shakhnarovich, G.: Feedforward semantic segmentation with zoom-out features. In: IEEE Conference on Computer Vision and Pattern Recognition, pp. 3376–3385 (2015)
8. Noh, H., Hong, S., Han, B.: Learning deconvolution network for semantic segmentation, pp. 1520–1528 (2015)

9. Chen, L.C., Papandreou, G., Kokkinos, I., Murphy, K., Yuille, A.L.: Semantic image segmentation with deep convolutional nets and fully connected crfs. Comput. Sci. **4**, 357–361 (2016)
10. Krizhevsky, A., Sutskever, I., Hinton, G.E.: Imagenet classification with deep convolutional neural networks. In: International Conference on Neural Information Processing Systems, pp. 1097–1105 (2012)
11. Simonyan, K., Zisserman, A.: Very deep convolutional networks for large-scale image recognition. Comput. Sci. (2014)
12. Szegedy, C., Liu, W., Jia, Y., Sermanet, P.: Going deeper with convolutions, pp. 1–9 (2014)
13. He, K., Zhang, X., Ren, S., Sun, J.: Deep residual learning for image recognition, pp. 770–778 (2015)
14. Hubel, D.H., Wiesel, T.N.: Receptive fields and functional architecture of monkey striate cortex. J. Physiol. **195**(1), 215–243 (1968)
15. Hubel, DH., Wiesel, TN.: Brain and Visual Perception: The Story of a 25-year Collaboration. Oxford University Press, Oxford, US, pp. 106 (2005). ISBN: 978-0-19-517618-6
16. Hubel, D.H., Wiesel, T.N.: Receptive fields of single neurones in the cat's striate cortex. J. Physiol. **148**(3), 574–591 (1959)
17. Fukushima, K.: Neocognitron. Scholarpedia **2**(1), 1717 (2007)
18. Fukushima, K.: Neocognitron: a self-organizing neural network model for a mechanism of pattern recognition unaffected by shift in position (PDF). Biol. Cybern. **36**(4), 193–202 (1980)
19. Ciresan, D., Meier, U., Schmidhuber, J.: Multi-column deep neural networks for image classification. In: 2012 IEEE Conference on Computer Vision and Pattern Recognition, pp. 3642–3649. IEEE, New York, NY (2012). arXiv:1202.2745. CiteSeerX 10.1.1.30.3283
20. LeCun, Y., Bengio, Y., Hinton, G.: Deep learning. Nature **521**(7553), 436–444 (2015). Bibcode: 2015Natur.521..436L
21. Weng, J., Ahuja, N., Huang, T.S.: Learning recognition and segmentation of 3-D objects from 2-D images. In: Proceedings of the 4th International Conference Computer Vision, pp. 121–128 (1993)
22. Schmidhuber, J.: Deep learning. Scholarpedia **10**(11), 1527–54 (2015). CiteSeerX 10.1.1.76.1541
23. Waibel, A.: Phoneme Recognition Using Time-Delay Neural Networks. Meeting of the Institute of Electrical, Information and Communication Engineers (IEICE), Tokyo, Japan (1987)
24. Denker, J.S., Gardner, W.R., Graf, H.P., Henderson, D., Howard, R.E., Hubbard, W., Jackel, L.D., Baird, H.S., Guyon, I.: Neural network recognizer for hand-written zip code digits, AT&T Bell Laboratories (1989)
25. LeCun, Y., Boser, B., Denker, Y., Henderson, D., Howard, R.E., Hubbard, W., Jackel, L.D.: Backpropagation Applied to Handwritten Zip Code Recognition. AT&T Bell Laboratories, Holmdel, NJ
26. Gao, J., Jiang, Q., Zhou, B., Chen, D.: Convolutional neural networks for computer-aided detection or diagnosis in medical image analysis: an overview. Math. Biosci. Eng. **16**(6), 6536–6561 (2019)
27. Shi, Z.H., Hao, H,. Zhao. M.H., et al.: A deep CNN based transfer learning method for false positive reduction, Multimed. Tools Appl **78** (2018)
28. Savitha, G., Jidesh, P.: Lung nodule identification and classification from distorted CT images for diagnosis and detection of lung cancer. In: Machine Intelligence and Signal Analysis, Springer, pp. 11–23 (2019)
29. Zhao, X.Z., Liu, L.Y., Qi, S., et al.: Agile convolutional neural network for pulmonary nodule classification using CT images. Int. J. Comput. Assist. Radiol. Surg. **13**, 585–595 (2018)
30. Dey, R., Lu, Z.J., Yi, H.: Diagnostic classification of lung nodules using 3D neural networks. In: 2018 IEEE 15th International Symposium on Biomedical Imaging (ISBI), pp. 774–778 (2018)
31. Nishio, M., Sugiyama, O., Yakami, M., et al.: Computer-aided diagnosis of lung nodule classification between benign nodule, primary lung cancer, and metastatic lung cancer at different image size using deep convolutional neural network with transfer learning. PLoS ONE **13**, e0200721 (2018)

32. Anton, D., Ramil, K., Adil, K., et al.: Large residual multiple view 3D CNN for false positive reduction in pulmonary nodule detection. In: 2017 IEEE Conference on Computational Intelligence in Bioinformatics and Computational Biology (CIBCB) (2017)

33. Ding, J., Li, A., Hu, Z.Q., et al.: Accurate pulmonary nodule detection in computed tomography images using deep convolutional neural networks. In: MICCAI 2017, pp. 559–567 (2017)

34. Dou, Q., Chen, H., Jin, Y.M., et al.: Automated pulmonary nodule detection via 3D Convnets with online sample filtering and hybrid-loss residual learning. In: MICCAI 2017, pp. 630–638 (2017)

35. Cheng, J.Z., Ni, D., Chou, Y.H., et al.: Computer-aided diagnosis with deep learning architecture: applications to breast lesions in us images and pulmonary nodules in CT scans. Sci. Rep. **6**, 24454 (2016)

36. Li, W., Cao, P., Zhao, D.Z., et al.: Pulmonary nodule classification with deep convolutional neural networks on computed tomography images. Comput. Math. Method. Med. **2016**, 1–7 (2016)

37. Dong, Y., Pan, Y., Zhang, J., Xu, W.: Learning to read chest X-ray images from 16000+ examples using CNN

38. Wang, X., Peng, Y., Lu, L., Lu, Z., Bagheri, M., Summers, RM.: Chestx-ray8: hospital-scale chest X-ray database and benchmarkson weakly-supervised classification and localization of common thorax diseases (19 July 2017)

39. Exploring the chestxray14 dataset: problems. https://lukeoakdenrayner.wordpress.com/2017/12/18/the-chestxray14-datasetproblems. Last accessed 2017/12/18

Age and Gender Prediction Using Convolutional Neural Network

Kunal Jain, Muskan Chawla, Anupma Gadhwal, Rachna Jain and Preeti Nagrath

Abstract Automatic age and gender classification has been widely used in a large amount of applications, particularly in human–computer interaction, biometrics, visual surveillance, electronic customer and commercial applications. Predicting age and gender of humans is very difficult and complicated. Since the increase in use of social media, autonomous prediction of age and gender has become extremely important. Nevertheless, performance of the existing methods on real-world images is still significantly lacking, especially when compared to the tremendous leaps in performance recently reported for the related task of face recognition. In this research paper, we have observed an increase in the performance on these tasks by using convolutional neural networks (CNN). We collected a dataset from 'Google' and 'IMFDB.' This dataset consisted of about 100,000 images of male and female of different age-groups (5–60). We created a CNN network adjoining the fully connected network using 'Keras' library. Following that, three convoluted layers were inserted; the first layer consisting of 64 neurons and window of 7×7, followed by a ReLU activation layer and a max-pooling layer of 2×2. Two more CNN layers of 100 and 64 neurons were inserted, followed by windows of 5×5 and 3×3, respectively, with activation function as ReLU. An output was received in matrix form which was flattened using a flatten layer. The layer was then fed to fully connected layers with 64 and 1 neurons each. The final output layer consisted of a single neuron with sigmoid activation function.

K. Jain (✉) · M. Chawla · A. Gadhwal · R. Jain · P. Nagrath
Bharati Vidyapeeth's College of Engineering, New Delhi, Delhi, India
e-mail: kjcdude97@gmail.com

M. Chawla
e-mail: chawla.muskan@yahoo.com

A. Gadhwal
e-mail: anupmagadhwal@gmail.com

R. Jain
e-mail: rachna.jain@bharatividyapeeth.edu

P. Nagrath
e-mail: preeti.nagrath@bharatividyapeeth.edu

© Springer Nature Singapore Pte Ltd. 2020
P. K. Singh et al. (eds.), *Proceedings of First International Conference on Computing, Communications, and Cyber-Security (IC4S 2019)*, Lecture Notes in Networks and Systems 121, https://doi.org/10.1007/978-981-15-3369-3_19

Keywords Convolutional neural network (CNN) · Dense layer · Pixels · Keras ·
ReLU · Sigmoid · Adam · Haar Cascades classifier · OpenCV · IMFDB

1 Introduction

The human face consists of features and information such as expression, gender and age. Humans are very advanced and can detect these features easily. For example, majority of humans can detect basic features like gender, age, race of a person by telling whether the person is male or female, if they are young or aged, just by seeing his/her face through their eyes. In this research paper, we have tried to minimize the gap between automatic face-recognizing ability of the computer and age and gender estimation methods. We obtained the desired results using a simple neural network architecture, with keeping the limited availability of age and gender labels in the existing datasets in our minds. Computer vision is the field of science that works on making a program that can make the computer see and describe our world, just like the human eyes work for humans. On the contrary, pattern recognition is the method of recognizing a pattern on basis of some algorithms. It provides identification, description and interpretation with the help of which computers can recognize and detect patterns such as shape, a handwritten word, environment or face of a human.

A pattern recognition system consists of three parts—preprocessing, feature extraction and classification. Automatic age and gender classification using facial features is one of the most difficult tasks as a person who is 40 years old might look like they are in their 20 s. Thus, getting sufficient and efficient data for age detection is another difficult task. Convolutional neural networks showed significant success in image detection. A convolutional neural network (CNN) consists of different layers, where each layer processes the output of its preceding layer and produces a strong, compact output. If the number of layers inside a CNN is large, it is called a deep network. In this research paper, a pretrained CNN has been used to predict the age and gender from unfiltered images of the face. The deep CNN has been initially trained on a large database for face recognition. At the moment, there is no database for age prediction.

In this research paper, we have observed an increase in the performance on these tasks by using convolutional neural networks (CNN). We collected a dataset from Google and IMFDB [1]. This dataset consisted of about 34,512 images of male and female of different ages and different ethnicities refer to Fig. 1, which shows how much variation occurs between the people. We created a CNN network, a fully connected network using 'Keras' library. Following that, three convoluted layers were inserted; the first layer consisted of 64 neurons and window of 7×7, followed by ReLU activation layer and a max-pooling layer of 2×2. Two more CNN layers of 100 and 64 neurons were inserted, followed by windows of 5×5 and 3×3, respectively, with activation functions as ReLU. An output was received in a matrix form which

Fig. 1 An example of change in facial characteristics in people [2]

was flattened using a flatten layer. The layer was then fed to fully connected layers with 64 and 1 neuron each. The final output layer consisted of a single neuron with sigmoid activation function.

2 Machine Learning

Machine learning revolves around the concept of making a computer program that improves the performance by automatically learning and adapting with experience. It is basically working on many hypotheses and then finding the best one that fits the observed data. It makes computers modify their action so that these actions get more and more accurate with experience, where accuracy is measured by how well the chosen action reflects the correct ones. Machine learning is a growing technology used to mine knowledge from data (known as data mining). Wherever data exists, things can be learned from it. Whenever there is an excess of data, the mechanics of learning must be automatic.

Machine learning technology is meant for automatic learning from voluminous datasets. Machine learning algorithms can be supervised or unsupervised. A data scientist or analyst works on supervised algorithms, and in unsupervised learning, we just have an input data. The various applications of machine learning are Web search engine, photo tagging applications, span detectors, etc.

3 Related Work and Data

Before describing the methodology, we briefly reviewed all the related methods that have been made in the field of facial recognition in [3]. A detailed survey of the gender and age prediction method can be found in [4]. More recently in [5], a hierarchical approach for automatic age estimation has been proposed and an analysis of how aging influences individual facial components has been made.

Deep CNN has additionally been successfully applied to applications including human pose estimation as in [6], face parsing technique is another example of achieving the required target as done in [7], facial keypoint detection can also be done just like in [8], speech recognition is another way as done in [9] and action classification is done in [10]. In the past, early methods for gender classification as in [11] have used a neural network trained on a small set of near-frontal face images. In [12], the combined 3D structure of the head and image intensities has been used for gender classification.

The dataset has been collected from Google and IMFDB [1], and it consists of tens of thousands of images of both male and female varying from every existing age-group. The dataset used contains pictures of the people from all ethnicities, with different gender, age-group, race and facial expressions. Earlier methods for age and gender predictions have been using size and proportions of the face of humans. They were limited to the young ages since the human face varies a lot in adulthood. Now, a method for age prediction has been designed that uses unfiltered facial images. Interesting approaches have been made in the field of facial recognition as done in [3]; we tried to use similar trade, but then, we thought why not use a predefined classifier for recognizing a face as that is not the objective of the research paper as done in [13]; some tried to detect the age using hierarchical classification [14]; some tried to work on unfiltered faces just like in [15]; others used the deep convolutional neural networks (DCNN) approach [16]; we based our research on these previous researches.

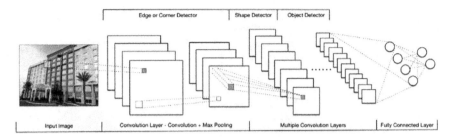

Fig. 2 Workflow of a convolutional neural network; the image is taken from [17]

4   Methodology

4.1   Convolutional Neural Network

Convolutional layer forms the most important building block of convolutional neural networks (CNN) Fig. 2 shows the working flow of a convolutional neural network. A convolution network has the ability to process large dataset, or in other words, a convolutional neural net can learn large amount of equations. The main task of a Convolutional Neural Network (CNN) is its ability to autonomously learn through a large number of filters; therefore, image and video processing usually consists of the convolutional neural network working on an image that comprises of a large number of the data points. Figure 3 shows the flowchart of the process followed. It is very important for machine learning since it increases the accuracy. Figure 4 shows the architecture of the convolutional neural network while processing an image.

5   Methods for Proposed Work

5.1   Data Collection

The dataset has been collected from Google and IMFDB [1] and consists of about 34,512 images of people from all ethnicities and ranging for all the existing age-groups.

5.2   Dataset Formation

Dataset formation is the critical step in every project. Data might be available in any form, raw and processed. There might be a possibility where there exists a data which might not satisfy our needs, or in other terms, it is not ready to be fed to

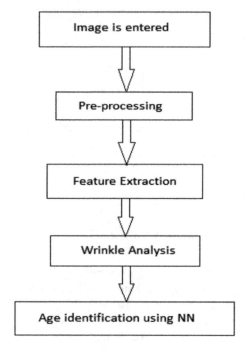

Fig. 3 Flowchart depicting the approach of the prediction model; this image is extracted from [18]

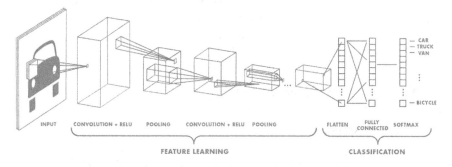

Fig. 4 Architecture of a convolutional neural network (CNN) taken from Google [19]

an algorithm. Therefore, it is always necessary to preprocess data before feeding the data to network. In machine learning or deep learning, data is everything. Our research is based on the supervised learning in which one feeds the labeled data to a network; basically, it is like teaching a child by showing him everything that this is one thing and that is another. In this project of age and gender detection, we tried to teach our neural network the difference between a male and a female. In order to do so, we created a labeled dataset; this is used for training and testing purposes of the neural net. In this project, the data of male and female has been stored in

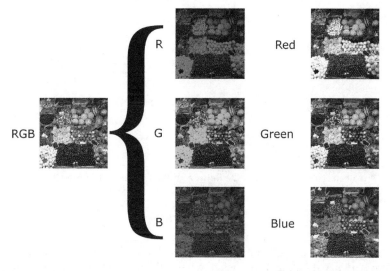

Fig. 5 Image depicting why grayscale image is used instead of color as grayscale in same in every color channel; this image is taken from [20]

different directories. The OS library has been used to travel through the directory to access the data. Since an image usually consists of three channels (red, green and blue which makes the whole spectrum of colors), which makes it hard to process, for example, consider an image of size 1080 × 2080, now for three channels, there will be three matrixes of this dimensions per image; therefore, the computing this large mathematical matrix value is very difficult; therefore to reduce the strain on the processor, all images are read in grayscale format as they are single channeled and resized to a matrix of 70 × 70 pixels. Figure 5 shows the difference between the RGB color format and the grayscale format. We could use any size of the image; it is not necessary to 70 × 70 image dimensions. We reduced the size and read the image in gray scale as for this particular project we do not need any color values and resizing the image doesn't compromise the number of features in an image hence we can do our work in small images and that too without straining our processor. For our project we labeled male as 0 and female as 1.

5.3   CNN Formation

A CNN network adjoining the fully connected network using Keras Library has been created. A convoluted layer has been inserted, with first layer containing 64 neurons and window of 7 × 7, followed by a ReLU activation (ReLU function stands for rectified linear unit which is nothing but a linear activation function that usually gets activated after some point the equation of the ReLU function is given in (1) and for the mathematical graph refer to Fig. 6).

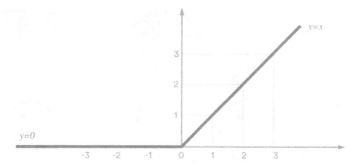

Fig. 6 A mathematical figure to show the ReLU activation function; the image is taken from [21]

$$R(z) = \max(0, z) \tag{1}$$

and max-pooling (a max-pooling layer extracts the maximum value from the values occurred in max-pooling layer window; this is the important step toward the prediction a clear picture of the max-pooling layer that can be understand by Fig. 7) layer of 2 × 2.

Along with that, two more CNN layers with 100 and 64 neurons and max-pooling layer of 5 × 5 and 3 × 3 have been inserted, respectively, with activation function ReLU. After receiving the output from CNN, the matrix output has been flattened using a flatten layer and fed to fully connected layers with 64 and 1 neuron each. The final output layer consists of a single neuron with sigmoid activation function (a sigmoid function has the 's'-shaped locus; most of the time, a logistic function is used as the sigmoid function. The main reason behind the use of the sigmoid function is that it gives the output between 0 and 1; hence, it eases the probability calculation). The equation of sigmoid function is given in (2), and the clear picture of the sigmoid function can be depicted from Fig. 8.

$$S(z) = \frac{1}{1 + e^{-z}} \tag{2}$$

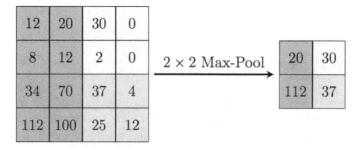

Fig. 7 A simple 2 × 2 max-pooling layer in action used to reduce the computational matrix; the above picture was taken from the [22]

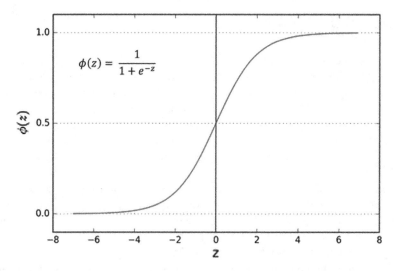

Fig. 8 Sigmoid function. As the locus of the graph is somewhat similar to '*s*,' hence the name sigmoid. The picture is taken from [23]

5.4 Training and Testing

Our model is based on the supervised learning. Therefore, our model is trained on the labeled dataset. The model has been trained on the dataset consisting of more than tens of thousands of pictures of people from different ethnicities and regions, with a validation split of 10% (this means 10% of the data is used to validate the result) and adam as the optimizer (an optimizer basically determines how the internal parameters of algorithms are set to optimize the output for the best a visual working of an adam optimizer with some other optimizers which are given in Fig. 9). We basically fed the image containing the only the face of every person in our dataset, and to do, this we used the help of Haar cascades algorithm. To train our network, we used features like mustache, beard, etc., to validate the gender.

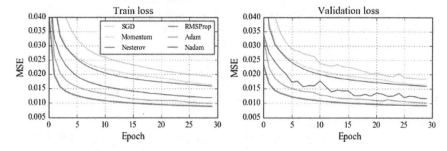

Fig. 9 Adam optimizer seems to works great with less error and moderate epochs therefore used in this project; this figure is taken from [24]

6 Implemented Work

The dataset has been collected from Google and IMFDB [1] and consists of more than tens of thousands of images of male and females of all ethnicities, ranging from all age-groups from 5 to 60. The dataset used contains pictures of people from all ethnicities, with different gender, age-group, race and facial expressions. Earlier methods for age and gender predictions have been using size and proportions of the face of the humans. They were limited to the young ages since the human face varies a lot in adulthood. Now, a method for age prediction has been designed that uses unfiltered facial images.

The model used is for the prediction of the gender of humans. As the prediction of the age is more complex and requires more features, a pretrained model for age detection has been used. The age and gender have been predicted on the live feed from the webcam. An OpenCV library has been used for this, and Haar cascades classifier has been used for face detection. Following this, the face was extracted from the image, turned it into grayscale format and resized it a 70×70 pixel matrix. Then, it was fed to our constructed model, and the output was obtained. Finally, the obtained results were wrapped onto the image frame using OpenCV library.

7 Results

For age and gender classification, we have measured and compared both the accuracy when the algorithm gives the exact age-group classification and when the algorithm is off by one adjacent age-group. This follows others who have done so in the past and reflects the uncertainty inherent to the task—facial features often change very little between oldest faces in one age class and the youngest faces of the subsequent class. Evidently, the proposed method outperforms the reported state of the art on both tasks with considerable gaps.

The proposed model has successfully predicted the age-group and gender of a person, by feeding image to the code via webcam with accuracy score of approximately 80%. Figure 10 shows the predicted age and the age-group of the people at the top of the bounding box, and Fig. 11 shows the age-group and gender as the output on the screen in real-time video. Figure 12 shows the bias value and the weights of the neuron connection.

8 Conclusion

Many people have tried methods to predict the age and gender in the past. Majorly, these methods focused on artificial and unnatural images that were taken in laboratories. Such images do not reflect the variations that we see in the human faces

Fig. 10 Output of the model with creation of bounding box figure with age and gender mentioned on the top of the box. The output is for the particular frame taken from the live feed

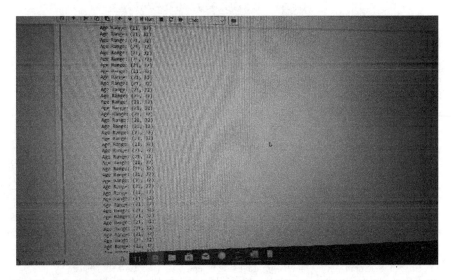

Fig. 11 Output image of the algorithm (printing the age and the gender) of every face detected in the frame

in the real world. Thus, to overcome this problem, we have used deep convolutional neural network (CNN) on the data available on Internet. We have tried to avoid over-fitting by redesigning the preexisting architecture due to limitation of data that is labeled. The size of the training data has been increased by inserting cropped part of

```
0B -> L1N0:  0.01451056357473135
0B -> L1N1:  0.03746987506747246
0B -> L1N2:  -0.0088733304291963 58
0B -> L1N3:  0.011707737110555172
0B -> L1N4:  -0.0120915193110 70442
0B -> L1N5:  0.014891318045556545
0B -> L1N6:  -0.02269248105585575
0B -> L1N7:  -0.0096962554 38029766
0B -> L1N8:  -0.0176525358110 66628
0B -> L1N9:  -0.0114759998 39603901
0B -> L1N10: -0.011685258708894253
0B -> L1N11: -0.01724427007138729
0B -> L1N12: -0.019513683393597603
0B -> L1N13: 0.032692670822143555
0B -> L1N14: 0.022855769842863083
0B -> L1N15: 0.035141684114933014
0B -> L1N16: 0.013138771057128906
0B -> L1N17: -0.00818723440170288
0B -> L1N18: 0.042872738093 13774
0B -> L1N19: -0.018101224675774574
```

Fig. 12 Different weights and biases of some neurons is the first layer of the model

the images. The method created has shown to outperform the existing methods. The use of CNN improved our detection results to a great extent.

References

1. Setty, S., Husain, M., Beham, P., Gudavalli, J., Kandasamy, M., Vaddi, R., Hemadri, V., Karure, J.C., Raju, R., Rajan, B., Kumar, V., Jawahar, C.V.: Indian movie face database: a benchmark for face recognition under wide variations. In: National Conference on Computer Vision, Pattern Recognition, Image Processing and Graphics (NCVPRIPG) (2013)
2. Medical daily picture. https://www.medicaldaily.com/how-sell-anything-anyone-hair-color-body-type-and-race-decide-consumer-behavior-327050
3. Emami, S., Suciu, V.: Facial recognition using OpenCV. J. Mob. Embed. Distrib. Syst
4. Fu, Y., Guo, G., Huang, T.S., Fu, Y., Guo, G., Huang, T.S.: IEEE Conference (2010)
5. Han, H., Otto, C., Liu, X., Jain, A.K.: Demographic estimation from face images: human vs. machine performance (2015)
6. Toshev, A., Szegedy, C.: Deeppose: human pose estimation via deep neural networks. In: IEEE Conference (2014)
7. Luo, P., Wang, X., Tang, X.: Hierarchical face parsing via deep learning. In: IEEE Conference Computer Vision
8. Sun, Y., Wang, X., Tang, X.: Deep convolutional network cascade for facial point detection. In: The Process Conference Computer Vision Pattern Recognition, pp. 3476–3483 (2015)
9. Graves, A., Mohammed, A.R., Hinton, G.: Speech recognition with deep recurrent neural networks. In: IEEE Conference on Acoutic, Speech and Signal Processing (ICASSP) (2013)

10. Karpathy, A., Toderici, G., Shetty, S., Leung, T., Sukthankar, R., Fei-Fei, L.: Large-scale video classification with convolutional neural netwoks. In: IEEE, pp. 1725–1732 (2014)
11. Golomb, B.A., Lawrence, D.T., Sejnowski, T.J.: Sexnet: a neural network identifies sex from human faces. In: Neural Information Processing Systems (1990)
12. O'toole, A.J., Vetter, T., Troje, N.F., Bulthoff, H.H. et al.: Sex classification is better with three-dimensional head structure than with image intensity information (1997)
13. Soo, s.: Object detection using haar-cascade classifier (2014)
14. Chao, S.E., Lee, Y.J., Lee, S.j., Park, K.R., Kim, J.: Age estimation using a hierarchial classifier based on global and local features, pattern recognition (2011)
15. Eidinger, E., Enbar, R., Hassner, T.: Age and gender estimation of unfiltered faces (2014)
16. Sun, Y., Wang, X., Tang, X.: Deep convolutional network cascade for facial point detection. In: The Process Conference Computer Vision Pattern Recognition, pp. 3476–3483 (2013)
17. Towardsdatascience.com: https://towardsdatascience.com/a-comprehensive-guide-to-convolutional-neural-networks-the-eli5-way-3bd2b1164a53
18. Thakur, S., Verma, L.: Age identification of facial images using neural network
19. Researchgate.net: https://www.researchgate.net/figure/Illustration-of-Convolutional-Neural-Network-CNN-Architecture_fig3_322477802
20. Wikipedia.org: https://upload.wikimedia.org/wikipedia/commons/3/33/Beyoglu_4671_tricolor.png
21. Medium.com: https://medium.com/@danqing/a-practical-guide-to-relu-b83ca804f1f7
22. computersciencewiki.org: https://computersciencewiki.org/index.php/File:MaxpoolSample2.png
23. Towardsdatascience.com: https://towardsdatascience.com/activation-functions-neural-networks-1cbd9f8d91d6
24. Towardsdatascience.com: https://towardsdatascience.com/adam-latest-trends-in-deep-learning-optimization-6be9a291375c

Vision-Based Human Emotion Recognition Using HOG-KLT Feature

R. Santhoshkumar and M. Kalaiselvi Geetha

Abstract Obviously, humans communicate and interact among each other through speech and body movement. It is also obvious that there is a close linkage between human emotion expressions and his body movements. This implies that emotion is an important aspect in the interaction and communication between people. Since the science of artificial intelligence (AI) is concerned with the automation of intelligent behavior. This paper aims to recognize the emotion of the human using histogram of orientation gradient (HOG) and Kanade–Lucas–Tomasi (KLT) HOG-KLT feature. The basic emotions used in this work are angry, joy, fear, sad and pride. The input videos are converted into gray frames. The HOG-KLT features are extracted from the sequences of frames. The emotions are recognized using support vector machine and random forest classifier. The GEMEP corpus dataset is used for this experiment.

Keywords Artificial intelligence · Histogram of orientation gradient (HOG) and Kanade–lucas–tomasi (KLT) · Emotion recognition · Support vector machine · Random forest classifier

1 Introduction

Recognizing emotion from human body movement is a challenging task. Human body language is a form of nonverbal cues that is used in human–human interactions almost entirely subconsciously. Moreover, body language can convey emotional state of a person in human communication such as angry, happy, fear, sad and neutral [1]. Body language can also enhance verbal communication. For example, hand and head movements give clue to the person you are interacting with that you understand the subject. The human head is a multi-signal communicative media with remarkable flexibility and specificity [2]. If a human face is not in front of camera and the human

R. Santhoshkumar (✉) · M. Kalaiselvi Geetha
Department of Computer Science and Engineering, Annamalai University, Annamalainagar, Tamilnadu, India
e-mail: santhoshkumar.aucse@gmail.com

M. Kalaiselvi Geetha
e-mail: geesiv@gmail.com

© Springer Nature Singapore Pte Ltd. 2020
P. K. Singh et al. (eds.), *Proceedings of First International Conference on Computing, Communications, and Cyber-Security (IC4S 2019)*, Lecture Notes in Networks and Systems 121, https://doi.org/10.1007/978-981-15-3369-3_20

is too far from camera, then it is the worst situation in emotion recognition problems. This paper aims to resolve these types of problems by using body movement of the person. In this case, body movements like head, hands, legs and center of body are considered as the major part for recognizing the emotion.

Another concern is that some emotion appearances are very close to each other like skeptical and frown emotions; this will make the recognition harder and forced the researchers to limit the classification to small number of emotions that concentrate on the basic six emotions or in other cases on maximum of eight emotions. Recognition of the emotions based on the face appearance is very concerned with the extracted face features. This process is called feature extraction [3]. The extracted features are then transformed into input data to any emotion recognizable system. It should be notes here that human emotions occur at different levels of intensity and at different moments in time. There is a difference between "a little bit sad" and "very sad" in their role of influencing behavior, facial expression and other emotions, even though both expressions are negative emotion. The objective of this paper is to recognize the emotion from the human body movements. The dataset used for this experiment is GEMEP corpus dataset. The angry, joy, fear, sad and pride are the emotions used for this experiment. Using HOG-KLT algorithm, the features are extracted from the sequence of frames. Finally, the emotions are recognized using SVM and random forest classifiers.

The following paper contains compressed literature survey in the second section. The proposed work is described in the third section. The experimental results are detailed in the fourth section. The conclusion and future scope are described in section five.

2 Related Works

Arunnehru et al. [4] developed an MI code for action recognition in video. Varghese et al. [5] detailed the recognition of human emotion using the proposed real-time emotion recognition system. Arunnehru et al. [6] proposed the motion projection profile (MPP) feature on motion information for automatic activity recognition from body movement. Glowinski et al. [7] describe a framework for behavior recognition from human upper body movements. The reduced amounts of visual information are used to analyze the affective behavior of body movements. Santhoshkumar et al. [8] proposed a different bin-level HOG for human emotion recognition system using human body movements. Arunnehru et al. [9] developed an action recognition system for real-time surveillance video using motion information from different images. Arunnehru et al. [10] proposed a gesture dynamic feature and recognized the human emotion supervised learning method. Piana et al. [11] proposed the sequences of 3D skeletons, the kinematic, geometrical and postural features which are extracted and given to the multi-class SVM classifier to categorize the human emotion. Karg et al. [12] summarize the survey on generation of such body movements and the state of the art on automatic recognition of emotion. The important characteristics such as the representation of affective state and the body movements are analyzed, and the use of information systems is discussed. Wang et al. [13] propose an advanced real-time system for human body movements to recognize emotions continuously. Fourat et al. [14] describe a system for recognition of emotion which depends on different actions, different expressions of emotions and low-level body cues from human body movement.

3 Proposed Methodology

3.1 Extraction of Histogram of Oriented Gradient Feature

Figure 1 demonstrates the proposed architecture for emotion recognition system. The HOG feature is defined as local object appearance and shape which can often be characterized rather well by the distribution of local intensity gradients of the corresponding gradient. The last line said that the description of the HOG method has been used in its higher form in scale invariant features transformation (SIFT), and it has been broadly demoralized in human detection [15]. The subsequent building of a 1D histogram whose concatenation supplies the feature vector from the HOG descriptor using gradient directions among the pixels in the cell. The image is to be analyzed as intensity function L. The image is further divided into cells of size 5×5 pixels with different histogram bins 9.

Compute the image gradient in both the x- and y-directions using Eqs. 1 and 2. Apply convolution to obtain the gradient image. Figure 2 shows the gradient image of x-direction and y-direction.

$$G_x = I^* D_x \tag{1}$$

$$G_y = I^* D_y \tag{2}$$

Equations 3 and 4 defined the gradient magnitude g and the gradient orientation θ used to compute for all the pixels in the block from the image gradients (Fig. 3).

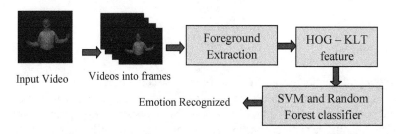

Fig. 1 Proposed architecture for emotion recognition system

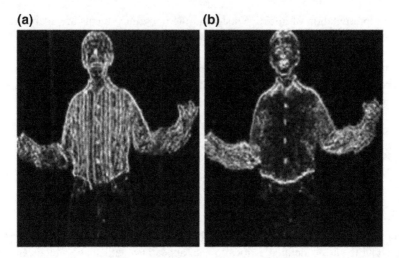

Fig. 2 **a** Gradient image of x-direction and **b** gradient image of y-direction

Fig. 3 Gradient magnitude
representation of image

$$|G| = \sqrt{G_x^2 + G_y^2} \qquad (3)$$

$$\theta = \tan^{-1}\left(\frac{G_x}{G_y}\right) \qquad (4)$$

Now, divide our image into cells and block. The cell is a rectangular region defined
by the number of pixels that belong in each cell. Image size is 200×150, and pixels
per cell are 5×5, $40 \times 30 = 1200$ cells. Now, construct histogram of oriented
gradient. Finally, each pixel contributes a weighted vote to the histogram; the weight
of the vote is simply the gradient magnitude $|G|$ at the given pixel. HOG feature is

extracted with different bin level 9, and the cell size is 5 × 5, and dimension is 225 for training and testing data. Figures 4 and 5 represent the histogram of angry and sad images, respectively.

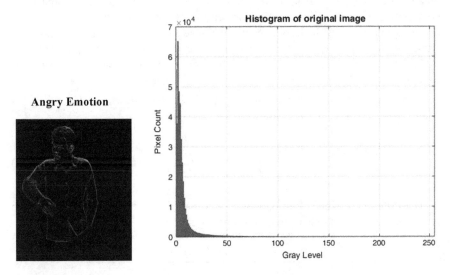

Fig. 4 Angry emotion and its histogram of gradient image

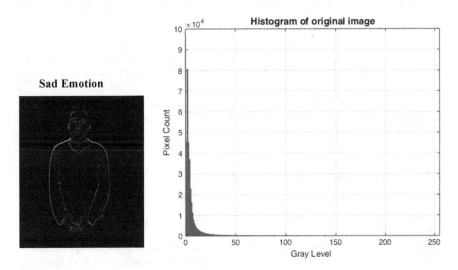

Fig. 5 Sad emotion and its histogram of gradient image

3.2 Kanade–Lucas–Tomasi (KLT) Feature

The KLT algorithm is described, respectively, in this section. The selected most important corner points from Harris corner detector are to be tracked using KLT. New computed corner points are connected to a newly detected corner, when old corner points are vanished. Wherever the image motion is high (most important corner points), the interest points between consecutive video frames are generated by KLT tracking algorithm [16, 17]. The Newton's method is used to minimize the sum of squared distances (SSD) within a tracking window.

To estimate the unknown d obtains the next linear system, where $H = \left[\frac{\partial I}{\partial x} \frac{\partial I}{\partial y}\right]$ is the image gradient vector at position x.

$$\left(\sum_w H^T H\right)(d) = \sum_W H^T \Delta I(Y, \Delta t) \tag{5}$$

The difference of the KLT equation is proposed by Tomasi, which uses both images symmetrically. This equation is derived in [18].

$$H = \left[\frac{\partial(I(*, t) + I(*, t + \Delta t))}{\partial x} \frac{\partial(I(*, t) + I(*, t + \Delta t))}{\partial y}\right] \tag{6}$$

The tracking points are selected by important corner points from Harris corner detector

$$c = \min\left(\mathrm{eig}\left(\sum_W \left[\frac{\partial I}{\partial x} \frac{\partial I}{\partial y}\right]^T \left[\frac{\partial I}{\partial x} \frac{\partial I}{\partial y}\right]\right)\right) \tag{7}$$

The most important key points are tracked using KLT. The extracted features are combined with the HOG feature of dimension 225. SVM and random forest classifier are used to classify the emotions.

3.3 Random Forest (RF) Classifier

A random forest classifier consists of number of trees with each grown by using some forms of randomization. The random forest uses random selection of features for splitting each node in each individual tree. The output can also be interpreted as a probabilistic measure. Random forest is the combination of tree predictions such that each tree depends on the value of a random vector sampled independently with the same distribution for all the trees in the forest. The random forest takes advantage of two powerful machine learning techniques: bagging and random feature selection. The theoretical and practical performance of ensemble classifiers is well documented in.

3.4   Support Vector Machine (SVM)

The support vector machine (SVM) is an important and efficient technique for classification in visual pattern recognition [15, 16]. The SVM is most extensively used in kernel learning algorithm. The elegant theory is used to separate two classes by large-margin hyperplanes. It cannot be extended easily to separate N mutually exclusive classes. The most popular "one-vs-others" approach is used for the multi-class problem where one class is separated from N classes. The classification task typically involves training and testing data.

The training data is separated by $(s_1, t_1), (s_2, t_2), \ldots, (s_n, t_n)$ into two classes, where $b_j \in \{+1, -1\}$ are the class labels and $s_j \in t_N$ contains n-dimensional feature vector. The goal of support vector machine is to develop a model which predicts target value from testing set. $w.s + b = 0$ is the hyperplane of binary classification, where $w \in R^N$. The two classes are separated by $b \in R$ [19]. $M = 2/\|w\|$ is the large margin as show in Fig. 6. The Lagrange multipliers α_i ($i = 1, \ldots, m$) are used to solve the minimization problem, where v and y are optimal values obtained from Eq. 8.

$$h(s) = \text{sgn}\left(\sum_{j=1}^{n} x_j b_j L(s_j, s) + y\right) \tag{8}$$

Fig. 6 Hyperplane of linear SVM

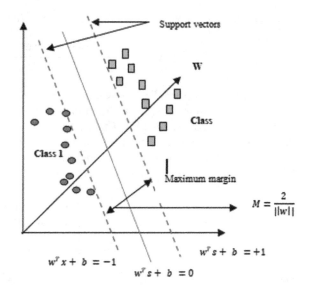

4 Results and Analysis

The angry, joy, fear, sad and pride emotions from GEMEP corpus dataset are used for this experiment. The open-source machine learning tool Weka is used for recognizing the emotion with random forest classifier. The experiments are conducted on Windows 10 operating system using MATLAB 2015b.

4.1 Performance Evaluation Metrics

The predicted versus actual classification can be charted in a table called confusion matrix.

Accuracy: overall how often is our model correct?

$$\text{Accuracy} = \frac{\text{truepositives} + \text{truenegatives}}{\text{totalexamples}} \tag{9}$$

Precision: when the model predicts positive, how often is it correct?

$$\text{Precision} = \frac{\text{trueppositives}}{\text{truepositives} + \text{falsepositives}} \tag{10}$$

Precision helps when the cost of false positives is high.

$$\text{Recall} = \frac{\text{trueppositives}}{\text{truepositives} + \text{falsenegatives}} \tag{11}$$

$F1$ score: $F1$ is an overall measure of a model's accuracy that combines precision and recall; in that weird way, that addition and multiplication just mix two ingredients to make a separate dish altogether. That is, a good $F1$ score means that you have low false positives and low false negatives, so you are correctly identifying real threats and you are not disturbed by false alarms. An $F1$ score is considered perfect when it is 1, while the model is a total failure when it is 0 (Fig. 7).

$$F1 - \text{score} = 2.\frac{\text{Precision. Recall}}{\text{Precision} + \text{Recall}} \tag{12}$$

Fig. 7 Confusion matrix

Actual Class	Predicted Class	
	True Positive (TP)	False Negative (FN)
	False Positive (FP)	True Negative (TN)

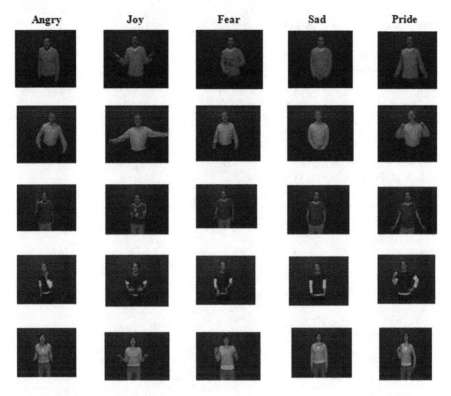

Fig. 8 Example frame of five emotions

4.2 GEMEP Dataset

The ten actors acted 18 affective states of emotional expression. Angry, fear, pride, joy and sad are the emotions used in this work. Each emotion in the video was acted by five male and five female. The height and width of the video are 720 × 576 and have 25 frames per second [20, 21] (Fig. 8).

4.3 Experimental Results on Random Forest Classifiers

The confusion matrix for the random forest and SVM results is shown in Tables 1 and 2, respectively.

The precision, recall and *F*-measure are the performance evaluation metrics. Random forest takes more time to learn and showed the good results of 95.9% accuracy for straight view GEMEP classifiers. The confusion matrix of random forest classifier and support vector machine is shown in 1 and 3. The performance measures of random forest classifier and support vector machine are shown in Tables 3 and 4, respectively (Figs. 9 and 10).

Table 1 Confusion matrix for random forest (RF)

	Angry	Joy	Fear	Sad	Pride
Angry	360	11	17	10	9
Joy	13	421	11	8	15
Fear	2	7	287	7	3
Sad	0	0	5	409	2
Pride	8	0	9	0	310

Table 2 Confusion matrix for SVM classifier

	Angry	Joy	Fear	Sad	Pride
Angry	329	32	22	4	22
Joy	26	385	16	11	30
Fear	8	16	254	13	15
Sad	7	8	15	379	7
Pride	12	14	20	7	274

Table 3 Performance measure of random forest (RF)

	Precession	Recall	F-Measure
Angry	93.1	82.2	87.3
Joy	87.8	89.1	88.4
Fear	83.8	91.5	87.5
Sad	91.5	97.8	96.4
Pride	89.9	90.2	90.1

Table 4 Performance measure of SVM

	Precession	Recall	F-Measure
Angry	86.1	81.2	81.4
Joy	85.7	79.4	83.5
Fear	78.9	81.4	80.2
Sad	91.5	82.6	91.5
Pride	79.4	90.3	79.3

5 Conclusion

This paper presented a novel method for emotion recognition using HOG-KLT feature. This experiment was conducted using GEMEP corpus dataset considering different emotions are angry, joy, fear, sad and pride. The performance of the proposed

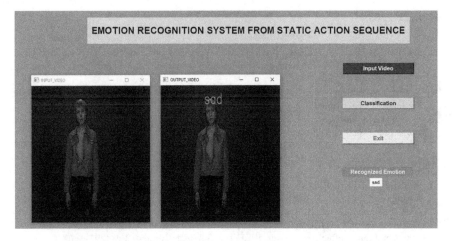

Fig. 9 Screenshot for sad emotion recognition

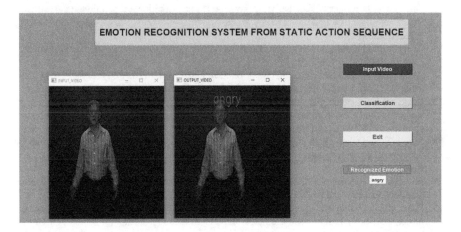

Fig. 10 Screenshot for angry emotion recognition

features was modeled using SVM with RBF and random forest. The proposed HOG-KLT feature performs better recognition of emotion with 95.9% in random forest classifier. It is observed from the experiments that the system could recognize sad, pride and fear emotions with high accuracy, and the system could not distinguish joy and angry with high accuracy and is of further interest. The future work will concentrate on emotion recognition applications for autism people (special child).

References

1. Akram, M., Zafar, I., Siddique Khan, W., Mushtaq, Z.: Facial expression recognition based on fuzzy logic. In: International Conference on Computer Vision Theory and Applications, VISAPP (2008)
2. Ashwini, A.V., Cherian, J.P., Kizhakkethottam, J.J.: Overview on emotion recognition system. In: International Conference on Soft-Computing and Network Security (2015)
3. Bänziger, T., Mortillaro, M., Scherer, K.R.: Introducing the geneva multimodal expression corpus for experimental research on emotion perception. Emotion **12**(5), 1161–1179 (2012)
4. Bänziger, T., Scherer, K.R.: Introducing the geneva multimodal emotion portrayal (GEMEP) corpus. In: Blueprint for affective computing: A sourcebook Oxford University Press, Oxford, pp. 271–294 (2010)
5. Birchfield, S.: Derivation of kanade-lucas-tomasi tracking equation. Unpubl. Notes (1997)
6. Chang, C.-C., Lin, C.-J.: LIBSVM: a library for support vector machines. ACM Trans. Intell. Syst. Technol. **2**, 1–27 (2011)
7. Glowinski, D., Dael, N., Camurri, A ., Volpe, G., Mortillaro, M., Scherer, K.: Toward a minimal representation of affective gestures. IEEE Trans. Affect. Comput. **2**(2) (2011)
8. Arunnehru, J., Kalaiselvi Geetha, M.: Automatic activity recognition for video surveillance. Int. J. Comput. Appl. **75**(9), 1–6 (2013)
9. Arunnehru, J., Kalaiselvi Geetha, M.: Automatic Human Emotion Recognition in Surveillance Video. Intelligent Techniques in Signal Processing for Multimedia Security, Springer-Verlag., 321–342 (2017)
10. Arunnehru, J., Kalaiselvi Geetha, M.: Behavior recognition in surveillance video using temporal features. In: 4th ICCCNT (2013)
11. Arunnehru, J., Kalaiselvi Geetha, M.: Motion intensity code for action recognition in video using PCA and SVM. Min. Intell. Knowl. Explor. **8284**, 70–81 (2013)
12. Lukas, B.D., Kanade, T.: An iterative image registration technique with an application to stereo vision. In: Proceedings of the International Joint Conference on Artificial Intelligence, pp. 674–679 (1981)
13. Karg, M., Samadani, A.-A., Gorbet, R., Kühnlenz, K., Hoey, J., Kuli, D.: Body movements for affective expression: a survey of automatic recognition and generation. IEEE Trans. Affect. Comput. **4**(4) (2013)
14. Miyamoto, Y., Uchida, Y.: Culture and mixed emotions: co-occurrence of positive and negative emotions in Japan and the United States. Am. Psychol. Assoc. (2010)
15. Piana, S., Stagliano, A., Odone, F., Verri, A., Camurri, A.: Real-time automatic emotion recognition from body gestures. In: Human-Computer Interaction. Computer Vision and Pattern Recognition (2014)
16. Tomasi, C., Kanade, T.: Detection and Tracking of Point Features. Technical Report CMU-CS-91132, Carnegie Mellon University (1991)
17. Wang, W., Enescu, V., Sahli, H.: Adaptive real-time emotion recognition from body movements. ACM Trans. Interact. Intell. Syst. **5**(4) (2015)
18. Fourat, N., Pelachaud, C.: Multi-level classification of emotional body expression. IEEE (2015)
19. Dalal, N., He, X.: Histograms of oriented gradients for human detection. In: International Conference on Computer Vision and Pattern Recognition, vol. 1, pp. 225–232. IEEE computer society press, San Diego, CA, 20–25 June (2005)
20. Prinzie, A., Van den Poel, D.: Random forests for multiclass classification: random multinomial logit. Expert. Syst. Appl. **34**(3), 1721–1732
21. Santhoshkumar, R., Kalaiselvi Geetha, M., Arunnehru, J.: SVM-KNN based emotion recognition of human in video using HOG feature and KLT tracking algorithm. Int. J. Pure Appl. Math. **117**(15) (2017)

An Analysis of Lung Tumor Classification Using SVM and ANN with GLCM Features

Vaibhavi Patel, Samkit Shah, Harshal Trivedi and Urja Naik

Abstract Lung cancer is dangerous when it comes to life of humans and it happened by growth of abnormal cells in lungs. Due to metastasis in closed tissues, it can be spread in other parts of the body. For detection and segmentation of tumors, various image processing methods are in use which can identify the tumor at different stages. Our proposed solution suggests to use k-means and EK-means clustering methods on various images of tumors where features are extracted by geometrical features and training is done by advance machine learning algorithms like artificial neural networks (ANN) and support vector machine (SVM) where it classifies the tumor into benign or malignant type and provides tumorous part as a result of segmentation. We have done feature extraction by further segmenting GLCM features into Haralick features and this is our achievement.

Keywords Computer tomography · Lung tumor · k-means and EK-means clustering · Support vector machine (SVM) · GLCM features · Artificial neural network (ANN)

V. Patel
Sardar Vallabhbhai Institute of Technology, Vasad, India
e-mail: 16mececv011@gmail.com

S. Shah
Government Engineering College, Gandhinagar, India
e-mail: Shahsamkit408@gmail.com

H. Trivedi
SoftVan, Ahmedabad, India
e-mail: Harshal@softvan.in

U. Naik (✉)
LD Engineering College, Ahmedabad, India
e-mail: Urjanaik98@gmail.com

© Springer Nature Singapore Pte Ltd. 2020
P. K. Singh et al. (eds.), *Proceedings of First International Conference on Computing, Communications, and Cyber-Security (IC4S 2019)*, Lecture Notes in Networks and Systems 121, https://doi.org/10.1007/978-981-15-3369-3_21

1 Introduction

Lung tumor is furthermore seen as sarcoma of the lung. Lung cancer in particular, is one of the major causes for cancer-related deaths worldwide [1]. Cancer patient has the smallest survival rate after the finding the tumor. The number of deaths increases every year. Survival from lung cancer is straightforwardly depending upon identification of tumor and at its discovery time. Sooner the cancer gets detected, the better the survival rate of the cancer patient [2]. Consequently, from this review, it is apparent that the lung malignancy is a genuine explanation behind death, what's more; its rate is expanding every year [3].

Three major reasons for lung cancer are tobacco consumption (85–90%), passive smoking (15–25%), and heredity and pollution (5–7%). As shown in Fig. 1, death rate at the age of 80 and above is the highest compare to other age groups.

Mid-section radio outlines can be taken to detect lung cancer and enlisted tomography (CT) analyzes [1]. CT lung image is mainly used for the detecting the lung tumor nodules. After taking CT image, the diagnosis process consists following stage shown in Fig. 2.

Image preprocessing—In these processes, it improvises the picture resolution via different methods. Pictures are cropped, normalized, and resized according to machine learning model requirement.

Fig. 1 Death rate per year due to lung tumor [1]

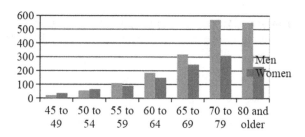

Fig. 2 Lung tumor detection and classification system [1]

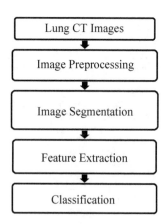

Image Segmentation—It concludes suspicious region from images via different clustering segmentation algorithms.

Feature Extraction—After segmentation, positive region will be extracted using various features like the perimeter, structure, size, area, tone, etc.

Classification—After identifying the cancerous nodule, classifies the tumor types [1, 4, 5].

This paper is structured as follows. We have worked on tumors type in second part of paper, while tumor classification methods are shown in third section. Section 4 discusses about approaches to detect lung tumor. In last section, we have analyzed performance using various metrics. Finally, conclusions and future enhancements are drawn. For the processing of vast amount of patients cancer data, cloud infrastructure plays an important role as a high performance compute engine [6, 7]. Nowadays, machine learning plays an important role for the lung cancer detection. Machine learning techniques like linear regression [8], support vector machine, artificial neural network, etc., are highly used in the medical domain.

Incorrect and/or late diagnosis are the two leading causes of cancer-related deaths. Lung cancer in particular, is one of the major causes for cancer-related deaths worldwide. So, this motivates us to make a system that is economical, swifter, and accurate that segments the lung cancer CT image and helps in better detection and classification of lung tumor. We used SVM classification technique which gave us more accuracy. Also, for feature extraction, Haralick features are used.

2 Tumor Types

Tumors will not always turn into cancer and it can affect both lungs due to spread of the infections. Unrestricted growth of abnormal cells is primary reason for tumor spread and it should be treated medically. Generally, cells that lie in the air passages are dangerous and due to metastasis and diverse body organisms, it will spread very fastly in body parts. Sarcoma that gets from epithelial cells is basic lung malignancies [1]. A lung nodule is round lesion that is mainly two types, cancerous or non-cancerous. It is illustrated in Fig. 3.

A tumor does not mean cancer. Malignant tumor can be injurious to human, whereas benign one is not cancerous. How the malignant tumor looks is shown in Fig. 4.

Fig. 3 Tumors types [5]

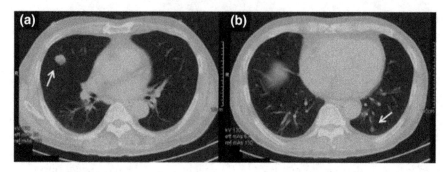

Fig. 4 Benign tumor [10]

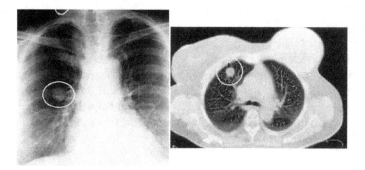

Fig. 5 Malignant tumor [10]

Harmful tumors can develop itself rapidly and try to search for the new region to spread. The unusual cells that frame a threatening tumor increase at a quicker rate. It is described in Fig. 5.

Lung harm is expanded, confined into two crucial sorts: Non-little cell lung disease and small cell lung malignancy, which is additionally subdivided into: adenocarcinoma, carcinoma, and squamous cell carcinomas [9].

3 Methodology

3.1 Segmentation Method

K-means clustering: *K*-means clustering is a technique to create image clusters by segregation. It focuses on segregating and grouping the pixels into cluster which have a nearest mean intensity value. The steps of the algorithm are shown below.

3.2   Feature Extraction

In the image highlight extraction, AI calculations gains from numerical picture information. There are three sorts of viewable signs individuals normally search for in an image: unearthly (normal tonal variety in different groups of noticeable wavelengths), con-printed (large-scale information reviewed from encompassing information), and textural. Textural information, or the spatial dispersion of tonal variety inside a band, is one of the most significant attributes utilized in distinguishing articles or districts of enthusiasm for a picture. Dark level co-event network (GLCM) is the framework-based highlights. It deals with lattice of recurrence which incorporates two pixels, isolated by specific vectors, happen in the pictures.

Co-occurrence matrix is defined as,

$$p_{ij}(\Delta x, \Delta y) = W Q(i, j|\Delta x, \Delta y) \tag{1}$$

where,

$$W = \frac{1}{(M - \Delta x)(N - \Delta y)} \quad Q\ (i, j|\Delta x, \Delta y) = \sum_{n=1}^{N-\Delta y} \sum_{m=1}^{M-\Delta x} A \tag{2}$$

where,

$$A = \begin{cases} 1 \text{ if } f(m, n) = i \text{ and } f(m + \Delta x, n + \Delta y) = j \\ 0 \qquad\qquad\qquad \text{otherwise} \end{cases}$$

It has size $N \times N$ (N = Number of gray-values), i.e., the rows and columns represent the set of possible pixel values. It is computed based on two parameters: relative distance between the pixel pair (d) and relative orientation/rotational angle (θ).

3.3   Classification of Extracted Tumor

(i)   Artificial Neural Network

Over AI in mental science, artificial neural networks (ANN) are a group about models pushed in the long run neural frameworks are utilized to overview or assess limits that may rely on an expansive entirety for data sources. Re-delivered neural frameworks need support everything considered showed as structures for interconnected "neurons" that trades information to one another. The affiliations require loads parameters that should be tuned in context of procedures, making neural systems versatile to loads other than skilled for taking in.

Fig. 6 Multi-layer ANN [5]

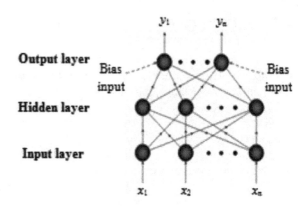

This multilayer perceptron neural system is an empowered forward neural system with many concealed layers. Cybenko Also Funahashi have turned out that the MLP organize with one stowed away layer need the proficiency should estimate whatever nonstop work up to specific exactness. Those request execution of the MLP compose will especially depend on upon those structures of the framework and arrangement calculations. Structural engineering about MLP organize is well illustrated in Fig. 6.

(ii) **Support Vector Machine (SVM)**

SVM recognizes it's separators in spatial space with various classes. In Fig. 3, Hyper planes and here would 2 classes x, o A, b. Hyperplanes fill in as separators between these classes, and partition the parts as indicated by different classes.

SVM is directed AI calculation which prepares on marked information. In our proposed strategy, SVM is prepared for 40 pictures of benign tumor and 40 of malignant tumor. The SVM classifier is sustained with highlights esteems and the help vectors are created from the preparation procedure. The given information is prepared and tried for 200 example pictures and the classifier is encouraged with two classes, for example, dangerous or benign. This is accurately described in the above Fig. 7.

4 Proposed Work

The proposed framework comprises four stages, to be specific, preprocess, segmentation, feature extraction, and classification. To start with, gathers the CT pictures and performs preprocessing activity for removing noise and improving the picture quality. After preprocessing, image classification is performed. Among the different techniques, k-means and EK-means clustering are utilized. EK-means has a few points of interest over other division strategies are given as the exact outcome. Following stage is the component determination, which gives the arrangement exactness. SVM calculation has a few points of interest over other order procedures. Figure 8 shows the flow diagram as mentioned.

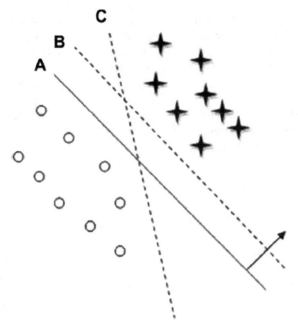

Fig. 7 Classification SVM [5]

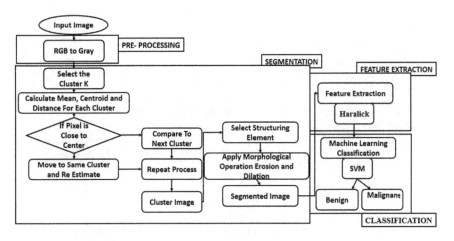

Fig. 8 Proposed workflow

4.1 Segmentation

The CT images are converted from RGB to gray and then are fed to k-means clustering algorithm. As the part of data preprocessing, gray scale images are fed and according to k-means, clusters are formed. Its prediction time is quite faster and is also a simple clustering method but contains false segmentation. Once done, the clustered images are fed to EK-means clustering techniques. It applies morphological operation on the images to extract the exact location of tumor. K-means and EK-means combinly perform as: Then, the cluster center k decides randomly and cluster distance is found for center and pixel values where shortest distance is chosen among all and transferred to cluster. Center values are recalculated again and same process is repeated again until the whole algorithm is converged [1]. This process is continued until all the centers are covered. Algorithmic step is given below:

Calculate the value of cluster mean m

$$M = \frac{\sum_{i=c(i)=k} x_i}{N_k}, \quad k = 1, 2, \ldots, K \tag{3}$$

Calculate the distance between cluster center and between each pixels

$$D(i) = \arg\min \|x_i - M_k\|^2 \quad i = 1, 2, \ldots, N \tag{4}$$

Repeat I, II steps until all mean value is coverage.
Algorithm performs following steps;

1. Set the number of clusters k
2. Choose random center of k number of clusters
3. Find mean and center of specific cluster
4. Calculate the square distance between pixel values to the center
5. Set point to nearest cluster.
6. Otherwise check for next cluster.
7. Recalculate the center values.
8. Repeat the process until the center does not move.
9. After k-means clustering, mark suspicious part of the cluster.
10. Apply morphological opening by reconstruction operation on desired cluster (cyst) as:

Pick a component, structural element and apply disintegration operation to expel foundation pixels in MATLAB using as:

$$Ie = imerode\ (cyst, SE) \tag{5}$$

Recreate the marker image, Ie, in light of cover cyst by imreconstruct code in MATLAB as:

$$RIe = imreconstruct(Ie, cyst); \tag{6}$$

Apply expansion operation on RIe for including pixels of item limits as

$$DRIe = imdilate(RIe, SE); \tag{7}$$

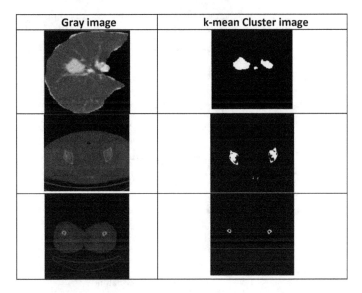

Fig. 9 Cluster images [2]

Apply reconstruction operation on complement of DRIe in light of supplement of RIe. As in step (b), by using regional maxima, the cyst area will be highlighted.

Figure 9 describes the gray scale image and the image is obtained by performing k-means clustering algorithm.

4.2 Feature Extraction

Computation of gray level co-occurrence matrix (GLCM)gives directional P_0, P_{45}, P_{90}, and P_{135} factors as shown in Fig. 10 and formulas as described in Table 1.

4.3 Tumor Classification

After calculating, GLCM feature output fed into the classifier. On the basis of image, the trained and test data are classified by ANN and SVM. The given figure shows classification of the segmented image after the GLCM feature extraction; benign means, it is non-cancerous tumor and malignant means cancerous tumor. The classification is done by the training and testing by ANN classifier.

The results in Fig. 11 show that the accuracy of the ANN classifier after the computation of the GLCM features is 78.50%.

The results illustrated in Fig. 12 show that the accuracy of the SVM classifier after the computation of the GLCM features is 87.50%.

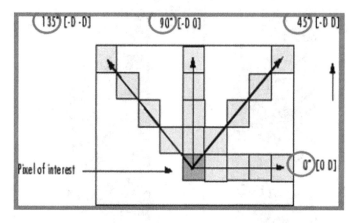

Fig. 10 Graph [1]

Table 1 Equations [2]

No	Features	Equation
1	Energy	$f_1 = \sum i \sum j \{p(i,j)\}^2$
2	Contrast	$f_2 = \sum_{n=0}^{Ng-1} n^2 \{\sum i \sum j \ p(i,j)\}$
3	Correlation	$f_3 = \frac{\sum i \sum j (i,j) p(i,j) - \mu x \mu y}{\sigma x \sigma y}$
4	Entropy	$f_9 = \sum i \sum j \ p(i,j) \ \log(p(i,j))$

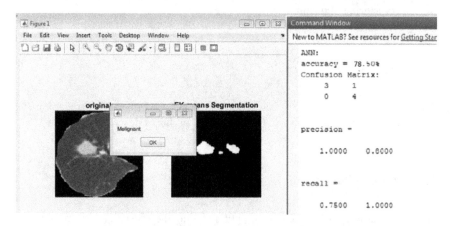

Fig. 11 Result of ANN classification

After taking results of support vector machine and back propagation technique, comparison between the methods is done using performance parameters which is shown in Table 2. Here, we check the measures based upon confusion matrix (An evaluation table that is often used to evaluate the performance of a classification model on a set of test data for which the true values are known).

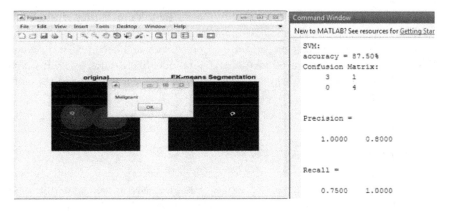

Fig. 12 Result of SVM classification

Table 2 Analysis

Classifier	Confusion matrix		Precision	Recall	Accuracy (%)
ANN	2	2	1.0	0.80	78.50
	0	4	0.75	1.00	
SVM	3	1	1.00	0.80	87.50
	0	4	0.7500	1.0000	

Table 3 Measures of lung tumor

Images	Accuracy (%)	Sensitivity (%)	Specificity (%)	Time (s)
T1	92.53	61.19	86.47	0.76
T2	90.23	62.17	88.58	0.58
T3	98.06	67.61	89.84	0.78
T4	91.23	71.91	84.25	0.83
T5	95.90	83.26	89.50	0.69

5 Conclusions and Future Work

In the field of medical sciences, segmentation is one of important features to find out malicious part which can provide ease to doctors to automatic diagnosis. It was found that several methods in our paper showed prominent results. We propose k-means and EK-means cluster-based segmentation method to extract the texture features (i.e., contrast, correlation, energy, homogeneity, etc.) from segmented image and the relevant features are used in SVM classifier to classify tumor types as whether benign and malignant and give a precision of about 87.50%. Future work can be extended by carrying with different feature extraction and advanced segmentation architectures which will detect segments more precisely and accurately. Various states of art models can also help to enhance the classification part also.

Next focus should be on using advanced segmentation architecture like U net and classification through advanced pretrained neural nets like Resnet, Mask R-CNN, YOLOv2, and so on. These ways, we can enhance our classification and get more accuracy in result.

References

1. Sangamithraa, P.B., Govindaraju, S.: Lung tumour detection and classification using EK-mean clustering. In: IEEE WiSP-NET (2016), pp. 2201–2204 (2016)
2. Nithila, E.E., Kumar, S.S.: Segmentation of lung nodule in CT data using active contour model and fuzzy C-mean clustering, pp. 2583–2586 (2016)
3. Wang, X-P., Zhang, W., Cui, Y.: Tumor segmentation in lung CT images based on support vector machine and improved level set. In: Tianjin University of Technology, Vol. 11, pp. 0396–9398 Springer, Berlin (2015)
4. Al-Tarawneh Mokhled, S.: Lung cancer detection using image processing techniques. Leonardo Electron. J. Pract. Technol. 234–236 (2012)
5. Dhaware, C.G., Wanjale, K.H.: Survey of image classification method in image processing. Int. J. Comput. Sci. Trends Technol. (IJCST) 114–115 (2016)
6. Bhatia, J., Govani, R., Bhavsar, M.: Software defined networking: from theory to practice. In: 2018 Fifth International Conference on Parallel, Distributed and Grid Computing (PDGC), pp. 789–794. Solan Himachal Pradesh, India (2018)
7. Bhatia, J., Patel, T., Trivedi, H., Majmudar, V.: HTV dynamic load balancing algorithm for virtual machine instances in cloud. In: 2012 International Symposium on Cloud and Services Computing, pp. 15–20. Mangalore (2012)
8. Jaykrushna, A., Patel, P., Trivedi H., Bhatia, J.: Linear regression assisted prediction based load balancer for cloud computing. In: 2018 IEEE Punecon, pp. 1–3. Pune, India (2018)
9. Solanki, E., Agrawal, A., Parmar, C.K.: A survey of current image segmentation techniques for detection of lung cancer. IJSRD Int. J. Sci. Res. Dev. 1–4 (2014)
10. Cabrera, J., Dionisio, A., Solano, G.: Lung cancer classification tool using microarray data and support vector machines, pp. 2–5 (2016)

Lane Detection Models in Autonomous Car

Kabindra Singh Bisht and Kamal Kumar

Abstract It has turned into an extraordinary subject to be explored for the better eventual fate of the vehicles. Path recognition is the issue of assessing the geometric structure of the path limits of a street on the pictures caught by a camera. To be a wise vehicle, path limit is vital data. There are numerous models and vehicles accessible till now, and yet at the same time, it is the point of research. How can get the completely self-sufficient vehicle utilizing any of the path-location procedures created yet. The edge at that point will be utilized to recognize conceivable path limit from the street. Use shading-based division to discover the path limit and utilize a quadratic capacity to approach it. This framework requests low computational power and memory prerequisites, and is vigorous within the sight of commotion, shadows, asphalt, and obstructions such like vehicles, bikes, and walkers conditions. The outcome pictures can be utilized as pre-prepared pictures for path following, street following, or impediment identification.

Keywords Lane detection · Road model · Behavior reflex methodologies · Direct perception approach · Mediated perception approaches

1 Introduction

In this paper, a road modular and color-based path discovery independent vehicle. It is imperative data for clever vehicle framework. In this framework, the undertakings of consequently determined vehicle incorporate street following and keeping inside the right path, keeping up a sheltered separation between vehicles, directing the speed of a vehicle as per traffic conditions and street attributes, moving crosswise over paths so as to overwhelm vehicles and maintain a strategic distance from deterrents, hunting down the right and most limited course to a goal. Smart vehicle frameworks offer a basic, protected, and productive activity while utilizing these savvy vehicles.

K. S. Bisht (✉) · K. Kumar
National Institute of Technology, Uttarakhand, Srinagar Garhwal, Uttarakhand, India
e-mail: kabindrabisht525@gmail.com

K. Kumar
e-mail: kamalkumar@nituk.ac.in

© Springer Nature Singapore Pte Ltd. 2020 285
P. K. Singh et al. (eds.), *Proceedings of First International Conference on Computing, Communications, and Cyber-Security (IC4S 2019)*, Lecture Notes in Networks and Systems 121, https://doi.org/10.1007/978-981-15-3369-3_22

Way acknowledgment is a well-examined zone of PC vision with applications in free vehicles and driver candidly strong system. This is to some degree in light of the way that, despite the obvious ease of finding white markings on a diminish road, it will, in general, be difficult to choose way markings on various sorts of road. These inconveniences rise up out of shadows, obstruction by various vehicles, changes in the road surfaces itself, and shifting sorts of way markings. Path recognition is a well-examined zone of PC vision with applications in self-sufficient vehicles and driver emotionally supportive network. This is to some extent in light of the fact that, in spite of the apparent straightforwardness of discovering white markings on a dim street, it tends to be hard to decide path markings on different sorts of street. These challenges emerge from shadows, impediment by different vehicles, changes in the street surfaces itself, and contrasting sorts of path markings. A path-location framework must almost certainly choose all way of markings from jumbled roadways. Since mistaken discoveries will create wrong guiding directions which may risk vehicle wellbeing, a vigorous and dependable calculation is a base prerequisite. Nonetheless, the incredible assortment of street conditions requires the utilization of complex vision calculations that requires costly equipment to execute as well as depends on numerous customizable parameters that are regularly decided as a matter of fact. Various pros have demonstrated way identifiers reliant on a wide grouping of methodology. The strategy regularly be used is to perceive the edges by various kind of channel. Numerous analysts have indicated path locators dependent on a wide assortment of numerous procedures. The system regularly be utilized is to distinguish the edges by different sort of channel and after that utilization, the Hough change to tit lines to these edges [1, 2]. A B Snake based path identification technique and following calculation are presented. The issues of identifying the two sides of path limits have been converged as the issue of distinguishing the midline of the path by utilizing the information of the point of view parallel lines. Some path discovery techniques depend on top view pictures. Some do street limit identification and following by ladar sensor or laser sensor [3, 4].

1.1 Property of Road

Path recognition shows every one of the markings which are drawn onto the path and the path limit is one sort of path checking [2]. Since the path stamping is one the street, the path checking straightforwardly from the street as opposed to looking through the entire street. Here, will give a short depiction about the street. The vast majority of the street is close to the shade of dull dim. Despite the fact that it is not actually dark, it is still in certain range. For RGB shading model, a pixel is said to be dark with the property:

$$Vr = Vg = Vb \qquad (1)$$

For genuine street, it is not actually the shade of dim when it is fixed, besmirched, cleaned, or with some other shading part because of the impression of light as described in Eq. 1. Be that as it may, the greater part of them is still in a range amid these tests.

For general road, the gray-level value of road is 100–130, and the difference between RGB is less than 25. Special case such like radiant day or blustery day, the parameter will change gently.

Radiant day, the dark estimation of street is 80–130 with shadow and 130–170 with daylight. The estimation of R will be 20–40 higher with daylight and the estimation of B will be likewise 20–40 higher with shadow.

Blustery day, the dark estimation of the street is 70–120 and the distinction between RGB is still under 25.

There for having property [2]:

$$|Vt - Vg| < th \qquad (2)$$

$$|Vg - Vb| < th \qquad (3)$$

$$|Vb - Vr| < th \qquad (4)$$

Value of th is found out during experiments as described in Eqs. 2, 3, and 4. The main reason that the road is not exactly gray in the sun light [2]. For example, take the color of road will be obviously different in the morning, afternoon, and evening. Backlight or headlight will also affect the results as described in Eqs. 2, 3, and 4.

Path stamping then again ought to be a difference to the street in shading. Despite the fact that can not ensure if the shade of the path stamping in each nation is all the equivalent, it ought to be a more splendid shading as opposed to the dull dim shade of street like white and orange red in Taiwan. Be that as it may, the street will get redder because of the reflection of the daylight. As per these analyses, the street is redder and the estimation of th is as yet appropriate on the off chance that it is set to be 40. With this esteem, the street will be recognized from orange red path checking even it is redder than expected because of the daylight. In spite of the fact that this will totally over recognize the street, it does not make any difference for misclassification except if. Also use lane marking class as shown in Fig. 1a. After processing: the color near gray is set to be black as shown in Fig. 1b [1].

2 Related Work

Most self-sufficient driving frameworks from industry today depend on intervened observation approaches. In PC vision, specialists have considered each assignment independently. Vehicle recognition and path identification are two key components of a self-governing driving framework. Run of the mill calculations yields jumping

Fig. 1 a Original image. **b** After processing: the color near gray is set to be black [1, 3]

boxes on distinguished autos and splines on recognized path markings. Be that as it may, these bouncing boxes what's more, splines are not the immediate affordance data used for driving. In this manner, a transformation is important which may result in additional commotion. Ordinary path discovery calculations.

For example, the one proposed in experience the ill effects of false identifications. Structures with unbending limits, for example, thruway guardrails or black top surface breaks, can be mis perceived as path markings. Indeed, even with great path identification results, basic data for vehicle restriction might miss [5, 6]. For example, given that just two path markings are generally identified dependably, it tends to be hard to decide whether a vehicle is driving on the left path or the correct path of a two path street [1].

To coordinate distinctive sources into a reliable world portrayal [7] proposed a probabilistic generative model that takes different location results as data sources and yields the design of the crossing point and traffic subtleties [8].

For conduct reflex methodologies, are the fundamental works that utilize a neural system to outline straightforwardly to directing points. All the more as of late, train an expansive intermittent neural system utilizing fortification learning approach. The system's capacity is equivalent to mapping the picture legitimately to the directing points, with the target to keep the vehicle on track. Thus, they utilize the video diversion TORCS for preparing and testing [9].

As far as profound learning for self-ruling driving, is a fruitful case of ConvNets-based conduct reflex approach as shown in Fig. 3. The creators propose a rough terrain driving robot DAVE that takes in a mapping from pictures to a human driver's guiding edges [2]. In the wake of preparing, the robot illustrates ability for deterrent evasion, proposes a rough terrain driving robot with self-administered learning capacity for long go vision. In their framework, a multi-layer convolutional organize is utilized to arrange a picture section as a navigable zone or not. For profundity map estimation, deep flow utilizes ConvNets to accomplish awesome outcomes for driving scene pictures on the KITTI dataset as shown in Fig. 3 [10, 11]. For picture highlights, profound adapting additionally shows huge improvement over hand-created highlights, for example, Essence. In this trials, will make an examination between educated ConvNet highlights and GIST for direct recognition in driving situations.

Mediated perception approaches [8] include different subsegments for perceiving driving pertinent articles, for example, paths, traffic signs, traffic lights, autos, people on foot, and so on [12]. The acknowledgment results are then joined into a reliable world portrayal of the vehicle's quick environment. To control the vehicle, an AI-based motor will consider the majority of this data previously settling on every choice. Since just a little part of the recognized articles are undoubtedly pertinent to driving decisions, this dimension of absolute scene comprehension may include superfluous intricacy to an effectively troublesome assignment. In contrast to other mechanical undertakings, driving a vehicle just requires controlling the course and the speed. This last yield space lives in an extremely low measurement, while interceded discernment figures a high-dimensional world portrayal, potentially including repetitive data. Rather than identifying a bouncing box of a vehicle and after that utilizing the bouncing box to appraise the separation to the vehicle, why not just anticipate the separation to a vehicle legitimately All things considered the individual sub undertakings engaged with intervened discernment are themselves considered open research inquiries in PC vision. Albeit intervened recognition includes the present best in class approaches for self-ruling driving, the vast majority of these frameworks need to depend on laser run discoverers, GPS, radar and exact maps of the earth to dependably parse protests in a scene. Expecting answers for some open difficulties for general scene understanding so as to settle the more straightforward vehicle controlling issue pointlessly expands the unpredictability furthermore, the expense of a framework.

Behavior reflex methodologies [8] build an immediate mapping from the tangible contribution to a driving activity. This thought dates back to the late 1980s when utilized a neural system to build an immediate mapping from a picture to controlling points. To get familiar with the model, a human drives the vehicle along the street while the framework records the pictures and directing points as the preparation information. In spite of the fact that this thought is exquisite, it can battle to manage traffic and entangled driving moves for a few reasons. Right off the bat, with different autos on the street, notwithstanding when the info pictures are comparable, unique human drivers may settle on totally unique choices which results in a not well-presented issue that is befuddling when preparing a regressor. For instance, with a vehicle straightforwardly ahead, one may pursue the vehicle, to pass the vehicle from the left, or to pass the vehicle from the right. At the point when every one of these situations exists in the preparation information, an AI model will experience issues choosing what to do given nearly the equivalent pictures. Also, the basic leadership for conduct reflex is too low level. The immediate mapping can not see a greater image of the circumstance. For instance, from the model's viewpoint, passing a vehicle and changing back to a path are only a succession of exceptionally low-dimension choices for turning the guiding wheel marginally one way and afterward the other way for some timeframe. This dimension of deliberation comes up short to catch what is truly going on, and it builds the trouble of the assignment superfluously. At long last, in light of the fact that the contribution to the model is the entire picture, the learning calculation must out which parts of the picture are important as shown in Fig. 3. Be that as it may, the dimension of supervision to prepare a conduct reflex model, for example, the directing point, might be too powerless to even think about forcing the calculation to become familiar with this basic data.

Portrayal that straightforwardly predicts the affordance for driving activities, rather than outwardly parsing the whole scene or indiscriminately mapping a picture to directing points.

Direct perception approach [8, 13] for self-governing driving—a third worldview that falls in the middle, intervened observation, and conduct reflex. To take in a mapping from a picture to a few significant affordance markers of the street circumstance, including the edge of the vehicle in respect to the street, the separation to the path markings, what's more, the separation to vehicles in the present and adjoining paths [2]. With this minimal yet significant affordance portrayal as observation yield, show that a very straightforward controller would then be able to settle on driving choices at a high level and drive the vehicle easily. Model is based upon the best in class profound convolutional neural network (ConvNet) system to naturally learn picture highlights for assessing affordance related to independent driving. To manufacture preparation set, inquire a human driver to play a vehicle hustling computer game TORCS for 12 h while recording the screen captures and the comparing names. Together with the basic controller that plan, model can make important expectations for affordance pointers and independently drive a vehicle in various tracks of the computer game, under various traffic conditions what's more, path designs [9, 12]. In the meantime, it appreciates a lot less complex structure than the common interceded recognition approach. Testing this framework on vehicle-mounted cell phone recordings and the KITTI dataset exhibits great real-world discernment too [10]. This immediate discernment approach gives a smaller, task explicit affordance portrayal for scene understanding in independent driving.

2.1 Mapping from an Image to Affordance

A best in class profound learning ConvNet as direct discernment model to delineate picture to the affordance markers [8]. In this paper, center is around parkway driving with various paths. From a sense of self driven perspective, the host vehicle just needs to concern the traffic in its present path and the two adjoining (left/right) paths when deciding. Along these lines, just need to demonstrate these three paths. Train a solitary ConvNet to deal with three path arrangements together: a street of one path, two paths, or three paths. Appeared are the average cases are managing. Every so often the vehicle needs to drive on path markings, and in such circumstances, just the paths on each side of the path stamping should be checked, as appeared.

Thruway driving activities can be sorted into two noteworthy types (following the path focus line) evolving paths or backing off to stay away from impacts with the first vehicles. To help these activities, characterize the framework to have two arrangements of portrayals under two facilitate frameworks: "in path framework" and "on checking framework." To accomplish two real capacities, path discernment and vehicle observation, three kinds of pointers have to speak to driving circumstances: heading point, the separation to the close by path markings, what's more, the separation to the previous vehicles. Altogether, 13 affordance pointers as this driving scene portrayal are shown. A total rundown of the affordance pointers is specified. They are the yield of the ConvNet as this affordance estimation and the contribution of the driving controller. The "in path framework" and "on stamping framework" are initiated under various conditions. To have a smooth change, characterize a covering region, where the two frameworks are dynamic. The format is appeared with the exception of heading edge, every one of the pointers may yield an inert state. There are two cases in which a marker will be inerted when the vehicle is driving in either the "in path framework" or "on stamping framework" and the other framework is deactivated, at that point, every one of the markers having a place with that framework is inert. when the vehicle is driving on limit paths (left most or right most path), and there is either no left path or no correct path, and at that point, the markers relating to the non existing nearby path are inert. As indicated by the pointers' esteem and dynamic/dormant express, the host vehicle can be precisely restricted out and about [13].

Complete list of affordance indicators in directed perception is representation as shown in Fig. 2 [14]

1: angle: angle between the car's heading and the tangent of the road "in lane system," when driving in the lane:
2: to Marking LL: distance to the left lane marking of the left lane
3: to Marking ML: distance to the left lane marking of the current lane
4: to Marking MR: distance to the right lane marking of the current lane
5: to Marking RR: distance to the right lane marking of the right lane
6: dist LL: distance to the preceding car in the left lane
7: dist MM: distance to the preceding car in the current lane
8: dist RR: distance to the preceding car in the right lane "on marking system," when driving on the lane marking.
9: to Marking L: distance to the left lane marking
10: to Marking M: distance to the central lane marking
11: to Marking R: distance to the right lane marking
12: dist L: distance to the preceding car in the left lane
13: dist R: distance to the preceding car in the right lane.

Controller logic as shown in Fig. 3 [4, 13–15]

1: while (in autonomous driving mode)

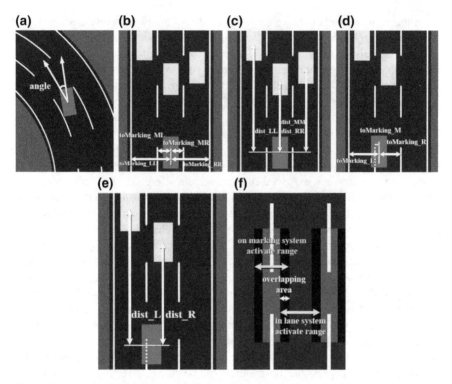

Fig. 2 Illustration of this affordance representation. **a** angle **b** in lane: to Marking **c** in lane: dist **d** on mark: to Marking **e** on marking: dist **f** overlapping area. A lane changing maneuver needs to traverse the in lane system and the on marking system. shows the designated overlapping area used to enable smooth transitions [4–6, 15]

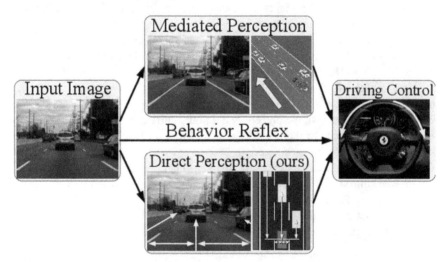

Fig. 3 Three paradigms for autonomous driving [8]

2:     ConvNet outputs affordance indicators
3:     check availability of both the left and right lanes
4:     if (approaching the preceding car in the same lane)
5:      if (left lane exists and available and lane changing allowable)
6:       left lane changing decision made
7:      else if (right lane exists and available and lane changing allowable)
8:        right lane changing decision made
9:      else
10:        slow down decision made
11:    if (normal driving)
12:      center line = center line of current lane
13:    else if (left/right lane changing)
14:        center line = center line of objective lane
15:    compute steering command
16:    compute desired speed
17:    compute acceleration/brake command based on desired speed.

2.2 Mapping from Affordance to Action

The directing control is figured utilizing the vehicle's position what's more, present, and the objective is to limit the hole between the vehicle's present position and the middle line of the path [8].

Characterizing dist focus as the separation to the middle line of the path, have [4, 6]:

$$steerCmd = C\_(angle - dist\ center/road\ width) \tag{5}$$

where C is a coefficient that varies under different driving conditions, and angle $\epsilon\ (-\pi, \pi)$ [4, 15]. When the car changes lanes, the center line switches from the current lane to the objective lane as described in Eq. 5. The pseudocode describing the logic of the driving controller is listed. At each time step, the framework figures desired speed. A controller influences the real speed to pursue the wanted speed by controlling the increasing speed/brake. The gauge wanted speed is 72 km/h. In the event that the vehicle is turning, an ideal speed drop is processed by the previous couple of directing edges. On the off chance that there is a former vehicle in short proximity and a hinder choice is made, the wanted speed is likewise controlled by the separation to the going before vehicle. To accomplish vehicle following conduct in such circumstances, actualize the ideal speed vehicle following model as [6, 14]:

$$v(t) = v.\ \max\left(1 - \exp\left(-\frac{c}{v.\ \max}\text{dist}(t) - d\right)\right) \tag{6}$$

where dist(t) is the distance to the preceding car, v.max is the largest allowable speed, c and d are coefficients to be calibrated [8]. With this implementation, the host car can achieve stable and smooth car following under a wide range of speeds and even make a full stop if necessary as described in Eq. 6.

3   Implementation

Immediate observation of ConvNet is based upon Caffe, what's more, utilize the standard AlexNet design. There are five convolutional layers pursued by four completely associated layers with yield measurements of 4096, 4096, 256, and 13, separately. Euclidian misfortune is utilized as the misfortune work. Since the 13 affordance pointers have different reaches, standardize them to the scope of. select 7 tracks and 22 traffic vehicles in TORCS, appeared and to produce the preparation set. Supplant the first street surface surfaces in TORCS with more than 30 tweaked black top surfaces of different path designs what's more, black top dimness levels. Additionally, program extraordinary driving practices for the traffic autos to make extraordinary traffic designs. Physically drive a vehicle on each track on numerous occasions to gather preparing information. While driving, the screen captures are all the while down examined to 280 × 210 what's more, put away in a database together with the ground truth marks. This information accumulation procedure can be effectively computerized by utilizing an AI vehicle [16–18]. However, when driving physically, Purposefully make extraordinary driving conditions (for example off the street, crash into different vehicles) to gather progressively compelling preparing tests, which makes the ConvNet all the more dominant and fundamentally decreases the preparation time. Altogether, gather 484,815 pictures for preparing. The preparing strategy is like preparing an AlexNet on ImageNet information [6]. The distinctions are: the info picture has a goal of 280 × 210 and is never again a square picture. Not utilizing any yields or a reflected form. Train model starting with no outside help. Pick an underlying learning rate of 0.01, also, every less clump comprises of 64 pictures arbitrarily chose from the preparation tests. After 140,000 cycles, stop the preparation procedure.

3.1   The Open Racing Vehicle Simulator Evaluation

Initially, assess this immediate discernment model on the TORCS driving amusement [18]. Inside the diversion, the ConvNet yield can be imagined and utilized by the controller to drive the host vehicle. To quantify the estimation precision of the affordance pointers, build a testing set comprising of tracks and vehicles excluded in the preparation set. In the elevated TORCS representation, treat the host vehicle as the reference object as shown in Fig. 4. As its vertical position is fixed, it moves on a level plane with a heading processed from edge. Traffic vehicles just move vertically.

Fig. 4 Examples of the 7 tracks used for training. Each track is customized to the configuration of one lane, two lane, and three lane with multiple asphalt darkness levels. The rest of the tracks are used in the testing set [9, 17, 19]

Do not picture the arch of the street, so the street ahead is continuously spoke to as a straight line. Both the estimation (void box) and the ground truth (strong box) are shown.

3.2 Qualitative Assessment

This framework can drive very well in TORCS with no crash. In some path evolving situations, the controller may somewhat overshoot; however, it rapidly recuperates to the ideal position of the goal path's middle. As found in the TORCS representation, the path discernment module is beautiful precise, and the vehicle discernment module is solid up to 30 ms away. In the scope of 30–60 m, the ConvNet yield ends up noisier. In a 280×210 picture, at the point when the traffic vehicle is more than 30 m away, it really shows up as a minor spot, which makes it trying for the system to appraise the separation. Be that as it may, in light of the fact that the speed of the host vehicle does not surpass 72 km/h in this tests, dependable vehicle observation inside 30 m can ensure palatable control quality in the diversion. To keep up smooth driving, this framework can endure moderate mistake in the pointer estimations. The vehicle is a ceaseless framework, and the controller is always revising its position. Indeed, even with some dissipated wrong estimations, the vehicle can at present drive easily with no impacts.

3.3 Comparison with Baselines

To quantitatively evaluate the performance of the TORCS-based direct perception ConvNet, compare it with three baseline methods. Refer to this model as "ConvNet full" in the following comparisons.

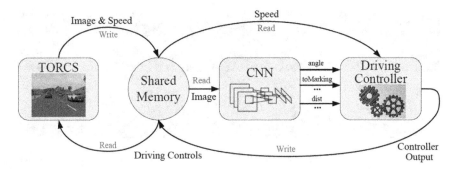

Fig. 5 System architecture the ConvNet processes the TORCS image and estimates 13 indicators for driving. Based on the indicators and the current speed of the car, a controller computes the driving commands which will be sent back to TORCS to drive the host car in it [6, 8]

Behavior Reflex ConvNet. The technique straightforwardly maps a picture to control utilizing a ConvNet. Train this model on the driving amusement by TORCS utilizing two settings:

(1) The preparation tests (more than 60,000 pictures) are altogether gathered while driving on a vacant track; the errand is to pursue the path as shown in Fig. 5.
(2) The preparation tests (more than 80,000 pictures) are gathered while driving in rush hour gridlock; the undertaking is to pursue the path, maintain a strategic distance from crashes by exchanging paths, and surpass moderate going before vehicles as shown in Fig. 5.

The video in this undertaking site appears as the regular execution. The conduct reflex framework can undoubtedly pursue void tracks. When testing on the same track where the preparation set is gathered, the prepared framework shows some ability at staying away from crashes by turning left or right. Be that as it may, the direction is inconsistent. The conduct is far not the same as a typical human driver also, is unusual the host vehicle slams into the former autos as often as possible.

Mediated Perception (lane detection): Run the Caltech path indicator on TORCS pictures. Since as it were two paths can be dependably recognized, map the directions of spline stay purposes of the best two recognized path markings to the path based-affordance pointers [7]. Train a framework made out of 8 support vector regression (SVR) and 6 bolster vector classification (SVC) models (utilizing libsvm) to execute the mapping (a vital advance for interceded observation approaches). The framework design is like the GIST-based framework (next area) delineated; however, without vehicle recognition [15]. Since the Caltech path identifier is a generally powerless standard, to make the assignment less complex, make an uncommon preparing set and testing set. Both the preparation set (2430 examples) also, testing set (2533 examples) are gathered from the equivalent track (not among the 7 preparing tracks for ConvNet) without traffic, and in a better picture goals of 640 × 480 [10, 15]. find that, notwithstanding when prepared and tried on a similar track, the Caltech path finder based framework still performs more regrettable than this model. Characterize this mistake metric as mean absolute mistake (MAE) between the affordance estimations and ground truth separations. An examination of the blunders for the two frameworks is appeared.

Directed Perception with GIST: Think about the handmade significance descriptor with the profound highlights learned by the ConvNet's convolutional layers in this model as shown in Figs. 6 and 7 [11]. A lot of 13 SVR and 6 SVC models are prepared to change over the GIST highlight to the 13 affordance pointers characterized in this framework. The strategy is shown. The GIST descriptor segments the picture into 4 × 4 portions. Since the ground territory spoken to by the lower 2 × 4 sections might be increasingly applicable to driving, attempt two unique settings in these tests: convert the entire GIST descriptor, also, convert the lower 2 × 4 fragments of GIST descriptor. Allude to these two baselines as "Substance entirety" also, "Significance half" separately.

Comparison with DPM-based baseline: Think about the execution of this KITTI-based ConvNet with the cutting edge DPM vehicle indicator (an intervened recognition approach). The DPM vehicle finder [10] is given and is upgraded for the KITTI dataset. Run the identifier on the full goals pictures and convert the bouncing boxes to separate estimations by anticipating the focal purpose of the lower edge to the ground plane (zero stature) utilizing the aligned camera model.

The projection is very exact given that the ground plane is level, which holds for most KITTI pictures as shown in Fig. 8 [8, 10]. DPM can identify different vehicles in the picture, and select the nearest ones (one on the host vehicle's abandoned, one to its right side, and one legitimately before it) to process the estimation blunder. Since

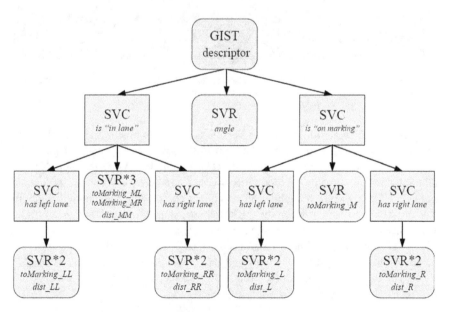

Fig. 6 GIST baseline. Procedure of mapping GIST descriptor to the 13 affordance indicators for driving using SVR and SVC [8]

(a) Autonomous driving in TROCS   (b) Testing on real video Testing the

Fig. 7 TORCS-based system [12, 19, 20]

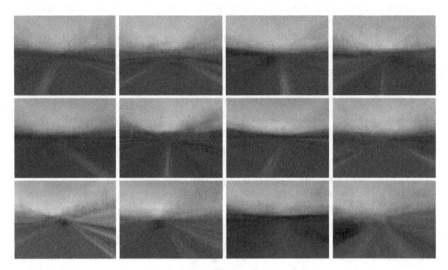

Fig. 8 Activation patterns of neurons. The neurons' activation patterns display strong correlations with the host car's heading, the location of lane markings, and traffic cars [7, 8, 10]

the pictures are taken while the host vehicle is driving, numerous pictures contain nearest autos that just somewhat show up in the left lower corner or right lower corner. DPM can not recognize these fractional vehicles, while the ConvNet can all the more likely handle such circumstances. To make the correlation reasonable, possibly tally mistakes when the nearest vehicles completely show up in the picture. The mistake is registered when the traffic vehicles appear inside 50 m ahead (in the y organize). At the point when there is no vehicle present, the ground truth is set as 50 m. Subsequently, if either model has a bogus positive, it will be punished. The mean absolute error (MAE) for the y and x facilitates, and the Euclidian separation d between the estimation, and the ground truth of the vehicle position are appeared A screen capture of the framework is appeared.

4   Visualization

To see how the ConvNet neurons react to the input pictures as shown in Fig. 8 can imagine the enactment designs. On a picture dataset of 21,100 examples, for each of the 4096 neurons in the main completely associated layer, pick the top 100 pictures from the dataset that enact the neuron the most and normal them to get an enactment design for this neuron. Along these lines, gain a thought of what this neuron gained from preparing. Demonstrates a few haphazardly chosen arrived at the midpoint of pictures. See that the neurons' enactment designs have solid relationship with the host vehicle's heading, the area of the path markings and the traffic autos. Therefore, trust the ConvNet has created task-specific highlights for driving. For a specific convolutional layer of the ConvNet, a reaction guide can be created by showing the most noteworthy esteem among all the channel reactions at every pixel. Since area data of articles in the first information picture is protected in the reaction map, realize where the striking areas of the picture are for the ConvNet when making estimations for the affordance pointers. Demonstrate the reaction maps of the fourth convolutional layer of the nearby extend ConvNet on an example of KITTI testing pictures as shown in Fig. 9. See that the ConvNet has solid reactions over the areas

Fig. 9 Response map of this Row (1–3) KITTI-based and Row (4–5) TORCS-based ConvNets. The ConvNets have strong responses over nearby cars and lane markings [8, 20]

of close by vehicles, which shows that it figures out how to "look" at these vehicles while assessing the separations. Likewise, demonstrate some reaction, n maps of this TORCS-based ConvNet in a similar as shown in Fig. 9. This ConvNet has exceptionally solid reactions over the areas of path markings.

5 Conclusions

In this paper, proposed a path discovery calculation. In contrast to the slope technique, the shading-based division strategy can undoubtedly dispose of the impact because of the daylight, shadow, asphalt, and impediments, similar to vehicles and people on foot. Utilizing the conceivable path checking sets additionally spare us a great deal of time so utilize technique to approach the path limit with high exactness without earlier data, while utilizing Hough change or other street display are substantially more muddled. This algorithm and technique review gives 70–80% improvements in self driving car.

A novel self-governing driving worldview dependent on direct discernment. This portrayal uses a profound ConvNet engineering to appraise the affordance for driving activities as opposed to parsing whole scenes (intervened observation approaches), or indiscriminately mapping an picture legitimately to driving directions (conduct reflex methodologies). Analyses demonstrate that this methodology can perform well in both virtual and genuine situations.

References

1. Humaidi, A.J., Fadhel, M.A.: Performance comparison for lane detection and tracking with two different techniques. In: Al Sadeq International Conference on Multidisciplinary in IT and Communication Science and Applications (AIC-MITCSA), pp. 1–6. IEEE, Iraq (May 2016)
2. Cheong, C.K..: Design of lane detection system based on color classification and edge clustering. In: Quality Electronic Design (ASQED) 2011 3rd Asia Symposium, pp. 266–271. IEEE, Malaysia (2011)
3. Wijesoma, W.S.: Road boundary detection and tracking using ladar sensing. Robot. Autom. IEEE Trans. Robot. Autom. 20(3), 456–464 (2004)
4. Streller, D., Furstenberg, K.: Vehicle and object models for robust tracking in traffic scenes using laser range images. In: Intelligent Transportation Systems 2002 Proceedings. The IEEE 5th International Conference on Intelligent Transportation Systems, pp. 118–123. IEEE, Singapore (2002)
5. Li, C., Wang, J.: A model based path planning algorithm for self driving cars in dynamic environment. In: 2015 Chinese Automation Congress (CAC), pp. 1123–1128. IEEE, China (Nov 2015)
6. Yadav, S., Patra, S.: Deep CNN with color lines model for unmarked road segmentation. In: Image Processing (ICIP) 2017 IEEE International Conference, pp. 585–589. IEEE, China (2017)
7. Ignatiev, K.V., Serykh, E.V.: Road marking recognition with computer vision system. In: Control in Technical Systems (CTS) 2017 IEEE II International Conference, pp. 165–167. IEEE, Russia (2017)

8. Chen, C.: Deepdriving learning affordance for direct perception in autonomous driving. In: The IEEE International Conference on Computer Vision (ICCV), pp. 2722–2730 (2015)
9. Liu, S., Lu, L.: Effective road lane detection and tracking method using line segment detector. In: Control Conference (CCC) 2018 37th Chinese, pp. 5222–5227. IEEE, China (2018)
10. Geiger, A.: Vision meets robotics: the KITTI dataset. In: The International Journal of Robotics Research, vol. 32 no. 11, pp. 1231–1237. Article first published online: 23 August 2013; Issue published: September 1 (2013)
11. Fan, Y., Zhang, W.: A robust lane boundaries detection algorithm based on gradient distribution features. In: Fuzzy Systems and Knowledge Discovery (FSKD) 2011 18th International Conference, vol. 3, pp. 1714–1718. IEEE, China (2011)
12. Cheng, H.Y.: Lane detection with moving vehicles in the traffic scenes. In: Intelligent Transportation Systems IEEE Transactions, vol. 7, no. 4, pp. 571–582 (2006)
13. Yim, Y.U.: Three feature based automatic lane detection algorithm (TFALDA) for autonomous driving. IEEE Transactions on Intelligent Transportation Systems, vol. 4, no. 4, pp. 219–225 (Dec 2003)
14. Krapukhina, N., Senchenko, R.: Intellectual simulation model for transport systems management and safety database. In: Transportation Information and Safety (ICTIS) 2017 4th International Conference, pp. 475–482. IEEE, Canada (2017)
15. Xu, Y., Jia, P.: Recognition of curvy lane lines on freeway and applications in intelligent vehicle. In: Image and Signal Processing (CISP) 2012 5th International Congress, pp. 926–930. IEEE, Piscataway, China (2012)
16. Li, Q.: Spring robot a prototype autonomous vehicle and its algorithms for lane detection. In: IEEE Transactions on Intelligent Transportation Systems, vol. 5, no. 4, pp. 300–308 (Dec 2004)
17. McCall, J.C.: Video based lane estimation and tracking for driver assistance: survey, system and evaluation. In: Intelligent Transportation Systems IEEE Transactions, vol. 7, no. 1, pp. 20–37 (2006)
18. Ozutemiz, K.B., Hacinecipoglu, A.: Self learning road detection with stereo vision. In: Signal Processing and Communications Applications Conference (SIU) 2013 21st, pp. 1–4. IEEE, Turkey (2013)
19. Kong, H.: General road detection from a single image. In: Image Processing IEEE Transactions, vol. 19, no. 8, pp. 2211–2220 (2010)
20. Mechat, N., Saadia, N.: Lane detection and tracking by monocular vision system in road vehicle. In: Image and Signal Processing (CISP) 2012 5th International Conference, pp. 1276–1282. IEEE, China (2012)

Various Noises in Medical Images and De-noising Techniques

Alamgeer Ali and Kamal Kumar

Abstract It has become a great topic to be researched for the better interpretation of the medical images. There are many de-noising techniques and filters available till now but still it is the topic of research. How, it is to get the fully de-noised image using any of the filtering techniques invented yet or using new idea of filtering. Wiener filtering is one of the most popular methods. Wavelets transform has become a very powerful tool for removing the noise from the images. Some of the noise cannot be ignored from the images while diagnosis. So, it is very important to remove the noise from such types of images. Numerous research papers have been published regarding this topic but still the research is going on to get fully de-noised medical images.

Keywords De-noising · Speckle noise · Rician noise · Salt and pepper noise · Poisson noise · N1-Means · Fast N1-Means · Total variation · Mean filter · Median filter

1 Introduction

There is a great improvement in the field of medical diagnosis after the emergence of the image acquisition techniques with the help of X-Ray, CT scan and MRI, etc. But, for achieving an accurate and efficient diagnosis, medical staff requires substantial quality medical images. As, we know that for getting an accurate diagnosis, the doctors or the medical staffs require good quality images of X-Ray, computed tomography, and MRI having very less noise present in it. Although, in the acquisition processes of the images, they may be corrupted by the disruptive noise which makes medical staff difficult to recognize the damage inside the body of the human being. Image de-noising is a vital image processing task.

A. Ali (✉) · K. Kumar
National Institute of Technology Uttarakhand, Srinagar, Pauri Garhwal, Uttarakhand, India
e-mail: alambulletin@gmail.com

K. Kumar
e-mail: kamalkumar@nituk.ac.in

© Springer Nature Singapore Pte Ltd. 2020
P. K. Singh et al. (eds.), *Proceedings of First International Conference on Computing, Communications, and Cyber-Security (IC4S 2019)*, Lecture Notes in Networks and Systems 121, https://doi.org/10.1007/978-981-15-3369-3_23

This paper is organized as follows: Sect. 2 gives an overview of the de-noising approaches for the different noises in the different types of images like ultrasound, MRI, computed tomography, scintigraphy and X-Ray data while the Sect. 3 discusses the methods for de-noising and image. Section 4 gives the idea of removing the speckle noise and Sect. 5 tells us about removing salt and pepper noise. Section 6 contains some idea of removing speckle noise while Sect. 7 is all about de-noising results for synthetic data. And the last Sect. 8 concludes the paper idea.

2 De-noising Techniques

Various approaches have been proposed for medical images de-noising field mainly NI-Means and Wavelets theory [1] which have been tested successfully to the different medical imaging data. This section describes the most widely used approaches for reducing noise from ultrasound images. It has been found that the most common noise like speckle noise is found in the ultrasound images and radar images. Rician noise is mostly present in MRI images. Poisson noise is mostly present in scintigraphy images. Now, the challenge is to remove the noise from the images with preserving as much as image characteristics because sometimes while removing the noise edges are not preserved. Due to which the interpretation of the experts may go wrong and the important information might be lost. Consequently, the dignity of the experts decreases and there is a risk for the patient. In today's modern era, most of the de-noising approaches are based on the wavelets threshold. In this approach, the image is decomposed into the co-efficient and then the noise-free co-efficient is filtered. While the acquisition of the ultrasound images and the radar images there is always a problem of the speckle noise. Speckle noise is common in these kinds of images. And our target is always to find out a way to remove the speckle noise from these images as maintaining the characteristics by the images is very important while de-noising it because it must not degrade its actual information otherwise it may be counterproductive. For example, we can easily loss tracking features used to follow fetus evolution during pregnancy.

Like ultrasound images, the MRI images are often affected by non-additive noise called Rician noise in case of magnetic imaging ranging data. Generally, MRI images are filtered by the threshold techniques. We can also use the strongest NL-means algorithm.

3 Methods for De-noising an Image

Mean filter is one of the most famous and effective technique for the removal of the noise from the medical images. This is the nonlinear technique which is also known as order static filtering in digital image processing. As the name suggests that there is something mean that is need to be calculated for the pixel intensities. An averaging

Table 1 Averaging of the pixels in mean filter

1/9	1/9	1/9
1/9	1/9	1/9
1/9	1/9	1/9

filter is used to remove the grain noise from the medical photographs like CT scan and MRI images. In this method, every pixel gets set to the average value of the intensities of the pixels in its neighborhood, and local variations caused by grains are reduced. But the disadvantage of using the mean filter is that it is not up to the mark in preserving the edges [2] due to which some information might get lost. So, in order to remove the drawbacks of the mean filter, a similar filter is used in place of the mean filter which is known as the median filter. In this filter, the same as the mean filter median is used. It means that every pixel is replaced by the median of the pixels in its neighborhood. In this filter, the edges are preserved while removing the noise from the images. The main objective of this is to reduce the noise from all types of images and we should be focused on the image characteristics mainly. Image characteristics must not change otherwise everything will go wrong. The Wiener filter is also another type of filter used frequently in de-noising of the medical images. The interpretation of the experts even may go wrong if the characteristics are changed (Table 1).

The MATLAB function which is used in the mean filter is imfilter (A, h), where A is the multidimensional array of the pixels with the filter value h. If A is an integer or logical array, imfilter truncates output elements that exceed the range of the given type and rounds fractional values. Note that, in the median filter, if the window has an odd number of entries, then median is very easy to define: it is just the middle element or the middle value after all the values in the window are sorted numerically in increasing order. The advantage of the median filter over the mean filter is that the median is more robust than the mean so a single very unrepresentative pixel will not affect the median value significantly. As, there is no new pixel created in case of median calculation of the pixels in its neighborhood, so the median filter does not create any unrealistic pixel value. Consequently, median filter is more edge preserved. Wiener filter is also a different method for filtering an image. It filters out the noise that has corrupted a signal. It uses statistical approach. Wiener filter is characterized as assumption, requirement and performance criteria. One is assumed to have knowledge of the spectral properties of the noise and the original signal. Wiener filter is to compute a statistical estimate of an unknown signal using a related signal as an input and filtering that known signal to produce the estimate as an output. For example, the known signal might consist of an unknown signal of interest that has been corrupted by additive noise. The Wiener filter can be used to filter out the noise from the corrupted signal to provide an estimate of the underlying signal of interest.

4 Results of Comparison of De-noising Algorithms

The implementation of various de-noising algorithm has been discussed in the previous section with the different filter using the MATLAB functions. Here, the images used were MRI images affected by Poisson noise, speckle noise, Gaussian noise and salt and pepper noise. Different filters like median filter, wiener filter and the Gaussian filter were used in order to remove the speckle noise from the different types of medical images. For this comparison, different 20 MRI images are taken for the experiment and on the different images the filters were applied to test the filter quality in de-noising the image. It has been a time taking process but the experiment was productive to compare the filters. From the above discussion, we come to know that in the case of wiener filter the value is minimum for speckle noise and the clarity is also given by the wiener filter. Figure 1a shows the speckle noise present in image and Fig. 1b represents the de-noised image.

Now, we will try to compare these three filters in case of the blurred noise and try to see the results in every case. Here, again we will use the three filters Wiener filter, Gaussian filter and the median filter and notice the result. Here, the results are somewhere different than the previous one, i.e., in case of speckle noise. In this case, we have noticed that the minimum value is given by the wiener filter in most of the cases as well as by the median filter but not given by the Gaussian filter in any of the test cases well.

But, it is very astonishing that the clarity is completely given by the Gaussian filter. Figure 2a shows the salt and pepper noise present in the image and Fig. 2b shows the filtered image. In the case of removing of the Poisson noise, the results were obtained for the different filters like median filter, wiener filter and the Gaussian filter. Here, the results are somewhere different than the previous one, i.e., in case of speckle noise and the blurred noise. In this case, we have noticed that the minimum value is given by the Gaussian filter in most of the cases as well as by the median

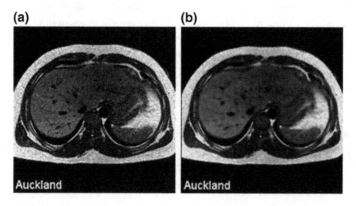

Fig. 1 **a** Speckle noise and **b** filtered image [3]

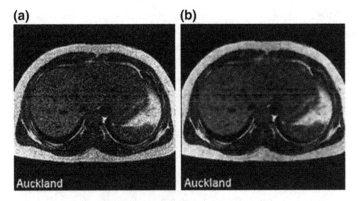

Fig. 2 **a** Salt and pepper noise and **b** filtered image [3]

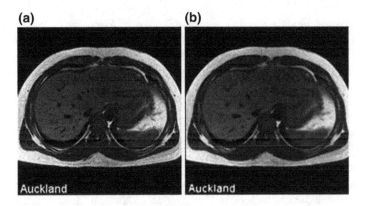

Fig. 3 **a** Poisson noise and **b** filtered image [3]

filter but not given by the Gaussian filter in any of the test cases. But the clarity is only given by the median filter. Figure 3 shows the Poisson noise present in it the filtered image as well.

In the case of removing of the salt and pepper noise, the results were obtained for the different filters like median filter, wiener filter and Gaussian filter. In the case of salt and pepper noise, median filter works better than any other filter because it gives the minimum value as well as the clarity is also given by the median filter.

5 Removing Salt and Pepper Noise

Customary middle channel does not take into thought for how picture qualities fluctuate from one area to another. It replaces each point in the picture by the middle of the relating neighborhood. Eventually, the versatile channel that is equipped for adjusting their conduct contingent upon the qualities of the picture in the region being shifted can deliver progressively powerful yield picture for a few input uproarious pictures. A versatile middle channel whose conduct changed dependent on measurable attributes of the picture inside the channel locale characterized by the $m \times n$ rectangular windows, Sxy is indicated in principle [4]. Same as middle channel, versatile middle channel likewise works in a rectangular window zone Sxy. Not at all like middle channel, in any case, the versatile middle channel changes the span of Sxy amid channel activity, contingent upon different conditions. The yield of the channel. Consider the following notation:

$Zmin$ = minimum intensity value in Sxy
$Zmax$ = maximum intensity value in Sxy
$Zmed$ = median of the intensity values in Sxy
Zxy = intensity value at coordinates (x, y)
$Smax$ = maximum allowed size of Sxy

The adaptive median filtering algorithm consists of two parts, denoted level A and level B.
Level A: If $Zmin < Zmed < Zmax$, go to level B
Else increase the window size
If window size < $Smax$, repeat level A
Else output $Zmed$
Level B: If $Zmin < Zxy < Zmax$,
Output Zxy
Else output $Zmed$.
Else yield Zdrug
Watching the calculation, the motivation behind dimension An is to decide whether the middle channel yield, $Zmed$, is an motivation (dark or white) or not. On the off chance that the condition $Zmin < Zmed < Zmax$ is valid, at that point Zdrug cannot be a motivation as indicated by the clamor hypothesis. For this situation, go to level B and test if the point in the focal point of the window, Zxy, is itself a drive. In the event that the condition $Zmin < Zxy < Zmax$ is valid, at that point Zxy cannot be a drive. For this situation, the calculation yields the unaltered pixel esteem, Zxy. By not evolving these "moderate dimension" focuses, contortion is decreased in the picture. On the off chance that the condition $Zmin < Zxy < Zmax$ is false, at that point either $Zxy = Zmin$ or $Zxy = Zmax$. In either case, the estimation of the pixel is an extraordinary esteem and the calculation yields the middle esteem $Zmed$, which we know from level An is not a clamor drive. With the help of this MATLAB code, we have tried to reduce the noise like salt and pepper noise with help of linear filter, median filtered and the adaptive filtered methodology. We have found that the performance of the Wiener filter is better than the mean filter and the median filter. MATLAB is a very important tool in the field of the image processing-related task. As it has been a big platform always so, reliability is somewhere better than the other platform to be used for this task. Wiener filter is better than the other filter in some aspects. It removes the blurring and additive noise present in the image and it is also very optimal in relation to the mean squared error where it minimizes the overall

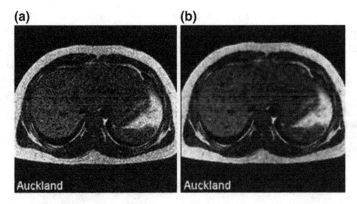

Fig. 4 a Salt and pepper and **b** filtered image [3]

mean square error in the operation of the filtering technique for noise removal. Salt and pepper noise is shown in Fig. 4a and filtered image is shown in Fig. 4b.

6 Removing Speckle Noise

There are distinctive kinds of commotion (noise) that shows up in pictures. Clamor may happen because of various factors such as while procuring pictures, catching pictures, changing pictures and packing pictures. Noises may have diverse sorts, and consequently, it is important to give distinctive techniques according to the kind of noise. Speckle noise contains high recurrence segments due to temporal development of organs as cerebrum, heart and so forth. So, it is important to give low pass channel to remove the high recurrence clamor. To expel spot commotion from pictures, till now numerous channels are used [5]. A few channels are great in visual elucidation, whereas some are great in smoothing abilities and clamor decrease. A few instances of such channels are mean, median, Lee, enhanced Frost, Wiener and gamma MAP channels [6]. A portion of these utilization window technique to evacuate dot clamor, called as piece [7]. This window size can extend from 3-by-3 to 33-by-33 yet it must be odd. To accomplish better outcome, window size ought to be littler. Speckle noise is a kind of noise that is taken very lightly but whole interpretation may go wrong while diagnoses. Speckle noise in conventional radar results from random fluctuations in the return signal from an object that is no bigger than a single image processing element. It increases the mean gray level of a local area. Speckle noise in SAR is generally more serious, causing difficulties for image interpretation. It is caused by coherent processing of backscattered signals from multiple distributed targets. If the problems of the speckle noise are removed then this helps a lot in data interpretation of the medical images. This is the major noise present in the medical images. It may happen because of the non-fitness of the machine which is used for the acquisition of the medical images (Fig. 5).

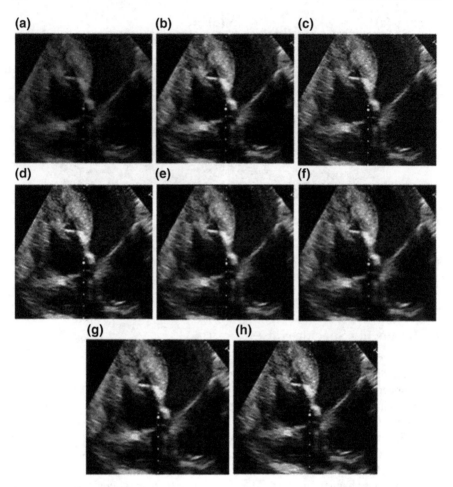

Fig. 5 Speckle noise removal

Scalar channels depend on the proportion of neighborhood measurements, which improves smoothing in homogenous locales of the pictures where dot is completely created and diminishes apparently in different districts of the picture so as to save the valuable subtleties of the picture [8]. Key sorts of scalar channels are mean channel and median channels examined in the next areas. It is basic and instinctive channel concocted by Pomalaza-Raez in 1984. It does not expel spot clamor at entire yet decreases at some expand. It chips away at normal premise that is the middle pixel is supplanted by the normal of the all pixels. Henceforth, this channel gives obscuring impact to the pictures, so it is least acceptable technique [9, 10] to expel dot commotion as it results in loss of subtleties. Median filter is nonlinear filter created by Pitas in 1990. It gives very preferred outcome over the mean channel. Here, the focused pixel is supplanted by the middle estimation everything being equal and subsequently delivers less obscuring. Due to this nature, it is utilized to diminish imprudent spot commotion. Favorable position is it protects the edges. Inconvenience is additional time required for calculation of the middle an incentive for arranging N pixels, the transient unpredictability is O ($N \log N$). Middle channel pursues calculation as pursues: (1) Take a 3×3 (or 5×5 and so forth) district based on the pixel (I, j). (2) Sort the force estimations of the pixels in the district into climbing request.

7   De-noising Results for Synthetic Data

Figure 6 demonstrates the got aftereffects of different de-noising calculations for the synthetic data. We have spotted the first engineered picture with $J = 0.4$ of commotion. For the total variation strategy, we halted at the twentieth emphasis, and for Pizurica's [11] methodology, we have set the K parameter at 4. This parameter controls the nature sharpness and the picture highlights. NI-Means and fast NI-Means calculations perform great visual outcomes since they consider all picture pixels. For sure, they supplant the de-noised picture. The picked window estimate is

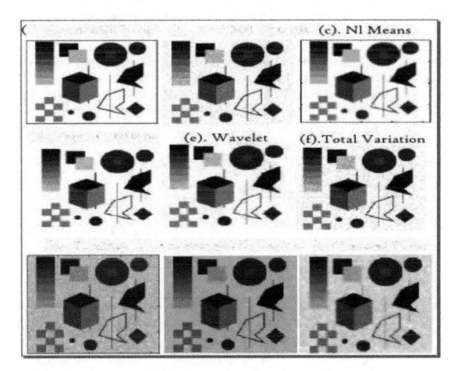

Fig. 6 Quality comparison of various de-noised images [8]

Fig. 7 From left to right: de-noised image by both Wiener and median filters (window used: 3 * 3)
[8]

5 * 5. Outwardly, we can see that NI-Means, fast NI-Means and wavelets Threshold approaches [12, 13], do not harm the power of every pixel in the loud picture by a weighted normal of all pixel powers (Fig. 7).

We note that bilateral channel, total variation [14], NI-Means and fast NL-Means approaches give higher MSSIM values which imply that they keep the picture structures unaltered in the wake of de-noising process. Anyway, the higher estimation of SNR [15] is coming about as it were with essential and faster NI-Means approaches (441,282 db for NL-implies and 252,186 db for fast NL-implies). Presently, we will degenerate the first picture by a few commotions levels so as to watch the clamor sway on the picture on one side, and the proficiency of commotion evacuation approaches in term of SNR on the opposite side. As indicated by Fig. 8 Nl-Means yields a huge SNR hole over other de-noising techniques. As far as RMSE value is concerned, Fig. 8 demonstrates that the littlest estimation of RMSE achieved by NI-Means approach.

We note that the NI-Means calculation gives a great MSSIM yield result. The MSSIM esteem stays great (nearer to 1) regardless of whether a significant clamor level adulterates the tried picture. In the following investigation, we include step by step different commotion levels to similar pictures than we de-noised just by N1-implies furthermore. We fluctuates the sigma between 0.2 and 0.7. The de-noised

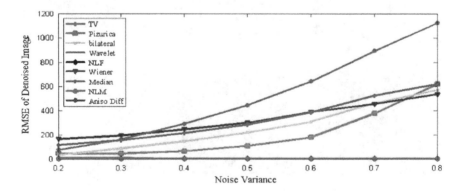

Fig. 8 Summarized results of de-noising performance [8]

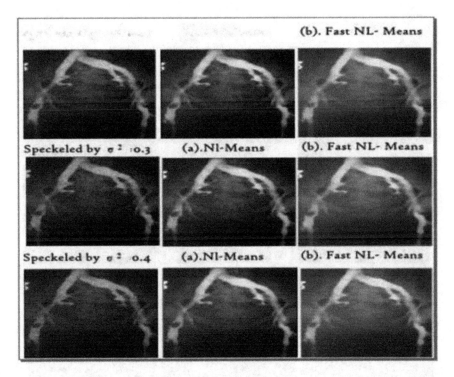

Fig. 9 Perceptual quality comparison of various de-speckled images with variance 0.2, 0.3 and 0.4 [8]

pictures are shown in Figs. 9 and 10. From these figures, we can take note of that NL-Means and quick Nl-Means approaches give commonly prevalent execution. However, it is not obvious to pass judgment on the execution of these approaches by just visual investigation, or even by SNR coming about since a specialist's learning is expected to affirm that point.

In both Figs. 9 and 10, it is tested for the different datasets. In Fig. 9, the different variance was taken like 0.2, 0.3 and 0.4 while in Fig. 10 it is tested for the variance value 0.5, 0.6 and 0.7. On the basis of the results, it would be told that which filter is better for the different types of noise present in the medical images. After the study of this, we will be able to find the best possible way for the reduction of noise from the medical images for the better interpretation of the images in which the sickness of the human is present. With the help of noise removal, the efficiency of diagnosis can be increased.

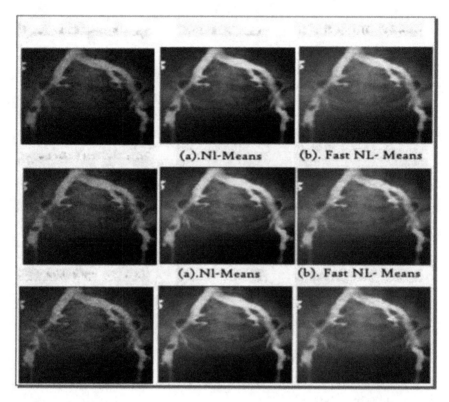

Fig. 10 Perceptual quality comparison of various de-speckled images with variance 0.5, 0.6 and 0.7 [8]

8 Conclusion

In the medical field, doctors face problems regarding the noise present in the medical images and they are in problem to recover the image after removing the noise from the images. Although, it is not certain to recommend any efficient method in which works for all medical images mobility as every method has its pros and cons. This paper gives a review of the de-noising filters used for the different MRI images and in every different noise the result noticed was different with respect to minimum value and the clarity. It infers that the determination of filters for expelling the noise from medicinal pictures depends on the sort of noise which is available in the picture and filter method which will be used.

References

1. Hathaliya, J., Tanwar, S., Tyagi, S., Kumar, N.: Securing electronics healthcare records in healthcare 4.0: a biometric-based approach. Comput. Electr. Eng. **76**, 398–410 (2019)
2. Perona, P., Malik, J.: Scale-space and edge detection using anisotropic diffusion. IEEE Trans. Pattern Anal. Mach. Intell. **12**, 629–639 (1990)
3. Kumar, N., Nachmani, M.: Noise removal and filtering techniques used in medical images. Orient. J. Comput. Sci. Technol. **10**(1), 103–113 (2017)
4. Pizurica, A., Philips, W., Lemahieu, I., Acheroy, M.: A versatile wavelet domain noise filtration technique for medical imaging. IEEE Trans. **22**, 323–331 (2003)
5. Tomasi, C., Manduchi, R.: Bilateral filter for gray and color images. In: IEEE Sixth International Conference on Computer Vision, Jan 1998, pp. 839–864
6. Kalra, M.K., Maher, M.M., D'Souza, R., Saini, S.: Multi detector computed tomography technology: current status and emerging developments. J. Comput. Assist. Tomogr. **28**, S2–S6 (2004)
7. Digenis, G.A., Sandefer, E.P., Page, R.C., Doll, W.J.: Gamma scintigraphy: an evolving technology in pharmaceutical formulation development-Part 1. Pharm. Sci. Technol. Today **I**, 100–108 (1998)
8. Amine, A., Rziza, M., Aboutajdine, D.: Noise reduction in medical images-comparison of noise removal algorithm, Oct 2012
9. Gupta, R., Tanwar, S., Tyagi, S., Kumar, N., Obaidat, M.S., Sadoun, B.: HaBiTs: blockchain based telesurgery framework for healthcare 4.0. In: International Conference on Computer, Information and Telecommunication Systems (IEEE CITS-2019), Beijing, China, 28–31 Aug 2019, pp. 6–10
10. Vora, J., Devmurari, P., Tanwar, S., Tyagi, S., Kumar, N., Obaidat, M.S.: Blind signatures based secured E-healthcare system. In: International Conference on Computer, Information and Telecommunication Systems (IEEE CITS-2018), Colmar, France, 11–13 July 2018, pp. 177–181
11. Pizurica, A., Wink, A.M., Vansteenkiste, E., Philips, W., Roedrdink, J.: A review of wavelet denoising in MRI and ultrasound brain imaging. Curr. Med. Imaging Rev., Bentham Sci. **2**, 247–260 (2006)
12. Bao, P., Zhang, L.: Noise reduction for magnetic resonance images via adaptive multiscale products thresholding. IEEE Trans. Med. Imaging **22**, 1089–1099 (2003). http://www.springer.com/lncs. Last accessed 2016/11/21
13. Lee, J.S.: Digital image smoothing and the sigma filter. Comput. Vis., Graph. Image Process. **24**, 255–269 (1983)
14. Rudin, L.I., Osher, S., Fatemi, E.: Nonlinear total variation noise removal algorithms. Phys. D: Nonlinear Phenom. **60**, 259–268 (1992)
15. Kumari, A., Tanwar, S., Tyagi, S., Kumar, N.: Fog computing for healthcare 4.0 environment: opportunities and challenges. Comput. Electr. Eng. **72**, 1–13 (2018)

Debunking Online Reputation Rumours Using Hybrid of Lexicon-Based and Machine Learning Techniques

M. P. S. Bhatia and Saurabh Raj Sangwan

Abstract The scale and scope of virality of a malicious rumour hurt the victim more as so many people see it quickly and it is also hard to respond to it. Quintessentially, reputation rumour is a tool for denigration bullying which can sully reputations and ruin the image of a person or an entity in public. This work is the primary study to demonstrate a correlation between denigration cyber-bullying and rumour. A novel *ReputeCheck* model is put forward to detect and debunk rumours sourced online on Twitter about global celebrities and world leaders/politicians. The model uses a lexicon to check for the presence of derogatory words and extracts message-based and user-account-based features to train a learning model. Supervised learning is used to train and test the model on a data set created from the Twitter post. A performance accuracy of 83.4% using support vector machine with Gaussian radial basis function kernel is observed. This study validates the vulnerabilities of denigration associated with a public figure or celebrity are the highest and rumour detection can be used as regulatory mechanism to inhibit the production, dissemination and impact of hateful messages online.

Keywords Rumours · Reputation · Social media · Denigration · Bullying

1 Introduction

With the wireless access to the Internet becoming omnipresent, recent years have witnessed a staunch progression in Web usage with new user-centric applications and media-rich services. Social media has been growing exponentially and has moved from a niche phenomenon to mass adoption. As per the recent statistics,[1] there are approximately 2 billion social media users worldwide with Facebook being the most popular social media platform and 91% of all social media users access social channels via mobile devices. People are no longer constrained by the local and

---

[1] https://www.oberlo.in/blog/social-media-marketing-statistics.

M. P. S. Bhatia · S. R. Sangwan (✉)
Division of Computer Engineering, Netaji Subhas University of Technology, New Delhi, India
e-mail: saurabhsangwan2610@gmail.com

© Springer Nature Singapore Pte Ltd. 2020
P. K. Singh et al. (eds.), *Proceedings of First International Conference on Computing, Communications, and Cyber-Security (IC4S 2019)*, Lecture Notes in Networks and Systems 121, https://doi.org/10.1007/978-981-15-3369-3_25

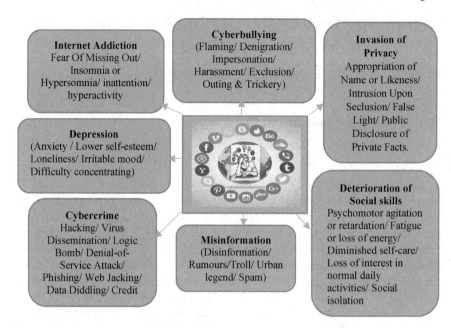

Fig. 1 Negative impact of social media

social boundaries and savour the increased connectivity with individuals around the world. The pros of using social media include seamless and fast transmission of information, communication, creative expression, collaboration, awareness and community building. On the flip side, the rampant proliferation of social networking in our day-to-day lives and the constant presence of screens and virtual access have a significant impact on user's emotional, physical and mental states. There are several studies which establish the fact that social media usage is addictive and has the capability to alter our neural pathways. Another disturbing trend is the amount of time people spend with their faces glued to a screen. A social media statistic uncovers that an average of 2 h and 22 min is spent per day per person on social networks and messaging.[2] Further, given the ease with which anyone anywhere can create an account and post, social media can be used as a weapon to affect and target people or communities/groups by means of bullying, fake news dissemination, spreading rumours and fraudulent activities. Figure 1 depicts the negative impacts of using social media.

With 90.4% of Millennials, 77.5% of Generation X and 48.2% of Baby Boomers as active social media users,[3] 'virtual living' tends to bend the social media impact much more heavily towards the negative side. Quite clearly, the global and pervasive reach of social media has in return given some unpremeditated consequences where people

---

[2]https://www.globalwebindex.com/reports/trends-18.

[3]https://www.emarketer.com.

have discovered illegal and unethical ways to use the socially connected virtual communities. Two of its most severe upshots are *cyber-bullying* where individuals find new means to bully one another over the Internet and *spread of misinformation* which includes disinformation, rumour, urban legend and troll to pollute the digital world. Cyber-bullying is defined by Smith et al. as an 'aggressive, intentional act carried out by a group or individual, using electronic forms of contact, repeatedly and over time against a victim who cannot easily defend himself or herself' [1]. Cyber-bullying involves harassment, threats or bullying through text messages or online. Victims may be children, teens or adults. The types of cyber-bullying include: flaming, outing and trickery, impersonation, denigration, exclusion, harassment, cyber-stalking. Denigration involves online 'dissing' or 'gossiping' about someone by writing and distributing vulgar, derogatory, cruel, mean or untrue rumours. Denigration is the most common bullying tactic involving public figures like celebrities and politicians where rumourous stories, pictures and videos are posted online to discredit and defame.

According to the social psychology, uncertainty, anxiety and mischief are the three driving forces behind the origin and spread of rumourous stories [2]. Further, the rumours can be categorized based on the motive and situation as follows [3, 4]:

- *Dread rumours*: convey fear about a potential negative event or threatening situations, i.e. something bad is going to happen.
- *Wish rumours*: relate a desired outcome, i.e. something good is going to happen.
- *Wedge-driving rumours*: divide people groups and serve to reinforce intergroup differences. The intent is to boost self-esteem when the rumour is about another group. These rumours, when spread strategically, can be used to improve the social status too.
- *Reputation rumours*: that can sully reputations ruin the image of a person or an entity in public.

Thus, the correlation between cyber-bullying and rumour as a critical digital threat is discernible. Quintessentially, reputation rumour is a tool for denigration bullying (Fig. 2). The scale and scope of virality of a malicious rumour hurt the victim more as so many people see it quickly and it is also hard to respond to it.

This work puts forward a model to detect and debunk online reputation rumours using a hybrid of lexicon and machine learning techniques. The proposed model *ReputeCheck* intends to uncover denigrate posts (reputation rumours) which stake and vilify the image of a person or an entity in public. A data set with known rumours sourced online on Twitter about global celebrities and world leaders/politicians is collected for evaluation and validation of the model. Lexical, message-based and user-account-based features are extracted to train the model using a support vector machine (SVM) classifier with Gaussian radial basis function (RBF) kernel. The performance is evaluated on the basis of accuracy.

The rest of the paper is organized as follows: The next section, Sect. 2, describes the related work followed by a detailed illustration of the proposed ReputeCheck model in Sect. 3. Section 4 gives the results, and finally Sect. 5 concludes the research conducted.

Fig. 2 Correlation between denigration bullying and rumour

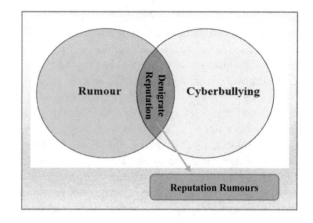

2 Related Work

The recent literature accounts the use of supervised machine learning and deep learning techniques for classifying hate speech, aggression, comment toxicity and bullying content on social forums [5, 6]. Dinakar et al. [7] construct a BullySpace, a common sense knowledge base that encodes particular knowledge about bullying situations and analyses the messages on Formspring (a social networking website) using AnalogySpace common sense reasoning technique. Hinduja et al. [8] explored the relationship between cyber-bullying and suicide among adolescents. The characteristic profile of wrongdoers and victims and conceivable strategies for its preventions are introduced in [9]. Till now, the majority of the work is devoted to text analysis for the most part. The work done in [10–13] basically is a research on cyber-aggression that has utilized text investigation approach on the comments. Yin et al. in [14] use contextual and sentiment features for analysing text and also noticed improved performance using these features for harassment detection achieving an accuracy of 50% on Kongregate, Myspace and Slashdot data set. A very similar work determined an accuracy of 80% in the detection of bullying comments on YouTube using textual context features [15, 16]. Recently, deep learning-based methods have been applied for cyber-bullying detection. Agrawal et al. [5] have experimented with four DNN-based models for cyber-bullying detection which include CNN, LSTM, BLSTM and BLSTM with attention on multiple social media platforms such as Formspring, Twitter and Wikipedia. Huang et al. [17] concatenated social network features with textual features for improved performance of cyber-bullying detection. Some rule-based classifications for the identification of bullies have also been done in [18] using Formspring data set. A new sentence-level filtering model was established by Xu et al. which semantically eliminates bullying words in texts by utilizing grammatical relations among words [19] on the YouTube data set. Works have been done where pictures are utilized for the discovery of cyber-bullying utilizing deep learning models like CNN and RNN or where semantic image features are utilized for identifying bullying [20].

Concurrently, rumour detection has been studied in the pertinent literature as a binary classification task with features extracted from textual comments, user profiles and information propagation patterns [21–29]. Among these feature types, the text-based rumour detection techniques have been prominently studied because the content stability and efficiency of textual feature extraction are higher as compared to modelling a user profile to detect a candidate rumour post. To the best of our knowledge, no correlation between denigration bullying and rumours has been reported in the literature. This research is the preliminary study which uncovers reputation rumour to identify denigration bullying. Consequently, a model is proposed which enables detection of online reputation rumours for timely intervention by moderators and further inhibiting the production and dissemination of inappropriate content to protect the victims.

3 The Proposed *ReputeCheck* Model

Denigration gossips and rumour stories are often outrageous, scandalous and absurd. As per the legal aspect of defamation or denigration, the plaintiff (victim) has a good case if he can prove that:

- the words were defamatory
- they referred to him and
- they were published to a third party.

Based on this, we build a model which intends to identify cases of denigration bullying by means of written defamatory posts in the form of online reputation rumours. The proposed *ReputeCheck* model is shown in Fig. 3.

The following subsections discuss the details.

3.1 Data Collection

We collect a data set with known rumours sourced online as tweets with mentions of global celebrities and world leaders/politicians. In this study, 2000 such comments are reviewed and identified manually. The balanced data set with 1000 'reputation rumours' and 1000 'non-reputation rumours' labels is created.

3.2 Pre-processing

Pre-processing is the process of cleaning and filtering the data to make it suitable for the feature extraction. The process includes:

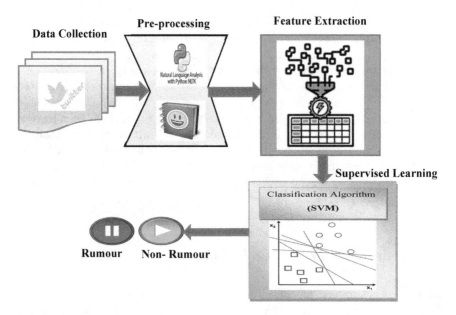

Fig. 3 Proposed *ReputeCheck* model

- Removing numeric and empty texts, URLs, mentions, hashtags, non-ASCII characters, stop words[4] and punctuations.
- Tokenization of comments is done using the TreebankWordTokenizer of Python Natural Language Toolkit (NLTK)[5] to filter the words, symbols and other elements called tokens [30]. The tokens are converted to lower case.
- Replacing slangs and emojis by their descriptive text using the *SMS Dictionary*[6] and *Emojipedia*,[7] respectively. As Internet is an informal way of communication, the use of slangs and emojis is a common practice. These may help understand the context and also intensify the emotion associated.
- Stemming to reduce the words to their root words using Porter stemmer.[8] Stemming enhances the likelihood of matching to the lexicon. Different users can use different terms for the same word, not every synonym word can be included to lexicon, it will increase the processing time, and stemming is crucial for the accuracy of the prediction model.

---

[4]University of Glasgow stop word list.

[5]https://www.nltk.org/.

[6]SMS Dictionary. *Vodacom Messaging*. Retrieved 16 March 2012.

[7]https://emojipedia.org/.

[8]Porter stemmer: http://snowball.tartarus.org/.

3.3 Feature Engineering

The feature vector for building the learning model is trained using a 7-tuple vector <w, i, p, e, m, d, v>, where for each post one binary feature is derived, such that:

- <*w: word*>: The presence or absence of defaming word using the derogatory lexicon base
- <*i: interrogatory*>: The presence or absence of a question mark
- <*p: polarity*>: The presence of negative sentiment polarity
- <*e: emblematic*>: The presence of capitalization or punctuations implemented as a binary feature
- <*m: metadata*>: The presence or absence of metadata (URL/media)
- <*d: date of account creation*>: Set to 1 if account creation is <30 days else 0
- <*v: verification status*>: Verified account or not implemented as a binary feature.

The details of feature extraction and method of feature value assignment to generate feature vector are as follows:

Derogatory Lexicon Base. Lexicons of negative words are necessary resources in extracting features of denigrate bullying based on the assumption that gossiping or defamatory messages usually contain specifically derogatory and vulgar words. The identification of the list of derogatory words contained in denigrated post is thus helpful for the automatic detection of denigrate bullying. Thus, this research makes use of a lexicon-based approach to find the presence or absence of derogatory word with a look-up in the list of English derogatory terms[9] and English vulgarities[10] taken from Wiktionary. The processed tokens from the comments are matched with the lexicon, and the feature value is set to true ('1') if the word is present in the lexicon base else it is set to '0'.

Interrogatory. The presence of derogatory or vulgar words may not necessarily exemplify denigrate bullying in the post. Sceptical or interrogative post triggers a dispute in claim. More specifically, a comment ending with a question mark indicates speculation which is not apparently associated with intimidation. Thus, if question mark/marks are present in the comment, the feature value is set to false ('0') else is set to '1'.

Polarity. Denigration and sentiment analysis are linked as clearly a derogatory post has negative sentiment. The polarity is determined using SentiWordNet (SWN) which is a subjective lexicon containing words with part-of-speech and 3 scores positive, negative and objective. Sum of positive, negative and neutral score sums to 1. The lexicon assigns single score to a word irrespective of the sense in which it is used. For each processed token from the comment, the positive and negative scores are extracted. A synset score is determined by subtracting the negative score from the positive. A comment score is calculated by summing the score of all words and averaging. Thus, if average comment score ≥ 0.5, it implies positive comment and

---

[9]https://en.wiktionary.org/wiki/Category:English_derogatory_terms.

[10]https://en.wiktionary.org/wiki/Category:English_vulgarities.

the feature is set to '0' else if the average comment score ≤ -0.5, it implies a negative comment and the feature value is set to '1'.

Punctuation and Capitalization. The presence of '!' or '...' is more often associated with non-rumours. The presence or absence of capitalization itself is also predictive of rumour. Thus, the feature value is set to '0' on absence else is set to '1'.

Metadata. The presence of URL or media, etc., adds authenticity to the post as these may support the authored information. Rumourous statements are not supported by such evident metadata. We define a binary feature that takes the value '0' when some evidential metadata is available with the tweet else it is set to '1'.

Account Creation Date. A young or a newly created account is more likely to be used for false propaganda/rumour. Thus, the feature value is set to '0' if the account was created more than 30 days before the tweet date else is set to '1'.

User Verification Status. Social verification status of user accounts helps to determine the genuineness of the information. That is, a verified account offers an at-a-glance assurance that the source of information is genuine, whereas an unverified account is very likely to be involved in spreading rumours. Thus, the feature value is set to '0' on verified account, else is set to '1'.

3.4 Classification

Classification in machine learning (ML) is the task of segregating instances of input into various classes based on previous training input [31, 32]. It is a supervised task, wherein a number of labelled training instances are given to the ML classifier as input. The classifier learns from the training sample. Then, the performance is tested on a sample of test instances. The data is run through the support vector machine (SVM) classifier with Gaussian radial basis function (RBF) kernel. Tenfold cross-validation method is used for evaluating the performance of the algorithm.

4 Results

The model used SVM with RBF kernel function. The results were evaluated and compared with the polynomial (POLY) and linear (LIN) function. The SVM-RBF kernel obtained the best performance accuracy of 83.4% followed by POLY and LIN kernels. A comparison of SVM (all kernel function variants) is given in Fig. 4.

The results were also evaluated for other supervised learning classifiers, namely Naïve Bayesian (NB), K-nearest neighbour (K-NN), decision tree (DT) and multi-layer perceptron (MLP) to validate the use of SVM (Fig. 5).

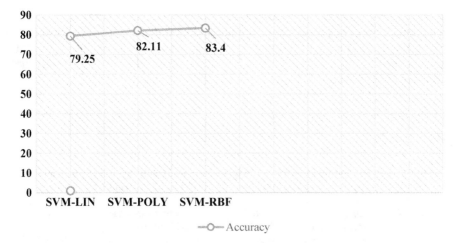

Fig. 4 Accuracy plots comparing SVM (variant kernel functions)

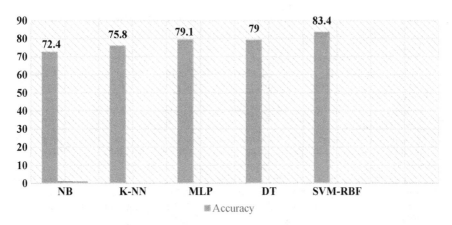

Fig. 5 Accuracy plots comparing with other supervised classifiers

5 Conclusion

With the ubiquitous Internet connectivity and bandwidth in abundance along with reduced cost of smart devices, the impact of computer-mediated technologies, like Twitter, Tumblr, Google+, Facebook, Instagram, Snapchat, YouTube, etc., is manifold and profound. Though social media facilitates digital experience and intense communications, it is often used as a weapon to affect and target people or communities/groups by means of frauds, rumours and bullying. This research presented a mechanism to counter denigration in online posts by debunking reputation rumours. A hybrid model using lexicon-based and supervised machine learning-based techniques was proposed which separates observations into two groups built on their characteristics. The preliminary results are motivating, and as possible future direction a holistic model with a detective mechanism for resolving online denigration can be contemplated. Also, deep learning techniques have proven capabilities in natural language processing tasks and their self-learning abilities can be used to improve the accuracy of classification.

References

1. Smith, P.K., Del Barrio, C., Tokunaga, R.S.: Definitions of bullying and cyberbullying: how useful are the terms. In: Principles of Cyberbullying Research: Definitions, Measures, and Methodology, pp. 26–40. Routledge, New York (2013)
2. Ting, C.S.W., Song, S.G.Z.: What lies beneath the truth: a literature review on fake news, false information. Report, Institute of Policy Studies, National University of Singapore (2017)
3. Difonzo, N., Bordia, P.: Rumor Psychology: Social and Organizational Approaches (2007). https://doi.org/10.1037/11503-000
4. Solove, D.: The Future of Reputation: Gossip, Rumor, and Privacy on the Internet. Yale University Press, New Haven (2007)
5. Agrawal, S., Awekar, A.: Deep learning for detecting cyberbullying across multiple social media platforms. In: European Conference on Information Retrieval. Springer, Cham (2018)
6. Kumar, A., Sachdeva, N.: Cyberbullying detection on social multimedia using soft computing techniques: a meta-analysis. Multimed. Tools Appl. 1–38 (2019)
7. Dinakar, K., Jones, B., Havasi, C., Lieberman, H., Picard, R.: Common sense reasoning for detection, prevention, and mitigation of cyberbullying. ACM Trans. Interact. Intell. Syst. 2(3), 18:1–18:30 (2012)
8. Hinduja, S., Patchin, J.W.: Bullying, cyberbullying, and suicide. Arch. Suicide Res. 14(3), 206–221 (2010)
9. Kokkinos, C.M., Antoniadou, N., Markos, A.: Cyberbullying: an investigation of the psychological profile of university student participants. J. Appl. Dev. Psychol. 35(3), 204–214 (2014)
10. Dadvar, M., de Jong, F.M.G., Ordelman, R.J.F., Trieschnigg, R.B.: Improved cyberbullying detection using gender information. In: Proceedings of the Twelfth Dutch-Belgian Information Retrieval Workshop (DIR 2012), Ghent, Belgium, pp. 23–25, Ghent, University of Ghent, Feb 2012
11. Nahar, V., Unankard, S., Li, X., Pang, C.: Sentiment analysis for effective detection of cyber bullying. In: Web Technologies and Applications, pp. 767–774. Springer, Berlin (2012)
12. Reynolds, K.K., Edwards, L.: Using machine learning to detect cyberbullying. In: Fourth International Conference on Machine Learning and Applications, vol. 2, pp. 241–244 (2011)
13. Ptaszynski, M., Dybala, P., Matsuba, T., Masui, F., Rzepka, R., Araki, K., Momouchi, Y.: In the service of online order tackling cyberbullying with machine learning and affect analysis (2010)
14. Yin D., et al.: Detection of harassment on web 2.0. In: Proceedings of the Content Analysis in the WEB 2, pp. 1–7 (2009)
15. Dinakar, K., Reichart, R., Lieberman, H.: Modeling the detection of textual cyberbullying. In: The Social Mobile Web (2011)
16. Marathe, S.S., Shirsat, K.P.: Approaches for Mining YouTube Videos Metadata in Cyberbullying detection. Int. J. Eng. Res. Technol. 4(05), 680–684 (2015)
17. Huang, Q., Singh, V.K., Atrey, P.K.: Cyber bullying detection using social and textual analysis. In: Proceedings of the 3rd International Workshop on Socially-Aware Multimedia, pp. 3–6. ACM (2014)

18. Reynolds, K., April, K., Lynne, E.: Using machine learning to detect cyberbullying. In: 2011 10th International Conference on Machine Learning and Applications and Workshops (ICMLA), vol. 2. IEEE (2011)
19. Xu, Z., Zhu, S.: Filtering offensive language in online communities using grammatical relations. In: Proceedings of the Seventh Annual Collaboration, Electronic Messaging, Anti-Abuse and Spam Conference (2010)
20. Zerr, S., et al.: Privacy-aware image classification and search. In: Proceedings of the 35th International ACM SIGIR Conference on Research and Development in Information Retrieval. ACM (2012)
21. Kumar, A., Sangwan, S.R., Nayyar, A.: Rumor veracity detection on twitter using particle swarm optimized shallow classifiers. Multimed. Tools Appl. (2019). https://doi.org/10.1007/s11042-019-7398-6
22. Kumar, A., Sangwan, S.R.: Rumor detection using machine learning techniques on social media. In: International Conference on Innovative Computing and Communications, pp. 213–221. Springer, Singapore (2019)
23. Yang, F., Liu, Y., Yu, X., Yang, M.: Automatic detection of rumor on Sina Weibo. In: Proceedings of the ACM SIGKDD Workshop on Mining Data Semantics, p. 13. ACM (2012)
24. Zubiaga, A., Geraldine, W.S.K., Maria, L., Rob, P.: PHEME dataset of rumours and non-rumours. figshare. Dataset (2016)
25. Ma, B., Lin, D., Cao, D.: Content representation for microblog rumor detection. In: Advances in Computational Intelligence Systems, pp. 245–251. Springer, Cham (2017)
26. Chen, T., Li, X., Yin, H., Zhang, J.: Call attention to rumors: deep attention based recurrent neural networks for early rumor detection. In: Pacific-Asia Conference on Knowledge Discovery and Data Mining, pp. 40–52. Springer, Cham (2018)
27. Nguyen, T.N., Li, C., Niederée, C.: On early-stage debunking rumors on twitter: leveraging the wisdom of weak learners. In: International Conference on Social Informatics, pp. 141–158. Springer, Cham (2017)
28. Wang, Z., Guo, Y., Li, Z., Tang, M., Qi, T., Wang, J.: Research on microblog rumor events detection via dynamic time series based GRU model. In: ICC 2019—2019 IEEE International Conference on Communications (ICC), Shanghai, China, pp. 1–6 (2019)
29. Bugueño, M., Sepulveda, G., Mendoza, M.: An empirical analysis of rumor detection on microblogs with recurrent neural networks. In: Meiselwitz, G. (ed.) Social Computing and Social Media. Design, Human Behavior and Analytics. HCII 2019. Lecture Notes in Computer Science, vol. 11578. Springer, Cham (2019)
30. Loper, E., Bird, S.: NLTK: the natural language toolkit. In: Proceedings of the ACL-02 Workshop on Effective Tools and Methodologies for Teaching Natural Language Processing and Computational Linguistics (2002)
31. Kumar, A., Dogra, P., Dabas, V.: Emotion analysis of Twitter using opinion mining. In: Eighth International Conference on Contemporary Computing (IC3), pp. 285–290. IEEE (2015)
32. Kumar, A., Khorwal, R.: Firefly algorithm for feature selection in sentiment analysis. In: Computational Intelligence in Data Mining, pp. 693–703. Springer, Singapore (2017)

A Novel Approach for Optimal Digital FIR Filter Design Using Hybrid Grey Wolf and Cuckoo Search Optimization

Suman Yadav, Manjeet Kumar, Richa Yadav and Ashwni Kumar

Abstract This paper represents designing of one-dimensional finite impulse filter (FIR) by computing the optimal filter coefficients in a way that the frequency response of the designed filter approximates ideal frequency response using proposed optimization algorithms. In this work, hybrid optimization of grey wolf and cuckoo search algorithm is considered to design a linear phase FIR low-pass, high-pass, band-pass and band-stop filters of 20th order. Analysis of simulation results of the designed filter is done, and various parameters are calculated such as maximum stopband ripple, maximum passband ripple and maximum attenuation in stop band. Results have been compared to recently published algorithms for optimization such as cat swarm optimization (CSO), particle swarm optimization (PSO), real-coded genetic algorithm (RCGA) and differential evolution (DE) to prove the superiority of the proposed algorithm.

Keywords Digital filter design · Filter response · Cat swarm optimization algorithm · Bio-inspired algorithms · Grey wolf algorithm · Cuckoo search algorithm

1 Introduction

Filtering plays a major role in digital signal processing, in different domains like biomedical image processing, communication system, satellite image processing, speech processing, audio signal processing, etc. Mainly filters are divided in two types that are FIR and infinite impulse response (IIR). They are used according to the requirement of the application. FIR filters are known for their linear phase

S. Yadav (✉)
ECE, Bharati Vidyapeeth College of Engineering, New Delhi 110063, India
e-mail: Suman.yadav@bharatividyapeeth.edu

M. Kumar
ECE, Bennett University, Greater Noida, Uttar Pradesh 201310, India

S. Yadav · R. Yadav · A. Kumar
ECE, Indira Gandhi Delhi Technical University for Women (IGDTUW), New Delhi 110006, India

© Springer Nature Singapore Pte Ltd. 2020
P. K. Singh et al. (eds.), *Proceedings of First International Conference on Computing, Communications, and Cyber-Security (IC4S 2019)*, Lecture Notes in Networks and Systems 121, https://doi.org/10.1007/978-981-15-3369-3_26

response and inherent stability, while IIR have an advantage of lesser filter coefficient, therefore less memory for the same order as needed to design FIR filter. So they are less computationally complex than FIR filter.

Still, the design and implementation of the digital filter is an interesting area for researchers as they are facing challenges to reduce ripples in passband and stopband, to get higher attenuation in stopband with smaller filter order as much as possible. They are trying to solve multi-objective problems to maintain a trade-off amongst multiple specifications using newly proposed nature-inspired algorithms. Earlier classical methods were used to design a digital filter such as windowing, frequency sampling and Park-McClellan. They have certain disadvantages as they are more likely to struck at local minima, not able to converge to global optimal solution. To avoid such conditions, various meta-heuristic and heuristic evolutionary algorithms replaced them with their ability to converge to global optima. There are many recently published algorithms available to the researcher which are based on meta-heuristic techniques to design and implement FIR filter such as genetic Algorithm (GA) [1], real-coded genetic algorithm (RCGA) [2], differential evolution (DE) [3], particle swarm optimization (PSO) [4], cuckoo search algorithm (CSA) [5], artificial bee colony algorithm (ABC) [6], gravitational search algorithm (GSA) [7], adaptive cuckoo search algorithm (ACSA) [8], flower pollination algorithm (FPA) [9], ant colony optimization (ACO) [10], grey wolf optimization (GWO) [11], cat swarm optimization (CSO) [7], bat algorithm (BA) [12], bacteria foraging algorithm (BFA) [7] and seeker optimization algorithm (SOA) [13] and their variants. Many hybrid variants of PSO and CSA have also been used.

A common problem encountered with above-mentioned algorithms is ripples in passband and stopband, convergence at an early stage, controlling parameters of these techniques. This work proposes a hybrid technique that includes grey wolf, and cuckoo search algorithm (HGWOCS) [14] provides better trade-off amongst exploration and exploitation of the search space. It also helps in reducing ripples in filter passband and attenuation is higher in the stopband.

The paper is written as follows, the problem statement of FIR filter design is specified in Sect. 2. Section 3 describes the methodology employed and pseudocode. Simulations and result are mentioned in Sects. 4 and 5, also signify the importance of proposed algorithm. Section 6 concludes the work and includes references.

2   Design Model of FIR

The response of FIR filters is specified in the frequency domain as follows:

$$B(z) = \sum_{p=0}^{p-1} b(p)z^{-p} \tag{1}$$

$$= b(0) + b(1)z^{-1} + \cdots + b(N-1)z^{N-1} \tag{2}$$

$b(p)$ are filter coefficients of the designed filter, whereas $p - 1$ is the order of the filter as defined in Eq. 1. The aim is to optimize these coefficients using hybrid grey wolf and cuckoo search algorithm to obtain $b(p)$ such that the deviation between the frequency response of the designed filter and the frequency response of the ideal filter is minimized.

For the symmetric condition of FIR filter, following condition of impulse response of filter needs to be satisfied:

$$b(p) = b(P - p) \quad \text{for } 0 \le p \le P \tag{3}$$

For the FIR filter to be of linear phase, the coefficients of the filter need to follow Eq. (2). Linear phase is the necessary property of a filter for not to loose any critical information while filtering a signal. Because of this condition, only half of the coefficients needs to be calculated and rest are duplicated to determine another half part of the impulse response.

Magnitude response of FIR filter

$$B(\omega) = \sum_{p=0}^{P-1} b(p)\mathrm{e}^{-j\omega p} \tag{4}$$

needs to be calculated and rest are duplicated to determine another half part of the impulse response. $B(\omega)$ is the real magnitude response of the estimated coefficients. It can be LP, HP, BP or BS filter depending on the values of $b(p)$.

Ideal FIR LP filter magnitude response:

$$B_{\mathrm{id}}(w) = \begin{cases} 1 \ \forall \omega \in [0, \omega_{\mathrm{c}}] \\ 0 \ \text{otherwise} \end{cases} \tag{5}$$

Ideal FIR HP filter magnitude response:

$$B_{\mathrm{id}}(w) = \begin{cases} 1 \ \forall \omega \in [0, \omega_{\mathrm{c}}] \\ 0 \ \text{otherwise} \end{cases} \tag{6}$$

Ideal FIR BP filter magnitude response:

$$B_{\mathrm{id}}(w) = \begin{cases} 1 \ \forall \omega \in [\omega_{\mathrm{l}}, \omega_{\mathrm{h}}] \\ 0 \ \text{otherwise} \end{cases} \tag{7}$$

Ideal FIR BS filter magnitude response:

$$B_{\mathrm{id}}(w) = \begin{cases} 0 \ \forall \omega \in [\omega_{\mathrm{l}}, \omega_{\mathrm{h}}] \\ 1 \ \text{otherwise} \end{cases} \tag{8}$$

$B_{id}(w)$ is defined as ideal magnitude response of FIR filter. For the ideal low-pass filter and ideal high-pass filter, ω_c is defined as the cut-off frequency. ω_l and ω_h are the lower and upper edge frequency of the band-pass and band-stop filters. There are a variety of error fitness functions available in literature like mean square error, absolute error difference, square root error, maximum error difference, etc. Selection of error fitness function depends on the problem identified. In this proposed work, mean absolute error is used as described below.

Mean absolute error (MAE)

$$A(\omega) = \sum_{\forall \omega} (|B_{id}(\omega)| - |B(\omega)|) \tag{9}$$

The purpose is to optimize the difference $A(\omega)$, i.e. the absolute difference between ideal frequency response and approximated frequency response of the FIR filter. $A(\omega)$ is the error fitness function for the optimization problem of designed filter.

3   Hybrid Grey Wolf and Cuckoo Search Optimization

In this proposed work, a hybrid combination of two optimization algorithms has been used. Grey wolf optimization is used for exploration and cuckoo search for exploitation in search space. Grey wolf optimization is based on hunting mechanism of grey wolves and mimics their hierarchy. Four types of wolves are there in the pack named as alpha, beta, delta, and omega according to their leadership hierarchy. Alpha wolf is the most dominant one or leader responsible for decision-making. It is the most powerful wolf in the pack. Beta wolf are the subordinates to the alphas, helps them in pack activity as an advisor. Omega comes at the lowest position in the hierarchy. They are like scapegoat, have to submit to all wolves. Delta wolves have to follow the instruction from alpha and beta, but they dominate omega.

Wolves live in a pack of 5–12. The procedure for hunting involves searching, encircling and attacking the prey. These steps are mathematically modelled as described in Eqs. (10–13) to design GWO algorithm and describes the encircling behaviour of the wolves. The social hierarchy in the GWO is designed such a way that the fittest solution is assumed to be alpha that is with minimum fitness. Best solution is followed by two more solutions with decreasing fitness function as beta and delta. These three solutions have a better idea of the position of prey. Rest of the wolves that is omega has to follow alpha, beta and gamma according to Eqs. (14 and 15) represent the hunting mechanism of the wolves.

$$\vec{D} = \left| \vec{C} . \vec{Y_p}(t) - \vec{Y}(t) \right| \tag{10}$$

$$\vec{Y}(t+1) = \vec{Y_p}(t) - \vec{A} . \vec{D} \tag{11}$$

$$\vec{A} = 2\vec{a}.\vec{r1} - \vec{a} \tag{12}$$

$$\vec{C} = 2.\vec{r_2} \tag{13}$$

Current iteration is indicated by t, $\vec{A}$ and $\vec{C}$ are coefficient vector. $\vec{Y_p}$ is the position of prey to be hunted and $\vec{Y}$ indicates the position of grey wolf. $\vec{D}$ represents the difference in distance between wolf and prey. $\vec{a}$ is linearly decremented from 2 to 0 as the iteration increases. $\vec{r_1}$ and $\vec{r_2}$ are random vectors in [0 1].

$$\vec{D_\alpha} = \left| \vec{C_1} \cdot \vec{Y_\alpha} - \vec{Y} \right|, \quad \vec{D_\beta} = \left| \vec{C_2} \cdot \vec{Y_\beta} - \vec{Y} \right|, \quad \vec{D_\gamma} = \left| \vec{C_2} \cdot \vec{Y_\gamma} - \vec{Y} \right| \tag{14}$$

$$\vec{Y_1} = \vec{Y_\alpha} - \vec{A_1} \cdot \left(\vec{D_\alpha} \right), \quad \vec{Y_2} = \vec{Y_\beta} - \vec{A_2} \cdot \left(\vec{D_\beta} \right), \quad \vec{Y_3} = \vec{Y_\gamma} - \vec{A_3} \cdot \left(\vec{D_\gamma} \right) \tag{15}$$

where Y_α, Y_β and Y_δ are the location of alpha, beta and gamma wolf. Wolves move towards the prey with a step size which is weighted by a constant and it increases the chance of stucking into local optima. So, cuckoo search optimization is used to update the current positions of the alpha, beta and gamma based on the best position obtained so far. Cuckoo search optimization mimics the parasitic behaviour of cuckoos in laying their eggs in other host bird nest that are fixed in number. Host bird takes care of the cuckoo egg as their own, but if it recognizes the cuckoo's egg it abandons its nest or either throws the cuckoo eggs away. Probability of identifying cuckoo egg is fixed between 0 and 1. It has some assumption:

1. Each cuckoo is allowed to lay one egg at a time and arbitrarily keep its egg in other host nest.
2. Only the nest with good quality of eggs is transferred to the next generation.
3. The next search space of pth cuckoo at time t is given as:

$$Y_p(t + 1) = Y_p(t) + \alpha * \text{levy}(\gamma) \tag{16}$$

$Y_p(t + 1)$ is the next search space, $Y_p(t)$ is previous search space and $\text{levy}(\gamma)$ describes the random step size through levy flight. α is a constant depends on the dimension of search space. It is assumed to be 1. CS optimality more relies on other habitat groups not only on time. After updating the positions by CS method, transfer will get back to GWO to calculate the average of three best positions again. Using this hybrid combination, local optima error is trade-off.

$$\vec{Y}(t + 1) = \frac{\vec{Y_1} + \vec{Y_2} + \vec{Y_3}}{3} \tag{17}$$

4. Psuedocode for design of FIR digital filter is shown in Fig. 1.

Pseudo Code

Initialize the population of wolf as search agent X_j as $j = 1,2,3\ldots\ldots$
Set parameter value of a, C and A
Fitness value of each search agent (wolf) is calculated.
Best three positions are selected as Y_α, Y_β and Y_δ with minimum fitness function
If l<(max iteration)
for every search agent
 Update the position of alpha, beta and gamma by CS based on best search agent
 Take the average of Y_α, Y_β Y_δ by GWO
End for
Update C, A and a.
Determine the fitness of every search agents.
Update Y_α, Y_β Y_δ using cuckoo search algorithm
L = l+1
End if
Output X_α

Fig. 1 FIR filter design using hybrid grey wolf cuckoo search optimization

4 Simulation Results

4.1 Low-Pass Filter

These simulations are done using MATLAB with Intel core i5 7th Generation 2.70 Ghz, 8 GB RAM to design 20th order FIR LP, HP, BP and BS filters. Specification for the FIR LPF is defined with cut-off frequency as $0.5 * \pi$. HGWOCS algorithm is run for 500 cycles to get the best result. Population size is taken as 30, whereas in all other algorithms, it is taken as 55 so less computation is done in the proposed algorithm. The range for the frequency components ω lies from $[0, \pi]$. The limits for the filter coefficients are taken from -1 to 1. Optimized coefficients are obtained by minimizing the error fitness function using HGWOCS and given in Table 1. Magnitude response based on these coefficients and its dB response is plotted in Fig. 2.

4.2 High-Pass Filter

FIR high-pass filter designed with certain specification like edge frequencies for the passband is taken as 0.5π. Order of the filter is set as 20. Limits for the filter coefficients are taken as -1 to $+1$. Owing to optimal filter coefficients given in Table 2, high-pass filter is designed as shown in Fig. 3.

Table 1 Optimized filter coefficients of FIR low-pass filter of 20th order

$b(p)$	HGWOCS	RGA [7]	PSO [7]	DE [7]	CSO [7]
$b(1) = b(21)$	−0.000286889	0.0206445080112550	0.0251167933552393	0.0270053999982491	0.0288986463975570
$b(2) = b(20)$	0.0135992	0.0487214131185106	0.0472192593000299	0.0472668667977926	0.0474313111424457
$b(3) = b(19)$	0.000242328	0.0058686011564964	0.0035462422723169	0.0053202042222841	0.0058656067648627
$b(4) = b(18)$	−0.0297009	−0.0409668653000227	−0.0400940472835999	−0.0389822948593733	−0.0357453079449558
$b(5) = b(17)$	−0.000172927	−0.0008635067800222	−0.0005204320672145	−0.0034522253386096	0.0013515901663467
$b(6) = b(16)$	0.0529637	0.0597960312655655	0.0609072077778672	0.0579468588721711	0.0596294544284159
$b(7) = b(15)$	0.000152428	−0.0014088428662974	−0.0017592407567773	−0.0020514005933964	0.0039075356955502
$b(8) = b(14)$	−0.0998583	−0.1031178347003011	−0.1036139949446693	−0.1027152676299155	−0.1031154183865900
$b(9) = b(13)$	−0.000203051	−0.0004406443820899	0.0006276230374221	0.0016929378017931	0.0029165885728897
$b(10) = b(12)$	0.316181	0.3176065126194617	0.3181190365486844	0.3197956762587685	0.3186922137395391
$b(11)$	0.500191	0.5000185389015569	0.5000185389015569	0.5000185389015569	0.5048214654662431

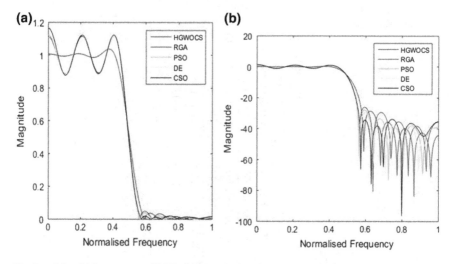

Fig. 2 **a** Magnitude response of LPF of 20th order and **b** magnitude response in dB scale

4.3   Band-Pass Filter

BPF has two edge frequencies one as passband 0.33π and other as stopband frequency 0.66π. HGWOCS generates the optimal filter coefficient given in Table 3. By minimizing the deviation between the desired and ideal frequency response of the filter, in designing FIR band-pass filter. The maximum and minimum range for the coefficients of filter is set as $+1$ and -1. Filter order is set as 20. Desired magnitude response and its dB response is plotted in Fig. 4.

4.4   Band-Stop Filter

Specifications to design BS filter are taken as follows: lower bound frequency 0.33π and upper bound frequency 0.66π for filter order 20. The limits of filter coefficients are set between -1 and $+1$. The optimized coefficient for band-stop FIR filter is given in Table 4, and using these coefficients, magnitude response and its dB plot are modelled in Fig. 5.

Table 2 Optimized filter coefficients of FIR high-pass filter of 20th order

$b(p)$	HGWOCS	RGA [7]	PSO [7]	DE [7]	CSO [7]
$b(1) = b(21)$	0.00757704	0.021731353545	0.025559145974814	0.029041921147266	0.027538232958190
$b(2) = b(20)$	0.013642	−0.048131602227058	−0.047413653181042	−0.045873202416582	−0.044380496079130
$b(3) = b(19)$	−0.00906027	0.006298189918824	0.005135430273491	0.002950561225606	0.003299313166652
$b(4) = b(18)$	−0.0294523	0.041895345956760	0.039988099089174	0.041311799862169	0.042933406737536
$b(5) = b(17)$	0.0105463	0.000879943669486	0.001405996354021	−0.000283997158910	0.000099323498881
$b(6) = b(16)$	0.0535474	−0.059027866591514	−0.060283192968605	−0.060002355552046	−0.058234539320593
$b(7) = b(15)$	−0.0105409	−0.000013559660394	0.000768613197325	−0.003921102337490	0.003181307723221
$b(8) = b(14)$	−0.1004630	0.104257677520726	0.105120739785348	0.106119151142982	0.102402693086120
$b(9) = b(13)$	0.0105959	0.003823743541217	0.001471927911810	−0.000565063060302	−0.002236399396214
$b(10) = b(12)$	0.316142	−0.316631427282300	−0.315471590838371	−0.320083906578923	−0.317925532944817
$b(11)$	−0.511346	0.499468012025621	0.499981461098444	0.499981461098444	0.499981461098444

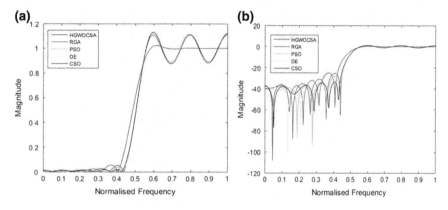

Fig. 3 **a** Magnitude response of HPF of 20th order and **b** magnitude response in dB scale

5   Analysis of Simulation Result

It can be seen from Table 5 that measure of stopband attenuation for FIR LP, HP, BP and BS filters designed with the respective algorithms are mentioned. For the FIR LPF, CSO has higher stopband attenuation (33.99) and HGWOCS is nearly equivalent to it (30.588) as compared to others like DE (29.53), PSO (28.03) and RGA (26.11). So, HGWOCS is efficient in attenuating the signal in stopband. For HPF, HGWOCS has got higher attenuation in stopband as desirable, whereas RGA is least amongst all. In the case of BPF, HGWOCS is not as efficient as for other filters in stopband. But for BSF has got higher capability of attenuating the signal in stopband as compared to all the other algorithms.

Table 6 shows that magnitude of ripples in passband for FIR LP, HP, BP and BS filters designed with given techniques. It is observed that filter designed with HGWOCS is superior to all other algorithms mentioned here. Magnitude of ripple in terms of ascending order is mentioned below:

For LPF: HGWOCS > CSO > PSO > DE > RGA
For HPF: HGWOCS > CSO > RGA > PSO > DE
For BPF: HGWOCS > CSO > PSO > RGA > DE
For BSF: HGWOCS > DE > PSO > RGA > CSO.

Table 3 Optimized filter coefficients of FIR band-pass filter of 20th order

$b(p)$	HGWOCS	RGA [7]	PSO [7]	DE [7]	CSO [7]
$b(1) = b(21)$	0.0310321	0.028502857104	0.024910907374264	0.0243157595656957	0.029219726125746
$b(2) = b(20)$	−0.00000854	−0.001893868108392	0.000092972958187	0.002788616388318	−0.002853754882048
$b(3) = b(19)$	−0.0506536	−0.076189026154460	−0.074535581888545	−0.075883731240533	−0.075276944714603
$b(4) = b(18)$	−0.000205944	0.000994123920259	−0.000579129089510	−0.003313368788351	0.003288541979801
$b(5) = b(17)$	0.000846892	0.053196793860741	0.058322287561503	0.056134992376798	0.051130186585096
$b(6) = b(16)$	0.000708793	−0.000639149080848	−0.000187613541059	0.000257826174027	0.001521235029526
$b(7) = b(15)$	0.128473	0.100057194730152	0.093164875388599	0.090912344142406	0.100018377612239
$b(8) = b(14)$	−0.000918319	0.001409980793664	0.001012723950710	0.002199187065772	−0.005673995620418
$b(9) = b(13)$	−0.271717	−0.299380312728113	−0.2968669179983546	−0.300934749358008	−0.298355272098757
$b(10) = b(12)$	0.000395847	−0.000752480372393	−0.00000392232750468	−0.001799401229551	0.002785499453246
$b(11)$	0.33439	0.400369877077545	0.400369877077545	0.400369877077545	0.400369877077545

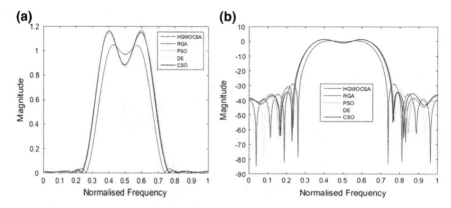

Fig. 4 **a** Magnitude response of BPF of 20th order and **b** magnitude response in dB scale

6 Conclusion

The proposed research aims in designing FIR LP, HP, BP and BS filters having minimum ripples in passband and higher attenuation in stopband based on HGWOCS optimization algorithm. Performance of this technique HGWOCS is compared with various other algorithms provided in literature for designing 20th order filter such as CSO, PSO, DE and RGA. It is found that there is improvement in both the objectives as HGWOCS has got improvement in reduction of ripples in passband with higher attenuation in stopband for designing FIR filters.

Table 4 Optimized filter coefficients of FIR band-stop filter of 20th order

b(p)	HGWOCS	RGA [7]	PSO [7]	DE [7]	CSO [7]
$b(1) = b(21)$	−0.0178001	0.0087652244382	0.005065078955931	0.0057381639377772	0.011453242233574
$b(2) = b(20)$	−0.00705384	0.0547969232249762	0.0544967166662981	0.053905628215447	0.0524374111749113
$b(3) = b(19)$	0.0517999	0.0017964199983890	0.005809988516188	0.002902448937586	0.011428069462394
$b(4) = b(18)$	0.0337037	0.048911654246731	0.051144048751957	0.049349878942931	0.047411160505444
$b(5) = b(17)$	−0.0624954	−0.054718457691943	−0.050663949788261	−0.050884656047053	−0.049098141901288
$b(6) = b(16)$	−0.0475653	−0.060963142228236	−0.062741465298722	−0.063088550820316	−0.065254829012881
$b(7) = b(15)$	0.0160424	0.004293459264617	−0.0000627184164445	0.004089341810059	−0.001020332055708
$b(8) = b(14)$	−0.0445172	−0.065342448643273	−0.068916923681426	−0.068023108311494	−0.068705355567975
$b(9) = b(13)$	0.0612498	0.300682045893488	0.297478557865240	0.296714629509332	0.296714629509332
$b(10) = b(12)$	0.566018	0.069036675664641	0.074390206250426	0.071701365941159	0.074218022294613
$b(11)$	−0.0996363	0.499582536276171	0.499582536276171	0.500000357523254	0.500000357523254

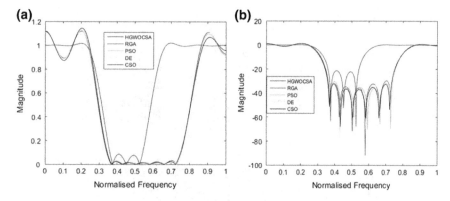

Fig. 5 **a** Magnitude response of BPF of 20th order and **b** magnitude response in dB scale

Table 5 Comparison of various algorithms in terms stopband attenuation for filter order 20

Filter type	Maximum stopband attenuation (dB)				
	HGWOCS	CSO [7]	DE [7]	PSO [7]	RGA [7]
LPF	30.588	33.99	29.53	28.03	26.11
HPF	35.8371	33.62	29.16	28.1	25.25
BPF	28.6866	34.47	32.58	32.03	30.8
BSF	34.1669	32.11	30.96	30.56	29.73

Table 6 Comparison of various algorithms in terms maximum passband ripple with filter order 20

Type of filter	Maximum passband ripple				
	HGWOCS	CSO [7]	DE [7]	PSO [7]	RGA [7]
LPF	1.036	1.124	1.360	1.230	1.650
HPF	1.023	1.100	1.130	1.120	1.110
BPF	1.047	1.144	1.161	1.150	1.160
BSF	1.086	1.140	1.101	1.106	1.108

References

1. Boudjelaba, K., Ros, F., Chikouche, D.: An efficient hybrid genetic algorithm to design finite impulse response filters. Expert Syst. Appl. **41**(13), 5917–5937 (2014)
2. Aggarwal, A., Rawat, T.K., Kumar, M., Upadhyay, D.K.: Optimal design of FIR high pass filter based on L1 error approximation using real coded genetic algorithm. Eng. Sci. Technol., Int. J. **18**(4), 594–602 (2015)
3. Reddy, K.S., Sahoo, S.K.: An approach for FIR filter coefficient optimization using differential evolution algorithm. AEU-Int. J. Electron. Commun. **69**(1), 101–108 (2015)
4. Kumar, M.: Fractional order FIR differentiator design using particle swarm optimization algorithm. Int. J. Numer. Model.: Electron. Netw., Devices Fields **32**(2) (2019)

5. Sharma, I., Kuldeep, B., Kumar, A., Singh, V.K.: Performance of swarm-based optimization techniques for designing digital FIR filter: a comparative study. Eng. Sci. Technol., Int. J. **19**(3), 1564–1572 (2016)
6. Karaboga, N.: A new design method based on artificial bee colony algorithm for digital IIR filters. J. Franklin Inst. **346**(4), 328–348 (2009)
7. Saha, S.K., Ghoshal, S.P., Kar, R., Mandal, D.: Cat swarm optimization algorithm for optimal linear phase FIR filter design. ISA Trans. **52**(6), 781–794 (2013)
8. Sarangi, S.K., Panda, R., Das, P.K., Abraham, A.: Design of optimal high pass and band stop FIR filters using adaptive Cuckoo search algorithm. Eng. Appl. Artif. Intell. **70**, 67–80 (2018)
9. Singh, S., Ashok, A., Rawat, T.R., Kumar, M.: Optimal IIR system identification using flower pollination algorithm. In: IEEE First International Conference on Power Electronics, Intelligent Control and Energy Systems, pp. 1–6, Delhi (2016)
10. Benhala, B.: Ant colony optimization for optimal low-pass butterworth filter design. WSEAS Trans. Circuits Syst. **13**, 313–318 (2014)
11. Mirjalili, S., Saremi, S., Mirjalili, S.M., Coelho, L.D.: Multi-objective grey wolf optimizer: a novel algorithm for multi-criterion optimization. Expert Syst. Appl. **47**, 106–109 (2016)
12. Kumar, M., Aggarwal, A., Rawat, T.K.: Bat algorithm: application to adaptive infinite impulse response system identification. Arab. J. Sci. Eng. **41**(9), 3587–3604 (2016)
13. Dai, C., Chen, W., Zhu, Y.: Seeker optimization algorithm for digital IIR filter design. IEEE Trans. Ind. Electron. **57**(5), 1710–1718 (2009)
14. Daniel, E., Anitha, J., Gnanaraj, J.: Optimum laplacian wavelet mask based medical image using hybrid cuckoo search–grey wolf optimization algorithm. Knowl.-Based Syst. **131**, 58–69 (2017)

Quasi-opposition-Based Multi-verse Optimization Algorithm for Feature Selection

Rahul Hans and Harjot Kaur

Abstract These days a category of algorithms known as metaheuristic algorithms is considered as more reliable for finding the solution of various optimization problems. Multi-verse optimization algorithm is one of the latest developed optimization algorithms, which mimics the behaviours of multi-verse theory in physics and impersonates the relationships amongst different universes. In machine learning, finding out the best feature subset from the original set containing the minimum number of features with maximum average accuracy (or at least remains same) is a difficult task and is considered as an optimization problem. This research proposes a quasi-oppositional learning-based multi-verse optimization algorithm which aims to improve the initial phase of setting up of random solutions along with quasi-opposite of a solution and considering the top best solutions from the set of initial solutions. The proposed algorithm has been assessed on twenty-three different benchmark functions, and furthermore, the binary variant of the proposed algorithm is developed for selecting the optimal set of features from various data sets. To gauge the performance of the proposed algorithms, various evaluation metrics are considered.

Keywords Multi-verse optimization · Quasi-opposition-based learning · Feature selection

1 Introduction

In the current era, optimization methods are playing an extremely significant role in finding the solutions to the problems related to engineering design and mathematics and other related areas, subject to conditions. Optimization techniques are the

R. Hans (✉)
DAV University, Jalandhar, Punjab, India
e-mail: rahulhans@gmail.com

R. Hans · H. Kaur
Guru Nanak Dev University, Regional Campus, Gurdaspur, Punjab, India
e-mail: harjotkaursohal@rediffmail.com

© Springer Nature Singapore Pte Ltd. 2020
P. K. Singh et al. (eds.), *Proceedings of First International Conference on Computing, Communications, and Cyber-Security (IC4S 2019)*, Lecture Notes in Networks and Systems 121, https://doi.org/10.1007/978-981-15-3369-3_27

methods where the finest solution to a problem is found by using random search mechanism. Recent times have witnessed the development of a class of optimization algorithms known as metaheuristic algorithms, which have gained the attention of various researchers. Metaheuristic class of algorithms performs a random search for the best solution of the problem to be optimized [1] and is considered to be more reliable when solving various optimization problems [2] that show the way to find good enough solutions to the complex problems in a rational time [3].

Optimization techniques fundamentally require maintaining the balance amongst two significant mechanisms: (i) exploring the search space for finding a good solution and (ii) exploiting the best solutions found during the process of exploration [4]. In other terms, the first step produces the different solutions on exploring the search space comprehensively, while the second step tries to find the good solution in the region where the best solution is found in the first step. Further, the two major classes of metaheuristics include exploration-oriented (population-based) like swarm intelligence and exploitation-oriented (single-solution-based) like simulated annealing [2].

Any individual measurable value of the procedure under observation [4] is defined as a feature. On the other hand, for different applications in machine learning, the data sets may contain a number of irrelevant features. Finding an optimal set of features from the original huge set of features is defined as an optimization problem.

In the current era, metaheuristic class of optimization techniques has been explored to find the solution of problems like feature selection. This paper puts forward an enhanced form of the multi-verse optimization algorithm using quasi-opposition strategy; performance of the continuous version of the proposed algorithm is evaluated on twenty-three benchmark functions. Furthermore, the continuous version of the proposed algorithm is converted to a binary version using the transfer function and is tested on various data sets for finding an optimal set of features.

1.1 Structuring of the Paper

The rest of the paper is structured as: Sect. 2 discusses some recent works accomplished in the field of metaheuristics including feature selection. Section 3 presents multi-verse optimization (MVO) algorithm and Sect. 4 discusses quasi-opposition-based multi-verse optimization and Sect. 5 briefly discusses experimental setup. Section 6 presents the performance analysis and results obtained by the improved version of MVO in comparison with other metaheuristic algorithms along with feature selection on various standard data sets. Finally, Sect. 7 presents some of the future works and conclusion of the proposed work.

2 Related Literature

Metaheuristics possess a much better capability than some of the usual optimization techniques to solve some of the complex and nonlinear problems, which have made these techniques very popular, and moreover, the features like flexibility have turned the attention of various researchers towards these techniques. As per the no free lunch theorem, no metaheuristic technique is suitably suited for solving every kind of the optimization problems [5]. Which is the primary reason, the researchers are trying to get better efficiency of optimization algorithms by altering some basic parameters or by hybridizing the algorithms; In [1] authors proposed an enhanced variant of sine cosine algorithm by incorporating the opposition-based strategy. Different benchmark functions and engineering problems were considered to validate the proposed algorithms. In [2], authors proposed simulated annealing-based whale optimization algorithm for selecting an optimal number of features, and authors used simulated annealing as a local search method. The proposed algorithm outperforms other algorithms when the algorithms were tested on various standard data sets for the feature selection problem. In [4], authors proposed binary multi-verse optimization algorithm by using four different transfer functions for the binarization of the algorithm, the authors implemented the proposed algorithms to find the solution of feature selection problem and results show the better performance of the proposed algorithms. In [6], authors proposed a quasi-opposition-based ant lion optimization by incorporating chaotic maps in it. Different benchmark functions were used to gauge the performance of the proposed algorithms. In [7], authors proposed, a customized position-updating equation with inertia weights to accelerate convergence, furthermore, to maintain the equilibrium in exploration and exploitation, a new nonlinear conversion parameter decreasing policy is used, to propose an enhanced version of sine cosine optimization. In [8], authors enhanced the particle swarm optimization algorithm by introducing a model of diverse particles groups. Different mathematical functions with different slopes were implemented to fine-tune the values of variables of the algorithm to impart diverse nature to particles. The results depict that PSO with diverse particle groups outperforms the other variants of PSO. In [9], authors proposed a superior version of GWO. A new position-updated equation is proposed for getting better exploration. Furthermore, to bring stability between exploration and exploitation, the authors introduced a nonlinear control parameter strategy. Twenty-three mathematical benchmark functions were considered to evaluate the proposed algorithms.

3 Multi-verse Optimization Algorithm

Multi-verse is one of the latest suppositions made by investigators in the field of physics [10]. The word multi-verse implies 'converse of a universe', which stands for the fact of the presence of another universe in addition to the existing. MVO algorithm

[11] is developed on the thoughts of three different suppositions, viz. white, black and wormholes. These suppositions were further mathematically mapped to put up the MVO algorithm.

For the mathematical mapping of the algorithm; the population is represented as U(set of possible solutions). Wherein each variable is assumed as $\text{\euro}_i^j$ (object) in the universe, the dimension is represented by dim and count of universes by N. The object $\text{\euro}_i^j$ is swapped by means of roulette wheel mechanism which picks universe U_k as in (1), where U_i is the ith universe, and the inflation rate of U_i normalized to form $N\text{Fit}(U_i)$.

The exploitation is performed by supposing that each U_i transfers the objects in the course of space randomly. The mathematical equations of the multi-verse optimization algorithms can be seen in (1) and (2). The wormhole existence probability (WEP) and travelling distance rate (TDR) are shown in (3) and (4). Furthermore, l and L define the current iteration and the total iterations correspondingly. p is allocated a default value of 6. In Eq. 3, the default lowest (MIN_{WEP}) and highest (MAX_{WEP}) values of WEP be 0.2 and 1 correspondingly.

$$\text{\euro}_i^j = \begin{cases} \text{\euro}_k^j, & \text{if } r1 < NF(U_i) \\ \text{\euro}_i^j, & \text{if } r1 > NF(U_i) \end{cases} \tag{1}$$

$$\text{\euro}_i^j = \begin{cases} \text{\euro}_j + \text{TDR} \times ((ub_j - lb_j) \times r4 + lb(j)) \geq 1 & r3 < 0.5 \quad r2 < \text{WEP} \\ \text{\euro}_j - \text{TDR} \times ((ub_j - lb_j) \times r4 + lb(j)) \geq 1 & r3 \geq 0.5 \; r2 < \text{WEP} \\ \text{\euro}_i^j & r2 > \text{WEP} \end{cases} \tag{2}$$

$$\text{WEP} = \text{MIN}_{\text{WEP}} + l \times \frac{\text{Max}_{\text{WEP}} - \text{MIN}_{\text{WEP}}}{L} \tag{3}$$

$$\text{TDR} = 1 - \frac{l^{1/p}}{L^{1/p}}. \tag{4}$$

4 Quasi-opposition-Based Multi-verse Optimization Algorithm

The first step in any optimization algorithm is to initialize random solutions which may be too far from the global optima and take more time to reach optimal value. To deal with this problem and to get enhanced convergence, authors in [12, 13] suggested to consider both the randomly produced solutions and their opposite solutions simultaneously and choose top best solutions from the set as initial solutions. Starting the search procedure with an initial solution set that consists of the better of the two sets aids to get better the convergence rate. Let 'a' be a real number in one-dimensional space lying within the range [ub, lb]. Then, the opposite of a number (op_a) is defined by (5)

$$op\_a = ub + lb - a \tag{5}$$

$$Qop\_a = rand\left(\left(\frac{(X + Y)}{2}\right), op\_a\right) \tag{6}$$

$$Qop\_a_j = rand\left(\left(\frac{(X_j + Y_j)}{2}\right), op\_a_j\right) \tag{7}$$

For one-dimensional search space, the quasi-opposite number (Qop_a), can be defined [6, 14] as a random number between $\frac{(ub+lb)}{2}$, and op_a, as in (6). Moreover, for an n-dimensional search space, the quasi-opposite point of 'a' is considered as in (7). The quasi-opposite point of a number has a superior probability to acquire a nearer position to the global optimum [6] and more competent to get convergence.

For feature selection, one of the mandatory requirements is to alter the continuous variant to its corresponding binary variant. The solutions in binary form are represented in the form of 0s and 1s. A transfer function describes the probability of altering a position vector's bits from 1 to 0 and opposite [15].

Algorithm 1. Pseudo code of QMVO Algorithm (MVO with Quasi-Opposition based Strategy)

1.	*Create U(population initialization of size N) and Initialize TDR, Best_universe and WEP, SU=Sorted universes;*
	NFit=Normalized inflation rate.
2.	**while the** *end criterion is not met*
	New_U=Union(U, Quasi_U) and select // *Select N Best Solutions from randomly*
	N best solutions from New_U to form U. // *initialized population and Quasi-*
	Evaluate the fitness of the universe // *Opposition of the population*
3.	**for** *each universe indexed by i*
	Update WEP and TDR ; index(Black_hole)= i;
	for *each object indexed by j*
	r1=rand(0,1)
4.	**if** *r1<NF(U₁)*
	index(White_hole)= RouletteWheelSelection(-NF);
	U(index(Black_hole),j)=SU(index(White_hole),j);
	end if
5.	*r2= rand(0,1)*
	if *r2<Wormhole_existance_probability*
	r3= rand(0,1), r4= rand(0,1)
	Update the position using equation (2)
	end if
6.	**end**
7.	**end**
8.	**end**

In order to deal with 0 and 1 in a binary search space, The MVO cannot perform updation of position by using (2) straightforwardly. Consequently, a different way to update the position is proposed by using the same (2) for transforming the position of the agent from 0 to 1 or from 1 to 0. In QMVO, the position of the agent is changed with the likelihood of position updation value $\in_i^j$ of the objects residing in the universe, having a dependency on r3. So, a modified sigmoidal function is implemented to map (2) values to probability values for updating the agent's position as shown in (8) and (9).

$$\epsilon_i^j = \begin{cases} 0, & \mathrm{trf1}(\epsilon_i^j) > \theta_1 \\ 1, & \text{otherwise} \end{cases} \tag{8}$$

$$\mathrm{trf1}(\epsilon_i^j) = \mathrm{sigmoid}(\epsilon_i^j) = \frac{1}{1 + e^{-2*\epsilon_i^j}} \tag{9}$$

As discussed in the above section the quasi-opposition of the point, the proposed algorithm applies the quasi-opposition strategy in multi-verse optimization at the population initialization step. In the proposed algorithm random population is initialized, along with it the quasi-opposition of the random population is also generated, the solutions in both the populations are evaluated for the fitness and the best solutions out of the two populations are considered for the next generation in the algorithm. Algorithm 1 depicts proposed QMVO in the Pseudo-code form.

5 Experimental Set-up

For experimental set-up, the configuration of the processor considered is Core 2 Duo, 2.00 GHz, and a RAM of 3.00 GB is considered. For the analysis of algorithms on various benchmark functions, the population size (search agents) is taken as 30 and number of iterations 300 and results have been taken on 30 different runs, and for feature selection, the results have been evaluated using 20 search agents and 100 iterations are taken and with 10 different runs.

5.1 Fitness Function

For the assessment of the dominating capability of the selected feature subset, a fitness function should be considered. The K-NN classifier [16]-based fitness function for the evaluation of the candidate solution is considered in (10). Where $\delta(\mathrm{Err})$ defines the rate of error in the process of classification:

$$\mathrm{Fitness} = W_1 \left| \frac{\mathrm{FSel}}{N} \right| + W_2 * \delta(\mathrm{Err}) \tag{10}$$

Moving ahead, features subset selected by an optimization algorithm is defined by |FSel| and initial feature set size is |N|, W_1 and W_2 are two weight factors considered defining the significance of the prominence of the classification and size of |FSel|, where $W_1 \in [0, 1]$ and $W_2 = (1 - W_1)$ adapted from [2]. For all data sets value, 5 is assigned to k.

Table 1 depicts different benchmark functions [17], and Table 2 shows the various benchmark data sets [18] considered to find the solution of feature selection problem.

Table 1 Various benchmark functions

Fun	Equation	lb	ub	dim
$F1$	$f(x) = \sum_{i=1}^{n} x_i^2$	-100	100	30
$F2$	$f(x) = \sum_{i=1}^{n} |x_i| + \prod_{i=1}^{n} |x_i|$	-10	10	30
$F3$	$f(x) = \sum_{i=1}^{n} \left(\sum_{j-1}^{i} x_i \right)^2$	-100	100	30
$F4$	$f(x) = \max_i \{|x_i|, 1 \leq i \leq n\}$	-100	100	30
$F5$	$f(x) = \sum_{i=1}^{n-1} \left[100(x_{i+1} - x_i^2)^2 + (x_i - 1)^2 \right]$	-30	30	30
$F6$	$f(x) = \sum_{i=1}^{n} (|x_i + 0.5|)^2$	-100	100	30
$F7$	$f(x) = \sum_{i=1}^{n} i x_i^4 + \text{random}[0, 1]$	-1.28	1.28	30
$F8$	$f(x) = \sum_{i=1}^{n} -x_i \sin\left(\sqrt{|x_i|} \right)$	-500	500	30
$F9$	$f(x) = \sum_{i=1}^{n} \left[x_i^2 - 10\cos(2\pi x_i) + 10 \right]$	-5.12	5.12	30
$F10$	$f(x) = -20\exp\left(-0.2\sqrt{\frac{1}{n} \sum_{i=1}^{n} x_i^2} \right) - \exp\left(\frac{1}{n} \sum_{i=1}^{n} \cos(2\pi x_i) \right) + 20 + e$	32	32	30
$F11$	$f(x) = \frac{1}{4000} \sum_{i=1}^{n} x_i^2 - \prod_{i=1}^{n} \cos\left(\frac{x_i}{\sqrt{i}} \right) + 1$	-600	600	30

(continued)

Table 1 (continued)

Fun	Equation	lb	ub	dim
F12	$f(x) = \frac{\pi}{n}\left\{10\sin^2(\pi y_1) + \sum_{1=1}^{n-1}(y_i-1)^2\left[1+10\sin^2(\pi y_{i+1})\right]\right.$ $\left. + (y_n-1)^2\right\} + \sum_{i=1}^{n} u(x_i, 10, 100, 4)U(x_i, a, k, m)$ $= \begin{cases} k(x_i-a)^m, & x_i > a \\ 0, & a \le x_i \le a \\ k(-x_i-a)^m, & x_i < -a \end{cases}$	-50	50	30
F13	$f(x) = 0.1\left\{\sin^2(3\pi x_1) + \sum_{i=1}^{n}\left(x_i-1^2\right)\left[1+\sin^2(3\pi x_i+1)\right]\right.$ $\left. + (x_n-1)^2\left[1+\sin^2(2\pi x_n)\right]\right\} + \sum_{i=1}^{n} u(x_i, 5, 100, 4)$	-50	50	30
F14	$f(x) = \left(\frac{1}{500} + \sum_{j=1}^{25}\frac{1}{j+\sum_{i=1}^{2}(x_i-a_{ij})^6}\right)^{-1}$	-65.536	65.536	2
F15	$f(x) = \left(\sum_{i=1}^{11}\left[a_i - \frac{x_1(b_i^2+b_ix_2)}{b_i^2+b_ix_3+x_4}\right]\right)^2$	5	5	4
F16	$f(x) = (4x_1^2 - 2.1x_1^4 + \frac{1}{3}x_1^6 + x_1x_2 - x_2^2 + 4x_2^4)$	-5	5	2
F17	$f(x) = \left(x_2 - \frac{5.1}{4\pi^2}x_1^2 + \frac{5}{\pi}x_1 - 6\right)^2 + 10\left(1-\frac{1}{8\pi}\right)\cos x_1 + 10$	$[-5, 0]$	$[10, 5]$	2
F18	$f(x) = \left[1 + (x_1+x_2+1)^2\left(19-14x_1+3x_1^2-14x_2+6x_1x_2+3x_2^2\right)\right]$ $\times\left[30 + (2x_1-3x_2)^2 \times \left(18-32x_1+12x_1^2+48x_2-36x_1x_2+27x_2^2\right)\right]$	-2	2	2
F19	$f(x) = -\sum_{i=1}^{4} c_i \exp\left(-\sum_{j=1}^{3} a_{ij}(x_j - p_{ij})^2\right)$	0	1	3

(continued)

Table 1 (continued)

Fun	Equation	lb	ub	dim
$F20$	$f(x) = -\sum_{i=1}^{4} c_i \exp\left(-\sum_{j=1}^{6} a_{ij}(x_j - p_{ij})^2\right)$	0	1	6
$F21$	$f(x) = -\sum_{i=1}^{5}\left[(X - a_i)(X - a_i)^T + C_i\right]^{-1}$	0	10	4
$F22$	$f(x) = -\sum_{i=1}^{7}\left[(X - a_i)(X - a_i)^T + C_i\right]^{-1}$	0	10	4
$F23$	$f(x) = -\sum_{i=1}^{10}\left[(X - a_i)(X - a_i)^T + C_i\right]^{-1}$	0	10	4

Table 2 Data sets
description with no. of
instances and attributes

S. No.	Dataset	Instances	Attributes
D_1	Zoo	101	16
D_2	Statlog	1000	20
D_3	Exactly	1000	13
D_4	Exactly2	1000	13
D_5	Heart	294	13
D_6	Vote	300	16
D_7	SPECT heart	267	22
D_8	Australian	690	14
D_9	Ionosphere	351	34
D_{10}	Water treatment	521	38
D_{11}	Wine	178	13

To quantify the performance of proposed algorithms on various benchmark functions, two measures have been considered as explained in the below section.

5.2 Evaluation Metrics

To quantify the performance of proposed algorithms on various benchmark functions, two different evaluation criteria are considered

(i) *Average fitness*—It is defined as the mean of fitness values obtained in different runs.
(ii) *Standard deviation*—It is explained as the deviation of the best solution, got after running an optimization algorithm in all runs.

For evaluation of feature selection, along with average fitness and standard deviation, authors have also considered

(i) *Mean classification accuracy*—The quantity of samples which are classified as correct from the full set of samples considered by an algorithm in all the runs.
(ii) *Average feature subset size*—It is defined as the number of features selected in all the runs.

6 Performance Analysis and Results

For the assessment of the proposed algorithms, a set of twenty-three different benchmark functions is taken into consideration [17], and along with it the comparison is done with some of the state-of-the-art metaheuristic algorithms, viz.

grey wolf optimization (GWO) [17], sine cosine algorithm (SCA), particle swarm optimization (PSO) and multi-verse optimization (MVO). The data in Table 3 shows that when algorithms are compared on the basis of average fitness and standard deviation, the proposed algorithm outperforms various other algorithms on some benchmark functions, which show the improvement capability of the algorithm up to some extent. When results were compared for feature selection on the basis of mean classification accuracy obtained, average feature subset size and average fitness values are obtained. Clearly, outperformance of the proposed algorithm is seen than the other algorithms taken for comparison, viz. SCA, GWO, PSO and MVO. The impact of incorporating quasi-opposition-based strategy can be seen in Table 3 when results were compared on various benchmark functions. Also, the role of the transfer function in the process of binarization can be seen from the results obtained in Table 4, Table 5 and graph in Fig. 1. Although the results got from QMVO on various criteria are better than the algorithms considered for comparison, the addition of the extra step of quasi-opposition-based population creation and evaluation for fitness brings overheads in terms of computational time complexity.

The major contribution of the research is twofold: (i) developing a quasi-opposite learning based multi-verse optimization algorithm and evaluating them for mean fitness and standard deviation on twenty-three benchmark functions; (ii) developing the binary version of proposed quasi-opposite learning based multi-verse optimization and applying the proposed algorithm for solving the feature selection problem. Also, the evaluation of the QMVO is done against some of the state-of-the-art algorithms in terms of various evaluation criteria.

7 Conclusions and Future Scope

This paper proposes a novel quasi-oppositional multi-verse optimization algorithm which has been tested on twenty-three benchmark functions, and also the performance of the binary variant of the proposed algorithm has been used for selecting an optimal set of features from various data sets. The results indicate that the proposed QMVO algorithm outperforms on eleven benchmark functions, and for feature selection, it outperforms all the other algorithms considered for comparison but the addition of quasi-opposition strategy and transfer function leads to increase in the time complexity.

In the future, work must be done to reduce the execution time for solving the feature selection problem for some real-world applications using metaheuristic algorithms. Furthermore, work must be done to use the opposition-based learning strategy with some other algorithms to improve the performance of the algorithms for solving various optimization problems including feature selection.

Table 3 Results on twenty-three benchmark functions

Fun	Criteria	GWO	SCA	PSO	MVO	QMVO
$F1$	Mean	4.18E−15	8.77E+01	1.29E−02	3.38E+00	3.22E−213
	Std_dev	4.48E−15	7.48E+01	1.65E−02	1.10E+00	0.00E+00
$F2$	Mean	1.98E−09	2.97E−01	5.86E−01	2.89E+01	1.53E−111
	Std_dev	1.02E−09	3.88E−01	4.27E−01	4.32E+01	4.40E−111
$F3$	Mean	7.01E−02	1.35E+04	1.81E+02	7.81E+02	8.40E−167
	Std_dev	2.34E−01	6.48E+03	7.08E+01	4.01E+02	0.00E+00
$F4$	Mean	9.06E−04	4.75E+01	1.82E+00	4.02E+00	3.69E−105
	Std_dev	6.55E−04	1.15E+01	6.33E−01	1.83E+00	2.02E−104
$F5$	Mean	2.75E+01	9.08E+05	1.63E+02	6.56E+02	2.90E+01
	Std_dev	5.80E−01	1.58E+06	2.01E+02	7.99E+02	2.22E−02
$F6$	Mean	9.68E−01	1.13E+02	1.06E−02	3.26E+00	2.08E+00
	Std_dev	4.66E−01	1.09E+02	1.33E−02	8.50E−01	4.60E−01
$F7$	Mean	3.34E−03	2.77E−01	1.79E−01	5.59E−02	7.74E−04
	Std_dev	1.44E−03	2.91E−01	7.47E−02	1.88E−02	4.91E−04
$F8$	Mean	−5.84E+03	−3.53E+03	−4.43E+03	−7.68E+03	−6.28E+03
	Std_dev	1.13E+03	2.11E+02	1.17E+03	7.56E+02	7.99E+02
$F9$	Mean	8.75E+00	7.75E+01	8.11E+01	7.75E+01	0.00E+00
	Std_dev	5.59E+00	5.39E+01	2.21E+01	5.39E+01	0.00E+00
$F10$	Mean	1.39E−08	1.66E+01	1.12E+00	2.32E+00	8.88E−16
	Std_dev	7.82E−09	6.74E+00	6.87E−01	7.25E−01	0.00E+00
$F11$	Mean	1.16E−02	3.09E+00	2.69E−02	1.02E+00	0.00E+00
	Std_dev	1.19E−02	3.25E+00	2.13E−02	2.71E−02	0.00E+00
$F12$	Mean	6.97E−02	9.89E+05	7.76E−02	3.19E+00	1.24E−01
	Std_dev	4.07E−02	2.07E+06	1.51E−01	1.29E+00	7.53E−02
$F13$	Mean	8.81E−01	2.27E+06	6.27E−02	4.55E−01	1.19E+00
	Std_dev	2.48E−01	3.76E+06	8.71E−02	3.23E−01	1.22E+00
$F14$	Mean	4.98E+00	2.26E+00	1.86E+00	1.03E+00	1.03E+00
	Std_dev	4.32E+00	1.88E+00	1.12E+00	1.81E−01	1.81E−01
$F15$	Mean	3.88E−03	1.11E−03	8.52E−04	5.60E−03	5.64E−04
	Std_dev	7.50E−03	3.87E−04	1.83E−04	1.28E−02	1.58E−04
$F16$	Mean	−1.03E+00	−1.03E+00	−1.03E+00	−1.03E+00	−1.03E+00
	Std_dev	7.89E−08	7.19E−05	5.30E−16	1.83E−06	1.45E−06
$F17$	Mean	3.98E−01	4.02E−01	3.98E−01	4.42E−01	3.98E−01
	Std_dev	1.67E−04	4.50E−03	0.00E+00	2.42E−01	2.18E−06
$F18$	Mean	5.70E+00	3.00E+00	3.00E+00	3.00E+00	3.00E+00
	Std_dev	1.48E+01	4.85E−04	2.81E−15	8.26E−06	1.12E−05

(continued)

Table 3 (continued)

Fun	Criteria	GWO	SCA	PSO	MVO	QMVO
$F19$	Mean	−3.86E+00	−3.85E+00	−3.86E+00	−3.86E+00	−3.86E+00
	Std_dev	1.61E−03	3.59E−03	2.43E−15	3.91E−06	6.75E−06
$F20$	Mean	−3.25E+00	−2.88E+00	−3.27E+00	−3.27E+00	−3.25E+00
	Std_dev	9.35E−02	3.15E−01	5.92E−02	6.31E−02	6.21E−02
$F21$	Mean	−8.80E+00	−2.56E+00	−6.49E+00	−6.96E+00	−1.02E+01
	Std_dev	2.54E+00	1.75E+00	3.56E+00	3.16E+00	8.68E−04
$F22$	Mean	−1.04E+01	−8.93E+00	−6.90E+00	−8.65E+00	−1.02E+01
	Std_dev	2.42E−03	2.77E+00	3.72E+00	3.04E+00	1.22E+00
$F23$	Mean	−9.78E+00	−3.39E+00	−9.65E+00	−8.97E+00	−1.01E+01
	Std_dev	2.31E+00	1.64E+00	2.34E+00	2.94E+00	1.70E+00

Table 4 Average feature subset size

	GWO	SCA	PSO	MVO	QMVO
D_1	10.3	11.3	10.5	11.1	8.1
D_2	11.3	10	12	12.8	7.5
D_3	10	8.8	9.9	9.6	7.2
D_4	6.7	5.1	6	5.6	5.1
D_5	5.2	6.4	5.8	7.7	5.1
D_6	8.7	9	8.1	9.6	4.2
D_7	11.1	12.8	12.4	15.9	10
D_8	6	6.9	6.2	6.5	4.3
D_9	13.9	19	18.9	21	9.1
D_{10}	19	21.4	20.7	25.9	15.3
D_{11}	7.4	9.3	9.1	8.2	6

Table 5 Average fitness values

	GWO	SCA	PSO	MVO	QMVO
D_1	0.087	0.076	0.063	0.067	0.054
D_2	0.286	0.281	0.293	0.282	0.264
D_3	0.302	0.255	0.282	0.266	0.137
D_4	0.249	0.252	0.247	0.249	0.233
D_5	0.197	0.184	0.186	0.17	0.163
D_6	0.062	0.062	0.074	0.052	0.044
D_7	0.284	0.255	0.274	0.251	0.229

(continued)

Table 5 (continued)

	GWO	SCA	PSO	MVO	QMVO
D_8	0.203	0.205	0.193	0.183	0.144
D_9	0.137	0.129	0.15	0.143	0.121
D_{10}	0.189	0.182	0.197	0.177	0.146
D_{11}	0.031	0.019	0.028	0.022	0.016

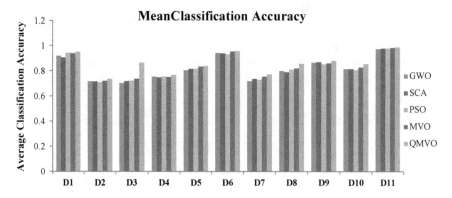

Fig. 1 Mean classification accuracy

References

1. Elaziz, M.A., Oliva, D., Xiong, S.: An improved opposition-based sine cosine algorithm for global optimization. Expert Syst. Appl. **90**, 484–500 (2017)
2. Mafarja, M.M., Mirjalili, S.: Hybrid whale optimization algorithm with simulated annealing for feature selection. Neurocomputing **260**, 302–312 (2017)
3. Yang, X.S., Deb, S., Fong, S.: Metaheuristic algorithms: optimal balance of intensification and diversification. Appl. Math. Inf. Sci. **8**(3), 977–983 (2014)
4. Hans, R., Kaur, H.: Binary multi-verse optimization (BMVO) approaches for feature selection. Int. J. Interact. Multimed. Artif. Intell. 1–16 (2019)
5. Wolpert, D.H., Macready, W.G.: No free lunch theorems for optimization. IEEE Trans. Evol. Comput. **1**(1), 67–82 (1997)
6. Saha, S., Mukherjee, V.: A novel quasi-oppositional chaotic antlion optimizer for global optimization. Appl. Intell. **48**(9), 2628–2660 (2018)
7. Long, W., Wu, T., Liang, X., Xu, S.: Solving high-dimensional global optimization problems using an improved sine cosine algorithm. Expert Syst. Appl. **123**, 108–126 (2019)
8. Mirjalili, S., Lewis, A., Sadiq, A.S.: Autonomous particles groups for particle swarm optimization. Arab. J. Sci. Eng. **39**(6), 4683–4697 (2014)
9. Long, W., Jiao, J., Liang, X., Tang, M.: An exploration-enhanced grey wolf optimizer to solve high-dimensional numerical optimization. Eng. Appl. Artif. Intell. **68**, 63–80 (2018)
10. Tegmark, M., Barrow, J.D., Davies, P.C., Harper Jr., C.L. (eds.): Science and Ultimate Reality. Cambridge University Press, Cambridge (2004)
11. Mirjalili, S., Mirjalili, S.M., Hatamlou, A.: Multi-verse optimizer: a nature-inspired algorithm for global optimization. Neural Comput. Appl. **27**(2), 495–513 (2016)
12. Tizhoosh, H.R.: Opposition-based learning: a new scheme for machine intelligence. In: International Conference on Computational Intelligence for Modelling, Control and

Automation and International Conference on Intelligent Agents, Web Technologies and Internet Commerce 2006, vol. 1, pp. 695–701. IEEE, Nov 2005

13. Tizhoosh, H.R.: Reinforcement learning based on actions and opposite actions. In: International Conference on Artificial Intelligence and Machine Learning, vol. 414 (2005)

14. Rahnamayan, S., Tizhoosh, H.R., Salama, M.M.: Quasi-oppositional differential evolution. In: 2007 IEEE Congress on Evolutionary Computation, pp. 2229–2236 (2007)

15. Mirjalili, S., Lewis, A.: S-shaped versus V-shaped transfer functions for binary particle swarm optimization. Swarm Evol. Comput. **9**, 1–14 (2013)

16. Altman, N.S.: An introduction to kernel and nearest-neighbor nonparametric regression. Am. Stat. **46**(3), 175–185 (1992)

17. Mirjalili, S., Mirjalili, S.M., Lewis, A.: Grey wolf optimizer. Adv. Eng. Softw. **1**(69), 46–61 (2014)

18. Asuncion, A., Newman, D.: UCI machine learning repository (2007)

Nonnegative Matrix Factorization Methods for Brain Tumor Segmentation in Magnetic Resonance Images

Harinder Kaur and Ram Singh

Abstract Automatic segmentation of objects in magnetic resonance images (MRI) is very challenging to provide reliable and automated computerized methods for clinical usage because this is a time-consuming process which requires more and more attention. MR imaging is the most usable technique of images to segment the tumor region from brain. By follow the novel framework of nonnegative matrix factorization (NNMF) along with fuzzy clustering and region growing, authors are able to do the tumor segmentation easily and accurately in this research work. By applying these methods, this paper represents the successful extraction of the whole tumor along with narcotic and edema from brain MR, T2, and FLAIR images. This proposed methodology is tested on the BRATS 2012 database with low-grade glioma (LGG) and high-grade glioma (HGG) brain tumor images. Moreover, along with NNMF, fuzzy c-means clustering and region growing, the algorithms of Dice, sensitivity, specificity, and Hausdorff are also evaluated to check the accuracy of the results.

Keywords Brain tumor segmentation · MRI · Nonnegative matrix factorization · Fuzzy c-means clustering · Region growing

1 Introduction

Brain is a part of our central nervous system [1]. Brain tumor starts with the abnormal growth of cells inside the brain. Brain tumor is very serious and also life threatening. The structure of brain tumor varied among the patients according to the size, location, etc. Brain tumor or glioma is a very common tumor with an expectancy of short life.

H. Kaur (✉) · R. Singh
Department of Computer Science and Engineering, Punjabi University Patiala, Patiala, Punjab 147002, India
e-mail: harinderkaur1994@gmail.com

R. Singh
e-mail: bhankharz@gmail.com

© Springer Nature Singapore Pte Ltd. 2020
P. K. Singh et al. (eds.), *Proceedings of First International Conference on Computing, Communications, and Cyber-Security (IC4S 2019)*, Lecture Notes in Networks and Systems 121, https://doi.org/10.1007/978-981-15-3369-3_28

Glioma is included as a part of the spinal or brain tumor which is occurred into the glial cells. Glioma has two types: low-grade glioma (LGG) and high-grade glioma (HGG). The detection of tumor is possible by doing the segmentation of an image into three regions like white matter, gray matter, and cerebrospinal fluid. There are various types of imaging techniques that are available to examine the tumor; but to detect the tumor from brain, multimodal MRI is widely used in medical field. Medical imaging technique that is MRI mostly depends on the computer technology which helps to scan the interior most part of the human being. For scanning the brain, MRI, magnetic fields, and radio frequency waves are used. The MRI scan is radiology technology which incorporates with the radio wave, magnetism, and computer to obtained scanned image of the body parts, in which radiations are used in the MRI scan; these are not harmful to human body. Magnet is the most important component for MRI scanning. It has some coils and wires from which electricity proceed. While capturing the MR image of body, these magnets are going all around the body part with hydrogen atoms. These atoms emit energy or radio frequency pulse [1], and then this energy is detected by the scanner. After scan the MR image, a digital image is produced which is stored into computer for doing further operations. The MRI scan provides a high-resolution image. FLAIR, T2, T1c, and T1 are channels of MRI which are needed to extract the various parts of the tumor [2]. These are also known as multimodal images of MRI. MRI scans take 20 min to 1 h for scanning any body part, and this is a limitation of the MRI. To remove this problem, gradients are exceeding in MRI scanning. Diagnosis of the brain tumor by using some techniques or tools can increase the chances of the tumor survival into our brain [3]. Automatic segmentation is very necessary to detect the brain tumor as compared to the manual segmentation which is very time-consuming [4] process. But the automatic segmentation is challenging task in medical field. Many types of glioma or tumor segmentation algorithms are available based on different types of properties [5]. Main purpose of the segmentation is dividing the image into various regions or sections based on the criteria [6]. Segmentation is an efficient technology to analyze various types of medical images. Glioma segmentation from brain is still challenging task. The proposed technique of segmentation which is nonnegative matrix factorization (NNMF) helps to analyze the data by using nonnegative elements like probabilities, spectra, and images. NNMF analyzes the datasets mathematically which is described using the matrices. In NNMF algorithm, nonnegative matrix V is factorized further into two nonnegative matrices that are W and H. NNMF has various applications, image recognition, and text classification and in recommender system (Fig. 1).

2 Literature Survey

Tamilselvan et al. [7] proposed the watershed and wavelet transform for segmenting the tumor region; wavelet helps to decompose an input image, and watershed

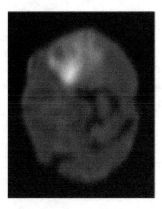

Fig. 1 MRI of brain

algorithm segments an image in detailed. Watershed algorithm includes the image gradients.

Joseph et al. [1] presented the k-means algorithms and morphological methods which help to remove the misclustered regions which are created by the k-means clustering algorithms while technology is implemented after de-noise an input image.

Veer et al. [8] proposed the techniques that are thresholding and watershed. Global thresholding technique extracts the tumor by applying only one value of threshold, and then watershed algorithm groups the image pixels by comparing intensity value.

Sara et al. [9] proposed the histogram to select threshold value for tumor segmentation. Histogram method includes the injection of contrast while avoiding the location of the image.

Kadkhodaei et al. [2] included the multimodal images for segment brain tumor by following supervoxel as well as saliency detection methodology. Finally, the classification is done by neural networks. All data are tested on the BRATS 2012.

Pereira et al. [10] proposed the convolutional neural networks (CNN) which contain 3 * 3 kernels. The proposed CNN methods are evaluated by BRATS and get the first position in online platform. The proposed methods are compared with some shallow architecture as resulted that it contains low performance while working on the number of feature maps.

Kapoor et al. [11] proposed the automatic approach of neural networking along with self-organized map (SOM). Wiener filter also helps to improve peak signal-to-noise ratio (PSNR) in this research paper.

Pezoulas et al. [12] proposed the skull stripping segmentation technique based on the normalized cut method which helps to remove brain skull. By removing the skull, histogram is applied to the image slices.

Nerurkar [13] proposed the methodology that is region growing and clustering. These are the accurate, efficient as well as fast algorithms. At the end by comparing both algorithms, it is concluded that region growing is better than k-means clustering in accuracy.

Avachar et al. [14] proposed the hardware implementation by using fuzzy algorithms along with the rough set algorithms on digital signal processor (DSP). Thresholding is also applied to remove algorithm complexity.

Remma Mathew and Antop [3] proposed Ostu's thresholding methodology which is used to detect tumor boundaries and to classify tumor detection. Feature extraction is done by discrete wavelet decomposition.

Milletari et al. [15] presented novel approach of CNN architecture with Hough voting technique to segment brain tumor. Hough voting is a patch-based methodology under CNN architecture.

Dou et al. [16] presented 3D convolutional architecture with 3D deeply supervised networks (DSN). These methods help to provide volumewise inferences and also include the Dice and Jaccard coefficients.

Baid et al. [17] proposed nonnegative matrix factorization (NMF) and fuzzy clustering for extract the image features. By Dice, sensitivity, specificity, and Jaccard algorithms, it gets accurate results for brain tumor segmentation on MRI BRATS 2012 database. In NMF coefficient, base matrix helps to create the data matrix. Dice shows the accuracy of 0.77 of the whole tumor and 0.81 of edema.

Goceri et al. [4] presented some hybrid algorithms to find out the skull location from an input image. Hybrid algorithms include Gaussian method, binary operations, and anatomy knowledge. Dilation and erosion operations also include in this paper.

Chen et al. [18], this paper used the voxelwise residual network for 3D MR brain image segmentation; it is basically a powerful tool for neuroimages and science. Voxelwise residual method is used for 2D images but this paper applies on 3D images.

Pardeep Kumar Reddy et al. [19] proposed k-means algorithms with triangular models. K-means helps in image enhancement or conversion of low-to-high resolution of image.

Dogra et al. [20] presented the graph cut method and also thresholding which are used to extract image region of interest. Graph cut involves for selecting the centroid values of an image.

Somasundaram et al. [21] proposed the region growing technique of segmentation in which each single pixel value called as a seed point. Erosion, dilation, and hole filling operations are also performed.

Dolz et al. [22] presented full convolutional neural network (FCNN) for brain segmentation; but most of the cases of this technology are avoided because of the computational cost along with memory recruitment. FCNN allows layers to apply on arbitrary sized input images.

Harshavardhan et al. [23] proposed skull stripping and filtering process to test the 15 images of the brain. The proposed methodology gives accurate segmentation performance with the perfect execution time.

Ezhilarasan et al. [24] proposed mean filter as well as morphological operations to segment T1-W MRI taking from IBSR. Binarization of image has done by the threshold value. Dice and Jaccard algorithms measure the similarity of results.

Guo et al. [25] presented the fuzzy clustering technique in this paper which helps to partition data into soft form. Homogeneous segments are obtained by applying fuzzy clustering approach. Then guided filter helps more and more to enhance all fuzzy clustering-based segmentation techniques.

Li et al. [5] included fuzzy c-means for ROI estimation and then extract the seed points using region growing. FCM is justifiable for rapid medical images segmentation. Final results show the accuracy in terms of Dice, PPV, and sensitivity that is 0.86, 0.90, and 0.84.

Zeinalkhani et al. [26] proposed the k-means clustering along with some genetic algorithm. K-means clustering converts an image into clusters with similar and dissimilar in nature easily because this is difficult to decide the cluster similarity and dissimilarity. Genetic algorithms are used for optimization. K-means along with genetic algorithm is computed for fifty replications, and together it gives more accurate results instead of applying only k-means for diagnosis of the brain tumor. Accuracy of k-means with GA is 96.99, and only k-means is 74.17.

Anwar et al. [6] proposed EMAP and k-means for segment LGG as well as HGG. EMAP is statistical method to find out posterior parameter, and k-means includes for extract tumor. EM algorithm has further two steps which are the most important part of this algorithm, i.e., E-step and M-step.

Ge et al. [27] proposed the novel CNN architecture and sensor fusion to enhance the overall performance. Moreover, this deep learning and multi-sensor fusion methods are used for glioma classification. CNN results of 83.73% accuracy, and sensor fusion further enhances it by 7%; final results are 90.87% accurate for glioma segmentation.

Vinoth et al. [28] proposed support vector machine (SVM) and CNN. Testing and training are the most important parts considered in this paper while working on neural networks. This paper also includes parameters like kurtosis, variance, and entropy, and by these parameters, stage and depth of tumor are identified. Authors proved that SVM is the best technique for classification.

Dobe et al. [29] presented a rough set of k-means clustering algorithm. Whole work was done by the rough set theory. As compared to traditional algorithm of k-means, in rough set theory in which boundary pixels contain equal probability, then it is associated with more than only cluster.

Nitta et al. [30] proposed the approach which is standard k-means clustering as well as dominant graylevel-based k-means. Selection of the centroid is done by first occurring 16 probabilities of dominant gray value rather than selecting the random initial center. Dominant graylevel k-means results are more accurate as compared to the standard k-means.

Yang et al. [31] proposed the multi-channel CNN inputs for feature extraction of image. Saliency detection is included in CNN for partition image by superpixel. Recognition rate of CNN model is 99.7%. Overall, work is done by 3 * 3 windows to attained primitive information.

Jemimma and Jacob Vetharaj [32] proposed the methodology which is watershed dynamic angle projection (WDAPP) and CNN; by applying it, authors get the accurate segmentation results. In this research paper, AP helps to extract some textured features of the brain, and CNN classified tumor or non-tumor regions.

Kumar et al. [33] proposed the techniques of this research paper that is k-means and morphological operations. Firstly, MRI scanned image is preprocessed and then conversion of it is done from RGB to grayscale. Whole work is done by using the software of Scilab.

Ma et al. [34] in this research paper explored the new methods which contain random forests and active contour and for glioma segmentation from multimodal MRI. It also is used in the feature representative learning techniques to extract local and textual information by applying random forests and kernels.

Nasiri et al. [35] proposed the methodology which is graphical model for MRI segmentation by label fusion techniques. In segmentation process, firstly, first slice has been segmented in MR image by the physician and then propagated it. They also compare method with convolutional Bayesian technique and state-of-the-art hepatic segmentation of tumor.

Lakshmi Narayan et al. [36] proposed genetic algorithms that are metaheuristic optimization technique and SVM. SVM classifier determines that image is abnormal or not. The performance attributes of this paper are PSNR, MSE, sensitivity, specificity, and accuracy.

3   Proposed Methodology

In this research paper, novel approach of NNMF, fuzzy c-means clustering, and region growing is the proposed methodology for the tumor segmentation in brain MR images. NNMF or NMF is a dimensionality reduction tool which is used in the machine learning along with the various types of applications like in the image processing and data mining. NNMF is basically group of multiple algorithms in linear algebra and multi-variant analysis. In NNMF, there are three matrices: V (data matrix), W (basic matrix), and H (coefficient matrix). The matrix V is factorized into the W and H matrices. In NNMF, all three matrices contain nonnegative elements. Why NNMF works on only nonnegative elements? There are mainly two types of causes behind it:

- In image processing field, the intensity images contain nonnegative values.
- In neurophysiology, firing rate of the visual perception neuron is also nonnegative.

NNMF is helpful to cluster the complex datasets, and these datasets are cluster with the extracting features. NNMF is part-based technique for the images. Steps are followed in this paper:

(i)   Preprocessing stage of MRI dataset
(ii)   Data matrix V, Construction
(iii) Data matrix V, Decomposition
(iv)   Segmentation of brain tumor by using fuzzy c-means clustering
(v)   Narcotic tumor segmentation by region growing.

3.1 Preprocessing Stage of MRI Dataset

Preprocessing is a very essential step to remove the noise from an image for skull stripping and also for normalization. The preprocessing operations have a direct impact on the tumor segmentation. FLAIR and T2 MR images are preprocessed in this research work by using biofilter to obtain a smooth image. It gives us a smooth image rather than affecting the necessary parts of the input image.

3.2 Data Matrix V, Construction

The construction of data matrix V is done by authors for a better segmentation because it gives us contextual information about the inter slices of MR images. Construction of this V matrix is done on $3 \times 3 \times 3$ window in which three types of slices are contained (previous slice, middle slice, and next slice), and each slice has nine pixels. So it represents by 27 pixels at the end.

3.3 Data Matrix V, Decomposition

For the decomposition of the data matrix V, nonnegative tensor decomposition (NTD) is proposed because decomposition is more important step to reduce the dimensionality of the input image. Size of data matrix V is inclining significantly because each pixel of image is presented in 27 pixels.

$$V \approx WH.$$

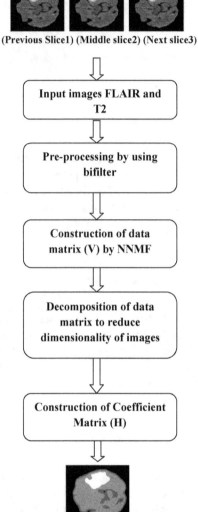

(Previous Slice1) (Middle slice2) (Next slice3)

Input images FLAIR and T2

Pre-processing by using bifilter

Construction of data matrix (V) by NNMF

Decomposition of data matrix to reduce dimensionality of images

Construction of Coefficient Matrix (H)

Segmentation of Matrix H, by Fuzzy c-mean algorithm for whole tumor (FLAIR image)

Apply region growing method for extract narcotic & edema (on T2 image)

3.4 Segmentation of Brain Tumor by Using Fuzzy C-Means Clustering

In the brain tumor segmentation process to segment the coefficient matrix H, algorithm of fuzzy c-means clustering is proposed. This algorithm does the segmentation by clustering the voxels. Fuzzy c-means algorithm is known as a soft partition of datasets. It is similar to the k-means clustering but it does not put any hard constraints on the pixel values. After applying it, whole tumor is segmented from FLAIR image of brain.

3.5 Narcotic Tumor Segmentation by Region Growing

Then by using the region growing methodology, brain tumor (edema as well as necrotic together) is segmented from the T2 image of brain accurately. Region growing method is processed after selecting an initial seed point from the input image. Then by using this initial seed, it examines the neighboring pixel values of initial seed point and the main purpose of this step is to grow the regions of image by comparing pixels (Fig. 2).

4 Results and Discussions

To evaluate the performance and accuracy of proposed technologies for the segmentation approach of brain tumor, all work is done on BRATS 2012 database. This contains various high-grade glioma and low-grade glioma brain images. In BRATS 2012 database T1, T1c, T2, and FLAIR images of MRI are available along with their ground truth images. In this research work, the segmentation of the whole tumor as well as narcotic is successfully done on the T2 and FLAIR images. By applying the proposed methods, obtaining outcomes are more accurate than other methodologies, and there is no any time complexity has occurred from the performance of these techniques. Authors include this methodology as the best methodology for brain tumor or glioma segmentation. Efficient and accurate results are attained for the extraction of the whole tumor as well as edema and narcotic tumor region from brain MRI. The proposed methods achieve Dice coefficient as 0.89 for whole tumor and 0.92 for edema and narcotic together. Experiment results of some other algorithms like sensitivity, specificity, and Hausdorff are also applied to determine the accuracy and represents in Table 1, Figs. 3, 4, 5, and 6.

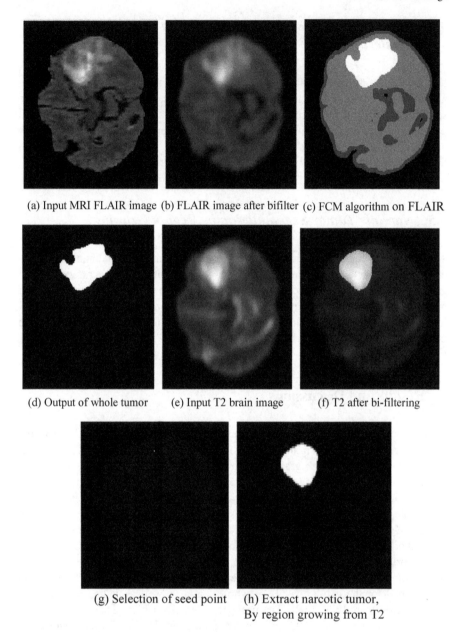

(a) Input MRI FLAIR image (b) FLAIR image after bifilter (c) FCM algorithm on FLAIR

(d) Output of whole tumor   (e) Input T2 brain image   (f) T2 after bi-filtering

(g) Selection of seed point   (h) Extract narcotic tumor,
                                By region growing from T2

Fig. 2 Brain tumor segmentation results

Table 1 Performance evaluation on BRATS training dataset

	Whole tumor	Narcotic tumor
Dice	0.89	0.92
Sensitivity	0.98	0.99
Specificity	0.93	0.87
Hausdorff	4.79	3.31

Fig. 3 Whole tumor accuracy (proposed method)

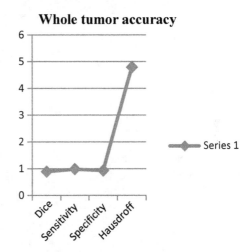

Fig. 4 Narcotic tumor accuracy (proposed method)

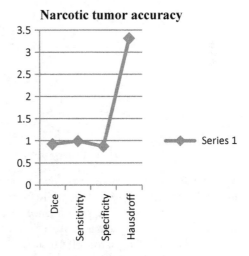

Fig. 5 Comparison of proposed and existing works

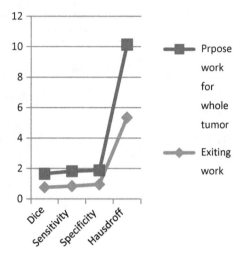

Fig. 6 Comparison of edima and narcotic data

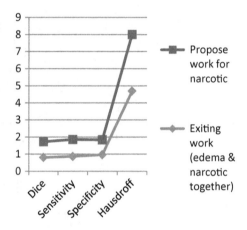

5 Conclusion

Segmentation of brain tumor has done by using the novel approach of nonnegative matrix factorization (NMF) along with a fuzzy c-means clustering and region growing segmentation method. The performance of the proposed method is evaluated on BRATS 2012 training dataset. Accuracy is defined regarding Dice, sensitivity, specification as well as Hausdorff. Accuracy is calculated to analyze the performance of proposed work. Results of research work illustrate that proposed methods outperform segmentation of brain tumor. $3 \times 3 \times 3$ Window gives the best results as compared to others.

In the future, researchers can improve the accuracy of segmentation results by using more reliable techniques or methods of segmentation on brain MRI for a better appearance of the tumor tissues or tumor boundaries in MRI volume datasets.

References

1. Joseph, R.P., Senthil Singh, C., Manikandan, M.: Brain tumor MRI image segmentation and detection in image processing. IJRTE **3**, 1–5 (2014)
2. Kadkhodaei, M., Samavi, S., Karimi, N., Mohaghegh, H., Soroushmmehr, S.M.R.: Automatic segmentation of multimodal brain tumor images based on classification of super-voxel. IEEE (2016)
3. Remma Mathew, A., Antop, B.: Tumor detection and classification of MRI brain image using wavelet transform and SVM. ICSPC **4**, 75–78 (2017)
4. Goceri, E., Songul, C.: Automated detection and extraction of skull from MR head images: preliminary results. IEEE (2017)
5. Li, Q., Gao, Z.: Glioma segmentation using a novel unified algorithm in multimodal MRI images. IEEE Access 1–9 (2018)
6. Anwar, S.M., et al.: Brain tumor segmentation on multi modal MRI scan using EMAP algorithm. IEEE (2018)
7. Tamilselvan, K.S., Murugesan, G., Gnanasekaran, B.: Brain tumor detection from clinical CT and MRI images using WT-FCM algorithm. IEEE (2013)
8. Veer, S.S., Patil, P.M.: An efficient method for segmentation and detection of brain tumor in MRI images. IRJET **2**, 912–916 (2015)
9. Sara, S., Yassine, S.T., Achraf, B., Ahmed, H.: New method of tumor extraction using a histogram study. IEEE (2015)
10. Pereira, S., Pinto, A., Alves, V., Silva, C.A.: Brain tumor segmentation using convolutional neural network in MRI image. IEEE Trans. Med. Imaging **35** (2016)
11. Kapoor, D., Kashyup, R.: segmentation of brain tumor from MRI using skull stripping and neural network. IJEDR **4**, 593–598 (2016)
12. Pezoulas, V.C., Pologiorgi, I., Seferlis, S., Giakos, G.C.: Tissue classification approach for brain tumor using MRI. IEEE (2017)
13. Nerurkar, S.N.: Brain tumor detection using image segmentation. IJERCSE **4**, 65–70 (2017)
14. Avachar, V., Mushrif, M., Dubey, Y.: Implementation of brain MRI image segmentation algorithm on DSP. IEEE, pp. 2066–2070 (2017)
15. Milletari, F., Ahmad, S., Kroll, C., Plate, A.: Hough-CNN: deep learning for segmentation of deep brain regions in MRI and ultrasound 92–102 (2017)
16. Dou, Q., Yu, L., Chen, H., Jin, Y., Yang, X., Qin, J., Heng, P.A.: 3D deeply supervised network for automated segmentation of volumetric medical images. Med. Image Anal. 40–54 (2017)
17. Baid, U.: Novel approach foe brain tumor segmentation with non negative matrix factorization. IEEE, pp. 101–105 (2017)
18. Chen, H., Dou, Q., Yu, L., Qin, J., Heng, P.A.: Voxresnet: deep voxelwise residual networks for brain segmentation from DMR images. NeuroImage 446–455 (2018)
19. Pardeep Kumar Reddy, R., Nagaraju, C.: Brain tumor MRI using gradient profile sharpness. IJANA **9**, 3557–3562 (2018)
20. Dogra, J., Prashar, N., Jain, S., Sood, M.: Improved methods for analyzing MRI brain images. IAEES **8**, 1–11 (2018)
21. Somasundaram, K., Helen Mercina, J., Magesh, S., Kalaiselvi, T.: Brain portion extraction scheme using region growing and morphological operation from MRI of human head scans. IJCSE **6**, 298–302 (2018)
22. Dolz, J., Desrosiers, C., Ayed, I.B.: 3D fully convolutional networks for sub cortical segmentation in MRI: a large-scale study. NeuroImage 456–470 (2018)

23. Harshavardhan, A., Babu, S., Venugopal, T.: An improved brain tumor segmentation method from MRI brain images
24. Ezhilarasan, K., Somasundaram, K., Kalaiselvi, T.: A simple method for automatic brain extraction from T1-W magnetic resonance images (MRI) of human head scans. IJCSE **6** (2018)
25. Guo, L., Chen, L., Philip Chen, C.L., Zhou, J.: Integrating guided filter into fuzzy clustering for noise image segmentation. Digit. Signal Process. 235–248 (2018)
26. Zeinalkhani, L., Alijamaat, A., Rostami, K.: Diagnosis of brain tumor using combination of k means clustering and genetic algorithms. IJMI (2018)
27. Ge, C., Gu, I.Y.H., Jakola, A.S., Yang, J.: Deep learning and multi-sensor for glioma classification using multi stream 2D convolutional networks. IEEE (2018)
28. Vinoth, R., Venkatesh, C.: Segmentation and detection of tumor in MRI images using CNN and SVM classification. IEEE (2018)
29. Dobe, O., Sarkar, A., Halder, A.: Rough K-means and morphological operation-based brain tumor extraction. Springer, Berlin (2019)
30. Nitta, G.R., Sravani, T., Nitta, S., Muthu, B.A.: Dominant gray level based K-means algorithm for MRI images. Health Technol. (2019)
31. Yang, A., Yang, X., Wu, W., Liu, H., Zhuansun, Y.: Research on feature extraction of tumor image based on convolutional neural network. IEEE Access (2019)
32. Jemimma, T.A., Jacob Vetharaj, Y.: Watershed algorithm based DAPP feature for brain tumor segmentation and classification. IEEE (2018)
33. Kumar, M., Sinha, A., Bansode, N.: Detection of brain tumor in MRI images by applying segmentation and area calculus method using SCILAB. IEEE (2018)
34. Ma, C., et al.: Concatenated and connected random forests with multiscale patch driven contour model for automated brain tumor segmentation of MR images. Trans. Med. Imaging (2018)
35. Nasiri, N., et al.: A controlled generative model for segmentation of liver tumors. ICEE (2019)
36. Lakshmi Narayan, T., et al.: An efficient optimization techniques to detect brain tumor from MRI images. ICSSIT (2018)

Constraint-Based Gujarati Parser Using LPP

Manisha Prajapati and Archit Yajnik

Abstract Integer linear programming has recently been used for decoding in a number of probabilistic models in order to enforce global constraints. However, in certain applications, such as non-projective dependency parsing and machine translation, the complete formulation of the decoding problem as an integer linear program renders solving intractable. We present an approach which solves the problem incrementally, thus we avoid creating intractable integer linear programs.

Keywords Gujarati sentences · LPP · LINGO

1 Introduction

Many inference algorithms require models to make strong assumptions of conditional independence between variables. For example, the Viterbi algorithm used for decoding in conditional random fields requires the model to be Markovian. Strong assumptions are also made in the case of [3] non-projective dependency parsing model. Here attachment decisions are made independently of one another [1]. However, often such assumptions cannot be justified. For example, in dependency parsing, if a subject has already been identified for a given verb, then the probability of attaching a second subject to the verb is zero. Similarly, if we find that one coordination argument is a noun, then the other argument cannot be a verb. Thus, decisions are often co-dependent. Integer linear programming (ILP) has recently been applied to inference in sequential constraint-based parsing. We apply this dependency parsing approach to Gujarati language's non-projective and projective nature and take the parser of [4] as a starting point for our model.

M. Prajapati
Gujarat Technological University, Ahmedabad, India
e-mail: purvansi263@gmail.com

A. Yajnik (✉)
Sikkim Manipal University, Gangtok, Sikkim, India
e-mail: archit.yajnik@gmail.com

© Springer Nature Singapore Pte Ltd. 2020   375
P. K. Singh et al. (eds.), *Proceedings of First International Conference on Computing,*
Communications, and Cyber-Security (IC4S 2019), Lecture Notes in Networks
and Systems 121, https://doi.org/10.1007/978-981-15-3369-3_29

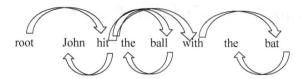

Fig. 1 Projective dependency tree (English)

Fig. 2 Non-projective dependency tree (English)

2 Dependency Parsing

Dependency graphs represent words and their relationship to syntactic modifiers using directed edges. Figure 1 shows a dependency graph for the sentence, "John hit the ball with the bat."

This example belongs to the special class of dependency graphs that only contain projective (also known as nested or non-crossing) edges, a projective graph is one that can be written with all words in a predefined linear order and all edges drawn on the plane above the sentence, with no edge crossing another. For non-projective graph, consider the sentence, "John saw a dog yesterday which was a Yorkshire Terrier" (Figs. 2 and 3).

3 Constraint-Based Parsing

3.1 Constraint-Based Parsing

Constraint-based parsing using integer programming has been successfully tried for Indian languages [1]. We used for Gujarati languages.
For example,

ધનશ્યામ   હાથથી   કેળું   ખાય છે.

(Ghanshyam eats Banana with his hand.)

A demand group (verb) is an element which makes demands for their karakas. These demands are satisfied by source group (noun). A source Group becomes a

Fig. 3 Dependency chart

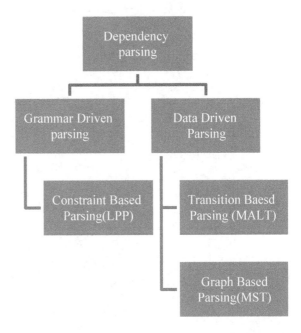

Table 1 Karaka charts for
ખા: (eat)

Karaka	Vibhakti	Presence
Karta	Ø	Mandatory
Karma	Ne or Ø	Mandatory
Karana	Thi or dvaaraa	Optional

potential candidate for a verb only after it satisfied the vibhakti specification as mentioned in the verb karaka frame (Table 1).

For a given sentences after the word group have been formed, karaka charts for the verb group are created and each of the noun group is tested against the karaka restriction in each karaka chart. When testing a noun group against a karaka restriction of a verb group, vibhakti information is checked, and it found satisfactory, the noun group becomes a candidate for the karaka of the verb group. This can be shown in the form of a constraint graph (Fig. 4).

3.1.1 Constraints

- A parse is a subgraph of constraints graph containing all the nodes the constraints graph and satisfying following condition.
- M1: For each of the mandatory karakas in a karaka chart for each demand group (verb group) there should be exactly one outgoing page labeled by the karaka from the demand group.

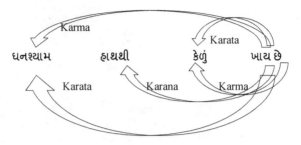

Fig. 4 Constraints graph

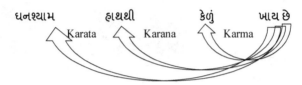

Fig. 5 Solution graph (corresponding to the meaning **ધનશ્યામ કેળું ખાય છે**, English: Ghanshyam Eats Banana)

Fig. 6 Another solution graph corresponding to the meaning **કેળું ખાય છે ધનશ્યામ**, English: Banana Eats Ghanshyam). Two Possible parse for the constraints graph

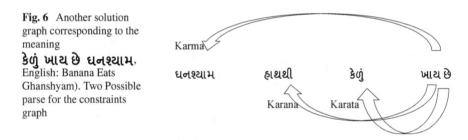

- M2: For each of the optional karakas in a karaka chart for each demand group, there should be at most one outgoing age labeled by the karaka from the demand group.
- M3: There should be exactly one incoming arc into each source group.
- If several subgraph of constraints graphs satisfy the above condition, it means that there are multiple parse and the sentences are ambiguous if not parse is called solution graph (Figs. 5 and 6).

3.2 Constraints Parser Using LPP

A constraint graph is converted into an integer programming problem by introducing a variable x for an arc from node i to j labeled by karaka k in the constraint graph such that for every arc there is a variable [2]. The variables take their values as 0 and

1. Equality and inequality constraints in the integer programming problem can be obtained from the condition as below:

(1) For each demand group i for each of its mandatory karakas k the following equality

must hold $O_{i,k} : \sum_j x_{i,k,j} = 1$

Not that $O_{i,k}$ stands for the equation form a given demand word i and karaka k thus

there will be as many equations as combination of i and k

(2) For each demand group i, for each of its optional karakas k the following inequality must hold $P_{i,k} : \sum_j x_{i,k,j} \leq 1$

(3) For each of the source group j, the following equality must hold

$$T_j : \sum_{i,k} x_{i,k,j} = 1$$

thus their will as many equation as there are source word. The cost function to be minimized is given as the sum of all variables.

Example 1 The projective sentence (refer Tables 2 and 3)

ધનશ્યામ	હાથથી	કેળું	ખાય છે.
k1	k3	k2	m.v.
a	b	c	A

Constraint C1:

$$O_{A,K1} : x_{A,K1,a} + x_{A,K1,c} = 1 \quad O_{A,K2} : x_{A,K2,a} + x_{A,K2,c} = 1$$

Constraint C2:

$$P_{A,K3} : x_{A,K3,b} \leq 1$$

Constraint C3:

$$T_a : x_{A,K1,a} + x_{A,K2,a} = 1$$
$$T_b : x_{A,K3,b} = 1$$
$$T_c : x_{A,K1,c} + x_{A,K2,c} = 1$$

Take $x_{A,K1,a} = y_1, x_{A,K1,c} = y_2, x_{A,K2,a} = y_3, x_{A,K2,c} = y_4, x_{A,K3,b} = y_5$
We get

$$O_{A,K1} : y_1 + y_2 = 1$$
$$O_{A,K2} : y_3 + y_4 = 1$$
$$P_{A,K3} : y_5 \leq 1$$

Table 2 POS tagging For Gujarati

No.	Category		Tag Name	Example
	Top Level	Sub-type		
1	Noun		N	
1.1		Common	NN	ચશ્માં, પેન
1.2		Proper	NNP	મોહન, રવિ.
1.3		Nloc	NST	ઉપર- નીચે
2.	Pronoun		PR	
2.1		Presonal	PRP	હું ,તું ,તે.
2.2		Reflexive	PRF	પોતે, જાતે, સ્વયં.
2.3		Relative	PRL	જે, તે, જ્યાં,
2.4		Raciprocal	PRC	અરસ-પરસ,પરસ્પર
2.5		Wh-word	PRQ	કોણ,ક્યારે, ક્યાં
2.6		Indefinite		કોઈ ,કંઈક, કશુંક
3	Demonstrative		DM	
3.1		Deictic	DMD	આ
3.2		Relative	DMR	જે , જેને.
3.3		Wh-word	DMQ	કોણ,શું, કેમ
3.4		Indefinite		કોઈ ,કંઈક, કશુંક
4	Verb		V	
4.1		Main	VM	ખાશે, ખાધું.
4.2		Auxiliary	VAUX	છે, હતું કર્યું,
5	Adjective		JJ	
6	Adverb		RB	
7	Postposition		PSP	
8	Conjuction		CC	

(continued)

Table 2 (continued)

No.	Category		Tag Name	Example
	Top Level	Sub-type		
8.1		Co-ordinator	CCD	અને, કે
8.2		Sub ordinator	CCS	તેથી, એવું, કારણ
9	Particles		RP	
9.1		Default	RPD	પણ,જા, તો
9.2		Interjection	INJ	હે!! .અરે!!, ઓ!!
9.3		Intensifier	INTF	બહુ, ધણું.
9.4		Negation	NEG	નહિ,ના.
10	Quantifiers		QT	
10.1		General	QTF	થોડું-ધણું
10.2		Cardinals	QTC	એક-બે-ત્રણ
10.3		Ordinals	QTO	પહેલું, બીજું
11	Residues		RD	
11.1		Foreign word	RDF	tv
11.2		Symbol	SYM	$, *,&
11.3		Punctuation	PUNC	, : ; { } ( ) .
11.4		Unknown	UNK	
11.5		Echowords	ECH	કામ-બામ, પાણી-બાણી

$$T_a : y_1 + y_3 = 1, \quad T_b : y_5 = 1, \quad T_c : y_2 + y_4 = 1$$

The cost function to be minimized is:

$$\text{Min} \, Z = y_1 + y_2 + y_3 + y_4 + y_5$$

Subject to,

$$y_1 + y_2 = 1$$
$$y_3 + y_4 = 1$$

Table 3 Dependency table

No.	Tag description			Example
1	K1	કર્તા (doer/agent/subject)	K-Labels	રામ બેઠો છે
2	K2	કર્મ (object/patient)		રામ દરરોજ એક સફરજન ખાય છે
3	K3	કર્ણ (instrument)		રામે ચપ્પુ થી સફરજન કાપ્યું
4	K4	સંપ્રદાન (recipient)		પ્રિયંકા પૂજા ને માટે ફૂલ લાવે છે
5	k4a	anubhava karta (experiencer)		મને રામ બુદ્ધિશાળી લાગે છે
6	K5	અપાદાન (source)		દીપ કંપાસમાંથી બોલપેન લે છે
7	K7t	સમય માં સ્થાન (location in time)		રામ દિલ્હી માં રહે છે
8	K7p	જગ્યા માં સ્થાન (location in space)		ટેબલ પર બુક છે
9	K7	અન્યત્ર સ્રોત (location elsewhere)		તે લોકો રાજનીતિ પર ચર્ચા કરી રહ્યા હતા
10	K*u	સમાનતા (similarity)		રાધા મીરા જેવી સુંદર છે
11	K1s	સંજ્ઞા પૂરક (noun complement)		રામ બુદ્ધિશાળી છે
12	K2p	ધ્યેય (Goal)		રામ ઘરે ગયો
13	K2g	gauna karma (secondary karma)		તે લોકો ગાંધીજીને બાપુ પણ કહે છે
14	Pk1	પ્રયોજક કર્તા (causer)		રામે બાળક ને ખાવાનું ખવડાવ્યુ
15	Mk1	મધ્યસ્થ કર્તા (madhystha karta)		નીમા એ આયા દ્વારા બાળક ને ખાવાનું ખવડાવ્યું
16	Jk1	પ્રાયોજિત કર્તા (causee)		નીમા એ આયા દ્વારા બાળક ને ખાવાનું ખવડાવ્યું
17	ras	સહયોગી (associative)	r-Labels	રામ તેના પિતાજી સાથે બજારમાં ગયો
18	rd	દિશા (direction)		સીતા ગામ ની તરફ જઈ રહી હતી
19	rh	કારણસર (cause)		મેં મોહન ને લીધે પુસ્તક ખરીદ્યુ
20	r6	માલિકીનું (possessive)		સન્માન ની લાગણી
21	r6v	(kA relation between a noun and a verb)		રામ ને એક દીકરી છે
22	rs	relation samanadhikaran		વાત એમ છે કે એ કાલે નહિ આવે
23	rsp	Relation for duratives		૧૯૯૦ થી લઇને ૨૦૦૦ સુધી ભારત ની પ્રગતિ તેજ રહી
24	rt	હેતુ (purpose)		મેં મોહન માટે પુસ્તક ખરીદ્યુ

(continued)

Table 3 (continued)

25	nmod	નામ ના સંશોધકો (noun modifier)	Modifier labels	ઝાડ પર બેઠેલી ચકલી ગીત ગાઈ રહી હતી
26	jjmod	વિશિષ્ટ (adjective modifier)		હલકી ભૂરી બુક
27	vmod	Verb modifier		તે ખાતો ખાતો ગયો
28	Adv	ક્રિયા વિષેશણ (adverb/manner)	(Other) non dependences	તે જલ્દી જલ્દી લખાવી રહ્યો હતો
29	ccof	સંબંધ સાથે જોડાયેલું (conjunction)		રામ ફળ ખાય છે અને સીતા દૂધ પીવે છે
30	pof	સબંધ નો ભાગ (part of complex predicates)		રામ રવિ ની રાહ જોઈ રહ્યો હતો
31	Fragof	Fragmen of		માઓવાદીઓના માણસો ની ધરપ કડ કરી દેવામાં આવી
32	nmod-relc	પ્રકાર ના સંજ્ઞા સંશોધક (noun modifier of thr type)		મારી બહેન જે દિલ્હી માં રહે છે તે કાલે આવી રહી છે
33	Jjmod-relc	પ્રકારના વિશિષ્ટ સંશોધક (adjective modifier of the relative)		મારી બહેન જે દિલ્હી માં રહે છે તે કાલે આવી રહી છે
34	Rbmod-relc	Adverb modifier of the relative		મારી બહેન જે દિલ્હી માં રહે છે તે કાલે આવી રહી છે

$$y_5 \leq 1$$
$$y_1 + y_3 = 1$$
$$y_5 = 1$$
$$y_2 + y_4 = 1$$

Result:

$$y_1 = 1, \quad y_4 = 1, \quad y_5 = 1$$

Same as Fig. 5. It is solved by two-phase simplex method

Example 2 Non-Projective sentence (Fig. 7)

રામે	ફળ	ખાઈને	મોહનને	રમકડું	આપ્યું.
k1	k2	V. Modi	k4	k2	M.V.
a	b	B	c	d	A

Constraint C1:

$$O_{A,K1} : x_{A,K1,a} + x_{A,K1,c} + x_{A,K1,d} + = 1$$
$$O_{A,K2} : x_{A,K2,a} + x_{A,K2,c} + + x_{A,K2,d} = 1$$
$$O_{B,K2} : x_{B,K2,b} = 1$$

Fig. 7 Constraint graph for
the non-projective sentences

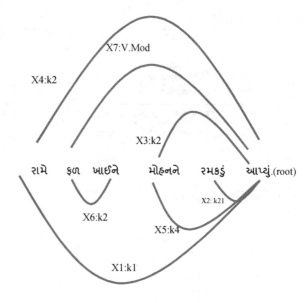

Constraint C2:

$$P_{A,K4} : x_{A,K4,c} \leq 1$$

Constraint C3:

$$T_a : x_{A,K1,a} = 1$$
$$T_b : x_{A,K2,b} + x_{B,K2,b} = 1$$
$$T_c : x_{A,K2,c} + x_{A,K4,c} = 1$$
$$T_d : x_{A,K2,d} = 1$$

Take $x_{A,K1,a} = y_1$, $x_{A,K1,c} = y_2$, $x_{A,K1,d} = y_3$, $x_{A,K2,a} = y_4$, $x_{A,K2,c} = y_5$, $x_{A,K2,d} = y_6$, $x_{A,K4,c} = y_7$, $x_{B,K2,b} = y_8$, $x_{A,K2,b} = y_9$
We get

$$O_{A,K1} : y_1 + y_2 + y_3 = 1$$
$$O_{A,K2} : y_4 + y_5 + y_6 = 1$$
$$O_{B,K2} : y_8 = 1$$
$$P_{A,K4} : y_7 \leq 1$$
$$T_a : y_1 = 1,$$
$$T_b : y_8 + y_9 = 1,$$
$$T_c : y_5 + y_7 = 1$$
$$T_d : y_6 = 1$$

Fig. 8 Solution of graph: parsing

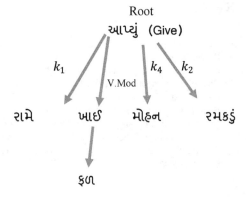

The cost function to be minimized is:

$$\text{Min } Z = y_1 + y_2 + y_3 + y_4 + y_5 + y_6 + y_7 + y_8 + y_9$$

Subject to,

$$y_1 + y_2 + y_3 = 1$$
$$y_4 + y_5 + y_6 = 1$$
$$y_8 = 1$$
$$y_7 \leq 1$$
$$y_1 = 1$$
$$y_8 + y_9 = 1$$
$$y_5 + y_7 = 1$$
$$y_6 = 1$$

Result: $y_1 = 1$, $y_6 = 1$, $y_7 = 1$, $y_8 = 1$ (Fig. 8).

4 Conclusion

In this research, we have shown how computational Paninian Gujarati grammar (for project and non-projective sentence) can be used for parsing word order languages. This paper based on translating Gujarati grammatical constraint to LPP constraints. A solution of these constraints produces a parse. The result is powerful and versatile parser in Gujarati language.

References

1. Bharati, A., Sangal, R.: Parsing free word order languages in Paniian Framework. In: ACL:93, Proceedings of Annual Meeting of Association of Computational Linguistics, Association of Computational Linguistics, New Jersey, USA (1993)
2. Bharati, A., Sangal, R., Papi Reddy, T.: A constraint based parser using integer programming. In: Proceedings of ICON-2002: International Conference on Natural Language Processing (2002)
3. Begum, R., Husain, S., Dhwaj, A., Sharma, D.M., Bai, L., Sangal, R.: Dependency annotation scheme for Indian languages. In: Proceedings of the Third International Joint Conference on Natural Language Processing (IJCNLP). Hyderabad, India (2008)
4. Shieber, S.M.: Evidence against the context-freeness of natural language. In: Linguistics and Philosophy, vol. 8, pp. 334–343 (1985)

A Review on Physiological Signals: Heart Rate Variability and Skin Conductance

Ankita Soni and Kirti Rawal

Abstract The activity of the autonomic nervous system (ANS) has been assessed with heart rate variability (HRV) and skin conductance (SC) or electrodermal activity (EDA) in many research works. From the last three decades, researchers have worked on the relationships between heart rate variability and skin conductance for different activities. The aim of this paper is to explain all possible methods which are used to analyze the variability of heart rate and skin conductance. HRV and SC are the physiological parameters that have been used for assessing body-related changes, stress, pain, depression, and in many different activities. There are many things left to know about their correlation in different activities and can be explored as a future scope.

Keywords Physiological parameters · Heart rate variability · Skin conductance · Time-domain and frequency-domain methods

1 Introduction

The heart rate variability is the standard parameter for analyzing the different changes in the body from the last three decades. Variation in cardiovascular activity and the autonomic nervous system can be detected by the HRV. Variability of heart rate is the parameter, reflecting the interaction of sympathetic and parasympathetic nervous system, which impacts the cardiovascular system as well as other organ systems [1].

Nowadays, skin conductance is in trend to assessing the skin sympathetic nervous activity. Skin conductance is also termed as electrodermal activity in which skin conductance level, skin conductance response are included. Electrodermal activity is regulated by sympathetic involvement of the sweat gland. A small change in response of sweat gland is reflected by the response of electrodermal activity [2].

A. Soni · K. Rawal (✉)
School of Electronics and Electrical Engineering, Lovely Professional University, Phagwara, Punjab, India
e-mail: kirti.20248@lpu.co.in

A. Soni
e-mail: er.ankita1512@gmail.com

© Springer Nature Singapore Pte Ltd. 2020
P. K. Singh et al. (eds.), *Proceedings of First International Conference on Computing, Communications, and Cyber-Security (IC4S 2019)*, Lecture Notes in Networks and Systems 121, https://doi.org/10.1007/978-981-15-3369-3_30

Skin behavior demonstrates the variety in the skin's electrical characteristics owing to sweat secretion from the sweat glands. These glands have three kinds: eccrine, apocrine, and apoeccrine. Emotional arousal can be evaluated by eccrine sweat gland because these glands are stimulated by sympathetic nerves that follow the psychological procedures [3]. In human body, the head/face, abdomen, arms/hands, and legs/feet are involved with different nerve bundles. For example, the nerves controlling the forehead and foot sweat glands differ from the nerves controlling finger sweat glands [4]. The Society for Psychophysiological Research recommended that the palms are as the preferred SC recording position [5]. The eccrine glands, however, are not only present in the palms and soles, but in decreased densities are effectively distributed throughout the body and it was estimated at the forehead, upper limbs, lower limbs and lastly the trunk [6, 7]. Therefore, in SC recording, body places other than traditional palmer recording places may indicate an emotional reaction [3].

The skin conductance is also recorded on different acupoints to evaluate the functioning of related organ [8] as a clinical practice for diagnostic purposes. Electrical current can be used as an electrical acupuncture at acupoint (or meridian point) of the body for generation of a little pulsation, this pulsation could affect heart rate and pulse rates [9]. It can be used for the monitoring of electrocardiogram and pulse rate variability to understand the mechanism of electrical acupuncture. It is primarily used by Chinese medical practice to treat pain, heart disease, and hypertension [10]. This Chinese medical practice also shows that stimulation of the particular point (specifically related to the heart) can slow down the heart rate by keeping the body in calm condition which decrease the skin conductance and further reduce the tension. Researchers found the correlation between heart rate and skin conductance for different mental activities such as anxiety during public speaking [11], interaction with stressful and benign content [12], social rejection [13], painful stimuli [14, 15], etc. Heart rate and skin conductance show similar trend in most of the cases, while in some cases, both the parameters showed negative correlations [16–18] and positive correlations [19–21] as well.

The skin conductance is the parameter for the measurement of electrical activity of skin which is used widely in the field of acupuncture therapy from the last two decades. The skin conductance parameter is also used for the measurement of the status of meridian points in Chinese medicine [10]. The status of an organ depends on the balance between the meridian energy. Yoshio Nakatani was the first person who developed the Ryodoraku theorem and measured the electrical activity of acupuncture points.

2   Literature Review

HRV is a principal parameter in case of diagnosing the cardiovascular autonomic dysfunction. Metelka [1] has explained that using short and long-term ECG recording, HRV can be assessed. It is one of the few techniques for diagnosing ambulatory cardiac autonomic neuropathy (CAN) monitoring progress, healing, and evaluating patient prognosis.

According to the European Cardiological Society Task Force et al., [22] there are different techniques used for the evaluation of heart rate variability. Time- and frequency-domain methods are used for HRV from the last many years. In these methods, different variables are used to record short-term and long-term variability that reflect the functioning and dysfunctioning of the autonomic nervous system and cardiovascular system.

It is also possible to assess the activity of the sympathetic nervous system by measuring the electrodermal activity of the skin. Posada-Quintero et al. [2] have explained the importance of the property of electrical conductivity of skin. Spectral analysis technique can be used for accessing the electrodermal activity of the skin. Skin conductance is now taken as a parameter for the evaluation of emotional arousal and cognitive stress. The skin conductance level and non-specific skin conductance responses (NS-SCRs) are the variable of time-domain methods and did not change during cognitive stress measurement. While the indices of frequency domain are EDASymp$_n$ (normalized sympathetic component of EDA) dependent on power spectral assessment and TVSymp (sympathetic component for time-varying assessment) dependent on time-frequency assessment showed significant change during mental pressure. Dooren et al. [3] explored that skin conductance is an outstanding measure among the most commonly utilized measures in psychophysiological research, including psychic excitement, and is generally estimated at the hand's fingers or the palms. Recording positions for palmar skin conductance be that as it may, not constantly favored for ambulatory recordings, all things considered. The sweat gland which is responsible for sweating in forehead is different from the gland that is responsible for sweating on foot [4].

Boucsein et al. [5] have mentioned that the palms are as the most preferred SC recording position according to the Society for Psychophysiological studies. However, Ménard et al. [6] and Saga et al. [7] have investigated that the eccrine sweat glands are not only present in the palms and soles, but are spread accurately throughout the entire body at reduced densities. The presence of these glands on the forehead, upper limbs, lower limbs, and the trunk in the body is estimated. Chamberlin et al. [8] have investigated that in the Chinese skin conductance is measured on acupoints for the diagnostic purposes. Shukla et al. [23] have explained that for the measurement of EDA, two variables should be measured such as SCL and NS-SCRs. Posada-Quintero et al. [24] have given the time-varying analysis of EDA during exercise. It is explained that when subjects performed low intensity and vigorous intensity exercise, then the upper frequency band which contain the EDA spectral power, shifted to higher frequencies compared to resting condition.

Norman et al. [25] discussed that the skin is a perfect medium for evaluating the electrodermal activities by the means of Galvanic skin response (GSR). Critchley [26] concluded that the skin is the biggest component of the body and the main organism environment interface. The autonomic sweat glands innervation is expressed in measurable modifications in surface skin conductance, called electrodermal activity (EDA). The skin responses can be measured in the range of micro Siemens (μS).

Christopoulos et al. [27] explained that emotional experiences can also be understood by measuring skin conductance.

Croft et al. [11] have explained the effect of anxiety on heart rate and skin conductance over public speaking task. Many people suffered with the problem of anxiety while they have faced the task regarding speaking in public and it affects their heart rate and the level of skin conductance. Raaijmakers et al. [12] investigated that both heart rate and skin conductance increased in stressful and benign situations.

Fujita et al. [28] and Schestatsky et al. [29] explored an interesting fact that in different physical activity which gives pain in the body have correlation between percentage falls in skin impedance (increased conductance). They evaluated the pain scores by a VAS scale and concluded that measurement of skin impedance may be a useful parameter for scaling of pain. Eriksson et al. [30] compared the pain profile of newborn infants and premature infants by using GSR.

Kyle et al. [14] and Eriksson et al. [30] have found positive correlation between Heart rate and pain stimuli, while Walker et al. [16] and Breau et al. [31] have found positive correlation between skin conductance and pain stimuli. Hence, the skin conductance [15, 30, 31] and heart rate [14, 32] could serve as a substitute measure of pain. For male subjects, skin behavior is researched to be more susceptible to pain detection, but when data are averaged over several stimuli, better predicator of pain was heart rate than skin conductance [33–37].

Tousignant-Laflamme et al. [20, 38] investigated that the gender plays a significant part in terms of pain prediction, cardiac activity, or autonomic responses. If the subjects have a pre-set mind, that the experiment will not cause any harm to them, but this perception could affect the outcomes of the experiment (heart rate and skin conductance). Tousignant-Laflamme et al. [39] have explained that male subjects showed considerable changes in heart rate as compared to female subjects with both heat pain and lower back pain. Some authors also investigated that the HR and SC are varied due to social rejection [13]. Andrianome et al. [40] compared the activity of the autonomic nervous system (ANS) between electro-hypersensitive (EHS) subjects and control individuals (healthy subjects) by using heart rate and skin conductance. The findings of this research showed that the variation in heart rate and skin conductance profile did not vary considerably between the electro-hypersensitive and control subjects.

Wang et al. [41] explained that ANS is controlled by parasympathetic and sympathetic systems and can be monitored and recorded by heart rate, GSR, and pupil size also. Autonomic nervous system is a regulatory part in human body, which regulates all fluctuations of arousal. Sarlo et al. [42] explored the effect of pleasant and unpleasant situations for different genders and concluded that the blood pressure shows significant changes as compared to heart rate and skin conductance. Rohrmann et al. [43] investigated that skin conductance is increased in women which implies strong feeling of disgust as compared to men. Bianchin et al. [44] examined that both genders show variability in heart rate and skin behavior with emotional responses.

Weng et al. [45] used Ryodoraku theorem to measure the skin electrical conductance at meridian and acupoints for determining the degree of pain. In their experimental procedure, they introduced a method of pulse stimulation for a specific duration on two acupoints and 12 meridian points on the patients of lower back pain, this treatment method is named as "Acupuncture Like Transcutaneous Electrical Nerve Stimulations (AL-TENS)". After AL-TENS treatment, it is concluded that the skin conductance is increased with decrease in pain.

Weng et al. [46] also used Ryodoraku theorem-based experiment for the treatment of upper back pain as well. Further, Hsu et al. [10] used skin conductance response, heart rate variability, and pulse rate variability to explore the impact of electrical acupuncture on heart meridian acupoint. The result of this study concluded that the SCR at acupoints near heart meridian were decreased which implies the body is in calm condition. Weng et al. [47] investigated that the skin conductance of meridians in obese people is lesser than the healthy people. But, when these obese people follow a weight reduction program than this conductance is increases with decrease in BMI.

Thus, it is clear from the literature survey that both skin conductance and heart rate are able to evaluate the functioning and dis-functioning of autonomic nervous system as well as cardiovascular system. Both parameters can be used for the diagnoses of different organs. From the literature survey, it is also clear that the skin conductance can be measured on different points of the body rather than the skin of palms because there are three types of sweat glands available in the body that are responsible for the electrical conductivity of the skin. Extensive summary of various important research papers on heart rate variability and skin conductance is tabulated in Table 1.

3 Methods

3.1 Heart Rate Variability Measurement

Two types of analysis usually performed for variation in heart rate are time-domain and analysis of the frequency domain.

Time-domain methods. Heart rate variability might be assessed by diverse techniques but time-domain measures are the simple way to evaluate the HRV. Continuous ECG recording gives each QRS complex and normal to normal intervals. The simple variables calculated in time-domain methods are the mean NN interval, the mean heart rate, the difference between the shortest and longest NN interval, the difference between night and day heart rate evaluations [22].

Statistical Measures. Measures of central tendency are known as statistical measures. There are many variables that come under statistical measurement of heart rate variability. Based on the duration, many variables are calculated, i.e., standard deviation of the NN interval (SDNN), standard deviation of the average NN interval for 5 min segments (SDANN), SDANN index, square root of the average square differences of successive NN intervals (RMSSD), number of interval greater than 50 ms in successive NN intervals (NN50), ratio of NN50 to total number of NN intervals (pNN50), mean of standard deviation of all NN intervals counted for 5 min

Table 1 Summary table on heart rate variability and skin conductance

Authors	Parameters	Techniques	Remarks
Task Force of the European Society of Cardiology [22]	Heart rate variability	Different linear and non-linear techniques used for HRV	Task Force of the European Society of Cardiology have given a detailed description on HRV, that it is a most promising marker evaluation of cardiac activity, autonomic nervous system activity, and increased sympathetic or reduced vagal activity
Bianchin et al. [44]	Heart rate, skin conductance, EMG, skin potential	ANOVA for statistical analysis	M. Bianchin et al. examined that both genders show variability in heart rate and skin conductance behavior with emotional responses
Dooren et al. [3]	Skin conductance on various points on body	ANOVA for statistical analysis	M. Van Dooren et al. explored that skin conductance is an outstanding measure among the most commonly utilized measures in psychophysiological research, including psychic excitement, and is generally estimated at the hand's fingers or the palms
Iffland et al. [13]	ECG (HRV) and EDA (SC)	ANOVA for statistical analysis	Authors have revealed in their research paper that social rejection evokes an immediate physiological reaction that can be evaluated by HRV and skin conductance

(continued)

Table 1 (continued)

Authors	Parameters	Techniques	Remarks
Ménard et al. [6]	HRV and skin conductance	Fast Fourier transform	HRV and skin conductance are very useful parameters for evaluation of emotional state whether it is joy, stress, disgust, etc.
Norman et al. [25]	Galvanic skin response (skin conductance)	Skin conductance properties	Galvanic skin response (GSR) is the measure of continuous dynamic variations of the electrical properties of skin
Broucqsault-Dedrie et al. [34]	Heart rate variability	Spearman correlation rank test for analysis purpose	Level of pain in sedated subjects can be assessed by autonomic nervous system activity and heart rate variability analysis is a good evaluation for this purpose
Chu et al. [32]	HRV and skin conductance	Feature extraction techniques	Y. Chu et al. have explained that the intensity of pain in patients can be evaluated with these both parameters (HRV and SC) by external electrical stimulation
Andrianome et al. [40]	HRV, skin conductance, respiration rate, blood pressure	Frequency-domain analysis for HRV and ANOVA for statistical analysis	Electro-hypersensitive population is affected by electromagnetic field that reflects their sensitivity to environmental intolerance and can be evaluated by heart rate variability and skin conductance

(continued)

Table 1 (continued)

Authors	Parameters	Techniques	Remarks
Ernst [48]	Heart rate variability	Different methods for HRV	Today, HRV is mostly used in cardiology and research purposes. G. Ernst has explained different short-term and long-term analysis methods for heart rate variability in this review article
Posada-Quintero et al. [2]	Electrodermal activity (EDA) or skin conductance	Computation of EDA in time and frequency domains	Posada et al. have examined that, when divers are at depth in water, the high pressure and low temperature alone can cause severe stress, challenging the human physiological control systems. This stress level can be evaluated by the analysis of skin conductance
Posada-Quintero et al. [24]	Electrodermal activity and heart rate variability	Time- and frequency-domain analysis for signal processing and ANOVA for statistical analysis	Authors have explained the both ECG and EDA are able to evaluate the sympathetic nervous system activity
Wang et al. [41]	Heart rate variability, GSR and pupil size	ANOVA for statistical analysis	Authors have explained that pupil size can also be parameter for evaluation of physiological arousal as HRV and GSR
Shukla et al. [23]	Electrodermal activity or skin conductance	Time and frequency domains	Author has explained electrodermal activity can indicate the physiological processes related to human cognition and emotion

(SDNN index), taking difference of adjacent NN interval and its SD (SDSD), etc. [22, 48].

Geometric Measures. Sample density distribution of NN intervals, sample density distribution of difference between adjacent NN intervals, Lorenz NN or R-R interval plot can also be converted into geometric structure, etc., and a straightforward formula is used to assess variation based on the resulting patterns of geometric or graphic characteristics. HRV triangular index methods are most extensively used method in which interpolation of R-R Interval histogram used. Other methods such as TINN and differential indexed are used for analyzing HRV [22, 48].

Frequency Domain. The frequency methods can understand by the analysis of power spectrum density (PSD) that provides nevertheless, vital data distributes power as a function of frequency. Two techniques categorized as parametric and nonparametric can be used to evaluate PSD. Non-parametric methods are simple to implement and have high processing speed, i.e., FFT. Parametric techniques facilitate the processing of smoother spectral parts, facilitate post-processing, and precise estimation of PSD can be carried out on tiny sample no. [1, 22], i.e., autoregressive methods (AR). Short-term and long-term (24 h) ECG recording can be evaluated by spectral analysis. The capacity to standardize testing circumstances is the benefit of short-term and long-term recordings. Long-term recording makes it possible to process spectral HRV parts that reflect long-term trends. For short-term recording, there are many variables can be evaluated, i.e., VLF, LF, HF, LF/HF, etc., and for long-term recording (suppose for 24 h) variables evaluated are total power, ULF, VLF, LF, HF, etc. All the methods used for the evaluation of heart rate variability are shown in Fig. 1.

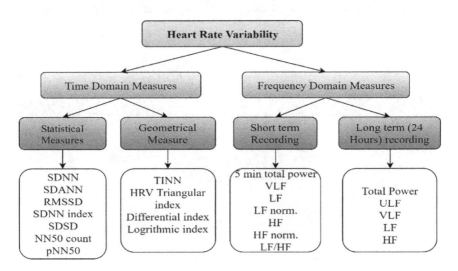

Fig. 1 Heart rate variability measurement

3.2 *Skin Conductance Measurement*

In the time domain, measurement of the EDA gives two parameters: First is the skin conductance level and second is the non-specific skin conductance responses [23]. SCL is measured in micro-Siemens (µS) and computed as a mean of several measurements which is taken during a specific period. The skin conductance responses are the quick short-term events contained by the EDA signals. The number of SCRs in a period of time is computed as the NS.SCRs index [24].

Recently, the power spectral density (PSD) and time-varying spectral analysis of the EDA signal have also been used to acquire the statistics about the spectral distribution of sympathetic arousal in the skin. Y. Shimomura et al. have evaluated the mental workload with the help of frequency-domain analysis of skin conductance. The frequency-domain analysis is very useful technique then traditionally used amplitude domain analysis because it can detect the small differences that are impossible to detect by the traditionally used methods [49].

In Fig. 2, the indices for the time-domain analysis are skin conductance level and non-specific skin conductance responses, while the indices of the frequency-domain analysis are EDASymp which is based on power spectral density and TVSymp which is based on time-frequency analysis [2]. The analysis of electrodermal activity depends on two constituents [50], which are tonic-level components (SCL) and phasic-level components (SCR) as mentioned in Fig. 2. The parameters of these biomedical signals, such as heart rate variability and skin conductance, can be stored through electronic health record in health care 4.0. This health care 4.0 system helps the doctors for analyzing the data of various subjects anywhere and at any time [51–54].

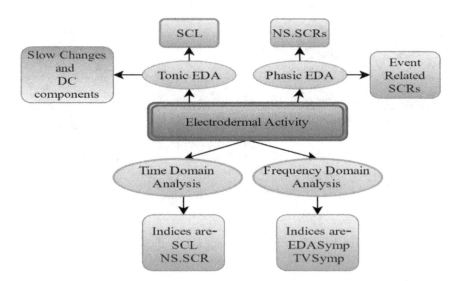

Fig. 2 Electrodermal activity measurement

4 Conclusion

Both physiological parameters, heart rate variability and skin conductance, are able to evaluate the changes occurred due to change in the physical level like exercise, jogging, rest, any physical activity, pleasant-unpleasant situations and mental level as well, i.e., disgust, crying baby video, stress, depression, arithmetic quiz, memory recall tasks, etc. According to different researchers, heart rate variability and skin conductance share positive as well as negative correlation that is completely dependent on the activity performed. On the basis of above literature, it can be concluded that there is a plenty of work is remaining to explore in this field. Some authors give contradictory results on finding the correlation of HRV and SC, which is the future scope for the new researcher to develop a reliable relation between them. Different linear and non-linear methods can be applied for the analysis purposes and new methods can be evolved further for the same.

References

1. Metelka, R.: Heart rate variability—current diagnosis of the cardiac autonomic neuropathy. A review. Biomed. Pap. **158**(3), 327–338 (2014)
2. Posada-Quintero, H.F., Florian, J.P., Orjuela-Cañón, A.D., Chon, K.H.: Electrodermal activity is sensitive to cognitive stress under water. Front. Physiol. **8**(Jan) (2018)
3. Dooren, M.V., Vries, J.J.G.G.J.D., Janssen, J.H.: Emotional sweating across the body: comparing 16 different skin conductance measurement locations. Physiol. Behav. **106**(2), 298–304 (2012)
4. Mauro, T.M.: Biology of eccrine and apocrine glands. Fitzpatricks Derm. Gen. Med. **7**, 929–935 (2012)
5. Boucsein, W., Fowles, D.C., Grimnes, S., Ben-Shakhar, G., Roth, W.T., Dawson, M.E., Filion, D.L.: Publication recommendations for electrodermal measurements. Psychophysiology **49**, 1017–1034 (2012)
6. Ménard, M., Richard, P., Hamdi, H., Daucé, B., Yamaguchi, T.: Emotion recognition based on heart rate and skin conductance. In: PhyCS—2nd International Conference on Physiological Computing Systems, Proceedings, 2015, pp. 26–32 (2015)
7. Saga, K.: Structure and function of human sweat glands studied with histochemistry and cytochemistry. Prog. Histochem. Cytochem. **37**(4), 323–386 (2002)
8. Chamberlin, S., Colbert, A.P., Larsen, A.: Skin conductance at 24 source (Yuan) acupoints in 8637 patients: influence of age, gender and time of day. JAMS J. Acupunct. Meridian Stud. **4**(1), 14–23 (2011)
9. Pearson, S., Colbert, A.P., McNames, J., Baumgartner, M., Hammerschlag, R.: Electrical skin impedance at acupuncture points. J. Altern. Complement. Med. **13**(4), 409–418 (2007)
10. Hsu, C.C., Weng, C.S., Liu, T.S., Tsai, Y.S., Chang, Y.H.: Effects of electrical acupuncture on acupoint BL15 evaluated in terms of heart rate variability, pulse rate variability and skin conductance response. Am. J. Chin. Med. **34**(1), 23–36 (2006)
11. Croft, R.J., Gonsalvez, C.J., Gander, J., Lechem, L., Barry, R.J.: Differential relations between heart rate and skin conductance, and public speaking anxiety. J. Behav. Ther. Exp. Psychiatry **35**(3), 259–271 (2004)
12. Raaijmakers, S.F., Steel, F.W., De Goede, M., Van Wouwe, N.C., Van Erp, J.B.F., Brouwer, A.M.: Heart rate variability and skin conductance biofeedback: a triple-blind randomized

controlled study. In: Proceedings of Humaine Association Conference on Affective Computing and Intelligent Interaction, ACII 2013, pp. 289–293 (2013)

13. Iffland, B., Sansen, L.M., Catani, C., Neuner, F.: Rapid heartbeat, but dry palms: reactions of heart rate and skin conductance levels to social rejection. Front. Psychol. **5**(Aug) (2014)

14. Kyle, B.N.I., McNeil, D.W.: Autonomic arousal and experimentally induced pain: a critical review of the literature. Pain Res. Manag.: J. Can. Pain Soc.—journal de la société canadienne pour le traitement de la douleur **19**(3), 159–167 (2014)

15. Dubé, A.A., Duquette, M., Roy, M., Lepore, F., Duncan, G., Rainville, P.: Brain activity associated with the electrodermal reactivity to acute heat pain. Neuroimage **45**(1), 169–180 (2009)

16. Walker, F.R., Thomson, A., Pfingst, K., Vlemincx, E., Aidman, E., Nalivaiko, E.: Habituation of the electrodermal response—a biological correlate of resilience? PLoS One **14**(1) (2019)

17. Rhudy, J.L., France, C.R., Bartley, E.J., McCabe, K.M., Williams, A.E.: Psychophysiological responses to pain: Further validation of the nociceptive flexion reflex (NFR) as a measure of nociception using multilevel modeling. Psychophysiology **46**(5), 939–948 (2009)

18. Tousignant-Laflamme, Y., Goffaux, P., Bourgault, P., Marchand, S.: Different autonomic responses to experimental pain in IBS patients and healthy controls. J. Clin. Gastroenterol. **40**(9), 814–820 (2006)

19. Rhudy, J.L., McCabe, K.M., Williams, A.E.: Affective modulation of autonomic reactions to noxious stimulation. Int. J. Psychophysiol. **63**(1), 105–109 (2007)

20. Tousignant-Laflamme, Y., Marchand, S.: Sex differences in cardiac and autonomic response to clinical and experimental pain in LBP patients. Eur. J. Pain **10**(7), 603–614 (2006)

21. Williams, A.E., Rhudy, J.L.: Emotional modulation of autonomic responses to painful trigeminal stimulation. Int. J. Psychophysiol. **71**(3), 242–247 (2009)

22. Camm, A.J., Malik, M., Bigger, J.T., Breithardt, G., Cerutti, S., Cohen, R.J., Coumel, P., Fallen, E.L., Kennedy, H.L., Kleiger, R.E., Lombardi, F.: Heart rate variability: standards of measurement, physiological interpretation and clinical use. Eur. Heart J. **17**(3), 354–381 (1996)

23. Shukla, J., Barreda-Angeles, M., Oliver, J., Nandi, G.C., Puig, D.: Feature extraction and selection for emotion recognition from electrodermal activity. IEEE Trans. Affect. Comput. (2019)

24. Posada-Quintero, H.F., Reljin, N., Mills, C., Mills, I., Florian, J.P., VanHeest, J.L., Chon, K.H.: Time-varying analysis of electrodermal activity during exercise. PLoS One **13**(6) (2018)

25. Norman, R., Mendolicchio, L., Mordeniz, C.: Galvanic skin response & its neurological correlates. J. Conscious. Explor. Res. **7**(7), 553–572 (2016)

26. Critchley, H.D.: Electrodermal responses: what happens in the brain. Neuroscientist **8**(2), 132–142 (2002)

27. Christopoulos, G.I., Uy, M.A., Yap, W.J.: The body and the brain: measuring skin conductance responses to understand the emotional experience. Organ. Res. Methods **22**(1), 394–420 (2019)

28. Fujita, T., Fujii, Y., Okada, S.F., Miyauchi, A., Takagi, Y.: Fall of skin impedance and bone and joint pain. J. Bone Miner. Metab. **19**(3), 175–179 (2001)

29. Schestatsky, P., Valls-Solé, J., Costa, J., León, L., Veciana, M., Chaves, M.L.: Skin autonomic reactivity to thermoalgesic stimuli. Clin. Auton. Res. **17**(6), 349–355 (2007)

30. Eriksson, M., Storm, H., Fremming, A., Schollin, J.: Skin conductance compared to a combined behavioural and physiological pain measure in new-born infants. Acta Paediatr. Int. J. Paediatr. **97**(1), 27–30 (2008)

31. Breau, L.M., McGrath, P.J., Camfield, C., Rosmus, C., Allen Finley, G.: Preliminary validation of an observational pain checklist for persons with cognitive impairments and inability to communicate verbally. Dev. Med. Child Neurol. **42**(9), 609–616 (2000)

32. Chu, Y., Zhao, X., Han, J., Su, Y.: Physiological signal-based method for measurement of pain intensity. Front. Neurosci. **11**(May) (2017)

33. Chapman, C.R., Nakamura, Y., Donaldson, G.W., Jacobson, R.C., Bradshaw, D.H., Flores, L., Chapman, C.N.: Sensory and affective dimensions of phasic pain are indistinguishable in the self-report and psychophysiology of normal laboratory subjects. J. Pain **2**(5), 279–294 (2001)

34. Broucqsault-Dédrie, C., Jonckheere, J.D., Jeanne, M., Nseir, S.: Measurement of heart rate variability to assess pain in sedated critically ill patients: a prospective observational study. PLoS One 11(1) (2016)
35. Aslaksen, P.M., Myrbakk, I.N., Høifødt, R.S., Flaten, M.A.: The effect of experimenter gender on autonomic and subjective responses to pain stimuli. Pain 129(3), 260–268 (2007)
36. Reeves, J.L., Radford-Graff, S.B., Shipman, D.: The effects of transcutaneous electrical nerve stimulation on experimental pain and sympathetic nervous system response. Pain Med. 5(2), 150–161 (2004)
37. Loggia, M.L., Juneau, M., Bushnell, M.C.: Autonomic responses to heat pain: heart rate, skin conductance, and their relation to verbal ratings and stimulus intensity. Pain 152(3), 592–598 (2011)
38. Tousignant-Laflamme, Y., Rainville, P., Marchand, S.: Establishing a link between heart rate and pain in healthy subjects: a gender effect. J. Pain 6(6), 341–347 (2005)
39. Tousignant-Laflamme, Y., Marchand, S.: Autonomic reactivity to pain throughout the menstrual cycle in healthy women. Clin. Auton. Res. 19(3), 167–173 (2009)
40. Andrianome, S., Gobert, J., Hugueville, L., Stéphan-Blanchard, E., Telliez, F., Selmaoui, B.: An assessment of the autonomic nervous system in the electrohypersensitive population: a heart rate variability and skin conductance study. J. Appl. Physiol. 123(5), 1055–1062 (2017)
41. Wang, C.A., Baird, T., Huang, J., Coutinho, J.D., Brien, D.C., Munoz, D.P.: Arousal effects on pupil size, heart rate, and skin conductance in an emotional face task. Front. Neurol. 9(2018)
42. Sarlo, M., Palomba, D., Buodo, G., Minghetti, R., Stegagno, L.: Blood pressure changes highlight gender differences in emotional reactivity to arousing pictures. Biol. Psychol. 70(3), 188–196 (2005)
43. Rohrmann, S., Hopp, H., Quirin, M.: Gender differences in psychophysiological responses to disgust. J. Psychophysiol. 22(2), 65–75 (2008)
44. Bianchin, M., Angrilli, A.: Gender differences in emotional responses: a psychophysiological study. Physiol. Behav. 105(4), 925–932 (2012)
45. Weng, C.S., Tsai, Y.S., Yang, C.Y.: Using electrical conductance as the evaluation parameter of pain in patients with low back pain treated by acupuncture-like TENS. Biomed. Eng. Appl. Basis Commun. 16(4), 205–212 (2004)
46. Weng, C.S., Tsai, Y.S., Shu, S.H., Chen, C.C., Sun, M.F.: The treatment of upper back pain by two modulated frequency modes of acupuncture-like TENS. J. Med. Biol. Eng. 25(1), 21–25 (2005)
47. Weng, C.S., Hung, Y.L., Shyu, L.Y., Chang, Y.H.: A study of electrical conductance of meridian in the obese during weight reduction. Am. J. Chin. Med. 32(3), 417–425 (2004)
48. Ernst, G.: Hidden signals-the history and methods of heart rate variability. Front. Public Health 5(2017)
49. Shimomura, Y., Yoda, T., Sugiura, K., Horiguchi, A., Iwanaga, K., Katsuura, T.: Use of frequency domain analysis of skin conductance for evaluation of mental workload. J. Physiol. Anthropol. 27(4), 173–177 (2008)
50. Braithwaite, J., Watson, D., Robert, J., Mickey, R.: A guide for analysing electrodermal activity (EDA) & skin conductance responses (SCRs) for psychological experiments. Psychophysiology 1017–1034 (2013)
51. Hathaliya, J.J., Tanwar, S., Tyagi, S., Kumar, N.: Securing electronics healthcare records in healthcare 4.0: a biometric-based approach. Comput. Electr. Eng. 76, 398–410 (2019)
52. Kumari, A., Tanwar, S., Tyagi, S., Kumar, N.: Fog computing for healthcare 4.0 environment: opportunities and challenges. Comput. Electr. Eng. 72, 1–13 (2018)
53. Vora, J., Devmurari, P., Tanwar, S., Tyagi, S., Kumar, N., Obaidat, M.S.: Blind signatures based secured E-healthcare system. In: International Conference on Computer, Information and Telecommunication Systems, 2018, pp. 177–181. IEEE CITS, Colmar, France, 2018
54. Gupta, R., Tanwar, S., Tyagi, S., Kumar, N., Obaidat, M.S., Sadoun, B.: Habits: block chain based telesurgery framework for healthcare 4.0. In: International Conference on Computer, Information and Telecommunication Systems, IEEE CITS-2019, pp. 6–10. 28–31 Aug 2019, Beijing, China

An Analysis on OCL/UML Constraints in E-commerce Application

Shikha Singh and Manuj Darbari

Abstract E-commerce is the most efficient and time-effective method to promote a product to worldwide customers. Its transaction may require submitting customer's personal information such as username, password, credit card numbers, and financial data. Unified Modeling Language (UML) or Object Constraint Language (OCL) was one of the standardized languages for software developers with visualizing the software system in e-commerce. In e-commerce, purpose security is the primary issue meant for transferring personal data, online payment, and user details. Due to the lack of such security, a significant portion of the customers is feeling uncomfortable to send their respective information over the Internet. This study investigated completely 40 different kinds of literature which are collected from standard publications. Each research paper analyzes different ideas to secure online shopping based on the UML/OCL. In the existing papers, some of the security drawbacks are pointed out and they are overcome by various languages with different software effectively is reviewed by using new research papers.

Keywords UML/OCL · E-commerce · Web security · Security threats

1 Introduction

E-commerce refers to "fundamental economic action including exchanging or purchasing and selling of merchandise and ventures." In other words, it is a value-based action of merchandise or service between sellers or merchants and buyers or customers explained in Sunitha and Philip [1]; as clients go into a shop, they observe the items, decide, and pay the amount for the given item. This type of commerce is referred to as conventional commerce where to satisfy the customer's requirements explored in [2] the shop and their needs to complete commercial exchanges and business capacities, for example, dealing with the store network [3], giving calculated

S. Singh (✉) · M. Darbari
BBD University, Lucknow, Uttar Pradesh, India
e-mail: shikhasingh2816@gmail.com

M. Darbari
e-mail: manujuma@gmail.com

© Springer Nature Singapore Pte Ltd. 2020
P. K. Singh et al. (eds.), *Proceedings of First International Conference on Computing, Communications, and Cyber-Security (IC4S 2019)*, Lecture Notes in Networks and Systems 121, https://doi.org/10.1007/978-981-15-3369-3_31

help, taking care of payments, etc. Through the growth of the World Wide Web (WWW), it significantly raises the knowledge [4] of e-commerce and also develops its possibility to cover various kinds of utilization. The OCL be a textual extension of the UML based on Huang et al. [5] mathematical rationale, and in this way, OCL is in the convention of other information-oriented formal specification languages [6]. The utilization of a common-use (texts) language such as "OCL" is necessary for representing a wide range of limitations that cannot be expressed by utilizing just the graphical [7] develops present by the theoretical modeling language such as UML. "OCL" can likewise be utilized as an inquiry language designed for important process or source system on UML models. It preserves some of the weaknesses of "UML" documentation like the absence of correctness [8]. In e-commerce applications, the exchange between consumers and business association hackers may destroy, change, or adjust different privileges of parties involved in exchange [9]. UML is the target language used by all state of the craftsmanship commitments dealing with the security at specification and design level in e-commerce applications. Some of the imperative applications [10] include marketing, taking orders, shipping, customer service, buying crude materials for the creation and supply.

1.1   Process Involved in E-commerce Applications

E-commerce is a nonstop sequence of information dispensation as well as a substitute, realizing a combined also incorporated data support of all members in the largely profitable exchange, suchlike their position of movement, sector, and nation. More formally it focuses on carefully enabled commercial exchanges between and among associations and people carried out in real time by means of telecommunication networks. E-commerce covers out open processes with customers, suppliers, and external partners. Applications for e-commerce from type "business–customer" are like business operations, supporting e-commerce from type "business–business," however, and are normally associated with retail trade, i.e., purchases by end users. The fundamental processes related to electronic commerce applications are listed below.

(1) Visit on Web site: By utilizing the Internet, the primary objective of this process is to pull in potential customers to the site. Customers in e-commerce are served by a Web site, i.e., customer e-shop is a member of the worldwide network. A user is a person whose objective is to pay by credit card or debit card and receive the products at home.

(2) To review electronic catalogs: Unlike conventional exchanging after the primary visit, the client began building their profile, i.e., attract conclusion about products interested at. The process then continues with the examination of electronic inventories containing data on individual products and services.

(3) Select goods and services: The e-commerce must have a well-developed management system that preserves the current status of the order (for the most

part through shopping basket), which contains data about the amount and price of goods.

(4) Insert an order: In this model, business–customer, the purchaser enters address information for delivery and shapes the method for payment. It can introduce extra requirements of the customer as a note (pressing the purchase as a blessing, time delivery and that's only the tip of the iceberg).

(5) The shaping of payment and delivery after receiving the data about the place of delivery and payment, e-shop determines the value of goods and delivery costs.

(6) Payment after ascertaining the value of merchandise and delivery costs presents charge payment from the buyer. Depending on the customer's desire credit, debit cards or money can be used. There is protection when transferring information over the network between the buyer and the online store so as to protect personal information and verify their authenticity.

(7) Order processing: After payment, the order is processed with particular products.

(8) Order execution: Then, the order is executed, and finally, we get the delivery of the product with the given time.

2 Overview of UML Model

Figure 1, UML is a demonstrating language utilized by programming developers. It is a graphical language for imagining, indicating, creating, and recording information about programming escalated frameworks, and it tends to be utilized

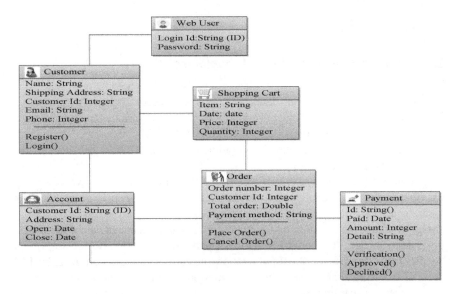

Fig. 1 Overview of UML

for displaying a framework free of a phase language. It can be utilized to create charts and give clients prepared to utilize expressive displaying models. Some UML tool assemblies produce program language code from UML. UML gives a standard strategy to compose a framework show, covering reasonable thoughts. With a comprehension of demonstrating, the utilization and usage of UML can make the product improvement process progressively effective. Web programming multifaceted nature of an application can be limited using diverse UML instruments.

2.1 UML Related with OCL

OCL is a relative of syntropy, a second-age object-situated examination, and structure technique. It supplements UML by giving articulations that have neither the ambiguities of typical language nor the characteristic inconvenience of using troublesome science. OCL is similarly a coarse language for graph-based models. OCL proclamations are built in four segments:

- A context that characterizes the restricted condition in which the state is real possession that speaks to a few qualities of the framework (e.g., if the context is a class, a property might be a trait).
- A property that symbolizes a number of the uniqueness of the situation (e.g., if the framework is a class, assets may be a quality).
- A process (e.g., calculation, set-oriented) that influences or succeeds possessions.
- Catchphrase (e.g., condition, then, else, and, or, not, implies) that are utilized to identify the provisional language.

2.2 Security Objectives in an E-commerce System

When security dangers are considered, this implies the acceptance of a security need in the system. The security objective is a property that describes the security need of a system, regularly expressed through the security objectives which are security characteristics of business assets. An e-commerce system, like any data system, has the accompanying security objectives.

Confidentiality For unauthorized disclosure, the information is protected. For example, lost privacy happens when the substance of correspondence or an organizer is disclosed.

Integrity means the information has not been unclear or damaged, which should be possible by mistake (e.g., broadcast fault) or pernicious intention (e.g., disrupt).

Availability describes the way to official people can contact information as well as systems inside a suitable age of instance. Reason for failure of accessibility might be assaults or instabilities of the system.

Responsibility: In the event that the responsibility of a structure is guaranteed, the members of a correspondence action can be sure that their correspondence partner is the one he or she professes to be. In this way, the correspondence partners can be detained responsible for their activities.

2.3 Security in E-commerce Using UML

The security problems of the e-commerce (processes and information) take great importance these days. The e-commerce is hence increasingly likely to be the subject of different disturbances, for example, congestions, malicious accesses, and assaults. A UML in security case represents a security service returned by the system for one or more performing artists. In the examination model, we defined security properties in e-commerce application by utilizing UML/OCL on the information.

3 Related Works

This part analyzed the different research papers used to analyze OCL on various Web applications such as E-commerce. The current research is taken from IEEE, ScienceDirect, Springer, and also some other journals. Table 1 discusses security in e-commerce applications using UML different techniques.

Oriol et al. [11], they used to outline the schema and the constraints; this work is self-governing of the language because it is based on a reason formalization of each "UML as well as OCL" since they may be broadly used inside the theoretical modeling network. The portion of "OCL" is utilized to describe the constriction which has the equal capability as relational algebra, and they additionally perceive a separation of it which presents some first-class possessions inside the modernize calculation method. The experiment is performed to describe the effectiveness of the progress.

Queralt et al. [12], the expressivity of UML patterns is commented with printed OCL imperatives implements the requirement for computerized view systems. The most important aim is to categorize a significant portion of the "OCL" language that ensures those homes. By this manner, conquer the limitation of the modern-day method while interpretation on one of this section.

Sunitha et al. [1], UML is achieved by using a hundred percentage automation in code technology system fashions will make a drastic development in software program enterprise. This model examines a way to enhance the code age from UML models, with the assistance of "OCL."

Li et al. [13], they recognized a position of relational reliability restriction bearing on the spatial, fleeting, and spatiotemporal homes of 3D spatiotemporal facts. They prolonged the "OCL" to handle spatiotemporal (ST) gadgets. ST–OCL is utilized to explain and document constraints. Constraints expressed through ST–OCL

Table 1 Analysis of security in e-commerce using UML different techniques

Author	Methodology	Merits and demerits
Chehida et al. [29]	To analyze the security necessities in e-trade using UML	1. Good security and efficiency using UML 2. Fails to implement in hardware configurations
Pathak et al. [30]	Proposed the utilization of security presentation flexibility expectation working method	1. Object-oriented modularization, basis system construction, and reorganization of a safe software model 2. Time requirement is more
Shuaibu et al. [31]	Risk analysis and threat modeling methods are proposed to secure e-commerce applications	Data is secured using risk analysis
Razzaq et al. [32]	Ontology engineering-based method is applied for designing and evaluating security system	More time-efficient
Khan et al. [33]	Proposed UML state machine diagram for improving security purpose	Same state machine diagrams may be used to expand test cases in a good way and to also include safety situations
Choshin et al. [34]	Proposed security of online payment using partial least squares (PLS-SEM)	More cost and time when the number of the customer is increased
Banerjee et al. [35]	Different safety risks of e-commerce transaction machine and also to provide an option to fill such deficiency of the machine. Use distinct tokens to recognize the consumer detail	The token is expired after single use; this minimizes the risk to the customer
Ozkaya et al. [36]	SAwUML structure explanation language is projected to develop UML for the elevated stage, precise condition of structural and behavioral intent	More compact than the SPIN model
El-Hajj et al. [37]	Framework for protecting the information flow in Web applications	Minimum developer attempt is necessary
Rodríguez et al. [38]	An expansion of UML 2.0 activity figures which will permit security necessities to be particular in industry methods	Lag in time

statements are estimate via the improved algorithms to recognize special topological associations among 3D spatiotemporal objects.

Hammad et al. [14], the "OCL" is generally utilized intended for identifying an extra restriction on representation. To assist practitioners as well as researchers to indicate "OCL" restriction, they intended and advanced an Internet-based device referred to as interactive OCL (iOCL) for interactively identify a restriction on a given version.

Büttner et al. [15], the OCL is a properly established component in representation-pushed manufacturing as well as associated modeling languages which include "UML" and Eclipse Modeling Framework (EMF) that helps object-oriented business improvement. Amid different concealed, "OCL" gives the formulation of sophistication invariants as well as the process convention in shape of pre- and submits conditions, plus aspect–impact-free question procedure.

Hoisl et al. [16], in current existence, UML-based Domain-Specific Modeling Languages (DSMLs) acknowledge happen to an acknowledged bit of leeway in model apprenticed development projects. Application of Web-based analysis amid MDD advisers and practitioners, they composed 80 able opinions on the accepted conveyance of manuscript and (re)using Architecture Account (DR) on UML-based DSMLs [17]. Data justification policies are normally utilized in mixture with statistical representation, e.g., OCL constraints that designate class member invariants in UML elegance fashions.

Aljumaily et al. [18], UML is one of the computer-aided software program engineering tools and its miles elegance figure, otherwise, that a specified magnificence figure is able to flawlessly convene the consumer's necessities. This examination focuses on the justification of class figure with black-box testing, a system utilized to test software program exclusive of focusing on the software program's accomplishment or shape.

Zaragoza et al. [19], the method explains software of cellular integration additives in public organization and e-commerce as a software improvement technique to basically incorporate the specific primary mechanism of the era right into a solitary net primarily-based explanation.

Vučković et al. [20], in order to come across the identification theft, there is a wanted to investigate customers' conduct on exceptional Web structures. This examines and analyzes e-commerce users' performance which will locate the individuality theft. Review facts are utilized with various factors which might be when it comes to identity theft occurrences.

Derdour et al. [21], they proposed a standard meta-modeling technique called Security Meta-model of Software Architecture (SMSA) for relating a software program machine as a compilation of additives that cooperate via refuge connectors. SMSA meta-version is formed as a UML SMSA profile. They develop UML dominant capability (meta-fashions and models) to outline security standards of SMSA (e.g., protection connectors, complex and area).

Gutiérrez et al. [6], the major motivation behind this paper is to offer the utility of the Process for Web Services Security (PWSSec) that created through the creators to genuine Web access-based contextual investigation. The manner by which security in between hierarchical data structures can be examined, planned, and actualized by utilizing applying PWSSec, which mixes a risk investigation and the board, alongside a safety design and a notably based strategy, is moreover appeared. Georg et al. [22], they proposed a strategy, based on aspect-oriented modeling (AOM), for joining security instruments in a utility. The usefulness of the product has portrayed the utilization of the essential adaptation, and the attacks are exact utilizing segments. The strike segment comprises of the essential model to acquire the abuse rendition. The security system, displayed as a security component, is made out of the main adaptation to accomplish the safety managed with the model.

Pathak et al. [23], UML-based establishments for model-driven structure and forward building of UML static models are investigated. The greater part of data on a Web server is openly legible and to be had to all clients; however, those clients have to never again be fit to trade the insights at the Web server. In this application, the requirement for security tests over the span of reads from disk resembles a misuse of CPU cycles.

Katchalov et al. [24], they present a technique which consolidates the model-driven methodology that utilized in secure MDD with the plan of utilitarian and security tests. They develop and assess new displaying rules that enable the modeler to effortlessly characterize such experiments during the demonstrating stage. They likewise actualize display change schedules to create runnable tests for the real execution of uses created with secure MDD.

Gorla et al. [25], they build up a general model to clarify the selection of business-to-business Internet business utilizing five business factors: outer condition, authoritative setting, chief's attributes, innovation setting, and hierarchical learning.

Li et al. [26], customized items and administrations in Internet business bring shoppers numerous experiences, yet in addition to trigger a progression of data security issues. Thinking about the limited reasonability of the amusement members, in this technique, they propose a transformative diversion model of security assurance among firms and buyers dependent on Internet business personalization.

Moebius et al. [27], they present a way to deal with these applications with the UML stretched out by a UML profile to tailor our models to security applications. They focus on demonstrating with UML and a few insights about the change of this model into the formal detail. They outline that this methodology on an electronic installment framework called Mondex.

Luhach et al. [28], Fast increments in data innovation additionally changed the current markets and changed them into e-markets (online business) from physical markets. SOA is outstanding among other methods to satisfy these necessities. This causes the security innovation usage of e-business extremely troublesome at other designing sciences. This paper talks about the significance of utilizing SOA in e-trade and distinguishes the blemishes in the current security investigation of e-business stages.

3.1 Existing Papers Analyzed with Security in E-commerce Using UML Different Techniques

3.1.1 Challenges in E-commerce

The hazard and the challenges of the belief that discourage the customer to participate in the e-commerce system are spoofing, interception of data, data alteration, denial of services, and overcharge.

Spoofing—The ease of replicating and creating the existing pages of a Web site makes it too easy to even consider creating duplicate sites that imagine being the first one published by different associations for leading fraudulent activities including the collection of personal data illegally.

Interception of sensitive data—When exchange data is transmitted through the Internet, hackers can intercept the transmission and get customers' sensitive data like credit card number, username, and secret word.

Data alteration—The content of an exchange may not exclusively be intercepted, yet in addition might be altered en route, either noxiously or accidentally. Customer names, credit card numbers, and sums sent through the Web all are vulnerable to such alteration.

Denial of services—The Web site can be changed by hackers with the goal that it declines service to the customers or may not work properly, for example, SYN flooding, DDoS, etc.

Overcharge—The fraudulent activities may contain incriminate the customers at a higher than the decided prices for the great or service ordered by the customer.

4 Review Analysis

This section explains the results of analyzed research papers about e-commerce applications in various fields. Research related to security in e-commerce applications based on the UML/OCL process by utilizing various techniques is analyzed and compared with other approaches. Figures and Table 2 explain the security analyzed

Table 2 Review analysis

E-commerce sites	Total respondents	No. of. respondents from the user	Percentage
	For security	For security	
Myntra	10	5	50
Snapdeal	10	2	20
Jabong	10	3	30
Amazon	10	1	10
Flipkart	10	0	0

with different languages with minimum time.

The above table explains the respondents from the different users and with security with each Web site. The considered e-commerce sites are Myntra, Snapdeal, Jabong, Amazon, Flipkart. For all e-commerce sites, the security of all the total respondents is 10 and the number of respondents is varied according to the users.

Figure 2 explains the response time analysis for various approaches such as object-oriented modeling approach (OOMA), case tool, aspect-oriented modeling (AOM), evolutionary game analysis (EGA) during online shopping by using UML/OCL. The x-axis represents various approaches described and the y-axis represents response time. Here, different authors explained different approaches for each response time analysis for each product purchased.

Figure 3 explains security analysis with various languages. The x-axis takes various languages as well as the y-axis takes security in percentage. The graph is plotted with various languages with security analysis. By using Extensible Markup

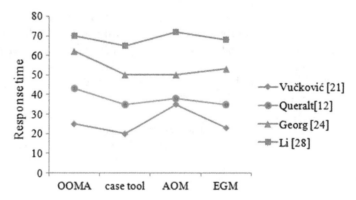

Fig. 2 Response time analysis

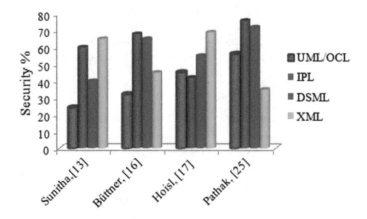

Fig. 3 Security analysis

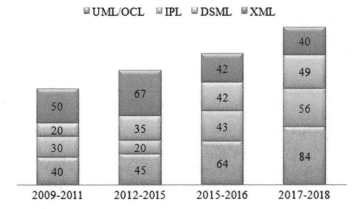

Fig. 4 Comparative analysis

Language (XML), security is 20 describes the 1st author, security is 32 explains the 2nd author, 45 describes for the 3rd author and 59 describes for the 4th author. Moreover, for Imperative Programming Languages (IPL) and Domain-Specific Modeling Language (DSML), the author describes different security levels.

Figure 4 explains the year-based analysis with various languages. Various languages such as UML/OCL, IPL, DSML, and XML were considered. In the year from 2009 to 2011, OCL achieves 40%, IPL achieves 30%, DSML achieves 20%, and XML achieves 50%. From 2012 to 2015, OCL reaches 45%, IPL reaches 20%, DSML reaches 35%, and XML reaches 76%. From 2017 to 2018, all languages give different values.

5   Conclusion

This paper reviews a clear explanation of the security in e-commerce applications for UML/OCL constraints. During online shopping, security is the major issue for transferring user details. UML is one of the efficient languages, and it is applicable to a large number of users in e-commerce applications. Nearly 15–40 research papers are analyzed related to the UML/OCL related to security in e-commerce applications. By using different software languages and approaches, the security is improved for online payment, user details, etc. The existing literature does not provide adequate results to analyze security issues with corresponding software, and the time requirement is a major concern. Achieving a better result by using various tool supports with code generation using UML diagrams.

References

1. Sunitha, E.V., Samuel, P.: Object constraint language for code generation from activity models. Inf. Softw. Technol. **103**, 92–111 (2018)
2. Cabot, Jordi, Pau, Raquel, Raventós, Ruth: From UML/OCL to SBVR specifications: a challenging transformation. Inf. Syst. **35**(4), 417–440 (2010)
3. Casola, V., De Benedictis, A., Rak, M., Villano, U.: Security-by-design in multi-cloud applications: an optimization approach. Inf. Sci. **454**, 344–362 (2018)
4. Sandhu, Ravi: Speculations on the science of web user security. Comput. Netw. **56**(18), 3891–3895 (2012)
5. Huang, Y.-W., Tsai, C.-H., Lin, T.-P., Huang, S.-K., Lee, D.T., Kuo, S.-Y.: A testing framework for Web application security assessment. Comput. Netw. **48**(5), 739–761 (2005)
6. Gutiérrez, C., Rosado, D.G., Fernández-Medina, E.: The practical application of a process for eliciting and designing security in web service systems. Inf. Softw. Technol. **51**(12), 1712–1738 (2009)
7. Herath, A., Al-Bastaki, Y., Herath, S.: Task-based interdisciplinary E-commerce course with UML sequence diagrams, algorithm transformations and spatial circuits to boost learning information security concepts. Int. J. Comput. Digit. Syst. **218**(1221), 1–9 (2013)
8. Gonzalez, R.M., Martin, M.V., Munoz-Arteaga, J., Garcia-Ruiz, M.A.: A measurement model for secure and usable e-commerce websites. In: Canadian Conference on Electrical and Computer Engineering, 2009, CCECE'09, pp. 77–82. IEEE (2009)
9. Agustin, J.L.H., Del Barco, P.C.: A model-driven approach to developing high-performance web applications. J. Syst. Softw. **86**(12), 3013–3023 (2013)
10. Woodside, M., Petriu, D.C., Petriu, D.B., Xu, J., Israr, T., Georg, G., France, R., Bieman, J.M., Houmb, S.H., Jürjens, J.: Performance analysis of security aspects by weaving scenarios extracted from UML models. J. Syst. Softw. **82**(1), 56–74 (2009)
11. Oriol, X., Teniente, E., Tort, A.: Computing repairs for constraint violations in UML/OCL conceptual schemas. Data Knowl. Eng. **99**, 39–58 (2015)
12. Queralt, A., Artale, A., Calvanese, D., Teniente, E.: OCL-lite: finite reasoning on UML/OCL conceptual schemas. Data Knowl. Eng. **73**, 1–22 (2012)
13. Li, J., Wong, D.W.S.: STModelViz: a 3D spatiotemporal GIS using a constraint-based approach. Comput. Environ. Urban Syst. **45**, 34–49 (2014)
14. Hammad, M., Yue, T., Wang, S., Ali, S., Nygård, J.F.: iOCL: an interactive tool for specifying, validating and evaluating OCL constraints. Sci. Comput. Program. **149**, 3–8 (2017)
15. Büttner, F., Gogolla, M.: On OCL-based imperative languages. Sci. Comput. Program. **92**, 162–178 (2014)
16. Hoisl, B., Sobernig, S., Strembeck, M.: Reusable and generic design decisions for developing UML-based domain-specific languages. Inf. Softw. Technol. **92**, 49–74 (2017)
17. Groenewegen, D.M., Visser, E.: Integration of data validation and user interface concerns in a DSL for web applications. Softw. Syst. Model. **12**(1), 35–52 (2013)
18. Aljumaily, H., Cuadra, D., Martínez, P.: Applying black-box testing to UML/OCL database models. Softw. Qual. J. **22**(2), 153–184 (2014)
19. Zaragoza, M.G., Kim, H.-K., Chung, Y.K.: Components of mobile integration in social business and E-commerce application. In: International Conference on Computational Science/Intelligence & Applied Informatics, pp. 59–68. Springer, Cham (2018)
20. Vučković, Z., Vukmirović, D., Milenković, M.J., Ristić, S., Prljić, K.: Analyzing of e-commerce user behavior to detect identity theft. Phys. A: Stat. Mech. Appl. **511**, 331–335 (2018)
21. Derdour, M., Alti, A., Gasmi, M., Roose, P.: Security architecture metamodel for model driven security. J. Innov. Digit. Ecosyst. **2**(1–2), 55–70 (2015)
22. Georg, G., Ray, I., Anastasakis, K., Bordbar, B., Toahchoodee, M., Houmb, S.H.: An aspect-oriented methodology for designing secure applications. Inf. Softw. Technol. **51**(5), 846–864 (2009)
23. Pathak, N., Sharma, G., Singh, B.M.: Towards designing of SPF based secure web application using UML 2.0. Int. J. Syst. Assur. Eng. Manag. **8**(1), 208–218 (2017)

24. Katchalov, K., Moebius, N., Stenzel, K., Borek, M., Reif, W.: Modeling test cases for security protocols with Secure MDD. Comput. Netw. **58**, 99–111 (2014)
25. Gorla, N., Chiravuri, A., Chinta, R.: Business-to-business e-commerce adoption: an empirical investigation of business factors. Inf. Syst. Front. **19**(3), 645–667 (2017)
26. Li, Y., Lu, X., Liu, B.: Evolutionary game analysis on e-commerce personalization and privacy protection. Wuhan Univ. J. Nat. Sci. **23**(1), 17–24 (2018)
27. Moebius, N., Haneberg, D., Reif, W., Schellhorn, G.: A modeling framework for the development of provably secure e-commerce applications. In: International Conference on Software Engineering Advances, 2007, ICSEA 2007, p. 8. IEEE (2007)
28. Luhach, A.K., Dwivedi, S.K., Jha, C.K.: Designing a logical security framework for e-commerce system based on soa. arXiv preprint arXiv:1407.2423 (2014)
29. Chehida, S., Rahmouni, M.K.: Security requirements analysis of web applications using UML. In: ICWIT, pp. 232–239 (2012)
30. Pathak, N., Singh, B.M., Sharma, G.: UML 2.0 based framework for the development of secure web application. Int. J. Inf. Technol. **9**(1), 101–109 (2017)
31. Shuaibu, B.M., Norwawi, N.M., Selamat, M.H., Al-Alwani, A.: Systematic review of web application security development model. Artif. Intell. Rev. **43**(2), 259–276 (2015)
32. Razzaq, A., Anwar, Z., Farooq Ahmad, H., Latif, K., Munir, F.: Ontology for attack detection: an intelligent approach to web application security. Comput. Secur. **45**, 124–146 (2014)
33. Khan, M.U.: Representing security specifications in UML state machine diagrams. Procedia Comput. Sci. **56**, 453–458 (2015)
34. Choshin, M., Ghaffari, A.: An investigation of the impact of effective factors on the success of e-commerce in small and medium-sized companies. Comput. Hum. Behav. **66**, 67–74 (2017)
35. Banerjee, S., Karforma, S.: On designing a secure E-commerce transaction management system—a UML based approach. J. Innov. Syst. Des. Eng. **2**(6), 102–108 (2012)
36. Ozkaya, M., Kose, M.A.: SAwUML–UML-based, contractual software architectures and their formal analysis using SPIN. Comput. Lang. Syst. Struct. **54**, 71–94 (2018)
37. El-Hajj, W., Brahim, G.B., Hajj, H., Safa, H., Adaimy, R.: Security-by-construction in web applications development via database annotations. Comput. Secur. **59**, 151–165 (2016)
38. Rodríguez, A., Fernández-Medina, E., Trujillo, J., Piattini, M.: Secure business process model specification through a UML 2.0 activity diagram profile. Decis. Support Syst. **51**(3), 446–465 (2011)

A Review on Crowdsourcing Models in Different Sectors

Akhil Bhatia and Ramesh Dharavath⑩

Abstract This paper systematically reviews association rules mining using crowd data in different sectors. Various sectors like health care, education, and tourism require human thinking to be harnessed to answer the queries that can be difficult for computers to answer. Traditional techniques, like Delphi, require experts in the domain which can make the system expensive, and results can be time taking. Crowdsourcing has emerged as a popular approach out of open innovation which can be used for problem solving, ideation, and even in software development because it enables us to make use of experience and background knowledge of crowd using any platform or system or generated forms. It involves a group of experts and non-experts with certain knowledge toward the problem which they solve together, and the solution is shared further. In association rule mining using crowd data, we outsource work to the crowd and mine association rules from their answers using different methods of aggregation. Crowdsourcing systems are beneficial when information required is not in a systematic manner or it is complex to aggregate. These systems can be utilized for many commercial and non-commercial platforms like Amazon Mechanical Turk (AMT) or Kaggle and they can be structured in different forms like crowdfunding or micro-tasking. We first review the current work of mining association rules using crowd data in different sectors. Then we propose suggestions in other sectors and problems which can be solved in those sectors. Finally, we conclude the limitations of the above framework in different sectors that have been covered in this paper.

Keywords Data mining · Crowdsourcing · Crowd mining

A. Bhatia · R. Dharavath (✉)
Indian Institute of Technology (Indian School of Mines), Dhanbad,
Jharkhand 826004, India
e-mail: drramesh@iitism.ac.in

A. Bhatia
e-mail: akhil.17KT000226@cse.iitism.ac.in

© Springer Nature Singapore Pte Ltd. 2020   415
P. K. Singh et al. (eds.), *Proceedings of First International Conference on Computing, Communications, and Cyber-Security (IC4S 2019)*, Lecture Notes in Networks and Systems 121, https://doi.org/10.1007/978-981-15-3369-3_32

Abbreviations

AMT   Amazon Mechanical Turk
AI   Artificial Intelligence

1   Introduction

In an era of digital transformation, data plays a key role as an enabler of immediate change and planning for the future in different sectors. The art of learning from data, i.e., statistics, deals with the collection of data, its metadata, and analysis which frequently leads us to a conclusion. Statistical data analysis holds a great importance in every area for results to be reliable. Different statistical methods serve different purposes according to different data structures and types of data it holds.

Data mining is a scientific discipline that takes the necessary tools from statistics. The methods of data mining lie in building models to detect patterns and relationships in data. It makes use of large databases and data sets which is not easy with the statistical model to analyze. Data mining has gained a great deal of importance due to a large amount of data in different applications belonging to various fields [1].

Many data mining tasks cannot be effectively solved by existing machine-only algorithms, such as rating things, choose the best out of some situations, and opinion mining [2]. Variously known as the wisdom of crowds or crowdsourcing defined as "the act of a company or institution taking a function once performed by employees and outsourcing it to an undefined (and generally large) network of people in the form of an open call. This can take the form of peer-production (when the job is performed collaboratively) but is also often undertaken by sole individuals (experts or novices). The crucial prerequisite is the use of the open call and large network of potential labours" (Howe 2006) [3].

Crowd users with their background knowledge and daily life experience learn to identify things that suit them best. Hence, if we make use of their gained experience by posing them, the right set of questions which in result can produce generalized trends using various aggregation methods. Crowdsourcing with data mining techniques forms the base of crowd mining [4].

However, it comes with few challenges in the form of budgetary constraints, data noise rule specialization or meaningless answers. The systems may need to perform data sanitization or denoising of raw data and also questions have to be simple to understand. If questions are complex, then answers may be meaningless or it can also lead to a meaningless rule which may not be generalized to the crowd [4].

The crowdsourcing concept has been applied in many applications like data cleaning, data labeling, image tagging, improving quality responses, association rule mining, etc. In crowdsourcing, requesters decompose the whole tasks into a small set of tasks [5]. These tasks are then pushed to the crowd by means of any platform, and workers will accomplish those tasks as part of any process or due to any reasons

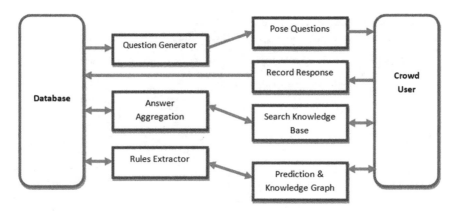

Fig. 1 Crowd mining system architecture

[6]. In some kind of scenario, it can help resolve the problems in an efficient way, producing positive results. Figure 1 denotes a generalized architecture for crowd mining systems.

As per ScienceDirect [7], there were 66 review articles in 2018 which make use of crowdsourcing as a term (accessed on September 16, 2019), while a search in Google Scholar threw about 18,800 articles since 2018. We collectively reviewed 13 articles for 11 different applications in six different sectors to show crowdsourcing work and how they can be extended. We present how it can be beneficial in different sectors, adding more benefits as compared to data mining techniques like the Delphi technique.

We also reviewed 12 articles including some of them from above for concluding key challenges in crowdsourcing. This paper also proposes other sectors for which crowdsourcing can be beneficial to produce results. We proposed a small overview that we can also cover Educational Data Mining (EDM) applications and IT industry small challenges integrating crowdsourcing with other advanced techniques. We also added more work which can be contributed to the healthcare sector and problem solving with the help of crowdsourcing.

The remainder of the paper is organized into four sections. Section 2 of the paper reviews the sectors in which association rule mining has been performed using crowd data. Section 3 of the paper proposes the sectors in which it can be used with problems it can solve efficiently. Finally, Sect. 4 provides the desired conclusions and limitations of using the above framework.

2 Review of Current Sectors Covered

Ghezzi et al. [3] presented significant research in this area. Applications and requirements to improve data quality were also added. Table 1 presents how

Table 1 Review of crowdsourcing models in various sectors

Related work	Year	Domain	Application	Conclusion
Amsterdamer et al. [8, 9]	2013	Health	Well-being-related practices	1. Support and confidence for rules generation have been on the basis of user interaction with the system 2. Question generation has been performed to maximize error reduction
McCoy et al. [10]	2012	Health	Problem medication	1. Association rule mining and crowdsourcing are complementary approaches 2. Summarization of data can improve clinical care
Chang et al. [11]	2015		Product design	1. Selection of best design for a product in an efficient way 2. Reduction of review load and capturing of good concepts for design
Al-Jumeily et al. [12]	2015	Education	Learning computer programming	Compared frameworks which make use of the intelligent adaptive system and crowdsourcing to enhance teaching process and programming skills
Abboura et al. [13]	2015		Matching dependencies (MDs)—data de-duplication	Proposed a hybrid approach (crowdsourcing algorithmic) for MDs generation
Someswar et al. [4]	2018	Health	Remedies for diseases	1. No use of AMT due to open-ended and close-ended questions 2. Attributes of crowd matters to improve rules learned

(continued)

Table 1 (continued)

Related work	Year	Domain	Application	Conclusion
Jiang et al. [14]	2018	Education	Education content creation	Use of a distributed and collaborative approach to enhance crowdsourced-based learning
Phuttharak et al. [15]	2018	IoT		Reviewed work on mobile crowdsourcing with the breakdown of the framework in detail
Cappa et al. [16]	2019	Stock market		Worked on models to detect the positive impact of crowdsourcing for stock market. They also produced implications of sectors that can improve and perform well
Taeihagh et al. [17]	2019		Policymaking	Identified types of crowdsourcing. On the basis of types, examined the use of crowdsourcing in policymaking on the basis of Hood's NATO model

crowdsourcing is changing a few sectors and enhancing the concept like Minet et al. [18] extended the work for farm sourcing.

3 Key Challenges in Crowdsourcing

The building of a crowdsourcing application relies on some important questions: "What do we achieve via this?", "Who is involved in this?", "Why are they involved?", and "How are they involved?" [15]. This brings us to challenges which we face at each and every step of crowdsourcing application. Every sector in which we build crowdsourcing solutions has different challenges at each step. Figure 2 shows a block diagram of key challenges in crowdsourcing applications. Four aspects of crowdsourcing applications with its challenges are as follows.

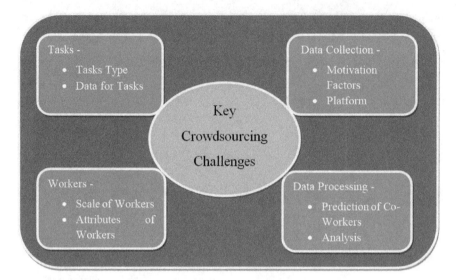

Fig. 2 Key Crowdsourcing challenges

3.1 Tasks

A task in a crowdsourcing framework is referred to as a small piece of work that needs human intervention (aided by machine computations) which involved humans and devices (e.g., mobile in the IoT sector) to collect data. Challenges related to tasks can be classified as follows.

3.1.1 Identifying the Task Type

There can be two types of tasks, i.e., human–companion device tasks or human intelligence tasks. Human–companion devices task is applied in personal data collection which could be done with the help of mobile devices like wearable devices [15]. It could be data regarding the environment or noise. Not all crowdsourcing frameworks make use of such data but basic attributes out of such tasks are required for excluding irrelevant results.

Human Intelligence Tasks require human intervention that is hard for a computer to do like face recognition or sentiment analysis. Mostly, frameworks utilize this task to get conclusions from experience and background knowledge of the worker. In many sectors, there are fewer attributes of data available from human–companion device tasks like basic remedies for diseases. Alimpertis et al. [19] made use of both types of tasks for application to monitor the network.

3.1.2 Identifying Data for Tasks

There can be types of data that have to be collected from tasks. This data can be the user's personal data, social data, experience data, etc. In many cases, the system cannot ask the user to share personal data so we need to identify certain attributes that are necessarily related to the sector in the framework.

Collection of different type of data from users requires a set of questions which have to be precise and evident. Identification of data is an important step to plan a set of questions according to fulfill the motive of that framework. Minet et al. [18] have reviewed the type of information to be collected in agriculture.

3.2 Workers

3.2.1 Identifying the Scale of Workers

The scale of workers indicates that the system needs to target individuals or a similar group of people who belong to the same social group or it could be a geographically broader task. This is completely dependent upon the sector for which we are developing the framework. For example, the healthcare sector needs data from individuals, IoT devices need geographical wide data, and social networking needs similar community data.

3.2.2 Attributes of Workers

Personal data from workers covers their attributes but they are important for systems to avoid irrelevant conclusions. This data can be collected by direct–indirect questions but always requires worker's consent for certain attributes.

For example, frameworks for education require a worker who is enough literate to give data for graduation or post-graduation-related questions (not academic). Frameworks for IoTs need not follow the above attribute as it can collect necessary attributes from mobile devices or sensors.

3.3 Data Collection

3.3.1 Motivation Factors

Encouraging workers to participate in applications for data contribution remains critical. There has to be some mechanism to identify motivating factors for a user to get involved with the system. This remains a challenge as every user cannot be pleased with payoffs or any other offer which is another reason why crowdsourcing is preferred over the Delphi technique [20].

Payoffs to the user make a strong effect to involve crowd but adding enjoyment-based motivation makes interaction and data collection more effective. Cost is an important factor in every framework to collect data from users. So in some cases, we cannot pay workers due to cost constraints although other factors can be added for a worker to collaborate. Li et al. [21] explained that there remains a trade-off among cost, quality, and latency which needs to be dealt with while collecting data for the system.

3.3.2 Platform to Collect Data

There has to be a medium to collect data that will preserve data, and also it has to be user-friendly. The framework which is being used for data collection is too complex then the users will not be able to contribute which will cause a lack of data for the system. It should be scalable and robust so that the collection of data does not face any halt. Minet et al. [18] identify various mediums to collect data for agriculture-related data mining using crowdsourcing and have also considered the attributes and background of workers.

3.4 Data Processing

3.4.1 Prediction of Co-worker

In frameworks for crowdsourcing, users cannot be compelled to answer every question. In such cases, we always give users a choice to skip questions. Here comes the need of prediction which has been skipped by any user on the basis of similarity.

For example, there two users X and Y who are answering on the same crowdsourcing platform. X has answered nine out of ten questions, and Y has answered five out of ten questions, so the similarity of users to predict unanswered questions can be done using co-answered questions. This can be done by using similarity measures like Pearson correlation, cosine similarity, etc. [22].

3.4.2 Analysis

Generating best results from a large set of crowd data required extensive analysis. Two main techniques that can be identified are basic and advanced approaches. Someswar et al. [4] made use of majority voting, weighted voting, and iterative methods in their framework, and a similar approach is also suggested by Li et al. [23]. McCoy et al. [10] used Spearman's rho, chi-square statistic, and logistic regression.

Chang et al. [11] used text mining, web mining, concept hierarchy, and concept clustering. Abboura et al. [13] used the Apriori algorithm to extend their algorithm for the solution.

There are many problems in which confidentiality of data related to problem and solution is kept at top priority like the finance industry. In such problems, crowdsourcing cannot be a methodology to solve it. In that case, we can look out for methodologies like open innovation or outsourcing. Crowdsourcing becomes a challenge when the problems are hard to explain or the abstract to form systems or questions, so open innovation might make more sense for such problems to bring creative solutions.

Some ideas come into existence with open innovation, and with time it brings the need for using crowdsourcing solutions to get at least one aspect of them. And sometimes crowdsourcing can be helpful in raw idea creation but then its refinement and shaping to produce effective results can be done with the usage of open innovation or outsourcing. In crowdsourcing and open innovation, we select a working solution that comes out of them, but in outsourcing, we select person/entity. The proposed sectors with their related applications are represented in Table 2.

However, there can be more sectors in which we can cover many applications of crowdsourcing models. A basic example that covers our location-based data and recommendations is Google Maps, which asks you reviews on the basis of your location data and visits. The use of location data or trajectory mining can be helpful for the tourism sector [28].

Table 2 Future work for crowdsourcing models in various sectors

Domain	Application
Education	1. Enhanced learning and teaching: using crowd users, i.e., students and teachers, the learning process and lecture deliverables can be improved 2. Career decision making: considering the attributes of crowd users, recommendations can be generated using generalized crowd data from a similar attribute group of users 3. Attrition in educational institutions: knowing the interests and similarities of data from crowd users, the attrition rate in institutions can be predicted or students can be recommended to opt for some other course work
IT	1. System setup challenges: there are many system challenges that students or interns face. These challenges include many first time errors or system setup issues with technologies they start to work on. Hence, to overcome these challenges, we can get data from the crowd like StackOverflow and present conclusive results [24]. Rahman et al. [25] made use of natural language processing (NLP) on StackOverflow and tried to covert search query into API classes which can be extended to suggest solutions for system setup

(continued)

Table 2 (continued)

Domain	Application
Planning and execution	There can be many instances where planning to a problem needs suggestions from people on whom the plan has to be implemented. In such cases, crowdsourcing platforms can be really helpful, for example, implementation of government schemes and the solution to problems related to water and pollution. Prime minister's office (PMO) gathered data via the mobile app for the speech of PM for August 15 Chaves et al. [26] have reviewed the work and use of crowd for urban planning. They reviewed the platforms and types of workers required for urban planning. Such applications can be bought and added by the government which can contribute to the infrastructure and healthcare sector of the country
Problem solving	Crowdsourcing is another form of open innovation which is helping providing solution to many AI problems. Kaggle [27] is one such platform that is making use of competitions to lure the creative and smartest crowd to solve AI problems The problem can be anything but such platforms can bring the best solution from the smartest crowd. It will attract good cost but the cost will be optimized for the problem which will be solved. For example, a tool for elder people to be safe from phishing attack which can save them from losing their hard-earned money just due to lack of technological knowledge
Healthcare	Crowdsourcing can be helpful in engaging difficult-to-reach people with data and discussion of critical diseases that they are aware of but less known to prevent it. Also in integration with social media, trends related to health can be generated. This can help various health agencies and hospitals to organize medical camps and vaccinations accordingly

4 Conclusions

Boim et al. [29] presented that asking the questions to the crowd which is not vague or too complex can result in better data quality. There can be chances of hiring redundant workers to complete tasks [22]. While applying the above methodology, the worker's reputation or his attributes can be taken into consideration to avoid inconsistent rules generation or better results. Particular models and patterns can be added to systems to avoid fraud.

However, certain systems can be extended to add recommendation systems using crowdsourcing techniques which can be helpful in domains like health care, education, and tourism. Aldhahri et al. [30] proposed the use of crowdsourcing for recommendation systems adding importance to workers, requesters and also the platform which is to be used to collect data from crowd users.

References

1. Kaya Keleş, M.: An overview: the impact of data mining applications on various sectors. Tehnički glasnik **11**(3), 128–132 (2017)
2. Chai, C., Fan, J., Li, G., Wang, J., Zheng, Y.: Crowd-powered data mining. arXiv preprint arXiv:1806.04968 (2018)
3. Ghezzi, A., Gabelloni, D., Martini, A., Natalicchio, A.: Crowdsourcing: a review and suggestions for future research. Int. J. Manag. Rev. **20**(2), 343–363 (2018)
4. Someswar, M., Bhattacharya, A.: MineAr: using crowd knowledge for mining association rules in the health domain. In: Proceedings of the ACM India Joint International Conference on Data Science and Management of Data, pp. 108–117. ACM (2018)
5. Karthika, K., Devi, R.D.: Crowdsourcing and its applications on data mining: a brief survey (2018)
6. Xintong, G., Hongzhi, W., Song, Y., Hong, G.: Brief survey of crowdsourcing for data mining. Expert Syst. Appl. **41**(17), 7987–7994 (2014)
7. ScienceDirect.com|Science, health and medical journals, full text articles and books, https://www.sciencedirect.com. Last accessed 4 Nov 2019
8. Amsterdamer, Y., Grossman, Y., Milo, T., Senellart, P.: Crowdminer: mining association rules from the crowd. Proc. VLDB Endow. **6**(12), 1250–1253 (2013)
9. Amsterdamer, Y., Grossman, Y., Milo, T., Senellart, P.: Crowd mining. In: Proceedings of the 2013 ACM SIGMOD International Conference on Management of Data, pp. 241–252. ACM (2013)
10. McCoy, A.B., Sittig, D.F., Wright, A.: Comparison of association rule mining and crowdsourcing for automated generation of a problem-medication knowledge base. In: 2012 IEEE Second International Conference on Healthcare Informatics, Imaging and Systems Biology, pp. 125–125. IEEE (2012)
11. Chang, D., Chen, C.H.: Product concept evaluation and selection using data mining and domain ontology in a crowdsourcing environment. Adv. Eng. Inform. **29**(4), 759–774 (2015)
12. Al-Jumeily, D., Hussain, A., Alghamdi, M., Dobbins, C., Lunn, J.: Educational crowdsourcing to support the learning of computer programming. Res. Pract. Technol. Enhanc. Learn. **10**(1), 13 (2015)
13. Abboura, A., Sahrl, S., Ouziri, M., Benbernou, S.: CrowdMD: Crowdsourcing-based approach for deduplication. In: 2015 IEEE International Conference on Big Data (Big Data), pp. 2621–2627. IEEE (2015)
14. Jiang, Y., Schlagwein, D., Benatallah, B.: A review on crowdsourcing for education: state of the art of literature and practice. In: PACIS, p. 180. AISeL (2018)
15. Phuttharak, J., Loke, S.W.: A review of mobile crowdsourcing architectures and challenges: toward crowd-empowered internet-of-things. IEEE Access **7**, 304–324 (2018)
16. Cappa, F., Oriani, R., Pinelli, M., De Massis, A.: When does crowdsourcing benefit firm stock market performance? Res. Policy **48**(9), 103825 (2019)
17. Taeihagh, A.: Crowdsourcing: a new tool for policy-making? Policy Sci. **50**(4), 629–647 (2019)
18. Minet, J., Curnel, Y., Gobin, A., Goffart, J.P., Melard, F., Tychon, B., Defourny, P.: Crowdsourcing for agricultural applications: a review of uses and opportunities for a farmsourcing approach. Comput. Electron. Agric. **142**, 126–138 (2017)
19. Alimpertis, E., Markopoulou, A., Irvine, U.C.: A system for crowdsourcing passive mobile network measurements. 14th USENIX NSDI, 17 (2018)
20. Flostrand, A.: Finding the future: crowdsourcing versus the Delphi technique. Bus. Horiz. **60**(2), 229–236 (2017)
21. Li, G., Zheng, Y., Fan, J., Wang, J., Cheng, R.: Crowdsourced data management: overview and challenges. In: Proceedings of the 2017 ACM International Conference on Management of Data, pp. 1711–1716. ACM (2017)
22. Thukral, R., Ramesh, D.: Ensemble similarity based collaborative filtering feedback: a recommender system scenario. In: 2018 International Conference on Advances in Computing, Communications and Informatics (ICACCI), pp. 2398–2402. IEEE (2018)

23. Li, G., Wang, J., Zheng, Y., Franklin, M.J.: Crowdsourced data management: a survey. IEEE Trans. Knowl. Data Eng. **28**(9), 2296–2319 (2016)
24. Al-batlaa, A., Abdullah-Al-Wadud, M., Hossain, M.A.: A review on recommending solutions for bugs using crowdsourcing. In: 2018 21st Saudi Computer Society National Computer Conference (NCC), pp. 1–4. IEEE (2018)
25. Rahman, M.M., Roy, C.K., Lo, D.: Rack: automatic API recommendation using crowdsourced knowledge. In: 2016 IEEE 23rd International Conference on Software Analysis, Evolution, and Reengineering (SANER), vol. 1, pp. 349–359. IEEE (2016)
26. Chaves, R., Schneider, D., Correia, A., Borges, M.R., Motta, C.: Understanding crowd work in online crowdsourcing platforms for urban planning: systematic review. In: 2019 IEEE 23rd International Conference on Computer Supported Cooperative Work in Design (CSCWD), pp. 273–278. IEEE (2019)
27. Kaggle: Your Home for Data Science, https://www.kaggle.com. Last accessed 4 Nov 2019
28. Basiri, A., Amirian, P., Winstanley, A., Moore, T.: Making tourist guidance systems more intelligent, adaptive and personalised using crowd sourced movement data. J. Ambient Intell. Humaniz. Comput. **9**(2), 413–427 (2018)
29. Boim, R., Greenshpan, O., Milo, T., Novgorodov, S., Polyzotis, N., Tan, W.C.: Asking the right questions in crowd data sourcing. In: 2012 IEEE 28th International Conference on Data Engineering, pp. 1261–1264. IEEE (2012)
30. Aldhahri, E., Shandilya, V., Shiva, S.: Towards an effective crowdsourcing recommendation system: a survey of the state-of-the-art. In: 2015 IEEE Symposium on Service-Oriented System Engineering, pp. 372–377. IEEE (2015)

An Efficient Provably Secure IBS Technique Using Integer Factorization Problem

Chandrashekhar Meshram and Mohammad S. Obaidat

Abstract Digital signatures are among the most fundamental primitives in cryptography, providing in an asymmetric setting, integrity, authenticity, and non-repudiation. An important technique for lightweight authentication is the identity-based signature (IBS). Recently, many attempts have been made to build IBS on the assumptions of the integer factorization problem (IFP), as they would stay secure in the future quantum era. In this article, we present an efficient provably secure IBS technique using IFP and show that it is protected against existential forgery and identity attack in the random oracle model. We also validate the outer performance of the technique in terms of security point of view and computational effectiveness. As it is intuitive, simple, and does not require paring, the resulting technique is of interest.

Keywords Digital signature · Public key cryptography · IBS · IFP · Random oracle

1 Introduction

In cryptographic methods, digital signatures are significantly promising. The digital signature ensures that the validity of the creator of a digital text and the truthfulness of the text, in spite of the fact that they likewise check the creator from withdrawing that real record later on. A signature on a specified bit string is legitimate in the event that it clears the related confirmation test that returns as inputs a confirmation key, an asserted signature, and a text. In the conventional signature technique, the

C. Meshram (✉)
Department of Mathematics, R.T.M. Nagpur University, Nagpur, Maharashtra, India
e-mail: cs_meshram@rediffmail.com

M. S. Obaidat
College of Computing and Informatics, University of Sharjah, Sharjah, UAE
e-mail: msobaidat@gmail.com

King Abdullah II School of Information Technology, The University of Jordan, Amman, Jordan

University of Science and Technology Beijing, Beijing, China

© Springer Nature Singapore Pte Ltd. 2020

427

P. K. Singh et al. (eds.), *Proceedings of First International Conference on Computing, Communications, and Cyber-Security (IC4S 2019)*, Lecture Notes in Networks and Systems 121, https://doi.org/10.1007/978-981-15-3369-3_33

confirmation key is a scientific object occupied at arbitrary from set. An external binding between the verification key and the signing entity is therefore needed. This coupling receives the sort of statement that a private key generator (PKG) produces.

Shamir [26] presented the idea of ID-based signature (IBS) technique in 1984. The idea is that the identity of the signer is used as the verification key. This allots through the necessity of an outer requisite. An identity can be somewhat string that records out the client, for example, such as open safety number or an IP address. The IBS technique has a disadvantage, viz. the client cannot construct the signing key by herself/himself. In its place, the PKG took the responsibility of transfer and distributing, via a trusted channel, the private key to every client, which is called key-escrowed problem, subsequently the PKG identifies the private keys (secret key) of the clients. An IBS technique or usually ID-based cryptographic (IBC) technique [15] does not need any endorsements to be substituted and thus can be favorable in excess of the conventional PKI-based frameworks in specific situations. A few IBS schemes using pairing [6, 10] have been presented in the writing after idea presented by Shamir in [26]. Recently, a few pairing-based developments were additionally presented [5, 8, 13, 22]. Numerous ID-based cryptographic technique and signature techniques [3, 4, 11, 16–21, 28] have been devised. However, in these strategies, the public key of individual client is not just some arbitrary number, as well as an identity is chosen either by the client or by the trustworthy parties. However, that marks the ID-based cryptography in current years.

Bellare et al. [1] gave the security verification or attacking model for a huge quantity of ID-based identification and signature techniques in 2009. Key matter here is a development that from one viewpoint clarifies how these plans are inferred. Furthermost is the standing IBS technique using Weil or Tate pairing [2], which frequently contains substantial calculation. Consequently, developing IBS techniques without exhausting pairing is of incredible enthusiasm for the field of ID-based cryptography. In spite of the fact that there are few techniques without exhausting pairings, there does not exist any provable protected IBS technique that depends on the IFP. Lee and Liao's [17] projected an alternative IBS technique using DLP without pairings. Chai et al. [7] presented IBS technique using quadratic residues without pairing. Furthermore, these techniques did not offer prescribed security verification and security discussion. Consequently, it is required to project secure IBS technique based on IFP.

As outlined above, inappropriately we initiate that standing IBS scheme cannot be regarded as protected. Consequently, our core impact in this article is to bridge this gap by recommending a provably efficient protected IBS technique using IFP. Our contemporary study is devoted to new creation of efficient protected IBS technique using IFP underneath the recognized security verification for existential forgery of N, the EF-ID-CPM in random oracle, utilizing the reversing model that has been presented by Pointcheval and Stern [23, 24]. This technique creates utilization of the enhanced Rabin-type digital signatures [25], where $\mathcal{N} = q_{\mathcal{P}}$ is a Williams integer, which have critical impact in the public keys of the unusual signer.

This paper is organized as follows. Required mathematical background is presented in Sect. 2. Our presented IBS technique using IFP is given in Sect. 3. Consistency of presented technique is explained in Sect. 4. Security investigation is discussed in Sect. 5. Execution examination of former IBS techniques is explained in Sect. 6. At last, Sect. 7 concludes and suggests future works.

2 Background Material

This section includes definitions of a couple of algorithms employed by our new technique, such as integer factorization problem, quadratic residuosity problem, Legendre symbol, and Jacobi symbol notations with few useful results to be used throughout this paper.

2.1 Integer Factorization Problem (IFP)

Definition 2.1 (*IFP*) Let $\mathcal{N}$ be a positive integer, find its prime factorization, i.e., write $\mathcal{N} = p_1^{e_1} \cdot p_2^{e_2} \cdot p_3^{e_3} \cdot p_4^{e_4} \dots p_t^{e_t}$ where the p_i are pairwise distinct primes and each $e_i \geq 1$.

2.2 Quadratic Residuosity Problem, Legendre Symbol, and Jacobi Symbol

Definition 2.2 Let $\mathcal{N}$ be an odd composite integer and $v \in J_{\mathcal{N}}$, decide whether or not a is a quadratic residue modulo $\mathcal{N}$.

Definition 2.3 For positive integer $\mathcal{N}$, an integer v is called a quadratic residue modulo $\mathcal{N}$ if $gcd(v, \mathcal{N}) = 1$ and $y^2 \equiv v \ (mod \ \mathcal{N})$ for some integer y, where y is called a square root of a modulo $\mathcal{N}$.

Definition 2.4 Let $\mathcal{N}$ be an odd prime number and c be a number with $\mathcal{N}$ is not divided c. c is called a quadratic residue modulo $\mathcal{N}$ if c is a square modulo $\mathcal{N}$, i.e., there is a number m such that $m^2 \equiv c \ mod \ \mathcal{N}$. If c is not a square modulo $\mathcal{N}$, i.e., there exists no such m, then c is called a quadratic non-residue modulo $\mathcal{N}$.

Let $\mathcal{N} = q \cdot p$ be a composite modulus, where q and p are two large prime numbers. Let $Q_{\mathcal{N}}$ denote the subgroup of squares in Z_N^*. Then, it is well known that $Q_{\mathcal{N}}$ is a cyclic group with order $\frac{\varphi(\mathcal{N})}{4} = \frac{(q-1)(p-1)}{4}$ [7].

Definition 2.5 Let q be an odd prime. The Legendre symbol of c is the quantity $\left(\frac{c}{q}\right)$ defined by

$$\left(\frac{c}{q}\right) = \begin{cases} 1, \text{ if } c \text{ is a quadratic residue modulo } q, \\ -1, \text{ if } c \text{ is a quadratic non residue modulo } q, \\ 0, \text{ if } q \text{ is not divide } c. \end{cases}$$

The Jacobi symbol $\left(\frac{c}{q}\right)$, a generalization of the Legendry symbol, will be used when the modulo $\mathcal{N}$ is a composite number instead of a prime number.

Definition 2.6 Let c and $\mathcal{N}$ be integers and $\mathcal{N}$ be odd and positive. Assume the factorization of $\mathcal{N}$ into primes is

$$\mathcal{N} = q_1^{e_1}.q_2^{e_2}.q_3^{e_3} \dots q_t^{e_t}$$

The Jacobi symbol $\left(\frac{c}{q}\right)$ is defined by the formula

$$\left(\frac{c}{q}\right) = (\frac{c}{q_1})^{e_1}(\frac{c}{q_2})^{e_2}(\frac{c}{q_3})^{e_3} \dots (\frac{c}{q_t})^{e_t}.$$

Theorem 2.1 [7] *Let v and n be integers that are positive and odd, then*

a. $\left(\frac{-1}{n}\right) = \begin{cases} 1, \text{ if } n \equiv 1 \ (mod \ 4), \\ -1, \text{ if } n \equiv 3 \ (mod \ 4). \end{cases}$

b. $\left(\frac{2}{n}\right) = \begin{cases} 1, \text{ if } n \equiv 1 \ or \ 7 \ (mod \ 8), \\ -1, \text{ if } n \equiv 3 \ or \ 5 \ (mod \ 8). \end{cases}$

c. $\left(\frac{v}{n}\right) = \begin{cases} \left(\frac{n}{v}\right), \text{ if } v \equiv 1 \ (mod \ 4) \ or \ n \equiv 1 \ (mod \ 4), \\ -\left(\frac{n}{v}\right), \text{ if } v \equiv 3 \ (mod \ 4) \ or \ n \equiv 3 \ (mod \ 4). \end{cases}$

Theorem 2.2 [7] *Let $c \in \mathcal{Q}_N$, $\mathcal{N} = q.p$, where q and p are two large prime numbers, satisfying $q = 2q' + 1$ and $p = 2p' + 1$, with q', p' themselves primes, then $c^{2d} \equiv c \ (mod \ \mathcal{N})$ where $d \equiv \frac{\mathcal{N} - q_1 - p_1 + 5}{8}$.*

3   An Efficient IBS Technique Based on IFP

In this area, we exhibit an IBS technique using IFP. An IBS technique is characterized by four pairs of algorithms. These algorithms are briefly explained below.

3.1 Setup

This algorithm will be endorsed by PKG by taking into consideration the security parameters ℓ as follows:

(1) Produce two huge arbitrary prime numbers $q \equiv 7 (mod\ 8)$ and $p \equiv 3 (mod\ 8)$ such that $\mathcal{N} = q * p$ satisfying $(q - 1)(p - 1) \geq 2^{\ell-1}$ and $qp < 2^{\ell}$.

(2) Choose $u \in Z_N^*$ satisfying Jacobi symbol $\left(\frac{u}{\mathcal{N}}\right) = -1$.

(3) Compute $d = \frac{\mathcal{N}-q-p+5}{8}$.

(4) Choose hash functions $\hbar_1: \{0, 1\}^* \rightarrow Z_N^*$, $\hbar_2 : \{0, 1\}^* \rightarrow \{0,1\}^{\ell}$ and $\hbar_3 : \{0, 1\}^* \rightarrow Q_N$, where $Q_N \subseteq Z_N^*$.

The public and private parameters of PKG are $\{\mathcal{N}, \hbar_1, \hbar_2, u, \ell\}$ and $\{q, p, d\}$

3.2 Extract

For a given identity ID, PKG calculates the resultant secrete key S_{ID} and two phases $(a_0, b_0) \in \{0, 1\}^2$ as follows:

(1) 1) $a_0 = \begin{cases} 0, & if\ \frac{\hbar_1(ID)}{\mathcal{N}} = 1, \\ 1, & if\ \frac{\hbar_1(ID)}{\mathcal{N}} = -1 \end{cases}$ and $b_0 = \begin{cases} 0, if\ \frac{2^{-a_0}\hbar_1(ID)}{p} = \frac{2^{-a_0}\hbar_1(ID)}{q} = 1, \\ 1, if\ \frac{2^{-a_0}\hbar_1(ID)}{p} = \frac{2^{-a_0}\hbar_1(ID)}{q} = -1 \end{cases}$

Note, let $\hbar_3(ID) = 2^{-a_0}\hbar_1(ID)(-1)^{b_0} (mod\ \mathcal{N})$, then $\hbar_3(ID) \in Q_N$.

(2) Then, compute S_{ID} as follows:

$$S_{ID} = (2^{-a_0}\hbar_1(ID))^d\ (mod\ \mathcal{N}).$$

Note that $S_{ID}^2 = 2^{-a_0}(\hbar_1(ID))(-1)^{b_0} (mod\ \mathcal{N}) = \hbar_3(ID)(mod\ \mathcal{N})$

Eventually, PKG proceeds safely S_{ID} and two phases (a_0, b_0) to the client with ID.

3.3 Signing

To sign a message $\mathfrak{M} \in \{0, 1\}^*$, a client

(1) Picks an arbitrary value $w \in Z_N^*$ and computes

$$\mathfrak{R} = w^2(mod\ \mathcal{N}), \gamma = \hbar_2(\mathfrak{R}, \mathfrak{M})\ and\ Y = S_{ID}^{\gamma}w\ (mod\ \mathcal{N})$$

Then, returned signature is $\sigma = (Y, \gamma, ID, a_0, b_0)$.

3.4   Verification

For a specified signature σ on the $\mathfrak{M}$ message, an authority might authenticate the validity by the subsequent procedures as follows:

(1)   Calculate $\hbar_3(ID) = 2^{-a_0}\hbar_1(ID)(-1)^{b_0}(mod\ \mathcal{N})$.
(2)   Check whether $\gamma = \hbar_2\left(\frac{\Upsilon^2}{\hbar_3{}^\gamma(ID)},\mathfrak{M}\right)$ or not.
      If there is an overhead situation (2), yield is valid, otherwise it is invalid.

4   Consistency

The rightness of the abovementioned technique can calculate

$$\Upsilon^2 = \left(wS_{ID}^\gamma\right)^2(mod\ \mathcal{N})$$
$$= w^2(S_{ID}^2)^\gamma(mod\ \mathcal{N})$$

$$= w^2\ \hbar_3{}^\gamma(ID)\ (mod\ \mathcal{N})$$
$$= \mathfrak{R}\ \hbar_3{}^\gamma(ID)(mod\ \mathcal{N}).$$

Since $\hbar_2\left(\frac{\Upsilon^2}{\hbar_3{}^\gamma(ID)},\mathfrak{M}\right) = \hbar_2(\mathfrak{R},\mathfrak{M}) = \gamma$.

5   Security Investigation

In this area, we describe the safety of IBS technique based on the hardness of IFP (it means difficulty in finding its factorization or solution) with offering security symbolization by describing an attacking model. We likewise demonstrate that the IBS technique is safe beside the random oracle under EF-ID-CPM attack, utilizing the rewinding technique familiarized by Pointcheval and Stern [23, 24].

Definition 5.1 We say that an IBS technique is $(t, q_\hbar, q_S, \epsilon)$-secure if for any t-time EF-ID-CPM—foe $\mathfrak{F}$ who creates at most $q_\hbar$ random oracle inquiries and q_S ordinary signature oracle inquiries. We have that $Adv_{IBS,A}^{attack}(1^\ell) < \epsilon$, where attack is EF-ID-CPM.

Theorem 5.1 *If an IFP is $\left(t', \epsilon'\right)$-hard, then our IBS technique is $(t, q_\hbar, q_S, \epsilon)$-protected against EF-ID-CPM attack, satisfying:*

$$\epsilon' \geq \frac{\left(2^{\ell-3}\epsilon - q_S(q_{\hbar_2} + q_S)\right)^2}{2^{2\ell-7}(q_{\hbar_2}+1)} - \frac{2^{\ell-3}\epsilon - q_S(q_{\hbar_2} + q_S)}{2^{\ell-5}}$$

and

$$t' = 2t + O\left(\ell^2(1 + \ell)\right)$$

where ℓ is a security parameter presented in Sect. 4.

Proof Assume that a $(t, q_{\hbar_2}, q_S, \epsilon)$-foe $\mathfrak{F}$ is against IBS technique, creating at most $q_{\hbar_2}$ random oracle inquiries and q_S regular signature oracle inquiries. Moreover, assume that it has at least ϵ advantage, and it keeps running in time at utmost t in the EF-ID-CPM game, which is described in [7].

We developed a probabilistic polynomial time (PPT) $\mathfrak{E}$ to crack IFP through non-negligible probability as follows:

$\mathfrak{E}$ will simulate the game with foe $\mathfrak{F}$:

(a) The simulator $\mathfrak{E}$ selects $u \in Z_N^*$ fulfilling Jacobi symbol $\left(\frac{u}{N}\right) = -1$ and sends (N, u) to $\mathfrak{F}$ as public parameter. Also, $\mathfrak{E}$ maintains three lists such as one signature record and two hash inquiries records.

(b) Then, simulator $\mathfrak{E}$ answers to foe $\mathfrak{F}$'s questions.

$\hbar_1$-**inquiries** $(q_{\hbar_1})$: To react every inquiry made by the foe's to the random oracle $\hbar_1$, simulator $\mathfrak{E}$ keeps up a $\hbar_1$-list. Every section in the $\hbar_1$-list of the structure $\langle ID, \hbar_1, s, a_0, b_0 \rangle$, where ID is the demanded identity, $s \in Z_N^*$, $\hbar_1$ is $\mathfrak{E}$'s answer and $(a_0, b_0) \in \{0, 1\}^2$ is within parameter as described below. At first, the list is unfilled. Simulator $\mathfrak{E}$ answers every question on ID as follows:

(1) If the question on ID, as of now, performs on $\hbar_1$-list in the tuples $\langle ID, \hbar_1, s, a_0, b_0 \rangle$, then simulator $\mathfrak{E}$ replies with $\hbar_1$ as the reply.

(2) If the question is another to the $\hbar_1$ oracle, simulator $\mathfrak{E}$ will choose an arbitrary integer $s \in Z_N^*$ and two arbitrary $(a_0, b_0) \in \{0, 1\}^2$, and reply by $\hbar_1 = \frac{s^2 2^{a_0}}{(-1)^{b_0}} \pmod{N}$ as the solution, then include the section $\langle ID, \hbar_1, s, a_0, b_0 \rangle$, to the $\hbar_1$-list.

$\hbar_2$-**inquiries** $(q_{\hbar_2})$: To react every inquiry made by the foe's to the random oracle $\hbar_2$, simulator $\mathfrak{E}$ keeps up a $\hbar_2$-list. Every section in the $\hbar_2$-list of the structure $\langle \mathfrak{R}, \mathfrak{M}, \hbar_2 \rangle$, where $(\mathfrak{R}, \mathfrak{M})$ is the demanded message, $\hbar_2$ is $\mathfrak{E}$'s answer. At first, the list is void. Simulator $\mathfrak{E}$ answers every inquiry on $(\mathfrak{R}, \mathfrak{M})$ as takes after:

(1) If the question on $(\mathfrak{R}, \mathfrak{M})$, as of now, performs on $\hbar_2$-list in the tuples $\langle \mathfrak{R}, \mathfrak{M}, \hbar_2 \rangle$, then simulator $\mathfrak{E}$ replies with $\hbar_2$ as the reply.

(2) If the question is another to the $\hbar_2$ oracle, simulator $\mathfrak{E}$ will choose an arbitrary number $\hbar_2 \in Z_N^*$ and reply $\hbar_2$ as the response. Then, it will include another section $\langle \mathfrak{R}, \mathfrak{M}, \hbar_2 \rangle$ to the $\hbar_2$-list.

Extraction inquiries (q_E): To respond to an extraction question on identity ID, simulator $\mathfrak{E}$ will choose $S_{ID} = s$, and in addition (a_0, b_0), as the response if ID previously occurs on the $\hbar_1$-list in the tuple of the $\langle ID, \hbar_1, s, a_0, b_0 \rangle$ type.

$\langle ID, \hbar_1, \mathit{s}, a_0, b_0 \rangle$. Moreover, simulator $\mathfrak{E}$ will include another section including identity ID, the identical route, as managing $\hbar_1$-inquiry, then return s and (a_0, b_0) as the response.

Signature inquiries (q_S): To respond to signature question on message $\mathfrak{M}$ with identity, ID can practically respond at arbitrary manner. Subsequently, simulator $\mathfrak{E}$ handles $\hbar_1$-list. Simulator $\mathfrak{E}$ initially calculates $\hbar_3(ID) = 2^{-a_0}(\hbar_1(ID))(-1)^{b_0}$

$(mod\ \mathcal{N})$, and then picks randomly $\Upsilon \in Z_N^*$, and $\gamma \in \{0, 1\}$, if $(\mathfrak{R}' = \frac{\Upsilon^2}{\hbar_3{}^\gamma(ID)}, \mathfrak{M})$ is in $\hbar_2$-list, $\mathfrak{E}$ information's is considered a disappointment, else, $\mathfrak{E}$ returns $\langle \Upsilon, \gamma, ID, a_0, b_0 \rangle$, as the signature, and $(\mathfrak{R}' = \frac{\Upsilon^2}{\hbar_3{}^\gamma(ID)}, \mathfrak{M}, \gamma)$ to the $\hbar_2$-list.

(c) Foe $\mathfrak{F}$ yields a fake signature $\langle \Upsilon^*, \gamma^*, ID^*, a_0, b_0 \rangle$ on particular message $\mathfrak{M}^*$, which has not been cross-examined on to signing oracle with ID^*. In addition, an identity ID^* ought not to ask about extraction inquiry oracle.

We apply the rewinding scheme [24] to factor $\mathcal{N}$. Now, simulator $\mathfrak{E}$ rearranges, $\mathfrak{F}$ with the identical arbitrary tape as the first run through, and tracks it once more. Subsequently, simulator $\mathfrak{E}$ has verified the record in the first run, and he/she can provide precisely the similar responses to all foe $\mathfrak{F}$'s questions before a $\hbar_2$ inquiry is tested which was formerly replied by γ^*. As foe $\mathfrak{F}$ requests for this $\hbar_2$ inquiry, simulator $\mathfrak{E}$ quits giving γ^* as the response yet chooses another number as the reaction at arbitrary. Give $\delta^* \in \{0, 1\}^*$ a chance to be the novel reaction. Remember, entirely the $\hbar_1$ questions in the next run are all the indistinguishable as in the previous run.

Formerly, if foe $\mathfrak{F}$ approaches up through another fake on $\mathfrak{M}^*$ message, simulator $\mathfrak{E}$ can find a couple of signatures $\langle \Upsilon^*, \gamma^*, ID^*, a_0, b_0 \rangle$ and $\langle \Upsilon^*, \delta^*, ID^*, a_0, b_0 \rangle$ satisfy

$$\frac{\Upsilon^{*2}}{\hbar_3{}^{\gamma^*}(ID^*)} \equiv \frac{\Upsilon'^2}{\hbar_3{}^{\delta^*}(ID^*)} \ (mod\ \mathcal{N})$$
$$\Rightarrow \hbar_3{}^{(\delta^* - \gamma^*)}(ID^*) \equiv \left(\frac{\Upsilon'}{\Upsilon^*}\right)^2 \ (mod\ \mathcal{N})$$

where $\hbar_3(ID^*) = 2^{-a_0}(\hbar_1(ID^*))(-1)^{b_0}(mod\mathcal{N})$.

Now, simulator $\mathfrak{E}$ examines in the $\hbar_1$-list to get the section $\langle ID^*, \hbar_1, \mathit{s}', a_0, b_0 \rangle$, and if $\mathit{s}' \not\equiv \pm \mathit{s}^2 (mod\ \mathcal{N})$, $\mathcal{N}$ can be factored by calculating capacity $Fac(\hbar_3(ID^*), \mathit{s}', \mathit{s}^2)$, which is presented in [7], else, simulator $\mathfrak{E}$ information's disappointment. Note, s was certain arbitrary square root of $\hbar_3(ID^*)$ picked by simulator $\mathfrak{E}$, which is free from the foe $\mathfrak{F}$'s perspective. In this manner, the probability that $\mathit{s}' \equiv \pm \mathit{s}^2 (mod\ \mathcal{N})$ is $\frac{1}{2}$.

Since simulator $\mathfrak{E}$ needs to trace foe $\mathfrak{F}$ double, and proceeds for various extra operations for factor $\mathcal{N}$, for example, gcd operations for division, exponentiations over modulo $\mathcal{N}$, and factorization of $\mathcal{N}$, the time simulator $\mathfrak{E}$ wants to factor $\mathcal{N}$ is expected as $t' = 2t + O(\ell^2(1 + \ell))$.

Assume that ϵ' be the probability that $\mathcal{N}$ could be calculated, ϵ^* be the probability that foe $\mathfrak{F}$ fakes a signature in a particular run, and ϵ be the probability that foe $\mathfrak{F}$

fakes a signature in the real attack. In the simulation algorithm of signature oracle, simulator $\mathfrak{E}$ selects Υ at arbitrary, which could bring about a crash of $\dfrac{\gamma^2}{\hbar_3{}^{\Upsilon}(ID)}$ along side a worth in the $\hbar_2$-list. Subsequently, the $\hbar_2$-list is full as per $\hbar_2$ and signature questions both, then the total probability is $\dfrac{(q_{\hbar_2} + q_S)}{|\mathcal{Q}_{\mathcal{N}}|}$, where $q_{\hbar_2}$ is the quantity of $\hbar_2$-inquiries and q_S is the signature inquiries. If the combination happens, simulation $\mathfrak{E}$ neglects to pretend. Therefore, the probability of foe $\mathfrak{F}$ in simulation algorithm will be decreased to

$$
\epsilon^* = \epsilon - \frac{q_S(q_{\hbar_2} + q_S)}{|\mathcal{Q}_{\mathcal{N}}|}
$$
$$
\geq \frac{2^{\ell-3}\epsilon - q_S(q_{\hbar_2} + q_S)}{2^{\ell-3}}.
$$

Let p_i be the probability that the fake depended on the i-th $\hbar_2$-inquiry in a solitary run. It can be calculated effectively as follows:

$$
\epsilon^* = \sum_{i=1}^{q_{\hbar_2}+1} p_i
$$

Assume that $p_{i,s}$ is the probability depended on the i-th $\hbar_2$-inquiry in a solitary keep running, for a given particular string s of length r, which decides the arbitrary type of foe $\mathfrak{F}$ and replies to every one of the inquiries. Thus,

$$
p_i = 2^{-r} \sum_{s \in \{0,1\}^r} p_{i,s}
$$

For a particular string s, the probability that a fake dependent on the i-th $\hbar_2$-inquiry in twice runs is $p_{i,s}(p_{i,s} - 1/2)$; later the reply of the i-th $\hbar_2$-question in the second run must be unique in relation to the main run. Assume that P_i is the probability that a fake dependent on the i-th $\hbar_2$-question in both runs. Then,

$$
\begin{aligned}
P_i &= 2^{-r} \sum_{s \in \{0,1\}^r} p_{i,s}\left(p_{i,s} - 1/2\right) \\
&= 2^{-r}\left(\sum_{s \in \{0,1\}^r} p_{i,s}^2 - \frac{1}{2}\sum_{s \in \{0,1\}^r} p_{i,s}\right) \\
&\geq \frac{2^{-r}(p_i 2^r)^2}{2^r} - \frac{p_i}{2} \\
&= p_i^2 - \frac{p_i}{2}
\end{aligned}
$$

Thus, the probability that foes yield two forgeries that are based on the same $\hbar_2$-inquiries in both runs is calculated as:

$$\sum_{i=1}^{q_{\hbar_2}+1} P_i \geq \sum_{i=1}^{q_{\hbar_2}+1} p_i^2 - \frac{1}{2} \sum_{i=1}^{q_{\hbar_2}+1} p_i$$

$$\geq \frac{\epsilon^{*2}}{q_{\hbar_2}+1} - \frac{\epsilon^*}{2}$$

$$\geq \frac{\left(\frac{2^{\ell-3}\epsilon - q_S(q_{\hbar_2}+q_S)}{2^{\ell-3}}\right)^2}{(q_{\hbar_2}+1)} - \frac{\left(\frac{2^{\ell-3}\epsilon - q_S(q_{\hbar_2}+q_S)}{2^{\ell-3}}\right)}{2}$$

$$= \frac{\left(2^{\ell-3}\epsilon - q_S(q_{\hbar_2}+q_S)\right)^2}{2^{2\ell-6}(q_{\hbar_2}+1)} - \frac{2^{\ell-3}\epsilon - q_S(q_{\hbar_2}+q_S)}{2^{\ell-4}}$$

Currently, we are almost near the outcome. Subsequently, yielding two fakes on the similar $\hbar_2$-inquiries, which implies that simulator $\mathfrak{E}$ has a most probability $\frac{1}{2}$ to the factor $\mathcal{N}$, as we have $\sum_{i=1}^{q_{\hbar_2}+1} \frac{P_i}{2}$. Thus, we have to divide the above probability by 2, then

$$\epsilon' \geq \frac{\left(2^{\ell-3}\epsilon - q_S(q_{\hbar_2}+q_S)\right)^2}{2^{2\ell-7}(q_{\hbar_2}+1)} - \frac{2^{\ell-3}\epsilon - q_S(q_{\hbar_2}+q_S)}{2^{\ell-5}}$$

6 Execution Examination of Other IBS Techniques

In this segment, we discuss nine most wide utilized IBS techniques and compare their exhibitions. These nine IBS techniques are Guillou et al.'s technique [12], Paterson's technique [22], Hess's technique [13], Cha-Cheon's technique [6], Chen et al.'s technique [8], Chai et al.'s technique [7], Tahat et al.'s technique [27], Chung et al.'s technique [9], and Ismail et al.'s technique [14]. These IBS techniques have diverse execution on server. The documentations utilized in calculation are as tracks: P = pairing operation, E = exponentiation in group, M = point multiplication in group or scalar, H = hashing, A = point addition in group, and $C(\theta)$ = computation cost of operation θ.

Table 1 Execution comparisons

IBS techniques	Signing algorithm	Verification algorithm	Total computational cost
Guillou et al. [12]	4E + 2M	4E + 2M	8E + 4M
Paterson [22]	3M + 2A + 2H	3P + 2E + M + 3H	3P + 2E + 4M + 5H + 2A
Hess [13]	P + E + A + 2M + H	2P + H + E	3P + 2E + A + 2M + 2H
Cha-Cheon [6]	2H + 2M + A	2P + H + M + A	2P + 3H + 3M + 2A
Chen et al. [8]	3H + 3M + A	4P + 3H + M + 2A	4P + 6H + 4M + 3A
Chai et al. [7]	$(l + 1)$E + M + H	2H + M + 2lE	$(3l + 1)$E + 2M + 3H
Tahat et al. [27]	4M + 2A + H	2A + H + 3M	7M + 4A + 2H
Chung et al. [9]	3A + 4M + H	A + H + 3M	4A + 7M + 2H
Ismail et al. [14]	2A + 2M + H	2M + 2A + H	4M + 4A + 2H
Our technique	H + M + 2E	2H + M + 2E	3H + 2M + 4E

6.1 Execution Examination on Signing Phase, Verification Phase, and Computational Cost

In this subsection, we clarified the execution, essential operations, and computation cost of proposed IBS technique and other nine IBS techniques as described below in Table 1.

As above outlined, it is obvious that our proposed IBS technique has better execution as compare to nine other competing techniques [6–9, 12–14, 22, 27] in signing stage and verification stage.

7 Conclusion and Future Work

In this article, we exhibited a structure to get productive provably safe IBS technique whose security can be broken to the complexity of the IFP. As we know, IFP is a principal intractable issue in cryptography. In addition, we indicated it is safe against EF-ID-CPM attack in random oracle. The new scheme has used enhanced Rabin-type signature schemes where $\mathcal{N} = q_p$, a Williams's integer which plays a vital part in listening to the public keys of the novel signer. The execution of our technique is better than other nine techniques. Consequently, our technique is safer, better entertainer, and having very low computation cost. Moreover, the subsequent techniques are among the most productive IBS techniques in random oracle; they benefit from current multi-CPU PCs. Because of the secure IBS technique using IFP developed in paper, it is easy to develop an ID-based partially blind signature technique for electronic cash. The partially blind signature technique-based IFP permits a signer to unequivocally incorporate a bit of basic data in a blind signature under some concurrence with a receiver, and it is crucial segment of electronic cash systems.

References

1. Bellare, M., Namprempre, C., Neven, G.: Security proofs for identity based identification and signature schemes. J. Cryptol. **22**, 1–61 (2009)
2. Boneh, D., Franklin, M.K.: Identity-based encryption from the Weil Pairing. In: Kilian, J. (ed.) CRYPTO 2001, LNCS, vol. 2193, pp. 213–229. Springer, Heidelberg (2001)
3. Boneh, D., Franklin, M.K.: Identity based encryption from the Weil pairing. SIAM J. Comput. **32**(3), 586–615 (2003)
4. Boneh, D., Canetti, R., Halevi, S., Katz, J.: Chosen-ciphertext security from identity-based encryption. SIAM J. Comput. **36**(5), 1301–1328 (2003)
5. Boneh, D., Lynn, B., Shacham, H.: Short signatures from the weil pairing. J. Cryptol. **17**(4), 297–319 (2004)
6. Cha, J., Cheon, J. H.: An identity-based signature from gap Diffie-Hellman groups. In: Desmedt, Y.G. (eds.) Public key cryptography—PKC 2003, LNCS, vol. 2567, pp. 18–30. Springer, Berlin, Heidelberg (2003)
7. Chai, Z., Cao, Z., Dong, L.: Identity-based signature scheme based on quadratic residues. Sci. China Ser. F: Inf. Sci. **50**, 373–380 (2007)
8. Chen, X., Zhang, F., Kim, K.: A new ID-based group signature scheme from bilinear pairings. In: Proceedings of International Workshop on Information Security Applications (WISA) 2003, LNCS, vol. 2908, pp. 585–592. Springer, Berlin, Heidelberg (2003)
9. Chung, Y.F., Huang, K.H., Lai, F., Chen, T.S.: ID-based digital signature scheme on the elliptic curve cryptosystem. Comput. Stan. Interfaces **29**, 601–604 (2007)
10. Dodis, Y., Katz, J., Xu, S., Yung, M.: Strong key-insulated signature schemes. In: Desmedt, Y.G. (eds.) Public key cryptography—PKC 2003, LNCS, vol. 2567, pp. 130–144. Springer, Berlin, Heidelberg (2003)
11. Gangishetti, R., Gorantla, M.C., Das, M.L., Saxena, A.: Threshold key issuing in identity-based cryptosystems. Comput. Stan. Interfaces **29**, 260–264 (2007)
12. Guillou, L.C., Quisquatar, J. J.: A "paradoxical" identity-based signature scheme resulting from zero-knowledge. In: CRYPTO '88 Proceedings of the 8th Annual International Cryptology Conference on Advances in Cryptology, LNCS, vol. 0403, pp. 216–231. Springer, Berlin, Heidelberg (1988)
13. Hess, F.: Efficient identity based signature schemes based on pairings. In: Nyberg, K., Heys, H. (eds.) Selected Areas in Cryptography. SAC 2002, LNCS, vol. 2595, pp. 310–324. Springer, Berlin, Heidelberg (2003)
14. Ismail, E.S., Wan-Daud, W.S.: ID-based signature scheme using elliptic curve cryptosystem. Appl. Math. Sci. **7**(73), 3615–3624 (2013)
15. Katz, J., Wang, N.: Efficiency improvements for signature schemes with tight security reductions. In: CCS '03 Proceedings of the 10th ACM Conference on Computer and Communications Security, pp. 155–164 (2003)
16. Kiltz, E., Vahlis, Y.: CCA2 secure IBE: standard model efficiency through authenticated symmetric encryption. In: Malkin, T. (ed.) Topics in Cryptology—CT-RSA 2008, LNCS, vol. 4964, pp. 221–239. Springer, Berlin, Heidelberg (2008)
17. Lee, W.C., Liao, K.C.: Constructing identity-based cryptosystems for discrete logarithm based cryptosystems. J. Netw. Comput. Appl. **22**, 191–199 (2004)
18. Lee, J., Chang, J., Lee, D.: Forgery attacks on Kang et al.'s identity-based strong designated verifier signature scheme and its improvement with security proof. Comput. Electr. Eng. 36, 948–954 (2010)

19. Meshram, C., Meshram, S., Zhang, M.: An ID-based cryptographic mechanisms based on GDLP and IFP. Inf. Process. Lett. **112**(19), 753–758 (2012)
20. Meshram, C., Meshram, S.: An identity-based cryptographic model for discrete logarithm and integer factoring based cryptosystem. Inf. Process. Lett. **113**(10–11), 375–380 (2013)
21. Meshram, C.: An efficient ID-based cryptographic encryption based on discrete logarithm problem and integer factorization problem. Inf. Process. Lett. **115**(2), 351–358 (2015)
22. Paterson, K.G.: IBS from pairings on elliptic curves. Cryptology ePrint Archive, Report 2002/004, http://eprint.iacr.org/2002/004 (2002)
23. Pointcheval, D., Stern, J.: Security proofs for signature schemes. In: Maurer, U. (ed.) Advances in Cryptology—EUROCRYPT '96, LNCS, vol. 1070, pp. 387–398. Springer, Berlin, Heidelberg (1996)
24. Pointcheval, D., Stern, J.: Security arguments for digital signatures and blind signatures. J. Cryptol. **13**, 361–396 (2000)
25. Rabin, M.: Digitalized signature and public key functions as intractable as factorization. MIT/LCS/TR-212, MIT Laboratory for computer science (1979)
26. Shamir, A.: Identity-based cryptosystems and signature schemes. In: Blakley, G.R., Chaum, D. (eds.) Advances in Cryptology. CRYPTO 1984, LNCS, vol. 196, pp. 47–53. Springer, Berlin, Heidelberg (1984)
27. Tahat, N., Ismail, E.S., Ahmad, F.B.: ID-based signature scheme using the conic curve over Zn on two hard problems. Int. J. Pure Appl. Math. **77**(3), 443–452 (2012)
28. Yu, J., Hao, R., Kong, F., Cheng, X., Fan, J., Chen, Y.: Forward-secure identity-based signature: security notions and construction. Inf. Sci. **181**, 648–660 (2011)

Data Analytics and Intelligent Learning

Advances in Single Image Super-Resolution: A Deep Learning Perspective

Karansingh Chauhan, Hrishikesh Patel, Ridham Dave, Jitendra Bhatia and Malaram Kumhar

Abstract With the increase in the resolution supported by displays and screens, the consumption of high-quality content such as 4 K resolution videos and images is increasing rapidly. To address the need for increasing fidelity, image super-resolution has been proposed to increase the resolution of images artificially. Image super-resolution produces a high-resolution image from a low-resolution image. The domain has seen outstanding growth in recent years due to the introduction of deep learning techniques. In this paper, we aim to give a quick review of the various approaches used, their merits and what inspired them. The highlight of this work is the inculcation of recently introduced layers and the review of non-traditional approaches such as the importance of the receptive field and perceptually oriented metrics. We also provide directions and recommendations to the community to extend the research done in this domain.

Keywords Single image super-resolution · Deep learning · Convolutional neural networks

K. Chauhan (✉) · R. Dave · J. Bhatia
Vishwakarma Government Engineering College, GTU, Ahmedabad, India
e-mail: c2karansingh@gmail.com

R. Dave
e-mail: ridhamdave5@gmail.com

J. Bhatia
e-mail: jitendrabbhatia@gmail.com

H. Patel · M. Kumhar
Institute of Technology, Nirma University, Ahmedabad, India
e-mail: hrishis.97@gmail.com

M. Kumhar
e-mail: malaram.kumhar@nirmauni.ac.in

© Springer Nature Singapore Pte Ltd. 2020
P. K. Singh et al. (eds.), *Proceedings of First International Conference on Computing, Communications, and Cyber-Security (IC4S 2019)*, Lecture Notes in Networks and Systems 121, https://doi.org/10.1007/978-981-15-3369-3_34

1 Introduction

Single image super-resolution is a domain in image processing that aims at recovering a high-resolution image from a single low-resolution image. It is considered as a classic problem in computer vision with various algorithms aiming at solving it ranging from conventional algorithms [1–4] to the deep learning-based algorithms [5, 6]. There is a wide applicability of image super-resolution in the domain of video surveillance under intelligent transportation system [7–9], security surveillance, etc.

Our work focuses on deep learning algorithms for image super-resolution and its subsequent architectural changes over time. We provide a compact overview of the currently existing methods and what were the motivations for each work. The architectural changes and variations of loss functions used in the proposed algorithms are highlighted in this paper. In addition to this, we review the performance metrics used to evaluate the networks and their shortcomings.

The main contributions of this review are threefold:

1. We reviewed and presented the recent enhancements in the deep learning-assisted super-resolution techniques with their comparison.
2. We provide an overview of the evaluation criteria used to quantify the performance of a super-resolution model.
3. We discuss current challenges faced in the image super-resolution domain and suggest future research directions.

The rest of this paper is organized as follows. Section 1.1 defines the image super-resolution as a domain of image processing. Section 2 covers the related work regarding the upscaling approaches that are used in various models, the various architectural structures used by the recent papers, the loss functions employed for the various architectures, and the performance metrics used to evaluate the previous models. Finally, we conclude our paper with future directions in Sect. 3.

1.1 Image Super-Resolution

In image super-resolution, the primary objective of the domain is to generate a high-resolution image from a given low-resolution image. Here, ϑ is defined as the degrading factor, which is responsible for the effect on the quality(noise level) and the resolution(scale) of the image. Mathematically, let I_{LR} be the low-resolution image and I_{HR} be the high-resolution image, and hence, low-resolution image can be expressed as:

$$I_{LR} = \vartheta(I_{HR}) \tag{1}$$

For an upscaling factor of S, I_{LR} has the dimensions of height H, width W and channels C as compared to I_{HR} which has the scaled dimensions of height $H * S$, width $W * S$ and channels C. These semantics will be utilized throughout the paper. In order to deduce the corresponding high-resolution image from the low-resolution image, considering Eq. 1, the super-resolution operator can be denoted as ϑ^{-1}. Therefore,

$$I_{SR} = \vartheta^{-1}(I_{LR}) \tag{2}$$

However, in practice, the I_{SR} does not contain all the features that the I_{HR} contains. The primary reason behind such variance is that during degradation some information is lost either due to the introduction of noise or size constraint at the time of downsizing the image. Therefore, Eq. 2 cannot be considered as an injective function. In other words, it is an underdetermined inverse problem, of which the solution is not unique.

2 Related Work

Over the past half a decade, various improvements have been proposed to improve image super-resolution. A comparison between these contributions is necessary to outline a clear direction in which research needs to be done. We address it in this section by selecting the most notable contributions and segment them in groups, as shown in Fig. 1. We also show the difference between the previous works, along with their advantages and disadvantages.

2.1 Upscaling Methods

Upscaling the input and generating the high-resolution output image is a key component of image super-resolution. Some approaches can utilize the feature maps from

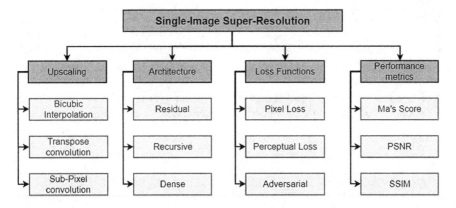

Fig. 1 Taxonomy

the low-resolution image to generate its high-resolution counterpart, whereas the more traditional methods cannot work with the feature maps but rather need to use the image as their input.

Bicubic Interpolation: This upscaling method is a traditional image scaling algorithm used to upscale a given image by performing conventional interpolation on the original pixel values. Given an input image, first, this technique takes the original pixel values onto an enlarged image and then fills the missing values through interpolation between the 16 pixels (4×4) surrounding it.

This method is employed by multiple models [10–14] for upscaling the low-resolution image prior to the task of passing it as an input to the network (CNN). The network then has to map the interpolated image to its high-resolution counterpart by learning the missing details. While this method is fast, it increases the network parameters without introducing any new image features.

Transpose Convolution Layer: This layer [15] (also referred to as the deconvolution layer) works in the opposite way as a convolution layer. Given some input feature map, a deconvolution layer attempts to predict the input image which was responsible for creating it.

To avoid increasing the network parameters by upscaling the image before feeding it to the network, first, the feature maps can be extracted from the low-resolution image space and then a transposed convolution layer can be used at the end of the network to generate the high-resolution image from the extracted features. In such a manner, the blocks of convolution layers do not have to process a large interpolated image, and thus reducing the computational complexity of the network. Apart from this, the deconvolution layer is a learning-based layer, whereas the bicubic interpolation is a more rigid algorithm. This results in better performance as seen in multiple models [16–19].

Although the proposed method does have some advantages over bicubic interpolation, it also causes checkerboard artefacts in the resulting image. This is due to the kernel size not being a multiple of the stride as seen in Fig. 2. In such cases, the output of multiple such convolutions overlaps on each other, which causes the infamous checkerboard artefact, as shown in Fig. 3.

Subpixel Convolution Layer: To alleviate the issues caused due to the transpose convolution layer as discussed above, Shi et al. [21] proposed a novel approach to reshuffle the pixels in the various feature maps in order to generate a single high-resolution image. This novel approach was organized as a single layer called the subpixel convolution.

For a low-resolution input image of dimension $H * W * C$ to be upscaled by a factor S, the sub-pixel convolution layer generates S^2 feature maps by using a

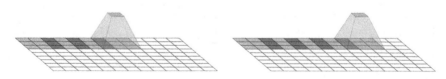

Fig. 2 Overlapping as the filter (3×3) is indivisible by stride (2)

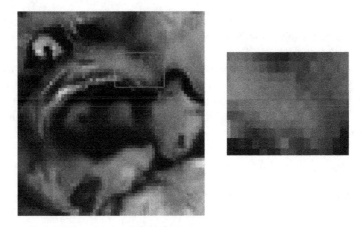

Fig. 3 An example of checkerboard artefacts [20]

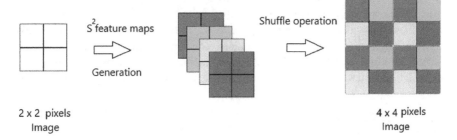

Fig. 4 Pixel shuffler applied for an upscaling factor of $\times 2$ ($s = 2$)

conventional convolutional layer. The output of the dimensions $H * W * S^2C$ is then reshuffled to generate an image of the dimensions $SH * SW * C$ which is the required super-resolution image. For the sake of clarity, an example of the pixel shuffler is shown in Fig. 4.

The sub-pixel convolution layer has a larger receptive field than the transpose convolution layer while also mitigating the appearance of checkerboard artefacts. Due to these reasons, it is widely adopted in multiple recent works [6, 21–24] for upscaling at the end of the network.

2.2 Network Architectures

Network architecture plays a crucial role in single image super-resolution. This subsection discusses the categories of network architecture with their positive and negative aspects.

Residual: Residual learning was first introduced by He et al. [25] after which their architecture was widely adopted for training deeper networks that can be trained comparatively faster.

In the context of image super-resolution, residual learning first aims to learn just the high-frequency image details and then combines those with the pre-existing low-frequency image details present in the LR image. The low-frequency details can be extracted through chained convolution layers. Mathematically, the residual can be represented as follows:

$$Residual = \{I_{HR} - \chi(w, h, c)\} \tag{3}$$

where χ represents features extracted by the convolution layers. By using residual learning, the network is relieved from the task of learning the redundant information already present in the I_{LR} image. Residual learning can be largely categorized into two, as mentioned below:

Global Residual Learning (GRL): GRL carries the input image to the end of the network by introducing a skip connection from the input to the final layer of the network as shown in Fig. 5a. Since I_{HR} and I_{LR} are highly correlated, it is intuitive to simply provide a shortcut for the same to reduce redundancy within the network.

Local Residual Learning (LRL): LRL carries the feature maps between various convolution layers by using skip connections as demonstrated in Fig. 5b. It is used to

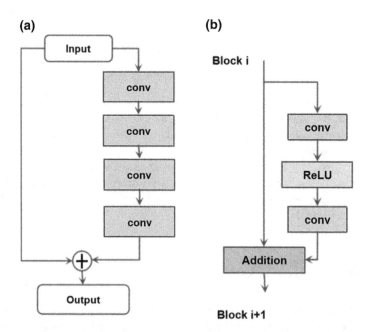

Fig. 5 **a** Global residual learning (VDSR [13]). **b** Local residual learning (EDSR [22])

tackle the loss of details due to convolutions in deep networks and to add the details from the previous layers to another layer which avoids the loss of feature maps.

Over the years, residual blocks have been used extensively in various models. SRGAN [6] used the ResNet [25] architecture in its generator, whereas Lim et al. [22] improved the ResNet architecture by removing the batch normalization layer in their network(EDSR). Residual learning is critical for training very deep networks as it can tackle the issues of slow convergence and degradation problems.

Recursive: While all the proposed techniques do provide state-of-the-art performance, their models are sub-optimal from memory and processing perspective. The primary reason behind this increase is that the networks get deeper; the model size increases with the increase in the number of parameters.

While this can potentially enhance the performance, the usability of such models becomes impractical in mobile systems where memory constraints are more severe.

To solve this issue, Kim et al. [12] proposed DRCN, which focuses on recursive blocks, as shown in Fig. 6b. Each recursive block contains chained convolution layers, and all such layers have the same neural weights. This weight sharing amongst the layers leads to a reduction in network parameters, ultimately leading to a more concise and compact model. DRRN [14] showed enhanced performance by intro-

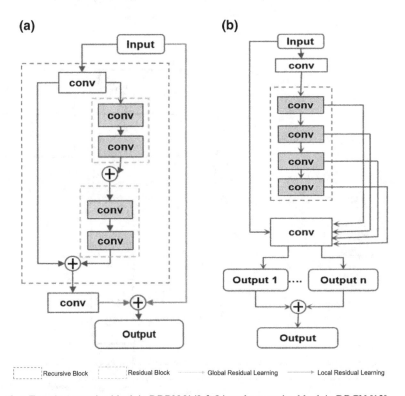

Fig. 6 **a** Two-tier recursive block in DRRN [14]. **b** Linearly recursive block in DRCN [12]

ducing multiple shared weights and using LRL within its network, as presented in
Fig. 6a. We have illustrated the weight sharing in Fig. 6a, b. The convolution layers
with the same colours share weights amongst them.

Dense: Dense connections were introduced by Huang et al. [26] in their work on
DenseNet. It was adopted for the context of super-resolution by Tong et al. [17], which
demonstrated superior performance than previous state-of-the-art results. Dense connections may essentially seem like repetitive residual blocks with more skip connections at first glance. However, conceptually, they have two fundamental differences.
Firstly, it contains skip connections from one convolutional layer to every succeeding
one within the block.

While residual blocks perform feature combination through element-wise addition, dense blocks simply concatenate the feature maps onto each other. Figure 7a,
b illustrates this difference. Second, The channel growth is not as high as residual
blocks. The skip connections encourage reusability of the previous feature maps
which can be lost in the network. This allows a given network to learn better inter-
pixel relationships. Due to this reason, dense connections are widely used in single
image super-resolution. Zhang et al. [23] improved on dense connections by increasing the channel growth rate, whereas ESRGAN [5] proposed a variant of dense
connections called residual-in-residual dense blocks (RRDB) which combines dense
blocks with residual skip connections.

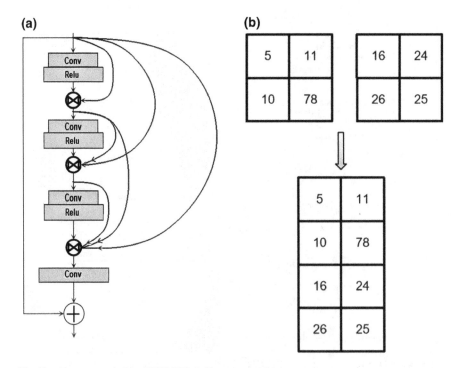

Fig. 7 **a** Dense connections RDN [23]. **b** Concat operation

2.3 Loss Functions

The loss function plays a key role in determining how the network is optimized. Different loss functions drive the neural weights to different values. Therefore, the selection of loss function plays a key role in any machine learning task. Various loss functions have been used over the previous years by different networks. We describe the most common loss functions and their behaviour in this subsection.

Pixel loss: Also known as mean square error (MSE), pixel loss compares the images pixel by pixel for differences. Mathematically, it is represented as

$$MSE = \frac{1}{H * W * C} \sum_{H*W*C} \{I_{HR}(H, W, C) - I_{SR}(H, W, C)\}^2 \qquad (4)$$

Pixel loss has been used extensively by almost all the models [11–13, 16] that have been put forward. PSNR, as discussed in Sect. 3.4, is a performance metric and since it relates inversely to MSE, optimizing for MSE automatically increases the PSNR of the super-resolution image. While this absolute comparison between the corresponding pixel may seem intuitive, it is restrictive for producing the image details and textures. This is because the above equation is similar to the mean spatial filter, which replaces a pixel with the mean of the 8 pixels filters surrounding it.

Perceptual loss: First introduced by Johnson et al. [18], perceptual loss adopts the use of a pre-trained model to compute the loss and to train the network. To achieve this, the SR and the HR image needs to be passed through the pre-trained model first. After this, the activations of a layer are extracted from the pre-trained model as a matrix. The matrices obtained by the activations due to HR and SR images are compared to compute the loss function. Mathematically, this can be represented as follows:

$$\frac{1}{H * W * C} * \sum_{H*W*C} \{\varphi^i_{HR}(H, W, C) - \varphi^i_{SR}(H, W, C)\}^2 \qquad (5)$$

where,

φ^i_{SR} is ith layer activation of pre-trained model on feeding I_{SR}.
φ^i_{HR} is ith layer activation of pre-trained model on feeding I_{HR}.

The intuition behind this approach is to leverage the training ability of certain pre-trained models used for image processing tasks to learn the semantic and perceptual qualities of images. Due to this property of such networks, the deeper layers can be utilized to determine the perceptual quality of images as these layers have been trained to extract specific features. An example of a frequently used network for this purpose is VGG19 [27]. In particular, the second layer of VGG's 5th block is popularly used for computing perceptual loss. It is a better-suited loss function than MSE since it aims at the perceptual features of the image rather than the pixel-wise comparison.

Adversarial loss: With the introduction of the use of GAN in the super-resolution field by Ledig et al. [6], a new loss function called adversarial loss was proposed. The proposed discriminator is trained on the same training dataset as the generator to distinguish between SR and HR image. The adversarial loss consists of the loss of the generator and the discriminator.

$$Adv_{Loss}^{Gen} = -log\,(Discrim\,(I_{SR}))\tag{6}$$

$$Adv_{Loss}^{Discrim} = -log\,(Discrim\,(I_{HR})) - log\,(1 - Discrim\,(I_{SR}))\tag{7}$$

By introducing adversarial loss, the generator is motivated to fool the discriminator by generating sharper details and better textures. At the same time, the discriminator is motivated to improve itself to differentiate the SR and HR image better. This min-max contest between the two components leads to an overall improvement in the perceptual quality of the image as each component motivates the other to improve itself. Recently, Jolicoeur-Martineau [28] proposed the relativistic discriminator, which showed enhanced performance as compared to the original discriminator by redefining the discriminator equation as follows:

$$D\,(\overline{x}) = sigmoid\left(C\,(x_r) - C\left(x_f\right)\right)\tag{8}$$

$$D\,(x) = sigmoid\,(C\,(x))\tag{9}$$

where
$C\,(x)$ is a function which assigns a score to an image regarding its fakeness or realness.

x_r is a real image, x_f is a fake image and $\overline{x} = (x_r\,,\,x_f)$.

2.4 Performance Metrics

SSIM: Structural similarity index (SSIM) [29] is a perceptually oriented criterion that measures how similar two images are based on their luminance and contrast. SSIM observes the dependency between spatially close pixels to determine the perceptual quality of the image. SSIM can be represented as:

$$SSIM = \frac{\left(2\mu_{I_{HR}}\mu_{I_{SR}} + O_1\right)\left(2\zeta_{I_{HR}}\zeta_{I_{SR}} + O_2\right)}{\left(\mu_{I_{HR}}^2 + \mu_{I_{SR}}^2 + O_1\right)\left(\sigma_{I_{HR}}^2 + \sigma_{I_{SR}}^2 + O_2\right)}\tag{10}$$

where μ = average, ρ = variance, ζ = covariance, $O_1 = (k_1 l_1)^2$, $O_2 = (k_2 l_2)^2$, k_1, k_2 are constants, and l_1, l_2 are dynamic range of pixel values.

The human eye is more perceptive to luminance in an image, and as can be seen, SSIM considers it as one of the factors in its formulation. Due to the above reasons, SSIM meets the performance metric criterion better than PSNR.

PSNR: Peak signal-to-noise ratio (PSNR) is frequently used to analyse the generated image quality for lossy image operations. Image super-resolution falls under the category of lossy image transformation since image features are lost when the high-resolution image is degraded to its low-resolution counterpart. Mathematically, PSNR is defined as:

$$PSNR = 10 * \log_{10} \left\{ \frac{L^2}{MSE} \right\} \tag{11}$$

However, this criteria have been deemed ineffective as they focus on pixel values at corresponding positions which have nothing to do with the perceptual quality of an image. Due to this, a higher PSNR value does not always correspond to better image quality in real scenes. But since many studies [11, 14, 16, 17] have used this as a performance metric, it is still used to compare the performance of various models.

Ma's score: In their work on evaluation criteria for image super-resolution, Ma et al. [30] propose a novel metric for measuring the performance of the various SR models. They conducted the largest subject study to collect subject scores (between 1 and 10) on the BSD data set [31].

Ma's score uses three classes of statistical details in both spatial and frequency domains to quantify artefacts generated in super-resolution images. The three classes being local frequency features, global frequency features and spatial frequency features. They also proposed a regression model that takes the three classes of features as stated above and uses them to regress over the subject study scores they conducted to give a perceptual index of the image. As their scoring method does not require the ground truth(original high-resolution image), it is considered a blind image quality assessment technique. Their work was also used to evaluate the perceptual quality of super-resolution images in the PIRM-SR 2018 challenge [32].

3   Conclusion and Future Direction

In our work, we have discussed the various architectural improvements, loss functions defined for model training, upscaling layers proposed as well as the performance metrics used to evaluate the various models.

Although the current state of image super-resolution is in a good position, there are still a lot of research opportunities to improve the domain. The lack of proper criteria to evaluate super-resolution models needs to be addressed by new blind methods such as Ma's score. We also encourage the community to further work on perceptual losses to make them more texture-oriented.

References

1. Zhang, K., Gao, X., Tao, D., Li, X.: Single image super-resolution with non-local means and steering kernel regression. IEEE Trans. Image Process. **21**(11), 4544–4556
2. Dong, W., Zhang, L., Shi, G., Wu, X.: Image deblurring and super-resolution by adaptive sparse domain selection and adaptive regularization. IEEE Trans. Image Process. **20**(7), 1838–1857 (2011)
3. He, H., Siu, W.-C.: Single image super-resolution using gaussian process regression. In: CVPR 2011, pp. 449–456. IEEE (2011)
4. Yang, J., Wright, J., Huang, T.S., Ma, Y.: Image super-resolution via sparse representation. IEEE Trans. Image Process. 19(11):2861–2873 (2010)
5. Wang, X., Yu, K., Wu, S., Gu, J., Liu, Y., Dong, C., Qiao, Y., Change Loy, C.: Esrgan: snhanced super-resolution generative adversarial networks. In: Proceedings of the European Conference on Computer Vision (ECCV), pages 0–0 (2018)
6. Ledig, C., Theis, L., Huszár, F., Caballero, J., Cunningham, A., Acosta, A., Aitken, A., Tejani, A., Totz, J., Wang, Z., et al.: Photo-realistic single image super-resolution using a generative adversarial network. In: Proceedings of the IEEE conference on computer vision and pattern recognition, pp. 4681–4690 (2017)
7. Bhatia, Jitendra, Dave, Ridham, Bhayani, Heta, Tanwar, Sudeep, Nayyar, Anand: Sdn-based real-time urban traffic analysis in vanet environment. Comput. Commun. **149**, 162–175 (2020)
8. Bhatia, Jitendra, Modi, Yash, Tanwar, Sudeep, Bhavsar, Madhuri: Software defined vehicular networks: a comprehensive review. Int. J. Commun. Syst. **32**(12), e4005 (2019)
9. Bhatia, J., Govani, R., Bhavsar, M.: Software defined networking: from theory to practice. In: 2018 Fifth International Conference on Parallel, Distributed and Grid Computing (PDGC), pp. 789–794 (2018)
10. Wang, Z., Liu, D., Yang, J., Han, W., Huang, T.: Deep networks for image super-resolution with sparse prior. In: Proceedings of the IEEE International Conference on Computer Vision, pp. 370–378 (2015)
11. Dong, C., Change Loy, C., He, K., Tang, X.: Image super-resolution using deep convolutional networks. IEEE Trans. Pattern Anal. Mach. Intell. 38(2):295–307 (2015)
12. Kim, J., Kwon Lee, J., Mu Lee, K.: Deeply-recursive convolutional network for image super-resolution. In: Proceedings of the IEEE Conference on Computer Vision and Pattern Recognition, pp. 1637–1645 (2016)
13. Kim, J., Kwon Lee, J., Mu Lee, K.: Accurate image super-resolution using very deep convolutional networks. In: Proceedings of the IEEE Conference on Computer Vision and Pattern Recognition, pp. 1646–1654 (2016)
14. Tai, Y., Yang, J., Liu, X.: Image super-resolution via deep recursive residual network. In: Proceedings of the IEEE Conference on Computer Vision and Pattern Recognition, pp. 3147–3155 (2017)
15. Zeiler, M.D., Krishnan, D., Taylor, G.W., Fergus, R.: Deconvolutional networks. In: Cvpr, vol. 10, pp. 7 (2010)
16. Mao, X., Shen, C., Yang, Y.-B.: Image restoration using very deep convolutional encoder-decoder networks with symmetric skip connections. In: Advances in Neural Information Processing Systems, pp. 2802–2810 (2016)
17. Tong, T., Li, G., Liu, X., Gao, Q.: Image super-resolution using dense skip connections. In: Proceedings of the IEEE International Conference on Computer Vision, pp. 4799–4807 (2017)
18. Johnson, J., Alahi, A., Fei-Fei, L.: Perceptual losses for real-time style transfer and super-resolution. In: European Conference on Computer Vision, pp. 694–711. Springer (2016)
19. Dong, C., Change Loy, C., Tang, X.: Accelerating the super-resolution convolutional neural network. In: European Conference on Computer Vision, pp. 391–407. Springer (2016)
20. Donahue, J., Krähenbühl, P., Darrell, T. : Adversarial Feature Learning. *arXiv preprint* arXiv:1605.09782 (2016)

21. Shi, W., Caballero, J., Huszár, F., Totz, J., Aitken, A.P., Bishop, R., Rueckert, D., Wang, Z.:
    Real-time single image and video super-resolution using an efficient sub-pixel convolutional
    neural network. In: Proceedings of the IEEE Conference on Computer Vision and Pattern
    Recognition, pp. 1874–1883 (2016)
22. Lim, B., Son, S., Kim, H., Nah, S., Lee, K.M.: Enhanced deep residual networks for single
    image super-resolution. In: Proceedings of the IEEE Conference on Computer Vision and
    Pattern Recognition Workshops, pp. 136–144 (2017)
23. Zhang, Y., Tian, Y., Kong, Y., Zhong, B., Fu, Y.: Residual dense network for image super-
    resolution. In: Proceedings of the IEEE Conference on Computer Vision and Pattern Recogni-
    tion, pp. 2472–2481 (2018)
24. Shamsolmoali, P., Li, X., Wang, R.: Single image resolution enhancement by efficient dilated
    densely connected residual network. Signal Process. Image Commun. 79, 13–23 (2019)
25. He, K., Zhang, X., Ren, S., Sun, J.: Deep residual learning for image recognition. In: Pro-
    ceedings of the IEEE Conference on Computer Vision and Pattern Recognition, pp. 770–778
    (2016)
26. Huang, G., Liu, Z., Van Der Maaten, L., Weinberger, K.Q.: Densely connected convolutional
    networks. In: Proceedings of the IEEE Conference on Computer Vision and Pattern Recogni-
    tion, pp. 4700–4708 (2017)
27. Simonyan, K., Zisserman, A.: Very deep convolutional networks for large-scale image recog-
    nition. *arXiv preprint* arXiv:1409.1556 (2014)
28. Jolicoeur-Martineau, A.: The relativistic discriminator: a key element missing from standard
    gan. *arXiv preprint* arXiv:1807.00734 (2018)
29. Wang, Z., Bovik, A.C., Sheikh, H.R., Simoncelli, E.P., et al.: Image quality assessment: from
    error visibility to structural similarity. IEEE Trans. Image Process. 13(4), 600–612 (2004)
30. Ma, C., Yang, C.-Y., Yang, X., Yang, M.-H.: Learning a no-reference quality metric for single-
    image super-resolution. Comput. Vis. Image Underst. 158, 1–16 (2017)
31. Martin, D., Fowlkes, C., Tal, D., Malik, J.:. A database of human segmented natural images
    and its application to evaluating segmentation algorithms and measuring ecological statistics.
    In: Proceedings of 8th Int'l Conference Computer Vision, vol. 2, pp. 416–423, July 2001
32. Blau, Y., Mechrez, R., Timofte, R., Michaeli, T., Zelnik-Manor, L.: The 2018 PIRM chal-
    lenge on perceptual image super-resolution. In: Proceedings of the European Conference on
    Computer Vision (ECCV), pp. 0–0 (2018)

Advancement of Text Summarization Using Machine Learning and Deep Learning: A Review

Rishi Kotadiya, Shivangi Bhatt and Uttam Chauhan

Abstract The rapid growth in the text available on the Internet in the variety of forms demands in-depth research for summarizing text automatically. The summarized form produced from one or multiple documents conveys the important message, which is significantly shorter than the original text. At the same time, summarizing the massive text collection exhibits several challenges. Besides the time complexity, the semantic similarity degree is one of the major issues in text summarization. The summarized text helps the user understand the large corpus much faster and with ease. In this paper, we reviewed several categories of text summarization. First, the techniques have been studied under the category of extractive and abstractive text summarization. Then, a review has been extended for text summarization using topic models. Furthermore, we incorporated machine learning aspects to summarize the large text. We also presented a comparison of techniques over numerous performance measures.

Keywords Text summarization · Text extraction · Test analysis · Machine learning · Deep learning · Statistical and linguistic model · Graph-based model · Topic modeling

1 Introduction

Various sources are generating data in different formats such as articles, blogs, and papers in huge quantity, and these data are increasing exponentially day by day. It is a highly time-consuming task to acquire the knowledge from these large bunch

R. Kotadiya (✉) · S. Bhatt · U. Chauhan
Vishwakarma Government Engineering College, Ahmedabad, Gujarat, India
e-mail: rishikotadiya581@gmail.com

S. Bhatt
e-mail: bshivangi47@gmail.com

U. Chauhan
e-mail: ug_chauhan@gtu.edu.in

© Springer Nature Singapore Pte Ltd. 2020
P. K. Singh et al. (eds.), *Proceedings of First International Conference on Computing, Communications, and Cyber-Security (IC4S 2019)*, Lecture Notes in Networks and Systems 121, https://doi.org/10.1007/978-981-15-3369-3_35

of the data. Hence, text summarization is a process that is utilized to summarize the available data and generate an overview that covers all the salient information. The process, automatic text summarization, is a text mining task that accepts a document as an input and generates the appropriate summary. It has a wide range of applications such as generating short reads, passage compaction, for instance, extracting the crux of sensitive reports such as accident reports produced by intelligent transport systems [1], etc.

However, the text summarization techniques can be classified on the basis of the number of input documents, that is, the summarization model can take both single and multiple documents as an input. Moreover, the summarization process can also be classified into query-based summarization approach that generates the summary revolving around a particular topic [2, 3] and generic summary that is used to get an overall idea of the data rather than information revolving around a particular topic [4]. The process can be classified based on the summary that can either be generated by extracting the sentences, or by constructing new sentences [5]. The sentence extraction process is called extractive text summarization [6–10], and the methodology in which new sentences are generated is known as abstractive text summarization [11–14]. Figure 1 represents the overview of architecture of text summarization.

On comparing the extractive and abstractive methods, the latter are quite challenging to implement because they generate new sentences on their own. Among various methods available for the abstractive text summarization, machine learning (ML) and deep learning (DL) play an important role. The reasons for the same are

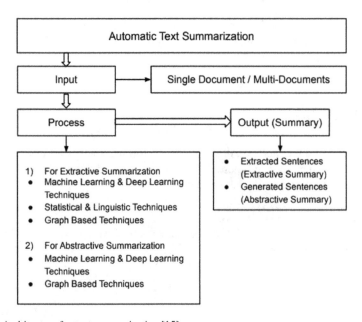

Fig. 1 Architecture for text summarization [15]

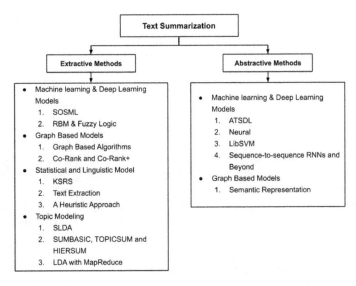

Fig. 2 Flow of methods

sentences can be generated easily with the help of ML and DL models. Also, different mechanisms can be integrated into these models to improve the quality of summary.

In this paper, we have discussed various methods of both extractive and abstractive summarization techniques incorporated to accomplish the task of generating a summary. We have distinguished the techniques based on the methods of data summarization and presented the comparison tables in each section that compares the accuracy of different techniques.

The rest of the paper is organized as follows: Sect. 2 is related to work done by various researchers. Sections 3 and 4 discuss the literature review of extractive and abstractive summarization methods, respectively, followed by their comparison table that compares various methods. Furthermore, Sects. 5 and 6 are the discussion and conclusion section with future directions, respectively. Figure 2 represents the flow of methods.

1.1 Role of ML and DL

Machine learning and deep learning models have been accommodated in various fields like data mining, artificial intelligence, machine translation, etc. One of these fields is text summarization. Notably, it has contributed significantly to the abstractive text summarization. For instance, neural networks, encoder-decoders, and many other models are integrated with different mechanisms such as attention mechanism and coverage mechanism. Further, these networks are used to generate the sentence according to the knowledge given to them. While in the extractive summarization, some of the models are used to assign the value to the sentences or to enhance the value of the score derived from the other methods [16].

2 Related Work

The use of text summarization techniques is becoming increasingly relevant and inevitable with growing number of digital records. There have been considerable studies that investigated various kinds of summaries, their generation techniques, and their evaluation methods.

Essentially, Nazari et al. have studied various types of summaries, such as abstractive, extractive, individual, and multi-documents [17]. They also outlined various overview methods including statistical, machine learning, semantic, and swarm intelligence methods concerning extractive overview.

Allahyari et al. have highlighted several extensively used extractive techniques for the generation of summary for both single and multi-documents. The authors practiced machine learning methods, frequency-driven techniques, topic representation methods, and graph-based approaches [18] .

Gambhir and Gupta have focused more on the latest extractive methods. Additionally, short overview of several abstract and multilingual methods has also been introduced [19]. Furthermore, the detailed discussion of both the evaluation methods, intrinsic and extrinsic, is also provided.

3 Extractive Methods

The extractive text summarization is a technique in which the input information is supplied to the model, and then, various logics are used to predict the significance of sentences. In this method, the users define the summary length (k) and scores are assigned to the sentences, based on the logic used. Lastly, top-ranking k sentences are used for generating the summary. Hence, this type of approach directly uses the sentences already present in the input data. Figure 3 represents the basic process of extractive summarization.

3.1 Using Machine Learning and Deep Learning

Various authors have suggested using ML and DL for summarization. Abdi et al. have presented three treatments to generate sentiment-oriented summary, and the model is called sentiment-oriented summarization of multi-documents (SOSML—sentiment-oriented summarization using machine learning) [6]. The first step in SOSML model is preprocessing followed by feature extraction. In this step, sentiment knowledge

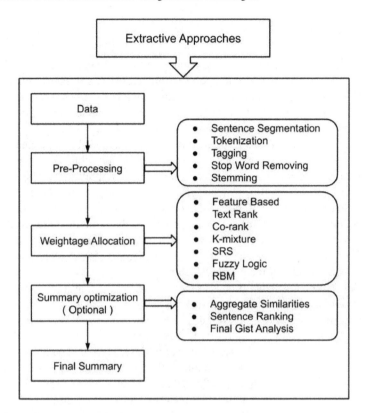

Fig. 3 Extractive text summarization

is used to assign a sentiment score and statistical and linguistic knowledge are used to assign feature scores and find the most important sentences. Further, the word-embedding model is used to extract vector representation of the words. Finally, classification is performed by determining the text polarity. Authors suggest that combination of information gain (IG) for sentiment score and support vector machine (SVM) for classification gives the best results among other combinations and methods.

Moreover, Shirwandkar et al. have used restricted Boltzmann machine (RBM) and fuzzy logic to enhance the sentence feature values [7] which is a hybrid version of method proposed in [20, 21]. This model uses the following steps: preprocessing, sentence feature extraction, RBM and fuzzy Logic, set operation, final summary. After preprocessing, nine features are used to find the sentence score. This sentence score and sentence feature matrix are given to the fuzzy logic and RBM, respectively. Further, the set operation is performed on the summary produced by the RBM and fuzzy logic in which common sentences are directly added to the final summary. The rest of the sentences are sorted according to position, and only half of them are added to the final summary. Lastly, summary sentences are arranged according to the original position they had in the main document.

3.2　Graph-Based Model

Thakkar et al. have broadly presented two algorithms that implements graph-based models [8]: graph-based ranking algorithm and shortest path algorithm. To apply the graph-based algorithms on the natural language texts, a token is recognized that most accurately defines the task and added as nodes to the graph. Afterward, the relations are found that joints text units which are used to plot edges between the vertices. This process is repeated until convergence and the value are used to sort the sentences which was received after the execution of the algorithm. However, frequent topic changes in the summary generated by this method might result into bewildering summary. Hence, the shortest path algorithm has been suggested as it generates the summary in a flow. There will be an edge between the sentences (nodes) that have at least one common word. Furthermore, if sentences share more similarity, then less cost will be assigned to the edge and vice versa. The summary is then generated by taking the shortest path that starts and ends with the very first sentence and the last sentence, respectively, of the original document.

Besides, Fang et al. have proposed the co-rank model that aims to incorporate the word–sentence relationship into the sentence graph for enhancing the sentence scoring [9]. In other words, co-rank uses the mutual feedback between words and sentences. To be specific, the mutual feedback here means that a sentence containing more words with a higher weight (i.e., keywords) deserves higher weight, and a word has a higher weight if it is included in important sentences. The summary generated may contain some redundancy for which they presented a redundancy elimination technique during the selection of top k sentences known as the co-rank+ model.

3.3　Statistical and Linguistic Approaches

The approach presented by Chandra et al. calculates the weights of the words or sentences, and according to the weight, the summary is generated [10]. The proposed approach consists of four main tasks: preprocessing, term weight determination, term relationship exploration, and summary generation. In this paper, semantic relation significance (SRS) is integrated with Katz's K-mixture probability model [22], which is named as K-mixture semantic relationship significance (KSRS) model where the K-mixture probabilistic model is used to enhance the quality of summaries. The main job of KSRS model is to determine the term weights in a statistical sense and recognize the term relationship to derive the SRS. Based on these SRS values, sentences are extracted and ranks are assigned to them. Finally, the summary is generated by electing top-ranked k sentences.

On the similar track, the paper by Afsharizadeh et al. suggested a method that is the extended version of the article submitted by Ahuja and Anand work, which includes many other features other than used by Ahuja and Anand [23]. This method incorporates text preprocessing, feature extraction, sentence scoring, and summary generation. Here, the task of feature extraction utilizes eleven features to assign feature values, and a score is assigned to each sentence based on a linear function of its feature values in the sentence scoring step. Finally, all of the sentences are ranked based on their scores, and the top-ranked sentences are selected to generate the summary [24].

On the other hand, Abujar et al. presented the model for Bengali text, which follows the extractive text summarization approach [25]. The process performs three steps. The first step is preprocessing followed by the recognition of superior sentences where the main leading sentences will be extracted from an original set of documents through word analysis and sentence analysis process because, words and sentences, both factors are equally important for preparing a qualitative summary [25]. Finally, filtering and improvement in the quality of the summary are carried out.

Table 1 shows the comparison between various extractive summarization techniques proposed by the scholars, the data sets used, the evaluation method. Recall-oriented understudy for gisting evaluation (ROUGE) [26] is a wildly known evaluation method for text summarization. It can be understood from Table 1 that the method proposed by Shirwandkar et al. [7] possesses the highest values of all the parameters evaluated using the ROGUE score. The summary generated for DUC 2007 corpus by using the technique proposed by Afsharizadeh et al. [24] can be considered as a less efficient method among others. However, these methods cannot be compared directly with each other as some factors might get ignored.

Table 1 Evaluation matrix of extractive techniques

Author	Dataset	Evaluation method	Values			Value
			Precision	Recall	F-measure	(If other)
Chandra et al. [10]	800 documents from reuters news	Not specified	78	78	78	–
Abujar et al. [25]	3 different bangla text	Not specified	–	–	–	87.9
Afsharizadeh et al. [24]	DUC 2007 corpus	ROUGE-1	36.12	36.79	36.44	–
Fang et al. [9]	10 Chinese documents and DUC02	ROUGE-1	–	–	–	61.15
Shirwandkar et al. [7]	Not specified	ROUGE	88	80	84	–
Abdi et al. [6]	DUC_2001, DUC_2002, the movie review dataset	Average ROUGE score (ARS)	–	–	–	38.59

3.4 Text Summarization Using Topic Modeling

The extraordinary enlargement has been made in the performance of techniques for text summarization [19, 27] and document classification [28, 29]. Latent Dirichlet allocation (LDA) is a probabilistic model that maps massive documents collection to a very small topic subspace [30, 31], and it works for unseen documents too.

Chang and Chien introduced an approach for summarizing multiple documents at the sentence level [32]. The proposed framework for sentence-based LDA (SLDA—sentence-based latent Dirichlet allocation) model has been applied for displaying corpus hierarchy with document summarization. They estimated the model parameters with the help of variational Bayes. The comparative matrix included VSM, LM, LDA, and SLDA, among which, the findings of the trials proved SLDA more precise, while the researchers foretold that the same model shall be practiced for the spoken documents also.

On a similar track, extractive summary over multiple documents was performed by Haghighi and Vanderwende [33]. The scholar used three techniques: SumBasic, Hire-Sum, TopicSum. The SumBasic derives the score based on how many high-frequency words a sentence comprises of [27]. TopicSum technique extracts summary with the help of the LDA-like topic model, while HireSum prepares the summary by considering more than one aspects of the documents. The evaluation based on the ROUGE score proved the HireSum technique more accurate compared to the other two considered for the experiments. However, the difference was fragile between HireSum and TopicSum summary.

In the research task presented by Nagwani, the scholars established LDA in a distributed or parallel environment for forming the set of topics where speeding up the process of summarizing the document collection is the prime objective [34]. The multi-document summarizer based on MapReduce architecture has been exhibited. The experiments were carried out for four nodes based on the MapReduce framework, and it was found that MapReduce architecture proved more scalability and reduced time complexity.

Table 2 depicts the experimental outcomes of various text summarization using LDA. The evaluation of the techniques was performed on ROUGH measures in addition to other methods, and the results were compared with the human-generated summary.

Table 2 Evaluation matrix of extractive techniques using topic modeling

Author	Dataset	Evaluation method	Values			Value
			Precision	Recall	F-measure	(If other)
Chang et al. [32]	DUC_2005	ROUGE-1	38.97	33.72	35.80	–
Haghighi et al. [33]	DUC_2007	ROUGE-1	–	–	–	43.1
Nagwani et al. [34]	Legal cases from FCA_2006-09	ROUGE-1	71.0	62.0	44.0	–

4 Abstractive Methods

Abstractive summarization techniques do not copy sentences from the original text, unlike the extractive summarization techniques. Nonetheless, these methods are a bit trickier to follow, as models are supposed to generate the summary sentences on their own. These methods can be implemented by machine learning, deep learning, graphs, and many more [11–14, 35].

4.1 Using Machine Learning and Deep Learning

Song et al. have proposed a long short-term memory (LSTM)-convolutional neural network (CNN)-based automatic text summarization framework (ATSDL—abstractive text summarization using deep learning) [11]. This model is provided with the fragments of the sentences or phrases to generate the summary sentences. After preprocessing, the phrase process takes place. During this phase, multiple order semantic parsing (MOSP) is used to generate a multiple order semantic tree (MOS tree) which is used here to create the phrases. After generating all the possible phrases, the phrase refinement is carried out, and a phrase combination is applied for a reduction of redundancy in the phrases. Finally, these phrases are provided to the LSTM-CNN model to generate the summary sentences.

Likewise, Esmaeilzadeh et al. have presented three models: baseline model, pointer-generator network, and transformers [35]. A sequence-to-sequence LSTM recurrent neural network (RNN) encoder–RNN–decoder architecture with attention is used as a baseline model. However, the output of this model had some limitations which have been overcome by using the other mentioned models. The second model pointer-generator network further consists of two mechanisms: the pointer-generator mechanism and the coverage mechanism. Moreover, the model proposed by Miyagishima et al. for machine translation is used to investigate performance on abstractive text summarization [36]. This model is auto-regressive at each step. However, researchers suggest that model 2 (LSTM encoder–decoder with attention, pointer-generator, and coverage mechanisms) gives the optimal output for the given data.

Day et al. have proposed the system architecture of deep learning for automatic text summarization [12]. The model preprocesses the raw data by encoding the tokens in a particular format for a library for support vector machine (LibSVM). Three sorts of candidate titles were generated by putting the data into three distinct models, namely a statistical model, a machine learning model, and a deep learning model. Term frequency–inverse document frequency (TF-IDF) method was used for statistical model. In the deep learning model, the seq2seq model developed by Keras was used for training in which the input layer is set as a character level.

Nallapati et al. have modeled the abstractive text summary process utilizing attentional encoder-recent neural networks [14]. Motivated by the concrete issues in the overall summary that the machine translation model does not adequately address. The investigates have suggested new designs and demonstrated that they offer further efficiency improvements. Several novel models for RNN-based encoder–decoder techniques are explained, such as RNN with attention and large vocabulary trick (LVT), utilizing feature-rich encoder to capture keywords, modeling out-of-vocabulary (OOV) words using switching generator-pointer as well as hierarchical attention model. They have also suggested a novel dataset to summarize the paper in several sentences.

4.2 Graph-Based Model

Liu et al. proposed the framework using abstract meaning representation (AMR) to generate summary that is primarily based on a graph-to-graph transformation that generates the summary by reducing the source graph [13]. Firstly, the input statements are parsed into individual AMR graphs. In the next step, these individual graphs are converted into the single-source graph and the subset of the nodes which are in the source graph is selected for the summarized AMR graph. In the end, a heuristic approach is used by the framework to generate the summary from the summarized AMR graph. In this framework, subgraph prediction is performed with the help of integer linear programming (ILP), and constraint is applied to the summary graph (which is measured by numbers of edges) to achieve the required compression rate.

Table 3 compares the multiple methods suggested by the researchers for abstractive text summarization approaches. It is interpretable from Table 3 that the summary generated for Newswire article from the English Gigaword corpus using the method proposed by Liu et al. is much more precise when evaluated using the ROUGE-1 method [13]. On the other hand, the technique utilized Nallapati et al. on the DUC corpus shows comparatively less value, among other methods [14].

5   Discussion

In extractive summarization, methods proposed by Shirwandkar et al. [7], which utilizes ML and DL concepts, show the highest result for ROUGE. It can be inferred from Table 1 that ML and DL have improved the accuracy of the extractive summarization method to a significant level. Besides that, in abstractive text summarization, graph-based model [13] has shown the best result, but Table 3 shows improvement in accuracy of the summarization using ML and DL models. As ML and DL are in the development phase, there is a room for improvement in the quality of summarized text. A different model can be integrated, or a mechanism can be added to the model to enhance the quality and remove the existing limitations. LDA forms the words

Table 3 Evaluation matrix of abstractive techniques

Author	Dataset	Evaluation method	Values			Value
			Precision	Recall	F-measure	(If other)
Song et al. [11]	CNN-Dailymail	ROUGE-1	–	–	–	34.9
Liu et al. [13]	Newswire article from the english gigaword corpus	ROUGE-1	51.2	40	44.7	–
Esmaeilzadeh et al. [35]	CNN-Dailymail	ROUGE-1	42.71	38.21	38.97	–
Day et al. [12]	50387 essays from the WOS database	ROUGE	–	–	–	–
Nallapati et al. [14]	DUC corpus	ROUGE-1				28.97
	Gigaword corpus		–	–	–	36.4
	CNN/Daily mail corpus					35.46

cluster based on bag-of-words concepts. However, linguistic characteristics such as synonyms and homonyms can be integrated with statistical inference. It may give better semantically coherent topics.

6 Conclusion

In this review article, we studied numerous techniques for summarizing an immense text collection. The gigantic size of the corpus is a widespread phenomenon as the drastic and omnidirectional development of social networking sites, microblogging sites, news forums, online shopping Web sites, etc. These data need to be summarized for knowledge gaining. We surveyed abstractive and extractive text summarization methods. For this, we considered a set of research work to portray the advancement in the field of text summarization. Furthermore, we compared the techniques employed by authors in a variety of research tasks. Under the category of extractive text summarization, we presented a review on text summarization using statistical inference methods. We also showed how machine learning plays its part when it comes to the summarization of massive text collection.

References

1. Bhatia, J., Modi, Y., Tanwar, S., Bhavsar, M.: Software defined vehicular networks: a comprehensive review. Int. J. Commun. Syst. 32(12), e4005 (2019)
2. Hasselqvist, J., Helmertz, N., Kågebäck, M.: Query-based abstractive summarization using neural networks. In: CoRR (2017). abs/1712.06100

3. Shah, S., Naiynar, S.A., Amutha, B.: Query based text summarization. ARPN J. Eng. Appl. Sci. **12**(19) (2006)
4. Kruengkrai, C., Jaruskulchai, C.: Generic text summarization using local and global properties of sentences. In: Proceedings IEEE/WIC International Conference on Web Intelligence (WI 2003), pp. 201–206 (2003)
5. Approaches to text summarization: An overview. https://www.kdnuggets.com/2019/01/approaches-text-summarization-overview.html. Last accessed 27 Aug 2019
6. Abdi, A., Shamsuddin, S.M., Hasan, S., Piran, J.: Machine learning-based multi-documents sentiment-oriented summarization using linguistic treatment. Expert Syst. Appl. **109**(20), 66–85 (2018)
7. Shirwandkar, N.S., Kulkarni, S.: Extractive text summarization using deep learning. In: 2018 Fourth International Conference on Computing Communication Control and Automation (IC-CUBEA), pp. 1–5. IEEE (2018)
8. Thakkar, K.S., Dharaskar, R.V., Chandak, M.B.: Graph-based algorithms for text summarization. In: 2010 3rd International Conference on Emerging Trends in Engineering and Technology, pp. 516–519. IEEE (2010)
9. Fang, C., Dejun, M., Deng, Z., Zhiang, W.: Word-sentence co-ranking for automatic extractive text summarization. Expert Syst. Appl. **72**, 189–195 (2017)
10. Chandra, M., Gupta, V., Paul, S.K.: A statistical approach for automatic text summarization by extraction. In: 2011 International Conference on Communication Systems and Network Technologies, pp. 268–271. IEEE (2011)
11. Song, S., Huang, H., Ruan, T.: Abstractive text summarization using lstm-cnn based deep learning. Multimedia Tools Appl. **78**(1), 857–875 (2019)
12. Day, M.-Y., Chen, C.-Y.: Artificial intelligence for automatic text summarization. In: 2018 IEEE International Conference on Information Reuse and Integration (IRI), pp. 478–484. IEEE (2018)
13. Liu, F., Flanigan, J., Thomson, S., Sadeh, N., Smith, N.A.: Toward abstractive summarization using semantic representations. In: CoRR (2018). abs/1805.10399
14. Nallapati, R., Zhou, B., Gulcehre, C., Xiang, B., et al.: Abstractive text summarization using sequence-to-sequence rnns and beyond. In: CoRR (2016). abs/1602.06023
15. Various text summarization approaches. https://www.researchgate.net/figure/Various-text-summarization-approaches_fig1_305628079. Last Accessed 7 Apr 2019
16. Text summarization using deep learning. https://towardsdatascience.com/text-summarization-using-deep-learning-6e379ed2e89c. Last Accessed 7 Apr 2019
17. Nazari, N., Mahdavi, M.A.: A survey on automatic text summarization. J. AI Data Min. **7**(1), 121–135 (2019)
18. Allahyari, M., Pouriyeh, S., Assefi, M., Safaei, S., Trippe, E.D., Gutierrez, J.B., Kochut, K.: Text summarization techniques: a brief survey. In: CoRR (2017). abs/1707.02268
19. Gambhir, M., Gupta, V.: Recent automatic text summarization techniques: a survey. Artif. Intell. Rev. **47**(1), 1–66 (2017)
20. Babar, S.A., Patil, P.D.: Improving performance of text summarization. Procedia Comput. Sci. **46**, 354–363 (2015)
21. Singh, S.P., Kumar, A., Mangal, A., Singhal, S.: Bilingual automatic text summarization using unsupervised deep learning. In: 2016 International Conference on Electrical, Electronics, and Optimization Techniques (ICEEOT), pp. 1195–1200. IEEE (2016)
22. Katz, S.M.: Distribution of content words and phrases in text and language modelling. Nat. Lang. Eng. **2**(1), 15–59 (1996)
23. Ahuja, R., Anand, W.: Multi-document text summarization using sentence extraction. Artif. Intell. Evol. Comput. Eng. Syst. **517**, 235–242 (2017)
24. Afsharizadeh, M., Ebrahimpour-Komleh, H., Bagheri, A.: Query-oriented text summarization using sentence extraction technique. In: 2018 4th International Conference on Web Research (ICWR), pp. 128–132. IEEE (2018)
25. Abujar, S., Hasan, M., Shahin, M.S.I., Hossain, S.A.: A heuristic approach of text summarization for bengali documentation. In: 2017 8th International Conference on Computing, Communication and Networking Technologies (ICCCNT), pp. 1–8. IEEE (2017)

26. Kanapala, A., Pal, S., Pamula, Rajendra: Text summarization from legal documents: a survey. Artif. Intell. Rev. **51**(3), 371–402 (2019)

27. Nenkova, A., McKeown, K.: A survey of text summarization techniques. In: Mining Text Data, pp. 43–76 (2012)

28. Li, Y.H., Jain, A.K.: Classification of text documents. The Comput. J. **41**(8), 537–546 (1998)

29. Britt, B.L., Berry, M.W., Browne, M., Merrell, M.A., Kolpack, J.: Document classification techniques for automated technology readiness level analysis. J. Am. Soc. Inf. Sci. Technol. **59**(4), 675–680 (2008)

30. Blei, D.M.: Probabilistic topic models. Commun. ACM **55**(4), 77–84 (2012)

31. Steyvers, M., Griffiths, Tom: Probabilistic topic models. Handbook of Latent Semantic Analysis **427**(7), 424–440 (2007)

32. Chang, Y.-L., Chien, J.-T.: Latent dirichlet learning for document summarization. In: 2009 IEEE International Conference on Acoustics, Speech and Signal Processing, pp. 1689–1692. IEEE (2009)

33. Haghighi, A., Vanderwende, L.: Exploring content models for multi-document summarization. In: Proceedings of Human Language Technologies: The 2009 Annual Conference of the North American Chapter of the Association for Computational Linguistics, pp. 362–370. Association for Computational Linguistics (2009)

34. Nagwani, N.K.: Summarizing large text collection using topic modeling and clustering based on mapreduce framework. J. Big Data **2**(1), 6 (2015)

35. Esmaeilzadeh, S., Peh, G.X., Xu, A.: Neural abstractive text summarization and fake news detection. In: CoRR (2019). abs/1904.00788

36. Miyagishima, K.J., Gruenert, U., Li, W.: Processing of s-cone signals in the inner plexiform layer of the mammalian retina. Vis. Neurosci. **31**(2), 153–163 (2014)

Pneumonia Detection Using Convolutional Neural Networks (CNNs)

V. Sirish Kaushik, Anand Nayyar, Gaurav Kataria and Rachna Jain

Abstract Pneumonia, an interstitial lung disease, is the leading cause of death in children under the age of five. It accounted for approximately 16% of the deaths of children under the age of five, killing around 880,000 children in 2016 according to a study conducted by UNICEF. Affected children were mostly less than two years old. Timely detection of pneumonia in children can help to fast-track the process of recovery. This paper presents convolutional neural network models to accurately detect pneumonic lungs from chest X-rays, which can be utilized in the real world by medical practitioners to treat pneumonia. Experimentation was conducted on Chest X-Ray Images (Pneumonia) dataset available on Kaggle. The first, second, third and fourth model consists of one, two, three and four convolutional layers, respectively. The first model achieves an accuracy of 89.74%, the second one reaches an accuracy of 85.26%, the third model achieves an accuracy of 92.31%, and lastly, the fourth model achieves an accuracy of 91.67%. Dropout regularization is employed in the second, third and fourth models to minimize overfitting in the fully connected layers. Furthermore, recall and F1 scores are calculated from the confusion matrix of each model for better evaluation.

Keywords Convolutional neural networks (CNNs) · Pneumonia detection · ReLU · Max-pooling · Forward and backward propagation

V. Sirish Kaushik (✉) · G. Kataria · R. Jain
Bharati Vidyapeeth's College of Engineering, New Delhi, Delhi, India
e-mail: shirishkaushik@gmail.com

G. Kataria
e-mail: gaurav.kataria2291999@gmail.com

R. Jain
e-mail: rachna.jain@bharatividyapeeth.edu

A. Nayyar
Graduate School, Duy Tan University, Da Nang, Vietnam
e-mail: anandnayyar@duytan.edu.vn

© Springer Nature Singapore Pte Ltd. 2020
P. K. Singh et al. (eds.), *Proceedings of First International Conference on Computing, Communications, and Cyber-Security (IC4S 2019)*, Lecture Notes in Networks and Systems 121, https://doi.org/10.1007/978-981-15-3369-3_36

1 Introduction

One of the major factors associated with pneumonia in children is indoor air pollution. Apart from this, under-nutrition, lack of safe water, sanitation and basic health facilities are also major factors. Pneumonia is an interstitial lung disease caused by bacteria, fungi or viruses. It accounted for approximately 16% of the 5.6 million under-five deaths, killing around 880,000 children in 2016 [1]. Affected victims were mostly less than two years old. Timely detection of pneumonia can help to prevent the deaths of children. This paper presents convolutional neural network models to accurately detect pneumonic lungs from chest X-rays, which can be utilized in the real world by medical practitioners to treat pneumonia [2]. These models have been trained to classify chest X-ray images into normal and pneumonia in a few seconds, hence serving the purpose of early detection of pneumonia. Although transfer learning models based on convolutional neural networks like AlexNet, ResNet50, InceptionV3, VGG16 and VGG19 are some of the most successful ImageNet dataset models with pre-trained weights, they were not trained on this dataset as the size of dataset taken for our research is not as extensive compared to ones which generally employ transfer learning [3]. Four classification models were built using CNN to detect pneumonia from chest X-ray images to help control this deadly infection in children and other age groups. Accuracy of the model is directly correlated with the size of the dataset, that is, the use of large datasets helps improve the accuracy of the model, but there is no direct correlation between the number of convolutional layers and the accuracy of the model.

To obtain the best results, a certain number of combinations of convolution layers, dense layers, dropouts and learning rates have to be trained by evaluating the models after each execution. Initially, simple models with one convolution layer were trained on the dataset, and thereafter, the complexities were increased to get the model that not only achieved desired accuracies but also outperformed other models in terms of recall and F1 scores. The objective of the paper is to develop CNN models from scratch which can classify and thus detect pneumonic patients from their chest X-rays with high validation accuracy, recall and F1 scores. Recall is often favored in medical imaging cases over other performance evaluating parameters, as it gives a measure of false negatives in the results. The number of false negatives in the result is very crucial in determining the real-world performance of models [4]. If a model achieves high accuracy but low recall values, it is termed as underperforming, inefficacious and even unsafe as higher false-negative values imply higher number of instances where the model is predicting a patient as normal, but in reality, the person is diseased. Hence, it would risk the patient's life. To prevent this, the focus would be only models with great recall values, decent accuracies and F1 scores [5].

The paper is organized into 5 sections: Sect. 1 introduces the subject of this research paper, addresses its importance and relevance, the purpose and motive to undertake this research work and the objective of the paper. Section 2 explores the work related to this field that has been accomplished till now. Section 3 explains the methodology of the paper, explaining the architecture of the models, flowchart and the dataset used to train and test the four models. Section 4 presents the results achieved by the various CNN models and compares the performance of each model using accuracy and loss graphs and confusion matrices. Section 5 provides a brief conclusion to the paper and delivers the best-suited model. Furthermore, the future scope of this research work has also been discussed. All the references which are cited in the paper have been listed in the end.

2 Related Work

Many researchers have tackled the problem of classifying images with high accuracy. Here are some citations related to our paper:

Rubin et al. [6] developed a CNN model to detect common thorax disease from frontal and lateral chest X-ray images. MIMIC-CXR dataset was used to perform large-scale automated recognition of these images. The dataset was split into training, testing and validation sets as 70%, 20% and 10%, respectively. Data augmentation and pixel normalization were used to improve overall performance. Their DualNet CNN model achieved an average AUC of 0.72 and 0.688 for PA and AP, respectively. A deep convolutional neural network to classify pulmonary tuberculosis was developed by Lakhani et al. [7]. Transfer learning models such as AlexNet and GoogleNet were also used to classify chest X-ray images. The dataset was split into training, testing and validation sets as 68%, 14.9% and 17.1%, respectively. Data augmentation and pre-processing techniques were employed to get the best performing model achieving an AUC of 0.99. Precision and recall of the model were 100 and 97.3%. An AG-CNN model was developed by Guan et al. [8] to recognize thorax disease. ChestX-ray14 dataset was used to detect thorax disease from chest X-ray images. Global and local branch attention-guided CNN was used for classification purposes. Their model was better than other models mentioned in their research paper, achieving an AUC of 0.868. A deep convolutional neural network model was developed by Rajpurkar et al. [9] to classify chest X-ray images into pneumonia and other 14 diseases. ChestX-ray14 dataset was used for training the model. They compared their ChXNet model (121 layered model) with practicing academic radiologists. Their ChXNet model achieved an F1 score (95% CI) of 0.435 outperforming radiologists which achieved an F1 score (95% CI) of 0.387.

A deep convolutional neural network model having five convolutional layers some followed by max-pooling layers, having three fully connected layers was trained by Krizhevsky et al. [10]. This network contained 60 million different parameters. By employing dropout, this model achieved a top-five error percent of 17%. Simonyan et al. [11] developed a highly accurate model employing multiple small kernel-sized filters to achieve top-five test accuracy 92.7%. This model was trained on the ImageNet dataset and submitted to the ILSVRC 2014 competition. A convolution neural network for classification and segmentation of brain tumor MRIs was developed by Xu et al. [12]. Multiple techniques such as data augmentation, feature selection and pooling techniques were employed in this model. The validation accuracy for classification achieved by this model is 97.5%, and validation accuracy of segmentation is 84%, 256 × 256 pixels sized frontal chest radiographs which were fed to a deep convolution neural network to detect abnormalities. A convolutional neural network with five convolution layers employing leaky ReLU, average pooling and three fully connected layers was developed by Anthimopoulos et al. [13] to detect interstitial lung disease patterns in a dataset containing 14,696 images belonging to seven different classes. This model achieved a classification accuracy of 85.5%. He et al. [14] developed a residual neural network (RNN) to classify images present in the ImageNet dataset. RNN introduced the concept of shortcut connections to tackle the problem of vanishing gradients. This model which was submitted to ILSVRC 2015 attained state-of-the-art classification accuracy. A transfer learning model, extension of AlexNet using data augmentation techniques, was developed by Glozman et al. [15]. This model was trained on ADNI database. Two neural network models were presented by Hemanth et al. [16] which are MCPN and MKNN. These models classified MRIs with high accuracies and tackled high convergence time period for Artificial Neural Networks.

3 Methodology

CNN models have been created from scratch and trained on Chest X-Ray Images (Pneumonia) dataset on Kaggle. Keras neural network library with TensorFlow backend has been used to implement the models. Dataset consists of 5216 training images, 624 testing images and 16 validation images. Data augmentation has been applied to achieve better results from the dataset. The four models have been trained on the training dataset, each with different number of convolutional layers. Each model was trained for 20 epochs, with training and testing batch sizes of 32 and 1, respectively. The following sub-headings further explain the above stages in depth.

3.1 CNN Architecture

CNN models are feed-forward networks with convolutional layers, pooling layers, flattening layers and fully connected layers employing suitable activation functions.

Convolutional layer. It is the building block of the CNNs. Convolution operation is done in mathematics to merge two functions [17]. In the CNN models, the input image is first converted into matrix form. Convolution filter is applied to the input matrix which slides over it, performing element-wise multiplication and storing the sum. This creates a feature map. 3 × 3 filter is generally employed to create 2D feature maps when images are black and white. Convolutions are performed in 3D when the input image is represented as a 3D matrix where the RGB color represents the third dimension. Several feature detectors are operated with the input matrix to generate a layer of feature maps which thus forms the convolutional layer.

Activation functions. All four models presented in this paper use two different activation functions, namely ReLU activation function and softmax activation function. The ReLU activation function stands for rectified linear function [18]. It is a nonlinear function that outputs zero when the input is negative and outputs one when the input is positive. The ReLU function is given by the following formula:

This type of activation function is broadly used in CNNs as it deals with the problem of vanishing gradients and is useful for increasing the nonlinearity of layers. ReLU activation function has many variants such as Noisy ReLUs, Leaky ReLUs and Parametric ReLUs. Advantages of ReLU over other activation functions are computational simplicity and representational sparsity. Softmax activation function is used in all four models presented in this paper. This broadly used activation function is employed in the last dense layer of all the four models [19]. This activation function normalizes inputs into a probability distribution. Categorical cross-entropy cost function is mostly used with this type of activation function.

Pooling layer. Convolutional layers are followed by pooling layers. The type of pooling layer used in all four models is max-pooling layers. The max-pooling layer having a dimension of 2×2 selects the maximum pixel intensity values from the window of the image currently covered by the kernel. Max-pooling is used to down sample images, hence reducing the dimensionality and complexity of the image [20]. Two other types of pooling layers can also be used which are general pooling and overlapping pooling. The models presented in this paper use max-pooling technique as it helps recognize salient features in the image.

Flattening layer and fully connected layers. After the input image passes through the convolutional layer and the pooling layer, it is fed into the flattening layer. This layer flattens out the input image into a column, further reducing its computational complexity. This is then fed into the fully connected layer/dense layer. The fully connected layer [21] has multiple layers, and every node in the first layer is connected to every node in the second layer. Each layer in the fully connected layer extracts features, and on this basis, the network makes a prediction [22, 23]. This process is known as forward propagation. After forward propagation, a cost function is calculated. It is a measure of performance of a neural network model. The cost function used in all four models is categorical cross-entropy. After the cost function is calculated, back propagation takes place. This process is repeated until the network achieves optimum performance. Adam optimization algorithm has been used in all four models.

Reducing overfitting. The first model exhibits substantial overfitting; hence, dropout technique was employed in the later models [24]. Dropout technique helps to reduce overfitting and tackles the problem of vanishing gradients. Dropout technique encourages each neuron to form its own individual representation of the input data. This technique on a random basis cuts connections between neurons in successive layers during the training process [25]. Learning rate of models was also modified, to reduce overfitting. Data augmentation technique can also be employed to reduce overfitting.

Algorithm of CNN classifiers. The algorithms used in the convolutional neural network classifiers have been explained in Figs. 1 and 2. Figure 3 shows the flowchart of the overall schema of research. The number of epochs for all the classifier models presented in this paper was fixed at 20 after training and testing several CNN models over the course of research. Classifier models trained for more number of epochs have showed overfitting. Several optimizer functions were also trained and studied. Adam optimizer function was finalized to be used for all classifiers after it gave the best results. Initially, a simple classifier model with convolutional layer of image size set to 64 * 64, 32 feature maps and employing ReLU activation function was trained. Fully connected dense layer with 128 perceptrons was utilized. To improve the result, the second classifier model was trained with one more convolutional layer of 64 feature maps for better feature extraction. The number of perceptrons in dense layer was also doubled to 256, so that better learning could be achieved. The third model was trained for three convolutional layers with 128 feature maps in third convolutional layer for more detailed feature extraction. Dense layer was kept unchanged. Dropout layer was introduced at 0.3, and learning rate of optimizer was

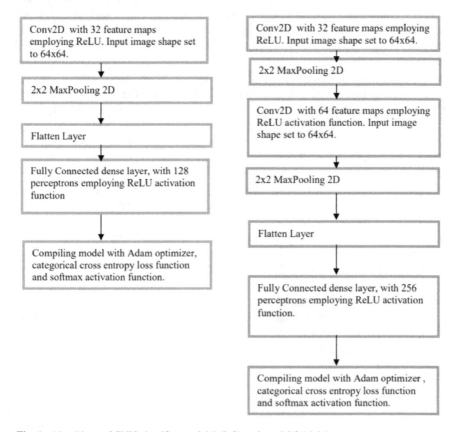

Fig. 1 Algorithms of CNN classifier model 1 (left) and model 2 (right)

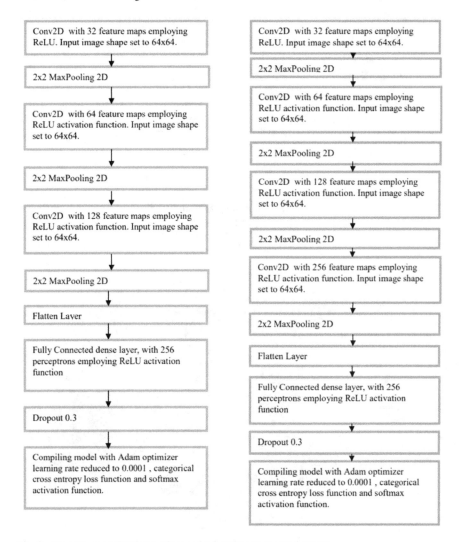

Fig. 2 Algorithms of CNN classifier model 3 (left) and model 4 (right)

lowered to 0.0001 to reduce the overfitting. The final fourth classifier model was trained for four convolutional layers with 256 feature maps in fourth convolutional layer. Dense layer, dropout layer and learning rate were kept same as third classifier model. The results have been summarized in the subsequent section of this paper.

Dataset. Chest X-Ray Images (Pneumonia) dataset of 1.16 GB size has been imported from Kaggle [26], with total of 5856 jpeg images split into Train, Test and Val folders each divided into category Pneumonia and Normal. Chest X-ray images (front and back) were selected from pediatric patients of one- to five-year olds from Guangzhou Women and Children's Medical Center, Guangzhou. Figure 4 provides

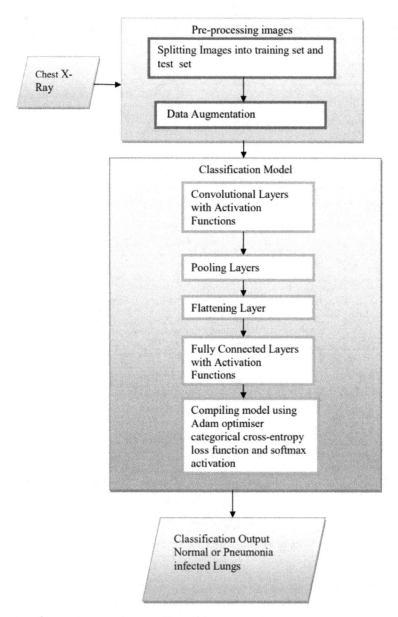

Fig. 3 Detailed schema of the experiment conducted

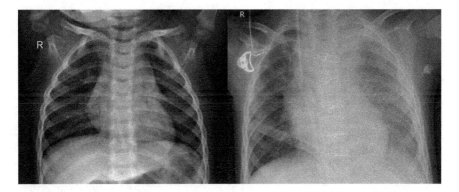

Fig. 4 Left image depicts normal lungs and right image depicts pneumonic lungs

the sample images from the dataset used during the research.

4 Experimental Results

To study the performance of each CNN classifier model, validation accuracy, recall and F1 score were evaluated as the performance measures [27, 28]. Accuracy and loss graphs were also studied. The confusion matrix was also computed for each model.

4.1 Comparison of Performance of Models

Figures 5 and 6 show the confusion matrices, accuracy graphs and loss graphs of all CNN classifier models. Table 1 and Figs. 5 and 6 show that classifier models 1 and 2 significantly underperformed compared to models 3 and 4. The accuracy graphs and loss graphs show overfitting. Accuracy, recall and F1 scores are also low. In addition to extra convolution layer, employing dropout and lowering the learning rate of optimizer in model 3 improved the performance considerably. It achieved the least overfitting along with highest accuracy and recall. Several attempts were made to better the performance by adding more convolutional layers and changing the parameters. Classifier model 4 with four convolutional layers showed good recall value and F1 score albeit with lower accuracy and higher overfitting compared to model 3. Thus, classifier model 3 performed the best among all CNN classifier models. In the following equations, tp = true positive, tn = true negative, fp = false positive and fn = false negative.

$$\text{Accuracy} = \frac{tp + tn}{tp + tn + fp + fn} \tag{1}$$

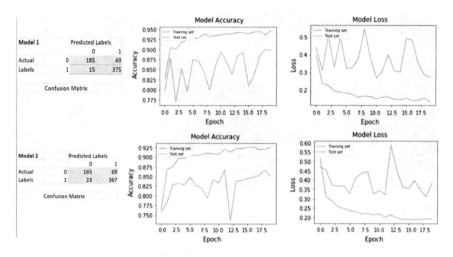

Fig. 5 Performance of classifier model 1 and model 2

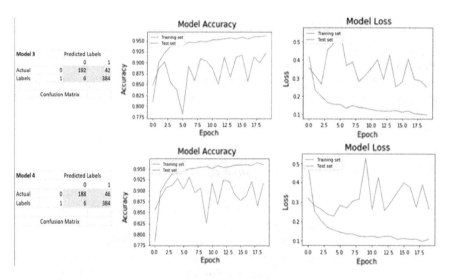

Fig. 6 Performance of classifier model 3 and model 4

$$Precision = tp/(tp + fp) \qquad (2)$$

Table 1 Performance comparison of different CNN models

Classifier model	Validation accuracy (%)	Validation loss (%)	Recall (%)	F1 score (%)
Model 1 (one conv.layer)	89.74	27.31	96	92
Model 2 (two conv.layers)	85.26	38.36	94	89
Model 3 (three conv.layers)	92.31	25.23	98	94
Model 4 (4 conv.layers)	91.67	26.61	98	94

$$\text{Recall} = tp/(tp + fn) \tag{3}$$

$$\text{F1 Score} = 2(\text{Precision} * \text{Recall})/(\text{Precision} + \text{Recall}) \tag{4}$$

5 Conclusion

The validation accuracy, recall and F1 score of CNN classifier model 3 with three convolutional layers are 92.31%, 98% and 94%, respectively, which are quite high compared to other models that were trained. CNN classifier model 4 with four convolutional layers also comes very close in performance with 91.67% validation accuracy, 98% recall and 94% F1 score. Both of these models have the same recall and F1 scores. The paper by Chakraborty [29] achieved the overall accuracy of 95.62% and recall of 95% trained on the same dataset. The paper by Liang [30] achieved recall of 96.7% on the same dataset. The models presented by us at best could achieve 92.31% accuracy which is lower, but 98% recall has been achieved. High recall values will ensure that the number of false-negative instances is lower, hence lowers the risk to the patient's life. Thus, it is concluded that CNN classifier model 3 and model 4 can, therefore, be effectively used by medical officers for diagnostic purposes for early detection of pneumonia in children as well as adults. A large number of X-ray images can be processed very quickly to provide highly precise diagnostic results, thus helping healthcare systems provide efficient patient care services and reduce mortality rates. These convolutional neural networks' models were successfully achieved by employing various methods of parameter tuning like adding dropout, changing learning rates, changing the batch size, number of epochs, adding more complex fully connected layers and changing various stochastic gradient optimizers [31].

In the future, it is hoped that transfer learning models would be trained on this dataset that would outperform these CNN models. It is intended that larger datasets will also be trained using the models presented in the paper. It is also expected that neural network models based on GAN [32], generative adversarial networks, would also be trained and compared with the existing models.

References

1. https://data.unicef.org/topic/child-health/pneumonia/. Accessed on 15 July 2019
2. Jaiswal, A.K., Tiwari, P., Kumar, S., Gupta, D., Khanna, A., Rodrigues, J.J.: Identifying pneumonia in chest x-rays: a deep learning approach. Measurement **145**, 511–518 (2019)
3. Kim, D.H., MacKinnon, T.: Artificial intelligence in fracture detection: transfer learning from deep convolutional neural networks. Clin. Radiol. **73**(5), 439–445 (2018)
4. Bernal, J., Kushibar, K., Asfaw, D.S., Valverde, S., Oliver, A., Martí, R., Lladó, X.: Deep convolutional neural networks for brain image analysis on magnetic resonance imaging: a review. Artif. Intell. Med. **95**, 64–81 (2019)
5. Arthur, F., Hossein, K.R.: Deep learning in medical image analysis: a third eye for doctors. J. Stomatology Oral Maxillofac. Surg.
6. Rubin, J., Sanghavi, D., Zhao, C., Lee, K., Qadir, A., Xu-Wilson, M.: Large Scale Automated Reading of Frontal and Lateral Chest X-Rays Using Dual Convolutional Neural Networks (2018). arXiv preprint arXiv:1804.07839
7. Lakhani, P., Sundaram, B.: Deep learning at chest radiography: automated classification of pulmonary tuberculosis by using convolutional neural networks. Radiology **284**(2), 574–582 (2017)
8. Guan, Q., Huang, Y., Zhong, Z., Zheng, Z., Zheng, L., Yang, Y.: Diagnose Like a Radiologist: Attention Guided Convolutional Neural Network for Thorax Disease Classification (2018). *arXiv preprint* arXiv:1801.09927
9. Rajpurkar, P., Irvin, J., Zhu, K., Yang, B., Mehta, H., Duan, T., Ding, D., Bagul, A., Langlotz, C., Shpanskaya, K., Lungren, M.P.: Chexnet: Radiologist-Level Pneumonia Detection on Chest X-rays with Deep Learning (2017). arXiv preprint arXiv:1711.05225
10. Krizhevsky, A., Sutskever, I., Hinton, G.E.: Imagenet classification with deep convolutional neural networks. In: Advances in Neural Information Processing Systems, pp. 1097–1105 (2012)
11. Simonyan, K., Zisserman, A.: Very Deep Convolutional Networks for Large-Scale Image Recognition (2014). arXiv preprint arXiv:1409.1556
12. Xu, Y., Jia, Z., Ai, Y., Zhang, F., Lai, M., Eric, I., Chang, C.: Deep convolutional activation features for large scale brain tumor histopathology image classification and segmentation. In: 2015 international conference on acoustics, speech and signal processing (ICASSP), pp. 947–951 (2015)
13. Anthimopoulos, M., Christodoulidis, S., Ebner, L., Christe, A., Mougiakakou, S.: Lung pattern classification for interstitial lung diseases using a deep convolutional neural network. IEEE Trans. Med. Imaging **35**(5), 1207–1216 (2016)
14. He, K., Zhang, X., Ren, S., Sun, J.: Deep residual learning for image recognition. In: Proceedings of the IEEE Conference on Computer Vision and Pattern Recognition, pp. 770–778 (2016)
15. Glozman, T., Liba, O.: Hidden Cues: Deep Learning for Alzheimer's Disease Classification CS331B project final report (2016)
16. Hemanth, D.J., Vijila, C.K.S., Selvakumar, A.I., Anitha, J.: Performance improved iteration-free artificial neural networks for abnormal magnetic resonance brain image classification. Neurocomputing **130**, 98–107 (2014)

17. Bi, X., Li, S., Xiao, B., Li, Y., Wang, G., Ma, X.: Computer aided Alzheimer's disease diagnosis by an unsupervised deep learning technology. Neurocomputing (2019)
18. Eckle, K., Schmidt-Hieber, J.: A comparison of deep networks with ReLU activation function and linear spline-type methods. Neural Netw. **110**, 232–242 (2019)
19. Ren, S., Jain, D.K., Guo, K., Xu, T., Chi, T.: Towards efficient medical lesion image super-resolution based on deep residual networks. Sig. Process. Image Commun. **75**, 1–10 (2019)
20. Zheng, Y., Iwana, B.K., Uchida, S.: Mining the displacement of max-pooling for text recognition. Pattern Recogn. **93**, 558–569 (2019)
21. Bhumika, P.S.S.S., Nayyar, P.A.: A review paper on algorithms used for text classification. Int. J. Appl. Innov. Eng. Manage. **3**(2), 90–99 (2013)
22. Kumar, A., Sangwan, S.R., Arora, A., Nayyar, A., Abdel-Basset, M.: Sarcasm detection using soft attention-based bidirectional long short-term memory model with convolution network. IEEE Access **7**, 23319–23328 (2019)
23. Saeed, F., Paul, A., Karthigaikumar, P., Nayyar, A.: Convolutional neural network based early fire detection. In: Multimedia Tools and Applications, pp. 1–17 (2019)
24. Kukkar, A., Mohana, R., Nayyar, A., Kim, J., Kang, B.G., Chilamkurti, N.: A novel deep-learning-based bug severity classification technique using convolutional neural networks and random forest with boosting. Sensors **19**(13), 2964 (2019)
25. Khan, S.H., Hayat, M., Porikli, F.: Regularization of deep neural networks with spectral dropout. Neural Netw. **110**, 82–90 (2019)
26. https://www.kaggle.com/paultimothymooney/chest-xray-pneumonia. Accessed on 15 July 2019
27. ALzubi, J.A., Bharathikannan, B., Tanwar, S., Manikandan, R., Khanna, A., Thaventhiran, C.: Boosted neural network ensemble classification for lung cancer disease diagnosis. Appl. Soft Comput. **80**, 579–591 (2019)
28. Vora, J., Tanwar, S., Polkowski, Z., Tyagi, S., Singh, P.K., Singh, Y.: Machine learning-based software effort estimation: an analysis. In: 11th International Conference on Electronics, computers and Artificial Intelligence (ECAI 2019), pp. 1–6, University of Pitesti, Pitesti, Romania, 27–29 June 2019
29. Chakraborty, S., Aich, S., Sim, J.S., Kim, H.C.: Detection of pneumonia from chest x-rays using a convolutional neural network architecture. In: International Conference on Future Information & Communication Engineering, vol. 11, no. 1, pp. 98–102 (2019)
30. Liang, G., Zheng, L.: A transfer learning method with deep residual network for pediatric pneumonia diagnosis. In: Computer Methods and Programs in Biomedicine (2019)
31. Du, S. S., Zhai, X., Poczos, B., Singh, A.: Gradient Descent Provably Optimizes Over-Parameterized Neural Networks (2018). arXiv preprint arXiv:1810.02054
32. Radford, A., Metz, L., Chintala, S.: Unsupervised Representation Learning with Deep Convolutional Generative Adversarial Networks (2015). arXiv preprint arXiv:1511.06434

A Machine Learning Algorithm for Classification, Analyzation and Prediction of Multimedia Messages in Social Networks

U. Tanuja, H. L. Gururaj and V. Janhavi

Abstract In the present-day scenario, the growth of social media gained much popularity around the world. A social media is a Website where people connect with each other. It is one of the most correlative media to communicate, share and exchange different types of information like text, image, audio, video…. It provides a platform to build relations among people to communicate and share their interests and activities, and with the enormous growth of social media, the users are spending more time for exchanging several contexts like text, image and video data. In this paper, the initiatory work is concerned with different types of classification algorithms used for analyzing audio and video data and also discusses about the machine learning of text and images in social networks.

Keywords Social networks · Convolution neural network (CNN) · Natural language processing (NLP) · Support vector machine (SVM) · Recurrent neural network (RNN)

1 Introduction

Social networks are the platform in which people or users build their own networks. Social networks also facilitate for creation and sharing information among each other. Social networks are important source of trending such as Facebook, WhatsApp, Twitter, LinkedIn and Instagram for communication. Nowadays, social networks are also more important in the business sectors [1]. Multimedia messages include text, audio, videos and images. Both large- and small-scale datasets can be used in machine learning (ML) algorithms [2]. All these multimedia messages are helpful for

U. Tanuja · H. L. Gururaj (✉) · V. Janhavi
Vidyavardhaka College of Engineering, Mysuru, Karnataka, India
e-mail: gururaj1711@gmail.com

U. Tanuja
e-mail: tanuumashankar715@gmail.com

V. Janhavi
e-mail: janhavi.v@vvce.ac.in

© Springer Nature Singapore Pte Ltd. 2020
P. K. Singh et al. (eds.), *Proceedings of First International Conference on Computing, Communications, and Cyber-Security (IC4S 2019)*, Lecture Notes in Networks and Systems 121, https://doi.org/10.1007/978-981-15-3369-3_37

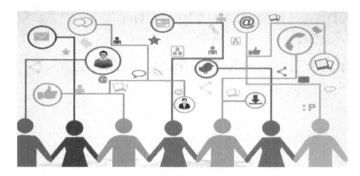

Fig. 1 Communication of multimedia messages in social networks

exchanging information like text messages, audios, videos and images among each other which creates network. As the number of media has increased, more people are enjoying to posting their experience through text and images and also express their opinions using audio and video.

The given Fig. 1 represents the communication of multimedia messages among different users. Social media provides different ways to communicate among each other.

- Social media like Twitter and Facebook created a sense of tenacity and share everything on the Internet.
- Social networks also provide updating and sharing contexts like text, audio, image and video with friends [3].
- It also enabled people inside perspective all over the world to share their stories.
- It also provides digital messages to share personally. Snapchat and Instagram provide smart filters.

In this paper, machine learning techniques are used for classification, analyzing and processing of multimedia messages like text, audio, image and video. Some methods of machine learning like support vector machine (SVM), recurrent neural networks (RNN), natural language processing (NLP) and CNN are used to process the multimedia contents.

The rest of this paper is collocated as follows. In Sect. 2, we present the text messages analysis using NLP method, and Sect. 3 represents various steps involving in NLP processing. Section 4 describes the classification of an image using CNN. Section 5 proposed a method for processing audio data using SVM. Section 6 depicts an algorithm for classification of video data using CNN, RNN and LSTM methods which are presented. Finally, Sect. 7 is concluded, and future research directions are proposed.

2 Text Analysis Using NLP

Social networking spaces such as Facebook, Instagram, WhatsApp, Twitter and Snapchat allow users to create their own personal profile and help to connect with each other. In social networks, there are new forms of communication between the users like text messages. For example, friend messages, postings, comments, tags… etc. Analyzing the text in social network with the help of natural language processing methods.

Text classification also helps in sentiment analysis, text categories and retrieving of information in social networks [4]. Machine learning is concerned with communication between humans and computers in natural language. Thus, the machine learning provides natural language processing (NLP) that helps computer to understand and manipulate human language [5].

Figure 2 depicts the different steps in natural language processing. Natural language processing can also be used for speech recognition, document summarization, question answering and machine translation. In social networks, texts like chatting, discussion forms, blogs and online reviews have been focused on NLP.

Fig. 3 represents that how text messages of social networks are processed using different analysis techniques of NLP method.

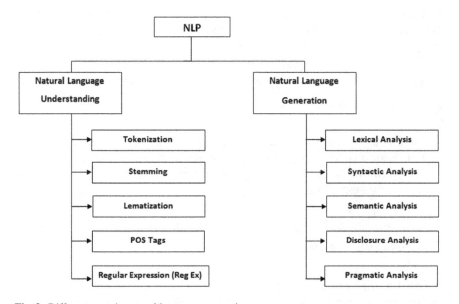

Fig. 2 Different steps in natural language processing

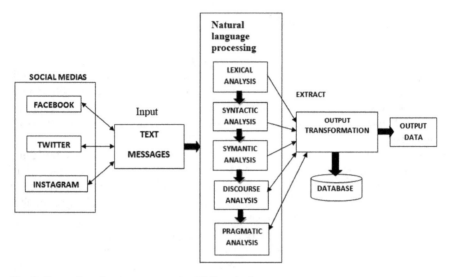

Fig. 3 Processing of text messages using NLP method

2.1 Lexical Analysis

In social networks, lexical analysis is responsible for analyzing and identifying the structure of words. Lexical analysis divides the whole sentence of texts into paragraph, sentences and words.

2.2 Syntactic Analysis

Syntactic analysis is also called as parsing or syntax analysis. Syntactic analysis in NLP is responsible for finding the syntactic structure of text/characters by analyzing its constituent words based on a given input (natural language). Figure 4 shows the example for syntactic analysis.

2.3 Semantic Analysis

In social media, texting is most important for communication; thus, the semantic analysis explicates the process of understanding the natural language of human's communication based on meaning and context. It involves reading all the words in context (natural language) to capture the meaning of any text. Semantic analysis also helps machines to understand languages automatically and in translating information.

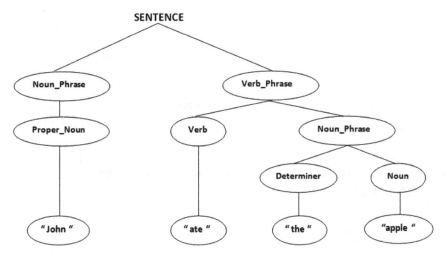

Fig. 4 Example of syntactic analysis

Semantic technology helps to find the data and separates the data from applications in the social networks [6]. It has a feature of text mining and comparison of data in the given text [7].

2.4 Discourse Analysis

Discourse analysis is responsible for processing of text or language analysis, which involves text and social interactions. Discourse means that it has a structured group of sentences. The sentence depends upon the meaning of the words and characters, and it also explicates the meaning of the next succeeding sentence.

2.5 Pragmatic Analysis

Pragmatic analysis involves the process of extracting information from text, and it also focuses on structured set of texts in social network and describes the actual meaning of that text. Pragmatic analysis involves deriving the aspects of natural language which requires the real-world knowledge.

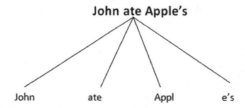

Fig. 5 Example of stemming

3 Various Steps Involving in the Natural Language Processing

3.1 Tokenization

In NLP, tokenization is the process of taking a text or group of text and splitting it up into individual words. Before processing of natural language in the computer, it is important to identify the words that constitute a string of words or characters. The meaning of text generally depends on the relation of character or words in that sentence. Different classes included in tokenization are simple tokenizer responsible for tokenizing raw text using characters. White space tokenizer is responsible for giving white spaces to tokenize the text. Tokenizers are responsible for converting raw text into separate tokens.

3.2 Stemming

In NLP, stemming is the process of semantic normalization, which reduces or chops off the words to their word root or derivational affixes to a common base form. A natural language processing (NLP) technique for stemming is applied to dissect each sentence and determines the semantic roles of significant words in the sentence [8]. It also helps to reduce the grammatical errors in the documents and also to reduce the ends of the words in a given text. Figure 5 represents the example of stemming—John ate Apple's.

3.3 Lemmatization

In NLP, lemmatization chops off the words to their base word, which are semantically correct lemmas. Lemmatization mutates root word with the help of vocabulary and the morphological analysis. Lemmatization also helps to remove the inflammation

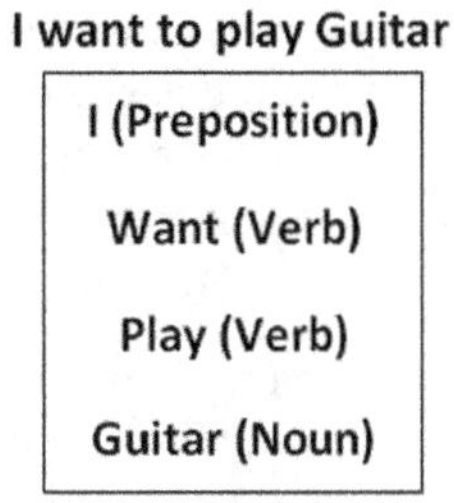

Fig. 6 Example of POS
tagging

I want to play Guitar

I (Preposition)
Want (Verb)
Play (Verb)
Guitar (Noun)

endings and return to the base form called lemma. Lemmatization is usually more advanced than stemming. Stemmer works only on a individual word without knowing the context.

3.4 POS Tagging

In NLP, POS tagging helps to recognize the grammatical mistakes of a given text or a word, and it also helps to identify whether the given word is a noun, pronoun, adjective, verb, adverb, etc. Based on the given context, POS tagging also helps in the relationships within the sentence and designates a corresponding tag to the word. Figure 6 depicts an example of POS tagging—I/want/to/play/Guitar.

3.5 Regular Expression (Regex)

In NLP, regular expression as a set of patterns matching commands helps to determine string or sentences in a large text data. These commands in regular expression help to match a digit and words per character of text which helps to handle group of texts. In regex, 'R' is helpful for parsing text data. Regular expression can be designed for metacharacters, sequences, quantifiers, character classes and POSIX character classes.

4 Image Processing Using Convolution Neural Networks

Nowadays, social media plays a vital role in marketing and communication between users and organization. Images and videos is one of the most important visual aspects that helps in communication and also today's marketing. People used to share their images and videos in social media. For analyzing and processing Image and Video with the help of convolution neural networks. Feature vectors are used to represent the low-level images represented in n dimensions [9]. Convolution neural network primarily helps to classify images. CNN algorithm takes image as an input to various objects in image and differentiates one from another. It has self-learning ability for image analysis, and it is also a popular algorithm [10]. CNN represents four important layers for processing of an image.

Figure 7 depicts the layers of a CNN.

(a) Convolution operation helps to extract the high-level features like edges, from a given input image. Convolution is also responsible for extracting the features (low-level features) like color, edged, etc. CNNs help to solve the inverse image problem by the trained sets of images and comparing of those images with the media in the social network [11]. In convolution neural network, filtering and optimization are used in trained image datasets [12].

Fig. 7 Different layers of convolution neural network

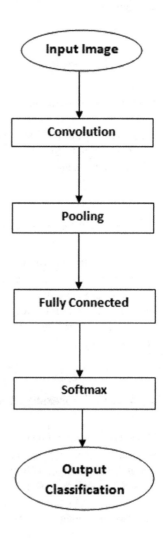

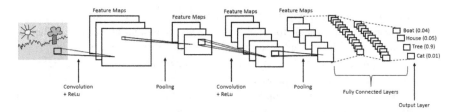

Fig. 8 Processing of an image using CNN

(b) Pooling layer is responsible for compressing the spatial size of the resultant convolved feature. Pooling layers help to reduce the computational power to process the data through dimensionality, and it also helps to extract features like positional and rotational to train the model. Max pooling helps to get the maximum value from the small position of the image. Average pooling helps to get the average values from the portion of an image.

(c) Fully connected layer (FC) in CNN represents the feature vectors of an input image. Fully connected layer helps to flatten the matrix of an image into vector and feeds that to fully connected neural network.

(d) SoftMax is the last layer in the CNN that is implemented through a neural network layer before the output layer. SoftMax is used to make the output more compatible with the image. SoftMax helps to aggregate the output of an image through neural network layer.

Figure 8 represents the processing of an image using different layers of convolution neural network. The advantage of CNN model is optimization of the trained datasets in the networks. If there is a low-level image, it will differentiate using linear and nonlinear filter methods [12]. To overcome with the retrieval of images, CNN low-level features are used to differentiate among them [13].

5   Audio Processing Using SVM

Nowadays, top social networks are taking up audio content and integrate in their platforms. People can share voice messages with each other. There are different formats of audio which are,

- Wav (Waveform Audio File) format.
- MPEG, OGG and PCD.
- WMA (Windows Media Audio) format.
- MP3 (MPEG-1 Audio Layer 3) format.
- MP4 (MPEG-1 Audio Layer 4) format.

Audio data is an important part of many new computers and multimedia applications in social networks. As social networks are growing rapidly, the amount of audio data is also increasing and demands for efficient computerized method.

Extraction and classification of audio are based on the sound context; extraction of an audio is very important for analyzing the audio context and finding relations between different things. Audio information refers to one of the most important sources of human perception. It is urgent to organize mass audio files in terms of the semantic description with rapid increasing of audio and video files [14]. Feature extraction of an audio is very important because data provided by audio cannot be understood by the model directly. Audio signal consists of three-dimensional signal axis represent time, amplitude and frequency.

5.1 Different Features for Audio Classification

Audio feature extraction is to compress sound signal to vector that helps to represent audio into meaningful information. Audio feature extraction helps to describe the respective audio signal with its descriptor, and these descriptors help to determine the audio signal characteristics based on the sound. [15]. Important feature, namely MFCC feature, is used for audio feature extraction.

Mel frequency cepstral co-efficient

Mel frequency cepstral co-efficient has a dominant feature which is widely used for the processing of audio and speech. MFCCs are short-term power spectrum based which has been proved that it has high effectiveness in recognizing the structure of music signals and frequency of audio signals. MFFCs improve the performance of audio classification. MFCCs also classify the audio with the help of the frequency scales [16]. MFCC helps to illustrate audio classifications with the help of linear cosine transform and nonlinear Fourier transform.

Audio processing in MFCCs has been applied for audio mining tasks, and MFCCs are based on the variation of human ears bandwidth with frequencies that capture the characteristics of speech and audio.

Figure 9 describes MFCC features, and firstly, the audio or sound signals are segmented, framed and windowed into short frames. Frames are calculated using fast Fourier transform (FFT) and that is converted into a Mel-scale filter outputs. A formula for calculating Mel-scale filters is.

$$
\text{Hm}(K) = \begin{cases} \frac{k-f(m-1)}{f(m)-f(m-1)} & K < f(M-1) \\ & f(m-1) \le k \le f(m) \\ \frac{k-f(m-1)}{f(m)-f(m-1)} & f(m) \le K \le f(M+1) \\ & k > f(m+1) \end{cases} \tag{1}
$$

where 'm' is the $\sum$number of filters and $f()$ is the list of $(m+2)$ Mel-spaced frequencies.

Next step, logarithm is applied to the Mel-scale filter output. Then, MFCCs are obtained by applying the discrete cosine transformation (DCT).

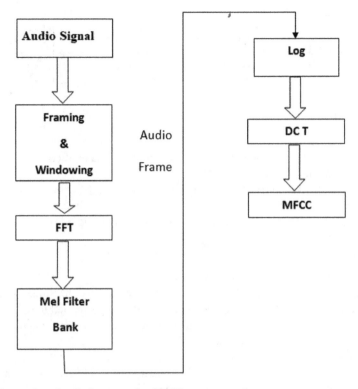

Fig. 9 Processing of audio features using MFCC

$$X_K = -\sum_{N=0}^{N=1} X_n \text{Cos}[\pi/N(n+1/2)k] \tag{2}$$

where $n = 0$, N is the number of audio or sound signal frames, and K is the number of segments.

DCT is applied to Mel filter bank to obtain MFCCs.

Support vector machine

In social networks, numerous audio files are sent or received by one another. In machine learning, support vector machine also known as supervised classification algorithm helps to categorize the different types of audio files. SVM can also be used for both regression and classification challenges.

'SVM is a machine learning method that works by looking for the best hyperplane (for classification) that separates the different labels. Hyperplane optimum can be found by measuring the margin/distance between hyperplane with data closest to each label [15]'.

SVMs include super vector machine methods for audio data classification detection, and it is based on the decision boundaries [16]. In this algorithm, we are firstly plotting each audio signal item as a point of n-dimensional space (when n

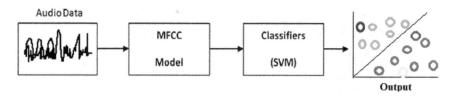

Fig. 10 Processing of audio data using MFCC and SVM classification model

is the number of signal features of audio data) with the value of coordinate and fed into MFCC model. Then, we need to perform classification of different audio data by finding the hyperplane or line that will make differentiation between two classes of audio data very well.

Figure 10 represents the processing of different audio data using classification model.

There are 2 types of support vector machine classifiers:

- *Linear support vector machine classifier*
  In linear support vector machine classifier, we assumed that the trained audio data is plotted in space, these audio data is separated by a gap called hyperplane, and the hyperplane is maximizing the distance to the nearest audio data point.
- *Nonlinear super vector machine classifier*
  In nonlinear super vector machine classifiers, sometimes our dataset is dispersed up to some extent. To solve this problem, nonlinear SVM classifiers and audio data sets are plotted in a higher dimension space.

6   Video Processing Using CNN, RNN and LSTM Method

Convolution neural network also helps to process the videos in social networks. A video is a sequence of images. Most existing video classification systems use only images. They classify images through convolutional neural networks (CNNs), which is a typical algorithm for image classification in machine learning and apply the results to video classification [17]. CNN is also used to identify the reduced datasets without any changes in the performance [18]. Video processing can also be categorized to speech and audio by removing the video analogous in the given video data [19]. In continuous video processing or classification, each frame is treated as discrete, and we will process these sequences of images using recurrent neural network (RNN) and convolution neural network which helps in classification of images as a frame.

Recurrent neural network helps to understand the context of video frames. For detecting spatial features in video, we use convolution neural network and for detecting temporal features in video. We use recurrent neural network.

Figure 11 represents fusion approach to combine spatial and temporal streams

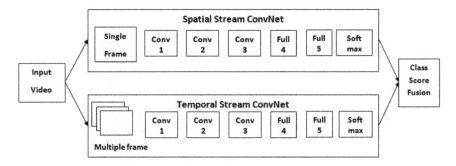

Fig. 11 Fusion approach for processing video data using CNN and RNN

using CNN and RNN. Temporal segment networks are responsible for using each segment as an input and final output of video clips is produced by combing segment stores. Classification of video data can be categorized into scenes, shots and also in the Ref. [19]. Pooling method allows us to get a representative feature from images. For each image extracted from the video, the feature vector matrices obtained through CNNs are transformed into one representative vector and then used for classification [17].

Long short-term memory (LSTM) is helpful for feedback connections that make it use general-purpose computers. LSTM is also responsible for processing each data points (such as images) and processing of sequence of data such as video. LSTM is suitable for classifying, processing and predictions based on data points.

As discussed earlier, the CNN and RNN are two stream approaches used for capturing the motion information in the videos. This is also not enough for analyzing the videos. LSTM demonstrates the effectiveness in capturing and actioning tasks in video.

Fig. 12 represents that two-layer LSTM network which is used for action recognition. Sequence or continuous online video are given to the CNN model, and the output from CNN model is fed to the two-layer LSTM as input. This algorithm helps to identify the cost estimation and limitations of existing technologies [20]. The outputs of both CNN and LSTM model are highly complementary. Nowadays, videos quality is becoming so viral world widely, and multimedia technologies are providing high definition and ultra-high definition videos [21].

7 Conclusion

Social networks are important and more convenient to communicate, share, exchange information like text, image, audio and video. In this paper, we leverage recently developed machine learning NLP method to processing text messages and various steps which involves processing of the texts in social networks. A novel approach for image classification using support vector machine was presented. The classification

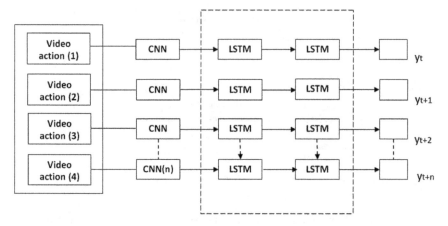

Fig. 12 Video action recognition using LSTM method

techniques like CNN and RNN used for audio and video data are discussed. The proposed approach helps in contributing to further generalization of classification and retrieval of audio and video data in social networks.

In future work, we would like to extend these classification algorithms for the effective usage of multimedia messages in social networks. These multimedia messages like text, audio, image and video data can be processed using different classification algorithms like logistic regression, stochastic gradient descent, K-nearest neighbors and neural networks. These algorithms also improve the performance in character recognition by filtering of unwanted text blocks, expression recognition and image detection in social networks.

References

1. Schiller, C., Meiren, T.: Enterprise Social Networks for Internal Communication and Collaboration. IEEE (2018). 978-1-5386-1469-3
2. ALzubi, J.A., Bharathikannan, B., Tanwar, S., Manikandan, R., Khanna, A., Thaventhiran, C.: Boosted neural network ensemble classification for lung cancer disease diagnosis. Appl. Soft Comput. (Elsevier) (2019)
3. Mou, W., Tzelepis, C., Mezaris, V., Gunes, H., Patras, I.: A deep generic to specific recognition model for group membership analysis using non-verbal cues. Image Vis. Comput. J. 81, 42–50 (2019). Elsevier
4. Cai, J., Li, J., Li, W., Wang, J.: Deep Learning Model Used In Text Classification, vol. 18. IEEE (2018). 978-1-7281-1536-8
5. Mukherjee, P., Santra, S., Bhowmick, S., Paul, A., Chatterjee, P., Deyasi, A.: Development of GUI for text-to-speech recognition using natural language processing. IEEE (2018). 978-1-5386-5550-4
6. Hassan, T., Hassan, S., Yar, M.A., Younas, W.: Semantic Analysis of Natural Language Software Requirement. IEEE (2016). 978-1-5090-2000-3
7. Merchant, K., Pande, Y.: NLP Based Latent Semantic Analysis for Legal Text Summarization. IEEE (2018). 978-1-5386-5314-2

8. Peng, T., Harris, I.G., Sawa, Y.: Detecting Phishing Attacks Using Natural Language Processing and Machine Learning. IEEE (2018) 0-7695-6360-0

9. Ritter, D., Rinderle-Ma, S.: Toward Application Integration with Multimedia Data. IEEE (2017).2325-6362

10. Sujit, S.J., Gabr, R.E., Coronado, I., Robinson, M., Datta, S., Narayana, P.A.: Automated Image Quality Evaluation of Structural Brain Magnetic Resonance Images using Deep Convolutional Neural Networks. IEEE (2018). 978-1-5386-8154-1

11. Gupta, H., Jin, K.H., Nguyen, H.Q., McCann, M.T., Unser, M.: CNN-Based Projected Gradient Descent for Consistent CT Image Reconstruction. IEEE (2018). 2832656

12. Liu, Z., Yan, W.Q., Yang, M.L.: Image Denoising Based on A CNN Model. IEEE (2018). 978-1-5386-6338-7

13. Liu, Y., Peng, Y., Hu, D., Li, D., Lim, K.-P., Ling, N.: Image retrieval using CNN and low-level feature fusion for crime scene investigation image database. APSIPA (2018). 978-988-14768-5-2

14. Rong, E.: Audio classification method based on machine learning. In: International Conference on Intelligent Transportation, Big Data & Smart City. IEEE (2017). 978-1-5090-6061-0

15. Ridoean, J.A., Sarno, R., Sunaryo, D.: Music mood classification using audio power and audio harmonicity based on MPEG-7 audio features and support vector machine. In: 3rd International Conference on Science in Information Technology. IEEE (2017). 978-1-5090-5864-8

16. Grama, L., Tuns, L., Rusu, C.: On the optimization of SVM kernel parameters for improving audio classification accuracy. In: 14th International Conference on Engineering of Modern Electric Systems (EMES). European union (2017). 978-1-5090-6073-3

17. Lee, J., Koh, Y., Yang, J.: A deep learning based video classification system using multimodality correlation approach. In: 17th International Conference on Control, Automation and Systems (ICCAS 2017). IEEE (2017). 18-21

18. Shankar, K., Lakshmanaprabu, S.K., Khanna, A., Tanwar, S., Rodrigues, J.J.P.C., Roy, N.R.: Alzheimer detection using group grey wolf optimization based features with convolutional classifier. Comput. Electr. Eng. **77**, 230–243 (2019)

19. Vinit Kaushik, R., Raghu, R., Maheshwar Reddy, L., Prasad, A., Prasanna S.: Ad analysis using machine learning. In: International Conference on Energy, Communication, Data Analytics and Soft Computing. IEEE (2017). 978-1-5386-1887-5

20. Vora, J., Tanwar, S., Polkowski, Z., Tyagi, S., Singh, P.K., Singh, Y.: Machine learning-based software effort estimation: an analysis. In: 11th International Conference on Electronics, computers and Artificial Intelligence (ECAI 2019), pp. 1–6. University of Pitesti, Pitesti, Romania, 27–29 June 2019

21. Liu, X., Li, Y., Liu, D., Wang, P., Yang, L.T: An Adaptive CU Size Decision Algorithm for HEVC Intra Prediction based on Complexity Classification using Machine Learning. IEEE (2017). 1051-8215

22. Tzelepis, C., Mezaris, V., Patras, I.: Linear maximum margin classifier for learning from uncertain data. IEEE Trans. Pattern Anal. Mach. Intell. **40**(12), 2948–2962 (2018). IEEE

Hope Enabler: A Novel Gamification-Based Approach to Enhance Classroom Engagement

Kavisha Duggal and Lovi Raj Gupta

Abstract Gamification is a two-state process that comprises enticement and engagement which are directly or indirectly related to hope. Hope is not just positive emotion, it is a motivational schema that consists of determination to achieve the target and build strategies that lead toward a goal. A unique concept of hope enabler is presented herein which is an innovative tool for hope, and it creates the next level of motivation for enticement and engagement. In this paper, the classroom teaching novel architecture of hope creation and hope support mechanism is formed. Factor analysis and other statistical tools are applied to the survey results to identify the gamified classroom learning factors/elements. The participation of each element would be differential, hence the present work is based on the survey in order to propose the level of participation by the virtue of the coefficients of each element identified. The classroom hope equation is formed which consists of four major elements: faculty conduct, teaching methodology, environment, and rewards. The complete theory of the gamification mechanism involves hope enabler, and even the D6 Process involves the atoms of hope and concludes the gamification methodology. When the new model of gamification is proposed by any organization to boost up productivity, there are hope enablers involved by the game designers to make the task more engaging, productive, and meaningful.

Keywords Classroom learning · Engagement · Enticement · Gamification · Gamified system · Hope · Rewards

K. Duggal (✉) · L. R. Gupta
Department of Computer Science and Engineering, Lovely Professional University, Punjab, India
e-mail: dhillon.kavisha89@gmail.com

L. R. Gupta
e-mail: lovi.gupta@lpu.co.in

© Springer Nature Singapore Pte Ltd. 2020
P. K. Singh et al. (eds.), *Proceedings of First International Conference on Computing, Communications, and Cyber-Security (IC4S 2019)*, Lecture Notes in Networks and Systems 121, https://doi.org/10.1007/978-981-15-3369-3_38

1   Introduction

The concept of gamification is defined with the use of game design elements in the non-game environment to encourage desired behavior and to enhance the participants' engagement [9]. Gamification is classified into twofold: structural gamification and content gamification [17]. Applications related to structural gamified scenarios frequently use levels, points, badges, achievements, leader boards, and progress bars to track individual performance. Content-based gamification is the usage of game thinking and game elements, in order to sort the content more game-like [17]. Without turning the content into a game, this type adds either challenges or story elements [19]. Increase in marketable deployment of "gamified" applications to huge spectators possibly assures innovation, inquiry-based on exciting lines, and data sources for human–computer interaction and game studies which indeed, gamification is gradually gathering the responsiveness of researchers [9].

Gamification is now trending, especially for the concerned community who are looking forward to embedding elements of a game into their services or products in order to motivate and enhance engagement of their customers or employees in a better way [12, 14]. There are gamification principles that increase engagement through hope [11]. Hope acts as a yearning for a goal congruent result, positively facilitates the relationships between the engagement and gamification principles [11]. The two major concepts of gamification—engagement and motivation—are directly proportional to hope. Students prefer to learn with the help of e-learning components. Adequate motivation and user engagement are expected when it comes to achieving a good score with the help of e-learning. It is possible to achieve adequate motivation with the deployment of gamification [23]. Enhancing community engagement motivates the users and in return contributes to a sustainable environment [3]. Gamification is a vast concept that is not at all about points, leader boards, and badges apart from this, and it includes dynamics such as self-expression and competition which play a major role to enhance the intrinsic motivation factors [22]. Rewards, recognition, and feedback are significant elements in gamifying a task [20].

Game mechanics are creating playful experiences and massive user engagement. But the factor of mixing game mechanics to achieve enhanced playful experience is missing [8]. Gamification taps into the simple desires and prerequisites of the user's impulses which revolve around the idea of achievement and status. Game mechanics that are used for gamification can be transformed conferring to the requirements [5]. This game mechanics like short-term goals, points, leader boards, badges, leveling, and onboarding can be attained with the highest level of enticement with the combination of hope (% of success) and motivation elements. In simple terms, it can be stated that a good employee hopes to get the best employee award, and motivation to become the best employee includes bonuses, status, and ownership.

Both game-based learning and gamification objective are to have fun-based learning and introducing values which are game-oriented [16] and will be accomplished by presenting mechanisms of games or creating a gamified learning process with the help of using games or gamification process as a part of learning. Gamification in education not only includes learners (students) but instructors also play a vital role

[6]. In this paper, various classroom teaching elements will be identified with the help of a survey that further enhances the gamified classroom learning. The elements introduced should nurture curiosity and intrigue effectiveness in gamified learning patterns [15].

When the concept of gamification is discussed, the main element which comes in mind is rewards. The continuous engagement and enticement of the students are not only based on extrinsic rewards (points, levels, badges, verbal appreciation, etc.), and it incorporates microelements of engagement like faculty conduct, teaching methodology, and environment which form an integral part of the gamified teaching–learning process. The concept of factor analysis is applied to the survey result, and various elements are identified.

2 Hope Enabler Mechanism

The small atoms of hope enjoin to build a mechanism of engagement termed as hope enabler. Hope acts as an imperious motivator as if there is something that one desires; it is the hope that motivates the individual in that direction. Hope plays a major factor that influences to take or avoid certain steps depending upon what one hopes to achieve in life. One of the most crucial aspects is to determine which game elements will be most meritoriously used in the learning context and will help us to attain learning objectives [10]. Hope enabler elements (x, y, z, k) will be derived from the survey analysis. The participation of each element would be differential, hence the present work is based on the survey in order to propose the level of participation by the virtue of the coefficients of each of these elements. $a_0, a_1, a_2, a_3 =$ coefficients.

2.1 Questionnaire for Faculty Based on the Hope Equation

- The processes which involve thinking and reasoning are more important than specific content in the curriculum.
- Inculcate positive social behavior among students, e.g., helping, sharing, and waiting.
- Desire to have 100% pass percentage.
- Today's classroom session will be full of live examples and will involve more student participation.
- Allow students to use cell phones in the classroom.
- Do you have a laugh and a talk with students (learn through play)?
- Creating a game environment while teaching in class.
- Environment of the classroom is based on scaffolding (handholding, feedback, slow yet steady).
- Well-defined levels for demonstrating the concepts to the students.
- Deployment of appropriate tools to make your teaching effective.
- Use verbal redirection for students who are disengaged.

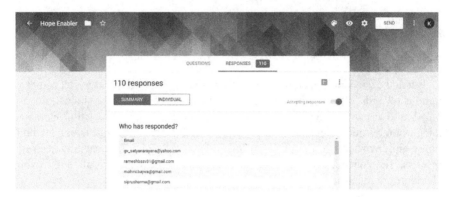

Fig. 1 Hope enabler survey response sheet

- Warn or threaten the students for their misbehavior in the classroom.
- Motivating the students by giving them positive feedback.
- Reward the achievers with incentives (e.g., stickers, badges).
- Do you follow the continuous reward system?
- Are you confident in your ability to encourage students socially, emotionally and inculcate problem-solving skills?

The survey is conducted on the methods adopted for classroom teaching, it consists of only 16 questions with the help of the feedback taken from the faculty, the values of x, y, z, k will be generated, and linear parametric modeling of the equation is done with the graphical representation. Link For the Google forms:—http://www.tinyurl.com/questionnaireforfaculty.

This survey has been filled by more than 100 faculties from different universities/colleges teaching in various domains. Only strongly agreed versions are taken to gauge the most acceptable outcomes of the survey.

2.2 Results of the Questionnaire

The questionnaire consists of 16 questions, and the response of each question is taken in the form of strongly agree, somewhat agree, neither agree nor disagree, somewhat disagree, and strongly disagree. The responses are collected through Google forms and are further recoded into the numeric number to be further used in the SPSS software to generate the valid results of each question consisting of frequency, percentage, and valid percentage according to the options mentioned in the survey.

Steps involved to generate following tables—open data in the SPSS → analyze → descriptive statistics → frequencies. Tables 1, 2, 3, 4, 5, 6, 7, 8, 9, 10, 11, 12, 13, 14, 15 and 16 consists of the results of survey question depicting the frequency and percentage of each attribute used in the survey form Fig. 1.

Table 1 Thinking and reasoning

		Frequency	Percent	Valid percent	Cumulative percent
Valid	Strongly disagree	1	0.9	0.9	0.9
	Somewhat disagree	1	0.9	0.9	1.8
	Neither agree nor disagree	4	3.6	3.6	5.5
	Somewhat agree	34	30.9	30.9	36.4
	Strongly agree	70	63.6	63.6	100
	Total	110	100	100	

Table 2 Inculcate positive social behavior

		Frequency	Percent	Valid percent	Cumulative percent
Valid	Somewhat disagree	4	3.6	3.6	3.6
	Neither agree nor disagree	3	2.7	2.7	6.4
	Somewhat agree	23	20.9	20.9	27.3
	Strongly agree	80	72.7	72.7	100.0
	Total	110	100.0	100.0	

Table 3 100% pass percentage

		Frequency	Percent	Valid percent	Cumulative percent
Valid	Strongly disagree	8	7.3	7.3	7.3
	Somewhat disagree	7	6.4	6.4	13.6
	Neither agree nor disagree	15	13.6	13.6	27.3
	Somewhat agree	37	33.6	33.6	60.9
	Strongly agree	43	39.1	39.1	100.0
	Total	110	100.0	100.0	

Table 4 Classroom session will be full of live examples

		Frequency	Percent	Valid percent	Cumulative percent
Valid	Somewhat disagree	3	2.7	2.7	2.7
	Neither agree nor disagree	3	2.7	2.7	5.5
	Somewhat agree	16	14.5	14.5	20.0
	Strongly agree	88	80.0	80.0	100.0
	Total	110	100.0	100.0	

Table 5 Laugh and a talk with students

		Frequency	Percent	Valid percent	Cumulative percent
Valid	Strongly disagree	1	0.9	0.9	0.9
	Neither agree nor disagree	5	4.5	4.5	5.5
	Somewhat agree	52	47.3	47.3	52.7
	Strongly agree	52	47.3	47.3	100.0
	Total	110	100.0	100.0	

Table 6 Use cellphones in the classroom

		Frequency	Percent	Valid percent	Cumulative percent
Valid	Strongly disagree	64	58.2	58.2	58.2
	Somewhat disagree	16	14.5	14.5	72.7
	Neither agree nor disagree	17	15.5	15.5	88.2
	Somewhat agree	11	10.0	10.0	98.2
	Strongly agree	2	1.8	1.8	100.0
	Total	110	100.0	100.0	

Table 7 Game environment

		Frequency	Percent	Valid percent	Cumulative percent
Valid	Strongly disagree	9	8.2	8.2	8.2
	Somewhat disagree	10	9.1	9.1	17.3
	Neither agree nor disagree	16	14.5	14.5	31.8
	Somewhat agree	37	33.6	33.6	65.5
	Strongly agree	38	34.5	34.5	100.0
	Total	110	100.0	100.0	

Table 8 Scaffolding

		Frequency	Percent	Valid percent	Cumulative percent
Valid	Strongly disagree	6	5.5	5.5	5.5
	Somewhat disagree	8	7.3	7.3	12.7
	Neither agree nor disagree	18	16.4	16.4	29.1
	Somewhat agree	55	50.0	50.0	79.1
	Strongly agree	23	20.9	20.9	100.0
	Total	110	100.0	100.0	

Table 9 Well-defined levels

		Frequency	Percent	Valid percent	Cumulative percent
Valid	Somewhat disagree	1	0.9	0.9	0.9
	Neither agree nor disagree	9	8.2	8.2	9.1
	Somewhat agree	24	21.8	21.8	30.9
	Strongly agree	76	69.1	69.1	100.0
	Total	110	100.0	100.0	

Table 10 Deployment of appropriate tools

		Frequency	Percent	Valid percent	Cumulative percent
Valid	Strongly disagree	1	0.9	0.9	0.9
	Somewhat disagree	1	0.9	0.9	1.8
	Neither agree nor disagree	3	2.7	2.7	4.5
	Somewhat agree	20	18.2	18.2	22.7
	Strongly agree	85	77.3	77.3	100.0
	Total	110	100.0	100.0	

Table 11 Use verbal redirection

		Frequency	Percent	Valid percent	Cumulative percent
Valid	Strongly disagree	1	0.9	0.9	0.9
	Somewhat disagree	4	3.6	3.6	4.5
	Neither agree nor disagree	13	11.8	11.8	16.4
	Somewhat agree	54	49.1	49.1	65.5
	Strongly agree	38	34.5	34.5	100.0
	Total	110	100.0	100.0	

2.3 Reliability Test on Dataset

To check the data reliability, Cronbach's Alpha method is used, and it helps to validate the data collected through the survey. The procedure used for validation makes the use of data more acceptable and checks the consistency of the received data. The SPSS software is used to perform this function. Steps applied in SPSS $\equiv$ analyze $\rightarrow$ scale $\rightarrow$ reliability analysis. Table 17 depicts that the collected data is valid and no values need to be excluded. Table 18 shows that data is highly acceptable as the value of Cronbach's Alpha is greater than 0.6. The Alpha value is 0.720 which indicates that items have relatively good internal consistency.

Table 12 Warn or threaten the students

		Frequency	Percent	Valid percent	Cumulative percent
Valid	Strongly disagree	9	8.2	8.2	8.2
	Somewhat disagree	11	10.0	10.0	18.2
	Neither agree nor disagree	23	20.9	20.9	39.1
	Somewhat agree	45	40.9	40.9	80.0
	Strongly agree	22	20.0	20.0	100.0
	Total	110	100.0	100.0	

Table 13 Motivating the students

		Frequency	Percent	Valid percent	Cumulative percent
Valid	Strongly disagree	2	1.8	1.8	1.8
	Somewhat disagree	1	0.9	0.9	2.7
	Neither agree nor disagree	3	2.7	2.7	5.5
	Somewhat agree	22	20.0	20.0	25.5
	Strongly agree	82	74.5	74.5	100.0
	Total	110	100.0	100.0	

Table 14 Rewards

		Frequency	Percent	Valid percent	Cumulative percent
Valid	Strongly disagree	1	0.9	0.9	0.9
	Somewhat disagree	3	2.7	2.7	3.6
	Neither agree nor disagree	12	10.9	10.9	14.5
	Somewhat agree	27	24.5	24.5	39.1
	Strongly agree	67	60.9	60.9	100.0
	Total	110	100.0	100.0	

Table 15 Continuous reward system

		Frequency	Percent	Valid percent	Cumulative percent
Valid	Strongly disagree	1	0.9	0.9	0.9
	Somewhat disagree	3	2.7	2.7	3.6
	Neither agree nor disagree	10	9.1	9.1	12.7
	Somewhat agree	47	42.7	42.7	55.5
	Strongly agree	49	44.5	44.5	100.0
	Total	110	100.0	100.0	

Table 16 Promote students emotionally and socially

		Frequency	Percent	Valid percent	Cumulative percent
Valid	Somewhat disagree	1	0.9	0.9	0.9
	Neither agree nor disagree	3	2.7	2.7	3.6
	Somewhat agree	27	24.5	24.5	28.2
	Strongly agree	79	71.8	71.8	100.0
	Total	110	100.0	100.0	

Table 17 Case processing summary

	N	%
Valid	110	100.0
Excluded	0	0.0
Total	110	100.0

Table 18 Reliability statistics

Cronbach's Alpha	N of items
0.720	16

3 Mathematical Build for Hope Enabler

A statistical method will be applied to the given data. The technique which is used to moderate a number of hefty variables into a scarer number of elements/factors is known as factor analysis. From all the variables, factor analysis abstracts extreme collective variance and situates them into a communal score [21]. The rotated component matrix helps us to identify which question loads on which component after rotation. Steps to perform factor analysis open data in SPSS → recode the values → analyze → dimension reduction → factor → descriptives → extraction → rotation → varmax. The rotated component matrix values signify that the classroom learning survey is further classified into four components or elements and relevance of each question with the defined four components as shown in Table 19. **Extraction method** is used as principal component analysis, and **rotation method** is used as varimax with Kaiser normalization.

3.1 Interpretation of the Rotated Above Matrices

Figure 2 is the interpretation of Table 19. The concept of rotation is involved which reduces the number of factors on which the variables show high loading. The 16 questions are further divided into four components/elements. For example, question "Are you confident in your ability to encourage students socially, emotionally and inculcate problem-solving skills?" has four different component values component $1 = 0.603$, component $2 = 0.413$, component $3 = 0.014$, and component $4 = -0.021$ as mentioned in Table 19. From the above value analysis, this question has high loading with component 1, so the rest three values are discarded, and similarly, the lower values are discarded for the rest 15 questions. Each question dependency is calculated, and from this question relation with element is generated.

The four elements formed are faculty conduct, teaching methodology, environment, and reward. Faculty conduct: A faculty's own ability or desire/belief is to make the classroom learning more effective and engaged to achieve goals. It includes observations as to what they do, how they react to a given situation; feedback loops as consistent or intermittent [4], reinforcement as actions → results, results → actions; Loss aversion as afraid of mistakes; confirmation bias as difficult to break the ice. **Teaching methodology**: content-based teaching with the help of real-life examples. Various new teaching pedagogies can be incorporated to enhance classroom learning [7]. **Environment**: The environment constraints always exist, and to apply gamification concepts there is a requisite of environment [1].There can be two types of environment: positive and negative. **Positive**: when one's basic qualities, i.e., academic background, work experiences, surroundings, player journey, etc., help to complete the work and boost his morale in the right direction. For example, a well-trained driver has to drive a car in a hilly area. **Negative**: what if one hopes for something that is beyond one's immediate control. A faculty wants to conduct a game-based learning

Table 19 Rotated component matrix

	Components			
	Faculty conduct	Teaching methodology	Environment	Rewards
The processes which involve thinking and reasoning are more important than specific content in curriculum	0.034	0.684	0.165	−0.114
Inculcate positive social behavior among students, e.g., helping, sharing, and waiting	0.595	−0.067	0.227	0.045
Desire to have 100% pass percentage	−0.021	0.591	0.040	0.285
Classroom session will be full of live examples and will involve more student participation	0.168	0.516	−0.013	0.089
Do you have a laugh and a talk with students (learn through play)?	0.233	0.290	0.542	−0.074
Allow students to use cellphones in the classroom	−0.304	−0.226	0.649	0.128
Creating a game environment while teaching in class	0.265	0.058	0.739	−0.079
Environment of the classroom is based on scaffolding (hand holding, feedback, slow yet steady)	0.248	0.239	0.528	0.070
Well-defined levels for demonstrating the concepts to the students	0.478	0.463	−0.203	0.155
Deployment of appropriate tools to make your teaching effective	0.575	0.261	0.051	0.016
Use verbal redirection for students who are disengaged	0.031	0.517	0.385	−0.013
Warn or threaten the students for their misbehavior in classroom	−0.029	0.085	−02.48	0.848
Motivating the students by giving them positive feedback	0.708	0.014	0.150	0.091
Reward the achievers with incentives (e.g., stickers, badges)	0.606	−0.062	0.180	0.454
Do you follow the continuous reward system?	0.414	0.138	−0.046	0.641
Are you confident in your ability to encourage students socially, emotionally and inculcate problem-solving skills?	0.603	0.413	0.014	−0.021

Rotated Component Matrix				
	Component			
	1	2	3	4
Are you confident in your ability to encourage students socially, emotionally and inculcate problem solving	.603			
Well defined levels for demonstrating the concepts to the students.	.478			
Reward the achievers with incentives (e.g. stickers, badges).	.606			
Motivating the students by giving them positive feedback.	.708			
Inculcate positive social behaviour among students. E.g helping, sharing, waiting.	.595			
Deployment of appropriate tools to make your teaching effective.	.575			
Warn or threaten the students for their misbehaviour in classroom.				.848
Do you follow the continuous reward system.				.641
Creating a game environment while teaching in class.			.739	
Allow students to use cellphones in the classroom.			.649	
Do you have a laugh and a talk with students (learn through play)?			.542	
Environment of the classroom is based on scaffolding (hand holding, feedback, slow yet steady).			.528	
The processes which involve thinking and reasoning are more important than specific content in curriculur		.684		
Desire to have 100 percent pass percentage.		.591		
Use verbal redirection for students who are disengaged.		.517		
Today's classroom session will be full of live examples and will involve more student participation.		.516		
	faculty Conduct	Teaching methodo	Environment	Reward

Fig. 2 Excel sheet for elements interpretation

session but the authorities of the concerned college do not approve. **Rewards**: They are like patrons which are given in recognition of services, efforts, and achievements [13]. Types of rewards are as follows: (1) tangible rewards—financial rewards [4] and intangible rewards–recognition rewards, (2) expected and unexpected rewards, and (3) contingency rewards as engagement contingent (To be awarded during the start of the task), completion contingent (To achieve award during finish of the task), and performance contingent (Reward if the task is accomplished well).

The rewards are one of the most important components of gamification (in the form of points, levels, badges, verbal appreciation) which acts as an extrinsic motivator for developing the perfect gamified learning environment. "The real accomplishment of gamified system is to raise feelings of achieving mastery, sovereignty, sense of belonging" with continuous engagement and enticement [11]. Rewards enhance more contribution of participants but a good framework leads to better quality contributors [18]. The major challenge is to check the joint variability of the elements to define the hope enabler to enhance the classroom engagement.

3.2 Computation of Regression Coefficients and Forming the Equation

Using least squares regression, develop a regression equation based on (1) faculty conduct (2) teaching methodology (3) environment (4) rewards. Four-Independent Variables Regression Independent Variables : $k = 4$ No Intercept, $e =$ error of tolerance Through SPSS software, regression is applied. Steps applied in SPSS are as follows: open data in SPSS $\rightarrow$ analyze $\rightarrow$ regression $\rightarrow$ linear $\rightarrow$ statistics. The ANOVA Table 20 represents the F-ratio which experiments whether the overall

Table 20 ANOVA

Model	Sum of squares	df	Mean square	F	Sig.	
1	Regression	4.460	4	1.115	24.367	0.000[a]
	Residual	0.549	12	0.046		
	Total	5.010[b]	16			

[a]Predictors of variables
[b]This total sum of squares is not corrected for the constant because the constant is zero for regression through the origin

model of regression is a virtuous fit for the data. Table 20 shows statistical data of independent variables as well as considerably envisage the dependent variable, $F = 24.367, p < 0.0005$.

$$H = a_0 x + a_1 y + a_2 z + a_3 k + e \tag{1}$$

To define the regression coefficients, we use the following equation [2].

$$a = (M'M)^{-1} M'N \tag{2}$$

N = Dependent variable in form of survey result (%Strongly agree)has been scaled on 0–1.
M = Each question dependency on four factors/elements. The data is collected from SPSS rotated matrix component which question load on which component after rotation Table 18.
From Table 21 we have N and M values:
Step 1: M'M = Matrix multiplication of M' and M Matrix M' dimensions = 4 * 16 and the values of the matrix are shown in Table 22.
M'M = The values of matrix multiplication M'M are mentioned in Table 23.
Step 2: (M'M) − 1 = Inverse of matrix multiplication will be performed. The values of inverse of matrix are mentioned in Table 24.
Step 3: M'N = Matrix multiplication. The matrix result is in the form of 4*1 as shown in Table 25.
Step 4: $a = (M'M)^{-1} M'N$, the matrix multiplication is performed, and the final result is mentioned in Table 26.
The values of a_0, a_1, a_2 have been obtained which will further help us to define the hope coefficients. The coefficient values are verified through SPSS software, and results of the coefficients are displayed in Table 27 where dependent variable is V5 and linear regression is through the origin.

Table 21 Dependency table

Q	N	M			
	Strongly agree (%) results	faculty conduct (M1)	Teaching methodology (M2)	Environment (M3)	Rewards (M4)
1	0.636	0.034	0.684	0.165	−0.114
2	0.721	0.595	−0.067	0.227	0.045
3	0.391	−0.021	0.591	0.04	0.285
4	0.8	0.168	0.516	−0.013	0.089
5	0.473	0.233	0.29	0.542	−0.074
6	0.18	−0.304	−0.226	0.649	0.128
7	0.345	0.265	0.058	0.739	−0.079
8	0.209	0.248	0.239	0.528	0.07
9	0.691	0.478	0.463	−0.203	0.155
10	0.773	0.575	0.261	0.051	0.016
11	0.345	0.031	0.517	0.385	−0.013
12	0.2	−0.029	0.085	−2.49	0.848
13	0.745	0.708	0.014	0.15	0.091
14	0.609	0.606	−0.062	0.18	0.454
15	0.445	0.414	0.138	−0.046	0.641
16	0.718	0.603	0.413	0.014	−0.021

The column B defines the values of the coefficients of defined elements. $a_0 = 0.82, a_1 = 0.66, a_2 = 0.10, a_3 = 0.35$.

$$H = a_0 x + a_1 y + a_2 z + a_3 k.$$

Hope equation for the classroom learning is representing the elements like faculty conduct, teaching methodology, environment, and rewards.

$$H = 0.8x + 0.7y + 0.1z + 0.3k.$$

Rewards act as one of the important elements of gamification but not the only element. Faculty conduct and teaching methodology act as one of the long term and imperious motivators which completes the hope enabler equation. The gamification provides companies with new platform to relate to customers enhance their employee productivity and innovation which can be easily achieved by hope microelements.

Table 22 M′ Matrix with values

	M′1	M′2	M′3	M′4	M′5	M′6	M′7	M′8	M′9	M′10	M′11	M′12	M′13	M′14	M′15	M′16
1	0.034	0.595	−0.021	0.168	0.233	−0.304	0.265	0.248	0.478	0.575	0.031	−0.029	0.708	0.606	0.414	0.603
2	0.684	−0.067	0.591	0.516	0.29	−0.226	0.058	0.239	0.463	0.261	0.517	0.085	0.014	−0.062	0.138	0.413
3	0.165	0.227	0.04	−0.013	0.542	0.649	0.739	0.528	−0.203	0.051	0.385	−2.49	0.15	0.18	−0.046	0.014
4	−0.114	0.045	0.285	0.089	−0.074	0.128	−0.079	0.07	0.155	0.016	−0.013	0.848	0.091	0.454	0.641	−0.021

Table 23 Result of matrix multiplication M′M

1	2.6267	0.93205	0.61453	0.6227
2	0.93205	2.0342	0.19123	0.28934
3	0.61453	0.19123	8.06974	−2.0584
4	0.6227	0.28934	−2.0584	1.50644

Table 24 Result of matrix multiplication $(M'M)^{-1}$

1	0.574473796	−0.1948590683	−0.1383867402	−0.3891263779
2	−0.1948590683	0.5887861393	−0.01138075389	−0.04809155444
3	−0.1383867402	−0.01138075389	0.2300604279	0.3737420858
4	−0.3891263779	−0.04809155444	0.3737420858	1.344579645

Table 25 Result of matrix multiplication M′N

1	3.06892
2	2.244716
3	0.775267
4	1.017025

Table 26 Final result

1	0.82339348925248777733
2	0.66598482153951469985
3	0.10685476612731086179
4	0.3528553136456018866

Table 27 Coefficients

Model	Unstandardized coefficients		Standardized coefficients	T	Sig
	B	Std. error	Beta		
V1	0.825	0.162	0.597	5.089	0.000
V2	0.666	0.164	0.424	4.058	0.002
V3	0.104	0.103	0.132	1.014	0.330
V4	0.348	0.248	0.191	1.404	0.186

4   Conclusion

In this paper, a novel model of hope enabler is being presented which is the next level of motivation to boost the engagement and enticement of the learner. The continuous engagement and enticement of the scholar are not only based on extrinsic rewards (points, levels, badges, verbal appreciation, etc.) but it also incorporates microelements of engagement like (1) faculty conduct (2) teaching methodology (3) environment (4) rewards. These elements are obtained by applying factor analysis on the survey results. The authenticity and reliability of data are thoroughly checked by the use of SPSS software. The microstructure of hope enabler actively takes part to form an impeccable gamified system to accomplish the goals of enticement and incessant engagement of the learner. The interpretation of survey clearly states that we need to bring a meticulously captured and revisited revolutionary change in our present education system by adopting gamified frameworks of teaching, involving rewards system beyond the marks, more lessons to be defined based on regular feedback, and embracing the new wave of technology in our teaching practices.

References

1. Akpolat, B.S., Slany, W.: Enhancing software engineering student team engagement in a high-intensity extreme programming course using gamification. In: 2014 IEEE 27th Conference on Software Engineering Education and Training (CSEE T), pp. 149–153, Apr 2014
2. Berman, H.: Regression coefficients, howpublished = https://stattrek.com/multiple-regression/regressioncoefficients.aspx?tutorial=reg, note = Accessed 2019
3. Bista, S.K., Nepal, S., Colineau, N., Paris, C.: Using gamification in an online community. In: 8th International Conference on Collaborative Computing: Networking, Applications and Worksharing (CollaborateCom), pp. 611–618, Oct 2012
4. Blagov, E., Simeonova, B., Bogolyubov, P.: Motivating the adoption and usage of corporate web 2.0 systems using fitness gamification practices. In: 2013 IEEE 15th Conference on Business Informatics, pp. 420–427, July 2013
5. Butgereit, L.: Gamifying a PhD taught module: a journey to phobos and deimos. In: 2015 IST-Africa Conference, pp. 1–9, May 2015
6. Casapicola, P., Polzl, M., Geiger, B.C.: A matlab-based magnitude response game for dsp education. In: 2014 IEEE International Conference on Acoustics, Speech and Signal Processing (ICASSP), pp. 2214–2218, May 2014
7. Chin, S.: Mobile technology and gamification: the future is now! In: 2014 Fourth International Conference on Digital Information and Communication Technology and its Applications (DICTAP), pp. 138–143, May 2014
8. Codish, D., Ravid, G.: Detecting playfulness in educational gamification through behavior patterns. IBM J. Res. Dev. **59**(6), 6:1–6:14 (2015)
9. Deterding, S., Dixon, D., Khaled, R., Nacke, L.: From game design elements to gamefulness: defining "gamification". In: Proceedings of the 15th International Academic MindTrek Conference: Envisioning Future Media Environments, MindTrek '11, pp. 9–15. ACM, New York, NY, USA (2011)
10. Dicheva, D., Irwin, K., Dichev, C., Talasila, S.: A course gamification platform supporting student motivation and engagement. In: 2014 International Conference on Web and Open Access to Learning (ICWOAL), pp. 1–4, Nov 2014

11. Eisingerich, A.B., Marchand, A., Fritze, M.P., Dong, L.: Hook vs. hope: How to enhance customer engagement through gamification. Int. J. Res. Mark. **36**(2), 200–215 (2019)
12. Erenli, K.: The impact of gamification—recommending education scenarios. Int. J. Emerg. Technol. Learn. (iJET) **8** (2013)
13. Hamzah, W.M.A.F.W., Ali, N.H., Mohd Saman, M.Y., Yusoff, M.H., Yacob, A.: Enhancement of the arcs model for gamification of learning. In: 2014 3rd International Conference on User Science and Engineering (i-USEr), pp. 287–291 (2014)
14. Herzig, P., Ameling, M., Schill, A.: A generic platform for enterprise gamification. In: Proceedings of the 2012 Joint Working IEEE/IFIP Conference on Software Architecture and European Conference on Software Architecture, WICSA-ECSA '12, pp. 219–223. IEEE Computer Society, Washington, DC, USA (2012)
15. Ibanez, M., Di-Serio, Delgado-Kloos, C.: Gamification for engaging computer science students in learning activities: a case study. IEEE Trans. Learn. Technol. **7**(3), 291–301 (2014)
16. Jayasinghe, U., Dharmaratne, A.: Game based learning vs. gamification from the higher education students' perspective. In: Proceedings of 2013 IEEE International Conference on Teaching, Assessment and Learning for Engineering (TALE), pp. 683–688, Aug 2013
17. Lamprinou, D., Paraskeva, F.: Gamification design framework based on sdt for student motivation. In: 2015 International Conference on Interactive Mobile Communication Technologies and Learning (IMCL), pp. 406–410, Nov 2015
18. Moccozet, L., Tardy, C., Opprecht, W., Leonard, M.: Gamification-based assessment of group work. In: 2013 International Conference on Interactive Collaborative Learning (ICL), pp. 171–179, Sep 2013
19. Ohno, A., Yamasaki, T., Tokiwa, K.: A discussion on introducing half-anonymity and gamification to improve students' motivation and engagement in classroom lectures. In: 2013 IEEE Region 10 Humanitarian Technology Conference, pp. 215–220, Aug 2013
20. Reid, E.F.: Crowdsourcing and gamification techniques in inspire (aqap online magazine). In: 2013 IEEE International Conference on Intelligence and Security Informatics, pp. 215–220, June 2013
21. Solutionsn, S.: Factor anaylsis. https://www.statisticssolutions.com/factor-analysis-sem-factor-analysis/. Accessed 2019
22. Suh, A., Wagner, C., Liu, L.: The effects of game dynamics on user engagement in gamified systems. In: 2015 48th Hawaii International Conference on System Sciences, pp. 672–681, Jan 2015
23. Wongso, O., Rosmansyah, Y., Bandung, Y.: Gamification framework model, based on social engagement in e-learning 2.0. In: 2014 2nd International Conference on Technology, Informatics, Management, Engineering Environment, pp. 10–14, Aug 2014

Analysis and Prediction of Stock Market Trends Using Deep Learning

Harshit Agarwal, Gaurav Jariwala and Akshit Shah

Abstract The paper proposes a progressive conclusion on the application of recurrent neural networks in stock price forecasting. We have also used random forest classifier to factor in the sudden fluctuations in stock prices which are derivatives of any abnormal events. Machine learning and deep learning strategies are being used by many quantitative hedge funds to increase their returns. Finance data belongs to time series data. A time series is a series of data points indexed in time. The nonlinearity and chaotic nature of the data can be combated using recurrent neural networks which are effective in tracing relationships between historical data and using it to predict new data. Historical data in this context is time series data from the past. It is one of the most important and the most valuable parts for speculating about future prices. Long short-term memory (LSTM) is capable of capturing the most important features from time series data and modelling its dependencies. Building a good and effective prediction system can help investors and traders to get a glimpse of the future direction of the stock and accordingly help them mitigate risk in their respective portfolios. The results obtained by our approach are accurate up to 97% for the values predicted using historic data and 67% for the trend prediction using news headlines.

Keywords Activation function · Back-propagation · Long-term short memory · Random forest classifier · Recurrent neural network · Sentimental analysis

H. Agarwal (✉) · G. Jariwala · A. Shah
Sarvajanik College of Engineering and Technology, Surat, India
e-mail: 9arshit@gmail.com

G. Jariwala
e-mail: gjariwala9@gmail.com

A. Shah
e-mail: shahakshit34@gmail.com

P. K. Singh et al. (eds.), *Proceedings of First International Conference on Computing, Communications, and Cyber-Security (IC4S 2019)*, Lecture Notes in Networks and Systems 121, https://doi.org/10.1007/978-981-15-3369-3_39

1   Introduction

1.1   Motivation

Stock market is a platform where buyers and sellers trade stocks driving the price either up or down. Stock prices change every day by market forces. These changes occur because of the law of supply and demand. If more people want to buy a stock (demand) than sell it (supply), the price moves up. Conversely, if more people wanted to sell a stock than buy it, there would be greater supply than demand, and the price would fall.

Understanding these mechanisms is easy. What is difficult to comprehend is what makes people buy a particular stock and sell another stock making the stock market highly volatile and unpredictable by nature. Therefore, investors are always taking risks in hopes of making a profit. People invest in the stock markets according to their risk appetite and expect profits in return of their investments. It is near-impossible to predict stock prices, owing to the volatility of factors that play a major role in the movement of prices. However, it is possible to make an estimate of future prices.

1.2   Contribution

Finance produces time series data. A time series is a series of data points indexed in time. Usually, a time series is a sequence taken at successive, equally spaced points in time: a sequence of discrete-time data. Time series data here refers to the indicators which trace the volatility in the prices of a stock on any particular day. These performance indicators are opening stock price ('Open'), closing price ('Close'), intra-day low price ('Low'), intra-day highest price ('High') and total volume of stocks traded during the day ('Volume').

Recurrent neural networks have been the most effective system to analyse sequential time series data and find relationships between historical data to forecast new data. Historical data in this context is time series data from the past. It is one of the most important and the most valuable parts for speculating about future prices.

Neural networks are interconnected data processing components that are composed of layers of artificial neurons (network nodes) that can process input and forward output to other nodes in the network. The nodes are connected by edges or weights that influence a signal's strength and the network's ultimate output.

Long short-term memory (LSTM) networks are a modified version of recurrent neural networks, which makes it easier to remember past data in memory. LSTM is well suited to classify, process and predict time series given time lags of unknown duration. It trains the model by using back-propagation.

Sudden massive fluctuations in the price which disrupts the pattern most of the time are derivatives of abnormal events. Accuracy of prediction depends on how well these factors are adjusted. We have used random forest classifier for the classification of stock's news headlines. Using these classifications, we generate labels for any particular news and then try adjusting the prices depended on the effect of the news on the stock price.

1.3 Organization

The remaining paper is organized as follows: Sect. 2 summarizes the related work done in this domain; Sect. 3 describes our approach and implementation; Sect. 4 describes the results obtained from the proposed system; and Sect. 5 summarizes the conclusion and future directions of the concept.

2 Related Work

Kar [1] has presented the application of artificial neural network to predict stock market indices. Several activation functions are implemented along with options for cross-validation sets achieving a best-case accuracy of 96% on their nifty stock index data set. The results obtained by them were fairly accurate but sudden variations in actual data make it impossible for them to predict the price for such fluctuations.

Barapatre et al. [2] have proposed a machine learning (ANN) artificial neural network model to predict the stock market price. The proposed algorithm integrates the back-propagation algorithm employed to train the ANN network model. They have used neural networks to predict the stock value by using historical details of the stock market value and applying a multilayer feed-forward network to solve the problem. The algorithm was tested on Tesla data set where 94% accuracy was achieved.

Lawrence [3] surveyed the application of neural networks to financial systems. He demonstrated how neural networks can be used to test the efficient-market hypothesis and how they outperform statistical and regression techniques in forecasting share prices.

Akita et al. [4] have proposed the application of deep learning models, Paragraph Vector and long short-term memory (LSTM) to financial time series forecasting. In the paper, they proposed an approach that converts newspaper articles into their distributed representations via Paragraph Vector and models the temporal effects of past events on opening prices about multiple companies with LSTM. In their approach, a recurrent network captured the changes of time series influence on stock price and the effectiveness of their approach can be evaluated from the experiments they conducted on market stimulation and the results showed that distributed representations of textual information are better than the numerical-data-only methods and bag-of-words-based methods.

3 Methods

The proposed system is a recurrent neural network which will be used to predict trend line and stock prices for the next day, from previous day data. The proposed system will have two modules, and the results of both modules will be compared to generate a final value. Keras library [5] is used to create the neural network, and TensorFlow library [6] is utilized for backend calculations.

All the values are normalized in the range of 0–1, before using as an input for the neural network. This is done so that high-value inputs do not dominate the final output. Also, inputs with high dimension will slow down the training process (Table 1).

3.1 Value Prediction Using Historical Data

Figure 1 is a representation of a simplified version of our proposed system. We are taking opening price, closing price, highest price, lowest price and volume of shares traded of a specific stock on a particular day to predict the next day's stock prices.

The output will be received in the form of the next day's open, close, high and low prices of a particular stock.

LSTM layers are being used with timesteps of 120 days. Four hidden layers are being used. Each layer contains 75 nodes where edges are initialized with random weights. The output layer contains four nodes which give us our four predicted values.

'Adam' optimizer is used for minimization of the loss function. It is better than the rest of the adaptive technique, and it rectifies every problem that is faced in other

Table 1 Terminologies used

Terms	Definitions
Min-max normalization	It is feature scaling method used to rescale the features of data in the range [0,1] $$x' = \frac{x - \min(x)}{\max(x) - \min(x)}$$
Long-term short memory	It is a neural network architecture which processes a sequence of data
Adam optimizer function	Its function is to update the weights of the neural network during training [7]
Mean squared error	It is a cost function used to reduce the difference between predicted and actual value. This function guides the optimizer function while training $$\text{MSE} = \frac{1}{n} \sum_{i=1}^{n} \left(Y_i - Y_i' \right)^2$$
Dropout rate	Rate at which nodes are dropped in each epoch to avoid overfitting

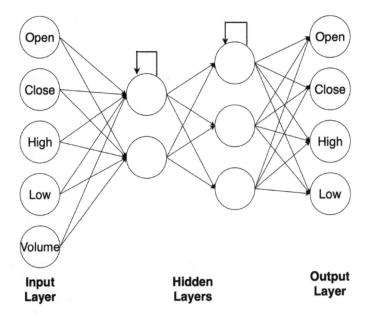

Fig. 1 Proposed system diagram

optimization techniques such as vanishing learning rate, slow convergence or high variance in the parameter updates which leads to fluctuating loss function [8].

After exploring other loss function, 'mean squared error' is found to be the most appropriate loss function for this system. Dense layer connection will be used with a dropout rate of 20% to avoid overfitting.

3.2 Trend Prediction Using News Headlines

This module will use sentiment analysis on news headlines for the stock whose next day's value is to be predicted. The news headlines are extracted from the web and the data is pre-processed and cleaned. After that, each headline is classified as positive, negative and neutral. Then, the feature extracting is done using N-Gram and finally the data is trained using random forest classifier as shown in Fig. 2.

The input for this module is the news headlines of the stock and labels to predict if the value of the stock will increase or decrease on the next day. The label for the current day is calculated by subtracting the open value of current day from the close value of the previous day, if the resulting value is negative then the label will be '1' and if the resulting value is positive or zero then the label will be '0', as shown in Eq. (1). '1' means the value of the stock has increased, and '0' means the value of the stock has decreased or is the same.

Fig. 2 Flow chart for
classification of news
headlines

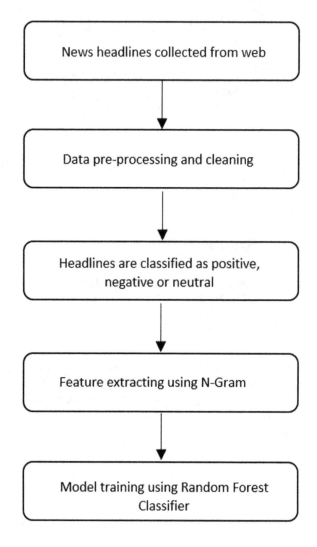

$$L_c = \begin{cases} 1 \text{ when } C_p - O_c < 0 \\ 0 \text{ when } C_p - O_c \geq 0 \end{cases} \tag{1}$$

where

L_c is label of current headline
C_p is previous day close value
O_c is current day open value

The headlines are processed using N-Gram to make a bag of words. N-Gram in simple words means a sequence of N words from the sentence. Then, using the random forest classifier model, headlines of the news are trained. A random forest is a meta-estimator that fits a number of decision tree classifiers on various sub-samples of the data set and averaging to improve the predictive accuracy and control overfitting [9]. This way the classifier can predict the labels of the news headlines.

4 Results

4.1 Value Prediction Using Historical Data

The RNN was trained on data set comprising of State Bank of India (SBIN) stock prices (open, high, low, close and volume) for a period of 10 years from 2 January 2009 to 9 July 2019.

Figure 3 Shows predicted prices for 41 days, from 9 July 2019 to 11 September 2019, where the last plotted point shows the predicted value for 11 September 2019. These graphs are plotted using matplotlib [10].

As shown in Fig. 3, all the predicted value lines follow the trend of actual values with accuracy of 97%.

Figure 3a shows the comparison of the actual open price for SBIN stock with the predicted price. Figure 3b shows the comparison of the actual close price for SBIN stock with the predicted price. Figure 3c shows the comparison of the actual high price for SBIN stock with the predicted price. Figure 3d shows the comparison of the actual low price for SBIN stock with the predicted price.

Figure 4 shows the comparison of the actual value of SBIN stock to predicted values from 11 July 2019 to 16 August 2019. Average accuracy of prediction is 97% using this system.

4.2 Trend Prediction Using News Headlines

The random forest classifier was trained on State Bank of India (SBIN) news headlines with the corresponding labels from 6 January 2009 to 21 November 2016. Figure 5 shows the sample of training data set.

The model classifies each word according to the respective label so when the headlines are provided to this model, it can predict the labels for the headlines accurately. Comparing the predicted label values with the actual label values of the test data set, we received an accuracy of 67.39%.

The precision we got for label '0' is 0.5 while the precision for label '1' was 0.67. Figure 6 shows the analysis of the sentimental analysis on the news headlines of the SBIN.

5 Conclusion and Future Work

The proposed system is accurately following the pattern of the stock prices where the predicted values are in close vicinity to the actual market values. We have used the data set of SBIN stock and achieved an accuracy of 97%, but the sudden fluctuations in the price of many stocks which are derived from abnormal events cannot be factored in the prediction system which is trained on past values of the respective stock prices. For such anomalies, sentimental analysis of the event is required. We have implemented the system by analysing the news which is available for any specific

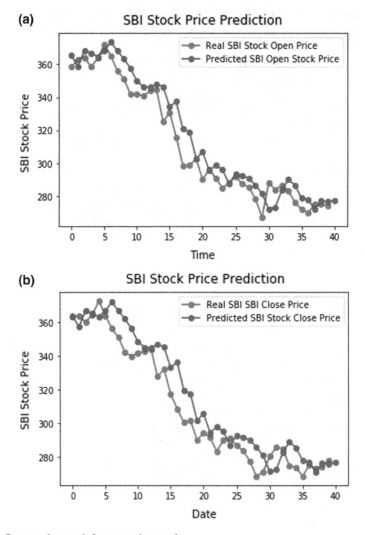

Fig. 3 Output value graph from neural network

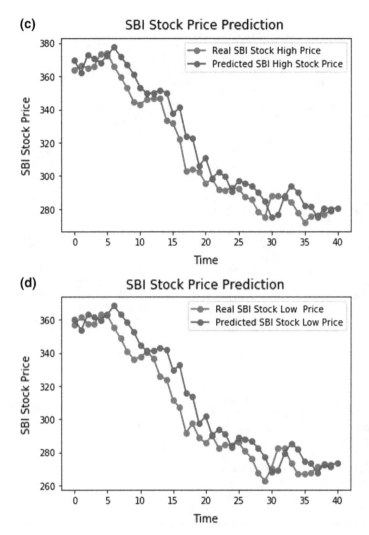

Fig. 3 (continued)

stock to reduce the deviation of the predicted prices from the actual prices. After adjusting the system and factoring in the sentimental analysis, we have achieved an accuracy of 67.39%. RNN is robust in modelling complex relationships between parameters but it is not good enough to find its application in live trading system.

For future work, adding more dimensionality to the data set can further improve the accuracy of prediction. Global trends and events can be used to further improve the prediction of the stock's direction for a longer timeline.

Date	Actual Values				Predicted Values			
	Open	High	Low	Close	Open	High	Low	Close
11/07/19	358.5	364	357	363.2	365.49	369.44	360.26	363.77
12/07/19	363	366.55	361.5	363.6	358.64	362.25	353.53	357.18
15/07/19	364.1	365	357.6	360.05	368.50	372.64	363.33	366.74
16/07/19	358.5	366	357.65	364.35	366.72	370.52	361.71	365.19
17/07/19	364.05	373.55	363.05	372.4	364.62	368.29	359.66	363.15
18/07/19	371.95	373.8	362.55	363.65	368.32	372.18	363.33	366.72
19/07/19	365.1	366.05	355.15	356	373.52	377.53	368.52	371.83
22/07/19	356	359.55	348.7	350.85	368.12	371.80	363.16	366.61
23/07/19	351.2	353	341	342.2	363.44	367.07	358.41	361.94
24/07/19	341.85	344.7	336.25	339.6	357.75	361.31	352.69	356.30
25/07/19	341.7	343.2	337.4	341.3	349.72	353.19	344.62	348.34
26/07/19	340.8	346.15	340.5	342.6	346.28	349.85	341.18	344.90
29/07/19	343.95	346.55	336.8	343.8	346.16	349.82	341.11	344.78
30/07/19	344.4	346.7	325.85	327.55	347.95	351.67	343.02	346.59
31/07/19	325.1	333.2	323.9	332.2	346.45	350.04	341.63	345.18
01/08/19	330.8	331.5	311.35	317.15	334.14	337.36	329.39	333.05
02/08/19	315.55	322.25	307.05	308.45	337.45	341.11	332.58	336.13
05/08/19	298.45	302.85	291.7	300.25	320.66	323.86	315.80	319.58
06/08/19	298.8	304.25	297.25	301.4	318.86	322.61	313.63	317.48
07/08/19	302.4	302.45	288.8	289.9	302.73	306.19	297.42	301.54
08/08/19	290	295.5	285.6	294.35	306.89	310.85	301.51	305.55
09/08/19	296.3	298	290.05	291.35	295.37	298.81	290.28	294.36
13/08/19	290.9	291.55	282.5	283.35	298.69	302.41	293.64	297.57
14/08/19	285.05	291.25	284.5	289.75	296.01	299.52	291.20	295.04
16/08/19	287.95	292.8	284.3	290.9	287.58	290.78	282.96	286.74

Fig. 4 Comparison table of actual prices and predicted prices of State Bank of India

Dates	Headlines	Label
06-01-2009	SBI Q3 PAT seen up at Rs 2364.2 cr: Religare	1
06-01-2009	RBI announces Indo-Nepal Remittance Scheme	1
08-01-2009	Simplex projects wins Rs 2000cr Housing Project in Libya	0
08-01-2009	SBI Q3 PAT seen up at Rs 2429.2 cr: M Oswal	0
08-01-2009	SBI Q3 PAT seen up at Rs 3415.2 cr: P Lilladher	0
08-01-2009	SBI Q3 net profit (stand) seen at Rs 2794 cr: Sharekhan	0
12-01-2009	SBI Q3 PAT seen up at Rs 2041.9 cr: IIFL	0
13-01-2009	DLF repays loans of about Rs 1,000 cr	1
19-01-2009	Buy SBI, target of Rs 1550: Karvy	0
19-01-2009	SBI hopes to cover one lakh villages next fiscal	0
22-01-2009	SBI Capital Mkts receives 'Bank of the Year' award	0
27-01-2009	Buy SBI, target of Rs 1605: Motilal Oswal	1
27-01-2009	Buy SBI at Rs 1,074: Sharekhan	1
27-01-2009	SBI wins 2 IBA Award	1
28-01-2009	Buy SBI, target of Rs 1447: Prabhudas Lilladher	1
28-01-2009	IL&FS arm opening office in London	1

Fig. 5 Sample data set to train the classifier

	precision	recall	f1-score	support
0	0.50	0.02	0.04	232
1	0.67	0.99	0.80	474
micro avg	0.67	0.67	0.67	706
macro avg	0.59	0.51	0.42	706
weighted avg	0.62	0.67	0.55	706

Fig. 6 Result summary

References

1. Kar, A.: Stock Prediction using Artificial Neural Networks. Department of Computer Science and Engineering, IIT Kanpur
2. Barapatre, O., Tete, E., Sahu, C.L., Kumar, D., Kshatriya, H.: Stock price prediction using artificial neural network. Int. J. Adv. Res. Ideas Innov. Technol. **4**(2), 916–922 (2018)
3. Lawrence, R.: Using Neural Networks to Forecast Stock Market Prices. Department of Computer Science, University of Manitoba (1997)
4. Akita, R., Yoshihara, A., Matsubara, T., Uehara, K.: Deep learning for stock prediction using numerical and textual information. In: IEEE/ACIS 15th International Conference on Computer and Information Science (ICIS), vol. 1, pp. 1–6 (2016)
5. Fran, C.: Keras, http://keras.io (2015)
6. Abadi, M., Agarwal, A., Barham, P., Brevdo, E., Chen, Z., Citro, C., Carrado, G. S., Da-vis, A., Dean, J., Devin, M., Ghemawat, S., Goodfellow, I., Harp, A., Irving, G., Isard, M., Jozefowicz, R., Jia, Y., Kaiser, L., Kudlur, M., Levenberg, J., Mané, D., Schuster, M., Monga, R., Moore, S., Murray, D., Olah, C., Shlens, J., Steiner, B., Sutskever, I., Talwar, K., Tucker, P., Vanhoucke, V., Vasudevan, V., Viégas, F., Vinyals, O., Warden, P., Wat-tenberg, M., Wicke, M., Yu, Y., Zheng, X.: TensorFlow: Large-scale Machine Learning on Heterogeneous Systems. http://tensorflow.org (2015)
7. Kingma, D., Ba, J.: Adam: A method for stochastic optimization. In: 3rd International Conference for Learning Representations, San Diego (2015)
8. Wallia, A.S.: Types of Optimization Algorithms used in Neural Networks and Ways to Optimize Gradient Descent. https://towardsdatascience.com
9. Pedregosa, F., Varoquaux, G., Gramfort, A., Michel, V., Thirion, B., Grisel, O., Blondel, M., Prettenhofer, P., Weiss, R., Dubourg, V., Vanderplas, J., Passos, A., Cournapeau, D., Brucher, M., Perrot, M., Duchesnay, É.: Scikit-learn: Machine learning in python. J. Mach. Learn. Res. **12**, 2825–2830 (2011)
10. Hunter, J.D.: Matplotlib: A 2D graphics environment. Comput. Sci. Eng. **9**(3), 90–95 (2007)

Prediction and Management of Diabetes

Nureni Ayofe Azeez, Michael Aja Okwe, Jonathan Oluranti, Sanjay Misra and Ravin Ahuja

Abstract Terminal illness like diabetes has become the major health issues common to every individual of all ages, groups and race in the world. The fast-growing rate of urbanization, increased social development, reaped financial growth leading to enormous wealth in the world today, alterations, modifications introduction of fatty foods, laziness and the revolution of sugary drinks in diets have brought about greater chances of developing certain diseases like diabetes. Diagnosis of this illness was basically based on the blood sugar level; but from this paper, we can see that it is also possible to perform some level of diagnosis from other symptoms such as body mass index and blood pressure. This is possible with the use of modern programming abilities. Fuzzy logic was used in the process of diagnosis to determine the possibility of a patient (user of the application) being diabetic. The application was also designed to recommend healthy dieting and lifestyle in accordance with the person correlation coefficient and the information provided by the user. The result obtained revealed that the solution is effective and efficient based on the metrics used for evaluation.

Keywords Real time · Diabetes · Management · Diagnosis · Solution

N. A. Azeez · M. A. Okwe
University of Lagos, Lagos, Nigeria
e-mail: nurayhn1@gmail.com

M. A. Okwe
e-mail: michaelokwe@yahoo.com

J. Oluranti · S. Misra (✉)
Covenant University, Ota, Nigeria
e-mail: sanjay.misra@covenantuniversity.edu.ng

J. Oluranti
e-mail: Jonathan.oluranti@covenantuniversity.edu.ng

R. Ahuja
Vishvakarma Skill University, Gurgaon, Haryana, India
e-mail: ravinahujadce@gmail.com

© Springer Nature Singapore Pte Ltd. 2020
P. K. Singh et al. (eds.), *Proceedings of First International Conference on Computing, Communications, and Cyber-Security (IC4S 2019)*, Lecture Notes in Networks and Systems 121, https://doi.org/10.1007/978-981-15-3369-3_40

1   Introduction

Effective management of people's health requires timely recognition and treatment, and hence, the need for a real-time application where the user can communicate with a doctor and get solution on time [1–3]. This paper aims at addressing and providing solution to the management of diabetes patients through an automated, real-time diabetes case-filing application. Today, a larger population of the people in the world suffers from poor health condition because of inappropriate diets. Poor dietary lifestyle has been a major contributor to the development of chronic diseases such as obesity, diabetes, and cardiovascular diseases [4, 5].

Findings in the field of medicine have shown that healthy eating/food strengthens the immune system hence preventing the person from all forms of illness and presenting people a greater chance of countering free radicals and warding off diseases [6–8]. It is important to know that healthy diets are dietary taken to develop the body cell and tissues for effective function [7].

With the adoption of information technology in modern medicine, medical procedures have been supported by technology advancement [9–11] to minimize the rapid growth of chronic diseases like diabetes [12–19]. This has helped the majority of the world population to gain access to medical intervention on time through the Web. Most people consume foods without considering their health state mostly because they were not properly guided due to the unavailability of medical experts [11].

The paper is aimed at providing relative information on all issues about diabetes by building a real-time solution that will leverage on the technological revolution in the medical field to deliver medical services aimed at the management and prevention of diabetes to all those that have access to the Internet. This paper presents a real-time diagnosis Web application for the management of diabetes.

2   Background and Related Work

Diabetes, often referred to as diabetes mellitus, describes a group of metabolic diseases in which the person has high blood glucose (blood sugar), either because insulin production is inadequate, or because the body's cells do not respond properly to insulin, or both [20, 21]. People with high blood glucose level always experience frequent urination, become increasingly thirsty and always hungry [21].

Diabetes, which has been recorded as one of the major causes of death in the world today, has been a major issue with the world health community. According to the World Health Organization (WHO), it is responsible for the death of about 1.6 million people above 18 years in 2016 and about 2.2 million deaths were attributable to high blood glucose in 2012 [22, 23]. They also stated that almost half of all deaths attributable to high blood glucose occur before the age of 70 years and that diabetes was the seventh leading cause of death in 2016 [22].

2.1 Types of Diabetes

Basically, there are four types of diabetes, namely Type 1 diabetes, Type 2 diabetes, gestational diabetes, and maturity-onset diabetes for the young.

2.1.1 Type 1 Diabetes

This type of diabetes amounts to approximately 10% of the total population of people living with diabetes. Basically, the body of someone with type 1 diabetes does not produce insulin. It is commonly referred to as insulin-dependent, juvenile or early-onset diabetes. Studies have shown that people usually develop type 1 diabetes before their 40th year, often in early adulthood/teenage years. Patients with type 1 diabetes are to receive insulin injections as long as they live and ensure that their blood glucose level is constantly monitored by carrying out regular blood tests and following a special diet [21].

2.1.2 Type 2 Diabetes

This type of diabetes amounts to approximately 90% of the total population of people living with diabetes. The body of the carrier of this form of diabetes does not produce sufficient insulin needed for proper functioning of the body or the body cells resist insulin either produced by the body or injected into the body. This form of diabetes is found amongst overweight and obese people [21].

2.1.3 Gestational Diabetes

This form of diabetes is found in women during pregnancy. The carrier of this form of diabetes has high level of glucose in their blood which limits the ability of their bodies to produce the needed amount of insulin which will transport all the glucose into their cells. This can be controlled by eating healthy food, exercise and some blood glucose-controlling medications [21].

2.1.4 Maturity-Onset Diabetes of the Young

This form of diabetes is very rare compared to the other three forms. It is an autosomal dominant inherited form of diabetes which is generally due to one of many single-gene mutations resulting in defects in insulin production by the carrier's body. This form of diabetes varies in age at presentation and in severity according to the specific gene defect; thus, we have about 13 subtypes of MODY. People leaving with this form of diabetes often can control it without using insulin.

In this paper, we seek to understand the major issues around diabetes in other to develop a real-time solution for prediction and management of diabetes.

3   Issues Considered in Automation Process

3.1   *Fuzzy-Based Diagnosis*

As part of the automation processes, the application of decision-making was built around fuzzy logic to determine the health status of the patient. With the combination of the user interface of the application and the knowledge base in the database, the diagnosis of the patients and the determination of the nature of the nutrition and lifestyle of the individual were possible [24]. This can be achieved by the following approaches:

information about the patient is entered through the frontend and stored in the database for processing.

The processed and some other raw data are then transferred to the knowledge base processor together with "If-Then" rules to enable the fuzzy operation to process them.

Fuzzy logic then processes the information to determine the health status of the patient. This process diagnoses and determines the classification of individual status.

To process ambiguous and vague information by the patient, the process of fuzzification and defuzzification was used. The defuzzification process converts the computations from the fuzzy process to crisp values using a centroid of gravity technique [25, 26]. Given as an aggregated membership function with Xi as centre of the membership function, the output value is determined with Eq. 1.

$$\text{CoG}(Y') = \left(\frac{\sum_{i=1}^{n} \mu Y(Xi)Xi}{\sum_{i=1}^{n} \mu Y(Xi)} \right) \tag{1}$$

Food Recommendation

The choice of food we eat/drink plays a very important role in our life. It helps in balancing our weight, boosting our immune system and ensuring healthy growth before we can carry out our daily activities. Over weighting has been established to be one of the major contributors/symptoms of diabetes. Overweighed or obese people have the highest risk of being diabetic. To find out if the patient is overweighed or obese, we shall calculate the body mass index using the equation below [27].

$$BMI = w(kg)/h^2(m) \tag{2}$$

where

w body weight in kilograms
h height in centimetres

SI unit is kg/m^2.

Table 1 Health risk associated with patient's BMI

Health risk	BMI (kg/m^2)
Risk of developing problems such as nutritional deficiency and osteoporosis	under 18.5
Low risk (healthy range)	18.5–23.0
Moderate risk of developing heart disease, high blood pressure, stroke, diabetes	23.1–27.5
High risk of developing heart disease, high blood pressure, stroke, diabetes	over 27.6

The above equation will help in the classification of the patient as underweight, normal overweight and obese which is demonstrated in the following notations.

$$\text{Cat} = \begin{cases} \text{BMI} < 18.5 \rightarrow \text{Underweight} \\ 18.5 \leq \text{BMI} \leq 24.9 \rightarrow \text{Normal} \\ 25.0 \leq \text{BMI} \leq 29.9 \rightarrow \text{Overweight} \\ \text{BMI} \geq 30.0 \rightarrow \text{Obese} \end{cases} \quad (3)$$

Table 1 shows the health risk as relates to the BMI calculation.

The main goal for the calculation of BMI is to help determine the risk of the patient and to proffer the best diet and other lifestyle for the patient. To recommend the best diet, we have to determine an approximated actual body weight for each patient with respect to their heights. To achieve that, we will use the "*Broca Index/formula*" as stated below which will profile the ideal body weight.

$$\text{Normal/ideal body weight (kg)} = (\text{Body Height (cm)} - 100) \quad (4)$$

This formula differs across gender

Men: Ideal Body Weight (kg) = [Height (cm) − 100] − ([Height (cm) − 100] × 10%)

Women: Ideal Body Weight (kg) = [Height (cm) − 100] + ([Height (cm) − 100] × 15%) [28].

After we have obtained an approximate ideal body weight, we can now determine the active level of the patient. This will help know the total amount of calories or energy the individual needs to carry out his/her daily activities. This active level is classified as sedentary, lightly active, active and very active [29, 30].

With the knowledge of the active level, diet for a diabetes patient can be recommended. In the case of diabetes, the basic kilocalorie sources are carbohydrate, protein and fat classified as macro-food nutrients. The recommendation of the intake of each of these macro-foods is based on the level of diabetes as study was done by the American Diabetes Association shows that low intake of carbohydrate keeps the blood glucose level normal [20]. Relationships between diabetes level and percentages of macro-nutrients in diabetic meals are given in Table 2.

These food items are measured in grams, and their calories are calculated to get the exact number of calories needed. Below is the equation for the conversion.

Table 2 Diabetes' level and macro-food nutrient percentage [20]

Diet	Carbohydrate (%)	Protein (%)	Fat (%)
Normal	60	20	20
Mild diabetes	56	23	21
Severe diabetes	50	25	24
Very severe diabetes	45	30	25

$$1 \text{ g of carbohydrate} = 4 \text{ kcal of carbohydrate}$$
$$1 \text{ g of protein} = 4 \text{ kcal of protein}$$
$$1 \text{ g of fat} = 9 \text{ kcal of fat} \tag{5}$$

With this, we proposed a diet with the ratio of breakfast, lunch and dinner to be 3:4:3. Other than this macro-food, trace and other food nutrients are under considerations.

3.2 Diet Personalization and Recommendation

In recommending diet for patients, the patient's ideal body weight will be calculated using Eq. 4. The solution from that will be used to determine the total energy required daily (ERD) in kilocalories by the patient. Together with the macro-food nutrients and the level of severity of the diagnosis of the diabetes, we can determine the quantity of carbohydrate, protein and fat in each of the patient's meal and calculate the proportion of food to be eaten as breakfast, lunch and dinner. Using a proposed diet ratio of breakfast, lunch and dinner to be 3:4:3, the total of the ratio is 10.

To determine the proportion of breakfast, lunch and dinner, Eq. 6 can be used.

$$\text{Meal} = (\text{Ratio of Meal}/\text{Total Ratio}) * \text{Total Kilocalories} \tag{6}$$

Having gotten the total amount of meal (breakfast, lunch and dinner), we can now calculate the amount of the three major food nutrients: carbohydrate, protein and fat for each of the meal by the application of the total proportion of each of the meal and the percentage of the suitable diabetic severity. This can be achieved using expression 7.

$$M = \begin{cases} Ca = \left(\frac{x1}{100}\right) * KclM \\ + \\ Pr = \left(\frac{x2}{100}\right) * KclM \\ + \\ Fa = \left(\frac{x3}{100}\right) * KclM \end{cases} \tag{7}$$

where

M	Meal (breakfast, lunch and dinner)
*X*1	Total percentage of carbohydrate macro-nutrient of the diabetic level of the individual
*X*2	Total percentage of protein macro-nutrient of the diabetic level of the individual
*X*3	Total percentage of fat macro-nutrient of the diabetic level of the individual
Ca	Total amount of carbohydrate in the meal
Pr	Total amount of protein in the meal
Fa	Total amount of fat in the meal
KlcM	Ratio of kilocalories of each of the macro-nutrient (carbohydrate, protein and fat) based on the total kilocalories per day.

This can be deduced from Eq. 6. The application of 7 in this work has led to the efficient recommendation and personalization of diet for patients that used the application.

3.3 Index Factors Used in Diagnosing Diabetes

To automate the process of diabetes diagnosis, some information about the patient were collected. The needed information is:

The person's age
Family health history
History of the person blood pressure
Sex of the person
Fitness
BMI and feeding/nutrition.

Each of them is assigned weight which was used in the calculation of the risk percentage of being diabetic. It is also right to say that being diabetic is a combination of many factors. In the implementation, we use the above indexes to determine the person risk of being diabetic.

4 Experimental Result and Evaluation

System Deployment—this application was developed with the concept of access control mechanism as there are different levels of users (Fig. 1).

Figure 2 provides information on the metrics considered for evaluation. The metrics are age, family history, blood pressure, waist size, sex, fitness, BMI and nutrition.

Fig. 1 Application home page

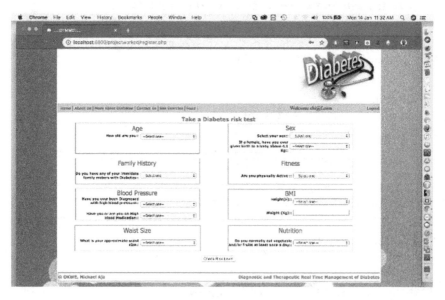

Fig. 2 Diabetes diagnosis page

It is important to evaluate the performance of the application. The application was evaluated against the standard of a secured health Web application. It was also checked to ensure that information is kept safe. The logic of the application reflects the true nature of the input which can be accessed via a network. Results from the application were measured against the risk standard of a diabetic person.

As part of diabetes diagnostic measures, users of the system were asked to provide data against each of the index listed. Each answer provided was weighted, and the values were used in determining the risk of being diabetic. The output is hereby presented according to the age category which is younger than 40, 41–50, 51–60 and 60 and above.

Table 3 indicates risk possibility of the users of the system who are below 40 years of being diabetic. Users entered some details and the system calculated the BMI, Broca Index, level of severity and others.

Figure 3 (graph) is the representation of the sum of the percentage of those being diabetic against sex. It shows that male younger than 40 have higher tendency of being diabetic than female.

Table 4 indicates risk possibility of the users of the system who are diabetic

Table 3 Information of users younger than 40

Id	BMI	Broca index	Calories per day	Level of severity	% of risk	Age	Sex
P1	21.50	58.50	1257.43	Normal	38.46	34	Male
P5	17.24	61.54	1061.25	Normal	30.77	34	Female
P6	20.69	61.54	1273.5	Mild normal	50.00	23	Female
P7	23.91	34.11	815.531	Normal	34.62	34	Female
P14	20.03	82.88	1660.32	Mild normal	50.00	34	Male
P16	55.04	34.11	1877.62	Mild normal	53.85	29	Male
P19	21.50	58.50	1257.43	Normal	38.46	34	Male
P23	17.24	61.54	1061.25	Normal	30.77	34	Female
P24	20.69	61.54	1273.5	Mild normal	50.00	23	Female
P25	23.91	34.11	815.531	Normal	34.62	34	Female
P32	20.03	82.88	1660.32	Mild normal	50.00	34	Male
P34	55.04	34.11	1877.62	Mild normal	53.85	29	Male
P37	21.50	58.50	1257.43	Normal	38.46	34	Male
P41	17.24	61.54	1061.25	Normal	30.77	34	Female
P42	20.69	61.54	1273.5	Mild normal	50.00	23	Female
P43	23.91	34.11	815.531	Normal	34.62	34	Female
P50	20.03	82.88	1660.32	Mild normal	50.00	34	Male
P52	55.04	34.11	1877.62	Mild normal	53.85	29	Male
P58	25.24	55.45	1399.73	Normal	33.33	33	Male

Fig. 3 Graph of sex against percentage of been diabetic of users below 40 years

between 41 and 50 years. The results were obtained after users entered some details, and the system calculated the BMI, Broca Index, level of severity and others.

Figure 4 is the graphical representation of the sum of the percentage of those diabetic against sex. It shows that male between 41 and 50 have higher tendency of being diabetic than female.

Table 5 indicates the risk possibility of the users of the system who are between 51 and 60 years being diabetic. Users entered some details and the system calculated the BMI, Broca Index, level of severity and others.

Figure 5 is the graphical representation of the sum of the percentage of users being diabetic against sex. It shows that female between 51 and 60 have higher tendency of being diabetic than male.

Table 6 indicates the risk possibility of the users of the system who are above 60 years being diabetic. Users entered some details and the system calculated the BMI, Broca Index, level of severity and others.

Figure 6 is the graphical representation of the sum of the percentage of users being

Table 4 Information of users between 41 and 50

Id	BMI	Broca index	Calories per day	Level of severity	% of risk	Age	Sex
P2	30.65	58.50	1793	Mild normal	50.00	46	Male
P4	21.12	76.78	1621.54	Normal	38.46	47	Male
P20	30.65	58.50	1793	Mild normal	50.00	46	Male
P22	21.12	76.78	1621.54	Normal	38.46	47	Male
P38	30.65	58.50	1793	Mild normal	50.00	46	Male
P40	21.12	76.78	1621.54	Normal	38.46	47	Male

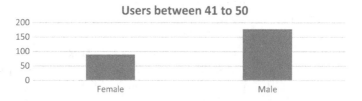

Fig. 4 Graph of sex against percentage of been diabetic of users between 41 and 50 years

Table 5 Information of users between 51 and 60

Id	BMI	Broca index	Calories per day	Level of severity	% of risk	Age	Sex
P3	47.31	37.16	1757.97	Very severe diabetic	80.77	54	Female
P9	18.22	73.74	1343.58	Mild normal	57.69	56	Male
P12	43.37	34.11	1479.33	Sever diabetic	73.08	57	Female
P15	37.25	55.45	2065.18	Very severe diabetic	80.77	53	Female
P17	37.55	61.54	2311.16	Mild normal	57.69	59	Female
P18	20.80	76.78	1596.98	Mild normal	46.15	55	Male
P21	47.31	37.16	1757.97	Very severe diabetic	80.77	54	Female
P27	18.22	73.74	1343.58	Mild normal	57.69	56	Male
P30	43.37	34.11	1479.33	Sever diabetic	73.08	57	Female
P33	37.25	55.45	2065.18	Very severe diabetic	80.77	53	Female
P35	37.55	61.54	2311.16	Mild normal	57.69	59	Female
P36	20.80	76.78	1596.98	Mild normal	46.15	55	Male
P39	47.31	37.16	1757.97	Very severe diabetic	80.77	54	Female
P45	18.22	73.74	1343.58	Mild normal	57.69	56	Male
P48	43.37	34.11	1479.33	Sever diabetic	73.08	57	Female
P51	37.25	55.45	2065.18	Very severe diabetic	80.77	53	Female
P53	37.55	61.54	2311.16	Mild normal	57.69	59	Female
P54	20.80	76.78	1596.98	Mild normal	46.15	55	Male
P59	18.22	73.74	1343.58	Mild normal	29.49	54	Female

Fig. 5 Graph of sex against percentage of been diabetic of users between 51 and 60 years

diabetic against sex. It shows that female above 60 have higher tendency of being diabetic than male.

The findings from those tables and figures (graphs) show that older people who are less active with heavyweight have higher risk of being diabetic. Also, we can

Table 6 Information of users above 60

Id	BMI	Broca index	Calories per day	Level of severity	% of risk	Age	Sex
P8	25.24	55.45	1399.73	Mild normal	50.00	67	Female
P10	17.24	61.54	1061.25	Normal	42.31	87	Male
P11	35.83	58.50	2095.71	Mild normal	61.54	90	Female
P13	17.24	61.54	1061.25	Mild normal	57.69	78	Female
P26	25.24	55.45	1399.73	Mild normal	50.00	67	Female
P28	17.24	61.54	1061.25	Normal	42.31	87	Male
P29	35.83	58.50	2095.71	Mild normal	61.54	90	Female
P31	17.24	61.54	1061.25	Mild normal	57.69	78	Female
P44	25.24	55.45	1399.73	Mild normal	50.00	67	Female
P46	17.24	61.54	1061.25	Normal	42.31	87	Male
P47	35.83	58.50	2095.71	Mild normal	61.54	90	Female
P49	17.24	61.54	1061.25	Mild normal	57.69	78	Female
P55	17.24	61.54	1061.25	Mild normal	44.87	74	Male
P56	20.69	61.54	1273.5	Very severe diabetic	75.03	87	Female
P57	23.91	34.11	815.531	Mild normal	37.18	93	Male

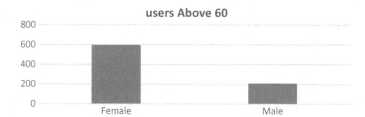

Fig. 6 Graph of sex against percentage of been diabetic of users above 60 years

deduce that younger male who not watch their diet and less active have higher risk of being diabetic than the female; but as they get older, the female risk becomes higher because of their less activity as well as the addition of weight and diet during pregnancy.

Figure 7 shows that in respect of the sex, the combination of BMI and age which increase the risk of been diabetic.

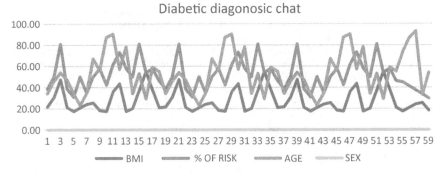

Fig. 7 Graph of 59 users

5 Conclusion and Recommendation

Diabetes mellitus is a serious health condition that can lead to untimely death. It is hard and expensive to maintain. It is one of the leading killer diseases in the world today. This application will go a long way to help in the reduction of this health issue and help the users to watch their blood sugar, monitor themselves through the risk assessment page and monitor their diets. The application also addresses the issue of access to e-medical personnel as all information inputted by the user can be accessed. This paper has shown that with the aid of a fuzzy logic, health conditions can be monitored and solutions can be proffered on time. Some of the limitations and constrains are access to large data for diabetic patients as most of the users were not ready to provide real information. Some users could not use the computer and some were not willing to provide information about their health status.

Acknowledgements The authors of this research appreciate the immense contribution of Covenant University Centre for Research, Innovation, and Discovery (CUCRID) for its support for this research.

References

1. Evert, A.B., Boucher, J.L., Cypress, M., Dunbar, S.A., Franz, M.J., Mayer-Davis, E.J., Yancy, W.S.: Nutrition therapy recommendations for the management of adults with diabetes. Diab. Care **37**(Supplement_1), S120–S143 (2013). https://doi.org/10.2337/dc14-s120
2. Shah, V.N., Garg, S.K.: Managing diabetes in the digital age. Clin. Diab. Endocrinol. **1**(1), 16 (2015). https://doi.org/10.1186/s40842-015-0016-2
3. Falvo, D., Holland, B.E.: Medical and psychosocial aspects of chronic illness and disability. Jones & Bartlett Learning, Burlington (2017)
4. Sharma, M., Majumdar, P.: Occupational lifestyle diseases: an emerging issue. Indian J. Occup. Environ. Med. **13**(3), 109–112 (2009)

5. Ley, S.H., Hamdy, O., Mohan, V., Hu, F.B.: Prevention and management of type 2 diabetes: dietary components and nutritional strategies. The Lancet **383**(9933), 1999–2007 (2014). https://doi.org/10.1016/s0140-6736(14)60613-9
6. Cattanach, N., Sheedy, R., Gill, S., Hughes, A.: Physical activity levels and patients' expectations of physical activity during acute general medical admission. Intern. Med. J. **44**(5), 501–504 (2014). https://doi.org/10.1111/imj.12411
7. Greenwood, D.A., Gee, P.M., Fatkin, K.J., Peeples, M.: A systematic review of reviews evaluating technology-enabled diabetes self-management education and support. J. Diab. Sci. Technol. **11**(5), 1015–1027 (2017)
8. Razavian, N., Blecker, S., Schmidt, A.M., Smith-McLallen, A., Nigam, S., Sontag, D.: Population-level prediction of type 2 diabetes from claims data and analysis of risk factors. Big Data **3**(4), 277–287 (2015)
9. Azeta, A.A., et al.: Preserving patient records with biometrics identification in e-health systems. In: Shukla, R., Agrawal, J., Sharma, S., Singh, Tomer G. (eds.) Data, Engineering and Applications, pp. 181–191. Springer, Singapore (2019)
10. Alade O.M., Sowunmi O.Y., Misra S., Maskeliūnas R., Damaševičius R.: A neural network based expert system for the diagnosis of diabetes mellitus. In: Antipova T., Rocha Á. (eds.) Information Technology Science. MOSITS 2017. Advances in Intelligent Systems and Computing, vol. 724. Springer, pp. 14–22 (2018)
11. Berndt, R.-D., Takenga, C., Preik, P., Kuehn, S., Berndt, L., Mayer, H., Schiel, R.: Impact of information technology on the therapy of type-1 diabetes: a case study of children and adolescents in Germany. J Personalized Med. **4**(2), 200–217 (2014)
12. Dunn, T.C., Xu, Y., Hayter, G., Ajjan, R.A.: Real-world flash glucose monitoring patterns and associations between self-monitoring frequency and glycaemic measures: a European analysis of over 60 million glucose tests. Diab. Res. Clin. Pract. **137**, 37–46 (2017)
13. Quemerais, M.A., Doron, M., Dutrech, F., Melki, V., Franc, S., Antonakios, M., et al.: Preliminary evaluation of a new semi-closed-loop insulin therapy system over the prandial period in adult patients with type 1 diabetes. J. Diab. Sci. Technol. **8**, 1177–1184 (2014)
14. Dandona, P.: Minimizing glycemic fluctuations in patients with type 2 diabetes: approaches and importance. Diab. Technol. Ther **19**, 498–506 (2017)
15. Shetty, D., Rit, K., Shaikh, S., Patil, N.: Diabetes disease prediction using data mining. In: International Conference on Innovations in Information, Embedded and Communication Systems (ICIIECS), pp. 1–5 (2017)
16. Singh, A., Halgamuge, M.N., Lakshmiganthan, R.: Impact of different data types on classifier performance of random forest, naive Bayes, and K-nearest neighbors algorithms. Int. J. Adv. Comput. Sci. Appl. **8**, 1–10 (2017)
17. Ahmed, T.M.: Using data mining to develop model for classifying diabetic patient control level based on historical medical records. J. Theor. Appl. Inf. Technol. **87**, 1–10 (2016)
18. Singh, D., Leavline, E.J., Baig, B.S.: Diabetes prediction using medical data. J. Comput. Intell. Bioinf. **10**, 1–8 (2017)
19. Wu, H., Yang, S., Huang, Z., He, J., Wang, X.: Type 2 diabetes mellitus prediction model based on data mining. Inf. Med. Unlocked **10**, 100–107 (2018)
20. American Diabetes Association: Diagnosis and classification of diabetes mellitus. Diab. Care **37**(Supplement_1), S81–S90 (2013). https://doi.org/10.2337/dc14-s081
21. Goldenberg, R., Punthakee, Z.: Definition, classification and diagnosis of diabetes, prediabetes and metabolic syndrome. Canadian J. Diab. **37**, S8–S11 (2013)
22. Organization W.H.: World health statistics 2016: monitoring health for the SDGs sustainable development goals. World Health Organization (2016)
23. Fagherazzi, G., Ravaud, P.: Digital diabetes: perspectives for diabetes prevention, management and research. Diab. Metab. (2018)
24. Rahmani Katigari, M., Ayatollahi, H., Malek, M., Kamkar Haghighi, M.: Fuzzy expert system for diagnosing diabetic neuropathy. World J. Diab. **8**(2), 80 (2017)
25. Lukmanto, R.B., Irwansyah, E.: The early detection of diabetes mellitus (DM) using fuzzy hierarchical model. Procedia Comput. Sci. **59**, 312–319 (2015)

26. Abdullah, A.A., Fadil, N.S., Khairunizam, W.: Development of fuzzy expert system for diagnosis of diabetes. In: 2018 International Conference on Computational Approach in Smart Systems Design and Applications (ICASSDA)
27. Abbasi, A., Juszczyk, D., van Jaarsveld, C.H.M., Gulliford, M.C.: Body mass index and incident type 1 and type 2 diabetes in children and young adults: a retrospective cohort study. J. Endocr. Soc. 1(5), 524–537 (2017)
28. Sheikh, M.A., Lund, E., Braaten, T.: The predictive effect of body mass index on type 2 diabetes in the Norwegian women and cancer study. Lipids Health Dis. 13, 164 (2014)
29. Edqvist, J., Rawshani, A., Adiels, M., Björck, L., Lind, M., Svensson, A.-M., Gudbjörnsdottir, S., Sattar, N., Rosengren, A.: BMI and mortality in patients with new-onset type 2 diabetes: a comparison with age- and sex-matched control subjects from the general population. Diab. Care 41(3), 485–493 (2018). https://doi.org/10.2337/dc17-1309
30. Su, Y., Ma, Y., Rao, W., Yang, G., Wang, S., Fu, Y., Kou, C.: Association between body mass index and diabetes in Northeastern China. Asia Pac. J. Pub. Health 28(6), 486–497 (2016)

Opinion Mining Techniques and Its Applications: A Review

Sonia

Abstract Opinion mining, generally referred to sentiment analysis, deals with the sentiments of people in terms of their opinions regarding any specific object, idea, topic or product. These opinions can be positive, negative or neutral. Opinion mining is gaining its importance in almost each and every domain such as product and services, financial services, health care or even politics. The task of opinion mining is divided into series of steps such as dataset acquisition, opinion identification, aspect extraction, classification, report summary and evaluation. Among these steps, aspect extraction is the most important one. This paper presents an overview of opinion mining and discusses aspect extraction methodologies proposed in the literature.

Keywords Opinion mining · Sentiment analysis · Opinion · Opinion holder

1 Introduction

Opinion mining [1, 2] is a synonym of sentiment analysis which is used to mine the data generated by the Web users in the form of reviews and comments. Opinion mining is a subfield of Web content mining that in turns comes under Web mining as shown in Fig. 1. Thus, it deals with Web data. This data majorly includes opinions, reviews posted on social media. The main objective of opinion mining is to mine these reviews and generate valuable information. The opinions on the Web are in the form of text and ratings and also pose some polarity, i.e., positive, negative or neutral. Moreover, this data is unstructured; therefore, opinion mining applies information retrieval [3], natural language processing [4] and text mining [5] techniques. Researchers also improved the results by applying various techniques of machine learning [6–9] and lexicon-based method [10, 11]. Opinion mining basically deals with the sentiment of people regarding any specific idea, topic or product. Opinion mining has significant importance in most of the areas like products, services, finance-related services, healthcare domain as well as in political domain. The next section describes the concept of Opinion Mining and its application areas.

Sonia (✉)
Delhi Technological University, Information and Technology, New Delhi, Delhi, India
e-mail: er.soniyachoudhary@gmail.com

© Springer Nature Singapore Pte Ltd. 2020 549
P. K. Singh et al. (eds.), *Proceedings of First International Conference on Computing, Communications, and Cyber-Security (IC4S 2019)*, Lecture Notes in Networks and Systems 121, https://doi.org/10.1007/978-981-15-3369-3_41

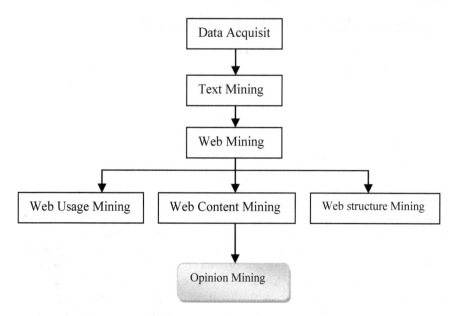

Fig. 1 Position of opinion mining

The organization of this paper is such that first opinion mining is described which includes the levels of opinion mining, process of opinion mining and then applications of opinion mining is elaborated. After that, paper discusses techniques of feature extraction in opinion mining then comes the conclusion.

2 Opinion Mining

Opinion is an idea, evaluation, belief judgment which any individual or a group of individuals hold for specific object item, idea, subject or topic which can be positive, negative or neutral. Opinions are used to take decisions these days by business houses on the basis of review which their customers express. It can be positive, negative or neutral on any object having features or attributes. It can be subjective or objective. An opinion has three parts:

- **Object or opinion target**: Opinion target can be a product, person, event, organization, or topic. It is an entity.
- **Opinion polarity**: Opinion polarity can be positive, negative or neutral.
- **Opinion holder**: Opinion Holder can be a person or an organization who holds the opinion.

Opinion mining. It is a task to mine a dataset D given to get evaluated which holds the opinion regarding any object then opinion mining is used to mine opinion which involves extracting the features of that object which are reviewed in each dataset

$d \in D$ to determine whether the reviews are negative, positive or neutral. Opinions are given on entities which can be anything like product, software, government schemes, topic, idea, etc. These opinions can be objective or it can be subjective.

2.1 Level of Sentiment/Opinion Mining [12]

Sentiment mining has been classified into three levels as in Fig. 2:

(1) *Document level* [13]: It is sentiment analysis which abstracts and focuses on the general information regarding polarity of the document which is not suitable for precise evaluations. At document-level opinion mining, opinions are mostly classified as positive and negative. It is not used where single review is used for sentiment on multiple topics/subjects.

(2) *Sentence level* [14]: This level of sentiment mining is used for sentence level to classify opinion. Here, each sentence is classified as positive, negative or neutral opinion. Opinion mining at sentence level is fine-grained analysis where each sentence of document is analyzed. Subjective classification which is a preprocessing step belongs to sentence-level opinion mining. Subjective classification classifies the sentence into subjective or objective sentences where objective sentence stands for factual information and subjective sentences represent subjective sentiment/opinion.

(3) *Feature or aspect level* [15, 10]: It is a fine-grained analysis of opinion expressed by the author. The document-level and sentence-level analyses are not potential enough to exactly describe the liking of the sentiment holders. It is also known as feature-based or aspect-based opinion mining. Aspect/feature-level analysis deals with opinion rather than going through the constructs of the language such as paragraph, clauses or sentences. User sometimes expresses their reviews on some specific feature or specific aspect of an entity instead of entity. An opinion can have multiple targets here where each target may have different opinions. Aspect/feature-based opinion mining provides structured summary of data from unstructured data as input. It provides summary of opinions about the entity and their features; thus, it can be used for qualitative as well as quantitative analysis.

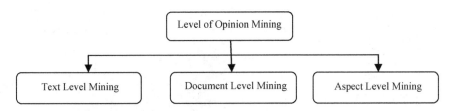

Fig. 2 Sentiment mining level

2.2 Opinion Mining Process

Opinion mining process mainly aim to uncover the opinion or sentiment in the document, its main objective is to discover the writer's attitude for different aspects of a problem. The opinion mining process is as in Fig. 3 where every element does some specific work as given below:

Data collection: Here, the first and foremost aim is to collect reliable dataset for opinion mining. The dataset can be obtained by any Web source; it can be Twitter (microblogs), Facebook (social networking sites), Weblogs, or Web sites (such as Amazon and Tripadvisor) having product reviews. Different techniques can be used to collect data from Web such as using Web scrapping [18], Web crawlers, etc.

Sentiment identification: In this phase, all the relevant comments or reviews are segregated from irrelevant or fake reviews. Here, opinion means the phrases which tend toward the sentiments or feeling about the services, product, person, etc. (desired entity).

Feature extraction: This phase is used to get information which is the most relevant from the dataset which we preprocessed using different techniques which this paper is going to summarize. It minimizes the dimensionality of the dataset by extracting only relevant and potential features which further improves to classify opinion.

Opinion classification: After sentiment identification and feature extraction which is considered as the preprocessing step, in this step, the opinions got classified.

Production summary: In production summary step, a summary of opinion results got produced which can be in any forms like text, charts, etc.

Evaluation: In this step, four parameters are used to evaluate the performance of opinion classification which are as follows:

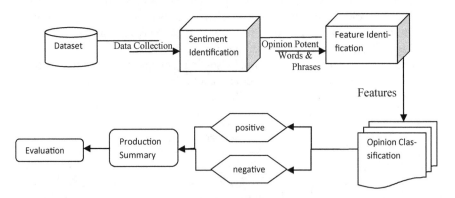

Fig. 3 Sentiment mining process [16, 17]

Precision, Recall, Accuracy and F-score

One of the most opinion mining objectives is to extract the feature of the object. Its significant application area is in product recommendation, product review summary, etc. In the next section, different techniques used for feature extraction in opinion mining are discussed.

2.3 Applications of Opinion Mining

Opinion mining is having vast application areas in different fields:

Commercial product areas [6, 19–21]: Opinion mining application in commercial product area is helpful for customers who want to purchase any product if he/she is having a small summary of the opinions of large number of users, for business organizations who can improve their product or services using reviews of the customers which would be helpful for marketing purpose, designing and manufacturing of the product. Opinion mining is helpful for advertising companies by recommending better product or service according to the reviews of the customers as well as the opinion mining will provide the idea regarding market demands. Important achievements in the commercial product areas are as follows [22]: products comparison, sentiments summarization, etc.

Product or service purchase: Opinion mining technique is helpful in taking decisions while purchasing products or services. Opinion mining is used to evaluate the opinions given by the customers in the form of reviews or comments and is helpful for comparing different competing brands. It provides structured summary to the customer for comparison by mining the unstructured text data.

Sentiment mining in the area of politics [23–27]: Social media users these days also give reviews on political, religious, cultural and social issues which give the insight about the mindset of people regarding any issue. This opinion mining helps politicians or government to form their policies well, and the opinion mining is also helpful in dealing any social, religious or cultural issues. One of the application areas which is in trend these days is political elections where a voter can take benefit of opinion mining in making decision to vote.

Sentiment analysis in area of finance (stock market) [28–30]: Opinion mining is helpful in providing information and proper knowledge to achieve long-term economic growth by allocating resources properly at national level. Opinion mining is helpful in taking better decisions in stock market.

Improving product or service quality: By using the Web sites like Amazon and C|Net [1, 12], RottenTomatoes.com [31] and IMDb [32] for product reviews, online companies can enhance the quality of the entity which can be product or service.

Marketing: Opinion mining-generated evaluation summary can be helpful for marketing [33]. As the summary generated by the opinion mining process is helpful in gaining the insight into the current ongoing trend of the market, it is helpful in gaining the public opinion trend toward the newly launched government plans and policies. All these results are contributing significantly toward intelligent marketing research [33].

Recommender system: Recommender system recommends any product or service by using opinion mining. Opinion of the customer can be gained and classified whether it is positive or negative; accordingly, recommender system recommends product or service [34].

Identification of 'flame': Using opinion mining, it is very easy to monitor social media, blogs, news articles, etc. Web content whether it can be any tweets, blogs, news or mails can easily got detected for any hate comments, harsh words, heated conversations or arrogant words [32] through opinion mining.

Spam detection: With the vast usage of Internet, it is very vulnerable to fake news or spam content anybody can put any spam content on Internet or Web. People usually write spam content which can confuse or misguide the people. Opinion mining and sentiment analysis can easily detect and specify the Web content into 'spam' or 'not a spam' content [2].

Policy making: Opinion mining can be very useful while making policies where the user's opinion can easily be mined regarding new policies to be launched and that information is useful in creating citizen-friendly policies.

3 Techniques of Feature Extraction in Opinion Mining

One of the most significant steps in opinion mining is feature extraction. It can be done by using supervised learning method where hidden Markov models [35] and conditional random field [36] are used for tagging the entity features. Even though it is good at some specific domain, it needs to get retrained when applied to some other domain. Therefore, to tackle such problems unsupervised NLP [37] was proposed for feature extraction of the opinion. Mining of syntactic patterns of features present in review documents is done for feature extraction. It does not extract invalid features because online reviews and comments are colloquial in nature. Association rule mining (ARM) [38] is proposed for frequently occurring or encountering features such as noun and noun phrases. One of the major drawbacks of this approach is that many invalid features also got extracted which are irrelevant. So, because of the encounter of invalid features mutual reinforcement clustering (MRC) approach [39] was developed. Here, for mining, the association between features and opinion words cooccurrence weight matrix was used. MRC is having low precision as it obtains good clusters on real-life views although MRC is potent of extracting infrequent features. An aspect ranking algorithm [40] came into existence where using probabilistic regression model, major product features got identified from customer reviews. But, its drawback is that it is totally focused on aspect ranking rather than the feature extraction. Latent Dirichlet allocation (LDA) [41] was proposed which mines the latent topics or aspects for feature extraction. Although it is good at extracting latent aspects in a review document, it is not that potent to identify specific feature terms.

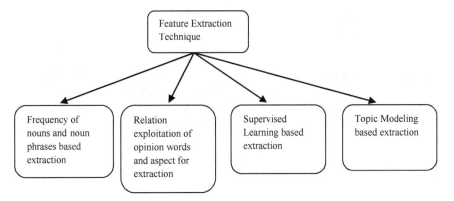

Fig. 4 Feature/aspect extraction classification

Feature extraction [42] can be broadly classified into four categories as shown in Fig. 4 and described follow:

- Frequency of nouns and noun phrases-based extraction [6, 12, 43, 44]
- Exploiting opinion and aspect relations-based extraction [45, 46]
- Supervised learning-based extraction [47, 48]
- Topic modeling-based extraction [49, 50].

3.1 Extraction Based on Frequency of Nouns and Noun Phrases

As this approach is easy yet it is significant due to its approach. When people write reviews or comments regarding any product, they use some words frequently to express their opinion [6]. In this method, POS tagger is used to determine noun and noun phrases [51] where tagger tags each word with its role in the sentence whether its noun, pronoun, adjective, etc. This method is further dependent on three approaches as follows:

- frequency-based mining and pruning,
- order based-filtering and
- similarity-based filtering.

3.2 Relation Exploitation of Opinion Words and Aspects for Extraction

This approach uses the features and opinion word relationship in the given opinion. The features which are not that frequent are extracted using this method. This approach [52] is focused on two important and basic issues to extract the features/aspects using opinion words as follows, and then, double propagation is used:

- Opinion Lexicon expansion,
- Opinion target extraction.

The features are extracted by using the syntactic relationship which creates link between review words and targets. Dependency parser is used to identify relations which can be further used for opinion lexicon expansion and extraction. This approach gives significant results.

3.3 Supervised Learning-Based Extraction

One of the significant techniques used for feature extraction is supervised learning where labeled data is used to generate aspects' model. Methods used here are as follows:

- support vector machine (SVM)
- conditional random field (CRF)
- hidden Markov model (HMM).

3.4 Extraction Based on Topic Modeling

Extraction based on topic modeling for aspect extraction is an unsupervised approach where it is assumed that any dataset given is having k hidden topics. Systematic approach is used to extract topics from the dataset by using statistical topic models [49, 50]. Hidden topics or features to get automatically extracted methods used here are as follows:

- latent semantic analysis (LSA)
- latent Dirichlet allocation (LDA).

These models are used at document-level opinion mining as both latent semantic analysis and latent Dirichlet allocation use the bag-of-words (BoW) technique.

4 Conclusion

In recent years, opinion mining is gaining its importance due to its vast application in various domains. The opinions expressed in the form of text are unstructured in nature; therefore, various efforts have been made to mine this unstructured data. This paper provides an overview of opinion mining and its application and also discussed one of the most prominent objectives, i.e., feature extraction.

References

1. Dave, K., Lawrence, S., Pennock, D.M.: Mining the peanut gallery: opinion extraction and semantic classification of product reviews. In: Proceedings of the 12th International ACM Conference on World Wide Web, pp. 519–528 (2003)
2. Liu, B., Zhang, L.: A survey of opinion mining and sentiment analysis. In: Aggarwal, C., Zhai, C. (eds.) Mining Text Data, pp. 415–463. Springer, Boston, MA (2012)
3. Scholer, F., Kelly, D., Carterette, B.: Information retrieval evaluation using test collections. Inf. Retr. J. 19(3), 225–229 (2016)
4. Daud, A., Khan, W., Che, D.: Urdu language processing: a survey. Artif. Intell. Rev. 47(3), 279–311 (2017)
5. Singh, J., Gupta, V.: A systematic review of text stemming techniques. Artif. Intell. Rev. 48(2), 157–217 (2017)
6. Jeyapriya, A., Selvi, K.: Extracting aspects and mining opinions in product reviews using supervised learning algorithm. In: 2015 2nd International Conference on Electronics and Communication Systems (ICECS), pp. 548–552. IEEE (2015)
7. Hajmohammadi, M.S., Ibrahim, R., Selamat, A.: Graph-based semi-supervised learning for cross-lingual sentiment classification. In: Guyen, N., Trawiński, B., Kosala, R. (eds.) Intelligent Information and Database Systems. ACIIDS 2015. Lecture Notes in Computer Science, vol. 9011, pp. 97–106. Springer, Cham (2015)
8. Habernal, I., Ptáček, T., Steinberger, J.: Supervised sentiment analysis in Czech social media. Inf. Process. Manag. 51(4), 532–546 (2015)
9. Li, G., Liu, F.: Sentiment analysis based on clustering: a framework in improving accuracy and recognizing neutral opinions. Appl. Intell. 40(3), 441–452 (2014)
10. Chinsha, T.C., Joseph, S.: A syntactic approach for aspect based opinion mining. In: IEEE International Conference on Semantic Computing (ICSC), pp. 24–31 (2015)
11. Velásquez, J.D.: Combining eye-tracking technologies with web usage mining for identifying Website Keyobjects. Eng. Appl. Artif. Intell. 26(5), 1469–1478 (2013)
12. Hu, M., Liu, B.: Mining and summarizing customer reviews. In: Proceedings of the Tenth ACM SIGKDD International Conference on Knowledge Discovery and Data Mining, pp. 168–177. ACM (2004)
13. Moraes, R., Valiati, J.F., Neto, W.P.G.: Document-level sentiment classification: an empirical comparison between SVM and ANN. Expert Syst. Appl. 40(2), 621–633 (2013)
14. Marcheggiani, D., Täckström, O., Esuli, A., Sebastiani, F.: Hierarchical multilabel conditional random fields for aspect-oriented opinion mining. In: de Rijke, M. et al. (eds.) Advances in Information Retrieval. ECIR 2014. Lecture Notes in Computer Science, vol. 8416, pp. 273–285. Springer, Cham (2014)
15. Xia, Y., Cambria, E., Hussain, A.: AspNet: aspect extraction by bootstrapping generalization and propagation using an aspect network. Cogn. Comput. 7(2), 241–253 (2015)
16. Hemmatian, F., Sohrabi, M.K.: A survey on classification techniques for opinion mining and sentiment analysis. In: Artificial Intelligence Review (2017)

17. Poobana, S., Sashi Rekha, K.: Opinion mining from text reviews using machine learning algorithm. Int. J. Innov. Res. Comput. Commun. Eng. **3**(3), 2320–9801 (2015)
18. Pandarachalil, R., Sendhilkumar, S., Mahalakshmi, G.S.: Twitter sentiment analysis for large-scale data: an unsupervised approach. Cogn. Comput. **7**(2), 254–262 (2015)
19. Chen, L., Wang, F., Qi, L., Liang, F.: Experiment on sentiment embedded comparison interface. Knowl. Based Syst. **64**, 44–58 (2014)
20. Marrese-Taylor, E., Velásquez, J.D., Bravo-Marquez, F.: A novel deterministic approach for aspect-based opinion mining in tourism products reviews. Expert Syst. Appl. **41**(17), 7764–7775 (2014)
21. Lu, T.J.: Semi-supervised microblog sentiment analysis using social relation and text similarity. In: 2015 International Conference on Big Data and Smart Computing (BigComp), pp. 194–201. IEEE (2015)
22. Tang, H., Tan, S., Cheng, X.: A survey on sentiment detection of reviews. Expert Syst. Appl. **36**(7), 10760–10773 (2009)
23. Tsagkalidou, K., Koutsonikola, V., Vakali, A., Kafetsios, K.: Emotional aware clustering on micro-blogging sources. In: D'Mello, S., Graesser, A., Schuller, B., Martin, J.C. (eds.) Affective Computing and Intelligent Interaction. ACII 2011. Lecture Notes in Computer Science, vol. 6974, pp. 387–396. Springer, Berlin (2011)
24. Unankard, S., Li, X., Sharaf, M., Zhong, J., Li, X.: Predicting elections from social networks based on subevent detection and sentiment analysis. In: Web Information Systems Engineering—WISE2014, pp. 1–16. Springer, Berlin (2014)
25. Kagan, V., Stevens, A., Subrahmanian, V.S.: Using twitter sentiment to forecast the 2013 Pakistani election and the 2014 Indian election. IEEE Intell. Syst. **1**, 2–5 (2015)
26. Mohammad, S.M., Zhu, X., Kiritchenko, S., Martin, J.: Sentiment, emotion, purpose, and style in electoral tweets. Inf. Process. Manag. **51**(4), 480–499 (2015)
27. Archambault, D., Greene, D., Cunningham, P.: Twitter Crowds: Techniques for Exploring Topic and Sentiment in Microblogging Data. Preprint. arXiv:1306.3839 (2013)
28. Bollen, J., Mao, H., Zeng, X.: Twitter mood predicts the stock market. J. Comput. Sci. **2**(1), 1–8 (2011)
29. Nofer, M., Hinz, O.: Using Twitter to predict the stock market. Bus. Inf. Syst. Eng. **57**(4), 229–242 (2015)
30. Bing, L., Chan, K.C., Ou, C.: Public sentiment analysis in Twitter data for prediction of a company's stock price movements. In: IEEE 11th International Conference on E-business Engineering (ICEBE), pp. 232–239 (2014)
31. Pang, B., Lee, L.: A sentimental education: sentiment analysis using subjectivity summarization based on minimum cuts. In: Proceedings of the 42nd Annual Meeting of the Association for Computational Linguistics, pp. 271–278 (2004)
32. Pang, B., Lee, L., Vaithyanathan, S.: Thumbs up? Sentiment classification using machine learning techniques. In: Proceedings of the Conference on Empirical Methods in Natural Language Processing (EMNLP2002), pp. 79–86 (2002)
33. Boiy, E., Hens, P., Deschacht, K., Moens, M.-F.: Automatic sentiment analysis in online text. In: Proceedings of the Conference on Electronic Publishing (ELPUB-2007), pp. 349–360 (2007)
34. Segaran, T.: Programming Collective Intelligence. O'Reilly Media, Inc., Sebastopol (2007)
35. Pang, B., Lee, L.: Opinion mining and sentiment analysis. Found. Trends Inf. Retrieval **2**(1–2), 1–135 (2008)
36. Jin, W., Ho, H.H.: A novel lexicalized HMM-based learning framework for web opinion mining. In: Proceedings of 26th Annual Int'l Conference on Machine Learning, pp. 465–472 (2009)
37. Jakob, N., Gurevych, I.: Extracting opinion targets in a single and cross-domain setting with conditional random fields. In: Proceedings of Conference on Empirical Methods in Natural Language Processing, pp. 1035–1045 (2010)
38. Qiu, G., Liu, B., Bu, J., Chen, C.: Opinion word expansion and target extraction through double propagation. Comput. Linguist. **37**, 9–27 (2011)
39. Hu, M., Liu, B.: Mining and summarizing customer reviews. In: Proceedings of 10th ACM SIGKDD Int'l Conference Knowledge Discovery and Data Mining, pp. 168–177 (2004)

40. Su, A., Xu, X., Guo, H., Guo, Z., Wu, X., Zhang, X., Swen, B., Su, Z.: Hidden sentiment association in chinese web opinion mining. In: Proceedings of 17th Int'l Conference on World Wide Web, pp. 959–968 (2008)
41. Yu, J., Zha, Z.-J., Wang, M., Chua, T.-S.: Aspect ranking: identifying important product aspects from online consumer reviews. In: Proceedings of 49th Annual Meeting of the Association for Computational Linguistics: Human Language Technologies, pp. 1496–1505 (2011)
42. Blei, D.M., Ng, A.Y., Jordan, M.I.: Latent dirichlet allocation. J. Mach. Learn. Res. **3**, 993–1022 (2003)
43. Rana, T.A., Cheah, Y.: Aspect extraction in sentiment analysis: comparative analysis and survey. Artif. Intell. Rev. **46**(4), 459–483 (2016)
44. Li, S., Zhou, L., Li, Y.: Improving aspect extraction by augmenting a frequency-based method with web based similarity measures. Inf. Process. Manag. **51**(1), 58–67 (2015)
45. Lv, Y., Liu, J., Chen, H., Mi, J., Liu, M., Zheng, Q.: Opinioned post detection in Sina Weibo. IEEE Access **5**, 7263–7271 (2017)
46. Qiu, G., Liu, B., Bu, J., Chen, C.: Opinion word expansion and target extraction through double propagation. Comput. Linguist. **37**(1), 9–27 (2011)
47. Wu, Y., Zhang, Q., Huang, X., Wu, L.: Phrase dependency parsing for opinion mining. In: Proceedings of the 2009 Conference on Empirical Methods in Natural Language Processing: volume 3, pp. 1533–1541. Association for Computational Linguistics (2009)
48. Jin, W., Ho, H.H., Srihari, R.K.: OpinionMiner: a novel machine learning system for web opinion mining and extraction. In: Proceedings of the 15th ACM SIGKDD İnternational Conference on Knowledge Discovery and Data Mining, pp. 1195–1204 (2009)
49. Yu, J., Zha, Z.J., Wang, M., Chua, T.S.: Aspect ranking: identifying important product aspects from online consumer reviews. In: Proceedings of the 49th Annual Meeting of the Association for Computational Linguistics: Human Language Technologies, vol. 1, pp. 1496–1505 (2011)
50. Mukherjee, A., Liu, B.: Aspect extraction through semi-supervised modeling. In: Proceedings of the 50th Annual Meeting of the Association for Computational Linguistics: Longpapers—volume 1. Association for Computational Linguistics, pp. 339–348 (2012)
51. Vulić, I., De Smet, W., Tang, J., Moens, M.F.: Probabilistic topic modeling in multilingual settings: an overview of its methodology and applications. Inf. Process. Manag. **51**(1), 111–147 (2015)
52. Wang, G., Zhang, Z., Sun, J., Yang, S., Larson, C.A.: POS-RS: a random subspace method for sentiment classification based on part-of-speech analysis. Inf. Process. Manag. **51**(4), 458–479 (2015)

Heart Disease Prediction Using Classification (Naive Bayes)

Akansh Gupta, Lokesh Kumar, Rachna Jain and Preeti Nagrath

Abstract This paper aims toward a greater idea and utilization of machine learning in the medical sector. In this paper, comparative performances of six classification models are presented, when used over the University of California Irvine's (UCI) Cleveland Heart Disease Records to predict coronary artery disease (CAD). At first, all the 13 provided independent features were used to build the models. On comparing the accuracy of models, it was found that K-nearest neighbors (KNN), support vector machine (SVM), and Naive Bayes have expected and better performances. Thereafter, feature selection is applied to improve prediction accuracy. The backward elimination method and filter method based on the Pearson correlation coefficient is used to choose major predicting features. The accuracy of models using all features and using features selected significantly enhanced the performance of Naive Bayes and random forest, while the other models did not perform as expected. Naive Bayes produced an accuracy of 88.16% on the test set thereafter.

Keywords Naive Bayes · Random forest · Classification model · Coronary artery disease · Cleveland dataset

1 Introduction

Heart diseases are common but serious health issue in most of the countries nowadays. The improvement in technology is making us physically less active and susceptible

A. Gupta (✉) · L. Kumar · R. Jain · P. Nagrath
Bharati Vidyapeeth's College of Engineering, New Delhi, Delhi, India
e-mail: akansh.gupta1298@gmail.com

L. Kumar
e-mail: krlokeshv99@gmail.com

R. Jain
e-mail: rachna.jain@bharatividyapeeth.edu

P. Nagrath
e-mail: preeti.nagrath@bharatividyapeeth.edu

© Springer Nature Singapore Pte Ltd. 2020
P. K. Singh et al. (eds.), *Proceedings of First International Conference on Computing, Communications, and Cyber-Security (IC4S 2019)*, Lecture Notes in Networks and Systems 121, https://doi.org/10.1007/978-981-15-3369-3_42

to diseases. They apply their intellect to work on machines, build them, enhance, and preserve their functions.

It is sometimes the case that a disease has different forms requiring various treatments. Like, among patients with angiographic coronary artery disease (CAD), patients with triple-vessel or left main CAD need surgical coronary artery revascularization, and on the other hand, patients with narrowing in single or double vessel may benefit from medical therapy [1]. Naturally, a physician would initially choose in case the patient has CAD, and, if so, then the patient needs surgery or not. Therefore, in the medical sector, machine learning algorithms can be used to recognize some severe diseases using the previous trends of symptoms.

Physical lethargy and malnutrition, overweight and corpulence, tobacco and substance abuse are among the dominant agents of cardiovascular disease as suggested by the Rochester's Medical Centre. Alwan [2] conferred the research in World Health Organization, where he elucidated the prevention of heart disease. The research showed that heart diseases are at the top among non-communicable disease which sums for 1/3 of mortality rate and 10% of the global disease trouble.

Also, it is not equitable to correlate the accuracy of models and resolve the finest since the operation depends upon data [3]. Numerous studies relate data mining and statistical concepts to crack prediction queries. The comparative studies have primarily considered a particular dataset or the same disposal of the dependent variable.

Anatomy of this paper is as follows: Sect. 2 will apprise you about the backdrop of heart disease prediction systems. Section 3 will expose you to the dataset used in this research, also data exploration and preprocessing techniques. Section 4 is composed with classification algorithms of machine learning used: KNN, SVM, logistic regression, decision tree, and random forest; feature extraction techniques will also be analyzed. Section 5 will include the Results and Discourse of the research conducted. The paper is accomplished in Sect. 6 by considering the future also.

The only encouragement of the research is to recommend the prediction system using the best classification model so it could help the medical experts to make successful decisions. The highest mortality rate in numerous countries is due to heart disease. Any successful treatment is always attributed by an accurate and precise diagnosis. Therefore, heart disease prediction systems based on machine learning algorithms assist in such cases to get the right results.

The evaluation of models was based on confusion matrix. Different feature extraction methods were used to select the most related attributes from given dataset to find an algorithm which performs best on the given dataset.

2 Background

Many researches have been accompanied on the prediction of CAD. At every moment, technology is leaping over what it had attained earlier. Things which seemed impossible before are now easy with the advancement of technology. Prediction of the disease using machine learning has become a common practice.

Considerable studies have appeared that were concentrated on heart disease analysis. Distinct methods have been tried to solve the given problem and they acquired high classification accuracies. Here are some citations:

Detrano et al. [4] had outcomes which achieved acceptable classification efficiency of almost 77% on administering logistic regression algorithm along derived discriminant. Zheng Yao [5] enforced an advanced model, based on C4.5 upgraded the efficacy of attribute selection and partitioning criteria. A study also revealed that the rules devised by the new model on C4.5 (R-C4.5) can give fair and appropriate explanations in healthcare fields. Resul Das [6] applied an approach that employs a statistical analytics application called SA software for diagnosing heart disease (ANN-based method). Imran Kurt et al. compared effectiveness of logistic regression, CART, and ANN for predicting CAD [7].

Jabbar et al. [8] employed 'Principal Component Analysis' as a medium to trim the dimensionality of a dataset and then used neural networks to produce better performance than traditional methods. John Gennari's [9] visionary CLASSIT[30] system of clustering exhibited an accuracy of 78.9% on UCI heart disease data. Alfeo Sabay's [10] model based on a neural network which worked upon fabricated data was capable to enhance the accuracy of CAD prediction to 96.7%. The 'Majority Voting Ensemble Method' used by Mechnovic yielded 87.37% for binary classification only outperformed by ANN [11].

3 Preliminaries of Research

3.1 Heart Disease Dataset

The following research has been performed using data from the heart disease database available at the UC Irvine repository [12]. This data has been available since 1988 and used by many researchers in heart disease prediction research because of its availability.

There are four available dataset, and out of these four sources, only the Cleveland dataset has been used in machine learning experiments mostly, due to its completeness of observations. Well, it was repeatedly mentioned in previous studies but still exploring the dataset during the data preprocessing stage showed it has the only six observations of missing data attributed with '?.'

The dataset has only 303 instances of the record with 14 attributes, 13 being the independent variable. The dependent variable had a multiclass distribution which was converted to a binary classified data due to the limitations of data provided. Results achieved for binary class were superior to that were achieved over five classes of dependent variable [9, 11].

The characterization of the Cleveland dataset [13] is given in Table 1, and

Table 1 Data type of attributes of Cleveland dataset

Age	Integer
Sex	Categorical
ChestPain	Categorical
RestBPS	Integer
Cholestrol	Integer
FBS	Categorical
MaxHR	Integer
RestECG	Categorical
Exang	Categorical
OldPeak	Float
Slope	Categorical
Nmajvess	Categorical
Thal	Categorical
ADS	Categorical

it represents the type of data (continuous int/float or categorical) and the brief information of attributes:

The missing values in some of the research were dropped, making the dataset shorter, with 297 instances instead of 303. Here, the missing values were replaced by nearest of the mean value of categorical variables. Two of thalassemia values were missing which are replaced by 6, i.e., fixed defect, and four Nmajvess missing values are replaced by 2, i.e., three major vessels are colored in fluoroscopy test. In the dependent variable, where narrowing in any major vessel was greater than 50% replaced by 1.

3.2 Exploration Using Graphs

Figure 1 shows that the patients who had asymptomatic type of pain had higher chances of artery disease.

Figure 2 depicts that patients that showed normal thalassemia had lesser chances of artery disease.

Figure 3 shows that patients who had exercise-induced angina have higher chances of artery disease.

Figure 4 shows that male patients have significantly higher chances of having artery disease.

Figure 5 shows patients those had higher ST depressions had higher chances of artery disease.

Figure 6 shows patients that achieved higher max. heart rates have lower chances of artery disease.

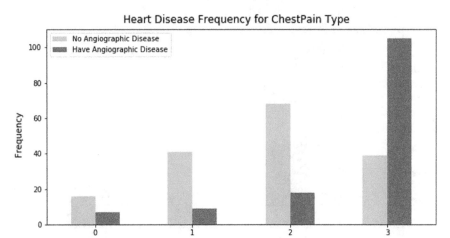

Fig. 1 Chest pain type versus ADS

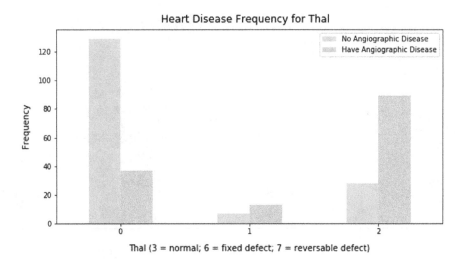

Fig. 2 Thalassemia frequency versus ADS

3.3 Data Preprocessing

The train to test ratio is taken as 3:1, and the model was also checked at a 3:2 ratio of train to test. This yielded that there were no significant changes in the performance of models, yet 3:1 train test splitting had way better results than the later.

Before, fitting the model to the training set encoding of multiclass categorical variables was done, and all categorical attributes were standardized, using the sklearn library.

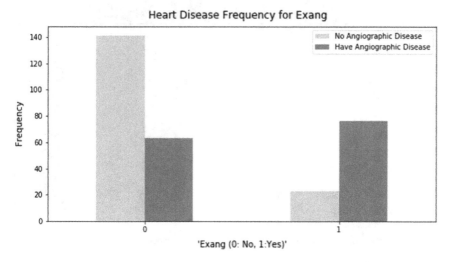

Fig. 3 Exercise-induced angina frequency versus ADS

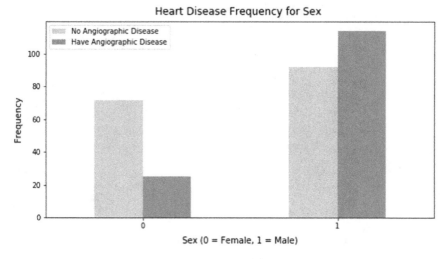

Fig. 4 Sex versus ADS

4 Methodology

4.1 Logistic Regression

Logistic regression is the simplest way to handle classification problems. In logistic regression, a sigmoid function is used instead of a linear function. [15] The value of sigmoid function varies between 0 and 1. It can be used for classification problem as

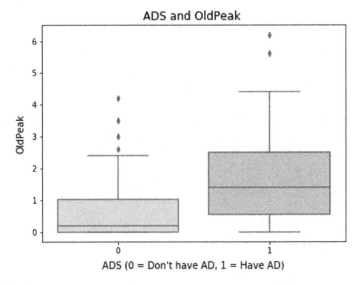

Fig. 5 OldPeak and ADS

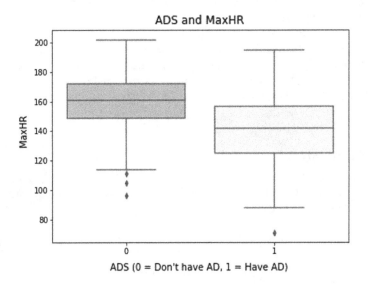

Fig. 6 ADS and MaxHR

when the value is greater than 0.5 it gives a label 1, otherwise 0. Logistic regression gives a linear classifier boundary. In this experiment, logistic regression gave an accuracy of 85.53% on Cleveland heart disease dataset.

4.2 K-Nearest Neighbors

K-Nearest neighbor algorithm is based on instance learning. Instance learning is also called lazy learning as the instances are not processed, they are just stored instead [25]. In this algorithm, a model is not learnt before instead when the test example is provided, it uses the stored examples to classify the class of the test example. This algorithm is based on distance metric. It finds the nearest neighbors depending on the value of k and looks at the value to predict the class of test example. The value of k was set to 5 for the experiment and the accuracy obtained on the Cleveland heart disease dataset was 88.16%. This also outperformed the other machine learning classification algorithms used.

4.3 Support Vector Machine

Support vector machine (SVM) algorithm is most effective for classification problems. It can be used for linear as well as nonlinear classifications depending on the different kernel functions. SVM is an alternative to Bayesian learning [27]. It depends on the support vectors chosen based on their distances. The points which are close to the decision boundary are known as support vectors. The greater the distance of the point from the margin, the higher is the confidence. This model completely depends upon the support vectors chosen and the distance metric. In this experiment, SVM algorithm was applied using default parameters and an accuracy of 88.16% was obtained. This algorithm outperformed all the other classification algorithms.

4.4 Naive Bayes

Naive Bayes algorithm is based on the application of Bayes law of probability. Bayes theorem is given by [19]:

$$P(A|B) = P(B|A) * P(A)/P(B) \tag{1}$$

where A and B are two events. When Bayes theorem is applied to classification problems, joint probability has to be calculated. It is very difficult to learn and represent joint probability as there are 2 to the power n possible combinations even if the features are Boolean [16]. This makes the calculation and storage intractable. So to simplify this, an assumption is made in Naive Bayes algorithm. The assumption is that all the features are independent to each other when the class of features is given. Conditional independence is assumed between attributes. This makes Naive Bayes a very simple classification algorithm. For n features, only the probability of $n - 1$ features needs to be calculated which can be computed easily.

In this experiment, using the Naive Bayes algorithm on Cleveland heart disease database, accuracy of 84.21% was obtained.

4.5 Decision Tree and Random Forest Classifier

A decision tree is a classifier in the form of a tree which has two types of nodes, decision nodes and leaf nodes [23]. The decision nodes specify a choice or test. The decision tree is just like any binary tree and can be easily followed to reach a leaf node. It can be used to solve problems of classification and regression, both. Generally, decision trees are used for classification problems. A decision tree can be constructed using two criteria. It can be based on entropy criteria or Gini index. A split is made at the internal node if the Gini index is low or the information gain is high. External nodes are also known as leafs. External node contains the value or the label also known as the target value. Random forest uses an army of trees and hence helps in avoiding overfitting in the tree [18, 28]. In decision tree and random forest, parameter 'criterion' was set to entropy, parameter 'max depth of tree' was set to 5, and parameter 'minimum samples leaf' was set to 8. In random forest algorithm, the combination of number of decision trees used was 800 (parameter number of estimators). The accuracies obtained were 82.89 and 86.84% on the dataset taken.

4.6 Feature Extraction

Figure 7 shows the heatmap of the Pearson correlation of all the attributes.

Filter method based on the Pearson correlation [21, 24] and backward elimination method was used to select the major predicting feature [26]. Using the results of both methods, 8 attributes come out to be major predictors or say most correlated to the dependent variables. Since these methods are based on linear relations and continuous variables, therefore, only Naive Bayes model was able to perform as per expectations.

Relevant Features: ChestPain, MaxHR, Exang, OldPeak, Slope, Nmajvess, Thal, Sex are the union of result achieved from both the methods.

5 Result and Discussions

To assess the performance [20] of various classification models, accuracy, recall, and precision were calculated. A comparison was also done by taking 25 and 40% as test data (tp: true positive, fp: false positive, tn: true negative, and fn: false negative) [14].

$$\text{Accuracy} = (tp + tn)/(tp + tn + fp + fn) \qquad (2)$$

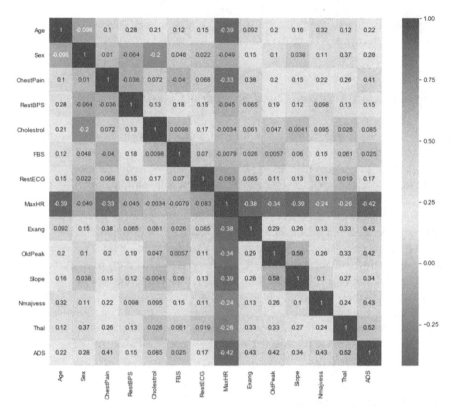

Fig. 7 Heatmap for correlation coefficients

Precision and recall [17] are as:

$$\text{Precision} = \text{tp}/(\text{tp} + \text{fp}) \tag{3}$$

$$\text{Recall} = \text{tp}/(\text{tp} + \text{fn}) \tag{4}$$

Comparison of models:

Table 2 compares the accuracies obtained by the algorithms. These are accuracy obtained before feature selection was applied. KNN and SVM outperform all other classification models.

Table 3 shows the precision and accuracies obtained by different models. It also tells about the number of instances which were classified as true (class-1) and false (class-0).

Table 2 displays the outcomes which were calculated for 25 and 40% of test data. Table 3 shows the results which were calculated for 25% of test data.

Table 4 shows performance of model after optimization, i.e., when only 8

Table 2 Comparison of different algorithms before optimization (25 and 40%)

Model	Accuracy (25%)	Accuracy (40%) (%)
Logistic regression	85.53 %	82.79 %
KNN	88.16 %	81.97 %
SVM	88.16 %	87.70 %
Naive Bayes	84.21 %	84.43 %
Decision tree	81.58 %	79.51 %
Random forest	86.84 %	85.25 %

Table 3 Calculated precision and recall of different models (25%)

Model	Precision	Recall	Predicted false	Predicted true
Logistic regression	85.5	85.5	39	37
KNN	88	88	39	37
SVM	88.5	88	39	37
Naive Bayes	84.5	84	39	37
Decision tree	81.5	81.5	39	37
Random forest	87	87	39	37

Table 4 Comparison of different algorithms after optimization (25%)

Model	Accuracy (%)
SVM	71.05
Naive Bayes	88.16
Logistic regression	89.47
KNN	76.32
Decision tree	81.89
Random forest tree	88.16

major attributes were used for prediction. Only logistic regression and Naive Bayes accuracy were increased significantly by approximately 5% and 4%, respectively.

6 Conclusion and Future Work

Many techniques can be used to predict and prevent coronary heart diseases including machine learning or deep learning techniques. Here, six classification algorithms are analyzed to predict CAD: Logistic regression, KNN, SVM [3], Naive Bayes, decision tree [3], and random forest. These algorithms are judged on the grounds of outcome accuracy. The data provided was of multiclass classification, and due to the small size of data and thirst to achieve better performance, data was converted as binary classified. This study showed that Naive Bayes model turned out to be the best classifier, in accordance with the data, for prediction. This model before feature extraction showed 46 incorrect cases (combining train and test) and only 36 incorrect cases with an accuracy of 88.16% thereafter.

There is certainly room for improvement in future. New features can be extracted using feature engineering. Deep learning [22] methods can also be applied to generate new features. PCA can also be used, instead of backward elimination for better results, to reduce the dimensionality of features and figure out informative features. Also, synthetic data, as used by Alfeo Sabay [10], can be a progressive step in disease prediction.

References

1. Detrano, R., Janosi, A., Steinbrunn, W., Pfisterer, M., Schmid, J. J., Meyer, M., Guppy, K.H., Abi-Mansour, P.: Algorithm to predict triple-vessel/left main coronary artery disease in patients without myocardial infarction. An international cross validation. Circulation **83**(5 Suppl), III89–96 (1991)
2. Alwan, A.: Global status report on noncommunicable diseases 2010. World Health Organization. Open J. Prev. Med. **5**(8) (2015)
3. Kumari, M., Godara, S.: Comparative study of data mining classification methods in cardiovascular disease prediction 1. Int. J. Comput. Sci. Technol. **2**, 304–308 (2011)
4. Detrano, R., Janosi, A., Steinbrunn, W., Pfisterer, M., Schmid, J.J., Sandhu, S., Guppy, K.H., Lee, S., Froelicher, V.: International application of a new probability algorithm for the diagnosis of coronary artery disease. Am. J. Cardiol. **64**(5), 304–310 (1989)
5. Yao, Z., Liu, P., Lei, L., Yin, J.: R-C4. 5 Decision tree model and its applications to health care dataset. In: Proceedings of ICSSSM'05. 2005 International Conference on Services Systems and Services Management, vol. 2, pp. 1099–1103. IEEE (2005)
6. Das, R., Turkoglu, I., Sengur, A.: Effective diagnosis of heart disease through neural networks ensembles. Expert Syst. Appl. **36**(4), 7675–7680 (2009)
7. Kurt, I., Ture, M., Kurum, A.T.: Comparing performances of logistic regression, classification and regression tree, and neural networks for predicting coronary artery disease. Expert Syst. Appl. **34**(1), 366–374 (2008)
8. Jabbar, M.A., Deekshatulu, B.L., Chandra, P.: Classification of heart disease using artificial neural network and feature subset selection. Glob. J. Comput. Sci. Technol. Neural Artif. Intell. **13**(3), 4–8 (2013)
9. Gennari, J.H., Langley, P., Fisher, D.: Models of incremental concept formation. Artif. Intell. **40**(1–3), 11–61 (1989)
10. Sabay, A., Harris, L., Bejugama, V., Jaceldo-Siegl, K.: Overcoming small data limitations in heart disease prediction by using surrogate data. SMU Data Sci. Rev. **1**(3), 12 (2018)
11. Mehanović, D., Mašetić, Z., Kečo, D.: Prediction of heart diseases using majority voting ensemble method. In: International Conference on Medical and Biological Engineering, pp. 491–498. Springer, Cham (2019)
12. Heart Disease Data Set, UCI Machine Learning Repository. http://archive.ics.uci.edu/ml/datasets/Heart+Disease
13. Detrano, R.: Heart Disease Data Set of Cleveland, V.A. Medical Center, Long Beach and Cleveland Clinic Foundation
14. Wikipedia: https://en.wikipedia.org/wiki/Precision_and_recall#cite_note-OlsonDelen-7

15. Chen, L., Cao, Q., Li, S., Ju, X.: Predicting heart attacks. Int. J. Comput. Appl. (0975–8887) **17**(8) (2011)
16. Chaki, D., Das, A., Zaber, M.I.: A comparison of three discrete methods for classification of heart disease data. Bangladesh J. Sci. Ind. Res. **50**(4), 293–296 (2015)
17. Wei, L., Altman, R.B.: An automated system for generating comparative disease profiles and making diagnoses. IEEE Trans. Neural Netw. **15**, 597 (2004)
18. Sen, S.K.: Predicting and diagnosing of heart disease using machine learning algorithms. Int. J. Eng. Comput. Sci. **6**(6) (2017)
19. Singh, Y.K., Sinha, N., Singh, S.K. Heart disease prediction system using random forest. In: International Conference on Advances in Computing and Data Sciences, pp. 613–623. Springer, Singapore (2016)
20. Basharat, I., Anjum, A.R., Fatima, M., Qamar, U., Khan, S.A.: A framework for classifying unstructured data of cardiac patients: a supervised learning approach. Framework **7**(2) (2016)
21. Hossain, J., FazlidaMohdSani, N., Mustapha, A., SurianiAffendey, L.: Using feature selection as accuracy benchmarking in clinical data mining. J. Comput. Sci. **9**(7), 883 (2013)
22. Chowdhury, D.R., Chatterjee, M., Samanta, R.K.: An artificial neural network model for neonatal disease diagnosis. Int. J. Artif. Intell. Expert Syst. (IJAE) **2**(3), 96–106 (2011)
23. Chavda, P., Bhavsar, H., Pithadia, Y., Kotecha, R.: Early Detection of Cardiac Disease Using Machine Learning. Available at SSRN 3370813 (2019)
24. Feature Selection with sklearn and Pandas. https://towardsdatascience.com/feature-selection-with-pandas-e3690ad8504b
25. Deekshatulu, B.L., Chandra, P.: Classification of heart disease using k-nearest neighbor and genetic algorithm. Procedia Technol. **10**, 85–94 (2013)
26. Jain, D., Singh, V.: Feature selection and classification systems for chronic disease prediction: a review. Egypt. Inf. J. **19**(3), 179–189 (2018)
27. Cortes, C., Vapnik, V.: Support-vector networks. Mach. Learn. **20**(3), 273–297 (1995). https://scikit-learn.org/stable/modules/generated/sklearn.ensemble.RandomForestClassifier.html
28. Aha, D., Kibler, D.: Instance-based prediction of heart-disease presence with the Cleveland database. University of California, 3(1), 3-2 (1988)

Scene Classification Using Deep Learning Technique

Aayushi Shah and Keyur Rana

Abstract In recent years, deep learning techniques have been of more focus due to its remarkable results, especially in computer vision. Computer vision involves processing and understanding image data which is well represented and processed by deep learning techniques. It has overcome the traditional methods of image classification where handcrafted features were used to classify images. With the advent of deep learning, researchers drew their focus on automated feature extraction for image classification. Deep learning is a promising approach to perform classification-oriented tasks. In this paper, we have focused on classification of classroom scenes using deep learning techniques by automatically extracting features and classifying the scenes into meaningful labels. This novel approach tends to ease the task of monitoring students in the classroom by automatically extracting information and analyzing it. At first, we introduce some of the traditional techniques and their limitations. We then introduce ResNet architecture of 152 layers pretrained with ImageNet data to accomplish our institute's classroom scene classification. Further, we demonstrate the results of our classroom data classified using ResNet architecture pretrained with ImageNet dataset.

Keywords Scene classification · Deep learning · ResNet · ImageNet · Transfer learning

1 Introduction

Scene classification [1] deals with classifying image data into defined labels. It belongs to computer vision [2] domain. Computer vision involves acquiring, analyzing, processing and understanding the given image. Classifying scene involves processing of the given image, understanding it and giving out classification results. It has numerous applications like retrieval of images/video for classification, remote

A. Shah · K. Rana (✉)
Sarvajanik College of Engineering and Technology, Surat, India
e-mail: keyur.rana@scet.ac.in

A. Shah
e-mail: aayushishah63@gmail.com

© Springer Nature Singapore Pte Ltd. 2020 575
P. K. Singh et al. (eds.), *Proceedings of First International Conference on Computing, Communications, and Cyber-Security (IC4S 2019)*, Lecture Notes in Networks and Systems 121, https://doi.org/10.1007/978-981-15-3369-3_43

sensing image recognition and so on. The task of classifying scenes can be done using image processing techniques also known as traditional techniques [3–5]. However, with image processing techniques, one has to handcraft the features of the scenes manually. For extracting features from images especially containing nonhomogeneous objects, it is possible to segment such images in too coarse or too fine way. Therefore, it may happen that the handcrafted features are not closely relevant to the subject of the image. This may affect the accuracy of scene classification. In a traditional way of classifying scenes, techniques like low-level image feature extraction and object detector-based feature extraction are used to extract features from the images. After features are extracted, some classification methods are used to classify the scenes into meaningful labels. This makes the traditional techniques a two-step approach to classify scenes. Another way to classify scenes is to adapt techniques that have automated feature extraction strategy and one-step classification. This points to deep learning [6] techniques that have evolved in the recent years. These techniques not only give good accuracy in terms of classification but also ensure appropriate automatic feature extraction from the scene images. This eradicates the limitation of traditional techniques which led to improper feature extraction. Therefore, it fits into our idea of classifying scenes with automatically extracted features. By using this approach, we need not worry about the number and type of features extracted as it extracts the relevant features by using its layered structure. Deep learning techniques allow us to classify scenes by providing feature extraction and classification in one go. The layered architecture in deep learning involves classification layer at the end that gives classification results. Deep learning is a subbranch of machine learning that deals with learning representation of given images by processing them in different layers and giving out results. It basically attempts to resemble human brain by a concept called neural networks [7]. It contains different architectures that have a different number of layers and differ in processing methods. Thereby, it is also known as convolution neural network, where convolution [8] acts as filters to extract features from the images. As a prerequisite, we need to learn the core concepts of deep learning and its architecture along with implementation environment. By using such technology, we opt to classify scenes related to classroom images by our novel approach. We apply scene classification in education domain where monitoring the behavior of the students and their seating in classroom is important. Nowadays, images from the CCTV camera are observed by a person continuously to determine discipline and smooth conduction of classes. However, it is a tedious job, so we require some automated system to do such task. We address issue of identifying student's behavior in a classroom so that appropriate actions can be taken. We classify the classroom scene images mainly into four categories, i.e., disciplined class, indisciplined class, empty class and partially empty class. This will surely help the supervisor to determine behavior in the classroom. To evaluate such task, we apply deep learning architecture, i.e., ResNet which is already trained with ImageNet data. ImageNet [9] data consist of 1000 different labels of more than 14 million images. It helps us to classify our classroom data which are comparable to ImageNet and the feature patterns help the architecture to do the classification task. This process is called transfer learning [10] which we apply to our work. Transfer learning is a process in which the architecture is trained with one kind of dataset which is comparable to required data to be trained. This architecture is trained with required data in the last few layers. The reason is that the first few layers will only extract basic features like points, line and edges which are basically same in each and every image data. However, the last few layers will extract the main features of the required data. Transfer learning greatly helps in training the architecture even if we have less number of trainable dataset.

The next section describes few methods that use deep learning concepts referred by us for our scene classification task. In the subsequent section, we discuss the related deep learning techniques for scene classification and relevant points that are included in our work. We define and discuss our proposed work and methodology with step by step explanation of each and every stage. Furthermore, with our novel approach, we evaluate our institute's classroom scenes classified into correct labels and demonstrate the results of individual labels along with respective output. In the very last section, we conclude our work along with future directions.

2 Related Work

To the best of our knowledge, there has been a very less focus on education domain where classroom scenes are classified. In this section, we discuss deep learning techniques used to classify scenes into meaningful labels. These techniques use automatic feature extraction by deep learning methods.

Yang et al. [11] proposed segmenting the scenes into regions for Indoor67 [12] scene dataset and classified into 67 different labels. King et al. [13] derived the concept of using residual nets [14] for places [15] dataset scene classification. The scenes are then classified using softmax classifier [16]. As VGG [17] has a plain simple and one-way architecture, it is not able to process the previous layer outputs which residual nets do. Wang et al. [18] proposed to extract fine feature contents from the places scene dataset and solved the problem of label ambiguity by combining ambiguous labels to super category. Xiao et al. [19] proposed improvised version of AlexNet [20] architecture by decomposing large kernels and strides of convolution into smaller ones. It classified SUN397 [21] dataset into meaningful labels. Herranz et al. [22] proposed hybrid dataset concept by combining ImageNet and places dataset. It extracted subregions larger to smaller features by scaling the scenes into different sizes. Table 1 describes the techniques discussed here along with important points that could be included in our research work. We show that we use the architecture and classifier based on our requirement for our education-based scene classification task. We also use confusion matrix [23] which is used as a measure to evaluate true and false results. Existing deep learning techniques for scene classification have been referred, and accordingly, we have used relevant approaches in our proposed work. It is mentioned in Table 1. Our proposed work is discussed in detail in the next section.

Table 1 Existing deep learning techniques for scene classification

Author	Brief summary	Included in our work?	Included points
Yang et al. [11]	Region proposal-based Indoor scene classification	No	–
King et al. [13]	Residual net (large network)-based places scene classification	Yes	– Softmax classifier – ResNet architecture
Wang et al. [15]	Imbibing knowledge of extra networks for large scale scene classification with multiple CNNs	Yes	– Confusion matrix
Xiao et al. [16]	Scene classification using improved Alexnet architecture to improve results	Yes	– Softmax classifier – Improvising architecture
Herranz et al. [19]	Scene classification by scaling scenes into different sizes for better recognition of objects inside the scene	Yes	– ImageNet dataset – Scaling (preprocessing)

3 Methodology

In this section, we discuss classroom scene classification for our institution using deep learning concepts. Hereby, we strive to provide efficient solution for classroom behavior by automated mechanism. Many approaches are available to classify scenes; however, our novel approach tends to provide classification of classroom scenes into meaningful labels to ease the task of the monitor. Human involvement is reduced by our approach as it provides automatic classification of various classroom scenes. We use ResNet architecture that is already trained with ImageNet dataset that contains more than 14 million images and classified into 1000 labels.

A small portion of ResNet architecture is shown in Fig. 1. It contains layers that learn from the previous layers and contains repetitive identical layers that make the architecture large. There is identity mapping (similarity mapping) in residual nets through skip connection. This identity does not contain parameters but adds output of the previous layer to the layer ahead. This enables input x and previous layer output $f(x)$ to be combined and become input to the next layer. We use transfer learning concept by replacing the last pooling layer of ResNet with new layers defined by us. In the beginning of the layers, the architecture will extract features like points, lines and edges of ImageNet dataset until intermediate level where it will extract objects and shapes of the same. After that in higher layers, the network will sum up

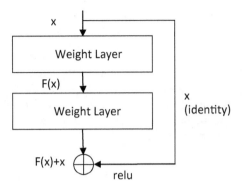

Fig. 1 ResNet architecture [14]

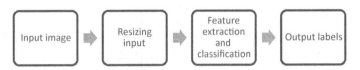

Fig. 2 General flow of scene classification

the features extracted in previous layers with features of new defined layers, of our classroom dataset. The general flow of scene classification is presented in Fig. 2.

3.1 Input Image

We take our classroom images by extracting them from CCTV videos. We tried to cover most of the classroom images with different orientations of benches and tables, different backgrounds and environments. Thus, the ResNet architecture gets trained with almost all kinds of classroom features. Along with our institute's classroom data, we also include classroom images from other institutes for more interpretation. We use transfer learning by taking the knowledge of ImageNet features which have about 1000 categories of labels of million images. With the basic features of ImageNet like points, line and edges, we need to make the architecture learn about our data. In this way, the architecture will learn in a better way even though we take around 500 training images having classroom scenes. Along with training data, we also need to validate whether our classroom images are trained well and they do not overfit. For that, we take 120 images to validate classroom data. We take more test images to know whether the network classifies scenes that are in a different background and orientations than train images. We tested more than 100 images for such task.

3.2   Resizing Input

We take our training data in the form of scene images extracted from videos and preprocess them according to the convolutional architecture requirement. We scale the classroom images into required size, i.e., 224 × 224. As the classroom images are in RGB format, we convert them into BGR format as required by the ResNet architecture. Preprocessing the input data makes sure that they are accepted by the ResNet architecture and processed accordingly.

3.3   Feature Extraction and Classification

In this stage, we take our preprocessed images and input them into ResNet architecture. We choose ResNet with 152-layer architecture which has residual nets. It is deep as it has 152 layers so gives better accuracy. Particularly, a residual unit consists of convolution layer, batch normalization [24], activation [25] and pool layers [26]. The convolution layer acts as a filter that convolves around images to extract features. Batch normalization normalizes the values of mean and variance by keeping their values 0 and optimum, respectively. This ensures that the features are learned properly. Activation layer contains certain activation function which activates the important features to be carried forward. ResNet has shortcut connections which lead to lesser parameters. Besides, it has identity layers that pass the input as it is to the last layer of a residual unit. So, it does not need to pass through all the in-between layers and can directly pass through identity connections. It can be seen in Fig. 1. In this way, it solves vanishing gradient problem [27] by preserving identity and learning all the important features. In the beginning, we load architecture of ResNet having definition of its layers. ResNet architecture is already pretrained with ImageNet having weights stored in an h5 file. It extracts thousands of features from ImageNet image data. Our classroom dataset is similar to part of ImageNet dataset from the various features point of view. Our data also contain common features like lines, edges, corners and shapes, objects along with some specific features like a room, an object called bench, chair and many more. The features extracted are stored in a file of numpy array in each folder, i.e., train, test and validation. It is important to specify that the weights are fixed for all the layers except the last fully connected dense layer. So, the network will get trained and weight updation will be done only according to the last layer. Thus, we get 2048 number of features of our own data as we made the network learn up to the last pooling layer with our own dataset. Thus, the dense layer will contain softmax classifier which will classify the scene dataset into four labels, i.e., disciplined class, indisciplined class, empty class and partially empty class. By training and validating the architecture, we got good accuracy and the network learned significantly. ResNet internally uses relu [28] activation as activation function and softmax classifier to classify scenes into different labels. We modified the architecture by decreasing its learning rate [29] from 0.1 to

Table 2 Statistics for ResNet152 architecture of our institute's classroom scene classification

Label	Disciplined class	Indisciplined class	Empty class	Partially empty class
Disciplined class	29	0	4	3
Indisciplined class	1	31	1	1
Empty class	0	0	36	0
Partially empty class	0	0	14	29

Table 3 Metric evaluation of scene classification with respect to labels

Metrics	Precision	Recall	F1-Score	Support
Disciplined class	0.97	0.81	0.88	36
Indisciplined class	1	0.91	0.95	34
Empty class	0.65	1	0.79	36
Partially empty class	0.88	0.67	0.76	43

0.01. We keep the epoch [30] and batch size [30] of images in normal range, i.e., 350 and 450, respectively. Epoch determines one iteration of all the training images at once. Basically, after giving test images to the trained ResNet architecture, we got 83% testing accuracy. Table 2 shows confusion matrix having label-wise scene classification. We analyze that 14 scenes of partially empty class are misclassified as empty class, besides this, almost every class has quite accurate classification. Table 3 shows metric evaluation of classroom scene classification with respect to labels. Precision [31] determines relevant results among the obtained results and recall [31] determines fraction of relevant results correctly retrieved over from total number of results. F1-score [32] is the harmonic mean of precision and recall. Support [32] is the total number of scenes pertaining to given label. As shown in Table 3, disciplined and indisciplined class images have high precision and recall with a good F1-score, i.e., nearer to 1. Empty class images have good recall; however, the precision is relatively low. Similarly, partially empty images have good precision but low recall. Hence, we achieve relatively low F1-score for empty class and partially empty class images. Overall, we achieve good scores for all the four labels (Fig. 3).

3.4 Output Labels

In the output, we run the test images randomly for which we got top four-label prediction. The test images of classroom scene dataset are shown in this section. We get good overall accuracy of 83% of our test scene images. As shown in Fig. 3, the disciplined class is misclassified as empty class. One of the reasons for this is that

Fig. 3 Actual: disciplined
class; result: empty class

Classifying   image: F:/resnet-classroom-
tl/dl/data/test/disciplined_class/screenshot_13.j
pg
    1. 0.89 empty_classroom
    2. 0.08 disciplined_class
    3. 0.02 partially_empty
    4. 0.01 indisciplined_class

the network learns the front benches features more than the other features. Figure 4
shows that the network again does the same and misclassifies the partially empty
class as an empty class. In other cases, i.e., from Figs. 5, 6, 7, 8, 9, and 10, the scenes
are classified accurately.

4   Conclusion and Future Work

We strive to benefit education domain by deriving an efficient solution. Monitoring
classrooms and behavior as well as seating of students can be determined by
performing classroom scene classification using advanced techniques. We use
automatic feature extraction-based deep learning approach to resolve such task.
ResNet architecture that is pretrained with ImageNet, when trained with our data
in the last few layers, works well and gives good results for training, testing and
validation scene sets. Additionally, the use of transfer learning concept enables
us to use less amount of training set and still achieves good testing accuracy of
83%. Although we attain such good accuracy, there are some challenges such as
determining the proper hyperparameters like learning rate and activations which
highly affect the accuracy of the scene classification. Such challenges need to

Fig. 4 Actual: partially empty; result: empty class

Classifying image: F:/resnet-classroom-tl/dl/data/test/partially_empty/cctv_cctv_Camera4_2018_1.jpg
1. 0.52 empty_classroom
2. 0.44 partially_empty
3. 0.03 disciplined_class
4. 0.01 indisciplined_class

Fig. 5 Actual: empty class; result: empty class

Classifying image: F:/resnet-classroom-tl/dl/data/test/empty_classroom/cctv_cctv_Camera3_2018_4.jpg
1. 0.88 empty_classroom
2. 0.07 indisciplined_class
3. 0.05 partially_empty
4. 0.00 disciplined_class

Fig. 6 Actual: partially empty; result: partially empty

Classifying image: F:/resnet-classroom-tl/dl/data/test/partially_empty/screenshot_6.jpg
1. 0.68 partially_empty
2. 0.30 empty_classroom
3. 0.01 discipined_class
4. 0.01 indiscipined_class

Fig. 7 Actual: indisciplined class; result: indisciplined class

Classifying image: F:/resnet-classroom-tl/dl/data/test/indiscipined_class/cctv_cctv_Camera4_2018_5.jpg
1. 0.92 indiscipined_class
2. 0.05 discipined_class
3. 0.02 partially_empty
4. 0.02 empty_classroom

Fig. 8 Actual: disciplined class; result: disciplined class

Classifying image: F:/resnet-classroom-tl/dl/data/test/discipined_class/screenshot_2.j pg
1. 0.83 discipined_class
2. 0.13 partially_empty
3. 0.04 empty_classroom
4. 0.00 indiscipined_class

Fig. 9 Actual: empty class; result: empty class

Classifying image: F:/resnet-classroom-tl/dl/data/test/empty_classroom/screenshot_6.j pg
1. 0.91 empty_classroom
2. 0.04 partially_empty
3. 0.04 indiscipined_class
4. 0.01 discipined_class

Fig. 10 Actual: empty class; result: empty class

Classifying image: F:/resnet-classroom-tl/dl/data/test/empty_classroom/47c4b7f04d.j
pg
1. 0.99 empty_classroom
2. 0.01 discipined_class
3. 0.00 partially_empty
4. 0.00 indicipined_class

be analyzed for better results. Determining number of layers of deep learning architecture required for our dataset is an open issue that requires certain analysis.

In the future, to deal with the aforementioned challenges and issues, we will certainly try with other variants of ResNet architecture having a different number of layers to analyze and compare the results of all the possibilities. We would also like to tune the hyperparameters (learning rate) to achieve better accuracy.

References

1. Dutt, R., Agrawal, P., Nayak, S.: Scene classification in images. (2005)
2. Singh, V., Girish, D., Ralescu, A.: Image understanding—a brief review of scene classification and recognition. In: Modern Artificial Intelligence and Cognitive Science, pp. 85–91 (2017)
3. Gorkani, M., Picard, R.: Texture orientation for sorting photos at a glance. In: International Conference on Pattern Recognition. Jerusalem, pp. 459–464 (1994)
4. Sande, K., Gevers, T., Snoek, C.: Evaluating color descriptors for object and scene recognition. IEEE Trans. Pattern Anal. Mach. Intell. **32**(9), 1582–1596 (2010)
5. Jia, L., Hao, S., Lim, Y., Fei, L.: Object as attributes for scene classification. In: European Conference on Computer Vision, vol. 6553, pp. 57–69. Springer, Berlin (2010)
6. Deep learning. https://en.wikipedia.org/wiki/Deep_learning. Last accessed 2018/08/18
7. A Beginner's Guide to Neural Networks and Deep Learning. https://skymind.ai/wiki/neural-network. Last accessed 2018/08/18

8. Chatterjee, S.: Different kinds of convolutional filters. https://www.saama.com/blog/different-kinds-convolutional-filters/. Last accessed 2018/08/18
9. ImageNet. http://www.image-net.org/. last accessed 2018/08/21
10. Lim, J., Salakhutdinov, R., Torralba, A.: Transfer learning by borrowing examples for multiclass object detection. In: Advances in Neural Information Processing Systems (2011)
11. Wang, Y., Wu, Y.: Scene Classification with Deep Convolutional Neural Networks. University of California
12. Indoor Scene Recognition. http://web.mit.edu/torralba/www/indoor.html. Last accessed 2018/08/24
13. King, J., Kishore, V., Ranalli, F.: Scene classification with convolutional neural networks. Cs231n.stanford.edu (2017)
14. He, K., Zhang, X., Ren, S., Sun, J.: Deep residual learning for image recognition. In: IEEE Conference on Computer Vision and Pattern Recognition, pp. 770–778 (2016)
15. Zhou, B., Khosla, A., Lapedriza, A., Torralba, A., Oliva, A.: Places: an image database for deep scene understanding. J. Vis. (2016)
16. Softmax Classifier. https://www.pyimagesearch.com/2016/09/12/softmax-classifiers-explained/. Last accessed 2018/03/08
17. Noh, H., Hong, S., Han, B.: Learning deconvolution network for semantic segmentation. In: Proceedings of the IEEE International Conference on Computer Vision, pp. 1520–1528 (2015)
18. Wang, L., Guo, S., Huang, W., Xiong, Y., Qiao, Y.: Knowledge guided disambiguation for large-scale scene classification with multi-resolution CNNs. IEEE Trans. Image Process. **26**, 2055–2068 (2017)
19. Xiao, L., Yan, Q.: Scene classification with improved AlexNet model. In: 12th International Conference on Intelligent Systems and Knowledge Engineering (2017)
20. Krizhevsky, A., Sutskever, I., Hinton, G.: ImageNet classification with deep convolutional neural networks. Commun. ACM **60**, 84–90 (2017)
21. Xiao, J., Ehinger, K., Hayes, J., Torralba, A., Oliva, A.: SUN database: exploring a large collection of scene categories. Int. J. Comput. Vision **119**, 3–22 (2016)
22. Herranz, L., Jiang, S., Li, X.: Scene recognition With CNNs: objects, scales and dataset bias. In: IEEE Conference on Computer Vision and Pattern Recognition, pp. 571–579 (2016)
23. Confusion Matrix. https://towardsdatascience.com/understanding-confusion-matrix-a9ad42dcfd62. Last accessed 2018/03/07
24. Jung, W., Jung, D., Kim, B., Lee, S., Rhee, W., Ho, J.: Restructuring batch normalization to accelerate CNN training. In: Proceedings of the 2nd SysML Conference, Palo Alto, CA, USA (2019)
25. Activation Function. https://towardsdatascience.com/activation-functions-and-its-types-which-is-better-a9a5310cc8f. Last accessed 2018/01/18
26. Pooling Layers. https://machinelearningmastery.com/pooling-layers-for-convolutional-neural-networks/. Last accessed 2018/01/18
27. Pascanu, R., Mikolov, T., Bengio, Y.: On the difficulty of training recurrent neural networks. Technical report, Cornell University (2013)
28. Relu—A Gentle Introduction to Rectified Linear Unit (Relu). https://machinelearningmastery.com/rectified-linear-activation-function-for-deep-learning-neural-networks/. Last accessed 2018/01/18
29. Optimization and Learning Rate. https://towardsdatascience.com/neural-network-optimization-algorithms-1a44c282f61d. Last accessed 2018/01/18
30. Batch size, Epoch, Learning Rate. https://stats.stackexchange.com/questions/360157/epoch-vs-iteration-in-cnn-training. Last accessed 2018/12/12
31. Precision and Recall. https://scikit-learn.org/stable/auto_examples/model_selection/plot_precision_recall.html. Last accessed: 2019/09/19
32. Confusion Matrix and Classification Report Support. https://stackoverflow.com/questions/30746460/how-to-interpret-scikits-learn-confusion-matrix-and-classification-report. Last accessed 2019/09/19

Multilingual Opinion Mining Movie Recommendation System Using RNN

Tarana Singh, Anand Nayyar and Arun Solanki

Abstract Twitter is a news and social networking site where people around the world post their blogs and share their feeling, point of view, and comments regarding any communication or about any latest movie, etc. Thus, Twitter generates a massive quantity of Twitter data every day. This data is real time, which is being used in the proposed work for implementing a "movie recommendation system." To enhance the performance of the framework, sentimental analysis is also being applied to the data. Nowadays, the recommendation system is also an essential tool for online businesses and used by various e-commerce sites, music applications, entertainment sites, etc. This work proposed a movie recommendation system for the movie domain which is developed using real-time multilingual tweets. These tweets are obtained from Twitter API using the LinqToTwitter Library. Sentimental analysis is also being performed on tweets. In this work, multilingual and real-time tweets are considered. These tweets are translated into the target language using Google Translate API. The proposed work used the Stanford library for preprocessing, and RNN is used for classifying the tweets. The tweets are classified as positive, negative, and neutral tweets. Preprocessing of the tweets is done to remove unwanted words, URLs, emoticons, etc. Finally, based on the classification, the movie is suggested to the user. This proposed work is better than the current practices as the implementation is being done on real-time tweets, and sentimental analysis is also being performed to get better results. This system is achieving 91.67% accuracy, 92% precision, 90.2% recall, and 90.98% f-measure.

Keywords Recurrent neural network · Artificial neural network · Natural language processing · Text categorization · Twitter API · The movie database

T. Singh (✉) · A. Solanki
Gautam Buddha University, Greater Noida, India
e-mail: taranasingh14@gmail.com

A. Solanki
e-mail: Ymca.arun@gmail.com

A. Nayyar
Duy Tan University, Da Nang, Vietnam
e-mail: anandnayyar@duytan.edu.vn

© Springer Nature Singapore Pte Ltd. 2020
P. K. Singh et al. (eds.), *Proceedings of First International Conference on Computing, Communications, and Cyber-Security (IC4S 2019)*, Lecture Notes in Networks and Systems 121, https://doi.org/10.1007/978-981-15-3369-3_44

589

1 Introduction

Recommendation systems are the subclass of the information filtering systems. These systems are used in a variety of areas, YouTube, Netflix, Amazon, Facebook, Twitter, Instagram, etc. [1–3]. Recommendation systems are basically of two kinds; for instance, content-based recommendation systems, these systems predict based on item properties [4–6]. Another sort of recommendation system is collaborative filtering [7–11]; these systems predict based on comparing the user's history [12–15]. Other than these two techniques, there is also the system named hybrid recommendation system. In this technique, both the collaborative and content-based filtering techniques are combined [16–19]. In this technique, content is used to infer ratings in case of the sparsity of ratings [20–22]. Both of these techniques are used in most recommendation systems at present. Netflix movie recommendation system is an example of a hybrid recommendation system [23, 24]. Sentiment analysis (SA) is also called opinion mining. Sentiment analysis is a logical mining of content which distinguishes and concentrates emotional data in source material. This will help in business to comprehend the social supposition of their image, item, or administration while checking on the Web discussions. However, investigation of online lifestreams is typically limited to the fundamental feeling or opinion analysis [25–28]. This is similar to digging the surface and omitting out the insights those need to be discovered. There are three approaches to sentimental analysis [29–33]. "Lexicon-based approach" of sentimental analysis is the unsupervised technique. Categorization is finished by looking at the highlights of a given content against conclusion vocabularies whose feelings are determined before the utilization [34–36]. The lexicon-based techniques to sentiment analysis are unsupervised learning because it does not require prior training to classify the data. [37–39]. "Machine learning approach" applicable to sentiment analysis mostly belongs to supervised classification. In this approach, there are two sets of documents, i.e., a training set and test set. The training set is used by an automatic classifier to learn the differentiating characteristics of materials, and a test set is used to check how well the classifier performs. Several techniques are Naïve Bayes, maximum entropy, support vector machine, etc. These techniques have achieved great success in sentiment analysis [40, 41]. The hybrid approach of sentimental analysis is the combination of both the above methods of sentimental analysis. Past research has shown that if the ML and LB both are combined, the performance of SA will improve [42].

2 Recurrent Neural Network

Recurrent neural network is a type of neural network that contains guided loops; these ends represent the promotion of activation for future entries in a sequence. Instead of accepting the same vector input X as a test example, an RNN can take the series of vector inputs $(x_1, x_2, x_3, \ldots, x_t)$ for arbitrary, where T is the value of the variable. When there is a single dimension in each x_t in our specific application, each x_t is a vector representation of a word, and $(x_1, \ldots, x_t))$ is a sequence of words in the movie review; it shares that RNN can imagine "unrolling" to copy to a network in each of all other copies of the same weight [43]. For each loop edge in the system, we connect the side to the same node in the x_t network, thereby creating a chain, which breaks any loop and gives us the standard forward neural network feed technique. First, calculate the value for it, the input gate, and the candidate value for the states of the memory cells at a time.

$$i_t = \sigma w_i x_t + U_i h_i + b_i \tag{1}$$

$$c_t = \tanh(w_i x_t + U_i h_t - 1 + bc) \tag{2}$$

Secondly, calculate the value of the function of activating the forget gate of the memory cells at time t:

$$f_t = \sigma w_f x_t + U_f h_t - 1 + b_f \tag{3}$$

Given the value of the input gate activation, the forget gate activation, and the candidate state value, now we can compute the memory cells' new state at time t:

$$c_t = i_t * c_t + f_t * c_t - 1 \tag{4}$$

With the new state of the memory cells, we can compute the latest value of the output gates and subsequently their outputs:

$$o_t = (w_o x_t + U_i h_t - 1) \tag{5}$$

Figure 1 shows the general architecture of RNN, also used in the proposed study [44–47]. This architecture used an embedded layer to break down the data into a sentence. These sentences are converted into words. In these words, the filter is applied and maps to output. In the proposed work, the main objective is to build a recommendation system for the movie domain. This system will use real-time Twitter data to generate predictions [48–50].

3 Research Paper Organization

The proposed work starts with Part I as introduction of the basics of recommendation system, sentimental analysis, and RNN. Part II discusses the latest work done by recent authors with the details of the techniques and tools used by different authors. Part III describes the architecture of the proposed recommendation system. Part IV shows the flowchart of the proposed method. Part V shows the algorithm of the

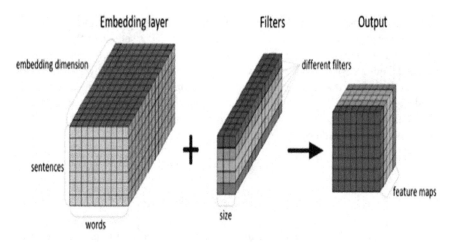

Fig. 1 RNN general architecture

proposed system. Part VI discusses the implementation of the proposed system with the help of the snapshot of the system while using the system. Part VII is presenting the results obtained in the proposed method. Part VIII is explaining the conclusion, and finally, in Part IX, the future work of the proposed system is stated. In the last part, X, in this paper, all the references are given.

4 Literature Review

Zhang et al. [36], the author, reviewed the preprocessing of tweets in detail. The author tackled various models as the results obtained by a solitary model may betray as they rely upon the data of that particular model. The author discussed that employing multiple models will assemble the exactness of the machine. The author used NB classifiers, which exhibits the most significant outcomes when appeared differently concerning SMO, SVM, and random forest. Rajput et al. [37] recommended that opinion word is utilized in many feeling characterization assignments. Positive sentiment phrases are being utilized to state any ideal conditions, while negative contemplations are being utilized to unwanted state conditions. There are some sentences and idioms, which are said to be as one as the lexicon of sentiment. There are three primary ways to deal with arranging or gathering rundown of sentiment words. The standard methodology takes too much time and is not utilized alone, and it is commonly joined with the other two computerized approaches because to keep mistakes away from mechanized techniques. Fradkin et al. [39], the author, proposed a recommendation system to foresee the user's review and information, dominatingly from massive collected data to suggest their preferences. The movie suggestion framework is a framework that helps users in characterizing users with common preferences. This framework (K-means cuckoo) has 0.68 MAE. Solanki et al. [49], the author, displayed a recommendation framework using "K-means clustering" and KNN, and these works for different estimations of "RMSE" are acquired. In this work, if authors are decreasing the no. of observations, the estimate of "RMSE" reduces. The most significant evaluation of "RMSE" got is 1.081648.

In the above discussion, there are various articles given by multiple writers who chip away at the sentimental analysis and recommendation system. In the existing systems, past researchers utilize the datasets from the given repositories, for instance, the GroupLens datasets, MovieLens datasets, and so forth; these datasets are the static datasets which are accessible on the Web on various Web sites. Previously, the authors took a shot at these available datasets to accomplish some degree of precision.

5 Architecture

Figure 2 demonstrates the working of the projected framework. It comprises three components, specifically the information component, a preparing component, and a yield component.

- **Input Module**: In this component, the client needs to give a contribution to the framework. The framework will pick the current date and area of the client. This data is additionally given to the next module, for example, the preprocessing; i.e., the output of the first module will be the input of the processing module.

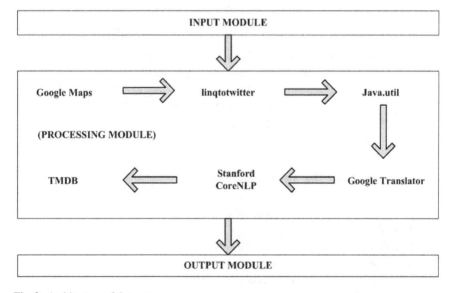

Fig. 2 Architecture of the system

- **Processing Module**: In the module, the framework procures the information from Twitter API, deciphers the multilingual information utilizing Google interpreter into the English language, and produces it accordingly.

  - **GoogleMap**: It is used to obtain the current geographical location of the user.
  - **LinqToTwitter**: It is used to link to the Twitter API to access the real-time Twitter data.
  - **Google Translate**: It is used to translate the other languages tweet into the target language, English.
  - **java.util**: It is used to access the property class to obtain the properties from the data and pass these properties to the model for training.
  - **Stanford NLP**: This library is used to train the model.
  - **TMDB**: This library is used to fetch the movie's data from the Web.

  This module will perform various operations like downloading the data, preprocessing the data, and finally applying the RNN classifier to categorize the tweets into positive, negative, and neutral categories. According to the input from the user, the system will generate a recommendation that has high positive reviews and will go to the next module that is an output module.
- **Output Module**: The yield module depicts the anticipated movie that the info client may like. The output of the processing module will be the input of this module.

6 Flowchart

Figure 3 demonstrates the stream outline of the movie recommendation framework. This chart reflects the procedure stream of the suggested framework, which portrays how the structure is functioning, how the system is managing the crude information, and how the framework predicts the beautiful motion pictures as per the contribution from the client.

7 Pseudocode of the Proposed System

Pseudocode of the proposed framework has the following steps:
Step 1: Input search text.
Step 2: System date and geographical location are automatically detected.
Step 3: Choose language.
Step 4: Search and download the data from Twitter.
Step 5: Translate the downloaded data using Google Translate.
Step 6: Preprocess the data using Stanford NLP.
Step 7: Apply RNN for the classification of the tweets.

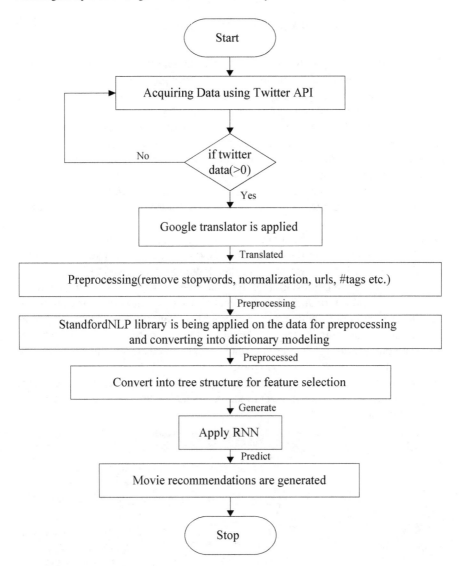

Fig. 3 Flowchart

Step 8: Movie recommendations are generated.
Step 9: Accuracy of the system is evaluated.

8   Implementation of the Proposed System

Step 1: In step 1, the proposed system search the query at client side to produce the recommendations. This framework will naturally pick the system date and geographical location using GoogleMap. The primary motivation behind the current geographic location of the client is that the proposed system can recognize in which geographical area the client is present so the framework can identify the most recent and mainstream movie. This encourages the context to produce the most appropriate recommendations to the user. Another part here is to choose the language. When we connect to the Twitter information at that point, there are the tweets in various languages like Hindi, Marathi, Spanish, and so forth recommended style to choose all language so the framework will obtain every tweet from the Twitter as presented in Fig. 4.

Step 2: After Step 1, users need to click on the search and download button, and the downloading of the twitter data will start. The downloaded information is stored in a.txt file and stored in the system for further operations as shown in Fig. 5a, b.

Step 3: In this step, the translation of the data is being performed using the Google Translator. As the information that is collected from Twitter is multilingual, that is why the conversion of the data into the target language (i.e., English) is essential. To translate the data into the targeted language, the user needs to click on the button "translate the data using Google Translate." Converted data is again stored in a.txt file for further operations as shown in Fig. 6a, b.

Step 4: The translated tweets are used to perform preprocessing. Preprocessing is done to eliminate the stop words, emoticons, hashtags, URLs, etc., and sentimental

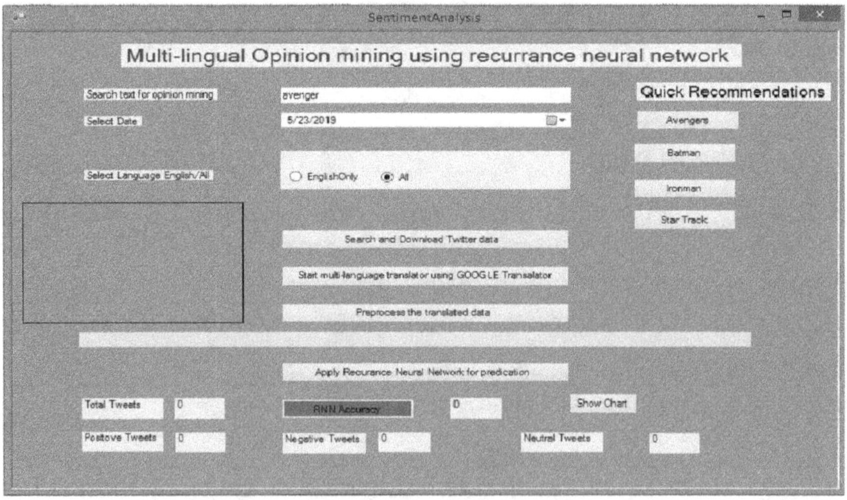

Fig. 4   Input data

(a)

(b)

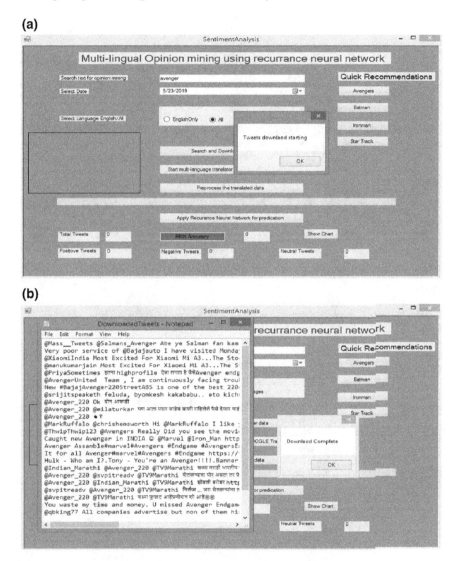

Fig. 5 **a** Tweet downloading. **b** Tweet downloading complete and .txt file of tweets

analysis is shown on the left data, which is without any unwanted words and characters. After this step is completed, we got the data on which the user can perform further tasks. After preprocessing, the processed tweets are stored in another.txt file, as shown in Fig. 7a, b. This file is used for further operations, i.e., sentimental analysis of the tweets.

Step 5: Preprocessing of the downloaded data is done in the above step using the Stanford NLP Library. Now, the classification of the preprocessed tweets is being done using RNN classification in Fig. 8. The tweets are classified into different

(a)

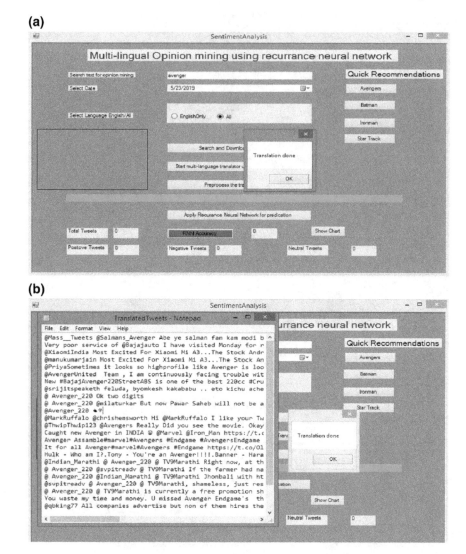

(b)

Fig. 6 **a** Translating the tweets. **b** Translated tweets

categories, namely positive tweets, negative tweets, and neutral tweets.

Step 6: In this step, based on the above classification, the recommendations to the user are being generated in the grid view. To produce the movie's recommendations in the system, the TMDB library is used. The user needs to enter the text, i.e., movie name in the searching textbox, and after all the above steps, the most prominent and exceedingly positive movie is predicted to the user by the system. The developer can manage the number of movies generated, which will be shown in the grid view. This is shown in Fig. 9.

(a)

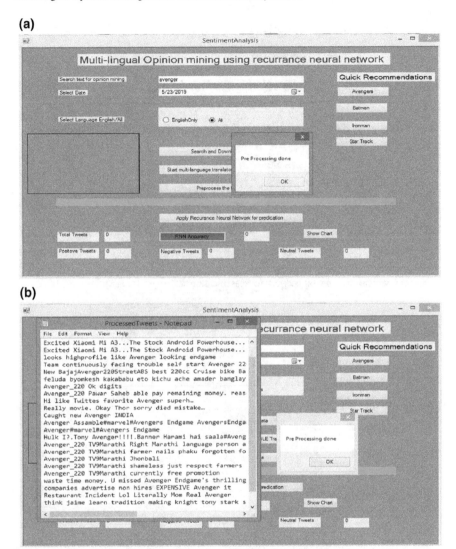

(b)

Fig. 7 **a** Preprocessing of tweets. **b** Preprocessed tweets

9 Result Analysis

In Fig. 10, the accuracy of the proposed framework is being represented, i.e., recommending the movies to the user according to the input data from the user with 91% accuracy.

Figure 11 shows the result where all the different positive, negative, and neutral tweets are being represented on the pie graph and all the values are in the %, so the pie chart is being utilized to describe the results of the proposed system.

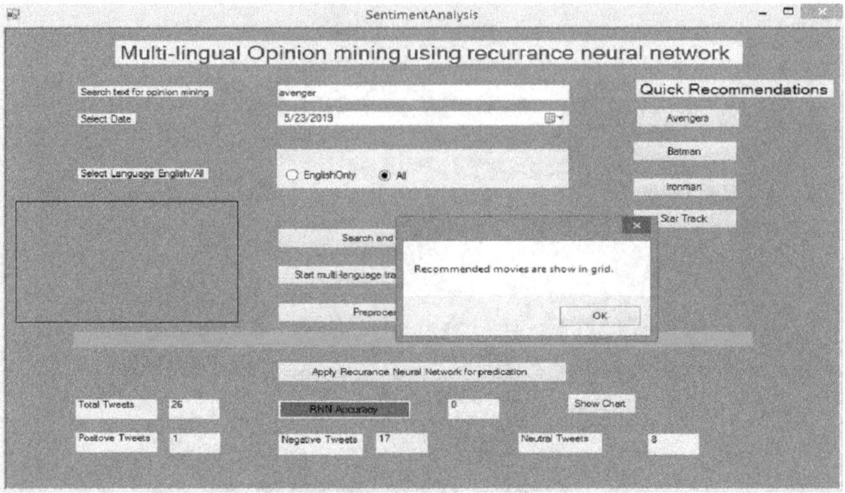

Fig. 8 Apply recurrence neural network for classification

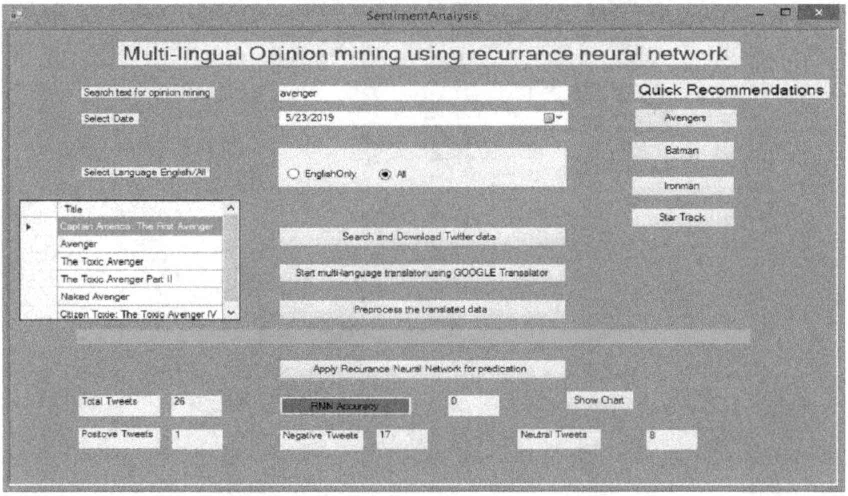

Fig. 9 Recommendations for the users

In Table 1, the result of the proposed system is represented with accuracy, precision, recall, and f-measures. The result of the proposed method is also compared with the results of the existing systems. The proposed work is improved than the already present work, as the proposed work is to obtain better outcomes [14, 27].

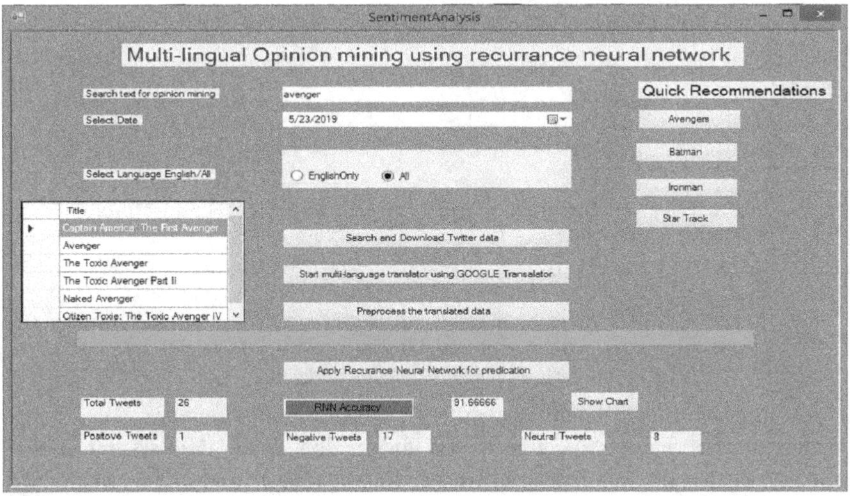

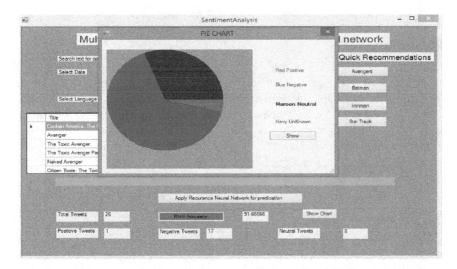

Fig. 10 Results of projected system

10 Conclusion

In this work, we have developed a movie recommendation framework using the TMDB database. RNN classification is applied for the sentimental analysis of the tweets; then, the movie recommendations are generated to the user based on the input given to the system by the user. In this proposed system, different operations are being done on the tweets like downloading of the information, interpretation

Fig. 11 PIE chart representation of results

Table 1 Result of performance measures for RNN and comparison with other classifiers

S. No.	Classifier	Dataset	Accuracy (%)	Precision (%)	Recall (%)	F-measures (%)
1	RNN (proposed work)	Twitter	91.67	92	90.2	90.98
2	Naïve Bayes (Soni)	Twitter	77.16	78.12	74.5	76.15
3	SVM (HailongZhang et al.)	Twitter	79.5	79	79	79

of the information, per-handling of the information, and afterward, the grouping of the data. For the classification of the information, the RNN classifier is being utilized, which is grouping the information with 91.67% accuracy. The discernments make it extraordinarily sure that the RNN classifier defeats each other classifier in anticipating the suspicions with the precision of 92%. This work has differentiated unmistakable request computations by getting the best results. There are various challenges for sentimental analysis heretofore. While attempting to counter this, this system used RNN classifiers which fall under AI frameworks to organize the ends did by slangs, incorrect spellings, emoticons, multilingual contentions, and different event of words and achieved a high precision. In this work, continuous Twitter information is being used utilizing different Twitter API keys to get to the report. In this way, the proposed framework is delivering the outcomes as indicated by the most recent discharge and fame of the movies in the specific geographical spam where the client is while scanning for the proposals.

11 Future Work

In the proposed work, the recommendation system is working only for the movie domain, but in the future, this system can be extended for the universal recommendation system, i.e., a system which can generate recommendations for other fields. In the future, these emoticons can be added in preprocessing to obtain the emotions of the users as these emoticons are also the part of the user's sentiments. So, in the future work, some techniques can be used for sentimental analysis.

References

1. Ziegler, C.N., McNee, S.M., Konstan, J.A., Lausen, G.: Improving recommendation lists through topic diversification. In: Proceedings of the 14th International Conference on World Wide Web, New York, NY, USA, vol. 2131, no. 34, pp. 222–234 (2005)

2. Gokulakrishnan, B., Priyanthan, P., Raghavan, T., Prasath, N., Perera, A.: Opinion mining and sentiment analysis on a twitter data stream. In: The International Conference on Advances in ICT for Emerging Regions, vol. 46, no. 12, pp. 182–188 (2012)
3. Singh, Y., Bhatia, P., Sangwan, O.: A review of studies on machine learning techniques. Int. J. Comput. Sci. Secur. 1(1), 70–84 (2007)
4. Tejeda, A.: A quality-based recommender system to disseminate information in a university digital library. Inf. Sci. 261(32), 52–69 (2014)
5. Fan, Y., Dong, L., Sun, X., Wang, D., Qin, W., Aizeng, C.: Research on auto-generating test-paper model based on spatial-temporal clustering analysis. In: Huang, D.S., Jo, K.H., Zhang, X.L. (eds.) Intelligent Computing Theories and Application, ICIC, Lecture Notes in Computer Science, vol. 10955, no. 2342, pp. 238–255. Springer, Cham (2018)
6. Luo, X., Zhou, M., Xia, Y., Zhu, Q.: An efficient non-negative matrix-factorization-based approach to collaborative filtering for recommender systems. IEEE Trans. Industr. Inf. 10(2), 1231–1245 (2014)
7. Kanta, V., Bharadwaj, K.: Enhancing recommendation quality of content-based, filtering through collaborative predictions and fuzzy similarity measures. Procedia Eng. 38(21), 939–942 (2012)
8. Lops, P., Gemmis, M., Semeraro, G.: Content-based recommender systems: state of the art and trends. In: Ricci, F., Rokach, L., Shapira, B., Kantor P. (eds.) Recommender Systems Handbook, vol. 2131, no. 34, pp. 2371–2384. Springer, Boston, MA (2011)
9. Hu, N., Bose, I., Koh, N.S., Liu, L.: Manipulation of online reviews: an analysis of ratings, readability, and sentiments. Decis Support Syst. 52(12), 674–684 (2012)
10. Jiang, J., Lu, J., Zhang, G., Long, G.: Scaling-up Item-based collaborative filtering recommendation algorithm based on Hadoop. In: 2013 IEEE World Congress on Services, vol. 52, no. 12, pp. 4–9 July 2013
11. Adeniyi, D., Wei, Z., Yongquan, Y.: Automated web usage data mining and recommendation system using k-nearest neighbor (KNN) classification method. Saudi Comput. Soc. King Saud Univ. 2152(120), 34–49 (2014)
12. Kataria, R., Verma, O.P.: An effective collaborative movie recommender system with cuckoo search. Egypt. Inform. J. 18(2), 105–112 (2019)
13. Xiao, P., Liangshan, S., Xiuran, L.: Improved collaborative filtering algorithm in the research and application of personalized movie recommendations. In: Fourth International Conference on Intelligent Systems Design and Engineering Applications, vol. 56, no. 6, pp. 401–414 (2013)
14. Munoz-Organero, M., Gustavo, A. González, R., Pedro, J., Delgado, C.: A Collaborative Recommender System Based on Space- Time Similarities, IEEE Pervasive Computing, vol. 2131, no. 34, pp. 232–246 (2010)
15. Czarnowski, I., Jdrzejowicz, P.: Data reduction algorithm for machine learning and data mining. In: Nguyen, N.T., Borzemski, L., Grzech, A., Ali, M. (eds.) New Frontiers in Applied Artificial Intelligence. IEA/AIE, Lecture Notes in Computer Science, vol. 5027, no. 2032, pp. 80–94. Springer, Berlin (2008)
16. Duarte, D., Stahl, N.: Machine learning: a concise overview. In: Said, A., Torra, V. (eds.) Data Science in Practice. Studies in Big Data, vol. 46, no. 12, pp. 95–115. Springer, Cham (2019)
17. Shamri, A.: Fuzzy-genetic approach to recommender systems based on a novel hybrid user model, expert systems with applications. In: 3rd International Conference on Computer Science and Information Technology, IEEE, vol. 1267, no. 104, pp. 300–313 (2008)
18. Jinming, H.: Application and research of collaborative filtering in e-commerce recommendation system. In: 3rd International Conference on Computer Science and Information Technology, vol. 1567, no. 204, pp. 338–352 (2010)
19. Jiang, Z., Zang, W., Liu, X.: Research of K-means clustering method based on DNA genetic algorithm and P system. In: Zu, Q., Hu, B. (eds.) Human-Centered Computing. HCC Lecture Notes in Computer Science, vol. 9567, no. 214, pp. 145–158. Springer, Cham (2016)
20. Yan, B., Chen, G.: AppJoy: personalized mobile application discovery. In: Proceedings of the 9th International Conference on Mobile Systems Applications and Services—MobiSys 11, vol. 2340, no. 123, pp. 11–25 (2011)

21. Davidsson, C., Moritz, S.: Utilizing implicit feedback and context to recommend mobile applications from first use. In: Proceedings of the Ca RR 2011, vol. 1240, no. 103, pp. 19–22. ACM Press, New York (2011)

22. Bilge, A., Kaleli, C., Yakut, I., Gunes, I., Polat, H.: A survey of privacy-preserving collaborative filtering schemes. Int. J. Softw. Eng. Knowl. Eng. **40**(13), 1085–1108 (2013)

23. Veselka, M., Schindler, K.: Mutual information estimation in higher dimensions: a speed-up of a k-nearest neighbor based estimator. In: Beliczynski B., Dzielinski A., Iwanowski M., Ribeiro B. (eds.) Adaptive and Natural Computing Algorithms. ICANNGA 2007. Lecture Notes in Computer Science, vol. 4431, no. 32, pp 2231–2245, Springer, Berlin (2007)

24. Calandrino, J.A., Kilzer, A., Narayanan. A., Felten, E.W., Shmatikov, V.: You might also like: privacy risks of collaborative filtering. In: Proceedings of the IEEE Symposium on Security and Privacy, vol. 1421, no. 26, pp. 231–246, Oakland, CA, USA (2011)

25. Soni, A.: Multi-lingual sentiment analysis of twitter data by using classification algorithms accepted to publish in IEEE, vol. 1238, no. 33, pp. 1022–1034 (2017)

26. Ahuja, R., Solanki, A.: Movie recommender system using K-Means clustering and K-Nearest Neighbor. In: Accepted for Publication in Confluence-2019: 9th International Conference on Cloud Computing, Data Science & Engineering, Amity University, Noida, vol. 1231, no. 21, pp. 25–38 (2019)

27. Ko, S.K.: A smart movie recommendation system. In: Smith, M.J., Salvendy, G. (eds.) Human Interface and the Management of Information. Interacting with Information. Human Interface, 2011. Lecture Notes in Computer Science, vol. 6771, no. 12, pp. 628–642. Springer, Berlin (2011)

28. Argamon, S., Bloom, K., Esuli, A., Sebastiani, F.: Automatically determining attitude type and force for sentiment analysis. In Human Language Technology. Challenges of the Information Society Springer, vol 223 issue 23, pp. 218–231, (2009)

29. Jinming, H.: Application and research of collaborative filtering in e-commerce recommendation system. In: 2010 3rd International Conference on Computer Science and Information Technology, vol. 40, no. 13, pp. 151–164 (2010)

30. Neri, F., Aliprandi, C., Capeci, F., Cuadros, M., Tomas.: Sentiment Analysis on Social Media. In: ACM International Conference on Advances in Social Networks Analysis and Mining, vol. 39, no. 20, pp. 324-338 (2012)

31. Xie, H.T., Meng, X.W.: A personalized information service model adapting to user requirement evolution. Acta Electron. Sin. **39**(3), 643–648 (2011)

32. Polat, H., Du, W.: Privacy preserving top n recommendation for distributed data. J. Am. Soc. Inf. Sci. Technol. **59**(7), 1093–1108 (2008)

33. Yakut, I., Polat, H.: Estimating NBC-based recommendations on arbitrarily partitioned data with privacy. Knowl. Based on Syst **36**(10), 2163–2178 (2012)

34. Okkalioglu, M., Koc, M., Polat, H.: On the discovery of fake binary ratings. In: Proceedings of the 30th Annual ACM Symposium on Applied Computing, ACM, USA, vol. 122. no. 20, pp. 901–907 (2015)

35. Wang, L.C., Meng, X.W., Zhang, Y.J.: A cognitive psychology-based approach to user preferences elicitation for mobile network services. Acta Electron. Sin. **39**(11), 2547–2553 (2011)

36. Zhang, H., Gan, W., Jiang, B.: Machine Learning and Lexicon based methods for sentiment classification: a survey. In: 11th Web Information System and Application Conference IEEE, vol. 132, no. 24, pp. 262–265 (2017)

37. Rajput, R., Solanki, A.: Real-time analysis of tweets using machine learning and semantic analysis. In: International Conference on Communication and Computing Systems (ICCCS-2016), Taylor and Francis, at Dronacharya College of Engineering, Gurgaon, 9–11 Sept, vol 138 issue 25, pp. 687–692, (2016)

38. Munoz-Organero, M., Ramíez-González, G.A., Munoz-Merino, P.J., Delgado, K.: A collaborative recommender system based on space-time similarities. IEEE Pervasive Comput. 2010, vol. 12 no. 5, pp. 1023–1039 (2017)

39. Fradkin, D., Muchnik, I.: Support vector machines for classification. In: Discrete Methods in Epidemiology, DIMACS Series in Discrete Mathematics and Theoretical Computer Science, vol. 70, no. 38, pp. 13–20 (2018)
40. Adomavicius, G., Tuzhilin, A.: Toward the next generation of recommender systems: a survey of the state-of-the-art and possible extensions. IEEE Trans. Knowl. Data Eng. **124**(10), 1051–1068 (2017)
41. Celma, O., Herrera, P.: A new approach to evaluating novel recommendations. In: Proceedings of the 2008 ACM Conference on Recommender Systems, vol. 512, no. 7, pp. 1028–1042, ACM, New York (2008)
42. Gamon, M.: Sentiment classification on customer feedback data: noisy Data, large feature vectors, and the role of linguistic analysis. In: Proceedings of the International Conference on Computational Linguistics (COLING), vol. 21, no. 15, pp. 841–847 (2004)
43. Kaur, N., Solanki, A.: Sentiment knowledge discovery in twitter using CoreNLP library. In: 8th International Conference on Cloud Computing, Data Science and Engineering (Confluence), vol. 345, no. 32, pp. 2342–2358 (2018)
44. Bell, R., Koren, Y., Volinsky, C.: Modeling relationships at multiple scales to improve the accuracy of large recommender systems. In: Proceedings of the 13th ACM SIGKDD International Conference on Knowledge Discovery and Data Mining, vol. 1221, no. 23, pp. 1234–1248. ACM, New York (2007)
45. Mishra, N., Chaturvedi, S., Mishra, V., Srivastava, R., Bargah, P.: Solving sparsity problem in rating-based movie recommendation system. In: Behera, H., Mohapatra, D. (eds.) Computational Intelligence in Data Mining. Advances in Intelligent Systems and Computing, vol. 556, no. 14, pp. 1231–1248. Springer, Singapore (2017)
46. Das, D., Chidananda, H.T., Sahoo, L.: Personalized movie recommendation system using twitter data. In: Pattnaik, P., Rautaray, S., Das, H., Nayak, J. (eds.) Progress in Computing, Analytics, and Networking. Advances in Intelligent Systems and Computing, vol. 710, no. 11, pp. 1232–1248. Springer, Singapore (2018)
47. Rajput, R., Solanki, A.: Review of sentimental analysis methods using lexicon based approach. Int. J. Comput. Sci. Mob. Comput. **5**(2), 159–166 (2016)
48. Sahoo, A., Pradhan, C., Barik, R., Dubey, H.: DeepReco: deep learning-based health recommender system using collaborative filtering. Computation **7**(2), 1283–1299 (2019)
49. Pandey, S., Solanki, A.: Music instrument recognition using deep convolutional neural networks. Int. J. Inf. Technol. **13**(3), 129–149 (2019)
50. Agarwal, A., Solanki, A.: An improved data clustering algorithm for outlier detection. Self-organology **3**(4), 121–139 (2016)

Analysis of Ex-YOLO Algorithm with Other Real-Time Algorithms for Emergency Vehicle Detection

Abhishek Baghel, Aprajita Srivastava, Aayushi Tyagi, Somya Goel and Preeti Nagrath

Abstract This paper highlights the topic of object recognition and its algorithms, used in detection of different objects. By using the concept of object recognition, emergency vehicles like ambulances, fire trucks and ambulances can be identified on the way so that a clear path can be made for them. This can be used in autonomous vehicles as an algorithm to identify emergency vehicles. This paper is an extension of a previous paper (Goel et al. in Intelligent communication, control and devices. Springer, Singapore [1]), highlighting present techniques related to object detection. In this paper, the focus is to develop a machine learning model that is built on YOLO algorithm and is able to recognize emergency vehicles. Model proposed is Extended-YOLO (Ex-YOLO), an extension of YOLO. It is basically an ML pipeline with two phases, one is the YOLO algorithm that creates bounding boxes across objects and the other would be the classifier that would detect emergency vehicles. The output from the YOLO algorithm is cropped out and then preprocessed. Image tensors are then passed to Phase II classifiers that are pretrained and classified the images to further categories. Classifiers present in Phase II of the pipeline work independently.

Keywords YOLO algorithm · Classifiers · Emergency vehicles · Autonomous driving

A. Baghel · A. Srivastava · A. Tyagi (✉) · S. Goel · P. Nagrath
Computer Science and Engineering (CSE), Bharati Vidyapeeth's College of Engineering, New Delhi, India
e-mail: aayushityagi2712@gmail.com

A. Baghel
e-mail: abhishekbaghel165@gmail.com

A. Srivastava
e-mail: aprajita4@gmail.com

S. Goel
e-mail: somyagoel77@gmail.com

P. Nagrath
e-mail: preeti.nagrath@bharatividyapeeth.edu

© Springer Nature Singapore Pte Ltd. 2020   607
P. K. Singh et al. (eds.), *Proceedings of First International Conference on Computing, Communications, and Cyber-Security (IC4S 2019)*, Lecture Notes in Networks and Systems 121, https://doi.org/10.1007/978-981-15-3369-3_45

1 Introduction

The use of autonomous cars is immense; it is a fast-growing field in which the machine has to deal with many different types of objects present on the road. The area of emergency vehicle detection is an application of object recognition where it is used to identify emergency vehicles among normal vehicles on the road. For example, suppose an ambulance is there behind the car, so if the car is able to detect it then it moves aside and gives space to ambulance. Similarly, other emergency vehicles such as police vehicle and fire station vehicles can also be detected so that people are able to deal with appropriate actions. This paper aims to give a model using which emergency vehicles can be detected and then handled accordingly. It finds its use mainly for autonomous vehicles where the vehicles solely rely on object detection and GPS for navigation.

This paper is an extension of a previous paper [1], highlighting present techniques related to object detection. Objective is to find an optimal way to identify and track emergency vehicles, which is fast and accurate. For the sake of this problem, we studied various present models that could serve our objective.

Out of all the models studied (region-based convolutional networks (R-CNN), Single Shot Detector (SDD) YOLO algorithm and Fast-YOLO algorithm), YOLO and Fast-YOLO are the only models which work in real time, which was a huge deciding factor for the nature of the problem statement. YOLO has a high precision rate with a medium FPS count, while Fast-YOLO is extremely fast, about 3 times faster than YOLO but had less precision for various classes. The frames per second in case of the YOLO algorithm is FPS = 45 and of Fast-YOLO is FPS = 155. Fast-YOLO is able to detect various objects in an image in a single shot. Hence, it is so fast.

Multiple objects inside bounding boxes along with class probabilities for the objects inside boxes are predicted by a single convolutional network simultaneously. YOLO directly optimizes detection performance and trains on full images. This model has various advantages over other models of object recognition [2].

- YOLO is fast. As frame detection is a regression problem, there is no need for a complicated pipeline. At test time, the neural network is simply run on a new image to predict detections. It allows processing streaming video in real time with latency of less than 25 ms.
- YOLO reasons about the image globally while predicting results. YOLO sees the entire image during training and test time, unlike sliding window and region proposal-based techniques.
- Generalizable representations and relationships of objects are learnt by YOLO.

So, to implement this, Ex-YOLO has two major processes with preprocessing at the beginning. The first phase is applying YOLO algorithm that creates bounding boxes around the objects and differentiates them from the environment. Next phase includes training the classifier by using python algorithms so that it is able to detect emergency vehicles.

2   Literature Survey

Yang et al. [3] have proposed a method that improves upon the correct YOLOv2 framework for the purpose of vehicle detection in a real-time manner. The method used to improve the framework includes optimization of the parameters present in the model including a expansion of the grid size as well as improving the number and sizes of anchors within the model. With this method, the accuracy rate came out to be 91.80%. Tao et al. [4] talk about the slow and inaccurate nature of the model when a classifier is executed on the features extracted from an object recognition algorithm. In this paper, the problem was overcome by optimizing the loss functions encountered in YOLO framework, which could increase the speed by 1.18 times the previous speed. Zhang et al. [5] used an end-to-end convolution network that is based on YOLOv2. Molchanov et al. [6] in their paper showed that YOLO algorithm showed really competitive results against the Fast-R-CNN algorithm, which is usually much faster in nature.

Grimnes et al. [7] in their paper create an object recognition model that is based on bounding box regression networks. They trained the model to correctly recognize the objects on the road such as cars, bikes, trucks and pedestrians. Tang et al. [8] investigate the use of YOLO algorithm for vehicle detection and explored other methods as well to achieve similar results. Du [9] talks about how the purpose of object recognition and the main priority is to accelerate the speed of the model so that the model can work in a real-time manner. This paper mentions YOLO, based upon the convolution neural network is a suitable option due to its high speed and accuracy. The advanced version of YOLO, YOLOv2 works even better and achieves an excellent trade-off between speed and accuracy as well as the ability to represent the whole image along with the object detector. Zhihuan et al. [10] state that deep convolutional neural networks (DCNNs) are a hotspot for object detection due to its extremely good ability to extract characteristics and the excellent outcomes given by it in the subject of computer vision. The paper talks about how YOLO model performs object recognition. YOLO model frames it as a regression problem. It used a single convolution neural network, which predicts the bounding boxes in an end-to-end manner, which makes it fast and highly accurate. Sentas et al. [11] in their paper created their own vehicle dataset with the help of an uncalibrated camera. They used Tiny-Yolo algorithm for real-time object recognition. This algorithm was followed by a support vector machine (SVM) classifier, which helped in classifying real-time streaming traffic video data. Chakraborty et al. [12] in their study show YOLO algorithm as well as deep convolutional neural network (DCNN) as a means to detect traffic congestion from camera images. To draw a comparison, they also used the support vector machine to understand the improvements that would come with using a deep learning algorithm, unlike SVM, which is a shallow algorithm. YOLO and DCNN achieved an accuracy rate of 91.5% and 90.2%, whereas SVM showed only 85.2% accuracy.

3 Proposed Methodology

There are three crucial steps in the development of Extended-YOLO (Ex-YOLO). The first step is the selection of the perfect algorithm for emergency vehicle detection, the next step deals with preprocessing the data to refine it, remove any noisy data and finally the pipeline is applied to help in better detection of vehicles effectively.

STEP 1: Algorithm used
Out of all the algorithms studied, YOLO algorithm is the only one which can work in real time, which was a huge deciding factor for the nature of the problem statement. The YOLO model [13] is a predictor and a classifier. It predicts the bounding boxes for each object and classifies the object into classes. This uses both regression and classification. YOLO does not have various preprocessing steps in the model; it is a single complete network that does object detection. Firstly, the model takes an input image. It is divided into an S × S grid. Each cell in the grid is responsible for identifying the image that center falls in the cell. A confidence score is given to each cell; it is a measure of the probability of an object being present in the cell. There are 24 convolution layers and two fully connected layers. Downsampling is done with 1 × 1 filters. Each cell of SxS grids outputs eight bounding boxes that may or may not contain an object along with the class of object if it is present. Another modification of this architecture is Fast-YOLO, which is lighter, containing fewer filters and namely nine convolutional layers. Fast-YOLO has higher computation speed of up to 155 FPS but has a higher error rate than YOLO.

YOLO algorithm has the following advantages over others:

- Fast and suitable for real-time processing as suggested by Fig. 2.
- A single network makes all the required predictions.
- YOLO is a network that is able to generalize. Other methods are outperformed when generalizing from natural images to other domains like artwork as shown in Fig. 1.
- The classifier is limited to a specified region using region proposal methods. YOLO develops boundaries to access the image set. YOLO shows a lesser number of false positives in the background (Fig. 2).

Architecture of YOLO [13] is depicted in Fig. 3.
Following graphs show accuracy vs. FPS of YOLO with other algorithms [14].
Architecture of YOLO algorithm is given below:
Ex-YOLO builds on top of YOLO algorithm as an ML pipeline.

STEP 2: Preprocessing
The real-time video input from camera is broken down to frame by frame images. Each image is preprocessed to form a fixed size tensor that is passed to the pipeline. The preprocessing step contains functions like dimension scaling of the image to a square, conversion of image to a tensor and normalization of the image tensor. Any ML model is as good as the data, and thus, the main objective of preprocessing is to enhance the relevant features of data and remove noise and irrelevant features.

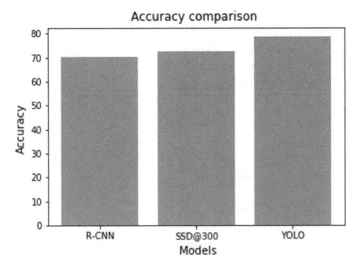

Fig. 1 Accuracy comparison of algorithms

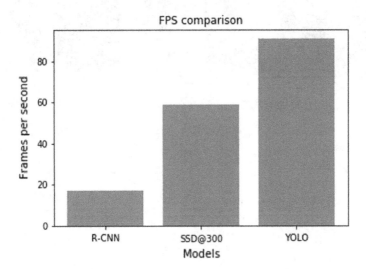

Fig. 2 FPS comparison of algorithms

Irrelevant features are the features that do not carry any relevant information which can be used to solve the problem. Noise can be due to error or negligence in capturing data. Some preprocessing steps as shown in Fig. 4 are:

- Image resizing
- Denoising
- Image segmentation

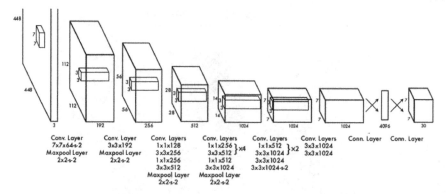

Fig. 3 Architecture of YOLO algorithm

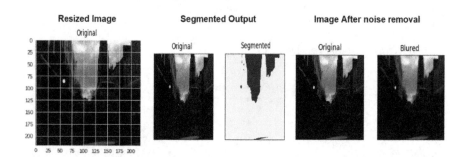

Fig. 4 Preprocessing of image

- Edge smoothing
- Image compression.

STEP 3: Machine Learning Pipeline

Our machine learning pipeline has two phases as shown in Fig. 5. These two phases are:

Phase I

Image tensor obtained after preprocessing acts as an input to this phase. In the first phase, the Fast-YOLO algorithm is applied to detect the position of various objects in the image as described earlier. The Fast-YOLO algorithm is tweaked to work fast and detect vehicles with precision so as to compensate for the delay produced by the second phase. This makes the model work in real time [15]. This step outputs a 7*7*30 tensor value containing the position of the objects relative to the image along with the confidence score for each object class detected.

This tensor is passed to an interface which crops the image of the vehicles detected in the previous step. It also includes a preprocessing computation on the image to change the dimensions of the image suitable for passing it to the next phase.

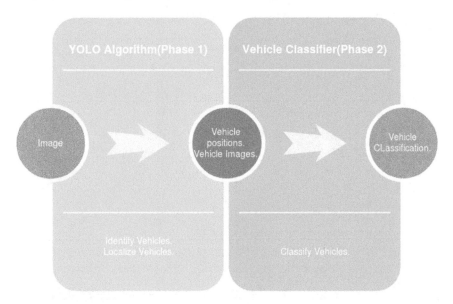

Fig. 5 Ex-YOLO pipeline

Phase II

The second phase of the pipeline is a pretrained CNN that is a simple vehicle classifier. It takes input an image of a vehicle and outputs a one-shot vector identifying the class of vehicle. The model classifies a vehicle into four classes, namely police, fire truck, ambulance and other. The second phase is completely independent of the first phase which makes it easy to maintain, update and debug. The interface between the two phases makes them work without error. The output layer consists of four nodes, each for one classification class [16]. It is a densely connected layer having a softmax activation function. After the vehicle is classified, the bounding box of the vehicle is made on the input image using the output of first phase along with class of object which is identified by the second phase. This is done in real time on every frame of the input video. Its architecture is shown in Fig. 6.

The layers used in the model are mentioned below:

1. Input: In this step, the system inputs an array of images. This layer is the entry point of data to the model. It defines the size of input data that is used for the model. In our case, the input is an image tensor of dimensions ($480 \times 480 \times 3$). The image must be converted to this input layer size before it is fed to the model.
2. Convolutional layer: This layer creates a convolution kernel that is convolved with the layer input to produce a tensor of outputs. It is used to extract useful features from an image that can be used to classify the image. Convolution functions are the heart of CNNs and are the reason why CNNs are light and accurate as compared to DNN [17].
3. Activation function: An activation function is a differentiable function that is applied to a layer to introduce nonlinearity in the model. It makes the features

Fig. 6 Flowchart explaining
the flow of the model

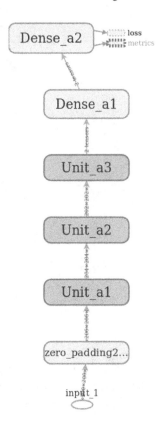

learned a nonlinear or complex function of input features and helps in forming an
efficient mapping between input and output. In our case, we use RectiLinear Unit
(ReLU) activation function. It is fast and makes the model converges quickly.

4. Zero padding: It is a pad applied to tensors to make their shape uniform and give
   the side feature equation weightage during convolution operation.
5. Batch normalization: Batch normalization reduces the amount by what the hidden
   unit values shift around (covariance shift). It normalizes the activations of the
   previous layer at each batch, i.e., applies a transformation that maintains the
   mean activation close to 0 and the activation standard deviation close to 1.
6. Flatten layer: It flattens the input and does not affect the batch size. It resizes in
   the input tensor to a single vector. For example, if the size of input tensor is $N
   \times M \times K$, then the output dimension after flatten would be $N*M*K \times 1$. It is
   usually used after the feature selection process is done before being connected
   to a fully connected layer described below.
7. Dense layer: Dense layer or fully connected layer is a layer in which each neuron
   is connected to every neuron of previous layer. This layer is quite heavy with
   a huge weight matrix. It is used for classification or regression after feature
   selection is done.

8. Max-pooling layer: Max-pooling is a downsampling applied to the model to reduce the number of features and choose only the most relevant ones for selection. For example, in case of object recognition, max-pooling selects the pixels which are the maximum lit up when an object is present and based on these features' classification is done. Max-pooling layers are the reason why CNN do not easily overfit [18].

9. Dropout: Dropout consists of randomly setting a fraction rate of input units to 0 at each update during training time, which helps prevent overfitting [19]. Dropout layer randomly "deactivates" some neurons of the layer during training to make them sort of independent to each other for making predictions. It forces the neurons to learn as much information as they can. While testing, the dropout layer is ignored.

Using the above layers, Phase II of the pipeline is created. The architecture of Phase II is as follows:

The Phase II classifier is made up of two types of blocks of computation.

1. Unit_a: It is a unit containing and convolution layer, followed by batch normalization to make it robust to different distributions. Then, "ReLU" activation is applied to obtain a nonlinear mapping function followed by a max-pool layer for downsampling. Its architecture is shown in Fig. 7.

2. Dense_a: It is a set of fully connected layer and a dropout layer attached to it to prevent overfitting. Its architecture is shown in Fig. 8.

Our model is made of three Unit_a blocks followed by two dense blocks as shown in Fig. 6. Dense_a2 outputs four values, each corresponding to the probability of each class, namely ambulance, fire truck, police vehicle and other.

Fig. 7 Flowchart explaining the layers in Unit_a

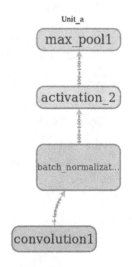

Fig. 8 Flowchart explaining
the layers in Unit_b

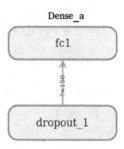

4 Experimental Setup

We trained our model on ImageNet 1000-class competition dataset using YOLOv2 to train detection networks. The model is tested on a hardware setup having an Intel i5 8th generation processor, 32 GB RAM, Nvidia Titan X graphic card. The images acquired from ImageNet were not of a unified size. We cropped the necessary images to maintain the aspect ratio of these images. The model is able to run at a stable 102 frames per second on the system, which makes it highly suitable for a real-time application. The training time was about 40 h average.

5 Results

Ex-YOLO is tested on a hardware setup having an Intel i5 8th generation processor, 32 GB RAM, Nvidia Titan X graphic card.

The model is trained on a custom-made dataset containing more than 6000 images of police cars, ambulances, fire trucks and normal cars. The images are taken from ImageNet and cropped according to need. Then, some preprocessing is done to get images with uniform aspect ratio. It is able to run at a stable 102 frames per second on the system as shown in Table 1, making it suitable for real-time application.

The training accuracy is mentioned in the following table:

Ex-YOLO is robust and can be easily trained to classify other sets of classes by simply changing the Phase II classifier.

Table 1 Training accuracy of datasets

S. No.	Dataset	Accuracy (%)
1	Custom-made dataset	91.6
2	VOC dataset	87

6   Conclusion

The paper aims to implement an emergency vehicle detection algorithm build on top of YOLO called Ex-YOLO. YOLO algorithm was considered suitable because of its better accuracy and frames per second value as compared with SDD and R-CNN. Considering the fact that YOLO algorithm gives real-time solution, it detects the objects present on the road by creating bounding boxes around them. After this, the classifier is used that classifies these objects as various emergency vehicles such as ambulances and police vehicles. Ex-YOLO has the training accuracy of 91.6% on the dataset used in this study. Hence, it can be relied upon for classifying objects accurately.

7   Future Scope

This method of using the lightweight individual network on top of a pretrained accurate and fast network can be heavily exploited in various applications where a deeper classification is required. The base model is not able to capture the complex features required to classify two similar objects. Hence, if aided by a light network whose only job is to classify those similar object, their accuracy would increase greatly.

References

1. Goel, S., Baghel, A., Srivastava, A., Tyagi, A., Nagrath, P.: Detection of emergency vehicles using modified Yolo Algorithm. In: Choudhury, S., Mishra, R., Mishra, R., Kumar, A. (eds.) Intelligent Communication, Control and Devices. Advances in Intelligent Systems and Computing, vol. 989. Springer, Singapore (2020)
2. https://towardsdatascience.com/yolo-you-only-look-once-real-time-object-detection-explained-492dc9230006
3. Yang, W., Zhang, J., Wang, H., Zhang, Z.: A vehicle real-time detection algorithm based on YOLOv2 framework. In: Real-Time Image and Video Processing 2018, vol. 10670, p. 106700N. International Society for Optics and Photonics (2018)
4. Tao, J., Wang, H., Zhang, X., Li, X., Yang, H.: An object detection system based on YOLO in the traffic scene. In: 2017 6th International Conference on Computer Science and Network Technology (ICCSNT), pp. 315–319. IEEE (2017)
5. Zhang, J., Huang, M., Jin, X., Li, X.: A real-time chinese traffic sign detection algorithm based on modified YOLOv2. Algorithms **10**(4), 127 (2017)
6. Molchanov, V.V., Vishnyakov, B.V., Vizilter, Y.V., Vishnyakova, O.V., Knyaz, V.A.: Pedestrian detection in video surveillance using fully convolutional YOLO neural network. In: Automated Visual Inspection and Machine Vision II, vol. 10334, p. 103340Q. International Society for Optics and Photonics (2017)
7. Grimnes, Ø.K. (2017). End-to-end steering angle prediction and object detection using convolutional neural networks (Master's thesis, NTNU)

8. Tang, T., Deng, Z., Zhou, S., Lei, L., Zou, H.: Fast vehicle detection in UAV images. In: 2017 International Workshop on Remote Sensing with Intelligent Processing (RSIP), pp. 1–5. IEEE (2017)

9. Chakraborty, P., Adu-Gyamfi, Y.O., Poddar, S., Ahsani, V., Sharma, A., Sarkar, S.: Traffic congestion detection from camera images using deep convolution neural networks. Transp. Res. Rec. **2672**(45), 222–231 (2018)

10. Zhihuan, W., Xiangning, C., Yongming, G., Yuntao, L.: Rapid target detection in high resolution remote sensing images using YOLO model. Int. Arch. Photogramm. Remote. Sens. Spat. Inf. Sci. **42**, 3 (2018)

11. Şentaş, A., Tashiev, İ., Küçükayvaz, F., Kul, S., Eken, S., Sayar, A., Becerikli, Y. Performance evaluation of support vector machine and convolutional neural network algorithms in real-time vehicle type classification. In: International Conference on Emerging Internetworking, Data & Web Technologies, pp. 934–943. Springer, Cham (2018)

12. Du, J.: Understanding of object detection based on CNN family and YOLO. In: Journal of Physics: Conference Series, vol. 1004, no. 1, p. 012029. IOP Publishing (2018)

13. Redmon, J., Divvala, S., Girshick, R., Farhadi, A.: You only look once: unified, real-time object detection. In: Proceedings of the IEEE Conference on Computer Vision and Pattern Recognition, pp. 779–788. (2016)

14. https://medium.com/@jonathan_hui/object-detection-speed-and-accuracy-comparison-faster-r-cnn-r-fcn-ssd-and-yolo-5425656ae359

15. Ghosh, R., Mishra, A., Orchard, G., Thakor, N.V.: Real-time object recognition and orientation estimation using an event-based camera and CNN. In: 2014 IEEE Biomedical Circuits and Systems Conference (BioCAS) Proceedings, pp. 544–547. IEEE (2014)

16. Yan, Z., Zhang, H., Piramuthu, R., Jagadeesh, V., DeCoste, D., Di, W., Yu, Y.: HD-CNN: hierarchical deep convolutional neural networks for large scale visual recognition. In: Proceedings of the IEEE International Conference on Computer Vision, pp. 2740–2748. (2015)

17. Wang, Y., Qiu, Y., Thai, T., Moore, K., Liu, H., Zheng, B.: A two-step convolutional neural network based computer-aided detection scheme for automatically segmenting adipose tissue volume depicting on CT images. Comput. Methods Programs Biomed. **144**, 97–104 (2017)

18. https://medium.com/@RaghavPrabhu/understanding-of-convolutional-neural-network-cnn-deep-learning-99760835f148

19. Srivastava, N., Hinton, G., Krizhevsky, A., Sutskever, I., Salakhutdinov, R.: Dropout: a simple way to prevent neural networks from overfitting. J. Mach. Learn. Res. **15**(1), 1929–1958 (2014)

Customization of Learning Environment Through Intelligent Management System

Prachi Jain, Anubhav Srivastava, Bramah Hazela, Vineet Singh and Pallavi Asthana

Abstract Current learning environment has expanded its domain immensely via online services. The benefits of those online services over a global network are multifold: No speed and time limitations in learning, place and pace no more impede the path to proficiency. The most essential segment of such development in online learning system is personalization and customization of contents according to the skills and competence of the individual user. The options for customization increase the performance of the system along with its accuracy for specific user. To customize the e-learning system and adapt it to be user specific, it can use the strategies to meet the requirements of the user. The adaptation can be shown in learning process or content or course. This paper scrutinizes the virtue of learning via integrative analysis to present a model for customization of e-learning using intelligent management system.

Keywords Customized learning · Predictability approaches · Intelligent system · Predictive model

P. Jain · A. Srivastava · B. Hazela (✉) · V. Singh · P. Asthana
Amity School of Engineering & Technology, Amity University Lucknow Campus, Lucknow, Uttar Pradesh, India
e-mail: bramahhazela77@gmail.com

P. Jain
e-mail: prachi.jainn0124@gmail.com

A. Srivastava
e-mail: anubhav763@gmail.com

V. Singh
e-mail: vsingh@lko.amity.edu

P. Asthana
e-mail: pallaviasthana2009@gmail.com

© Springer Nature Singapore Pte Ltd. 2020
P. K. Singh et al. (eds.), *Proceedings of First International Conference on Computing, Communications, and Cyber-Security (IC4S 2019)*, Lecture Notes in Networks and Systems 121, https://doi.org/10.1007/978-981-15-3369-3_46

1 Introduction

Online platform of learning has revolutionized the education system all over. It has already filled the gap between the educators and the learner, where anybody can be a student and can choose any course online to learn in a hassle-free manner.

> A click of a mouse button provides any student anywhere with unprecedented opportunities to learn. So, if a child in Grand Junction wants to master Japanese, it's possible online. If a budding artist in Five Points wants to study the masterpieces of the Louvre, it's possible online. If a future Stephen Hawking in La Junta wants to study Gravitational Entropy with the man himself, it's possible online. If military parents want continuity in their children's education throughout frequent moves to serve our country, then it's possible online.
>
> Rod Paige, US Secretary of Education, 2002.

With the evolution of learning, the infinity of learning has changed. Learning is no longer an educator's issue; it is now a parallel important learners' issue. Online education platforms like Coursera, edX, Udemy, NPTEL, etc. have provided online courses for the learners. By providing massive open online course (MOOC) services, they are creating parallel changes between physical and virtual mediums for the enrichment of learners. Given everything online is a boon for a novice, but having it all from many sources can sometimes prove to be a bane for that someone who clearly has brief knowledge about the topic. It affects the decision-making ability as to which course would be more beneficial to learn. There are many learning Web sites with each providing a specialization in a particular domain and many other courses related to it. In present times, learning strategy is at its peak but it lacks the recommender system, predictability and learning mechanism concept for online learning. To make e-learning more accurate, we need to customize it according to the learner's demand and needs. According to Klašnja-Milićević A., personalization of the content according to the user profile in e-learning system is important and proposes major standards in e-learning along with its specifications [1]. According to Bousbia N., other important factors to adaptation are learning style, learning path and interactions of the learner with the course provider [2].

In this paper, authors propose to use an intelligent system to create learner's customized environment for the specialization in courses related to their online search. The proposal is to improve the decision-making ability of the learners using machine learning technique. The key challenge in this project is to know the user, sort the available courses from different sources and select few specific courses that suit the user online search. In this paper, we have done a rigorous review on recommender system and proposed intelligent recommender system based on user's online search.

2 Background

With the development of Web and its speed, alongside, the availability of electronic mediums like PCs, mobiles, PCs at sensible cost virtual learning flourished. This enabled all learners across the globe to exploit Web-based education and learning condition. So now whenever a student needs to know anything in any area, he just demands to the specialist organization and supplier at that point. It furnishes the student with the essential assets.

Every student is an individual and requests explicit administrations for courses. MOOC administrations are summed up for a gathering of individuals who need to take in some explicit course. We propose here to take this present virtual learning condition somewhat further. We propose to make a redid interface which would enable the students to make their very own shell of learning condition, where we would get to the necessities of the individual student and utilize diverse consistency ways to deal with and show some best source way for him, where he would then be able to settle on a decision most appropriate for him.

Lot of research work has been done to explore the infrastructure to improve the learning environment through customization, personalized learning methodologies, etc. which help to increase the entire e-learning environment in a more proficient manner. Research work [3, 4] proposed the concept of an auto-generating platform that helps non-computing background trainers to explore e-learning without focusing on e-learning interface. The proposed platform was capable enough for developing an interface by adopting systematic scientific methodologies. Work suggested in [5] discusses the incorporation of e-learning with network technologies for data transfer in real time exchange. In the paper [6], author discussed the adaptive model of learning that helps MOOC providers to plan learning strategies according to the level of understating of the learners. Researchers discussed group gray wolf optimization techniques for precise learning environment. This technique is used to increase the performance of a machine learning algorithm and provide a significant solution to our proposed model. This technique will be beneficial for the learners by providing a better search solution [7].

Research work [8] discussed the importance of e-learning by making the connection between educational theories and their practical implementation. It is inclined toward the method of delivery and types of methodologies like descriptive, open challenge, practical, review, design, etc. for courses provided by the MOOC. In [9], researchers highlighted the artificial intelligent technique incorporated with neural network that would work on both small and big data sets by minimizing the classification time during searching. Here, authors also discussed the experimental results and proposed approach with minimum comparison time and a higher rate of accuracy.

Authors have proposed the system named PEDAL-NG that encourages customized learning on the basis of learner's knowledge and maintain the learning style by improving the operational learning environment. It focuses on prior knowledge, past experiences, learner skills, learning style and interest [10]. Research has also discussed the customized approach for accurate prediction with the help of machine learning techniques and big data approaches [11]. Once our proposed model started dealing with huge amount of data, it will require a proper methodology through which we can handle the learners' data. In research work [12], the authors have discussed the estimation effort by using the machine learning approaches to figure out the most accurate amount of effort to develop the software which surely helps us in our model to calculate the right amount of effort while implementing the model.

3   Methodologies

In this work, we propose a framework for customization of online course for a student where they will be able to find the specific course as per their requirement in the given domain. Here, the best-suited course will be provided to the learner by applying intelligent system techniques and then applying it to prioritize them in the customized learning environment that would also help students to concentrate on the topic they specifically want to learn and remove those topics that appear similar but less relevant. Here is the proposed heuristic-based model as shown in Fig. 1.

3.1   Data Collection

In the present e-learning environment, multiple massive open online course (MOOC) providers are available, having their own collection of data respective of courses provided by them. In this proposed model, we are taking all the information related to the courses from the available MOOC and store it in a single place. This will act as a training data set which will help to train the algorithm and heuristic model. The training data would include students' feedback of MOOC courses, learner's aptitude, qualification, area of interest, his current knowledge base, learning style, etc.

3.2   Preprocessing

After the collection of raw data that includes (i) information extraction: It will extract the fruitful information from the raw data; (ii) information classification: It will classify data in a meaningful manner; and (iii) information evaluation: All the extracted and classified data is properly arranged.

3.3   Sampling

15After preprocessing, from the larger population, some samples are collected to overcome the problems like accessing larger data set, destructive observations, accuracy issues, etc. Two important components of sampling are feature extraction and dimensionality reduction. Feature extraction is to extract relevant feature from the preprocessed data, and dimensionality reduction is to remove overlapping features.

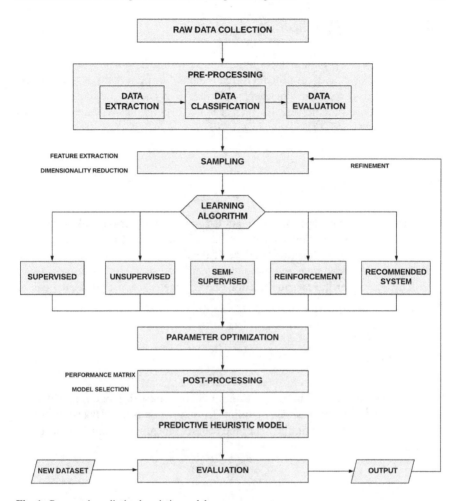

Fig. 1 Proposed predictive heuristic model

Sampling strategies include random oversampling and undersampling, informed undersampling, synthetic data generation, adaptive synthetic sampling and sampling using data cleaning technique as stated in [13].

3.4 Learning Algorithm

Different learning mechanisms like supervised, unsupervised, semi-supervised, reinforcement and recommended system are based on result obtained from input sample data set. [14].

3.5 Parameter Optimization

The basic purpose of parameter optimization is to find optimal hyperparameters that would work for predictive model formation. For this purpose, two strategies are commonly used: grid search which searches exhaustively across the chosen hyperparameter subset and random search which searches hyperparameter randomly rather than exhaustively [15].

3.6 Post-processing

In the post-processing segment, two main tasks will be performing like creating a performance matrix and model selection. In performance matrix, we chose metric to evaluate and compare the performance of the different learning algorithms. For this, we use different parameters—conclusion matrix, accuracy, precision, sensitivity, specificity, etc. After the metric selection and parameter evaluation, model can be decided.

3.7 Data Set

New data set is provided by the learner to the predictive heuristic model, created by the above procedure of preprocessing, sampling, learning technique and post-processing. This model produces the optimal list of the courses meeting the requirements of the learner.

3.8 Predictive Heuristic Model

Heuristic approach is a technique that works for quick decision, especially for complex data. The decisions made by heuristic approach are fast, do not incorporate exhaustive search and produce results that may not be accurate but are near to optimal. Predictive heuristic model would provide the list of courses that would work well with the learner requirements. Table 1 contains various training and prediction complexities of different machine learning algorithms.

M is equal to the number of training data samples, N is equal to the number of selected features, M_{tree} is the number of trees, M_{sv} is the number of support vector, M_{li} is the number of neurons in layer i in the described neural network [16].

Table 1 Training and prediction complexities

S. No.	Algorithm	Training complexity	Prediction complexity
1.	Decision tree	$O(M^2N)$	$O(N)$
2.	Random forest regression	$O(M^2NM_{tree})$	$O(NM_{tree})$
3.	Random forest classification	$O(M_2NM_{tree})$	$O(NM_{tree})$
4.	Extremely random tree	$O(MNM_{tree})$	$O(MNM_{tree})$
5.	Gradient boosting	$O(MNM_{tree})$	$O(NM_{tree})$
6.	Linear regression	$O(N^2M + N^3)$	$O(N)$
7.	Support vector machine	$O(M^2N + M^3)$	$O(M_{sv}N)$

3.9 Evaluation

The data set provided by the learner is extracted, classified and evaluated, and after feature selection, it is processed and given to predictive heuristic model to produce optimal result.

3.10 Outcome

As a result of the above processing when the data set is passed to the predictive model, a personalized list of courses customized as per the learner's requirements is produced in a single page leaving the final call of decision on the learner to choose the course that best suits him. This data set then acts as test data and is passed to the sampling stage of the process to refine the predictive heuristic model.

3.11 Canvas of Proposed Business Model

See Fig. 2.

SEGMENTS	KEY ACTIVITIES	VALUE PROPORTION	LEARNER RELATIONSHIP	KEY FEATURES
o MOOC providers o learners	o Manage MOOC network. o Manage learner network. o Promotions and marketing. o Product development and technical assessment.	o MOOC providers ▪ Fare indexing ▪ Open marketing ▪ 24*7 support o Learners ▪ Quick access ▪ Get desired outcome ▪ Secure payment ▪ 24*7 support	o Query system o Review system o Social media **KEY RESOURCES** o Technocrats & policy makers. o Strong technical set-up & database	o Learners o MOOC providers o Content providers o Recruiters o Investors o Media firms

ACCESSING MODES	REVENUE ANALYSIS
o Website and Apps	o Cost per access and Advertising

Fig. 2 Canvas of proposed business model

4 Research Questions and Answers

4.1 What is the Model All About?

In this model, a customized learning environment has been created for the learners by knowing their needs. It will have modules about demands, choices, knowledge background, aptitude to know the student and a recommender system to help the student in decision making. This model would be beneficial for both learner and course manager to create the courses that are preferred by learner or to update such courses for better visibility.

4.2 How Will it Benefit the Learner?

In this model, the authors propose to make a personalized environment with a robotic touch for the student that would teach him and make a choice list of courses and its sources for him leaving the final call of decision on him. It will follow steps to know the student by using a questionnaire, by conducting a simple test to know his aptitude or using grades for the same and then use machine learning to develop a model that promises to be customer specific. After this, run a query to search in Internet, and those courses best suited for him will be listed and along with it a set of sources associated with each course.

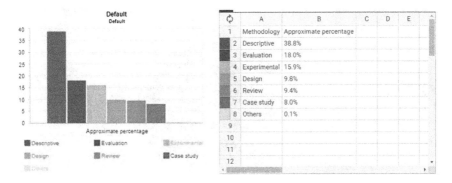

Fig. 3 Methodology of learning

4.3 What are the Approaches Applied to Increase the Efficiency of the Model?

It uses "heuristic" approach where many different paths are explored and success of each is found out. The one closest to the desired result is chosen, and other paths are dropped. The other method intelligent system uses is "operations research" where we have used mathematical tools, mathematical models and statistics to analyze the complex real-world problems to improve performance that would assist in decision making. By applying heuristic approach to learning, this model will provide the different methodologies of learning like descriptive, evaluation type, experimental type, design, review, case study, etc. Figure 3 shows the statistics of different methodologies of learning, data from the source [17].

4.4 What is the Unique Selling Point of this Model?

This intelligent system used would know about the current state of the learner and make a knowledge base to guide him accurately to enhance his skills and knowledge which in turn will boost the knowledge base of the system and so the learner's skills and so on. The model is designed in such a manner that the request query of the learner will first be processed at the intelligent system data set and then at MOOC database. So, the result of same query will be different for different learners making it learner specific, hence creating a personalized environment for the learner. Figure 4 shows the data flow diagram of the proposed model.

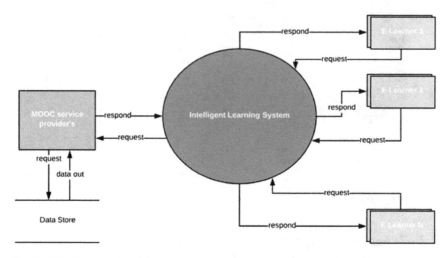

Fig. 4 DFD of proposed model

4.5 *What are the Main Paybacks of this Model as a Business Model?*

It will introduce new dimension of learning that customizes the learning environment and increases the learning rate of the learners, giving a hardcore impact on learning environment like from online learning to mobile learning. In this model, intelligent system, classical machine learning approach and modern deep learning approach are being used. Although these algorithms will produce accurate and efficient results as per the learner's requirement, still there are some prominent factors that need to be considered for better evaluation [18].

5 Result and Discussion: Performance, Advantages, Limitations and Future Scope

On the basis of these five factors like credentialing, course diversity, course features, social feature and partner institutions, we are scrutinizing the top three most progressive MOOCs Coursera, edX and FutureLearn. The number of users and revenue generated by these MOOCs are shown in Fig. 5 [19–21].

Here in Fig. 6, we are showing the subjectwise analysis of last four years' growth. On the basis of individual growth rate of massive open online course provider "*Coursera*," we can predict the growth rate of our proposed business model and that will be more than 88% approximately [22, 23].

In Fig. 7, we can see that every year MOOC is growing at an accelerating rate.

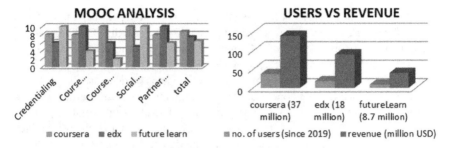

Fig. 5 MOOC analysis and learner versus revenue comparison

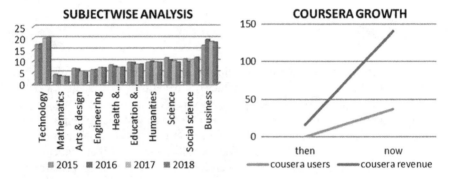

Fig. 6 Subjectwise analysis and MOOC growth rate

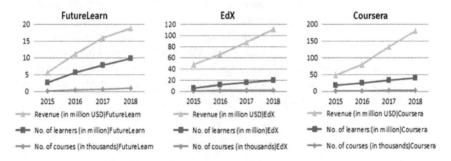

Fig. 7 Progressive growth rate of MOOCs (FutureLearn, edX, Coursera)

Considering the topmost progressive MOOCs: Coursera, edX, FutureLearn, an analysis has been made. FutureLearn of Europe has moved up in the number of courses at 200 courses added per year, whereas the number of learners increased with 0.35 million learners per year and revenue increased at a rate of 0.475 million USD per year. edX too has shown great progress by 1550 average increase in courses per year, 3.25 million average increase in learners per year and 5 million average increase in revenue per year from 2015. Coursera increased at 1525 number of courses on an average per year, 12 million new registrations on an average per year and 27.5 million USD revenue in an average per year. Considering growth of MOOCs in relation to the number of courses, the MOOCs have increased at a rate of 2000 courses per year from 2012. With drastic increase in MOOC services, the users are bound to get confused as to which suits him the most, and our model is made to use the prediction algorithm to channelize the courses that best fits the interest of the learner. The efficiency of learners and MOOCs will only increase. The efficiency of all MOOCs is supposed to increase by at least 20% at start and will increase with time by intelligent system. All the statistical data are taken from the MOOC report presented by "Class Central" [24–26].

6 Conclusion

This paper exhibits giving learning condition to a distinctive diverse kind of students. It chiefly centers around how a wise framework benefits student and open online course suppliers. An astute framework utilizes various types of methodologies from managed figuring out how to a recommender framework that can be helpful for the students either by performing quality examination or by performing community-oriented separating and spotlights on every single student and their dimension of comprehension, and by analyzing that dimension it gives the most appropriate reward to them. Predictive analysis based on proposed model would help the student for better learning experience. Albeit here, machine learning can be utilized as a lively apparatus to enhance the learning result of the students by means of making a framework and anticipating most appropriate courses to them. In vast majority of the cases, it will anticipate effectively just about 70–80% yet at the same time it relies upon some different elements like information about the conduct of students like learning style, inspiration, environment, understanding dimension and so on, here a framework endeavoring to foresee the learning prerequisite of the students and continue spurring itself to perform better.

References

1. Klašnja-Milićević, A., et al.: Introduction to e-learning systems. In: E-Learning Systems, pp. 3–17. Springer, Cham (2017)
2. Bousbia, N., Labat, J-M., Balla, A.: Detection of learning styles from learner's browsing behavior during e-learning activities. In: International Conference on Intelligent Tutoring Systems, Montreal, QC, Canada, 23–27 June 2008, pp. 740–742
3. Wang, Z., Wang, X., Wang, X.: Research and implementation of Web-based e-learning course auto-generating platform. In: International Conference on Technologies for E-Learning and Digital Entertainment, Nanjing, China, 25–27 July 2008, pp. 70–76
4. Bitzer, D.L., Braunfeld, P.G., Lichtenberger, W.W.: PLATO II: A multiple-student, computer-controlled, automatic teaching device. In: Programmed Learning and Computer-based Instruction, pp. 205–216, 1962

5. Yang, Y., et al.: An overview on mobile e-learning research of domestic and foreign. In: International Conference on Web-Based Learning, Jinhua, China, 20–22 August, 2008, pp. 521–528

6. Qu, Y., Wang, C., Zhong, L.: The research and discussion of web-based adaptive learning model and strategy. In: International Conference on Hybrid Learning and Education, pp. 412–420, 2009. ISBN 978-3-642-03697-2

7. Shankar, K., et al.: Alzheimer detection using Group Grey Wolf Optimization based features with convolutional classifier. Comput. Electr. Eng. **77**, 230–243 (2019)

8. Fernández-Manjón, B., et al., eds. Computers and Education: E-learning, from theory to practice. Springer Science & Business Media, 2007. ISBN 978-1-4020-4914-9

9. ALzubi, J.A., et al.: Boosted neural network ensemble classification for lung cancer disease diagnosis. Appl. Soft Comput. **80**, 579–591 (2019)

10. Takhirov, N., et al.: Organizing learning objects for personalized eLearning services. In: International Conference on Theory and Practice of Digital Libraries, Corfu, Greece, vol. 5714, pp. 384–387, 22 Sept–2 Oct 2009

11. Verma, J.P., et al.: Evaluation of pattern based customized approach for stock market trend prediction with big data and machine learning techniques. Int. J. Bus. Anal. (IJBAN) **6**(3), 1–15 (2019)

12. Lu, F., Li, X., et al.: Research on personalized e-learning system using fuzzy set based clustering algorithm." In: International Conference on Computational Science, Beijing, China, LNCS vol. 4489, pp. 587–590, 27–30 May 2007

13. Goel, G., et al.: Evaluation of sampling methods for learning from imbalanced data. In: Proceedings of International Conference on Intelligent Computing, 28–31 July 2013, Nanning, China, pp. 392–401

14. Sullivan, R.: Machine-Learning Techniques. In: Introduction to Data Mining for the Life Sciences, pp. 1–31. Humana Press, Jan 2012. ISBN-978-1-59745-290-8

15. Feurer, M., Hutter, F.: Hyperparameter optimization. In: Automated Machine Learning, pp. 3–33. Springer, Cham, May 2019. ISBN 978-030-05318-5

16. https://www.thekerneltrip.com/machine/learning/computational-complexity-learning-algorithms/

17. https://www.igi-global.com/book/intelligent-systems-operations/37306

18. http://openaccess.uoc.edu/webapps/o2/bitstream/10609/75705/6/ELR_Report_2017.pdf

19. https://www.classcentral.com/report/futurelearn-2018-year-review/

20. https://www.reviews.com/mooc-platforms/

21. https://www.classcentral.com/report/mooc-stats-2018/

22. https://www.classcentral.com/report/mooc-providers-list/

23. https://www.classcentral.com/report/coursera-2018-year-review/

24. https://www.classcentral.com/report/moocs-2015-stats/

25. https://www.classcentral.com/report/mooc-stats-2017/

26. https://www.classcentral.com/report/mooc-stats-2016

Classification of Comparison-Based Analysis Between Rattle and Weka Tool on the Mental Health Disorder Problem Dataset Using Random Forest Tree Classifier

Vivek Kumar and Ajay Kumar

Abstract Nowadays the major problem is related to the mind disorder or we can say it mental health problem. This type of problem is basically observed in the old aged people, but in the present time, it is also observed in the youth age people. It is very important to deal with such type of problems because as the time passes the treatment goes complex. Depression is the main cause of mental health problem. This paper uses the data mining technique for the help of diagnosis. We use two data mining tools, namely Rattle and Weka, for the classification of geriatrics-mental-health dataset, and we use the decision tree classifier for the classification purpose.

Keywords Rattle · Weka · Random forest tree

1 Introduction

As the clinical data is growing day by day due to the new diseases coming in the environment and due to the number of records are stored because the today the data is entered in the databases manually and with the electronics devices as well. This data is very huge in nature so it can not be handled or can not be analyzed efficiently manually; so with the help of data mining, researcher can work such type of data and get the desired result. Data mining is the technique by which we can extract the desired pattern from the huge type of data. This pattern is very useful in the diagnosis of diseases. Data mining has a vital area and it applies number of fields like neural network, pattern recognition, artificial intelligent (AI), high-performance computing, database management system (DBMS), data visualization and many more.

Mental health disorder is the major problem, and if a person is suffering from these diseases, then he/she can not enjoy life in a proper manner. Today the persons are busy in their life, they do not have time for their health, and they do not go for the medical checkup on a regular basis. And if they are affected by any health-related

V. Kumar (✉) · A. Kumar
Meerut Institute of Engineering and Technology, Meerut, UP, India
e-mail: vivek.kumar@miet.ac.in

A. Kumar
e-mail: ajay.kumar@miet.ac.in

© Springer Nature Singapore Pte Ltd. 2020　　　　　　　　　　　　　　　　633
P. K. Singh et al. (eds.), *Proceedings of First International Conference on Computing,
Communications, and Cyber-Security (IC4S 2019)*, Lecture Notes in Networks
and Systems 121, https://doi.org/10.1007/978-981-15-3369-3_47

problem and they are not going for the better treatment on time, then their body will response worse. Depression is the cause in the mental health problem; basically, depression may be due to the stress. The stress may be coming from the workload and may be from social life or personal life. And due to depression, one can not do the work or enjoy life in a proper way, because there are some changes occurred in the body.

Mental health problem is mainly happened in the old age because it may be that the person is in this age not having a proper attention from their family, may be that they cannot do the physical work because of weakness in the body, may be that there is vision problem, may be that these people have hearing problem, may be that there is mobility problem, may be that high tension factor is in the mind, may be that blood pressure and diabetes are the factors and may be that they are having heart-related problem. But it is no so that mental health problem arises in the old age only but it is also happened in the middle age group and children also. We have to care about our changes in the body, organs and activities.

2 Literature Survey

In this section, the pre-existing work related to the mental health disorder is reviewed.

Ojeme et al. [1] explain that depression is a very serious cause and one should not avoid it because detection of depression is very difficult thing in the mental health disorder. By then, depression body is also affecting physically. In the paper, work is done on the dataset which is based on depression and physical illness in Nigeria. Multi-dimensional Bayesian network classification approach on the data is used. Class-bridge MBC predictive model performs the best result on the dataset.

Sau et al. [2] discuss that depression is mainly found in the old aged persons because of change in body organs. In the old age, brain is not working properly. Some factors related to the mental health in the old age are cision problem, mobility problem, hearing problem and sleeping problem. In this paper, ANN model is used on the data which are collected from number of healthcare centers. In the dataset having 30 instances, trained ANN model is applied and this model gives the correct result 18 non-depressed and 12 depressed with kappa statics $= 1$.

Daimi et al. [3] explain about the depression problem in the brain. The paper has the data related to the depression patient, and then, data mining technique is applied on it. Weka tool is used for the analysis, and the classification algorithm used C4.5.

Patel et al. [4] discuss the mental disorder problem related to depression. It is too difficult to detect it, and if the timely treatment is not done, then its face may be worst. For the improvement purpose, machine learning can be used on such type of disease data. In this paper, depression background, imaging and number of machine learning methods are discussed. Cross-validation, feature reduction methods and machine learning methods are performed. And create the best model on the basis of depression and non-depression patients.

Orabi et al. [5] tell that in the natural language processing work a deep neural network architecture. The data is used from the Twitter social media for the architecture for the unstructured text data. The work focuses on pointing out the persons who are affected by the mental problem which is depression, and the classification is done on the data which is coming from their social post.

Hathaliya et al. [7] tell with the exponential increase in the usage of Healthcare 4.0-based diagnostics systems, the patients' records are now stored in electronic health record (EHR) repository that is easily accessible by doctors from any location. Database repository is an open channel which requires security protocols, and one of them is biometric-based authentication scheme to ensure secure access from anywhere and is designed with Automated Validation of Internet Security Protocols and Applications (AVISPA) tool. Kumari A. et al. [8] explain Industry 4.0 also known as the fourth revolution is a new era dealing with technologies like robotics, automation and artificial intelligence (AI). The adoption of robots in industries worldwide is on high rise and provides better quality product with higher accuracy in less time. Industry 4.0 is also providing an environment rich enough for big data analysis and self-correcting procedures. This all will result in improved productivity and meeting customer expectations at the fullest way possible.

Gupta et al. [9] define the existing state-of-the-art schemes handling the security of EHRs have resulted in data being generally inaccessible to patients. Blockchain technology resolves the aforementioned issues because it shares the data in a decentralized and transactional fashion and can be leveraged in the healthcare sector to maintain the balance between privacy and accessibility of EHRs. This provides secure and efficient access to medical data by patients, providers and third parties while preserving the patient private information with all security measures.

Vora et al. [10] show the e-health cloud paradigm has evolved from the exchange and enhanced sharing of valuable information between various medical institutions, hospital systems and respective care providers. It reduces the cost and making of efficient process. It preserves privacy of the health record and identity of a patient and helps them to get rid of information loss fear. Here, the main motive is to provide an approach to preserve the identity of an individual.

In this paper, four types of models are used which is based on the neural network and they are CNNWithMax, MultiChannelCNN, MultiChannelPoolingCNN and BiLSTM (context-aware attention). CNNWithMax, MultiChannelCNN and MultiChannelPoolingCNN are based on the CNN, and the BiLSTM (context-aware attention) is based on the RNN. CLPsych2015 dataset is used for the analysis for test generalization ability on the Bell and detection of depression, and the result shows that the RNN model is not so good but CNN models give the best performance on the data.

3 Methodology

In this section, the methodology used in the paper is described.

3.1 Collection of Data

Some medical diseases are very dangerous for humans, and the proper information can save the life of ones. For the classification analysis, we have taken data from the Kaggle which is a machine learning repository and the dataset is related to the geriatrics-mental-health. There are 60 instances and 14 input variables and 1 is target variable. Table 1 describes the attribute name and its data type.

3.2 Data Mining

As per the definition by Jiawei Han in his book Data Mining: Concepts and Techniques [6], the data mining is extraction of interesting, non-trivial, implicit, previously unknown and potentially useful patterns or knowledge from huge amount of data. Data mining has been used in various areas like healthcare, business intelligence, financial trade analysis and network intrusion detection.

Table 1 Data description

Attribute name	Data type
Age	Numeric
Gender	Numeric
Living spouse	Numeric
Family type	Numeric
Education	Numeric
Occupation	Numeric
Substance abuse	Numeric
Personal income	Numeric
DM	Numeric
HTN	Numeric
Hearing problem	Numeric
Vision problem	Numeric
Mobility problem	Numeric
Sleep problem	Numeric
Health	Category/nominal

There are a lot of hidden information resided in a huge amount of data, and by using the data mining, we can get valuable and meaningful information from that. This is very helpful in generating the pattern on the extract information, and this pattern is very important for the analysis and for the last result. As we can understand that the data is coming from the different locations and from the various devices and of different types of format and such type of data cannot be handled by the human manually, the data mining is the way by which it can be done in a very easy way.

3.2.1 Classifier Used

As in the data mining tool, there are a number of classifier available but for our analysis we are using random forest tree classifier.

When the input variables are of hundreds or in thousands, then the random forest tree is the best option. In the big dataset, it shows the importance of the variables in the classification. Today in machine learning, the random forest tree is very helpful. For the Weka, we used tenfold cross-validation, and for the Rattle, we us number of tree is 500 and variables are taken 3.

4 Result and Analysis

We have collected data from the Kaggle (machine learning repository), and the dataset name is geriatrics.csv. This dataset has 60 observations, 14 input variables and 1 target variable, and we are using here two data mining tools, namely Rattle and the Weka, to classify our data, and using random forest tree model for the analysis.

4.1 Random Forest Tree Using Rattle

In the very first step, we have to load our dataset in the Rattle which is in.csv file format and the name of the dataset is geriatrics.csv. In this dataset, 14 input variables are of numeric data type and 1 target is categorical data type. Figure 1 shows the image after loading the dataset into Rattle. Below we summarize the dataset. The data is limited to the training dataset.

Data frame: crs*dataset*[crstrain, c(crs*input, crs*risk, crs*target*)]

42 observations and 15 variables	Maximum NAs: 0
Levels	Storage
Age	Integer

(continued)

(continued)

42 observations and 15 variables	Maximum NAs: 0
Gender	Integer
Living_spouse	Integer
Family_Type	Integer
Education	Integer
Occupation	Integer
Substance_abuse	Integer
Personal_income	Integer
DM	Integer
HTN	Integer
Hearing_problem	Integer
Vision_problem	Integer
Mobility_problem	Integer
Sleep_problem	Integer
Health	Integer
Variable	—Levels—
Health	—Disease, normal—

For the simple distribution of our data, consider Figs. 2 and 3. Below the first and third quartiles refer to the first and third quartiles, indicating that 25% of the observations have values of that variable which are less than or greater than (respectively) the value listed.

Below in the table indicate that first quartiles and third quartiles show that twenty-five percentage have the variables' value which are ¡ Or ¿ (less than Or greater than) the listed values.

For the random forest tree model, we have to click on the model tab in Rattle and click on the forest radio button or we can code in the R console as well. The code is shown in Fig. 3.

```
randomForest(formula = Health.,
data = crsdataset[crstrain, c(crsinput, crstarget)],
ntree = 500, mtry = 3, importance = TRUE, replace = FALSE, na.action =
randomForest::na.roughfix)
```

Type of random forest: classification number of trees: 500
Number of variables tried at each split: 3 (Fig. 4).

	Disease	Normal
Disease	16	2
Normal	1	23

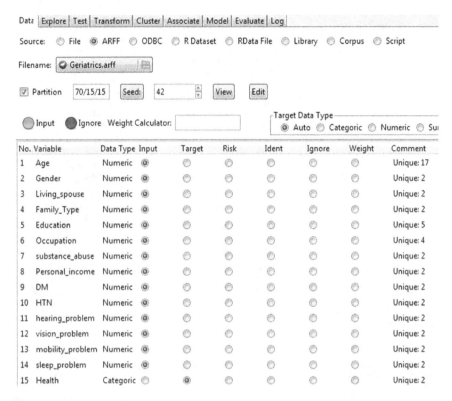

Fig. 1 Loading dataset image

Figure 5 shows the relative importance of variables, which tells that which variable is playing the most important role.

Now the confusion matrix related to this observation and Table 2 show the observation comes from the confusion matrix. Confusion matrix:

4.2 Experiment Using Weka

Weka is also a data mining tool, and we are using our dataset in the file extension ARFF. So first load the dataset into the Weka, then it will show that there are 15 attributes in the left side of Weka window. If we click on any attribute, then Weka will show its property. We have 15 attributes that is shown in Table 1. Now after loading the dataset geriatrics.arff, the pre-process tab will give us some observation related to the attributes used which is shown in Table 3. If we want to visualize all the attributes, then click on visualize all; then Fig. 6 shows result. Now go to the

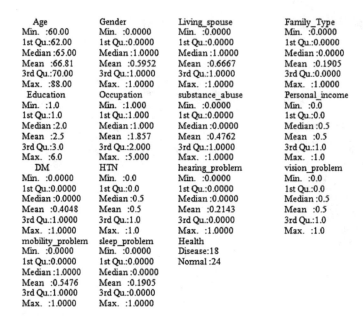

```
Age                Gender             Living_spouse       Family_Type
Min.   :60.00      Min.   :0.0000     Min.   :0.0000      Min.   :0.0000
1st Qu.:62.00      1st Qu.:0.0000     1st Qu.:0.0000      1st Qu.:0.0000
Median :65.00      Median :1.0000     Median :1.0000      Median :0.0000
Mean   :66.81      Mean   :0.5952     Mean   :0.6667      Mean   :0.1905
3rd Qu.:70.00      3rd Qu.:1.0000     3rd Qu.:1.0000      3rd Qu.:0.0000
Max.   :88.00      Max.   :1.0000     Max.   :1.0000      Max.   :1.0000
Education          Occupation         substance_abuse     Personal_income
Min.   :1.0        Min.   :1.000      Min.   :0.0000      Min.   :0.0
1st Qu.:1.0        1st Qu.:1.000      1st Qu.:0.0000      1st Qu.:0.0
Median :2.0        Median :1.000      Median :0.0000      Median :0.5
Mean   :2.5        Mean   :1.857      Mean   :0.4762      Mean   :0.5
3rd Qu.:3.0        3rd Qu.:2.000      3rd Qu.:1.0000      3rd Qu.:1.0
Max.   :6.0        Max.   :5.000      Max.   :1.0000      Max.   :1.0
DM                 HTN                hearing_problem     vision_problem
Min.   :0.0000     Min.   :0.0        Min.   :0.0000      Min.   :0.0
1st Qu.:0.0000     1st Qu.:0.0        1st Qu.:0.0000      1st Qu.:0.0
Median :0.0000     Median :0.5        Median :0.0000      Median :0.5
Mean   :0.4048     Mean   :0.5        Mean   :0.2143      Mean   :0.5
3rd Qu.:1.0000     3rd Qu.:1.0        3rd Qu.:0.0000      3rd Qu.:1.0
Max.   :1.0000     Max.   :1.0        Max.   :1.0000      Max.   :1.0
mobility_problem   sleep_problem      Health
Min.   :0.0000     Min.   :0.0000     Disease:18
1st Qu.:0.0000     1st Qu.:0.0000     Normal :24
Median :1.0000     Median :0.0000
Mean   :0.5476     Mean   :0.1905
3rd Qu.:1.0000     3rd Qu.:0.0000
Max.   :1.0000     Max.   :1.0000
```

Fig. 2 First quartiles and third quartiles

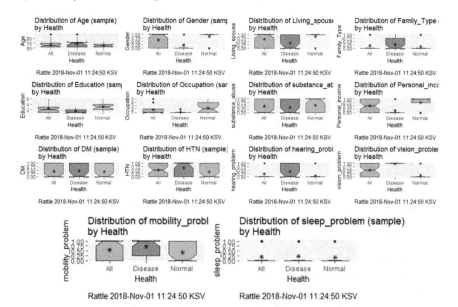

Fig. 3 Distribution of data

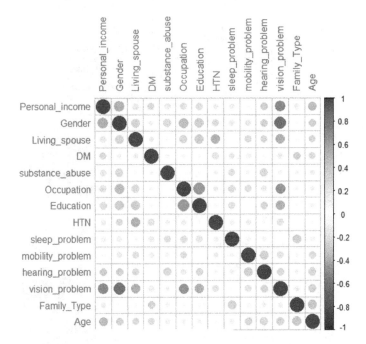

Fig. 4 Distribution of data

classify tab and click on the choose button, then a drop-down list is shown and click on tree then select random forest from the list. Take tenfold cross-validation for the classification. Summary of this classifier is

Correctly classified instances	50	83.3333%
Incorrectly classified instances	10	16.6667%
Kappa statistic	0.6606	
Mean absolute error	0.2242	
Root mean squared error	0.3131	
Relative absolute error		45.4831%
Root relative squared error		62.9843%
Total number of instances	60	

The detailed accuracy by the classification is shown in Table 4. The above observation is coming from the confusion matrix

a	b	<–classified as
29	5	—a = Normal
5	21	—b = Disease

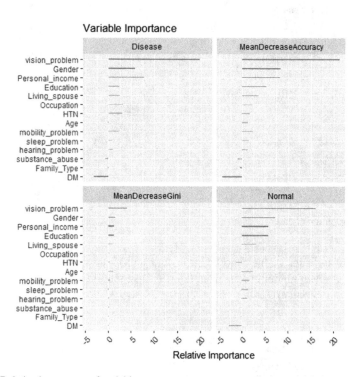

Fig. 5 Relative importance of variable

Table 2 Observation from the confusion matrix

Observations	Values
True positive rate (normal)	0.95833
False positive rate(normal)	0.11111
True negative rate (disease)	0.88888
False negative rate (disease)	0.41666
Accuracy	0.9285
Precision	0.92
Recall	0.95833
F-Measure	0.93877
ROC area	0.85

Table 3 Weka data processing

Name	Type	Missing	Distinct	Unique	Minimum	Maximum	Mean	Standard deviation
Age	Numeric	0	17	5	60	88	66.5	5.832
Gender	Numeric	0	2	0	0	1	0.533	0.503
Living spouse	Numeric	0	2	0	0	1	0.7	0.462
Family type	Numeric	0	2	0	0	1	0.183	0.39
Education	Numeric	0	5	0	1	6	2.5	1.501
Occupation	Numeric	0	4	0	1	5	1.883	1.391
Substance abuse	Numeric	0	2	0	0	1	0.55	0.502
Personal income	Numeric	0	2	0	0	1	0.55	0.502
DM	Numeric	0	2	0	0	1	0.483	0.504
HTN	Numeric	0	2	0	0	1	0.6	0.494
Hearing problem	Numeric	0	2	0	0	1	0.2	0.403
Vision problem	Numeric	0	2	0	0	1	0.483	0.504
Mobility problem	Numeric	0	2	0	0	1	0.55	0.502
Sleep problem	Numeric	0 2	0	0	1	0.217	0.415	

Name	Type	Missing	Distinct	Unique	Label	Count	Weight
Health	Nominal	0	2	0	Normal	34	34.0
					Disease	26	26.0

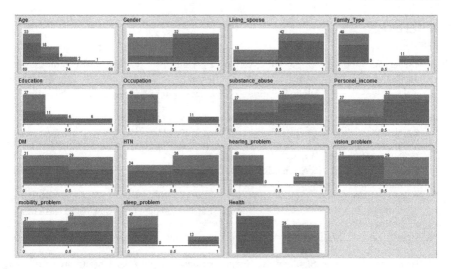

Fig. 6 Loading dataset image

Table 4 Detailed accuracy by class

True positive rate	False positive rate	Precision	Recall	F-Measure	MCC	ROC area	PRC area	Class
0.853	0.192	0.853	0.853	0.853	0.661	0.946	0.963	Normal
0.808	0.147	0.808	0.808	0.808	0.661	0.946	0.933	Disease
0.833	0.173	0.833	0.833	0.833	0.661	0.946	0.950	

5 Conclusion and Future Work

We have used mental health disorder-related dataset for the classification purpose, it has class name health, and it has two inputs for the diagnosis purpose: One is normal, and the second is disease. If a person is normal, then he/she do not have any mind disorder, and if he/she has disease, then there is mind disorder and they need treatment; for this, we have used two data mining tools, namely Rattle and Weka. Random forest tree classifier is used for the classification of our data, and we observe that the accuracy of Rattle tool is coming 92.85% and from the Weka tool it is 83.33%. So from the result, we can say that Rattle tool is giving good result than Weka for our data.

We have taken random forest tree classifier and apply on two data mining tools as discussed above. So we can also take more than one data mining classifier and check the result that which one tool is giving the best result and helping the medical centers for improving the pattern. Clustering and neural network method can also be applied to such type of data for enhancing the result.

References

1. Ojeme, B., Mbogho, A.: Selecting learning algorithms for simultaneous identification of depression and comorbid disorders. In: 20th International Conference on Knowledge Based and Intelligent Information and Engineering Systems, Procedia. Comput. Sci. vol. 96, pp. 1294–1303 (Aug 2016)
2. Sau, A., Bhakta, I.: Artificial neural network (ANN) model to predict depression among geriatric population at a slum in Kolkata, India. J. Clin. Diagn. Res. 11(5), 1–4 (2017)
3. Daimi, K., Banitaan, S.: Using data mining to predict possible future depression cases. Int. J. Public Health Sci. (IJPHS) 3(4), 231–240 (2014)
4. Patel, M.J., Khalaf, A., Aizenstein, H.J.: Studying depression using imaging and machine learning methods. NeuroImage Clin. 10(6), 115–123 (2016)
5. Orabi, A.H., Buddhitha, P., Orabi, M.H., Inkpen, D.: Deep learning for depression detection of twitter users. In: Proceedings of the Fifth Workshop on Computational Linguistics and Clinical Psychology: From Keyboard to Clinic, New Orleans, Louisiana, 5 June 2018, pp. 88–97
6. Han, J., Kamber, M.: Data Mining: Concepts and Techniques. Morgan Kaufmann Publishers, 2nd edn, (2006)
7. Hathaliya, J., Tanwar, S., Tyagi, S., Kumar, N.: Securing electronics healthcare records in healthcare 4.0: a biometric-based approach. Comput. Electr. Eng. 76(5), 398–410 (2019)

8. Kumari, A., Tanwar, S., Tyagi, S., Kumar, N.: Fog computing for healthcare 4.0 environment: opportunities and challenges. Comput. Electr. Eng. **72**, 1–13 (2018)
9. Gupta, R., Tanwar, S., Tyagi, S., Kumar, N., Obaidat, M.S., Sadoun Habits, B.: Blockchain based telesurgery Framework for Healthcare 4.0. In: International Conference on Computer, Information and Telecommunication Systems (IEEE CITS-2019), Beijing, China, pp. 6–10 (Aug 2019)
10. Vora, J., Devmurari, P., Tanwar, S., Tyagi, S., Kumar, N., Obaidat, M.S.: Blind signatures based secured e-healthcare system. In: International Conference on Computer, Information and Telecommunication Systems (IEEE CITS-2018), Colmar, France, 11–13 July 2018, pp. 177–181

Evaluation of Students' Performance Based on Teaching Method Using LMS

Faisal Bawazir Muchtar, Mosleh Hmoud Al-Adhaileh,
Pradeep Kumar Singh, Minah Japang, Zeti Darleena Eri, Hazlina Haron
and Farkhana Muchtar

Abstract The traditional teaching approach in Malaysia is claimed to have caused boredom and potentially reduces students learning performance, and their abilities have been decreasing over the years, leading them into confusion and frustration. Through exhaustive observations made, it was found that a lot of higher institutions of education in Malaysia have not yet adapted learning management system (LMS) to improve students' learning performance and interest. Based on the current and existing work reviewed, none of the universities evaluated the effectiveness of the

F. B. Muchtar
School of Education, Faculty of Social Sciences & Huminiites, Universiti Teknologi Malaysia, 81310 Skudai, Johor, Malaysia
e-mail: fbwzeer@gmail.com

M. H. Al-Adhaileh
Deanship of e-Learning and Distance Education, King Faisal University, Al-Hofuf, Al-Ahsa, Kingdom of Saudi Arabia
e-mail: mdaileh@kfu.edu.sa

P. K. Singh
Department of CSE & IT, Jaypee University of IT, 17334 Waknaghat, Solan, Himachal Pradesh, India
e-mail: pradeep_84cs@yahoo.com

M. Japang
Labuan Faculty of International Finance, Universiti Malaysia Sabah, Labuan International Campus. Jalan Sg Pagar, 87000 Federal Territory Labuan, Malaysia
e-mail: mina1511@ums.edu.my

Z. D. Eri
Faculty of Computer and Mathematical Sciences, Universiti Teknologi MARA, Kampus Kuala Terengganu, 21080 Kuala Terengganu, Terengganu, Malaysia
e-mail: zetid415@uitm.edu.my

H. Haron
Segi College Subang Jaya, Persiaran Kewajipan, USJ1, 47600 Subang Jaya, Selangor, Malaysia
e-mail: hazlinaharon@segi.edu.my

F. Muchtar (✉)
School of Computing, Faculty Engineering, Universiti Teknologi Malaysia, 81310 Skudai, Johor, Malaysia
e-mail: farkhana@gmail.com

© Springer Nature Singapore Pte Ltd. 2020 647
P. K. Singh et al. (eds.), *Proceedings of First International Conference on Computing, Communications, and Cyber-Security (IC4S 2019)*, Lecture Notes in Networks and Systems 121, https://doi.org/10.1007/978-981-15-3369-3_48

LMS despite claims of successes. From the information gathered, previous works have overlooked the significance of evaluating effectiveness of using LMS, and therefore, this research takes the opportunity to look into untapped issues. For this research, the sample population to measure the effectiveness of using LMS at the university level is taken from students in pre-MUET classes who will be sitting for MUET examination in University Putra Malaysia (UPM).

Keywords E-learning · Computer-based learning · Online learning · Learning management system · LMS

1 Introduction

The adaptation of the Internet has changed and revolutionized how we think, use and learn information [1, 2]. In the year 2003, a survey regarding online learning was conducted, and it was found that 81% of all the universities and higher institutions in USA provided either a complete online learning mode or a hybrid (blended) mode in almost all their courses [3]. Additionally, in the survey, 34% of the universities and higher institutions in USA offer at least one or more complete online degree courses(s). In fact, authors in [4] predicted that within seven years from the time of his writing, 50% of all university students in USA will pursue their course online and actively uses technology on a daily basis.

In another study conducted across six regions in Europe, authors in [5] found that learning management system worked effectively in European learning institutions using either commercially of self-developed learning management system (LMS) software.

The schools in the district of Baltimore in USA on the other hand implemented district-wide LMS that unite teachers and students alike across the district using T3 systems or locally dubbed as 'The Teacher Student Support System'. Further, Hill [6] in his article stated that the district of Baltimore uses LMS to post 27,000 classes online for 83,000 students and 62,000 teachers.

The successes of online learning are without a doubt attributed to the availability of learning management system (LMS) as its platform [5], where scalability, security, interoperability, stability and security are its key features evaluated in LMS [7]. The learning management system (LMS) through its efficient use of materials for learning has the ability to provide immediate online feedback and assessment facilitates for both teaching and learning environments [8], where communications can also be improved through e-mails, online discussions and forum within the learning management system platform [9]. Furthermore, students can determine their own pace of learning, according to their own personal needs, and thus, the learning process can become an engaging experience [10].

It is to be noted that LMS has been widely used with proven effectiveness toward increasing the students' performance in various higher learning institutions and universities abroad. In developed countries, LMS has also been successfully adapted in schools, with similar proven effectiveness.

However, based on current and existing work in Malaysian context, it was found that the use of LMS is still underrated and lacks implementation in both higher education and school levels in Malaysia. Therefore, this research tries to fill the areas of this untapped issue, that is to demonstrate and calculate the performance of student who utilized e-learning materials based on LMS and compare the results with the standard traditional teaching method.

To the researcher's knowledge, until present, there has not been any research or existing work that calculates the effectiveness of using LMS and compared them to the traditional teaching method at higher learning institutions level or school level in Malaysia. Therefore, this research is so far and the only research that looks into this issue.

It is to be noted that this research does not seek to replace the current existing teaching method used in Malaysian schools; rather, it serves as an alternative tool for the purpose of increasing the students' performance.

2　Literature Review

This section will describe literature review from three angles, namely description of LMS, literature review on the teaching aide apparatus previously used and related work that uses LMS. From the literature review, we further describe out proposed idea for this research.

2.1　Learning Management System

For better understanding regarding LMS, this section is divided into three sub-section, namely overview of LMS, definition of LMS and characteristics of LMS.

2.2　Overview of LMS

To allow better understanding between learning management system (LMS) features and other features such as e-learning or learning content management system (LCMS), this paper will use the analogy of two Perodua Kancil cars in a showroom. Both the cars have the same color, the same wheel and made of the same materials. In fact, both the cars do the same things, that is, to take a person from point A to point B, run on petrol, and both cars carry a maximum of four passengers. Looking again, however, the second Kancil is fitted with a Ferrari engine.

Supposing that, the owner of the second Kancil wishes to place the Ferrari engine into a 12 seater multi-purpose vehicle (MPV), it is still practical to do so as there is adequate torque power in the engine to move the vehicle. However, to place the Kancil's engine into a 12 seater MPV (multi-purpose vehicle), vehicle is definitely impractical. Additionally, it is also possible to extend other advanced technology such as 'Turbo' technology into the second Kancil, as the engine was designed and explicitly built to allow for such incorporation. The first Kancil, however, has limited ability for any advanced incorporation of any advanced technology.

Similar with the use of this analogy with the LMS, system, it can be said that the first Kancil is akin to the LMS. It is powerful and able to cater large amount of users. It is also possible and practical to extend its purpose and capability as the core technology was designed for incorporations of other technologies in mind. The second Kancil on the other hand is akin to other definitions other than LMS such as virtual learning environment (VLE) and learning content management system (LCMS), although its implementation will to some extent serve its purpose and limitations and constraints apply with this option.

In short, using this analogy to clear the confusion of many that tends to generalize the specific term of learning management system (LMS) with other features such as e-learning or learning content management system (LCMS), it can be deduced that although the technologies embedded in virtual learning environment (VLE), learning course management system (LCMS) and e-learning may share the purpose and function, the distinction of learning management system (LMS) lies in its ability and features which is not possible to be implemented or incorporated in features such as learning content management system (LCMS) or e-learning due to its limiting nature in the design of the core technology itself.

2.3 Definition of LMS

While it can be understood that the development of computers and related technology related to education takes place very rapidly, one common mistake that seemed to be at rampant is the overgeneralization of specific terms at the expense of computer literature developments. For example, authors in [11] stated that learning management system (LMS) is also known as virtual learning environment (VLE). The mass public on the other hand may comprehend that virtual learning environment (VLE) is similar to learning course management system (LCMS) and further similar to virtual learning or e-learning.

This overgeneralization in the end often resulted in the creation of a generic term rounded by many as learning management system (LMS). Therefore, without a good, firm understanding and ability to differentiate the terms and its specification, it is unfortunately easy to entangle oneself in the web of confusion that later will result in inappropriate allocation of the end product (software), especially in determining its compliances within the learning management system (LMS) specification. Understandably, however, the terms can in fact literally provide the same interpretation to many.

This paper acknowledges the intertwining definitions by various scholars; nonetheless, this paper seeks to find a definition that is reliable and able to provide a clear distinct characterization and specification for each of the terms mentioned. Provided in this section is the definition, characteristics, features and elements that clearly separate e-learning, virtual learning system (VLE), learning content management system (LCMS) and learning management system (LMS).

E-learning (electronic learning): According to authors in [12], e-learning term is only been used in 1983, coined by White (1983) 13] in a journal article titled 'Synthesis of Research on Electronic Learning.' Before that, e-learning is known with several different names such as computer-assisted instruction (CAI), computer-based education (CBE) and computer-assisted learning (CAL). The definition of e-learning also evolves according to the availability of the current technology in computer such as multimedia, network, Internet and smart phone technology. In short, authors in [12] are defining e-learning as learning process, that is using computer technology in general during class session regardless students and teacher are available in the class physically or virtually.

Virtual learning system (VLE): *'All in one solution that can online learning for an organization. It includes features of learning management system (LMS) for courses within the learning environment, but it is not able to track courses not created in the particular learning environment'* [14].

Learning content management system (LCMS): *'A centralized software application or set of applications facilitate and streamline the process of designing, testing, approving and posting e-learning content, usually on Web pages'* [15].

Learning management system (LMS): *'Software automates the administration of training. The LMS registers users, tracks courses in a catalog, records data from learners, and provides reports to management. An LMS is typically designed to handle courses by multiple publishers and providers. It usually does not include its own authoring capabilities; instead, it focuses on managing courses created by a variety of other sources'* [15].

2.4   Characteristics of LMS

At this juncture, it is possible to see although the terms are somehow related; conceptual differences existed within specified terms. Although intertwining understanding of the definitions to some extent appears harmless at this point of discussion, the effects of this issue present itself the development stages, namely when learning contents are added into software made for a particular feature (thus determining it constraints and advantages pending in the embedded core technology available in the feature).

Emphasizing educational LMS in particular, the following 'general' characteristics was given by authors in [16], that is, the lessons are linked to the objectives of the instruction, incorporated into the lessons are standardized curriculum, the course extends to several other grades consistently, it collects results of students performance, and lessons are based on individual students learning progress.

Authors in [15] additionally 'recommends' the following functioning LMS; provides administration tools which enable the management of user registrations, profiles, roles, curricula, certification paths, tutor assignments, content, internal budgets, user payments and chargebacks, scheduling for learners, instructors and classroom; the provision of content that involves three aspects that is the medium (classroom, online), the method (instructor-led, self-paced) and learners; develop content that includes authoring, maintenance and storage; ability to integrate into third-party software; able to assess learners competency and skills acquisition; assessment for authoring is provided and supported; adheres to SCORM and AICC standards that allows content to be imported regardless of what the authoring system (features) that it was created in; configured to function within existing systems and internal processes and security features such as encryption and passwords are enabled.

Further discussion of this section on characteristics of LMS requires special attention which will be discussed in Sect. 2.5.4 (misconception of LMS used in schools in Malaysia).

2.5   Apparatus of Teaching Aide Used in Previous Works

The method of teaching has come a long way, always shifting and continuously evolving. Thus, in this section, we shall describe the method of teaching according to the era years since it was first introduced in Malaysia, namely the 1970s, 1980s, 1990s, and 2000s.

2.5.1   Apparatus of Teaching Aide Used in the Era of 1970s

Back in the 1970s and 1980s, the use of radio and television programs such as TV Pendidikan (TVP) and Radio Pendidikan (RP) received exiting response from Malaysian students whenever the teachers used it as additional learning tool, complementing face-to-face teaching interaction with the teacher. Teachers were provided with a time schedule of the programs beforehand. However, most schools never did practice a unified time schedule system, making it difficult for the teachers to incorporate this form of teaching into their classroom teaching.

2.5.2 Apparatus of Teaching Aide Used in the Era of 1980s

Next, the advancement in recording technology in the 1980s comes with the introduction of VCR, VCD and DVD. This particular method of teaching was used slightly more extensively by teachers as it allows playbacks of materials according to as and when the teacher requires it. Although it gave teachers a slight form of control, materials differ from actual classroom teaching, occasionally repeating and to some extent quite dull the students. Its positive effect to the teaching and learning process was vague. Blackboards, books and exercise books were the dominating, preferred and irreplaceable medium to teach, to record progress and to keep the entire student activities throughout the entire course of instruction for teachers at the time.

2.5.3 Apparatus of Teaching Aide Used in the Era of 1990s

The 1990s introduced new teaching aids and teachers quickly adapt to the use of overhead projectors (OHP) and screens. At this point, teachers began to realize how teaching methods and teaching aids can be used to optimize the students' learning process, although at that time, the understanding of most teachers was largely on how it can be used to minimize repeating process to write and rewrite on the whiteboard and how time can be used to focus on the students' learning process. Nonetheless, it served as a preliminary round on the successes of incorporating technology with the teacher's teaching methods and the student's learning process.

2.5.4 Apparatus of Teaching Aide Used in the Era of the 2000s

Today, the phase of teaching evolution enters a new revolution with the utilization of computers and the Internet. The computers' flexible ability to incorporate elements such as audio, video and document transfers is becoming a popular method of instructional mode in the education arena [17], and it changes how the students learn and how the teachers teach using their method of instruction [18]. In fact, according to authors in [19], everything can be incorporated with the use of computers from video content right down to hand phones that have been used in the classroom to support and enhance the learning process of the students today.

Although the apparatus before the introduction of computers such as television and VCR serves a common purpose with computers today is to aid the students learning process through a particular method of instruction, because of its versatility and adaptability, computers deserve a rather distinctive recognition. Depending on its use, it is vital to distinguish the elements that a particular method of instruction has to offer and the importance of its characteristics in education [20] as well as the roles that computers fill in the students learning process.

Two major distinctions of purposes and roles of computers in the learning process were categorized by authors in [21], whether the students learn 'from' computers (as tutors to increase knowledge) or learn 'with' computers as a tool (to develop higher-order thinking and creativity skills).

Authors in [22] stated that where learning 'from' computers takes place and discrete educational software (DES) is involved. Incorporated in DES are terms such as computer-based instruction (CBI), computer-assisted instruction (CAI) and computer-assisted learning (CAL) that authors in [23] described it as drill and practice programs, more sophisticated tutorials and more individualized instruction, respectively. These software applications are the most available computer software in the market and the most extensively used educational technology in schools today. Along with word processing software such as Words, Lotus 123 and Word Perfect, these applications have been used in the classroom for over two decades [24].

The use of discrete educational software (DES) enjoys a significant recognition as a method of learning until educators start realizing the potential of learning 'with' computers. Partially, this actual realization was due to the decrease of computer prices that greatly attributes to the widespread availability of computers both in schools and students' home that also allows students to perform, as individuals, these computer-based instruction.

In the case, language learning among early discoveries was how computers can be utilized in reading programs for language learners. Complementing to available L2 learning theories at the time, authors in [25] applied schema theories as its basis to evaluate reading software programs. He developed the criteria and questions on how to evaluate reading software, namely interactive ability of the reading software (such as its flexibility to respond to students' errors and the ability to differentiate between major and minor errors), information processing (such as the use of text as the base for its activities and the availability to use prediction and problem-solving strategies), background knowledge (such as building schemata through pre-reading activities) and general software construction and its implementation.

Authors in [26] implemented another computer program to determine the entry score of ESL students' reading speed and to find out if improvements can be made with the use of computer programs. The study indicated significant improvements on targeted learners' reading speed although the program developed did little to improve the students' comprehension of the text read.

A similar study with contrasting effect, however, was conducted by authors in [27], who found that computer-assisted reading instruction did improve reading skills and knowledge improvement in mathematics and science.

In another study, authors in [28] in a related study found that computer software can handle a range of activities and perform tasks with speed and accuracy. Software enables the students to check their work almost immediately and allowed maneuverability to gradually move from one stage to another according to their level and ability. Further, when mistakes were made, the program was able to drill, simulate and even explain the errors made in such a way that students found it very easy to understand. In addition, various other studies conducted have also indicated similar importance of applying computer programs in ESL learning [29].

Authors in [30], on the other hand, investigated how tools, techniques and application of computers can be used to engage children learners to read, write, invent, solve problems and explore the world. Their study refers computer technology used in the learning process as 'media' in which four distinct 'media' were distinguished and focused, namely 'media for enquiry' (such as data modeling, spreadsheets and access to online databases), 'media for communication' (such as e-mail, conferencing, simulation and tutorials), 'media for construction' (such as robotics and computer-aided design and control systems) and 'media for expression' (such as interactive video, animation and music composition). This stage marks the point where computers, documents, programs, media and the Internet integrates and thus began an era of adopting another form of learning—known as e-learning.

The process of realizing the importance of computers as the students learn 'with' technology as a mean of enhancing the students reasoning and problem-solving abilities, therefore, gave birth to the origin of the overall framework of learning management system, known as LMS which incorporates another term, integrated learning system (ILS).

Compared to discrete educational software (DES), integrated learning system (ILS) offers functionality beyond instructional content in teaching and learning as it includes content such as overall management, ability to track and provide personalized attention and instruction to the students as well as the integration that incorporates itself across the system [16, 31, 32].

The discussion in this section provides historical background of the shifting methods used in education and provides historical developments in the use of technology as well as the terms required for our understanding on the basis of which LMS was founded. The next section looks at areas where LMS was given due recognition.

2.6　Related Work that Used LMS

This section will discuss existing work that proclaimed itself using LMS as a teaching method both at the university and the school levels.

2.6.1　The Use of LMS in Universities Abroad

In Nordic countries, for example, Sweden, Finland, Czech Republic, etc., learning management system (LMS) is widely used in higher education. 'It is not easy to find Nordic institution without experiences with LMS' [33].

Authors in [34] also reported similar tendencies for LMS in UK and Ireland and mentioned that these two countries have adapted very extensive implementation of e-learning via LMS both at degree and diploma levels.

2.6.2   The Use of LMS at Higher Educational Level in Malaysia

At the national level, however, universities in Malaysia successfully implemented the LMS as University Malaya pioneered the first use of LMS in the country dubbed 'COL' (course online) at the time in 1998. Multimedia University (MMU), International Medical University (IMU) and University Tun Abdul Razak (UNITAR) later implemented LMS in 1999 [35].

In addition to the universities already mentioned, authors in [36] further added Universiti Sains Malaysia (USM), Universiti Kebangsaan Malaysia (UKM), Universiti Putra Malaysia (UPM) and Universiti Technology Mara (UiTM), as evidence of successes, where LMS played its role.

Based on a study conducted by Su Lih Teng in 2007 [37], it was found that e-learning has become one of the most popular forms of teaching and learning in the field of education. The outcome of this study has initiated UTM to develop its e-learning program that utilizes Moodle-based LMS in 2004. The purpose of the mentioned study is to survey the influence of information technology in relation to gender, type of courses and the attitudes toward e-learning in UTM.

Further, in another study by Ahmad Rafee Che Kassim and Mazlan Samsudin [38] in the same year, they presented the aspects of construction, implementation and the management and development of e-learning and IAB's step by step measures that were implemented. The research was conducted using action research and observations which were made by selected lecturers. The results from the observation indicated heightened quality of teaching and learning by the lecturers and students in IAB. The project was conducted in detail and received support from all relevant parties in Institute Aminuddin Baki.

Later in 2009, UPM introduced a fully implemented LMS in the country known as Putra LMS that integrates five previously self-managed LMS in respective faculties, namely Faculty of Engineering, Faculty of Educational Studies, Faculty of Economy and Management as well as the Faculty of Modern Languages (UPM, 2009) [39], thus proving LMS adaptability of extending its features as mentioned in Sect. 2.3 (Definition of LMS).

Although the successes on the implementation of LMS in Malaysian universities gained widespread acknowledgment (effort from providers), the acceptance of LMS to Malaysia's general public and schools deserves a second look nonetheless. Here, discrepancies existed between the providers and the receivers.

Despite the government's initiative for Malaysian Super Corridor (MSC) in 1996, Malaysia's ranking in Internet penetration (number of users) today is bypassed by other competing countries at the time. Malaysia was rather slow to changes in the ICT industry, and its usage is under-utilized [40].

A survey was conducted in 2004 that included 5779 respondents (4625 were students, 977 were university faculty members, 102 were policy-makers and 75 came from the industry) found that Malaysian public is only 'moderately' ready for any form of online learning with a mean scale of 5.5 out of 10 [41].

Further, the survey also indicated a mean of 4.76 (the lowest in the study) for its environmental readiness, that is the country's population readiness as a whole in terms of government policy, role of mass media and proficiency in the English language. The same study also indicated that learners are more ready for online learning (with a mean of 6.33) when compared to the perception of the lecturers (with a mean of 5.73).

2.6.3   The Use of LMS in Schools Abroad

Hill (2009) in his interview with one of the K-12 teachers in the Baltimore district in USA mentioned that 'in a typical algebra classes, if you ask a question you may get one or two hands, but by asking the question in a discussion board or chat feature associated with LMS, you may get 100 hits within 15 min.' In the same report, another teacher also stated that the 'anecdotal accounts from teachers in his districts indicate that the environment in the classroom of teachers who use LMS is far superior to what is reported in non-LMS participating classrooms.' Mentioned earlier in this article, Hill [6] in his article stated that the district of Baltimore uses LMS to post 27,000 classes online for 83,000 students and 62,000 teachers.

In Ontario and Quebec in Canada, LMS is provided at provincial level by the Ministry of Education (MOE), and there have also been discussions to incorporate LMS in schools at a national level. Across Ontario province alone, it was estimated that 25,000 students took online courses set up by the Canadian Ministry of Education [42].

In the same report, Powell and Patrick stated that in Singapore 75% of all its schools adapted LMS. The figure works out to about 400,000 students, and the Ministry of Education aimed to incorporate LMS in its entire primary, secondary and junior colleges by the end of 2006.

2.6.4   Misconception of LMS Used at School Level in Malaysia

There are many claims that Malaysian schools implemented learning management system (LMS) as mentioned in the previous work; however, this section aims to highlight the fact that learning management system (LMS) is never applied in any Malaysian school despite the numerous claims by various parties involved.

For example, a simple search on Google pertaining to smart schools in Malaysia will return uncountable metadata with the tag learning management system (LMS) as the subheading. Extensive search, however, reveals that none has complied with the actual definition of learning management system (LMS) discussed in Sect. 2.2, Definition of LMS.

Visits to the schools' official portals revealed static information that serves to provide the school's organizational chart, examination dates and general messages. Some teachers included lesson plans as its content. However, much of the links are broken. Structure and organization of content are limited to individual schools and administrators with extremely limited features.

In short, misconception existed within the Malaysian school context that likened the process involved in learning content management system (LCMS) as a similar concept that is available within the LMS infrastructure, which it definitely is not.

LMS characteristics do not limit any of its functionality such as in learning content management system (LCMS). In fact, based on literature review, LMS is an open ended and very systematic in nature with almost no limitations on the system whatsoever. In fact, any system be it, virtual learning environment (VLE), learning content management system (LCMS) and e-learning can be incorporated into the LMS structure, however, not vice versa.

This misconception emerged largely due to confusion pertaining to the exact definition that differentiates LMS, which later escalates the process of generalization of this specific term. The assumption at large is that, when the computer and the Internet are combined with education, LMS comes into effect. The misconception especially can be noted especially in Malaysian smart schools where computers and the Internet are made available.

On a worldwide scale, in terms of readiness for any form of online learning, Malaysia is currently ranked at number 36th among 65 countries that includes countries mentioned earlier such as USA *(ranked 2nd)*, *Sweden (ranked 2nd, tie with USA)*, *Singapore (ranked 6th)*, *UK (ranked 7th)*, *Finland (ranked 10th)*, *Canada (ranked 13th)*, Ireland *(ranked 21st)*, *and Czech Republic (ranked 31st)* [43]. Based on this fact, to increase the public readiness to accept the implementation of LMS and to rectify the misconception of this particular teaching aide, this paper proposes an idea to implement LMS at an early stage at the school level. The concept of this idea will be discussed in detail in the next section.

2.7 Quality of Online Teaching Versus Traditional Teaching

Authors in [2] in their study found that respondents generally agreed that the quality of online education will improve in the future. Sixty percent of the respondents expected that the quality of online courses would be identical to traditional instruction by the year 2006 (see Fig. 1: Expected Quality of Online Versus Traditional Education).

Additionally, majority of the respondents predicted that the quality of online courses would be superior to (47%) or the same as (39%) that of traditional instruction by 2013. Only 8% predicted that the quality of online courses would be inferior in 2013.

Nonetheless, there are also predictions from the respondents that learning outcomes of online students would be either the same as (39%) or superior to (42%) those of traditional method students by 2013.

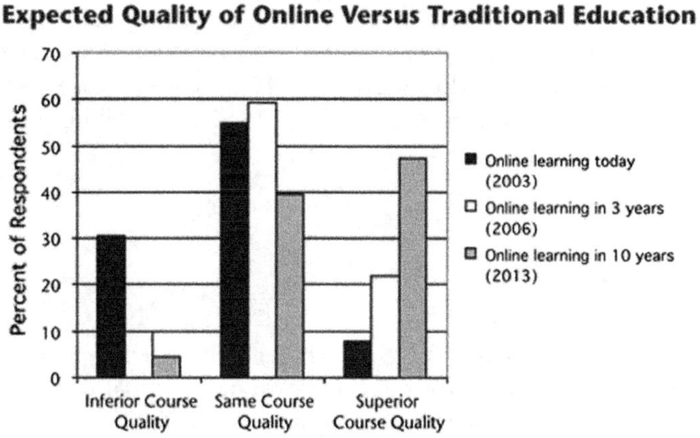

Fig. 1 Expected quality of online versus traditional education [2]

2.8 Evaluation Method to Measure Quality of Online Learning

Evaluation is an important part of ensuring the quality of online courses and programs. Table 1 summarizes respondents' predictions about future trends concerning the evaluation of online learning.

Authors in [44] in their study questioned respondents on what would they perceive the most effective measurement that will be effective to determine the quality of online education in the coming decade. The results obtained indicated that 43.8% of the respondent answered that a 'comparison of online student achievement with that of students in face-to-face classroom settings' would be the most effective. This

Table 1 Predictions of how the quality of online learning will be measured [44]

Predictions about how the quality of online learning will be measured		
Response options	Number of respondents	Response rate (%)
Comparison of student achievement with those in live or face-to-face classroom settings	237	43.8
Student performance in simulated tasks of real-world activities	80	14.8
Student course evaluations	47	8.7
Course completion rates	36	6.6
Course interactivity ratings and evaluations	24	4.4
Other	24	4.4
Student placement into jobs	23	4.3
Student satisfaction questionnaires	17	3.1
Computer log data of student usage and activity	1	0.2

was followed by students 'performance in simulated tasks of real-world activities' (14.8%), 'student course evaluations' (8.7%), 'course completion rates' (6.6%), 'course interactivity rating and evaluation' (4.4%), 'student placement into job' (4.3%), 'student satisfaction questionnaire' (3.1%) and 'computer log data of student usage and activity' (0.2%).

3 Design and Implementation

Learning management Systems (LMS) or virtual learning environments (VLE) is an application developed to assist students' learning in virtual environment. This project was made possible with the cooperation with Putra LMS, MUET students in University Putra Malaysia and the researchers for this paper.

Therefore, it is necessary that the subtopics included in this article provide a brief description about Putra LMS, the features available in Putra LMS and the development slots that were created in the e-learning process for UPM students undertaking MUET course.

Here, the prototype that we named teaching method using LMS (TMUL) will be constructed in this research. To ease discussions in later sections, TMUT will be used as an acronym that stands for teaching method using traditional approach.

3.1 Putra LMS

Putra LMS is the latest UPM learning management system deployed to facilitate all aspects of e-learning activities in the university. To enter into Putra LMS, type in http://lms.upm.edu.my at the browser address (see Fig. 2).

Fig. 2 Putra learning management system

3.2   Slots Created for MUET Course Based on Putra LMS

The slots created for MUET course based on Putra LMS are based on the features as the following: Slot-Assignments, Slot-Quizzes, Slot-Forum, Slot-Online Chat Room, Slot-Glossary, Slot-E-mail: Slot-Blogs, Slot-Helpdesk: Slot-Attendance List, Slot-MarkSheet.

Slot-Assignments: This particular slot allows students to send in their assignment. It also allows students to view deadlines for projects.

Slot-Quizzes: The slot for quiz is important to determine whether objectives set to the students were achieved or vice versa. The quiz provided in the e-learning materials includes all the subtopics. In this quiz, 10 objective questions were provided. The quiz of choice for most of the users was multiple-choice questions. Students need to answer the quiz according to the time stipulated.

Slot-Forum: The forum slot provides a venue for the students to discuss and provide input particularly the topics of the current study. Students can view all discussions, ideas and input from their course mate. Additionally, students can also create their own discussion topic. This in turn encourages the students to learn via social constuctivism.

Slot-Online Chat Room: Chatting allows individual students to chat online with their course mate. Further, this slot allows teacher and students to interact with each other.

Slot-Glossary: In this particular slot, users have the opportunity to search for unknown meaning of words in their learning process for the MUET course. This is to ensure that users understand important words and definition relating to the subject. Hence, it provides avenues for better and concrete understanding. Students need to place the cursor on the unknown words and the meaning of the words will be displayed in the section provided.

Slot-E-mail: The e-mail slot allows students to send electronic mails to teachers and students. Unlike public access e-mails and yahoo, this slot provides a directory for easy access with course mates and lecturers.

Slot-Blogs: This is a place for blogger to share their knowledge and experience. The student can read the article and make comments.

Slot-Helpdesk: To enable user-friendly e-learning materials based on Putra LMS, the 'helpdes'k' section provides students with of the objective for each learning process, guides on how to use included softwares and materials, references as well as links to other related Web pages and other related information.

Slot-Attendance List: Attendance list allows lecturers and the students to check their attendance digitally.

Slot-MarkSheet: Mark sheets contain the quiz marks for the quiz conducted throughout the course.

4 Result and Analysis

This section pays close attention to the MUET students related to this research. The students were given ample opportunity to familiarize themselves with the Putra LMS. After a window period of 2 months of getting acquainted with the Putra LMS, the students were given an evaluation questionnaire to complete.

4.1 Result and Analysis

In this section, the outcome of the result analysis would be discussed in detail.

4.1.1 Quality of Teaching Method Using LMS Versus Traditional Education

This research utilizes 'questionnaires' in its approach to obtain information pertaining to the usage of LMS programs from students who took MUET courses in University Putra Malaysia. The list of questions forwarded to the students in the questionnaire for this paper is listed in Table 2.

Table 2 Result from questionnaire survey completed by UPM student

No.	Question	Agree	Disagree
1	Do you think that using the Putra LMS for this MUET course helped you to learn more compared to traditional methods of learning alone? N = 24	24	0
2	Which slot of the Putra LMS did you find most interesting and helpful compared to traditional teaching method? (you may pick more than one):		
	Slot-course material/theory	24	0
	Slot-posting assignment	24	0
	Slot-quizzes	19	4
	Slot-forum	18	4
	Slot-online chat room	17	7
	Slot-glossary	16	8
	Slot-e-mail	16	8
	Slot-blogs	15	9
	Slot-helpdesk	19	5
	Slot-attendance list	20	9
	Slot-marksheet	24	0

From the data obtained in the questionnaire survey in Table 2, it was found that majority of the student (100%) utilized Putra LMS to download notes and other learning materials provided by the lecturers. Additionally, the students also utilized Putra LMS to send and produce assignments as well as to obtain information and be reminded of due dates, etc. (100%). All of the students indicated that LMS facilitated them to check on their carry forward marks and examination results (100%).

The students were also very enthusiastic about online quizzes and online exercises supplied (79.1%) and indicated similar liking for the helpdesk feature that provides information and external links (66.7%). The glossary slot that made it possible for the students to check unknown definition and meaning of word with one-click access also received favorable response (66.7%).

Further, the results from this questionnaire survey also indicated that the students felt comfortable to interact and communicate with lecturers, fellow course mates and networks of friends who registered their profiles in Putra LMS through (66.7%) e-mails, online chat rooms (70.8%), forums (75%) and blogs (62.5%).

It is worth mentioning that majority of the students interviewed stated how easy it was to utilize Putra LMS and remarked that 'with one click, all the information that is required can be obtained.' Additionally, the students also found it to be an advantage, especially for shy students to interact with lecturers, course mates and other students in Putra LMS network with the possibility of using nicknames in some features of the system.

4.1.2 Descriptive Result of TMUL and TMUT

To demonstrate that the results of TMUL were better than TMUT based on the set of questionnaire survey in Table 3, a comparison based on mean scores of both results is provided. Graphical comparison chart is also provided to support the analysis of the mean scores.

In this research, a lower mean score illustrated a better result. Thus, if the mean score for TMUL was much lower than the mean score of TMUT, it can be concluded then that TMUL had performed better than TMUT.

From the comparison of mean scores for both results, it was found that the mean score for TMUL was 8.83 and the mean score for TMUT was 2.04. This indicated that the results obtained from TMUL were better than TMUT.

To support the results obtained by comparing the mean scores, graphical chart is used and portrayed in Fig. 3.

Graphical analysis from Fig. 3 indicated that the result of TMUL had always been lower than the result of TMUT. This proved that the performance of TMUL

Table 3 Descriptive static result of set of questionnaire survey from student of UPM	Description	Type	Value (N)	Mean
	Score	TMUL	24	8.83
		TMUT	24	2.04

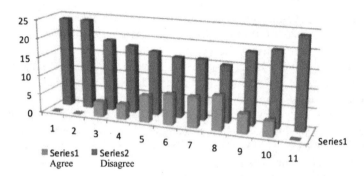

Fig. 3 Bar chart—set of questionnaire survey versus scores from student of UPM

was better when compared to the performance of TMUT for all the data slots gathered and captured from the questionnaire survey listed in Table 2.

4.2 Overall Conclusion

The performance evaluation illustrated in this article was based on the results captured in questionnaire survey filled by UPM students who took MUET courses in Second Semester year 2017/2018.

It is interesting to note that the findings obtained from the questionnaire survey showed that students engaged very positively to the use of teaching method using LMS compared to standard traditional teaching method.

The result obtained from each of the questions in the questionnaire survey conducted proved that the usage of teaching method using LMS can indeed increase and contribute toward their performance to learn English subject particularly for those sitting for MUET paper.

5 Summary

From the study conducted, it can be concluded that teaching method through the use of e-learning made possible in LMS provided significant advantages compared to disadvantages.

Among its many advantages, LMS includes a certain kind of bonus for the students to utilize the facilities as it was provided in a systematic and efficient manner. Teaching method utilizing LMS-based e-learning process provided a new alternative form of teaching and learning that is meaningful and fun in its implementation. It can also be regarded as an approach with vast potential to make students understand and upgrade their performance through guided inquiry.

Realizing that the use of e-learning materials based in LMS can assist the process of teaching and learning, and it is imperative to note that education through LMS is gaining widespread acceptance in global arena. It is, therefore, very important that all institution of higher learning to take the initiative and highlight its effectiveness.

Further, higher institution also needs to demonstrate its effectiveness by demonstrating that students who utilized LMS as learning method do fare above traditional teaching method alone.

Even though this research is considered to be a small scale, it provided the necessary preliminary areas to venture, especially into a new dimension of topic in field of education. For future work, it is hoped that this research can be extended in order for it to contribute even more toward education in Malaysia.

References

1. Lipschultz, J.H.: Visualizing social network influence: measurement and case studies. Contemporary Research Methods and Data Analytics in the News Industry pp. 208–232. https://doi.org/10.4018/978-1-4666-8580-2.ch012. https://www.igi-global.com/chapter/visualizing-social-network-influence/132916 (2015)
2. Lipschultz, J.H.: social media communication: concepts, practices, data, law and ethics. Taylor & Francis (2017)
3. Allen, I.E., Seaman, J.: Entering the mainstream: the quality and extent of online education in the United States, 2003 and 2004. Sloan Consortium (2004). https://eric.ed.gov/?id=ED53006
4. Horton, W.K.: Designing Web-based training: how to teach anyone anything anywhere anytime, vol. 1. Wiley, New York, NY (2000)
5. Paulsen, M.F.: Experiences with learning management systems in 113 European Institutions. J. Edu. Tech. Soc. 6(4), 134–148 (2003). https://www.jstor.org/stable/jeductechsoci.6.4.134
6. Hill, E.E.: Pioneering the use of learning management systems in K-12 education. Dist. Learn. 6(2), 47–51 (2009)
7. Hall, J.L.: Assessing learning management systems (2002). https://www.chieflearningofficer.com/2002/12/11/assessing-learning-management-systems/
8. Breen, L., Cohen, L., Chang, P.: Teaching and learning online for the first time: student and coordinator perspectives. ECU Publications Pre. 2011 (2003). https://ro.ecu.edu.au/ecuworks/3589
9. Beard, L.A., Harper, C.: Student perceptions of online versus on campus instruction. Edu. 122(4) (2002)
10. Zhang, D., Zhao, J.L., Zhou, L., Nunamaker Jr, J.F.: Can e-learning replace classroom learning? Commun. ACM 47(5), 75–79 (2004)
11. Rienties, B., Giesbers, B., Lygo-Baker, S., Ma, H.W.S., Rees, R.: Why some teachers easily learn to use a new virtual learning environment: a technology acceptance perspective. Interact. Learn. Environ. 24(3), 539–552 (2016). https://doi.org/10.1080/10494820.2014.881394
12. Aparicio, M., Bacao, F., Oliveira, T.: An e-Learning theoretical framework. J. Educ. Tech. Soc. 19(1), 292–307 (2016). https://www.jstor.org/stable/jeductechsoci.19.1.292
13. White, M.A.: Synthesis of research on electronic learning. Educ. Leadersh. 40(8), 13–15 (1983). https://www.learntechlib.org/p/134860/
14. Hall, B.: White paper brandon hall research center. Brandon Hall Excellence in Learning Awards (2009)
15. Glossary—Learning Circuits—ASTD (2012). https://web.archive.org/web/20120224174841 http://www.astd.org/LC/glossary.htm

16. Bailey, G.D.: Wanted: a road map for understanding integrated learning systems. Educ. Tech. **32**(9), 3–5 (1992)
17. Khan, M.A., Omrane, A.: A theoretical analysis of factors influencing students decision to use learning technologies in the context of institutions of higher education. Adv. Soc. Sci. Res. J. **4**(1) (2017)
18. der Kleij, F.M.V., Feskens, R.C.W., Eggen, T.J.H.M.: Effects of feedback in a computer-based learning environment on students' learning outcomes: a meta-analysis. Rev. Educ. Res. **85**(4), 475–511 (2015). https://doi.org/10.3102/0034654314564881
19. Tlili, A., Essalmi, F., Jemni, M., Kinshuk, Chen, N.S.: Role of personality in computer based learning. Comput. Hum. Behav. **64**, 805–813 (2016). https://doi.org/10.1016/j.chb.2016.07.043. http://www.sciencedirect.com/science/article/pii/S0747563216305337
20. Ostanina-Olszewska, J.: Modern technology in language learning and teaching. Linguodidactica (22), 153–164 (2018). https://www.ceeol.com/search/article-detail?id=766584
21. Goldie, J.G.S.: Connectivism: a knowledge learning theory for the digital age? Med. Teach. **38**(10), 1064–1069 (2016). https://doi.org/10.3109/0142159x.2016.1173661
22. Ringsta, C., Kelley, L.: The learning return on our educational technology investment: a review of findings from research. WestEd (2002). https://eric.ed.gov/?id=ED462924
23. Murphy, R., Penuel, W.R., Means, B., Korbak, C., Whaley, A., Allen, J.E.: E-DESK: a review of recent evidence on the effectiveness of discrete educational software. Technical Report SRI Project 11063, SRI International (2002)
24. Parr, J.M., Fung, I.: A review of the literature on computer-assisted learning, particularly integrated learning systems, and outcomes with respect to literacy and numeracy. Technology Republic Ministry of Education New Zealand, Wellington, New Zealand (2016). Retrieved 20 Nov 2016
25. Becker, H.J., Ravitz, J.L., Wong, Y.: Teacher and teacher-directed student use of computers and software. Teaching, learning, and computing: 1998 National Survey. Report #3 (1999). https://eric.ed.gov/?id=ED437927
26. Preisinger, R., Sargeant, K., Weibel, K.: An evaluation of reading software according to schema theory. Tech. Rep. Eric Doc. Reprod. Serv. No. ED 298761 (1988)
27. Van Laere, E., Agirdag, O., van Braak, J.: Supporting science learning in linguistically diverse classrooms: factors related to the use of bilingual content in a computer-based learning environment. Comput. Hum. Behav. **57**, 428–441 (2016). https://doi.org/10.1016/j.chb.2015.12.056. http://www.sciencedirect.com/science/article/pii/S0747563215303277
28. Skryabin, M., Zhang, J., Liu, L., Zhang, D.: How the ICT development level and usage influence student achievement in reading, mathematics, and science. Comp. Educ. **85**, 49–58 (2015). https://doi.org/10.1016/j.compedu.2015.02.004. http://www.sciencedirect.com/science/article/pii/S0360131515000457
29. Rahimi, M., Allahyari, A.: Effects of multimedia learning combined with strategy-based instruction on vocabulary learning and strategy use. Sage Open **9**(2), (2019). https://doi.org/10.1177/2158244019844081
30. Chatterjee, P., Mishra, D., Padhi, L.K., Ojha, J., Al-Absi, A.A., Sain, M.: Digital story-telling: a methodology of Web based learning of teaching of folklore studies. In: 2019 21st International Conference on Advanced Communication Technology (ICACT), pp. 573–578 (2019). https://doi.org/10.23919/icact.2019.8702047
31. Becker, H.J.: A Model for improving the performance of integrated learning systems: mixed individualized/group/whole class lessons, cooperative learning, and organizing time for teacher-led remediation of small groups. Educ. Tech. **32**(9), 6–15 (1992). https://www.jstor.org/stable/44427613
32. Szabo, M.: CMI Theory and practice: historical roots of learning management systems. In: Association for the Advancement of Computing in Education (AACE), pp. 929–936 (2002). https://www.learntechlib.org/primary/p/15322/
33. Paulsen, M.F.: An analysis of online education and learning management systems in the Nordic countries. Online J. Distance Learn. Adm. **5**(3), 1–12 (2002)

34. Keegan, D.: The use of learning management systems in North Western Europe. Web-education systems in Europe, Zentrales Institut für Fernstudienforschung pp. 58–81 (2002)
35. Kaur, K., Zoraini Wati, A.: An assessment of e-learning readiness at Open University Malaysia (2004)
36. Soekartawi: Constraints in implementing 'E-Learning' using WebCT: lessons from the SEAMEO regional open learning center. Malays. Online J. Instr. Technol. **2**(2), 97–105 (2005)
37. Su, L.T.: Kemahiran Teknologi Maklumat di Kalangan Pelajar Fakulti Pendidikan, UTM dan Hubungannya Dengan Sikap Terhadap E-Pembelajaran. Master's thesis, Universiti Teknologi Malaysia, Faculty of Education (2007). http://eprints.utm.my/id/eprint/6587/
38. Rafee, A., Kassim, C., Lew, Y.L., Hafizi, S.: Aplikasi E-Pembelajaran Dalam Program Latihan di Institut Aminuddin Baki. Jurnal Kajian Tindakan: Pengurusan Dan Kepimpinan Pendidikan (2009–2010) pp. 63–82 (2011)
39. Juferi, N.E., Ahmad, K.N.: UPM launches putra future classroom (2018). https://www.thestar.com.my/news/education/2018/02/18/ upm-launches-putra-future-classroom
40. Adni, N.D.A.: Pembangunan bahan E-Pembelajaran berasaskan moodle berta-juk 'SETS'dan'TRIGONOMETRY II'matematik tingkatan empat. Ph.D. thesis, Universiti Teknologi Malaysia (2008)
41. Abas, Z.W.: E-Learning in Malaysia: moving forward in open distance learning. Int. J. E-Learning **8**(4), 527–537 (2009). https://www.learntechlib.org/primary/p/30506/
42. Powell, A., Patrick, S.: An international perspective of k-12 online learning: a summary of the 2006 NACOL international e-learning survey. N. Am Counc. Online Learn. (2006)
43. Adams, D., Sumintono, B., Mohamed, A., Noor, N.S.M.: E-learning readiness among students of diverse backgrounds in a leading Malaysian higher education institution. Malays. J. Learn. Instruc. **15**(2), 227–256 (2018)
44. Kim, K.J., Bonk, C.J.: The future of online teaching and learning in higher education. Educ. Q. **29**(4), 22–30 (2006)

Enhancing Stock Prices Forecasting System Outputs Through Genetic Algorithms Refinement of Rules-Lists

Abraham Ayegba Alfa, Ibraheem Olatunji Yusuf, Sanjay Misra and Ravin Ahuja

Abstract The intent of stock market was to amass capital in an economy and distribute of same to high-yielding return ventures. Recently, stock markets are considered the foremost meeting point of information such as macroeconomic, national and investor. There are significant mechanisms for measuring future progress in the economy and the markets. Studies have revealed that fuzzy logic control (FLC)-based forecasting models rely on the composition of rules-lists, which are often redundant due to poor mapping of their antecedents and conditions to the consequents. This paper introduced a process of refining the rules-lists with the use of genetic algorithm. A refined rules-list was constructed for FLC rules base after the removal of inherent redundancy. To evaluate the proposed enhanced FLC model, the inputs and output variables were opening, highest and closing prices of Dangote Cement Company Shares, respectively. The outcomes showed that rules-lists of the enhanced FLC were shortened to five (5) rules as against the nine (9) rules in the human expert system. Also, the forecasts of enhanced FLC constructed with refined rules-lists were better than those FLC built with human expert system on the basis of mean square error (MSE) and mean absolute percentage error (MAPE) calculated. In the case of MSE, forecasts improved from 24.898% to 75.102%. Similarly, MAPE forecasts accuracy improved from 32.424% to 67.576% for the enhanced FLC against FLC.

A. A. Alfa
Kogi State College of Education, Ankpa, Nigeria
e-mail: abrahamsalfa@gmail.com

I. O. Yusuf
Federal University of Technology, Minna, Nigeria
e-mail: yiomoore@yahoo.com

S. Misra (✉)
Covenant University, Otta, Nigeria
e-mail: sanjay.misra@covenantuniversity.edu.ng

R. Ahuja
Vishvakarma Skill University, Gurgaon, Haryana, India
e-mail: ravinahujadce@gmail.com

© Springer Nature Singapore Pte Ltd. 2020
P. K. Singh et al. (eds.), *Proceedings of First International Conference on Computing, Communications, and Cyber-Security (IC4S 2019)*, Lecture Notes in Networks and Systems 121, https://doi.org/10.1007/978-981-15-3369-3_49

Keywords FLC · Rules-Lists · Rule base · Enhanced FLC · Genetic algorithm · Refine · Forecast

1 Introduction

The discussions about whether returns on stock can be accurately forecasted are still on-going subject in several financial research and asset pricing. In 1990, Lo and MacKinnlay opined that the methods of predicting stock returns might just involve snooping of data. It is a widely held notion that certain economic and financial factors can largely portray the discrepancy of the returns on stock which could be used to effectively predict trends for a stock return. Certain studies have revealed that linear models can be used to forecast stock returns alongside other financial indicators (such as earnings yield), or special economic progression components (such as rate of inflation, growth of industrial production, exchange and interest rate) [1].

Though, there are many financial organizations and corporate individuals utilizing this approach to predict profit [2]. The most basic concept for forecasting stock return is the capital asset pricing model (CAPM) which considered the surplus stock returns as being linked to the systematic risk of the market. In fact, economic indicators are reflective of the total economic conditions and provide signals about the health status of the nation's economy [3].

Nowadays, the accuracy of forecasts has utilized many optimizations algorithms in searching for the best possible solution in space. There are prospects of utilizing genetic algorithm operations to improve forecast performance accuracy of fuzzy logic controls through parameters optimizations.

In this paper, a genetic algorithm is used to refine the fuzzy logic control rules-lists in order to enhance the predictive outcomes of stock prices. The paper is organized as follows. The literature review and background are provided in Sect. 2. The methodology, results and conclusion are in Sects. 3, 4 and 5, respectively.

2 Literature Review

Stock market prediction attempts to ascertain the future value of company stock. Whenever the accurate forecast can be arrived for future stock price, there is the propensity of amassing significant returns on it. Recently, the trends in stock prices are regulated by efficient-market theories of random assumptions [4]. These techniques are discussed in the next subsections.

2.1 Computer-Assisted Reasoning

Generally, the computer system became useful in the process of making decisions validly and accurately on the basis of the dataset available to it, which is the commencement of the artificial intelligence reasoning era. These include:

Decision tree is a field of artificial intelligence derived from pattern recognition and computational learning theory, which is capable of learning and producing forecasts from data. In decision tree learning problems, the machine learning approaches are used to construct a decision tree in order to support certain decision-making processes. To construct a decision tree for a given task, a decision tree learning algorithm carefully chooses specific attributes to be tested at each node for each level of the tree.

Artificial neural networks are born out of the biological neural systems composed of highly sophisticated webs of interconnected neurons, which respond to external stimuli outside of the system. A typical ANN has specific structures and methods of learning. The perceptron is one of the foremost ANN models designed to classify patterns by a supervised learning technique [5]. It includes a processing unit that accepts a set of inputs having unique values (called a bias) and incremental values of 1. The connection of input and processing unit makes use of specially assigned weights. In general, the training dataset is continuously offered for the network to obtain network output, which is same or nearest to the desired output [5].

Statistical quantities can be used to estimate relationships and predict possible trends in stock markets investments returns through the analysis of two or more variables over a period of time. This kind of method is popularly known as time series. Statistical inferences have been derived to support decision-making processes as follows:

Linear Regression Analysis. It is used to estimate the stock activities given by the general Eq. 1:

$$y = a + bx. \tag{1}$$

a and b are calculated by Eqs. 2 and 3.

$$b = \left(n \sum xy - n \sum x \sum y \right) \Big/ \left(n \sum x^2 - \left(\sum x \right)^2 \right) \tag{2}$$

and,

$$a = Y - bx \tag{3}$$

where

x and y are mean values computed,
n is the number of recorded observations,
In fact, the linear regression is referred to as finest linear unbiased estimator.

Coefficient of Determination. The formulation and computation of coefficient of determination are given by Eq. 4:

$$r^2 = \left[\left(\left(n\sum xy - n\sum x \sum y \right)^2 \right) \Big/ \left(n\sum x^2 - \left(\sum x\right)^2 . n\sum y^2 - \left(n\sum y\right)^2 \right) \right] \qquad (4)$$

Multiple Regressions. It is used to predict stock as follows: from Eq. 1, given that $Y = a + bX$; then $V = a + bZ$, where Y can be represented as V, which is the value of stock shares associated with variables such as incomes, competition, unemployment rates, inflation rates and exchange rate. And the dependent variable Z is not a single independent variable. X is number of different independent variables x_i. Thereafter, Eq. 1 can be rewritten as shown in Eq. 5:

$$y = a + b_1 x_1 + b_2 x_2 + \cdots + b_n x_n. \qquad (5)$$

In the same vain, by substituting the various dependent and independent variables (such as cost of advert, stock value, share price, unemployment, inflation and competition) in Eq. 5 gives Eq. 6 as follows [6]:

$$V = a + b_1 D + b_2 P + b_3 U + b_4 I + b_5 C \ldots + b_n x_n. \qquad (6)$$

where

D the advert cost,
V the stock value,
P the share price,
U the unemployment,
I inflation,
C competition

2.2 Fuzzy Logic Controls

Fuzzy logic controls can be constructed and built to solve complex problems. This can be achieved with fuzzy inference systems as follows:

1. Determination of input variables ranges with corresponding names.
2. Determination of output variables ranges and corresponding names.
3. Generate the fuzzy membership functions degree for inputs and output variables.
4. Build the fuzzy rules for operating the FIS.
5. Assign strengths for action and execution processes to the rules and defuzzification.

An entire fuzzy logic control system is made up of the following [7]:

- Rule editor,

- FIS editor,
- Membership functions,
- Rule viewer,
- Surface viewer.

Defuzzification. It uses a fuzzy set (the cumulative output fuzzy set) as an input with a single number matching output [8]. In FLCs, there are fewer values, rules and decision process needed as compared to other AI approaches. Other benefits of FLCs include:

1. Natural language (linguistic) variables are used as against numerals similar to human thinking method.
2. There is flexibility in terms of the number of observed variables evaluated.
3. It adopts simple design processes and suitable for solving complex problems.
4. The method of knowledge representation and acquisition is simple.
5. There are fewer rules incorporating huge complexity for better robustness.
6. It connects the output to input without the full grasp of the nature of the underlying variables, which increases the accuracy and stability of the system.

Limitations of Fuzzy Logic Controls. The following are the limitations of fuzzy logic controls: [9]

1. It cannot be easily modelled.
2. The rule-based concept employed by fuzzy logic makes it less favourable over mathematical precision or crisp system and linear models.
3. It is easier to design and prototype, but requiring several simulations and fine-tuning before releases.

2.3 Genetic Algorithm Search

Genetic algorithm (GA) is a stochastic all-inclusive search technique that simulates the characteristics of natural sciences evolution such as crossover, mutation and selection. GA models the artificial principal of Darwinians survival of fittest theory, while the genetic task is abstracted from nature to build a strong tool useful for effectively determining best solutions to multifaceted real-life problems [9, 10]. GAs are simple and potent from the calculation standpoint, because they reduce the hypotheses in search space rather than restraining it [11]. GA can operate on linear and nonlinear problems articulated in continuous and discrete search space. According to [12], the objective of GA is to continuously produce fitter offspring after successive generations of simple genetic procedures. Majority of genetic procedures progressively point the search in the direction of a global best solution as illustrated in Fig. 1.

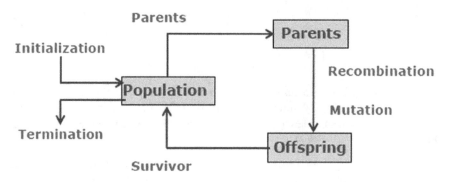

Fig. 1 Simple genetic algorithm procedure [13]

2.4 Related Studies

There are several works available in literature on prediction for stock market [11, 14–20]. The use of fuzzy inference in stock market decision-making process was explored by [14]. Consequent upon this, four indicators were selected for technical analysis including: relative strength index, moving average convergence, and on-balance volume and stochastic oscillator. Each indictor was combined with trading rules to produce fuzzy rules and membership functions that support the fuzzy inference system. The results support three main decisions such as buy, sell or hold. Though, this model's efficiency depends partly on individuals' trading expertise and other factors.

A translated Nigeria stock market prices (NSMP) prediction method gave rise to increased accuracy over untranslated NSMP using error backpropagation algorithm (or regression analysis) according to a study conducted by [15]. The forecasts were conducted on the basis of prevailing foreign exchange rate, and NSMP showed increased accuracy and consistency. However, other models such as fuzzy logic, genetic algorithm and rough sets [15] could improve translated NSMP forecasts when incorporated.

Autonomously, genetic algorithm and fuzzy-neural networks were deployed for stock price index forecasts [11]. Mean square error, mean relative error, mean square percentage error and root mean square error were used to evaluate and compare the outcomes. The results revealed that hybridized model of fuzzy-neural and genetic algorithm offered more accurate predictions and enhanced approximations of stock price index when put side-by-side with neural network model [11].

A genetic fuzzy system (GFS)-based expert system for stock price forecasts was introduced by [13]. The rule base extraction and database turning of the GFS model were used to predict stock price daily. The GA automation of GFS model offered veritable information patterns through descriptive rule inference. The model was implemented for IBM stock price (having input as open and close, and next day close price as output), which provided the best outcomes when compared to ANN, ARIMA and HMM using MAPE. The GFS had a smaller number of input variables through step-by-step regression technique. But, the training rule base for fuzzy expert system was constructed with the Pittsburgh technique.

Historical stock prices were considered helpful in analysing the stock prices movements in future [16, 17]. One model for stock price prediction is a two-layered reasoning [17]. The model employs domain knowledge realized from technical analysis carried out at the first layer of reasoning to monitor a second layer of reasoning with machine learning. Again, the model is accompanied with a money management strategy, which preserves historical successes of forecasts carried out to support the decision on the quantity of capital to be invested on the basis of future outlooks.

Again, fuzzy logic and genetic algorithm were combined to control the process of procurement for a firm. Wiecek [18] considered the influence of exterior factors such as lead time and demand uncertainty, in developing time-variable membership function parameters in order to model outcome values of fuzzy sets. Genetic algorithm was introduced to optimize the fuzzy rule base composed of four criteria of a firm. The outcomes revealed that inventory status was influenced largely by high demand and lead time, which further impacted on the inventory control system.

3 Methodology

The initial rules-lists generated by human expert system for FLC on the basis of DangoteCem Company's opening price, highest price and closing price dataset collected whose pseudocode is presented in Table 1.

The human expert's rules-lists are used to derive the indices for genetic algorithm procedures are presented in Table 2.

From Table 2, the rules-lists combines the indices of the two inputs and one output for the nine rules derived from the human expert. The input index of 3, 2 and 1 represent high, medium and low degrees. The output index of 3, 2 and 1 represent rises, stable and falls degrees. These valued rules-lists variables are required for the subsequent optimization processes on the RB.

3.1 FLC Rules-Lists Derivation

In the GA operations, each fuzzy rule index is used to generate the corresponding chromosome's allele. The individual having the best fitness in preceding population (that is, the minimum value of fitness function) is used to construct fuzzy rules-lists as illustrated in step 1–6.

Step I. Encoding of chromosomes was achieved through a row-wise combination of each input fuzzy set which is an integer number depicting membership values (1–3) presented in Table 2. These codes are used to generate chromosomes presented in Table 3.

Table 1 Pseudocode of initial fuzzy rules-lists generation

INPUT: Opening price, Highest price **OUTPUT: Closing price**

1	**INITIALISATION: INPUT** membership functions: low, medium, high.
2	**INITIALISATION: OUTPUT** membership functions: low, medium, high.
3	**CREATE RULE LIST COUNT <= 9**
4	WHEN RULE LIST COUNT = 1
5	IF Opening Price {High} AND Highest Price {High} THEN Closing Price {Rises}
6	WHEN RULE LISTS COUNT = 2
7	IF Opening Price {High} AND Highest Price {Medium} THEN
8	Closing Price {Stable}
9	WHEN RULE LISTS COUNT = 3
10	IF Opening Price {High} AND Highest Price {Low} THEN
11	Closing Share Price {Rises}
12	WHEN RULE LISTS COUNT = 4
13	IF Opening Price {Medium} AND Highest Price {High} THEN
14	Closing Share Price {Rises}
15	WHEN RULE LISTS COUNT = 5
16	IF Opening Price {Medium} AND Highest Price {Medium} THEN
17	Closing Share Price {Stable}
18	WHEN RULE LISTS COUNT = 6
19	IF Opening Price {Medium} AND Highest Price {Low} THEN
20	Closing Share Price {Stable}
21	WHEN RULE LISTS COUNT = 7
22	IF Opening Price {Low} AND Highest Price {High} THEN
23	Closing Share Price {Rises}
24	WHEN RULE LISTS COUNT = 8
25	IF Opening Price {Low} AND Highest Price {Medium} THEN
26	Closing Share Price {Stable}
27	WHEN RULE LISTS COUNT = 9
28	IF Opening Price {Low} AND Highest Price {Low} THEN
29	Closing Share Price {Falls}
30	END IF
31	**TERMINATE** Rules generation operation
32	**RETURN** Rules base

Table 2 Initial fuzzy rules-lists

Rules-list	Input 1	Input 2	Output
1	3	3	3
2	3	2	2
3	3	1	3
4	2	3	3
5	2	2	2
6	2	1	2
7	1	3	3
8	1	2	2
9	1	1	1

Table 3 Initial fuzzy rules-lists

Rules-list	Input 1	Input 2	Output
1	3	3	3
2	3	2	2
3	3	1	3
4	2	3	3
5	2	2	2
6	2	1	2
7	1	3	3
8	1	2	2
9	1	1	1

In Table 3, the row-wise collection of the alleles encodes the chromosome [333231232221131211].

Step II. The initial population for the chromosomes ($N_{POP} = 30$) are constructed randomly for $N = 18$.

Step III. The mean square error (MSE) fitness function is chosen for genetic procedure is given by Eq. 7.

$$X_j = \frac{1}{n} \sum_{i=1}^{n} \left(X_i - \hat{X}_i \right)^2. \tag{7}$$

where,

X_j mean square error (MSE),
X_i actual value of ith training data of RB coded in jth chromosome (X_j),
$\hat{X}_i$ predicted value of ith training data of RB coded in jth chromosome (X_j),
X_j jth chromosome, and
n the training dataset quantity.

The operation of GA is similar to a typical search/minimization problem because it requires the best solution set at smallest errors.

Step IV. The new population of ($N_{POP}-1$) chromosomes is created by means of genetic operation using selected parameters of iterations, mutation rate and crossover.

Step V. The best rule set of present population with minimum MSE was used to build the FLC rules-lists. Again, the best fitness chromosome generated is used to build the next population ($N_{POP}-1$).

Step VI. The genetic operations are terminated whenever the population generation reached 30. Otherwise, go back to step III. The enhanced rules-lists are realized from these genetic procedures.

4 Results Presentation

The genetic procedure involves selection, mutation and crossover using the start chromosome string [3332312322211131211] to generate a new chromosome string [231313311133233122] at minimum fitness function of 0.00041142 and shortest time of 14.14773s. The new rules-lists generated for the FLC after the genetic procedure is given as follows:

R1.   (Opening_Price==Medium)   and   (Highest_Price==High)   => (Closing_Price=Rise)

R2.   (Opening_Price==Low)   and   (Highest_Price==High)   => (Closing_Price=Stable)

R3.   (Opening_Price==Low)   and   (Highest_Price==High)   => (Closing_Price=Rise)

R4.   (Opening_Price==High)   and   (Highest_Price==Low)   => (Closing_Price=Rise)

R5. (Opening_Price==Low) and (Highest_Price==Low) => (Closing_Price=Stable)

R6.   (Opening_Price==High)   and   (Highest_Price==High)   => (Closing_Price=Stable)

R7.   (Opening_Price==Medium)   and   (Highest_Price==High)   => (Closing_Price=Rise)

R8.   (Opening_Price==High)   and   (Highest_Price==Low)   => (Closing_Price=Stable)

R9. (Opening_Price==Medium) and (Highest_Price==Medium) => (Closing_Price = Fall)

In computational logic system, all unique sets of inputs (or antecedents) are expected to produce unique outputs (or consequents). For that reason, rules (R2, R3, R4 and R8) are rejected due to:

1. Duplication of antecedents,
2. The matching consequents for these antecedents are inconsistent,
3. The inability of the rules to contribute validly to decision-making.

Upon the removal of redundancy in the FLC rules-list, resultant FLC is enhanced with five rules-lists using genetic algorithm procedure as follows:

R1.   (Opening_Price==Medium)   &   (Highest_Price==High)   => (Closing_Price=Rise)

R5. (Opening_Price==Low) & (Highest_Price==Low) => (Closing_Price=Stable)

R6. (Opening_Price==High) & (Highest_Price==High) => (Closing_Price=Stable)

R7.   (Opening_Price==Medium)   &   (Highest_Price==High)   => (Closing_Price=Rise)

R9.   (Opening_Price==Medium)   &   (Highest_Price==Medium)   => (Closing_Price=Fall)

Table 4 Evaluation of FLC against enhanced FLC RB rules-lists

Model	MSE (%)	MAPE (%)
FLC	75.102	67.576
Enhanced FLC	24.898	32.424

The FLC is evaluated for the rules-lists offered by the GA procedure, and those of the enhanced rules-lists to forecast the closing price of Dangote Cement Company Shares as presented in Table 4.

In Table 4, the FLC built with enhanced rules-lists offered the best forecasts with minimal errors. There are large errors resulting from rules-lists generated by human expert system as used in FLC. The genetic procedure introduced into FLC increased the accuracy of predictions by 24.898 to 75.102% using the fitness function (MSE) and the MAPE. These results show that FLC rules-lists accuracy improved significantly when fewer number of rules make up the rules base. Again, it is found that only valid rules-lists are to be incorporated in the RBs in order to support appropriate decisions making process about future stock prices trends. This can be attained through effective redundancy removal processes such as genetic algorithm procedure proposed in this paper.

5 Conclusion

This paper found that the outcomes of enhanced FLC model improved significantly over FLC by 24.898 to 75.102% for MSE. In the case of MAPE calculated, enhanced FLC outperformed FLC model by 23.424 to 67.576%. The main reason is removal of redundancy in the rules-lists offered by the human expert system rules-lists. Again, human experts' system rules-lists require further enhancement processes in order to support and elicit accurate and appropriate decisions standpoints about stock prices movements. Also, the enhanced FLC rule base offered by GA procedure decreased rules-lists from 9 to 5, which minimized losses and time consumptions. In future works, there is a need to deploy other optimization procedures to eliminate redundancy in rules-lists of FLC RBs in order to support improved decision-making processes in fields such as medicine, manufacturing and transportation.

Acknowledgements We would like to acknowledge the sponsorship and support provided by Covenant University through the Centre for Research, Innovation and Discovery (CUCRID).

References

1. Sun, C.: Stock market returns predictability: does volatility matter? Unpublished Ph.D. Thesis, University of Nottingham, Nottingham, United Kingdom, pp. 1–180 (2008)
2. Feng, H.M., Chou, H.C.: Evolutionary fuzzy stock prediction system design and its application to the Taiwan stock index. Int. J. Innov. Comput. Inf. Control. **8**(9), 6173–6190 (2012)
3. Karlsson, M., Orselius, H.: Economic and business cycle indicator: accuracy, reliability and consistency of swedish Indicators. Unpublished Masters' Thesis, Department of Business Administration, Jonkoping University, Sweden, pp. 1–64 (2014)
4. Taylor, S.J.: Modeling Financial Time Series. No. 2nd, New York: World Scientific Publishing (2008)
5. Jabbari, E., Falhi, Z.: Prediction of stock returns using financial ratios based on historical cost, compared with adjusted prices (accounting for inflation) with neural network approach. Indian J. Fundam. Appl. Life Sci. **4**(4), 2231–6345 (2014)
6. Naik, R.L., Manjula, B., Ramesh, D., Murthy, B.S., Sarma, S.S.V.N.: Prediction of stock market index using neural networks: an empirical study of BSE. Eur. J. Bus. Manag. **4**(12), 60–71 (2012)
7. MATLab: Fuzzy Logic Toolbox; User's Guide. R2013a: The MathWorks, Inc. (2013)
8. Kayacan, E.: Interval type-2 fuzzy logic systems: theory and design. Unpublished Ph.D. Thesis, Department of Electrical and Electronic Engineering, Bogazici University, Istanbul, Turkey, pp. 1–149 (2011)
9. Chen, C., Li, M., Sui, J., Wei, K., Pei, Q.: A genetic algorithm-optimized fuzzy logic controller to avoid rear-end collisions. J. Adv Transp. vol. 50, no. 8, pp. 1735-1753 (2016)
10. Tahmasebi, P., Hezarkhani, A.: A hybrid neural networks-fuzzy logic-genetic algorithm for grade estimation. Elsevier J. Comput. Geosci. **42**, 18–27 (2012)
11. Delnavaz, B.: Forecasting of the stock price index by using fuzzy-neural network and genetic algorithms. J. Appl. Sci. Agric. **9**(9), 109–117 (2014)
12. Yefimochkin, O.: Fundamental: using macroeconomic indicators and genetic algorithms in stock market forecasting. Unpublished Master's Thesis, Department of Computer Engineering, The Technical University of Lisbon, Portugal, pp. 1–120 (2011)
13. Hadavandi, E., Shavandi, H., Ghanbari, A.A.: Genetic fuzzy expert system for stock price forecasting. In: Proceedings of 7th IEEE International Conference on Fuzzy Systems and Knowledge Discovery, Yantai, China, pp. 41–44 (2010)
14. Acheme, D.J., Vincent, O.R., Folorunso, O., Olusola, O.I.: A predictive stock market technical analysis using fuzzy logic. Comput. Inf. Sci. **7**(3), 1–17 (2014)
15. Akinwale, A.T., Arogundade, O.T., Adekoya, A.F.: Translated Nigeria stock market prices using artificial neural network for effective prediction. J. Theor. Appl. Inf. Technol. **1**, 36–43 (2009)
16. Floreano, D., Mattiussi, C.: Bio-inspired artificial intelligence. The MIT Press, Cambridge, Massachusetts (2008)
17. Larsen, J.I.: Predicting stock prices using technical analysis and machine learning. Unpublished M.Sc. Thesis, Department of Computer and Information Science, Norwegian University of Science and Technology, Trondheim, Norway, pp. 1–82 (2010)
18. Wiecek, P.: Intelligent approach to inventory control in logistics under uncertainty conditions. In: Elsevier 12th Conference on Transport Engineering, vol. 18, pp. 164–171 (2016)
19. Alhassan, J., Misra, S.: Using a weightless neural network to forecast stock prices: a case study of Nigerian stock exchange. Sci. Res. Essay **6**(14), 2934–2940 (2011)
20. Alhassan, J.K., Misra, S., Ogwueleka, F., Inyiama, H.C.: Forecasting Nigeria foreign exchange using artificial neural network. J. Sci. Technol. Math. Educ. **9**(1), 47–56 (2012)

A Novel Approach for Multi-pitch Detection with Gender Recognition

Rahul Kumar Singh, Pardeep Singh, Navdeep Kumar and Rajeev Tiwari

Abstract An audio signal can be estimated for its frequency ad saliencies of pitches sounds using multiple pitch estimation for a short-time frame. The focus of this paper is to solve the difficulty for predicting the fundamental frequencies of a signal comprising various harmonically related sinusoidal signals. The framework proposed in this paper can be used to detect single as well as multi-F0 along with gender identification. The performance of the approach is evaluated on 135 mixture audio signals which compromise of human speech of different genders. The successful evaluation was performed for a number of human in the complex mixture speech signal and their respective genders. It is based on the iterative approach in which frame-by-frame analysis is carried out for multi-F0 detection. The algorithm effectively provides optimal results in classifying gender identification and person calculation.

Keywords Multiple pitch estimation · Gender identification · Frequency domain · Fast Fourier transform

1 Introduction

Pitch is a perceptive characteristic of a sound. Pitch has a very important acoustical feature in speech analysis and plays a vital role in many applications

R. K. Singh (✉) · P. Singh · R. Tiwari
School of Computer Science, University of Petroleum and Energy Studies, Dehradun 248007, India
e-mail: rk.singh@ddn.upes.ac.in

P. Singh
e-mail: pardeep.singh@ddn.upes.ac.in

R. Tiwari
e-mail: rajeev.tiwari@ddn.upes.ac.in

N. Kumar
Becton Dickinson India Pvt. Ltd., 160101 Chandigarh, India
e-mail: navdeep.cb@gmail.com

© Springer Nature Singapore Pte Ltd. 2020
P. K. Singh et al. (eds.), *Proceedings of First International Conference on Computing, Communications, and Cyber-Security (IC4S 2019)*, Lecture Notes in Networks and Systems 121, https://doi.org/10.1007/978-981-15-3369-3_50

like speech recognition, prosody analysis, speaker identification and computational auditory scene analysis (CASA). Pitch perception is a very complex process, and determination of pitch from a single source is easy as compared to multi-sources or of polyphonic signals. The difficulty of calculating the essential frequency or pitch of periodic waveforms arises in several forms of application. The same has received prominent attention of researchers in the recent era. For instance, numerous audio and speech problems especially depend on the primary forming of an estimate on the pitch or pitches including problems. Estimating the fundamental frequency in an audio signal plays a vital role in gender identification. There are many effective and accurate proposed algorithms on single-source pitch determination and detection. But the multi-pitch real-life scenario occurs regularly than single-pitch case, and often also in speech processing. It is tough to correctly estimate the various pitches of a mixed signal. Multi-pitch assessment has potential application in speech sources separation, speech recognition and speech enhancement. The analysis of harmonic sound mixtures such as multi-pitches is considered as a basic problem in processing audio signal. The academicians are working on retrieving music information from music transcripts, source separation [1], melody extraction [2], etc. The same can be used in speech processing as it is useful in multi-talker speech recognition [3], prosody analysis [4] and solving the cocktail party problem [5]. Each approach either considers the music or speech but does not cover both in the same. It is observed that dealing with multiple frequencies in mixed periodic signals is way very tough than single fundamental frequency estimation. The research on estimating multiple pitches in a sound started with separating co-channel speech signals [6]. The approach provided a motivation to identify the multi-pitches in polyphonic music, which leads to the investigation in the area of multiple fundamental frequency estimations, which required achieving the automatic transcription of duets [7]. The research than focused towards more complex compositions in music for the identification of fundamental frequencies of various instrumental voices was taken into consideration [8].

Various researchers focused on iterative approaches to estimate the fundamental frequency of predominant periodic signal for identifying multiple pitches. The identified frequency is then removed in order to estimate the next fundamental frequency. This process is iterated for all harmonic sources in the signal to identify all the frequency in a signal. The ambiguity and uncertainty in a harmonic model lead to octave errors. This points towards estimating the fundamental frequencies in a harmonic signal as a noticeable task for processing audio signal. The problem of intrinsic ambiguity can be resolved by taking into consideration the smoothness of a spectral envelope [9].

The problem of uncertainty in a harmonic model diversifies when multiple pitches are taken into consideration. The spectral overlap of various frequencies can be resolved by following two below mentioned approaches:

i.   Iterative approaches: In this approach, we recursively calculate the dominant pitch and eliminate its harmonics from the mixture;
ii. Joint approaches: In this approach, we concurrently calculate all the fundamental frequencies by improving a joint criterion.

The iterative approaches are fast in estimating the multiple frequencies but multiple errors occur at each iteration as the result of spectral overlap between partials belonging to different sources. On the other hand, the method discussed in second category do not lead to errors while estimating the frequencies but demands more computations as they estimate large dimensions which nearly equals o the number of harmonic components in an audio signal.

The collection of sinusoids, each containing amplitude, phase and frequency (also referred as harmonics), results in periodic signals, and frequency of these sinusoids is an integer multiple of a fundamental frequency. These signals also contain some observation noises which corrupt the signal. Estimating fundamental frequencies from these corrupted signals is known as pitch estimation. Concept of multi-pitch estimation came into picture when the signals contain many such corrupted periodic signals. So the area of interest moves towards estimating multiple pitches of periodic signals in noise or from corrupted signals. The problem for identifying sound mixtures is resolved by proposing an approach multi-pitch estimation (MPE). In this approach, the relationship between pitches is considered at each and every frame level. Compared with existing methods, the proposed approach has the following advantages:

(a) Unsupervised: No training is required for the source models to identify the multiple pitches.
(b) General: This approach considers both music and speech for identification of multiple pitches.

The motivation of the paper is to acquire original audio mixture and provide an output of estimated pitches for each time frame. It also provides the detail of a number of sources and their gender in audio signal.

The rest of the article is planned as follows: in Sect. 2, related work has been covered. Section 3 describes problem formulation and then describing the algorithm to solve the problem is covered in Sect. 4. Section 5 shows the experimental work done on speech and its results. Section 6 summarizes the main conclusions and future scope.

2 Related Work

The aim of the recent work is forming the pitch estimation using second-order statistics [10, 11]. For multi-F0 source signals, comprising numerous harmonically related signals, these methods calculate each of the present pitch signals separately forming different forms of iterative estimation schemes, typically demanding a priori information of both the number of sources and the model order of each of the sources. This research effort examines a new system for calculating the fundamental frequencies of a signal with multiple pitches, without conveying any prior information of either the number of sources present or their number of harmonics.

Tolonen et al. [12] presented the model for computing multiple pitches from speech signal by diving it into two channels namely low channel and high channel. This is followed by autocorrelation method. This is used for real-time scenario. In 2001, Klapuri [13] proposed a new model after implementation of three models for multi-pitch detection in speech signals. This work contains spectral smoothness evaluation. The speech corpus being used varies from one to six speeches in an audio wave. The error rate reduces as we reduce the number of sounds in one audio wave.

Wu et al. [14] used HMM for detection of pitch tracks being framed. The robust algorithm is proposed for speech recognition algorithm for noisy speech. Klapuri [9] in his next paper performed another experiment in which he calculates harmonicity and spectral smoothness. This is repetitive process in which one sound is being detected first. This sound is then removed from a complex mixture of sounds in signal. Now, the residual signal is undergone through the same process.

Abeysekera et al. [15] proposed a new technique for the same using bispectrum that means two-dimensional frequency log. This undergoes detection of one signal and removing it from the testing signal. Same process is being performed for residual signal. Removal of two-dimensional signals is comparatively easy and convenient. Christensen et al. [16] introduced some solutions for separation of speech signals from music.

Zhang [17] introduced an algorithm based on weighted summary correlogram. Accuracy measures are observed better. Klapuri [18] performed better accuracy with fundamental frequency detection and removal of sound one by one iteratively. Badeau et al. [19] proposed new expectation maximization algorithm for the same. It also separates the overlapped harmonic spectra obtained from speech signals.

Vincent et al. [20] proposed models for time-varying amplitude speech signals. They used their basic model as non-negative matrix factorization. Harmonicity and spectral smoothness constraints are taken into account for music signals and estimation of pitches in those signals. The proposed adaptive spectral decomposition model reveals that the basis spectra estimated by this model are having a finite structure. In addition, the spectral envelope is adapted to the observed data; this model is implemented on piano and woodwind data. The experimental results explain that the constrained NMF procedure is ample better than the unconstrained NMF or NMF under harmonicity constraint alone. This model sets the benchmark for multiple pitch estimation through NMF algorithms. However, Benetos et al. [21] also worked on time-varying multi-pitch detection in speech signals. They used HMM for speech corpus containing MIDI database. Koretz et al. [22] used maximum a posteriori probability algorithm for multiple pitch detection. Adalbjornsson et al. [23] worked on block sparsity. They introduced an alternative algorithm for the detection of multiple pitches in speech signal. The work is going on a similar concept and new techniques have been introduced. This is further carried on to better accuracy and size of data set.

Wohlmayr et al. [24] proposed a methodology to estimate the stream pitches of simultaneous talkers by using factorial hidden Markov approach. Initial training of system is done by using the isolated recordings mixtures of talkers. This is the basic limitation of the systems where initial training on sources is not feasible.

Recently an unsupervised approach for estimating the stream pitches is proposed by Hu and Wang [25]. The approach focuses on separating the signals of two simultaneous talkers. The approach is tested only for speech and its applicability for other audio signals such as music is not tested. In 2011, Christensen et al. [26] The task of high-resolution pitch estimation is motivation of research of this paper. Already proposed systems such as the classical comb filtering, maximum likelihood approaches and others based on optimal filtering are extended for an unknown number of harmonics. This is also known as model order which is based on the posteriori principle. So, this proposed method of estimating orders and the fundamental frequency is applicable to those situations also where there is no prior knowledge of the model order. Also, a computationally efficient order-recursive execution that is ample quicker than a direct execution has been proposed.

Zhang et al. [27] extend the multi-pitch estimation to a level of multi-channel. A new estimator is proposed which deals with estimating fundamental frequency as well as DOA of multiple sources. Subspace analysis along with time/space model is used in this estimator. This estimator deals with real signals having simulated anechoic array recording. This estimator shows better performance even under adverse conditions.

Giacobello et al. [28] overview the various linear predictors for speech investigation and coding. In this paper, sparsity is introduced into the linear prediction environment. These sparse linear predictors perform extra effective decoupling between the pitch harmonics and the spectral envelope. This leads to the uncorrupted predictors of the pitch excitation. Properties such as shift invariance and pitch invariance are also offered by these predictors. A more synergistic new approach is proposed in this paper to encode a speech segment along with a compact representation. This approach also reduces the size and cost of computations required.

Nilsen et al. [29] taken both the real- and complex-valued periodic signals along with additive noise as input in order to derive the probability model. In this paper, the prior knowledge is curved into an observation model and prior distributions which are further turned into g-prior which is the more convenient prior. This is done using approximation on signal-to-noise ratio (SNR) and the number of observations. In this paper, the posterior distribution is also derived from the fundamental frequency. Comparison between various approximations is also done in this paper. Result of this comparison concluded that the BIC approximation is worse than the other approximations. In 2013, Nilsen et al. [30] estimated both the DOA and the pitch of harmonic source using ULA. This technique reveals that if harmonic structure is taken into account, the estimation done for DOA is more accurate. Also, the use of multiple sensors increases the accuracy of pitch estimation. Two estimators named as NLS and aNLS are proposed in this paper. This method gives better results as compared to other methods in terms of mean squared error. Performance is also enhanced through this method, and this method is also applicable on real-life signals.

Duan et al. [31] proposed constrained-clustering approach in this paper. This approach works on harmonic sound sources ad perform the streaming of multiple pitches. Estimation of pitches is done in time frames using multi-pitch estimation (MPE) algorithm. No pre-training is required for systems following this approach, and this approach is applicable to both music and speech signals. A new cepstrum named as uniform discrete cepstrum (UDC) is proposed, which presents the timbre of sound sources. Table 1 describes the techniques, approaches and outcomes in the early research.

Table 1 Summary of research work

Name of author	Algorithms used	Input	Output
Walmsley et al. [32]	Polyphonic pitch tracking using joint Bayesian estimation	Musical signal	Parameters based on frequency variation
Klapuri [13]	Spectral smoothness principle	Polyphonic, multi-instrumental music and mixtures of simultaneous speakers	Segregate harmonic signals
Klapuri [9]	Based on harmonicity and spectral smoothness	Acoustic input signal sampled at 44.1 kHz rate and quantized to 16-bit precision.	Clean signal and noisy signal with different SNRs
Abeysekera [15]	Using frequency-lag domain and bispectrum	Multi-pitch input spectrum	Autocorrelation channels of 32–120 filter bank outputs.
Zhang et al. [17]	Based on weighted summary correlogram	Multi-pitch input signal	Estimation of closest pitch frequency
Bedeau et al. [19]	Expectation maximization algorithm	Musical chord	Estimate successive single-pitch and spectral envelope.
Vincent et al. [20]	Adaptive harmonic spectral decomposition	Multi-pitched musical audio, piano recordings, narrow band spectra	Estimates the model parameters and performance of pitch
Benelos et al. [21]	Using harmonic envelope estimation	Musical signal and recordings with sampling rate 8 kHz	Estimation of log-frequency spectral envelope.
Koretz et al. [22]	Based on maximum a posteriori probability	Monophonic and polyphonic signals, music and synthetic signals	Determine different harmonic sources present
Huang et al. [33]	Based on multi-length windows harmonic model	Harmonics of two concurrent speech signals	Estimation of prominent pitch and autocorrelation
Adalbjornsson [23]	Using block sparsity	Multiple harmonically related sinusoidal signals	Estimate pitch frequencies of two different signals

3 Problem Statement

Algorithms proposed so far in pitch analysis mainly works on the concept of detecting pitches from a music corpus or mixture signals containing various pitches of different instruments or humans. So far there is no such algorithm which itself identifies the number of person from the mixture signals. We consider the problem in our research work and now we proposed an algorithm which identifies a number of pitches from the mixture signal. Along with this, our system also performs the task of gender identification of those pitches.

There are previously many algorithms proposed in the field of multi-pitch detection. All the algorithms determine multi-F0 but most of them work on music notes or music transcription. We propose a recursive algorithm which is used for estimating fundamental frequencies and number of harmonics of a multi-pitch complex signal based on the harmonic summation method. Determining the number of harmonics is also done using the same algorithm and it continues to extract sources until the pitch detection criteria fails.

Range of pitch filter considered in our algorithm is 100–400 Hz.

4 Proposed Approach

An iterative approach for frame-by-frame analysis has been considered for multi-F0 detection.

Firstly, the person detection is explained for two speakers at dissimilar signal-to-signal ratio (SSR) stages, and then explanation is extended for multiple persons in speech signal. The excellence of the parted output signals is also considered by the proposed method. After finding the single-talk regions by using double-talk detector, the SCSS work is only to isolate the mixture segments.

Diversified signal with N samples $y \in R^N$ is taken. These are collected of up to J speaker signals as $y = \sum_{j=1}^{J} s(\varphi_j) + e$. The matrix transpose is represented by T, $j \in [I, J]$ represents the number of signals in the mixed signal. The jth signal is represented with $s(\varphi_j) \in R^N$. This jth signal is characterized by parameter vector φ_j and $e \in R^N$ and depicts the noise signal. All these are incorporated in the model.

Sinusoidal modelling is used in this proposed approach for modelling the jth speaker signal in the mixture as a parametric feature vector φ_j. This is composed of sinusoidal parameters such as phase vectors, amplitude and frequency. This system uses $K = 3$ candidate models each represented by M_k, for explaining the mixed signal, y. M_0, M_1 and M_2 are used to designate noise-only, single-talk and double-talk, respectively.

Parameter vector $\varnothing_k$ with L_k sinusoids is used to describe these models. So, basically, the suggested method explains the succeeding difficulty: specified the mixed signal, hand-picked the model that is the supreme likely. Three models considered for y, which are [34]:

Case 1 M_0: $y = e$,
Case 2 M_1: $y = s(\psi_j) + e$ for j (belongs to) $\in [1, 2]$,
Case 3 M_2: $y = s(\psi_1) + s(\psi_1) + e$,

Case 3 equation $s(\psi_1) + (\psi_2)$ shows estimation for the mixed signal, Case 2 $s(\psi_j)$ with $j \in [1, 2]$ Represents the jth signal modelled. ψ_j depicts the parameter set.

Now, the posterior probabilities of M_k with $k \in Z_k = \{0, 1, 2\}$ is evaluated by system. The system calculation of the utmost possible hypothesis is denoted by M_k and is achieved as:

$$\hat{M}_k = \arg \max_{M_k : k \in Z_K} \left\{ \int_{\theta_k}^{k} p(y|\theta_k, M_k) p(\theta_k|M_k) d\theta_k \dots \right. \tag{1}$$

Equation (1) represents a convoluted nonlinear maximization problem which occurs for the used models. But in this proposed system, the different criterion is used in its place of numerical integration for the estimation of marginal density in (1).

Asymptotic criterion is defined as:

$$\hat{M}_k = \arg \max_{M_k : k \in Z_K} \{-\ln p(y|\hat{\theta}k, M_k) + pc\} \tag{2}$$

In Eq. (2), pc is model-dependent penalty of the criterion, $\hat{\theta}_k$ is an estimate of θ_k for the kth model M_k and $-\ln p(y | \theta_k, M_k)$ Log-likelihood term gained from an approximation of (1). Multiple hypothesis method when there are more than two persons speaking at one time.

Now, the target is to determine $-\ln p(y | \theta_k, M_k)$ for each of the three primary candidate models M_k with $k \in Z_k = \{0, 1, 2\}$. Again sinusoidal modelling is used for modelling the speaker signals in the mixture. Let $s_i(\psi_j)$ be the jth speaker signal with $j \in [1, 2]$ for the ith frequency band which is modelled by the parametric vector ψ_j. It is assumed in this system that the signal modelling error, e, has a Gaussian distribution.

Another assumption made by this system is that the modelling error sub-band signal e_i is white in each ith frequency band. Due to sub-band decomposition and the independence assumption for all frequency bands, it is also assumed that e_i is independent from one band to another.

Based on this assumption, it is clear that the likelihood function for all bands for each class M_k is given by

$$P(e|\sigma^2) = \prod_{i=1}^{Q} p(e_i|\sigma_i^2) = \frac{1}{(2\pi)^{\frac{N}{2}} \prod_{i=1}^{Q} \sigma_i} \exp\left(-\frac{1}{2} \sum_{i=1}^{Q} \frac{e_i^T e_i}{\sigma_i^2}\right) \tag{3}$$

In Eq. (3), a total number of frequency bands represented by Q, variance due to the modelling error signal in ith band represented by σ_i and band is represented by e_i.

For single speaker class, $M1$, the modelling error at the ith frequency band is given by $\widehat{e_i} = y_i - s_i(\hat{\psi}_j)$, and for the mixed class, $M2$, the estimated error is defined as $\widehat{e_i} = y_i - s_i(\hat{\psi}_1) - s_i(\hat{\psi}_2)$. The noise appraised for the ith frequency band as a coloured noise is not fitted by $M2$. The condition for sinusoids composed of unknown amplitudes and frequencies reduces to

$$\hat{M}_k = \arg \max_{M_k : k \in Z_K} \left\{ \frac{N}{2} \sum_{i=1}^{Q} \ln \sigma_i^2 + \frac{5L_k}{2} \ln N \right\} \tag{4}$$

where
$\sigma_i^2 = \frac{1}{N} \widehat{e_i^T} \, \hat{e}_i$ Shows the estimated variance
ith represents the frequency band and depicts the number of sinusoids.

In the mixture class $M2$, A mixture estimate is required to replace $s(\hat{\psi}_1) + s(\hat{\psi}_1))$, so that the best pair of $\{\hat{\psi}_1$ and $\hat{\psi}_2\}$ can be extracted from the speaker models of the primary speakers. The minimum mean square error (MMSE) calculator is used for the mixture magnitude spectrum for finding out the joint best states in the speaker models. When we combined these models, it describes the magnitude spectrum for the experiential mixture, y. The noise model, M0, is also included in this as one of the examined models by setting $y = \hat{e}$ and locating the number of sinusoids equal to zero ($L_k = 0$).

The estimated noise variance is given by

$$\sigma_i^2 = \frac{1}{N} y_i^T y_i.$$

Finally, using the estimated value for σ_i depending on each possible class of M_k with $k \in Z_k = \{0, 1, 2\}$, the best model, as a result, is the one which yields high log-likelihood and low model order, which is achieved in (4).

4.1 Harmonic Model

The voiced speech signal is defined through various models; one of them is harmonic model. According to this model, the sum of sinusoids with time-varying amplitudes, frequencies and phase is equal to the voices speech signal [35].

This can be explained as follows:

$$x(m) = \sum_{r=1}^{R} A_r \cos(2\pi F_r m + \theta_{0r})$$

where

R harmonic order

A_R instantaneous amplitude

F_r frequency of the rth harmonic component, approximates to $r F_0$

θ_{0r} initial phase

F_0 fundamental frequency, also called pitch.

4.2 Harmonic Structure

Initially, the acoustic input signal is taken into account for calculating fast Fourier transform. This transformation is used to convert time/space domain signal to complex frequency domain signal as output. Now, this complex signal is divided into window frames of length 25 ms. In our system, sampling is done at 8000 kHz and quantized to 32-bit float precision. Width of each window function is considered to be 16,384 samples.

4.3 Harmonic Structure Subtraction

Suggested method is applied on each and every frame in order to detect the dominant pitch in mixture signals. Amplitude of harmonic partials of the detected pitch is estimated. After this, detected pitch is removed from the frequency domain of selected frame and the algorithm is iteratively applied on the residual signals.

4.4 When to Stop the Algorithm

Iterative execution of harmonic subtraction step will lead to a stage where the pitch in residual signals is nearly same. So, no peak can be detected with the algorithm. At that time, algorithm can stop working on that frame.

5 Experimental Set-up and Results

Figure 1 represents the flow of the proposed approach. Performance of proposed method is evaluated on attested set of 135 mixed sounds. These 135 sounds consist of mixture of both male and female voice speech. Sampling of these speech signals is done at 8000 kHz originally. The FFT point is set to 16,384 in order to accomplish a high frequency resolution. Frame length is set to 25 ms.

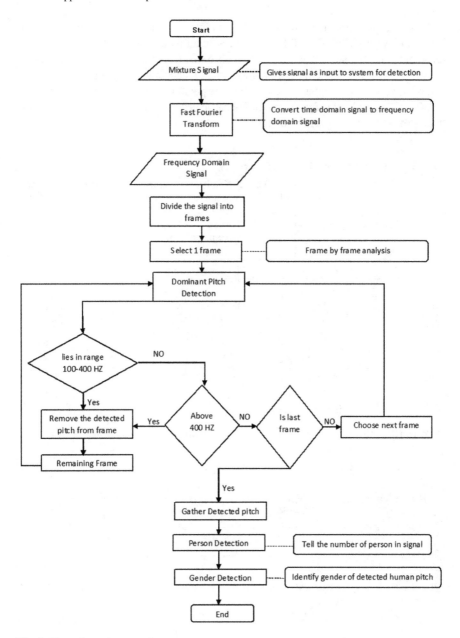

Fig. 1 Flow chart of proposed system

Table 2 represents the performance evaluation of our test set and Fig. 2. presents the accuracy graph based on Table 2. Groups contain multiple genders where male is represented by M and female is by F. It is observed from the results that as a number of persons increase in the mixture audio signal, its accuracy is going down for both person as well as for gender detection. For one person or two persons, i.e. Group G0, G1, G2, G3 and G4, it shows more than 90% accuracy in person detection and more than 85% accuracy in gender detection.

Table 2 Accuracy rates for mixture audio signals

Group	Number of humans in mixed audio signal	Gender	Number of test cases	Numbers of times right person predicted	Numbers of times right Gender predicted	Accuracy rate for person detection	Accuracy rate for gender detection
G0	1	M	15	14	15	93.33	100
G1	1	F	15	15	15	100	100
G2	2	M, M	15	14	13	93.33	86.66
G3	2	F, M	15	15	14	100	93.33
G4	2	F, F	15	14	13	93.33	86.66
G5	3	M, M, M	15	12	11	80	73.33
G6	3	M, M, F	15	13	12	86.66	80
G7	3	M, F, F	15	13	13	86.66	86.66
G8	3	F, F, F	15	13	11	86.66	73.33

M male; *F* female

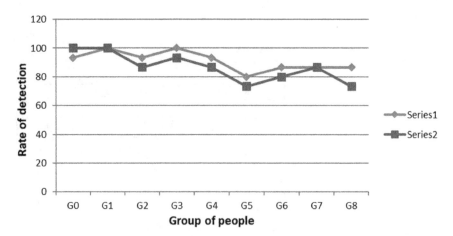

Fig. 2 Accuracy graph. Series1 = accuracy rate for person detection and Series2 = accuracy rate for gender detection

Gender detection accuracy for a single person is near about 100%. When three persons are involved, i.e. group G5, G6, G7 and G8, accuracy level drops down to 80% for person detection and 73% to gender detection. It is also seen that it shows good results when mixture is composed of different genders. Accuracy rate for similar gender is lower than that of different gender.

6 Conclusion and Future Scope

The experimental results demonstrate a novel technique to predict many F0 in mixed signals and gender identification (male and female) of those F0s. The main scope of our idea or method is to determine the predominant pitch after that subtract it from specified signal to acquire the residual signal for subsequent iteration. Our method results consider the number of peoples in the signal along with their femininity like male and female. The experiments result of our method performs fine for human-voiced signal and intelligent to recognize their femininity with a good accuracy rate. It shows higher accuracy for different gender mixture signal than similar gender mixture signal. However, it only works for three persons' mixture signal. For greater number, it is accuracy level drops down significantly. In future, we will try to improve our person detection system so that it will be able to detect more than three persons and try to improve our accuracy level for more than three persons.

References

1. Duan, Z., Zhang, Y., Zhang, C., Shi, Z.: Unsupervised single-channel music source separation by average harmonic structure modeling. IEEE Trans. Audio Speech Lang. Process. **16**(4), 766–778 (2008)
2. Han, J. Chen, C.-W.: Improving melody extraction using probabilistic latent component analysis. In: Proceedings of IEEE International Conference on Acoustics, Speech and Signal Processing (ICASSP), pp. 33–36 (2011)
3. Cooke, M., Hershey, J.R., Rennie, S.: Monaural speech separation and recognition challenge. Comput. Speech Lang. **24**, 1–15 (2010)
4. Jiang, D.-N., Zhang, W., Shen, L.-Q., Cai, L.-H.: Prosody analysis and modeling for emotional speech synthesis. In: Proceedings of IEEE International Conference on Acoustics, Speech, and Signal Processing (ICASSP), pp. 281–284 (2005)
5. Cherry, E.C.: Some experiments on the recognition of speech, with one and two ears. J. Acoust. Soc. Amer. **25**, 975–979 (1953)
6. Shields, J.V.C.: Separation of added speech signals by digital comb filtering. Dissertation, Ph.D. thesis Massachusetts Institute of Technology, Cambridge (1970)
7. Moorer, J.A.: On the transcription of musical sound by computer. Comput. Music J. 32–38 (1977)
8. Chafe, C., Jaffe, D.: Source separation and note identification in polyphonic music. In: Proceedings of IEEE International Conference on Acoustics, Speech, and Signal Processing, vol. 11, pp. 1289–1292 (1986)

9. Klapuri, A.P.: Multiple fundamental frequency estimation based on harmonicity and spectral smoothness. IEEE Trans. Speech Audio Process. 11(3), (2003) 119, 214

10. Christensen, M.G., Stoica, P., Jakobsson, A., Jensen, S.H.: Multi-pitch estimation. Sig. Process. 88(4), 972–983 (2008)

11. Zhou, Z., So, H.C., Chan, F.K.W.: Optimally weighted music algorithm for frequency estimation of real harmonic sinusoids. In IEEE International Conference on Acoustics, Speech and Signal Processing, Kyoto, Japan, 25–30 Mar 2012

12. Karjalainen, M., Tolonen, T.: A computationally efficient multipitch analysis model. IEEE Trans. Speech Audio Process. 8(6), (20000

13. Klapuri, A.P.: Multipitch extimation and sound separation by the spectral smoothness principle. IEEE (2001)

14. Wu, M., Wang, D., Brown, G.J.: A multipitch tracking algorithm for noisy speech. IEEE Trans. Speech Audio Process. 11(3), (2003)

15. Abeysekera, S.S.: Multiple pitch estimation of poly-phonic audio signals in a frequency-lag domain using the bispectrum. IEEE (2004)

16. Christensen, M.G., Stoica, P., Jakobsson, A., Jensen, S.H.: The multi pitch estimation problem: some new solution. IEEE (2007)

17. Zhang, X., Liu, W., Li, P., Xu, B.: Multipitch detection based on weighted summary correlogram. National Laboratory of Pattern Recognition, Beijing

18. Klapuri, A.: Multipitch analysis of polyphonic music and speech signals using an auditory model. IEEE Trans. Speech Audio Process. 16(2), (2008)

19. Badeau, R., Emiya, V., David, B.: Expectation maximization algorithm for multi pitch estimation and separation of overlapping harmonic spectra. IEEE (2009)

20. Vincent, E., Bertin, N. Badeau, R.: Adaptive harmonic spectral decomposition for multiple pitch estimation. IEEE Trans. Speech Audio Process. 18(3), (2010)

21. Benetos, E., Dixon, S.: Joint multi-pitch detection using harmonic envelope estimation for polyphonic music transcription. IEEE J. Sel. Top. Signal Proces. 5(6), (2011)

22. Koretz, A., Tabrikian, J.: Maximum a posteriori probability multiple-pitch tracking using the harmonic model. IEEE Trans. Speech Audio Process. 19(7), (2011)

23. Adalbjornsson, S.I., Jakobsson, A., Christensen, M.G.: Estimating multiple pitches using block sparsity. IEEE (2013)

24. Wohlmayr, M., Stark, M., Pernkopf, F.: A probabilistic interaction model for multipitch tracking with factorial hidden markov models. IEEE Trans. Audio Speech Lang. Process. 19(4), 799–810 (2011)

25. Hu, K., Wang, D.: An unsupervised approach to cochannel speech separation. IEEE Trans. Audio Speech Lang. Process. 21(1), 122–131 (2013)

26. Christensen, M.G., Højvang, J.L., Jakobsson, A., Jensen, S.H.: Joint fundamental frequency and order estimation using optimal filtering. EURASIP J. Adv. Signal Process. (2011)

27. Zhang, J.X., Christensen, M.G., Jensen, S.H., Moonen, M.: Joint DOA and multi-pitch estimation based on subspace techniques. EURASIP J. Adv. Signal Process. (2012)

28. Giacobello, D., Christensen, M.G., Murthi, M.N., Jensen, S.H., Moonen, M.: Sparse linear prediction and its applications to speech processing. IEEE Trans. Audio Speech Lang. Process. 20(5), (2012)

29. Nielsen, J.K., Christensen, M.G., Jensen, S.H.: Default bayesian estimation of the fundamental frequency. IEEE Trans. Audio Speech Lang. Process. 21(3), (2013)

30. Jensen, J.R., Christensen, M.G., Jensen, S.H.: Nonlinear least squares methods for joint DOA and pitch estimation. IEEE Trans. Audio Speech Lang. Process. 21(5), (2013)

31. Duan, Z., Han, J., Pardo, B.: Multi-pitch streaming of harmonic sound mixtures. IEEE/ACM Trans. Audio Speech Lang. Process. 22(1), (2014)

32. Walmsley, P.J., Godsill, S.J., Rayner, P.J.W.: Polyphonic pitch tracking using joint Bayesian estimation of multiple frame parameters. In: Proceedings of 1999 IFIM Workshop on Applications & Signal Processing 10 Audio and Acoustics, 1999

33. Huang, Q., Wang, D.: Multi-pitch estimation for speech mixture based on multi-length windows harmonic model. In: Proceedings of IJCCSO, 2011

34. Vincent, E.: Musical source separation using time-frequency source priors. IEEE Trans. Audio Speech Lang. Process. **14**(1), 91–98 (2006)
35. Macon, M.W., Link, L.J.: A singing voice synthesis system based on sinusoidal modeling. In: Proceedings of ICASSP, vol. 1, pp. 435–438 (1997)

Malaria Parasite Classification Employing Chan–Vese Algorithm and SVM for Healthcare

Pragya, Pooja Khanna and Sachin Kumar

Abstract Malaria a contagious disease that spreads through Anopheles mosquito bite, it's deathly infection which if not treated in time can take lives and become an epidemic. According to WHO, it is already one of the major reasons of death in tropical and sub-tropical regions across the globe. Early detection and diagnosis of malaria can help to prevent lot of casualties; however, gold test till date is by microscopy. Different species of plasmodium parasite have evolved over generations, five main that can be listed are: *P. falciparum*, *P. vivax*, *P. ovale*, *P. malariae*, and *P. knowlesi*, these parasites affect human adversely. Deadliest among these is *P. falciparum* which contributes to majority of deaths and has also developed resistances to various drugs employed to prevent the infection. Accurate evaluation of parasite density and, hence, identification of infection till date are done by an expert; which is an exhaustive process and lack accuracy. Paper presents an automated algorithm based on SVM for accurate and early determination of parasite density and infection in red blood cells, the algorithm was designed for automated detection and was tested on sample of 100 images obtained from online UCI repository. An efficiency of 91% percent was achieved on sample infected images utilizing SVM classifier.

Keywords Malaria · RBC · Chan–Vese algorithm · UCI · SVM

Pragya
Department of Chemistry, MVPG College, Lucknow University, Lucknow, India
e-mail: dr.pragya2011@gmail.com

P. Khanna · S. Kumar (✉)
Amity University, Lucknow Campus, Lucknow, India
e-mail: skumar3@lko.amity.edu

P. Khanna
e-mail: pkhanna@lko.amity.edu

© Springer Nature Singapore Pte Ltd. 2020
P. K. Singh et al. (eds.), *Proceedings of First International Conference on Computing, Communications, and Cyber-Security (IC4S 2019)*, Lecture Notes in Networks and Systems 121, https://doi.org/10.1007/978-981-15-3369-3_51

1 Introduction

WHO has been constantly maintaining statistics and suggesting preventive measures to be adopted to reduce number of malaria infected cases, though the total number of reported infected cases reduced from an 262 million an estimated figure globally in 2000 to 214 million in 2015, a decline of 18%. The number of malaria deaths globally fell from an estimated 839,000 in 2000 to 438,000 in 2017, a fall of 48%. Largely, deaths in 2017 are estimated to have taken place in e WHO African Region (88%), followed by WHO South-East Asia Region. WHO has set a global target of reducing the mortality rates by less than 40% by 2020 as compared to cases reported in 2015, by 75% in 2025 and by 90% in 2030, further WHO has set a target of completely eradicating malaria infection from at least ten countries in 2020, by at least twenty countries in 2025 and at least thirty-five countries by 2030 as compared to cases reported in 2015 and to also ensure that malaria infection does not re-establishes itself in countries in which it has been already eradicated. The common cause that leads to malaria is plasmodium parasite. Every year millions of cases are recorded across the globe and several people amongst them loose their lives, and Fig. 1 depicts a pictorial representation of casualties due to malaria infection across the globe, the data has obtained from Institute of Health Metrics and Evaluation. Female Anapholes mosquitos are accountable for spreading plasmodium parasite, these generally infect during night hours usually between dusk and dawn. Parasite primarily responsible plasmodium is of varied types but mainly five of them are responsible for causing malaria infection in humans [1, 2]. Figure 2 depicts five main malaria causing parasites.

- *Plasmodium falciparum*—It is the common parasite found in Africa also responsible for most of the deaths.

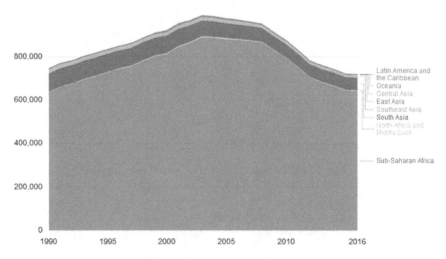

Fig. 1 Malaria causalities reported worldwide

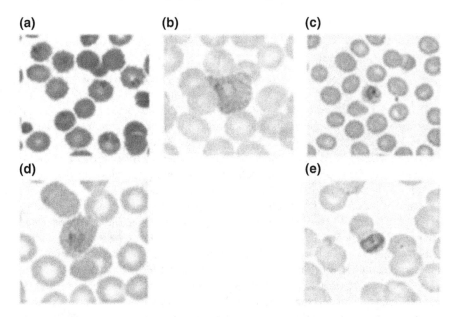

Fig. 2 Different parasites causing malaria infection **a** *Plasmodium falciparum* **b** *P. vivax* **c** *P. knowlesi* **d** *P. ovale* **e** *P. malariae*

- *Plasmodium vivax*—found in South America and Asia, the symptoms caused by this parasite are milder but it retains in liver for about three years.
- *Plasmodium ovale*—it is not so common, usually occurring in western parts of Africa. The parasite persists for good amount of years in liver not displaying any signs.
- *Plasmodium malariae*—this is fairly rare and generally found in parts of Africa only.
- *Plasmodium knowlesi*—Generally found in southeast parts of Asia and is also very rare.

Once the parasite enters the blood streams of a person, it passes on in liver, before again entering the veins and contaminating red blood cells, and it infects the liver. In red blood cells, parasite grows and multiplies. Infected cells burst at every 48–72 h, whenever this event happens, person experience fever, sweating, and chills [3].

World Malaria report published in 2017 estimated 216 million cases of malaria by 91 countries in 2016. Out of complete proportion of malaria infected population, the share of India is about 6 and 7% deaths happen in India. [4, 5]. Paper presents an automated technique for early and accurate detection of malaria infection; the article is organized as: Sect. 2 reviews about potential work done in the area in the form of literature survey, Sect. 3 proposes a model for the algorithm, Sect. 4 presents the technique for parasite plasmodium detection, and finally Sect. 4 presents result and conclusion.

2   Literature Survey

World Health Organization is consistently issuing standards and guideline to be followed which leads for compete eradication of the disease. Malaria if not detected and treated early with proper medication can prove fatal and may cause death. Anand V. K. Chhaniwal utilized digital holographic microscopy to detect malaria parasite. They suggested a technique that utilized digital holographic microscopy for quantitative analysis of cells that evaluates thickness of the cell from direct form. The model proposed the application of digital holographic interferometric microscopy (DHIM) with an aim to focus on numerical method to perform identification of parasites of malaria present in red blood cells in an automated procedure [6].

Isha Suvalka and Ankit Sanadhya in their paper suggested the pattern recognition technique to identify parasite by detecting its curved structure. The type of parasite is determined by defining specific ranges in the program. The range is defined numerically by number of pixels expressed as area for the suspect of infection in the image undergone staining procedure; as is depicted in Table 1 [7].

M. S. Suryawanshi and V. V. Dixit in 2013 proposed an advanced algorithm for parasite detection by cell segmentation. The model proposed involved several steps like binarization of image through Poisson distribution, theoretically based on minimized error thresholding [8]. Further, P. A. Pattanaik and Tripti Swarankar suggested three-phase pattern identification of machine vision through kernel detection and Kalman process of filtration to identify infection arising due to malaria. The usage of kernel-based identification with careful pixel data ensures that the proposed procedure efficiently identifies and restricts the target infected by parasite, this method was achieved via thin blood cell image [9]. Adedeji olugboja and Zenghui wang in 2017 employed various machine learning classifiers to identify malaria parasite. They opine that the traditional method of microscopy method had major flaws like time consumption and was not able to determine the density of the parasite. Watershed segmentation method was employed to obtain plasmodium contaminated and not contaminated erythrocytes [10].

S. D. Bias and S. Kareem in 2017 proposed low complex and efficient novel algorithm for segregating edges, particularly for malaria infection in thin blood smears, the algorithm works on fuzzy logic utilizing dynamic thresholding via histogram analysis followed by morphological operations for removing noise [11]. D. Kaur and G. K. Walia in 2017 proposed an ant colony optimization algorithm to identify edges parasites, it is was used maintain pheromone matrix which gives the data regarding

Table 1 Ranges that define type of parasite [7]	Parasite categories	Range
	P. malariae	6–7
	P. ovale	0.60–0.75
	P. vivux	1.2–2
	P. falcipurum	0.40–0.60

edges at every pixel position in blood smear sample, related to number of ants movement. Intensity level changes help in predicting motion of ants [12]. C. Mehanian, M. Jaiswal, C. Delahunt et al. proposed a system vision algorithm that utilizes deep learning to segregate parasites in standardized thick films of blood [13]. J. Soyemi, E. Adebiyi et al. in 2017 proposed an algorithm for protein interaction network (PIN) for DEG at red blood condition was identified, segregated, and extracted via protein interaction reserve maintained in database. Experiment segregated sixty four protein complexes utilizing Molecular Complex Detection (MCODE) technique in Cytoscape. Enrichment of the identified protein complexes in functional form displayed functions related to rRNA working, ribosome biogenesis, RNA processes for metabolic activity, cellular based process, nucleic and process on metabolic activity and others that are active in the RBCs and could be vulnerable to infection by *P. falciparum* [14]. E. V. Haryanto, M. Y. Mashor et al. in 2017 proposed an algorithm that utilized comparison of image based on pigment color space and color space to obtain optimum color channel to segregate plasmodium features from Giemsa stained sample images of blood cell which are infected with parasites suspected of malaria [15]. J. H. Brenas, M. Sadnan, A. Manir et al. in 2017 proposed a method for implementation of design utilizing the interoperability based on semantics and different phases of development for analytics environment for malaria parasites, having a sole aim of increasing efficiency and interoperability based on semantics for parasite surveillance of malaria and in order to keep the authenticity of sample value across various diversified scales dynamically [16]. D. Biben, M. S. Nair and P. Punitha proposed pre-trained proto-type utilizing deep belief network to segregate using classification algorithms 4100 peripheral images of blood smear into infected or non-infected parasite class. Proto-type presented based on DBN was pre-trained by a stack of restricted Boltzmann algorithms utilizing divergence algorithm. In order to pre-train DBN, vector features were segregated from sample images and variables that were visible belonging to deep belief network were initialized. Combined feature containing information of color and texture were utilized as vector feature [17].

3　Proposed Model

The technique employed presents integrated analysis model for comprehensive analysis of blood samples to identify and evaluate density of parasites causing malaria. The implementation of proposed system model is achieved through processes involving pre-processing steps, segmentation, stained object detection, RBC count, extraction of features, comparison, and classification as depicted in Fig. 3. The algorithm must be able to operate in an unverified environment and must be sufficiently sensitive to detect parasites at every stage especially at the initial stages of their life. MATLAB computational environment has been utilized to develop proposed algorithm. Major steps involved can be summarized as:

Image acquisition: Sample infected images were acquired from UCI online data repository.

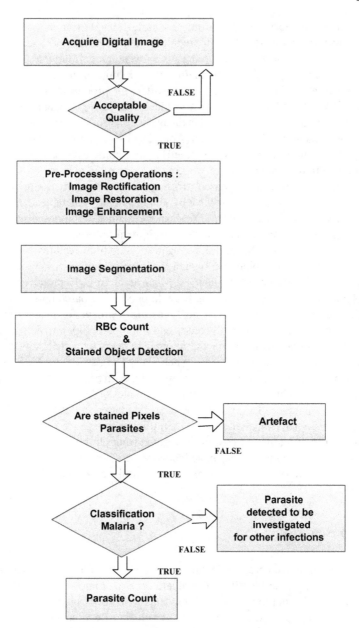

Fig. 3 Generalized block diagram of proposed model

Image preprocessing: Primary aim of preprocessing is to rectify degradation of image and improve contrast of target regions. The sample infected images with malaria parasites obtained may be of poor quality due noise present or improper contrast or uneven staining, and they are required to be improved through the process of restoration and enhancement, further objective that is covered is to achieve visual evaluation by increasing the contrast of the image. Infected sample image has its contrast improved through histogram equalization operated locally; this is done so as to make the distinction of RBC and parasite easy [18, 19].

Image segmentation: Various algorithms for segmentation have utilized for segregating red blood cells and infected parasites, like threshold-based and region-based segmentation, depending upon features to be extracted. The threshold value is selected from the histogram of the image by the medical support as the quality of image may vary depending upon the operating conditions. The categorization process for detecting infected parasite requires that individual cells in the slides are automatically isolated for analysis, thresholding achieves above process. The cells are then segregated from the background with the help of morphological operations. Finally, region of interest segmented from microscopic images is compared to standard parameters of malaria parasites. The common method proposed for this purpose is Chan–Vese et al. [20], the sole aim behind implementing this algorithm is to segregate image into areas containing useful information, and case is clearly depicted in Fig. 4, displaying separate color scale for RBC and artefacts against the background.

Stained object detection: Segregation and identification of stained objects through pixels values can be interpreted as a binary problem dealing with two-class classification. Take two classes as x_s *defined* for stained and x_{ns} *defined* for non-stained class. A classifier based on Bayesian theory with a feature defined as $rgb = \{r, g, b\}$ color space vector can be put in equation form as follow:

$$rgb \in x_s \quad \text{if} \frac{p(rgb|x_s)}{p(rgb|x_{ns})} >= \theta \tag{1}$$

Fig. 4 RBC's with boundaries segregated through Chan-Vese algorithm

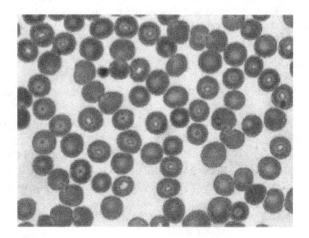

Here, $p(rgb|x_i)$ ($i \in \{s, ns\}$) depicts the class of conditional probability density function. θ is defined as the likelihood ratio for threshold. It directly gives the cost which is dependent on application for decision and primarily probabilities which *are priori for different* classes $P(w_s)$, $P(w_{ns})$ cases as they were determinable easily. PDF has been evaluated with the help of training, set of images already available in database. Stained regions forming objects in the set are manually segregated and labelled *one after one* by means of thresholding, and this is achieved through small program modules written specially for the said task, this facilitates viewing of output of the image while finalizing threshold values of the two variables. Procedure adopted for formulation of histograms is described as:

1. Stained $H_s(r, g, b)$ and non-stained $H_{ns}(r, g, b)$ categories both are set to zero value. Every repeated value of (r, g, b) will increase the counter by one for specific (r, g, b) to build the histograms:

$$H_s(r, g, b) = H_s(r, g, b) + 1 \quad \text{if } rgb \in w_s \tag{2}$$

$$H_{ns}(r, g, b) = H_{ns}(r, g, b) + 1 \quad \text{if } rgb \in w_{ns} \tag{3}$$

2. Histogram so obtained is normalized utilizing total occurrence of (r, g, b) given as (N_s, N_{ns}), respectively

$$P(rgb|ws) = Hs(r, g, b)/Ns \tag{4}$$

$$P(rgb|wns) = Hns(r, g, b)/Nns \tag{5}$$

Feature Extraction: Features play a primary role in segregating normal and red blood cells with infection. The objective of feature identification is to convert the sample test values to abridged set of indicators that segregate pertinent data against the sample test data in the form of information about infected and benign RBC's, the same is achieved and is depicted in Fig. 4.

Chan–Vese algorithm is utilized to segregate the sample infected data; algorithm differentiates between RBC from rest of the background. But owing to biconcave structure of RBC, the pallor located at center takes same feature as the background, as depicted in Fig. 5 [21].

Later image is converted to binary format for though thresholding. Fig. 5 shows the result of the threshold-based segmentation, and the given sample image is thresholded at a reference point 104.03. The threshold value is selected from the histogram of the image by the medical support as the quality of image may vary depending upon the operating conditions. The categorization process for RBC's requires that individual cells in the slides are automatically isolated for analysis, thresholding achieves the above process. The cells are then segregated from the background with the help opening.

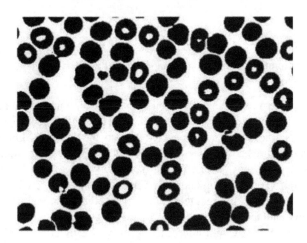

Fig. 5 Binary image of RBC's

Identification process would need elimination of the central pallor as essential requirement, to achieve this, 'hole filling' algorithm is implemented, given by equations mentioned below and is depicted in Fig. 6 [22].

$$X_k = (X_{k-1} + B) \cap A^c \qquad (6)$$

where: $k = 1, 2, 3, \ldots$

Stop the iteration if $X_k = X_{k-1}$

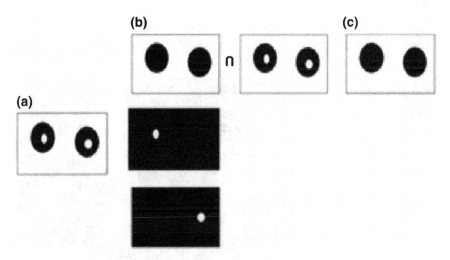

Fig. 6 Algorithm for hole filling **a** sample image having holes **b** connected neighbors **c** sample image with filled holes

where

X_t: represents an array of zero's
B: Structuring element
A: Sample image

Algorithm performing filling of holes in the sample digital binary image takes out the largest component that is connected between all the neighbors that are connected. Analysis of components that are connected extracts the data regarding connectivity of pixels in a two-dimensional data in form of image by labeling connected pixels that have similar intensity levels. Figure 6a depicts image with holes, extraction of all connected pixels or components is achieved and is depicted in Fig. 6b and intersection with the background, that is the largest component that is connected for any given sample image was done with original image is depicted in Fig. 6c.

Sample infected image with hole filled is shadowed by erosion utilizing a disk structuring element having radius of 5 pixels values; the operated sample data image obtained after performing hole filling and eroding as depicted in Fig. 7. Artifacts are minimized through erosion which removes random and isolated pixels [6, 8].

Parasite Classification: Stained objects on slides consist of diversified components present in blood sample along with malaria parasites if infected, such as platelets and WBCs. Therefore, the detection process further requires processing in order to segment the infection causing parasites.

Evaluation of pattern spectrum function $(G_\Lambda(Y))$ area on negative scale of image gray portion provides an excellent approximation of mean value of area (A_μ) covered by RBC and radius [16]. Once the stained object is determined in the form of pixel values, infinite morphological reconstruction is applied [15] utilizing the pixels having gray values near to stained group S as the indicators and the gray level on negative axis of image to approximate the region of cell that has the stained group. Resultant binary object S_b is marked for extraction as it represents one of the features. Pixels with similar color scale as stained are grouped as single entity and are merged. During the process, non-parasites present can be identified by comparing

Fig. 7 Binary image after removing spurious boundaries

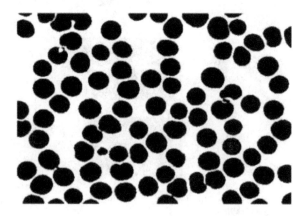

size of cell and heuristics (Location—background and foreground) can be eliminated at this point.

These heuristics apply only in the case of WBC's which are large in size and platelets that are clearly submerged and surrounded against background. The present model passes all objects through feature extraction step and then classes are assigned to them as infected and non-infected by malaria parasite with help of SVM classifier. SVM was utilized to segregate two classes determined after the analysis of blood smear as malaria parasite infected and benign, SVM achieves this by creating an separating hyperplane between two classes of data taking into consideration features chosen for identification of parasites. Generally, a kernel function (K) is utilized for same. Hyperplane separating the two classes is based on features obtained from stained object detection and size and shape of RBC's. In equation form, hyperplane is represented as:

$$w \cdot y + a = 0, \tag{7}$$

where: w gives normal to hyperplane; a gives the perpendicular distance to hyperplane from origin. Feature vector lying on left side of the plane is represented by $+1$ class, marked as infected by malaria parasite and is given by:

$$w \cdot y_i + a \geq +1 \text{ for } y_i = +1 \tag{8}$$

and feature vector that lies other side represent class -1 and are marked as benign [12, 13].

$$w \cdot y_i + a \leq +1 \text{ for } y_i = -1 \tag{9}$$

The resultant is independent of the heuristics [23–25].

RBC Count: Binary image obtained after processing with infected RBC's identified can be utilized to obtain the density of the parasites in region of interest, this can be estimated through dividing total region occupied by RBCs from the area of single RBC and the resultant so obtained is rounded to the nearest integer value to approximate the parasite density. Hough transform is utilized for segregating of sample infected image to extract shapes like ellipse, straight line, or circle already defined with data image. Hough transform technique is designed here to identify circles within an image. The circles so identified has been classified as infected depending on feature parameters already defined, this is depicted in Fig. 8 [26–29].

Figure 9 depicts integrated graphical user interface designed for the algorithm explained above, as depicted sample images processed are subjected segmentation, feature extraction, stained object detection and parasite classification. Interface has been designed on MATLAB. The interface implemented is able to identify parasite density as well as infection. The model was tested on a sample of more than 100 images with an accuracy of 91% for predicting level of infection. Figures 9 and 10 depict two simultaneous cases of parasite detected and not detected, respectively.

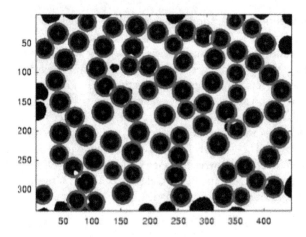

Fig. 8 Circles with various radii are detected

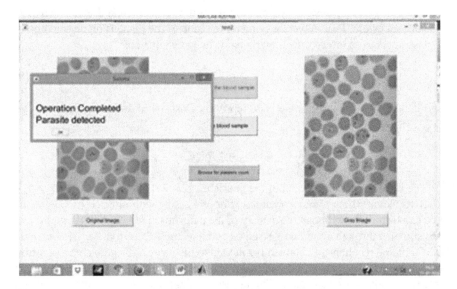

Fig. 9 Integrated graphics user Interface-Infection detected

4 Conclusion

Paper presents an automated algorithm based on SVM for accurate and early determination of parasite density and infection in red blood cells. Features are extracted from the pre-processed images through stained object detection and shape and color of RBC's as compared to standard parameter values. Plasmodium parasites are primary parasite responsible for malaria infection, which contributes a largely in mortality rate. Detection system has been implemented for Plasmoduim parasite. The presence

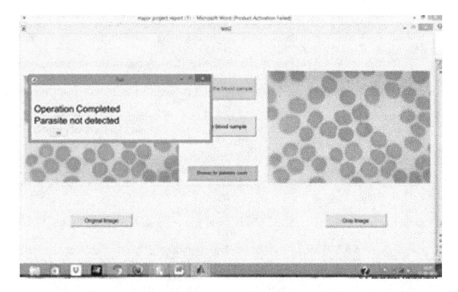

Fig. 10 Integrated graphics user Interface-Infection not detected

of the malaria parasite was detected by analyzing the features extracted from the image. Also a graphical user interface is created to ease the process of identification and estimation of malaria parasite. The results obtained from algorithm embedded in integrated user interface were compared with the results of traditional technique of microscopy. Efficiency achieved is at par and sensitivity of the model proposed is better than most of the traditional techniques. The algorithm was designed for automated detection and was tested on sample of 100 images obtained from online UCI repository. An efficiency of 91% percent was achieved on sample infected images utilizing SVM classifier.

References

1. Makkapati, V., Rao, R.: Segmentation of malaria parasites in peripheral blood smear images. In: ICASSP, IEEE International Conference on Acoustics, Speech and Signal Processing—Proceedings, pp. 1361–1364 (2009). https://doi.org/10.1109/icassp.2009.4959845
2. Zou, L., Chen, J., Zhang, J., Garsia, N.: Image processing algorithms to automate the most of all diagnosis of malaria on thin blood smears. In: 2010 International Conference on Digital Image Computing: Techniques and Applications
3. Kareem, S., Morling, R.C., Kale, I.: A novel method to count the red blood cells in thin blood films. In: 2011 IEEE International Symposium of Circuits and Systems (ISCAS), pp. 1021–1024 (2011)
4. https://www.who.int/malaria/media/world-malaria-report-2017
5. Rakshit, P., Bhowmik, K.: Detection of presence of parasites in human RBC in case of diagnosing malaria using image processing. In: (ICIIP-2013), pp. 329–334. https://doi.org/10.1109/iciip.2013.6707610

6. Anand, A., Chhaniwal, V.K., Patel, N.R., Javidi, B.: Automatic identification of malaria-infected RBC with digital holographic microscopy using correlation algorithms. IEEE Photonics J. **4**(5), 1456–1464 (2012)
7. Suwalka, I., Sanadhya, A., Mathur, A., Chouhan, M.S.: Identify malaria parasite using pattern recognition technique. In: 2012 International Conference on Computing, Communication and Applications (ICCCA), Dindigul, Tamilnadu, 2012, pp. 1–4 (2012). https://doi.org/10.1109/ICCCA.2012.6179129
8. Suryawanshi, M.S., Dixit, P.V.V.: Comparative study of malaria parasite detection using euclidean distance classifier & SVM. Int. J. Adv. Res. Comput. Eng. (IJARCET) **2**(11), 2994–2997 (2013)
9. Pattanaik, P.A., Swarnkar, T., Sheet, D.: Object detection technique for malaria parasite in thin blood smear images. In: 2017 IEEE International Conference on Bioinformatics and Biomedicine (BIBM), Kansas City, MO, 2017, pp. 2120–2123. https://doi.org/10.1109/bibm.2017.8217986
10. Olugboja, A., Wang, Z.: Malaria parasite detection using different machine learning classifier. In: International Conference on Machine Learning and Cybernetics Ningbo, China, 9–12 July 2017
11. Bias, S.D., Reni, S.K., Kale, I.: A novel fuzzy logic inspired edge detection technique for analysis of malaria infected microscopic thin blood images. In: 2017 IEEE Life Sciences Conference (LSC), Sydney, NSW, 2017, pp. 262–265. https://doi.org/10.1109/lsc.2017.8268193
12. Kaur, D., Walia, G.K.: Detection of malaria parasites using ant colony optimization. In: 4th IEEE International Conference on Signal Processing, Computing and Control (ISPCC 2K17), 21–23 Sept 2017, Solan, India
13. Mehanian, C., et al.: Computer-automated malaria diagnosis and quantitation using convolutional neural networks. In: 2017 IEEE International Conference on Computer Vision Workshops (ICCVW), Venice, 2017, pp. 116–125. https://doi.org/10.1109/iccvw.2017.22
14. Soyemi, J., Isewon, I., Oyelade, J., Adebiyi, E.: Functional enrichment of human protein complexes in malaria parasites (2017). IEEE. 978-1-5090-4642-3/17/$31.00
15. Haryanto, S.E.V., Mashor, M.Y., Nasir, A.S.A., Jaafar, H.: Malaria parasite detection with histogram color space method in Giemsa-stained blood cell images. In: 2017 5th International Conference on Cyber and IT Service Management (CITSM), Denpasar, 2017, pp. 1–4. https://doi.org/10.1109/citsm.2017.8089291
16. Brenas, J.H., Al-Manir, M.S., Baker, C.J.O., Shaban-Nejad, A.: A malaria analytics framework to support evolution and interoperability of global health surveillance systems. IEEE Access **5**, 21605–21619 (2017). https://doi.org/10.1109/access.2017.2761232
17. Bibin, D., Nair, M.S., Punitha, P.: Malaria parasite detection from peripheral blood smear images using deep belief networks. IEEE Access **5**, 9099–9108 (2017). https://doi.org/10.1109/access.2017.2705642
18. Shet, N.R., Sampathila, N.: An image processing approach for screening of malaria. Canar. Eng. Coll. Mangalore, pp. 395–399, 2015
19. Ghate, D.A., Jadhav, P.C.: Automatic Detection of Malaria Parasite from Blood Images. Int. J. Adv. Comput. Technol. **1**(3), 66–71 (2012)
20. Tek, F.B., Dempster, A.G., Kale, I.: A colour normalization method for giemsa-stained blood cell images. In: Proceedings of Sinyal Isleme ve Iletisim Uygulamalari, Apr 2006
21. Di Ruberto, C., Dempster, A., Khan, S., Jarra, B.: Automatic thresholding of infected blood images using granulometry and regional extrema. In: ICPR, pp. 3445–3448, 2000
22. Dong, Y., Jiang, Z., Shen, H., Pan, W.D.: Classification accuracies of malaria infected cells using deep convolutional neural networks based on decompressed images, 2017. IEEE. 978-1-5386-1539-3/17
23. Tek, F.B., Dempster, A.G., Kale, I.: Blood cell segmentation using minimum area watershed and circle radon transformations. In: Proceedings of International Symposium on Mathematical Morphology, vol. 1, pp. 739–752, Apr 2005
24. Špringl, V.: Automatic malaria diagnosis through microscopy imaging. Czech Tech. Univ. Prague Fac. Electr. Engeneering (2009)

25. Gonzalez, R.C., Woods, R.E., Eddins, S.L.: Digital image processing using MATLAB. Prentice Hall (2001)
26. Iancu, D.A.: Eye detection using variants of the hough transform. Final project in CS475 Computational Vision and Biological perception (2004)
27. Verm, A., Scholar, M.T., Lal, C., Kumar, S.: Image segmentation: review paper. Int. J. Educ. Sci. Res. Rev. **3**(2), (2016)
28. Jones, M.J., Rehg, J.M.: Statistical color models with application to skin detection. Int. J. Comput. Vision **46**(1), 81–96 (2002)
29. Webb. Statistical pattern recognition, 2nd edn. J Wiley and Sons Inc., New York, USA (2002)

A Novel Budget-based Collaborative Recommendation System for Tourism (BCRS)

Sourabh Garg, Diksha Mohnani, Dhwani Sachwani, Maitri Savani and Nirali Nanavati

Abstract It is customary for people to visit and rate a place, hotel, or restaurant in online traveling apps. The travel history of people reflects the type of places, hotels, or restaurants they like and indicates their future likings. In most traveling apps, the app tries to recommend the places, hotels, and restaurants based on ratings. These apps use filters that do not take the consideration of user preferences and budget under which a user would want to explore the places, hotels or restaurants of a particular unknown destination. In this paper, we have developed a generative model that will not only take into consideration individual liking but also connects it with the budget in which they want to explore a destination. Our proposed model is Budget based Collaborative Recommendation System (BCRS). The BCRS is intuitive and highly interpretable. We have applied this model on real-world datasets, and the results show the advantage of uncovering user's budgets for the recommendation of places, hotel, or restaurants which will be a boon for tourism as it will increase the reachability of users to many places and user would able to explore places of their preferences in their budget.

Keywords Budget-based recommendation system · Collaborative filtering · Tourism

S. Garg (✉) · D. Mohnani · D. Sachwani · M. Savani · N. Nanavati
Sarvajanik College of Engineering and Technology, Dr R K Desai Marg, Opp. Mission Hospital, Athwalines, Athwa, Surat, Gujarat, India
e-mail: saurabh.g23arg@gmail.com

D. Mohnani
e-mail: dikshamohnani@gmail.com

D. Sachwani
e-mail: dhwanisach53@gmail.com

M. Savani
e-mail: savanimaitri41@gmail.com

N. Nanavati
e-mail: nirali.nanavati@scet.ac.in

© Springer Nature Singapore Pte Ltd. 2020 713
P. K. Singh et al. (eds.), *Proceedings of First International Conference on Computing, Communications, and Cyber-Security (IC4S 2019)*, Lecture Notes in Networks and Systems 121, https://doi.org/10.1007/978-981-15-3369-3_52

1 Introduction

In today's world, the amount of information on the web and the number of users interacting with the internet are increasing day by day. The information available on the internet is useful and necessary for users who wish to plan for a trip to some unknown destination. To plan their trip, tourists search for information about the places, hotels, and sightseeing associated with that particular unknown destination. However, the list of possibilities offered by traveling apps is pretty vast. This long list of options is very complex and time consuming for tourists to select the one that fits best with their needs. Nowadays, tourist wants to avoid agents (middle-men) to make their own choices. They want to make bookings and payments for themselves directly. But because of this time consuming and complex task, normally most of the tourists choose the best available package option provided by famous traveling apps and have to take whatever it offers which might skip certain options of their preferences [1].

With the growth of the online tourism market, there is an urgent need for an intelligent recommendation of places, sightseeing, and restaurants for the customers. Most of the online traveling apps today show results filtered according to the rating and the number of users who have rated them. However, most of the traveling apps do not consider recommending items to the users according to their interests and also the budget which is a very essential parameter one should consider while recommending places to users. According to [2], users allocate their budgets by considering both the "whether to spend" and the "how much to spend" decisions to maximize the utility function. Hence, the tourist's decision implies a unified preference structure for different places. In this paper, we propose a model called Budget based Collaborative Recommendation System (BCRS) for tourism which will recommend places considering the user's budget also. With the understanding of how users' budget affects their purchasing behaviors, we have provided a more intelligent recommendation system. To the best of our knowledge, this is the first work that considers modeling users' budget in a generative way for recommending places according to user preferences.

Why try to model a recommendation system on the user's budget in tourism? Price can play a very important role in affecting user's decision of purchasing anything around the world whether it is a cell phone or booking a 5-star hotel. As pointed out by Du and Kamakura [2] and Kooti et al. [3], users usually have finite budgets and they will assign more budgets to the places, hotels, or restaurants that they prefer the most. So, to understand how the users will divide their budgets across different categories like sightseeing, hotels, or restaurants is of great practical importance. Moreover, some users can allocate more budget for a particular attribute like some users will be willing to pay more for the hotel than other attributes. Thus, all these phenomenon demonstrate the need of a recommendation system that will not only recommend items but also takes care of budget in tourism so that more and more people would get the chance of exploring a particular city at their budget. The existing tourism apps do not consider this important feature and do not target user preferences.

How the tourism industry will get affected through our proposed recommendation model? In today's world, tourism is an important economic driver. Embedding budget into tourism-based recommender system will not only benefit travelers but will also boost up the tourism industry which will have further implications on economy such as greater employment opportunities, better infrastructure, and increased spending in the local community. The growth of tourism sector will prove to be a boon for the nation's economy.

Hence, in this paper, we have proposed a new generative model named Budget based Collaborative Recommendation System (BCRS) to connect the collaborative recommendation system with budget and user preferences as derived from his/her past travel visits. Hence, we summarize our contribution as follows:

- We have proposed a generative model which will recommend places to users according to their budget.
- We have proposed an extension of matrix factorization technique with the SVD decomposition model for collaborative recommendation system to which we have added budget as one of the parameters.
- The experimental results reveal the advantage of the proposed model compared with state-of-the-art budget unaware recommendation models in tourism.

The rest of the paper is organized as follows. We discuss related work in Sect. 2. Further, we give an overview of the background in Sect. 3. These include details about collaborative filtering, matrix factorization, and web scraping. In Sect. 4, the problem statement is defined, while in Sect. 5, we have proposed our algorithm. We explain the evaluation methodology, the datasets used, and the tools in Sect. 6. Further, in Sect. 7, we show the comparative experimental results and we give our conclusion in Sect. 8.

2 Related Work

Recommender systems have drawn more attention from the machine learning community since the emergence of e-commerce. Various approaches have been proposed to provide better product recommendations. Among them, collaborative filtering is a leading technique that tries to recommend a user with products by analyzing similar user records [4–6].

Many collaborative filtering researchers have recognized the problem of sparseness, i.e., many values in the rating matrix are null since all users do not rate all items. Computing distances between users is complicated by the fact that the number of items users have rated in common is not constant. An alternative to inserting global means for null values or significance weighting is singular value decomposition (SVD) [7], which reduces the dimensionality of the rating matrix and identifies latent factors in the data.

Understanding how budgets affect users' purchasing decisions is an important topic in the field of economics. However, all these works are taken from an economic perspective. Based on our analysis above, uncovering users' budgets can also be an important issue in the field of recommender systems. However, few works consider incorporating budget into recommender systems. To the best of our knowledge, this paper is the first one that works on incorporating users' preferences and budget into tourism recommender systems in a generative way.

3 Background

Here, in this section, we have defined the modules which are the basis of our model that are collaborative filtering, matrix factorization, and web scraping.

3.1 Web Scraping

Web scraping is a technique employed to extract specific data from websites where the extracted data is saved in the files or on database in a tabular format. Data displayed on the website can only be viewed by a web browser. They do not offer any functionality to save a copy of this data for personal use [8].

Web scraping is the technique of automating this process so that instead of manually copying the data from websites, the web scraping software will perform the same task within a few seconds. Thus, this technique is useful to extract real-time data from the internet such as the price of places, sightseeing, and restaurants from the website so that the user will get the up-to-date price of the places [9].

In our proposed work for web scraping, we have used a tool called Selenium [10, 11]. Selenium is a Web application testing framework that supports a wide variety of browsers and platforms including Java, .Net, Ruby, Python, and others. The Selenium IDE is a popular tool for browser automation, mostly because of its software testing application. Also, Web scraping techniques for tough dynamic websites may be implemented with this IDE along with the Selenium remote control server. ChromeDriver 2.42 is used for fetching the value from websites. This chromeDriver is used by Selenium which creates a virtual chrome via which permissible information can be extracted.

3.2 Recommendation System

A recommendation system is a computer program that filters the data using different algorithms and recommends the most relevant items to users [12]. There are two types of recommendation systems:

- Content-based recommendation system: This method is based on the description of an item and a profile of the user's preferred choices. This recommendation system recommends products which are similar to the ones that a user has liked in the past.
- Collaborative recommendation system: This method is usually based on collecting and analyzing information on the user's behaviors, their activities, or preferences and predicting what they will like based on the similarity with other users.

3.2.1 Collaborative Filtering Recommendation System

Collaborative filtering is a recommendation system technique that takes "User Behavior" in consideration for recommending items. Let us understand with an example (as shown in Fig. 1), say if a person A likes oranges, apples, and bananas and a person B likes apples, watermelons, and bananas, then we can say that they have almost similar fruit choices. Based on this, we can say with some certainty that person A would like watermelon and person B would like orange. Thus, we will recommend watermelon to person A and orange to person B. In this way, the collaborative filtering for the recommendation system works.

3.3 Matrix Factorization

Users give feedback for places they visit in the form of ratings, and this collection of feedback can be represented in a form of a matrix, where each row represents a user and each column represents a destination. The matrix will be sparse since

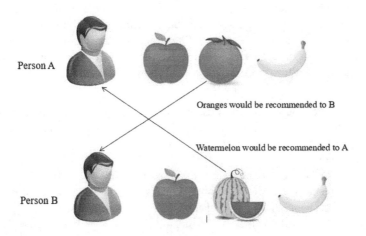

Fig. 1 Collaborative filtering recommendation system

not everyone has visited every destination (we all have different preferences when it comes to visiting places) [13, 14].

One strength of matrix factorization is the fact that it can incorporate implicit feedback, information that is not directly given but can be derived by analyzing user behavior. Using this strength, we can estimate if a user is going to like a place that he/she never visited. And if the estimated rating is high, we can recommend that place to the user. Below is how matrix factorization works for predicting ratings:

So, based on each latent feature, all the missing ratings in the R matrix will be filled using the predicted rating value ($r_{(ui)}$) which will be obtained through gradient descent algorithm.

4   Problem Statement

In this paper, we assume there are U users and V places. For matrix factorization, the latent factor will be denoted by K. We will make a UXV user-destination matrix whose values are ratings which are denoted by R and the matrix is denoted by A. The budget allocated by user for a place is denoted by B. So, after constructing matrix A and applying stochastic gradient algorithm, we will denote an observable matrix A' in which for a user Ui, we will have a row of predicted rating and then we will recommend to user, places having higher predicted rating and which satisfy the budget B.

5   Proposed Approach

In BRCS, we have tried to uncover how the users rate a place and how the places are recommended to users based on their budget. Unlike the traditional collaborative recommendation system, this system will recommend places to users which will not only satisfy user's liking but also will satisfy the budget set by the user. So, the generative process of BRCS is as follows (as shown in Fig. 2) [12]:

1. For all users U and places V:
   We will fetch the ratings from users past travel records. Here, if a user has not explored a place (or is yet to visit a place), we will denote the rating for that place as 0.
2. Initial matrix A:
   We will make a UXV matrix where the user (mainly user-id) will be row and places (place id) will be a column. Thus, if a user has explored a place and rating is available in the user's history, then we will assign that to that particular A ($U_{(i)}XV_{(j)}$) position otherwise as mentioned above, we will assign 0 if the user is yet to visit that place.

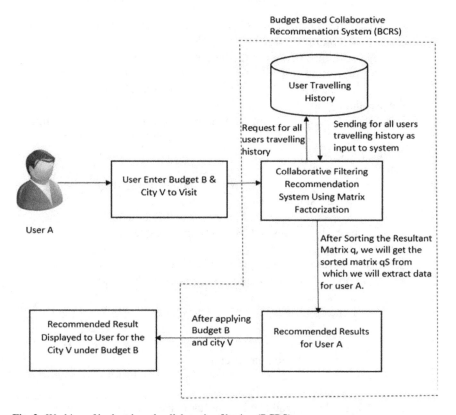

Fig. 2 Working of budget-based collaborative filtering (BCRS)

3. Latent factor k:
   We have to find a set of features that can define how a user rates the places. These are called latent features. We need to find a way to extract the most important latent features from the existing features. Thus, here we will denote the latent feature as k which will help us in removing noise from the data. Thus here we will be taking the category under which a place falls as a latent feature. For example, sightseeing may be categorized as a museum or as a theme park and so on.

5.1 Implementation Details

Following are the steps that our recommender system will follow in order to recommend places to a user.

- Let us say a user $U_{(i)}$ sets budget B for a city $V_{(j)}$.

1. First of all with the help of library sklearn.model_selection, we will divide the database result into 90% train-set and 10% test-set.
2. A matrix A is formed which consists of user (user_id) as row, places (place_id) as column, and rating as their value with the help of train-set.
3. Let us define a class MF which will return the predicted matrix. So, following is the pseudocode for it.
   i. Initialize the matrix A, number of latent feature k, alpha (learning rate for stochastic gradient descent [15]) and beta (regularization bias term).
   ii. Initializing user-feature and place-feature matrix from matrix A.
      self.P = np.normal.random(scale = 1./self.k,
                                  size= (self.num_users, self.k))
      self.Q = np.normal.random(scale = 1./self.k,
                                  size= (self.num_items, self.k))
   iii. Initializing the bias terms and list of training sample in a list named S.
      #Initializing the bias terms
      self.b_u = np.zeros(self.num_users) #for user
      self.b_i = np.zeros(self.num_items) #for items
      #List of training sample
      self.b = np.mean(self.R[np.where(self.R != 0)])
      self.S = [
            (i, j, self.R[i, j])
            for i in range(self.num_users)
            for j in range(self.num_items)
            if self.R[i, j] > 0]
   iv. Applying Stochastic gradient descent for a given number of iteration iter.
      T = []
      for i in range(iter):
            np.random.shuffle(self.S)
            self.sgd()
            mse = self.mse()
            t.append((i,mse))
   v. The function mse() (mean square error)is defined as follows:
      error += pow(self.R[x, y] - predicted[x, y], 2)
      error = np.sqrt(error)
   vi. The function sgd() is defined as follows:
      for i,j,r in self.S:
            prediction = self.get_rating(i, j)
            e = (r - prediction)
            self.b_u[i] += self.alpha * (e - self.beta * self.b_u[i])
            self.b_i[j] += self.alpha * (e - self.beta * self.b_i[j])
            self.P[i,:] += self.alpha * (e * self.Q[j,:] -
                                          self.beta * self.P[i,:])
            self.Q[j,:] += self.alpha * (e * self.P[i,:] -
                                          self.beta * self.Q[j,:])

vii. For getting the prediction as mentioned in step 6 we will define a function get_rating():

def get_rating(self, i, j):
    prediction = self.b + self.b_u[i] + self.b_i[j] +
                self.P[i, :].dot(self.Q[j, :].T)

4. Now, we will pass the matrix A to the object O of class MF().

    q = MF(A,K=8,alpha = 0.001,beta = 0.01,iterations = 200)

Here, we will get k from the database and we will keep alpha, beta, and iterations as constants.

5. After getting the matrix B in q, we will sort the matrix in descending order.

6. When the step 5 gets over, we will get the sorted matrix qS. Thus, now we will pass the budget and the column which do not satisfy it will be mentioned its value as NA.

7. Thus, for the user $U_{(i)}$, we will extract the places from matrix qS with the help of rows. So, the first 10 places will be recommended to the user $U_{(i)}$ (excluding the column with NA).

6   Performance Evaluation

In this section, we give the details of the methodology of evaluation, the tools, and the datasets used. We also discuss the methodology with the help of a sample test application.

6.1   Metrics

The metrics which we have used for performance evaluation are [12]:

1. Recall: Recall can be defined as the proportion of items that a user actually likes to total recommended items [16]. It is given as

$$\text{Recall} = \frac{t_p}{t_p + t_n} \tag{1}$$

Here, t_p stands for true positive and represents the number of items recommended to a user that he/she likes and $t_p + t_n$ represents the total number of recommended items where f_n stands for false negative. If a user likes three items out of the total five recommended items that the engine decided to show, then the recall is 0.6.

2. Root mean square error (RMSE): It measures the error in the predicted ratings [17, 18]

$$RMSE = \sqrt{\frac{\sum_{i=1}^{N} Predicted_i - Actual^i}{N}} \qquad (2)$$

Here, predicted is the rating predicted for the item I by the model and actual is the original rating for item I. If a user has given a rating of 5 to an item I and we predicted the rating as 4, then RMSE is 1.

We have used recall as an evaluation metric and RMSE as an accuracy metric.

6.2 Datasets

We performed our experiments on a dataset with 60 users. Thus, we modeled our BCRS by randomly dividing the data into train-set and test-set. The aim is to predict the rating of all the places for each user and recommend the places to the user according to the similarity among different users and also considering the budget which they entered. Here, we had a total of 39 places for which we had scrapped the information from websites through selenium.

7 Result and Analysis

7.1 Performance Results

After testing our recommender system BRCS on the test-set, following results were obtained for recall and RMSE. We evaluated the result for 10 different iterations (as shown in Table 1) where the ratio of train-set to test-set was (9:1).

Table 1 Result of RMSE and Recall after running BCRS model for recommendation

Iteration	RMSE	Recall (out of 100 in %)
1	1.67	75.01
2	0.53	82.18
3	1.47	90.47
4	0.96	85.00
5	1.44	84.44
6	0.79	90.00
7	0.90	81.00
8	1.09	90.27
9	1.21	82.00
10	0.84	85.67

7.2 Empirical Analysis

From the above table, we observed that the recall ranges from a minimum of 75% to a maximum of 90% which depicts that on an average 8 recommendations out of the top 10 recommended are actually liked by the tourist (user).

Further, we observed that RMSE ranges from a minimum of 0.53 to a maximum of 1.67. Thus, our algorithm has performed better as it not only recommends user the places according to his interests as depicted by his previous visits but also takes into consideration the budget allocated by the user.

8 Conclusion and Scope of Future Work

Thus, in this paper, we have worked on improving the state-of-the-art for collaborative filtering in the form of BCRS in tourism which would be advantageous for tourists as they can now get rid of exhaustive searches and can make their packages based on the recommendation provided that fits in their budget. Our primary contributions are:

- We have made a generative model which will recommend places to user according to their budget and preferences, considering user's travel history.
- We have proposed an extension of matrix factorization technique with the SVD decomposition model for collaborative recommendation system in which we have added budget as an additional parameter.
- We further have provided the graphical and empirical analysis of our scheme and it shows that it can perform better after taking into account the budget of the user.

So, our scheme could further be applied to lots of other scenario and applications where budget as a parameter can play a very important role in determining future likings of users.

References

1. Kzaz, L., Dakhchoune, D., Dahab, D.: Tourism recommender systems: an overview of recommendation approaches. Int. J. Comput. Appl. **180**, 9–13 (2018)
2. Du, R.Y., Kamakura, W.A.: Where did all that money go? Understanding how consumers allocate their consumption budget. J. Mark. **72**, 109–131 (2008)
3. Kooti, F., et al.: Portrait of an online shopper: understanding and predicting consumer behavior. In: Proceedings of the Ninth ACM International Conference on Web Search and Data Mining, San Francisco, USA, pp. 205–214 (2016)
4. Koren, Y., Bell, R.: In: Ricci, F., Rokach, L., Shapira, B., Kantor, P.B. (eds.) Recommender Systems Handbook, pp. 77–118. Springer, Boston, MA (2015)
5. Goldberg, D., Nichols, D., Oki, B.M., Terry, D.: Using collaborative filtering to weave an information tapestry. Commun. ACM **35**, 61–71 (1992)

6. Koren, Y.: Factorization meets the neighborhood: a multifaceted collaborative filtering model. In: Proceedings of the 14th ACM SIGKDD International Conference on Knowledge Discovery and Data Mining, Las Vegas, Nevada, USA, pp. 426–434 (2008)
7. Sarwar, B., Karypis, G., Konstan, J., Riedl, J.: Application of dimensionality reduction in recommender system—a case study. Tech. Rep. (Department of Computer Science, University of Minnesota, Minneapolis, 2000)
8. SysNucleus. WebHarvy Web Scraper. https://www.webharvy.com/articles/what-is-web-scraping.html (2019)
9. Vargiu, E., Urru, M.: Exploiting web scraping in a collaborative filtering-based approach to web advertising. Artif. Intell. Res. **2**, 44–54 (2013)
10. Singla, S., Kaur, H.: Selenium keyword driven automation testing framework. IJARCSSE **4** (2014). ISSN: 2277 128X
11. Bandi, A.: Web Scraping Using Selenium | Python. https://towardsdatascience.com/web-scraping-using-selenium-python-8a60f4cf40ab (2019). Oct 2018
12. Pulkit, S.: Comprehensive guide to build recommendation engine from scratch. https://www.analyticsvidhya.com/blog/2018/06//comprehensive-guide-recommendation-engine-python/ (2018). Accessed 06 Nov 2019
13. Seo, J.D.: [Paper summary] Matrix factorization techniques for recommender systems. https://towardsdatascience.com/paper-summary-matrix-systems-82d1a7ace74 (2019). Nov 2018
14. Mavridis, A.: Matrix factorization techniques for recommender systems (2017)
15. Bottou, L.: In: Neural Networks: Tricks of the Trade, pp. 421–436. Springer, Heidelberg (2012)
16. Herlocker, J.L., Konstan, J.A., Terveen, L.G., Riedl, J.T.: Evaluating collaborative filtering recommender systems. ACM Trans. Inf. Syst. (TOIS) **22**, 5–53 (2004)
17. Cremonesi, P., Koren, Y., Turrin, R.: Performance of recommender algorithms on top-n recommendation tasks. In: Proceedings of the Fourth ACM Conference on Recommender Systems, Barcelona, Spain, pp. 39–46 (2010)
18. Willmott, C.J., Matsuura, K.: Advantages of the mean absolute error (MAE) over the root mean square error (RMSE) in assessing average model performance. Clim. Res. **30**, 79–82 (2005)

Context Annotated Graph and Fuzzy Similarity Based Document Descriptor

Akshi Kumar, Geetanjali Garg, Rishabh Kumar and Mahima Chugh

Abstract The document descriptors are widely researched and developed in the NLP literature. This study deals with the task of designing context-aware document descriptors, which can be used for information retrieval (IR) based on generic queries. We propose a graph-based data structure that preserves contextual as well as structural information of the document. We also design a novel metric for calculating fuzzy similarity between two vectors. Three different types of contexts, collocational words co-occurring frequently with the keyword in the given paper (C1), collocational words co-occurring with the keyword in the defined corpus (C2), and words synonymous with the keywords in the WordNet lexical database (C3) were used. Each document in a corpus of research papers was represented with a context annotated graph (CAG). These graphs were vectorized, and similarity between paper vector and the query vector was calculated by using fuzzy similarity metric. We compare the results of generic search queries using average weighted precision (AWP) and discounted cumulative gain (DCG) performance metric. The descriptor with contextual information coupled with fuzzy similarity algorithm (AWP score of 0.915 and DCG score of 0.922) outperforms baseline of non-contextual, cosine similarity-based results (AWP score of 0.752 and DCG score of 0.796). We also concluded that contextual information gives better result with fuzzy similarity (AWP score of 0.915 and DCG score of 0.922) than with cosine similarity (AWP score of 0.865 and DCG score of 0.912).

Keywords Natural language processing · Fuzzy similarity · Context-aware graph · Information retrieval · Descriptors

A. Kumar · G. Garg (✉) · R. Kumar · M. Chugh
Department of Computer Science and Engineering, Delhi Technological University, Delhi, India
e-mail: geetanjali.garg@dtu.ac.in

A. Kumar
e-mail: akshikumar@dce.ac.in

R. Kumar
e-mail: rishabhkr1729@gmail.com

M. Chugh
e-mail: kosmikoffe601@gmail.com

1 Introduction

Development of document descriptors is an active area of research in the NLP literature. In the most trivial sense, documents can be represented by bag-of-words (BOW) matrix or a term frequency (TF) matrix. However, these representations are not space efficient and do not consider word orderings or inter-word relationships. Some methods also use machine learning algorithms to learn vector representations of the document. In this work, we explore and validate the use of contextual information for vectorizing the document and also representing the knowledge not contained in the documents itself but included in the context. The use of context has been applied in the domain of computer vision where location or date of an image can be used as a context which provides more features for automatic face recognition and person detection [1]. Similarly, in textual information, there can be an implied context in a sentence. For example, in the English language sentence, 'I cast my vote yesterday', there is contextual information indicating that elections took place yesterday. This information, which is not directly written, provides useful semantic knowledge for the IR task.

We collected the metadata of research papers published from 1992 to 2018 from the Kaggle dataset [2]. Firstly, the papers were converted into JSON format and preprocessed to eliminate stop words and non-alphabetic characters. The lemmatization of the remaining words was then carried out while preserving their position. The document descriptor for research papers was developed using three types of contexts. These contexts are defined as

1. **C1**: These are the words in the neighbourhood of a keyword in the given document. The collocational words around a keyword in a window of size $2n$ were used. These words were present in position ranging in $\pm n$ words from the keyword. The results for $n = 2$ and $n = 5$ are calculated and compared. These collocational features around keyword have also been used in other studies [3–5].
2. **C2**: These are the words in the neighbourhood of keyword in the defined corpus. Wikipedia was used for this purpose. It provides collocational words which generally co-occur with keyword but might not be co-occurring in the given document. This context provides real-world knowledge to descriptors. Similar to C1, we have a window of size $2n$. Gabrilovich and Markovitch [6] have also used Wikipedia as a knowledge repository in their work.
3. **C3**: WordNet is a freely available lexical database which contains sets of words having similar meanings defined as synsets. We find words synonymous with the keyword in the WordNet, if present, were found and then used as the third type of context.

In this work, we also design a method for computing the similarity between the document vector and the query vector using fuzzy logic. The key points addressed in this study are

- **G1**: Comparing the performance of IR system when the context of the document is also used as opposed to only keywords.

- **G2**: Comparing the performance of IR system when using different types of contexts as defined above.
- **G3**: Comparing the performance of IR system when different values of window sizes for C1 and C2 are used.
- **G4**: Comparing the performance of IR system when the fuzzy similarity metric is used for ranking the queries as opposed to cosine similarity.

The rest of the work is organized as follows: Sect. 2 discusses the literature review of context-based learning, document descriptors and fuzzy similarity algorithms. Section 3 describes the algorithms in detail, while Sect. 4 discusses the implementation. Results and their theoretical implications are listed in Sect. 5. Finally, Sect. 6 describes conclusion and future work.

2　Related Work

2.1　Context-Aware Document Descriptors

Different disciplines and domains have defined and used context in their study, yet there is an ample scope for using it for document descriptors. Baltrunas and Ricci [7] used the context, like the weather in which a user gave the rating for their context-based splitting of items. He et al. [3] used global context words in title and abstract and local context words in a window around a keyword for the context-aware citation recommendation. Stroppa et al. [8] used context-informed features for translation from Italian to English and Chinese to English, where context was defined as features surrounding target word to be translated. Lu et al. [9] used the social context of a user like the number of followers for prediction of quality of reviews. The context of user defined by the user's most recent 'n' queries was defined and used for song recommendation by Hariri et al. [10]. Chapman et al. [11] researched clinical text and used contextual information to derive whether symptoms are acute or chronic. For context-aware spelling correction, part-of-speech tag and words surrounding target word were used as features by Carlson et al. [4] and Hagenau et al. [12]. Apart from context-based learning, several studies have also attempted to automatically describe documents. Torki [13] used covariance of embedding of all the words in a document to develop a covariance-based document descriptor. Cigarrán et al. [14] used noun phrases as document descriptors using formal concept analysis for document retrieval. Garofalakis et al. [15] developed XTRACT algorithm to infer document type descriptor (DTD) for XML documents. Keyword extraction from documents is itself a well-researched problem in the NLP literature. Few of the studies are listed in this section. Kim and Kan [16] selected candidates for key phrases using regular expressions. Rose et al. [17] designed and tested rapid automatic keyword extraction (RAKE) algorithm. Forsyth [18] empirically compared nine methods of finding features from text for describing documents. Litvak and Last [19] represented semantic information of text and Web documents in a graph-based entity capturing structural document features. Ohsawa et al. [20] in their work KeyGraph extracted keywords by representing terms with high co-occurrences as a cluster. Leskovec et al. [21] summarized documents by representing semantic structure as a graph and using a subgraph to obtain the summary. A number of other studies have worked on semantic graphs and networks of textual documents [22–26].

2.2 Fuzzy Similarity

Fuzzy similarity algorithms and matrices have been used in several disciplines in recent past. Chen et al. [27] developed a novel algorithm for fingerprint matching based on a fuzzy similarity measure. Begum et al. [28] applied fuzzy similarity matching on case-based reasoning system in the psycho–physiological domain. Tolias et al. [29] introduced a new family of fuzzy similarity indices. Lee [30] proposed an algorithm for the transitive closure of a fuzzy similarity matrix. Fuzzy logic helps to create a non-binary relationship between objects. It is more representative of real-world scenarios where an object cannot be classified into a single class but is part of multiple different categories.

3 Proposed Framework

The proposed framework comprises three novel algorithms as described below:

3.1 Algorithm for Making Context-Aware Graph

Pseudocode for developing context annotated graph from keywords, context words and their frequencies is as follows.

$for\ each\ unique\ keyword\ K_i:$
  $n = new\ node(K_i)$
  $mark\ n\ as\ Type\ I\ node$
  $n.weight = occuranceFrequency(K_i)$
    $for\ each\ context\ C_j\ associated\ with\ K_i:$
      $m = newnode(C_j)$
      $mark\ m\ as\ Type\ II\ node$
      $m.weight = occuranceFreuency(C_j)$
      $m.type = ContextType(C_j)$
      $e_{ij} = createDirectedEdge(n, m)$
      $e_{ij}.weight = cooccuranceFreuency(K_i, C_j)$

$for\ each\ pair\ of\ Type\ I\ node\ n_i\ and\ n_j:$

if n_j is directly after n_i in a sentence:
$e_{ij} = createDirectedEdge(n_i, n_j)$
$e_{ij}.weight = cooccuranceFreuency(n_i, n_j)$

Here, weights of keyword nodes and context word nodes are marked by their frequencies in the research paper. The directed edges capture the structural relationship of words within a sentence. A directed edge is added between a keyword node and its context word node and between keyword nodes '*i*' and '*j*' if keyword '*j*' comes immediately after the keyword '*i*' in a sentence. Their co-occurrence frequency marks the edge weights. CAG captures the entire structure of the document, preserving inter-keywords relationship and gives them weights depending on their frequencies. The keywords and context words are lemmatized, and only one node is created for each unique word. The CAG eliminates the need to study *n*-grams, thus simplifying the computations.

Algorithm for Vectorization of Graph

for each prominent node n_i of graph:
create orthogonal dimension D_i
create a vector component $\widehat{D_i}$ for the graph
asign magnitude$|x_i| = n_i.weight$ for $\widehat{D_i}$

The second algorithm discusses the conversion of the graph into a vector so that fuzzy similarity can be calculated between documents or documents and query. We extract the most prominent nodes from the graph using the Hyperlink-Induced Topic Search (HITS) algorithm. It is described in detail in Sect. 4.5. For each prominent node, we created a new dimension of the vector and assigned the weight of the node as the magnitude of the vector in the newly created dimension.

3.2 Algorithm for Fuzzy Similarity

Let $\vec{D}_1 = x_1\widehat{K_1} + x_2\widehat{K_2} + \cdots + x_n\widehat{K_n}$ be a vectorized document
for each path P_i between K_i and K_j:

$$RelationshipWeight_{ij}+ = \left(\frac{1}{k * e^{PathLength-k}}\right)$$

Let $\vec{Q}_1 = y_1\widehat{K_1} + y_2\widehat{K_2} + \cdots + y_k\widehat{K_k}$ be a vectorized query

$$FuzzySimilarity = \frac{\sum_{i=1}^{k}\sum_{j=1}^{n} y_i x_j * RelationshipWeight_{ij}}{|\vec{D}_1||\vec{Q}_1|\sum_{i=1}^{k}\sum_{j=1}^{n} RelationshipWeight_{ij}}$$

The vectors created in above step do not have orthonormal dimensions since words in a document have many types of relationships between them. Due to this reason, traditional similarity algorithms like cosine similarity perform poorly as compared to the proposed fuzzy similarity algorithm. The fuzzy similarity computes a relationship score for each pair of dimensions found above. It assigns the score based on the

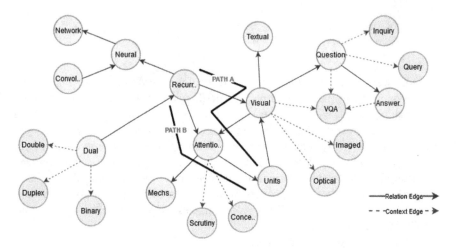

Fig. 1 Context annotated graph for sample text

inter-keyword relationship between nodes and how far apart the nodes are present in CAG, thereby taking into account the structure of the document. This score helps us to determine how strongly two keywords are related and is used to provide a fuzzy component while calculating scores of 'traditionally unrelated' dimensions. The constant 'k' is the sensitivity of the algorithm for far away nodes. We chose a value of $k = 2$ in our implementation.

Figure 1 shows how the above three algorithms were used for the creation of CAG and calculation of fuzzy similarity. The blue nodes are keywords, and red nodes are context words. The edges of the graph capture the structural information of the document. The nodes named 'recurrent' and 'units' are part of most prominent nodes in the graph, and hence, they form dimensions of the vectorized document. While calculating fuzzy similarity, there exists two paths (A and B) of path length 3 and 2, respectively, between these nodes, which is used to give a relationship weight to the dimensions.

4 Implementation

4.1 Data Collection and Preprocessing

A corpus of research papers from 1992 to 2018 was collected from the Kaggle dataset [2]. In the text, all the numeric characters, punctuations and stop words were removed. The content was tokenized into words using Natural Language Toolkit (NLTK) library in Python. Inflectional endings were also removed from the word using WordNet Lemmatizer followed by part-of-speech tagging. Thus, after preprocessing, every research paper can be represented as a list of lemmatized nouns written in the content of the paper. The structure of the paper was preserved in the process.

4.2 PDF to JSON Conversion

Science Parse [31] is an open-source JAVA and Scala-based project which uses supervised machine learning models to convert Portable Document Format (PDF) of research papers in structured JavaScript Object Notation (JSON) format.

4.3 Keyword Extraction

We calculate the TF-IDF score was calculated for each word 'w' in each paper 'p' using the following equations:

$$\text{TF}(w, p) = \frac{\text{Count}(w, p)}{\text{Count}(w, p)} \tag{1}$$

$$\text{IDF}(w) = \ln \frac{N_p}{N_w} \tag{2}$$

$$\text{TF-IDF}(w, p) = \text{TF}(w, p) * \text{IDF}(w) \tag{3}$$

Here, Count(w, p) gives the count of the word 'w' in the paper 'p'. N_p is the total number of papers, and N_w is the number of papers containing the word 'w'. This score is high for the words frequently occurring in the specific paper only, and IDF term lowers the score for words which are frequent in maximum documents. The top ten keywords with the highest TF-IDF scores for each paper were derived.

4.4 Context Extraction

We extracted three different types of contextual information [32] for the selected keywords:

- **C1**: The words around keywords in $\pm n$ window size of the keyword whose TF-IDF score is exceeding a threshold value $t = 0.02$.
- **C2**: It captures the real-world knowledge by extracting contextual information from the Wikipedia dump. We included the contextual words around $\pm n$ window size of the keywords whose TF-IDF score was above threshold.
- **C3**: It captures the language-specific features of the keyword. We include the words having similar semantics as the keyword.

4.5   Hyperlink-Induced Topic Search (HITS)

Kleinberg [33] developed HITS algorithm for extracting useful information from a graph or network entities. Popularly, this algorithm is used for getting relevant pages from the World Wide Web network. We used this algorithm for extracting most important and relevant keyword nodes and context nodes from the CAG. This algorithm works for directed graph taking into account the in links and out links of the nodes.

5   Results and Analysis

Valcarce et al. [34] studied the problem of evaluating IR metrics [35, 36] and concluded precision to be most robust and discounted cumulative gain providing the highest discriminative power. The following metrics were used to assess the result.

5.1   Average Weighted Precision (AWP)

Precision is defined as the fraction of the search results which are relevant.

$$\text{Precision} = \frac{\text{True Relevant}}{\text{True Relevant} + \text{False Relevant}} \tag{4}$$

However, as the search results were ranked according to maximum fuzzy similarity with the query, precision should be weighted more for the top-ranked results. We define precision (k) as the precision for top 'k' search results [37]. The average weighted precision, AWP(k) was calculated by taking a weighted average of precision (1) to precision (k) with the weight of precision (i) as $(k - i)/k$. This calculates precision by giving more weight to the top search results of the query.

$$\text{AWP}(k) = \sum_{i=1}^{k} \frac{k - i}{i} * \text{Precision}(k) \tag{5}$$

5.2   Discounted Cumulative Gain (DCG)

DCG [38] measures the relevance of search results while giving logarithmic weight to positions of search results. For 'k' positions, we aggregate the relevance with the

Table 1 AWP and DCG metrics using cosine similarity for ranking results

Model (using cosine similarity)	AWP	DCG
Baseline: keywords only	0.752	0.796
Keywords + C1 (WS = 2)	0.804	0.895
Keywords + C1 (WS = 5)	0.851	0.901
Keywords + C2 (WS = 2)	0.794	0.875
Keywords + C2 (WS = 5)	0.832	0.897
Keywords + C3	0.772	0.809
Keywords + C1 (WS = 5) + C2 (WS = 5) + C3	0.865	0.912

Table 2 AWP and DCG metrics using fuzzy similarity for ranking results

Model (using fuzzy similarity)	AWP	DCG
Baseline: keywords only	0.705	0.728
Keywords + C1 (WS = 2)	0.892	0.910
Keywords + C1 (WS = 5)	0.751	0.881
Keywords + C2 (WS = 2)	0.850	0.877
Keywords + C2 (WS = 5)	0.855	0.891
Keywords + C3	0.780	0.813
Keywords + C1 (WS = 2) + C2 (WS = 5) + C3	0.915	0.922

query of all search results and penalize it with logarithmic factor proportional to its position.

$$\mathrm{DCG}\,(k) = \sum_{i=1}^{k} \frac{\mathrm{relevance}_i}{\log_2(i+1)} \tag{6}$$

The relevance of search results for the queries was annotated for both metrics (0 or 1 for AWP and [0, 1] for DCG). The baseline was taken as the output of queries using only keywords from documents and cosine similarity. The AWP and DCG metrics were calculated for baseline as well as contexts with different window sizes (WS) for both cosine similarity and fuzzy similarity. The results are given in Tables 1 and 2. These are the average values of metrics which were derived using several queries (2k+).

5.3 Theoretical Significance of Results

Tables 1 and 2 both clearly show that all the descriptors built using context annotated graphs outperform trivial baseline irrespective of the type of similarity algorithm used. The context C1 is better than other contexts for both types of similarity algorithms. It performs better with a smaller window size while using fuzzy similarity and with a larger window size with cosine similarity, whereas C2 contributes more with larger window size. As C2 extracts words from Wikipedia corpus, having a large window size will include more words co-occurring with the target keyword. However, increasing window size too much may result in the addition of noise. As synsets in WordNet database do not contain many technical words occurring in scientific literature, C3 did not contribute much in the improvement of search results. For non-scientific documents, C3 might perform better. A combination of keywords and all context words (C1 with window size 2 and C2 with window size 5) gives the best results for both similarity algorithms. The significant improvement over baseline shows that existing cosine similarity or rule-based approaches of finding document similarity fail significantly in capturing inter-keyword relationships. This is the reason that fuzzy similarity outperforms existing schemes. The use of contextual information enhances the real-world knowledge of our descriptors, which is another area that traditional approaches might ignore.

The work is contributing to scientific literature in several ways:

- Definition and validation of different types of context for NLP literature.
- A novel algorithm for calculating fuzzy similarity, as opposed to traditional similarity metric.
- Usage of directed edges for capturing inter-word structural relationship in a sentence instead of n-grams.

6 Conclusion and Future Work

In this work, a document descriptor based on context-aware graph and fuzzy similarity has been presented. A novel metric for fuzzy similarity was also developed. This metric outperforms cosine similarity-based measures for information retrieval. A corpus of research papers from varied domains was collected, and the PDF format was converted to JSON. These papers were preprocessed, tokenized and lemmatized, and top ten keywords were extracted using TF-IDF scores. The three defined contexts C1, C2 and C3 were extracted and represented as nodes in the graph. The most prominent nodes of the graphs were vectorized, and similarity to the query was calculated using the fuzzy similarity metric. It has been found that document descriptors made using context-aware graph are better for representing the research papers. The descriptor with contextual information coupled with fuzzy similarity algorithm outperforms baseline of non-contextual, cosine similarity-based results. For future works, several new contexts like the context of the user from his past queries and social context from the user's profile can be included with the query. More language-specific relations like antonyms, super-subordinate relations, etc., can be used as context. The descriptor can be extended to other types of documents as well as documents in other languages.

References

1. O'Hare, N., Smeaton, A.F.: Context-aware person identification in personal photo collections. IEEE Trans. Multimed. **11**(2), 220–228 (2009)
2. https://www.kaggle.com/neelshah18/arxivdataset
3. He, Q., Pei, J., Kifer, D., Mitra, P., Giles, L.: Context aware citation recommendation. In: Proceedings of the 19th International Conference on World Wide Web, Raleigh, NC, USA, pp. 421–430. ACM (2010)
4. Carlson, A.J., Rosen, J., Roth, D.: Scaling up context sensitive text correction. In: IAAI, pp. 45–50 (2001)
5. Golding, A.R., Schabes, Y.: Combining trigram-based and feature-based methods for context sensitive spelling correction. In: Proceedings of the 34th Annual Meeting on Association for Computational Linguistics, Santa Cruz, CA, pp. 71–78. Association for Computational Linguistics (1996)
6. Gabrilovich, E., Markovitch, S.: Overcoming the brittleness bottleneck using Wikipedia: enhancing text categorization with encyclopedic knowledge. In: AAAI, Boston, MA, vol. 6, pp. 1301–1306 (2006)
7. Baltrunas, L., Ricci, F.: Context-based splitting of item ratings in collaborative filtering. In: Proceedings of the Third ACM Conference on Recommender Systems, New York, USA, pp. 245–248. ACM (2009)
8. Stroppa, N., Van den Bosch, A., Way, A.: Exploiting source similarity for SMT using context-informed features (2007)
9. Lu, Y., Tsaparas, P., Ntoulas, A., Polanyi, L.: Exploiting social context for review quality prediction. In: Proceedings of the 19th International Conference on World Wide Web, Raleigh, NC, USA, pp. 691–700. ACM (2010)
10. Hariri, N., Mobasher, B., Burke, R.: Context-aware music recommendation based on latent topic sequential patterns. In: Proceedings of the Sixth ACM Conference on Recommender Systems, Dublin, Ireland, pp. 131–138. ACM (2012)
11. Chapman, W.W., Chu, D., Dowling, J.N.: Context: an algorithm for identifying contextual features from clinical text. In: Proceedings of the Workshop on Bio NLP 2007: Biological, Translational, and Clinical Language Processing, pp. 81–88. Association for Computational Linguistics (2007)
12. Hagenau, M., Liebmann, M., Hedwig, M., Neumann, D.: Automated news reading: stock price prediction based on financial news using context-specific features. In: 2012 45th Hawaii International Conference on System Sciences, Maui, HI, USA, pp. 1040–1049. IEEE (2012)
13. Torki, M.: A document descriptor using covariance of word vectors. In: Proceedings of the 56th Annual Meeting of the Association for Computational Linguistics, vol. 2. Short Papers, pp. 527–532 (2018)
14. Cigarrán, J.M., Peñas, A., Gonzalo, J., Verdejo, F.: Automatic selection of noun phrases as document descriptors in an FCA-based information retrieval system. In: International Conference on Formal Concept Analysis, Lens, France, pp. 49–63. Springer (2005)
15. Garofalakis, M., Gionis, A., Rastogi, R., Seshadri, S., Shim, K.: Xtract: learning document type descriptors from xml document collections. Data Min. Knowl. Discov. **7**(1), 23–56 (2003)
16. Kim, S.N., Kan, M.-Y.: Re-examining automatic keyphrase extraction approaches in scientific articles. In: Proceedings of the Workshop on Multiword Expressions: Identification, Interpretation, Disambiguation and Applications, pp. 9–16. Association for Computational Linguistics (2009)
17. Rose, S., Engel, D., Cramer, N., Cowley, W.: Automatic keyword extraction from individual documents. Text Mining: Applications and Theory, pp. 1–20. Wiley, Chichester, UK (2010)
18. Forsyth, R.S.: Deriving document descriptors from data. In: Emotion, Creativity and Art, Perm, Russia, pp. 1–28 (1997)
19. Litvak, M., Last, M.: Graph-based keyword extraction for single-document summarization. In: Proceedings of the Workshop on Multi-source Multilingual Information Extraction and Summarization, pp. 17–24. Association for Computational Linguistics (2008)

20. Ohsawa, Y., Benson, N.E., Yachida, M.: KeyGraph: automatic indexing by co-occurrence graph based on building construction metaphor. In: Proceedings IEEE International Forum on Research and Technology Advances in Digital Libraries-ADL'98, pp. 12–18. IEEE (1998)
21. Leskovec, J., Grobelnik, M., Milic-Frayling, N.: Learning sub-structures of document semantic graphs for document summarization. In: Link KDD Workshop, pp. 133–138 (2004)
22. Moawad, I.F., Aref, M.: Semantic graph reduction approach for abstractive text summarization. In: 2012 Seventh International Conference on Computer Engineering & Systems (ICCES), pp. 132–138. IEEE (2012)
23. Sowa, J.F.: Semantic Networks (1987)
24. Yih, S.W.-t., Chang, M.-W., He, X., Gao, J.: Semantic parsing via staged query graph generation: question answering with knowledge base. In: Proceedings of the 53rd Annual Meeting of the Association for Computational Linguistics and the 7th International Joint Conference on Natural Language Processing, vol. 1. Long Papers, pp. 1321–1331. Association for Computational Linguistics (2015)
25. Alvarez, M.A., Lim, S.J.: A graph modeling of semantic similarity between words. In: International Conference on Semantic Computing (ICSC 2007), pp. 355–362. IEEE (2007)
26. Tsatsaronis, G., Varlamis, I., Nørvåg, K.: Semanticrank: ranking keywords and sentences using semantic graphs. In: Proceedings of the 23rd International Conference on Computational Linguistics, pp. 1074–1082. Association for Computational Linguistics (2010)
27. Chen, X.J., Tian, J., Yang, X., Tian, J., et al.: A new algorithm for distorted fingerprints matching based on normalized fuzzy similarity measure. IEEE Trans. Image Process. **15**, 767–776 (2006)
28. Begum, S., Ahmed, M.U., Funk, P., Xiong, N., Von Schéele, B.: A case-based decision support system for individual stress diagnosis using fuzzy similarity matching. Comput. Intell. **25**(3), 180–195 (2009)
29. Tolias, Y.A., Panas, S.M., Tsoukalas, L.H.: Generalized fuzzy indices for similarity matching. Fuzzy Sets Syst. **120**(2), 255–270 (2001)
30. Lee, H.-S.: An optimal algorithm for computing the max–min transitive closure of a fuzzy similarity matrix. Fuzzy Sets Syst. **123**(1), 129–136 (2001)
31. https://github.com/allenai/science-parse
32. Kumar, A., Garg, G.: Systematic literature review on context-based sentiment analysis in social multimedia. Multimed. Tools Appl. 1–32 (2019)
33. Kleinberg, J.M.: Authoritative sources in a hyperlinked environment. J. ACM (JACM) **46**(5), 604–632 (1999)
34. Valcarce, D., Bellogín, A., Parapar, J., Castells, P.: On the robustness and discriminative power of information retrieval metrics for top-n recommendation. In: Proceedings of the 12th ACM Conference on Recommender Systems, pp. 260–268. ACM (2018)
35. Bhatia, M.P.S., Kumar, A.: Paradigm shifts: from pre-web information systems to recent web-based contextual information retrieval. Webology **7**(1) (2010)
36. Bhatia, M.P.S., Khalid, A.K.: Contextual proximity based term-weighting for improved web information retrieval. In: International Conference on Knowledge Science, Engineering and Management, pp. 267–278. Springer, Berlin, Heidelberg (2007)

37. Tunkelang, D.: Evaluating Search: Measure It (2017)
38. J'arvelin, K., Kekäläinen, J.: Cumulated gain-based evaluation of IR techniques. ACM Trans. Inf. Syst. (TOIS) **20**(4), 422–446 (2002)

An Efficient Image Compression Technique for Big Data Based on Deep Convolutional Autoencoders

Shefali Arora and M. P. S. Bhatia

Abstract Big organizations and enterprises store a large amount of data and images comprise a huge portion of it. Thus, compression of image data becomes important to remove redundancy as well as to make file size manageable. Autoencoders can be used for image compression in such scenarios. Deep autoencoders train on a large set of images to compress them and represent using fewer codes. In this paper, we build deep convolutional autoencoders and use them to reduce dimensionality of images present in the CIFAR-10 dataset and MNIST dataset. We train the autoencoder using commonly used loss functions and quantize images to achieve a superior identification performance. The compressed image vectors are later decompressed to obtain a near approximation of the original images.

Keywords Compression · Autoencoders · Dimensionality reduction · Big data

1 Introduction

Big data refers to a large amount of data which maybe in structured or unstructured format. This data is complex and is in huge volume, so it becomes difficult to process as compared to traditional processing techniques. The main challenges faced in processing big data are [1]:

- Storage issues
- Transport issues
- Management issues
- Information sharing
- Relevant data searching.

S. Arora (✉) · M. P. S. Bhatia
Division of Computer Engineering, Netaji Subhas Institute of Technology, Delhi, India
e-mail: arorashef@gmail.com

M. P. S. Bhatia
e-mail: bhatia.mps@gmail.com

© Springer Nature Singapore Pte Ltd. 2020
P. K. Singh et al. (eds.), *Proceedings of First International Conference on Computing, Communications, and Cyber-Security (IC4S 2019)*, Lecture Notes in Networks and Systems 121, https://doi.org/10.1007/978-981-15-3369-3_54

Compression is a technique which helps in reduction of data to save space and transmission time. Compression helps to reduce file size, save storage space, increase transfer speed and allow real-time transfer of data [2].

In many cases, group of images stored in a file system are almost identical. For example, images in a medical database, biometric templates, etc. These images have similar pixel intensities. It is useful to reduce their size while storing or analysing them using inter-image encoding techniques. It is important to extract relevant features from images when a large set of images is analysed collectively, while extraction, quantification of relevant information have been an important area of research in image processing.

This challenge is also applicable to applications involving the use of biometric identification. With the enrolment of a huge number of users, it is very important to manage a large amount of data efficiently. Conventional image compression methods were designed so as to maintain the image quality, size and resolution. However, as the size of image data increases, it becomes impractical to store and compute it on a single resource. A large bottleneck is created by using many analytics tools involving distributed storage, transmission and storage of a huge amount of data. Furthermore, image quality and resolution can vary to a vast extent when databases are mined for the task of image retrieval. In this paper, we improvise a deep convolutional autoencoder framework to compress images, which would help to provide efficient and scalable processing of images without any loss in performance.

The use of autoencoders has been used in dimensionality reduction of images, generative model learning and compact representation. They can be effective in extracting compressed codes from images with minimum amount of loss. Thus, they help to achieve better performance as compared to conventional image compression standards including JPEG and JPEG2000 [3].

Another advantage of deep learning is that this approach is quicker when it comes to diverse image formats. Thus, deep learning-based image compression methods are more efficient and generic as compared to traditional codec techniques. Deep autoencoders are one of the techniques that encode inter-image pixels [4]. In a set of similar images, significant amount of inter-image redundancy is present. The similar images have similar pixel intensities, comparable histograms and edge distributions. The correlation between them can be calculated using product-moment correlation coefficient or Pearson's. The correlation between these images is high, which is due to the presence of inter-image redundancy. This redundancy is so-called set redundancy. This redundancy can be used to improve image compression ratio.

We make use of deep convolutional autoencoders to compress images present in CIFAR-10 and MNIST datasets.

The rest of this paper is organized as follows. Section 2 presents the work done in the field of deep learning and image compression. Section 3 presents the proposed autoencoder architecture which includes the design of the CAE network architecture and quantization. Section 4 summarizes the experimental results and compares the performance of our technique with traditional compression methods. Section 5 concludes the paper.

2 Related Work

Various image compression techniques [5] have been used by researchers to reduce image file size in frameworks using big data. Some of the popular techniques are shown in Fig. 1.

Huffman coding is a lossless compression algorithm in which pixels are assigned codes based on their frequency of occurrence. Thus, pixels occurring more frequently will have shorter codes. The codes are stored on a binary tree. An experiment was performed to compress a grey-scale image using Huffman coding algorithm. The result gave a compression ratio of 0.8456 [6]. Authors [7] presented the use of autoencoders and compared the use with PCA and JPEG compression techniques. Images were fed to a nonlinear encoding layer. The encoding layer was compressed, and the input pixels were converted into a compressed feature vector. After the encoding was complete, the compressed vector was fed into a decoder which reversed the process of encoding and gave reconstructed input images. The difference between the reconstruction and input images was regarded as error to further minimize the difference. Further, k-means clustering was used to find quantization values to encode the activation values. On comparison with JPEG and PCA techniques, deep autoencoders surpassed JPEG in high compression and reduced quality scheme. Also, deep autoencoder showed better nonlinear representation of the input image than that of PCA, and hence, deep autoencoder had better reconstruction quality.

Authors observed that deep autoencoder was able to find statistical regularities in a specific domain of images which was not possible by JPEG [8]. Authors [9] also demonstrated the use of autoencoder to reduce the dimensionality of data. This was done using Restricted Boltzmann machines (RBMs) to pre-train the neural network, and each layer consisted of feature detectors. Later, the RBMs were unrolled to get all the layers of neural network, and backpropagation was done for fine-tuning of weights. Efficient hierarchial autoencoders have been used by authors [10] to predict human motion.

Deep autoencoder is efficient in representing and generalizing nonlinear structures in data. Authors [11] have demonstrated the use of large-scale unsupervised learning algorithms for building high-level features which includes the use of stack of RBMs

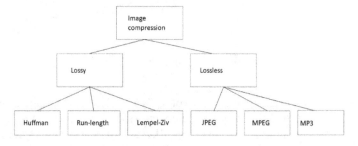

Fig. 1 Image compression techniques

to construct autoencoders. They used unlabelled data to create high-level feature detectors. They trained a nine-layer autoencoder, with sparse layers and applied local contrast normalization on a large dataset of images. They trained the network using asynchronous stochastic gradient descent (SGD) on a cluster of 1000 machines (16,000 cores) for three days. In a nutshell, they show that deep autoencoders perform well as feature detectors in case of images.

An autoencoder is a neural network with encoder $g_e(\theta)$ which reconstructs an image X to $\hat{X}$ as the new representation Y. When autoencoders are used for image compression, the representation is also quantized as:

$$\hat{Y} = Q(Y) \tag{1}$$

Autoencoder comprises convolutional layers and nonlinear operators [12]. Given image matrix $n = h \times w \times n_c$, $Y \in R_n$. The encoder g_e followed by normalization, parameterized by α_e is denoted as $g_e(\cdot; \theta, \alpha_e)$. Similarly, the decoder followed by normalization, parameterized by α_d is denoted as $g_d(\cdot; \alpha_d, \phi)$.

$$\min_{\theta, \alpha_e, \alpha_d, \sigma} E\left[||X - g_d(Q(g_e(X; \theta, \alpha_e)); \alpha_d, \phi)||_F^2\right] + \gamma \sum_{i=1}^{n_c} H_i \tag{2}$$

$\hat{p}_i$ is the probability of the ith quantized feature of the feature map. θ and ϕ can be calculated using gradient-based methods. The expectation E is approximated over a set of training images. The quantization step size can be explicitly learned for each feature map of Y.

$$H_i = -\frac{1}{n} \sum_{j=1}^{n} \log_2(\hat{p}(\hat{y}_{ij})), \quad \gamma \in R_+^* \tag{3}$$

Due to quantization, the implicit function is minimized in step sizes δ_i where $i = 1$ to n_c so as to make it an explicit function of δ_i for $i = 1$ to n_c. Removing normalizations and finding continuous uniform distribution of support $\{-0.5\delta_i, 0.5\delta_i\}$, we get

$$\min_{\theta, \alpha_e, \alpha_d, \sigma} E[||X - g_d((g_e(X; \theta + E; \phi)))||_F^2] + \gamma \sum_{i=1}^{n_c} h_i \tag{4}$$

$$h_i = -\log_2(\delta_i) - \frac{1}{n} \sum_{j=1}^{n} \log_2(\hat{p}_i(y_{ij} + \epsilon_{ij})) \tag{5}$$

The ith matrix of E contains n realizations of $\{\epsilon_{ij}\}_{j=1,2,...,n}$ of E_i, E_i being a continuous random variable of probability density function. However, we cannot learn δ_i for $i = 1$ to n_c as the minimized function is not differential with respect to δ_i for $i = 1$ to n_c. This is followed by change in variable $E_i = \delta_i T$ where T is a random variable, given continuous uniform distribution of support $[-0.5, 0.5]$.

$$\min_{\theta, \alpha_e, \alpha_d, \sigma} E[||X - g_d((g_e(X; \theta + \Delta \odot T; \phi)))||_F^2] + \gamma \sum_{i=1}^{n_c} h_i \quad (6)$$

$$h_i = -\log_2(\delta_i) - \frac{1}{n} \sum_{j=1}^{n} \log_2(\widehat{p_i}(y_{ij} + \delta_i T_{ij})) \quad (7)$$

The ith matrix of T contains n realizations $\{T_{ij}\}$ $j = 1, \ldots, n$ of T. All the coefficients in the ith matrix of $\Delta \in R_n$ are equal to δ_i.

Balle et al. [13] gave an image compression algorithm consisting of a nonlinear transform and a uniform quantizer. The nonlinear transform consisted of convolutional neural networks, consisting of a convolutional layer, downsampling and nonlinearity. Unlike the traditional nonlinear activation functions, this approach used a generalized divisive normalization transform. The optimization was done using a distortion function, $D + \lambda R$, assuming uniform quantization. The authors approximate quantization noise with uniform noise; this is associated with smoothing of the discrete probability mass function of the obtained coefficients with a box filter. One of the limitations of this model is that it requires retraining of the model for every value of λ.

Toderici et al. [14] make the encoding and decoding process progressive, i.e. after first encoding and decoding, a residue is computed with respect to the original image. Later, the residue is further encoded and decoded, to find difference in residue with respect to the previous patches. This scheme is applied on 32×32 pixel patches. A binarizer is used, and the model is based on recurrent neural networks. Several architectures derived from the well-known LSTM networks were used to provide the best results [15].

In the work [13], authors propose the main difference in the way of dealing with quantization. The transformations consist of single linear layer and contrast gain control, whereas in the work proposed by authors [14], the network tries to lessen the amount of distortion. This takes more time to perform the compression.

Gregor et al. [16] explored image compression using variational autoencoders with recurrent encoders and decoders for compression of small images. The lower bound of log likelihood $p(y|x)$ is minimized in this process, which acts as the encoder, and $q(x|y)$ which represents the decoder.

$$-E_{p(y|x)} = \left[\log \frac{q(y)q(x|y)}{p(y|x)}\right] \quad (8)$$

While Gregor et al. [16] used a Gaussian distribution for the encoder, Balle et al. assumed it to be uniform, i.e. $p(y|x) = f(x) + u$. The objective function on assuming likelihood with fixed variance $p(y|x) = N(x|g(y), \sigma^2 I)$ is

$$E_u\left[-\log q(f(x)+u) + \frac{1}{2\sigma^2}||x - g(f(x)+u||^2\right] + C \tag{9}$$

Here, C represents the encoder's negative entropy and normalization constant of the Gaussian likelihood.

2.1 Contribution

- We simplify the architecture of the convolutional autoencoder used and compare the impact of loss functions on the quality of images.
- We propose the use of quantized ReLu function, which performs round quantization on images.
- We implement the use of such a framework on different datasets, proving that autoencoders help to achieve with negligible loss in image quality.

3 Methodology

3.1 Proposed Framework

The process of compression on Hadoop framework is shown in Fig. 2:

1. Store the images folder on a local file system.
2. Use image compression algorithm to compress the images.
3. Using Hadoop shell commands; transfer the compressed images to Hadoop file system (HDFS).
4. Get the output files from the Hadoop file system to local file system to read conveniently.

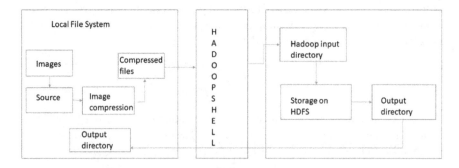

Fig. 2 Image compression architecture in big data framework

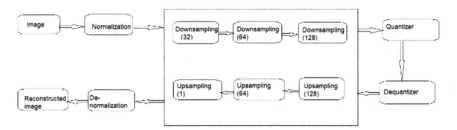

Fig. 3 Proposed architecture

To obtain the compressed representation of the input images, convolutional autoencoders are used. Thus, downsampling/upsampling operations are required during the encoding and decoding process. It is observed that a good resolution of images is obtained by convolving images and then upsampling them, rather than doing consecutive downsampling operations.

We use 3×3 filters on images. N_i denotes the number of filters used. By using max pooling operation, we can obtain downsampled featured maps. This is followed by convolving filters and upsampling of the feature maps obtained.

Due to its non-differentiable property, quantizer cannot be integrated into the convolutional autoencoder's optimization process. Thus, smooth approximations were proposed for the same. In our method, we apply round-based quantization using quantized ReLu activation function. The proposed autoencoder is shown in Fig. 3.

As the pre-processing steps before the CAE design, the images are normalized by mapping of values between 0 and 1. The convolutional autoencoder can be considered as an encoder $y = g_e(x)$ and a synthesis transform with the decoder function, $z = g_d(y)$, where x represents the original image, y represents the reconstructed image, and z represents the compressed data. Optimized parameters are θ and φ. Downsampling and upsampling operations are required to encode and decode an image using this framework. A good resolution of the image is obtained after the process of upsampling. The number of filters used in each step is shown in the architecture. The loss functions used for the convolutional autoencoder are mean squared error and contrastive loss function, with margin set as 0.01.

Along with mean squared error loss, we also explore the use of contrastive loss in image compression.

Contrastive loss is used to learn discriminant features from images in tasks involving images. When a pair of images is given as input to a model, y equals 1 if the two images are similar and 0 otherwise. The loss function for a single pair is

$$yd^2 + (1 - y)\max(\text{margin} - d, 0)^2 \tag{10}$$

where d is the Euclidean distance between the two image features (suppose their features are f1 and f2): $d = \|\text{f1} - \text{f2}\|_2$. The margin term is used to "tighten" the constraint.

In this work, Adam is used to optimize the CAE model with a batch size of 32 and number of epochs set to 20. The learning rate was kept at a fixed value of 0.001, and the momentum was set as 0.7.

3.2 Quantization

ReLU is a popular activation function used in feedforward neural networks with the representation:

$$\sigma(x) = \max(0, x) \tag{11}$$

Each ReLu unit connects to units in the next layer. Other versions of ReLu like leaky ReLu and PReLu can be used by increasing the number of units and weights by a constant factor [17]. If the number of distinct weight values is finite, it is represented as λ ($\lambda \in Z+$ and $\lambda \geq 2$), for both linear quantization and nonlinear quantization.

For linear quantization, as there is no loss in generality, values can be given as $\{-1, 1\lambda, 2\lambda, ..., \lambda - 1\lambda\}$, with uniform spaces in these values. For nonlinear quantization, these values are not constrained to any specific values. Only $\log(\lambda)_2$ bits are needed to encode the index, i.e. the bit-width is $\log(\lambda)$.

The quantization of a neural network can be done stochastically as

$$x^b = \{+1 \text{ with probability } p, = \sigma(x),$$
$$- 1 \text{ with probability } 1 - p\} \tag{12}$$

where σ is hard sigmoid function defined as

$$\sigma(x) = \text{clip}\left(\frac{x+1}{2}, 0, 1\right) = \max\left(0, \min\left(1, \frac{x+1}{2}\right)\right) \tag{13}$$

Given the number of bits per pixel as λ, the quantization is performed as

$$x = \text{round\_through}\left(\frac{\sigma(x) * 2^\lambda}{2^\lambda} - 1, 0, 1 - \frac{1}{2^\lambda - 1}\right) \tag{14}$$

where round_through is defined as the round of the function value in parenthesis.

4 Results

4.1 Experiments and Datasets Used

The experiments have been performed on CloudxLab, which provides a framework for storing and accessing files over a Hadoop cluster.

We apply the image compression technique using deep convolutional autoencoders on the following dataset:

We use CIFAR10 database consisting of 60,000 32 × 32 colour images in ten classes, with 6000 images per class. There are 50,000 training images and 10,000 test images. We also repeat the results on MNIST dataset, which is a database of handwritten datasets, with 60,000 training examples and 10,000 testing examples. In order to evaluate the efficiency of the proposed autoencoder, the rate is measured in terms of bits per pixel (bpp), and the quality of the reconstructed images is measured in terms of PSNR [18], which stands for peak signal-to-noise ratio.

4.2 Analysis

We compare the results achieve don the two datasets based on the following parameters:

We compare the results achieved on the datasets using our proposed convolutional autoencoder with results achieved using JPEG and JPEG2000 image compression standards.

We compare the use of two different loss prediction techniques for images, i.e. root-mean-square error loss and contrastive loss.

It is observed that the efficiency of convolutional autoencoder is better as compared to the results achieved using JPEG and JPEG2000 standards. This is because after training and optimizing the network, the PSNR metric is better for reconstructed images using CAE. Examples of reconstructed patches and quantized images using convolutional autoencoder are shown in Fig. 4.

The comparison of loss prediction techniques has been done on the CIFAR-10 database, the results of which show mean squared error loss to be a better measure as compared to contrastive loss for this task. This is shown in Table 1.

Fig. 4 CIFAR-10 images: **a** images after compression; **b** quantized images

Table 1 Comparing loss values for CIFAR-10 dataset

Loss	Time per step (ms)	Value
Mean squared error	13	0.0023
Contrastive loss	12	0.2874

Table 2 PSNR for CIFAR-10 dataset

Bits per pixel	PSNR
0.5	31.728
1	38.359
2	40.864

Table 3 PSNR for MNIST dataset

Bits per pixel	PSNR
0.5	240.512
1	249.149
2	252.100

Corresponding to the number of bits per pixel, the PSNR metric observed for CIFAR-10 and MNIST datasets has been tabulated in Tables 2 and 3.

The comparison of results achieved using proposed CAE has been done with the results achieved using JPEG and JPEG2000 standards has been visualized in Figs. 5

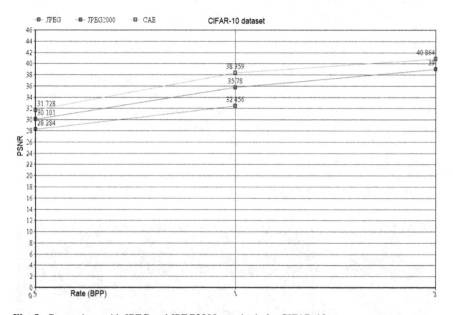

Fig. 5 Comparison with JPEG and JPEG2000 standards for CIFAR-10

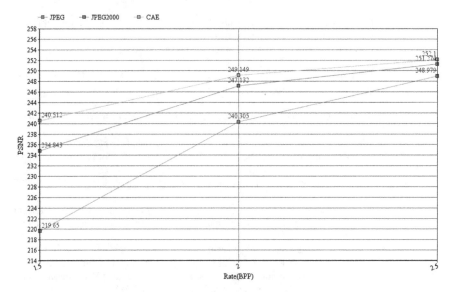

Fig. 6 Comparison with JPEG and JPEG2000 standards for MNIST

and 6 for CIFAR-10 and MNIST, respectively.

5 Conclusion and Future Work

From the experiment conducted, it is concluded that image compression is important in big data frameworks, when storing a large amount of data. The use of convolutional autoencoders is better as compared to standard image compression algorithms and can be applied in big data frameworks for reducing redundancy and improving the image storage capacity. Convolutional autoencoders with many hidden downsampling and upsampling layers are used to extract features from the image, which give a better representation and also minimal loss in performance.

In the future, we would like to work on variational autoencoders and investigate the use of autoencoders on more real-world datasets.

References

1. Bonde, P., Barahate, S.: Data compression techniques for big data. Int. J. Emerg. Technol. Innov. Res. **4**, 193–198 (2017)
2. Angelov, P., Manolopoulos, Y., Papadopoulos, A.: Special issue of Big Data Research Journal on "Big data and neural networks". Big Data Res. **11**, 3–4 (2018)
3. Oh, T., Besar, R.: Medical image compression using JPEG-2000 and JPEG: a comparison study. J. Mech. Med. Biol. **02**, 313–328 (2002)

4. Bhatt, G., Jha, P., Raman, B.: Representation learning using step-based deep multi-modal autoencoders. Pattern Recogn. **95**, 12–23 (2019)
5. Babu, T., Murty, M., Subrahmanya, S.: Compression Schemes for Mining Large Datasets, 1st edn, p. 197. Springer, London (2016)
6. Ebrahami, T., Santa Cruz, D., Christopoulos, D., Askelof, J., Larrson, M.: JPEG2000 still image coding versus other standards. In: SPIE International Symposium, San Diego, USA, pp. 1–9 (2000)
7. Atreya, A., O'Shea, D.: Novel Lossy Compression Algorithms with Stacked Autoencoders, pp. 1–9. Stanford University (2009)
8. O'Shea, T., Karra, K., Clancy, T.: Learning to communicate: channel auto-encoders, domain specific regularizers, and attention. In: 2016 IEEE International Symposium on Signal Processing and Information Technology (ISSPIT), Limassol, pp. 223–228 (2019)
9. Hinton, G.: Reducing the dimensionality of data with neural networks. Science **313**, 504–507 (2006)
10. Li, Y., Wang, Z., Yang, X., Wang, M., Poiana, S., Chaudhry, E., Zhang, J.: Efficient convolutional hierarchical autoencoder for human motion prediction. Vis. Comput. **35**, 1143–1156 (2019)
11. Ng, A., Ng, Q.: Building high-level features using large scale unsupervised learning. In: 29th International Conference on Machine Learning, Edinburgh, Scotland, UK, pp. 2–5 (2009)
12. Wen, T., Zhang, Z.: Deep convolution neural network and autoencoders-based unsupervised feature learning of EEG signals. IEEE Access **6**, 25399–25410 (2018)
13. Balle, J., Laparra, V., Simoncelli, E.: End to-end optimized image compression. In: International Conference on Learning Representations (ICLR), pp. 1–27 (2017)
14. Toderici, G., O'Malley, S.M., Hwang, S.J.: Variable rate image compression with recurrent neural networks. arXiv:1511.06085, pp. 1–12 (2015)
15. Siddeq, M., Rodrigues, M.: A novel high-frequency encoding algorithm for image compression. EURASIP J. Adv. Signal Process. **26**, 1–17 (2017)
16. Gregor, K., Danihelka, I., Graves, A., Wierstra, D..: Towards conceptual compression: arXiv: 1601.06759, pp. 1–14 (2016)
17. Hubara, I., Courbariaux, M., Soudry, D., El-Yaniv, R., Bengio, Y.: Quantized neural networks: training neural networks with low precision weights and activations. J. Mach. Learn. Res. **18**, 1–30 (2018)
18. Lee, J.: PSNR analysis of ultrasound images for follow-up of hepatocellular carcinoma. J. Korean Soc. Radiol. **9**, 263–267 (2015)

Security and Privacy Issues

Designing a Transformational Model for Decentralization of Electronic Health Record Using Blockchain

Monga Suhasini and Dilbag Singh

Abstract Electronic Health Records management in healthcare system is one of the perplexing systems which should be secured, decentralized, and authentic, and any minute error in these systems can be all around expensive both monetarily and socially. The data residing in existing EHR systems is centralized in healthcare institutions or hospitals in databases on servers inside the institutions or hospitals. These systems may suffer from different issues like no distributed access of EHRs, interoperability issues like handling different formats of data and their exchange, security and recovery issues like any loophole in the system can lead to cyber-attacks. The blockchain innovation empowers the usage of profoundly secure and protection saving decentralized frameworks where exchanges are not under the influence of any outsider associations. Utilizing the blockchain innovation information put in a fixed compartment called block conveyed over the system in an undeniable and mutable way. Data security and protection are improved by the blockchain innovation in which information are scrambled and disseminated over the whole system. This paper provides an insight into the existing health record management systems and their challenges. Also this paper introduces blockchain technology, its architecture, and impacts in the healthcare industry. In addition, a model for the generation of the unique identity of the patient as Unique Patient Identifier (UPID) is proposed to have the single patient identification record for universal access using blockchain technology.

Keywords Blockchain · Healthcare · Security · Decentralized access · EHRs

M. Suhasini (✉) · D. Singh
Department of Computer Science and Applications, Chaudhary Devi Lal University,
Sirsa, Haryana, India
e-mail: suhasini.monga@gmail.com

D. Singh
e-mail: dscdlu7@gmail.com

© Springer Nature Singapore Pte Ltd. 2020   753
P. K. Singh et al. (eds.), *Proceedings of First International Conference on Computing,
Communications, and Cyber-Security (IC4S 2019)*, Lecture Notes in Networks
and Systems 121, https://doi.org/10.1007/978-981-15-3369-3_55

1 Introduction

In current scenario and in future data and correspondence advancements for health care will be universal in the everyday lives of patients. An innovation as a decentralized and secure digital technology, i.e. blockchain is now laying its impact in other fields than financial sector [1]. The combination of electronic procedures with versatile and implanted gadgets can possibly improve access, effectiveness, and nature of customized care. While the usage of these digital physical frameworks can significantly profit society all in all, the security and protection of every patient must be guaranteed [2]. Although advancements in digital technologies and storage capabilities healthcare industry has transitioned from paper-based health records to Electronic Medical Records (EMRs) and Electronic Health Records (EHRs), yet patients have to maintain their records in hard copies. There is no provision for maintaining a single universal copy of health record for patient. There is only centralized access of data internal to the particular hospital or health institution. Often used interchangeably with the similar term Electronic Medical Record systems (EMR), Electronic Health Records refer to a single medical practice's methodical collection of patient health information stored in digital format. With these EHRs, healthcare providers are now able to more easily share and access their medical records through network-connected, enterprise-wide record management platforms on a centralized server in a hospital or an institution. The transition to Electronic Medical Records (EMRs) from paper-based health records has introduced various benefits to healthcare industry yet EMRs had their own challenges and issues.

Centralized EHRs have overcome various issues of EMRs like voluminous data handling digitally using technologies like Big Data and Cloud, but the security and decentralized access of crucial health records stored over a distributed framework is still a challenge for healthcare IT. Healthcare system needs an advancement of major structure changes for Electronic Health Records (EHRs). There is a basic requirement for such advancement, as transition from centralized access and storage to decentralized and secure access of EHRs anywhere and anytime. Centralized EHRs were never intended to oversee multi-institutional, lifetime medicinal records [3]. A public blockchain network is totally open and anyone will be part of and participate within the network. Bitcoin is one among the most important public blockchain networks in production these days, whereas private/permissioned blockchains use blockchain-based platforms running on personal Cloud infrastructure. A non-public blockchain network needs to be authenticated by the initiator or the participants involved in the network having admin rights.

1.1 Contributions

Based on the study and analysis, following are the major contributions of this paper:

- Identification of issues and challenges in centralized EHRs and an overview of blockchain technology through a comprehensive literature review of existing systems. Also the role of blockchain in healthcare and its applications is presented.
- Proposing a decentralized model for Universal Patient Identifier (UPI) using blockchain technology as a solution to address the issues and challenges in centralized EHRs.
- Comparative analysis of centralized EHR and blockchain-based EHR.

1.2 Issues and Challenges in Centralized EHRs

Access of health records: The biggest challenge in current EHRs is the data that is saved on the centralized database in the healthcare institution or hospital which is not accessible outside. Some institutions or hospitals are using advanced EHRs in which the data is stored on a Cloud but for that also no provision of universal copy of the digital health record is maintained.

Security and confidentiality: Healthcare industry faces various security breaches and loss of confidentiality of critical health information. Various cyber-attacks, data theft, and hacking incidents can lead to disastrous situations.

Interoperability: Sharing and exchange of heterogeneous data in different formats are difficult in present systems. Nationwide interoperability is a challenge for centralized systems.

Single point of failure: As the data is being stored and accessed from a centralized server means having a single point of failure can cause the entire system to fail.

Paper-based medical documentation: Since there are lot of issues in recovery, transparency, and efficiency in centralized systems, maintenance of medical records and patient's history still relied on paper or hard copies.

Integrity and availability: Data integrity is another challenge as there is duplication of a single patient record in various healthcare institutions and hospitals. Because of load on the single centralized server, lack of availability of up-to-date information poses difficult situations in patient care.

1.3 Overview of Blockchain Technology

Blockchain may be considered as a distributed ledger in which the transactions that are in the commit mode are added to form a list of blocks. Each transaction is referred as block that is added in a chain-like structure forming a blockchain. Blockchain is enabled by integration of many core technologies like cryptographic hash, digital signature, and distributed consensus mechanism [3]. A list of blocks is created that is sequential. Every block in the list is referring to its parent block using the pointer as hash key. Child blocks are also saved in the blockchain using forward-referencing hashed keys. A genesis block is the first block of the blockchain which has no parent block. Every block consists of block header and body comprising of data (transactions), hash key, timestamp, parent block key, etc. A combination of secret and public key is held by every participant in the network. The block information or data is signed using the secret key. Every user within the network can access these digital signed blocks that are distributed throughout the network. There are two phases of digital signature process: the signing phase and the verification phase. A process in which every node in the blockchain takes part in the validation of block information is called consensus determination. Every node can participate in consensus process in a public blockchain, but in a consortium blockchain, only few special nodes can validate the blocks. In a private blockchain, the validation process for consensus determination can be done by a centralized node only [3].

Intranet term could be the term used for private blockchain and Internet could be the term used for public blockchain. Each user having distinctive identity would be categorized as permissioned blockchain user, whereas anyone can participate in a permissionless blockchain. Users in permissionless blockchain do not need permission from any decentralized authority or central administrator to leave the network [4]. There are few important features of blockchain such as collaboration, flexibility resilience, and distributed verification [5]. Blockchain-driven information is replicated on multiple computers across several geographic areas being accessed by different users. In today's technological world, a trivial and remarkable invention after internet is blockchain technology [6]. Blockchain facilitates trust and validates identity while not intermediate third parties, enabling a web useful. TCP/IP has been for the exchange of knowledge, and blockchain is for the exchange useful. The experts declared it well, blockchain "the trust machine" [7].

2 Literature Review

Knirsch et al. (2019) present the center ideas of blockchain innovation and explore a genuine world use case from the vitality area, where clients exchange parts of their photovoltaic power plant through a blockchain. This does include blockchain innovation, yet additionally requires client communication. Along these lines, a completely custom, private, and also permissioned blockchain is actualized without any preparation. Author assesses the requirement for blockchain innovation inside this utilization case, just as the ideal properties of the framework [1].

Brogan et al. (2018) expound the way electronic ledger in a distributed manner which can assume a key job in progressing Electronic Health Records, by guaranteeing credibility and uprightness of information produced by wearable gadgets. Additionally exhibited that the confirmed informing module can be utilized to safely share, store and recover encoded movement information utilizing sealed appropriated record [2].

Zheng et al. (2018) give an outline of blockchain and its component functioning. Comparison of consensus algorithms is done, and applications of blockchain in various industries are listed. At last, the future trends and innovative use cases are also elaborated. Although the technology is new and trending, it also has some limitations like scalability which could be a hindrance when blockchain is implemented for a very large-scale system [3].

Matthew N. O. et al. (2018) introduce blockchain technology, technical aspects, and types. It also discusses the various applications in healthcare industry like record management, clinical drugs, drug management, and security concern. Author has discussed the main benefits of blockchain, i.e., interoperability and data security. At last, author concluded with various challenges of blockchain technology [4].

Efanov et al. (2018) illustrate the all-pervasive nature of blockchain technology. Three phases or generations of the block chain development are highlighted in this research: Blockchain 1.0 as digital currency, Blockchain 2.0 as digital economy, and Blockcchain 3.0 as digital society. This paper abridges unmistakable use instances of the blockchain innovation, including cryptographic money, IoT, communication among machines, Electronic Health Records, unique identity generation, smart communities, and contracts [8].

Peck et al. (2017) elicited the IEEE initiatives and support for the blockchain technology to be implemented in various sectors of public services like education, energy healthcare, etc. Also author explained the important features of blockchain in various domains like every feature could act as a basis for the application of blockchain technology. Paper was concluded with various challenges of technology [5].

3 Research Methodology

Research methodology comprises procedures or techniques which are used to identify, select, process, and analyze the data for the establishment of theories, deducing facts, and generating conclusions. This research paper follows an exploratory method of research identifying the issues and challenges in the existing centralized Electronic Health Records management systems and the finding the ways to propose solutions with the help of blockchain along with an overview of blockchain technology and its architecture followed by a comprehensive review. At last, a transformational model along with the information flow is proposed for generation of Universal Patient Identifier (UPI) for decentralization of EHRs using blockchain. Research concludes with a qualitative comparison of centralized EHRs with the proposed blockchain-based EHRs through Table 1.

Table 1 Comparison of centralized EHRs and blockchain-based EHRs

Issues	Centralized EHR	Blockchain-based EHR
Access of health records	Internal to healthcare institution/hospital	Universal access through a hash key
Patient identification record	Multiple copies of health records of a single patient	Single patient identification record through Unique Patient Identifier
Vulnerable to data loss	More vulnerable to data loss as a single point of failure in central server and whole data is lost	Data loss is minimal as data is stored on distributed nodes in distributed databases
Privacy and confidentiality	Difficult to maintain	Cryptographic hash key is generated to every patient to maintain privacy and keep data confidential
Interoperability	Very difficult	Easy nationwide interoperability

4 Role of Blockchain in Healthcare

An important source of healthcare intelligence is healthcare data. Healthcare data sharing is important to create an efficient healthcare system to have qualitative services of health care. Healthcare data is a crucial and personal asset that must be under ownership and control of patient not only in authority of the concerned healthcare system that might have security risks that could put data at risk. A research proposes an application called Healthcare Data Gateway (HGD) using blockchain to control, own, securely access, and share information as per patient's authentication without privacy violation that provides a new prospective in improving the intelligence and operations of healthcare system including security and privacy of patient data [9].

Sharing of data in the modern scenario of healthcare industry in a blockchain based architecture takes out the information duplicity and irregularity problem of the conventional framework. It likewise kills the need of agents for data and agreement sharing. Healthcare industry has encountered various uprisings, and earlier manual records were kept for patients and were more doctor-driven then a reasonably innovation driven of keeping electronic records, be that as it may, not the patient driven. Focused on patient-driven data also transformed the healthcare sector but may not proficient to deal with ongoing large medicinal information [10]. In this new era of healthcare, awareness among the people has risen to the level where the major concern is health and people are using smart healthcare devices such as IoT devices. IoT devices that help to collect healthcare data for the analysis and monitor through wearables, smart devices such as healthband, pacemaker, and blood pressure measurement [11]. Today, the new healthcare industry may have expelled the previously mentioned obstructions up to some extent, but the solution to the above-mentioned issues and applications that blockchain can provide is still in its conception stage.

There is a massive extent of blockchain innovation in health care. This is upheld by a report that expresses the worldwide blockchain advancement innovation in healthcare services that were esteemed at roughly $34.47 million that will see a precarious ascent in 2024 with income near $1415.59 million. Also, the CAGR development will be around 70.45% between 2018 and 2024. There may be several applications of blockchain in health care [12]:

- Single Patient Identification Record for Universal Access
- Health Insurance and Claims Settlements
- Supply chain management for counterfeit drug tracking.

BC can improve interoperability over a worldwide market, disposing of framework limits and geographic restrictions. Other than interoperability and security, BC holds the guarantee to join the dissimilar social insurance forms, lessen costs, improve regulatory consistence, improve patient experience, giving medicinal services at lower expenses, and independent observing and preventive upkeep of restorative gadgets. It will accelerate the R&D cycle and time to market of new medications [4]. Sharing of data in the modern scenario of healthcare industry in a blockchain organize takes out the information duplicity and irregularity problem of the conventional framework. It likewise kills the need of agents for data and agreement sharing. Healthcare industry has encountered various uprisings, earlier manual records were kept for patients and those were only handled by the doctors than an innovation driven approach of keeping electronic records. Focused on patient driven data also transformed the healthcare sector but may not proficient to deal with ongoing large medicinal information [10]. Today, the new healthcare industry may have expelled the previously mentioned obstructions up to some extent, but the solution to the above-mentioned issues and applications that blockchain can provide is still in its conception stage.

5   Proposed Transformational Model for Decentralization of Electronic Health Record

This section proposes a model based on blockchain technology to resolve the issue of decentralized access of Electronic Health Records and also maintaining a universal copy of patient identification of health record through Universal Patient Identifier (UPI). Every individual is registered and allocated a UPID for the access of health records internal or external to the healthcare institution/hospital. There will be no separate copies of a patient record in different hospitals and healthcare institutions. Proposed model will have two modules: Generation of UPID and block creation.

Fig. 1 Generation of the
Universal Patient Identifier
(UPI)

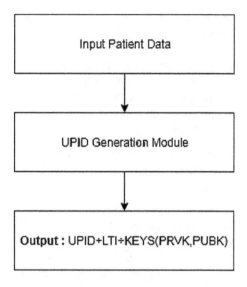

5.1 Generation of Universal Patient Identifier (UPI)

Once the data is entered for the first time, a new UPID will be generated for new patient registered. Also, two keys, private and public (PRVK, PUBK), will be generated for secure sharing of data on blockchain. This unique key for patient will act as a primary key in the distributed database. This identity or key will only be used for accessing patient records anywhere internal/external to the healthcare institution/hospital. But there may be a situation where the patient is not having knowledge to access digital records or not having enough resources. To resolve these kinds of situations, biometric details of the patient: left thumb impression (LTI) will be scanned and uploaded in the system while entering the information regarding the patient the system. Figure 1 illustrates the complete flow of process of UPID generation starting from entering the patient data to the UPID module for registration, and the output of the module is Unique Patient Identifier (UPID), left thumb impression (LTI) and two keys: private (PRVK) and public (PUBK). This output will act as an input for the next module of Block Creation.

5.2 Creation of a New Block

Once the UPID is generated, a new block will be created and appended to the existing blockchain. The structure of the new block will have UPID, LTI, Keys (PRVK, PUBK). Also every block will have a unique hash code address of its own and of previous block also.

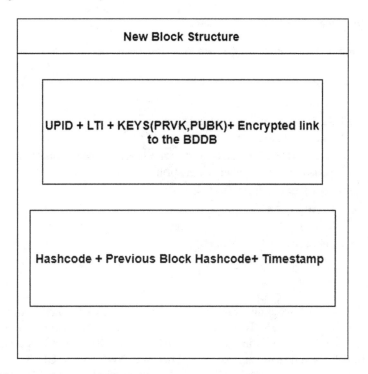

Fig. 2 Structure of the new block

It will also include the time of creation of the block as timestamp. This new block will be created whenever a new record is created and will have an encrypted link to the entire health record of patient which will be saved in the blockchain distributed Database (BDDB). Figure 2 pictorially represents the new block structure having the above-mentioned attributes.

5.3 Information Flow Process

Information flows in a sequential manner from input of patient details, creation of the new block then incorporating the details by appending the new block to the blockchain and saving an encrypted link of the health record in the BDDB. At last the allocation of Universal Patient Identifier (UPI) to the patient. Figure 3 represents the algorithm including various steps for the functioning of the modules in the proposed model.

The data will be input to the Patient API (Application Interface) in the registration form. Left thumb will scanned and uploaded for LTI. Once all the data is entered, the information will be registered, a pair of keys will be generated and a new UPID will be generated. Once the generation of UPID is done, a new block will be created

Algorithm: Registration of a New Patient

1. **Input:** Patient Registration Data. Scan and upload Left Thumb Impression

2. **Processing:**
   2.1 Register()
   2.2 Keys(PRKV,PUBK) = Generate Keys()
   2.3 UPID = Create UPID()
   2.4 NewBlock = Create Block(Hashcode + Prev. Hashcode + Timestamp + UPID +LTI + Keys + Encrypted Link to BDDB)
   2.5 For n blocks in Blockchain
       Append (NewBlock)
       End for
3. **Output:** Send UPID(Email, SMS)

Fig. 3 Algorithm for the proposed model

having block structure comprising hash code, previous block hash code, timestamp, UPID, LTI, and Keys (PRVK, PUBK). It also consists of the encrypted link of the health record in the blockchain distributed database (BDDB). The entire state is saved in the BDDB for future use and access in a decentralized fashion. UPID is assigned to the patient through an email/SMS. Now the patient will be identified through this UPID for Electronic Health Record access. A flowchart has been depicted through Fig. 4 for the above-mentioned processing.

6 Comparison of Centralized EHRs to Blockchain-Based EHRs

A transformation from centralized EHRs to blockchain-based EHRs may resolve all the challenges and issues. A blockchain-based global healthcare system will enhance the management of health data in terms of accessibility without networking constraints as data will be stored on the distributed ledger blocks on the network which can be accessed in any hospital using a UPID and joint consensus mechanism of patient and other entities. Data can be maintained by both patients and other entities involved in the system.

A comparative analysis between both systems is illustrated through Table 1.

Fig. 4 Flowchart for
information flow process in
the proposed model

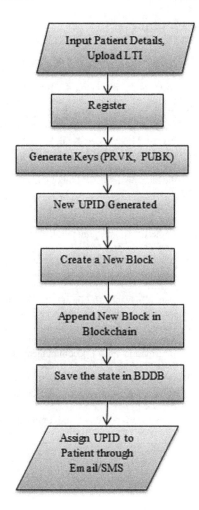

7 Conclusion and Future Work

In this paper, a model is proposed based on blockchain technology for management
of Electronic Health Records in a secure and decentralized manner that can play
a vital role in assisting healthcare administration for a full implementation of an
efficient blockchain-based healthcare system. This paper covers and provides an
effective solution toward all the issues of security, privacy, and single patient identity
for a health record which is stored in various healthcare institutions/hospitals having
centralized access and multiple copies.

References

1. Knirsch, F., Unterweger, A., Engel, D.: Implementing a blockchain from scratch: why, how, and what we learned. EURASIP J. Inf. Secur. **2019**, 2–14 (2019)
2. Brogan, J., Baskaran, I., Ramachandran, N.: Authenticating health activity data using distributed ledger technologies. Comput. Struct. Biotechnol. J. **16**, 257–266 (2018)
3. Zheng, Z., Xie, S., Dai, H.-N., Chen, X.: Blockchain challenges and opportunities: a survey. Int. J. Web Grid Serv. **14**(4), 352–375 (2018)
4. Sadiku, M.N.O., Eze, K.G., Musa, S.M.: Block chain technology in healthcare. Int. J. Adv. Sci. Res. Eng. (IJASRE) **4**, 154–159 (2018). E-ISSN: 2454-8006
5. Peck, M., Tonti, W.R., Stavrou, A., Rupe, J.W., Rong, C., Kostyk, T.: Reinforcing the links of the blockchain. IEEE Future Directions Blockchain Initiative White Paper, pp. 1–16 (2017)
6. Azaria, A., Ekblaw, A., Vieira, T., Lippman, A.: MedRec: using blockchain for medical data access and permission management. In: 2nd International Conference on Open and Big Data, pp. 25–30 (2016)
7. Bethesda: Blockchain: the chain of trust and its potential to transform healthcare. IBM Ideation Challenge—Use of Blockchain in Health IT and Health Related Research, pp. 1–12 (2016)
8. Efanov, D., Roschin, P.: The all-pervasiveness of blockchain technology. In: 8th Annual International Conference on Biologically Inspired Cognitive Architectures, BICA, pp. 116–121 (2018)
9. Yue, X., Wang, H., Jin, D., Li, M., Jiang, W.: Healthcare data gateways: found healthcare intelligence on blockchain with novel privacy risk control. J. Med. Syst. **40**(218), 1–8 (2016)
10. Gupta, R., Tanwar, S., Tyagi, S., Kumar, N., Obaidat, M.S.: HaBiTs: blockchain-based telesurgery framework for healthcare 4.0. In: International Conference on Computer, Information and Telecommunication Systems (CITS) (2019). https://doi.org/10.1109/cits.2019.8862127
11. Gupta, R., Tanwar, S., Tyagi, S., Kumar, N.: Tactile internet and its applications in 5G era: a comprehensive review. Int. J. Commun. Syst. **32**(14), e3981 (2019). https://doi.org/10.1002/dac.3981
12. https://www.peerbits.com/blog/blockchain-technology-on-healthcare-industry.html. Accessed 09 Aug 2019

Awareness Learning Analysis of Malware and Ransomware in Bitcoin

Garima Jain and Nisha Rani

Abstract Ransomware is a subset of malware in which it typically enables cyberextortion for financial gain. It is a serious and growing cyberthreat that often affects an individual's privacy. It has recently made headlines for border attacks on businesses. It results in loss of sensitive information, disruption, regular operations, and damage to an organization's reputation. It spreads through phishing emails that contain malicious attachments or through drive-by downloading. It occurs when a user visits an infected website unknowingly and then malware is downloaded and installed without the acknowledgment of the user. This paper provides a brief of ransomware's history and as well as preventive measures and solutions to network security, data privacy, and menace of ransomware that challenge computer.

Keywords Ransomware · Malware · Cryptovirus · Crypto-Trojan · Cryptoworm · Cybercrime

1 Introduction

Ransomware is a sort of malware to extort cash from a computer person by infecting and taking control of the victim's device, or other files or documents stored on it. Large corporations like Microsoft and Google are designing and building software products which can save your documents without any difficulty. Using these, files can be stored online and are accessible through any device [1]. The attacker encrypts the files in such a way that they are not accessible by the user. After assurance of decrypting the files, he demands a ransom amount in the form of Bitcoins. It spreads through electronic mail attachments, inflamed packages, and website. A ransomware [2] also can be known as a crypto-Trojan, cryptovirus, or cryptoworm. The ransomware which first blocked a display screen and the charge window traumatic a payment became first seen in Russia 2009, earlier than which ransomware become approximately

G. Jain (✉) · N. Rani
Swami Vivekanad Subharti University, Meerut, UP 250005, India
e-mail: jaingarima2011@gmail.com

N. Rani
e-mail: nishadaksh1996@gmail.com

© Springer Nature Singapore Pte Ltd. 2020
P. K. Singh et al. (eds.), *Proceedings of First International Conference on Computing, Communications, and Cyber-Security (IC4S 2019)*, Lecture Notes in Networks and Systems 121, https://doi.org/10.1007/978-981-15-3369-3_56

scrambling statistics and traumatic an installment forget entry to. This attack has infected thousands of computers worldwide traumatizing people, compelling them to pay ransom amount. As ransomware assaults multiplied extra than 1500% on 2017 contrasted with the past a long term, the ransomware threat stood out as virtually newsworthy as the maximum eminent malware glides after focused assaults in 2017.

1.1 Types of Ransomware Attack

There are two major types of this attack:

- Encryption of documents/folders—once the documents are encrypted, it will block the access to data demanding a huge amount of ransom. And, once the time is elapsed, the data is gone forever. This form of ransomware [3] is referred to as 'scareware.' For example, cryptoLocker.
- Locks the display—it displays a full display screen photograph that blocks all other home windows and demands money. Private documents are not encrypted. This type of ransomware which is explained in Fig. 1 is known as 'Winlocker.'
- Master boot document (MBR) ransomware—a section in the hard drive of the computer that enables the operating system to boot is known as master boot record. The attack [4] infects computer's MBR which interferes with the everyday boot procedure and demands ransom thereafter.

Fig. 1 Locker ransomware [3]

1.2 Biggest Ransomware of Attack

The latest ransomware versions to upward thrust is Jigsaw, an advanced version that has some unmistakable highlights. One imperative detail of Jigsaw is its utilization of sensational effects for social designing—the payoff display consists of an image of a manikin, Billy, from the noticed film association, along the headings for paying the fee in Bitcoin. In any case, what in reality separates Jigsaw as a progressed ransomware version is that instead of offering a direction of activities for paying the payment, Jigsaw basically erases more information every hour the fee is not paid. "reality be instructed, inside the occasion you endeavor to stop the method or restart the framework, Jigsaw will erase 1000 documents, alongside those traces proscribing the sports that the consumer can take to try and recoup their records without paying the charge." One of the major ransomware variations to target Apple OS X is moreover advanced in 2016. KeRanger is a ransomware horse targeting maximum part affected clients usage of the transmission application, however motivated round 6500 pcs inside a day and a half gain with a provoke response [5], and KeRanger turned into expelled from transmission the day following its revelation. All matters considered, 2016 changed into a pennant 12 months for ransomware attacks, with reviews from mid-2017 comparing that ransomware were given cybercriminals a mixture of $1 billion for the year [6, 7].

1.3 Ransomware Working

In ransomware attack, the Trojan spreads via network, frequently through email attachments from rogue senders. However, it is explained in Fig. 2 that ransomware cycle works for the famous 'WannaCry Worm' travels from computer to computer without user's interrelation. The contamination chain [2] involves a three-step procedure:

- URL Redirection
- Getting on malicious page
- CAPTCHA code verification.

1.4 Strategies for Fighting Ransomware Attack

In order to fight against the ransomware attack, following points will be helpful:

- Backing up the data locally as well as online regularly. Backup of the data must be taken on a secondary device so that it can be retrieved.
- Segmentation of the network will be helpful in defending against this attack. Whole network security will not be compromised due to dynamic control access.

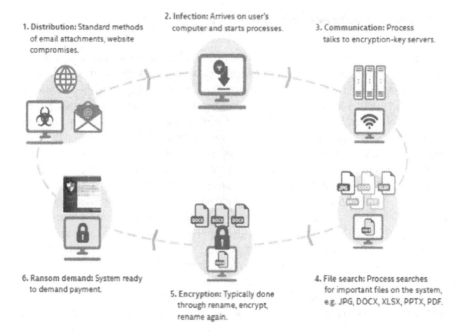

Fig. 2 Ransomware life cycle [5]

- Installing anti-ransomware software will be more secure. These software help in prevention of this attack. Intrusions are detected on time, thus eliminating those from the system.
- Updating the system is also important as the security updates are installed by an operating system.
- Scheduled security checks must be executed on the devices. Otherwise, the device may become vulnerable to ransomware attack.

2 Related Work

In 1966, fatal weakness was spotted by Adam L. Young and Moti Yung in 'AIDS' Trojan. An idea of using public key cryptography was then introduced for the attacks. That is why it was mentioned as 'cryptoviral extortion.' A combination of crypto-Trojan, cryptovirus, or cryptoworm encrypts the data. And, the victim has to pay commanded ransom in order to get the access of encrypted data.

2.1 Evolution of Ransomware

In 1989, Joseph L. Popp created the first ransomware which was named AIDS Trojan. Symmetric cryptography is the main component of Trojan. In the beginning, it was easy to overcome this attack but it became more powerful within a few decades [8]. Two categories of ransomware attack remain persistent: crypto- and locker-based. In crypto-based ransomware attack, data is being encrypted by the attacker whereas in locker-based attack, the victim is locked out of the device [9]. This attack is mostly seen in the android-based devices. According to the recent reports of ransomware attacks, major attacks are reported in the previous years. Rise in the ransom demands is also seen. As per the latest information, average ransom demand is from $300 to $500. And sometimes, the attacker doubles the ransom amount or destroys the file, if the deadline passes. The Bitcoin term has gathered a public growing interested throughout the ages, undaunted by incalculable hacks and tricks. In this way, it is significant for users to completely comprehend the suggestions and impediments of the Bitcoin framework. Various studies have completely analyzed the flawed arrangements for protection inborn in the Bitcoin framework [10]. First analysis on privacy in the field of Bitcoin framework done by Reid and Harrigan where they had an option to ascribe outer recognizing data to address, utilizing an early portrayal of the Bitcoin exchange organize [11]. In the latest scenario, author Androulaki et al., the Bitcoin surroundings in both the genuine Bitcoin condition and a recreation of Bitcoin in a college setting, demonstrated that conduct-based and exchange-based grouping methods could successfully deanonymize up to 40% of Bitcoin clients in the field of simulation [12]. The entire analysis of transaction graph is discussed by Ron and Shamir and looks at how user spends Bitcoins, in this scenario how Bitcoins are dispersed among various users, and means by which user secures their protection in the Bitcoin framework [13]. Similar to Androulaki et al. to classify the Bitcoin wallets, owners have discussed the how user spends Bitcoins, in this scenario how Bitcoins are dispersed among various users, and means by which users secure their protection in the Bitcoin framework. Meiklejohn et al. explored discussed about clustering heuristics in same ongoing area in Bitcoin wallets [14]. On analyzing the user behavior on Bitcoin, the performance is measured by Ron and Shamir about the block-chain data, rather than user information for trying to deanonymize. The assumption about the user character behavior is described by multiple addresses belong to the same user for the order characterization [15]. In [16] Ober et alter empirically study Bitcoin has a global property that transaction graph had at the time of evolution on 6th January since the Bitcoin creation. The Bitcoin address which is distinguishable in same they called as used address, the same is used to make the payment also present in some transaction. The introduction about specification language CTPL i.e.., Computation Tree Predicate Logic is discussed in [17] the malware variants has discussed that using a dataset the total malware are thirteen on windows malware as a set of worms between 2002 and 2004. There is the link mentioned on the display which the Bitcoin victim is expected to pay into, also a paytm link is given on Web site which also sometime displays the address.

The unique ransomware [18] address for many ransomware families generate unique address on paying victims for automate identification, where other various victims uses multiple addresses (e.g., WannaCry and CryptoDefense).

3   Research Design and Method

As we have discussed about ransomware attack, attack is classified into five stages: setup, communicating head office, handshake and keys, encryption, and extortion. There are five steps that are explained below:

Setup: That is step one of the ransomware setup techniques. Once a victim's computer is infected, the cryptoransomware installs itself. It then attacks inside the windows registry so that it can be started mechanically each time the computer boots up.

Communicating head office: Before attacking your device, it contacts a server which is operated by cybercriminals.

Handshake and keys: On the third stage of attack, the ransomware customer and server authenticate each other by using a technique called 'handshake.' The server then generates cryptographic keys. One key is stored on the victim's computer, and the second one is stored on the criminals' server.

Encryption: The step four is encryption that explains that as soon as the cryptographic keys are set up, the ransomware attack in your pc starts.

Extortion: At last, the ransomware displays a screen showing a time restriction to pay the ransom and the attacker will decrypt the data. But if time given is lapses, then the whole of yours will be deleted forever. A normal ransom amount degrees from $300 to $500. It needs to be paid in Bitcoins or other electronic bills that are untraceable.

The Anatomy of Ransomware Attack

Ransomware is not a new phenomenon, but its effects are starting to be felt more widely and more deeply than ever before. Ransomware continues to be always a major risk to organizations of all sizes. Ransomware continues to develop, and more erudite modifications are being introduced all the time, offering a better mode of encryption. Detecting a ransomware attack before encryption begins is very difficult to perform [19]. One of the most communal ways of infecting ransomware in a system is through an ill-advised click. However, if you know what to look for, it is possible to identify an infection before encryption even starts. Ransomware is known to be highly automated, such that beyond the 'distribution stage,' most of the ransomware methods runs autonomously without demanding communication with a C2 (command and control) for receiving any instructions. Instead, Ransomware executable file contains all the rationality which is required to hijack a system. The given step in a whole process can happen right under your muzzle [20]. If it is decided not to pay for the attack, the data may vanish endlessly.

Step 1: Identify an inclined network the usage of state-of-the-art tools to hit upon and probe networks for lax security protocols, unpatched software, or single-factor authentication.

Step 2: Scrape user key off the shady Web. In this stage, the malicious cipher is transferred and code execution begins from this stage.

Step 3: Use a third-party site to confirm the stolen password. The attacker from this step will 'own' the system.

Step 4: Obscure their location by logging in via global proxies. First the malware will scan your local mainframe to find documents to encrypt.

Step 5: Step pull down your registered data, encrypt it, and spread it through the block-chain in data centers across the worldwide. Local file encryption can occur in minutes; however, system file encryption and the process can take several hours.

Step 6: Demand thousands of bucks for the safe return of your data and cripple your day-to-day actions in the meantime. Hence, in this step, the attack is demanding payment.

Finally, the anatomy of ransomware attack explained in Fig. 3 for detection of ransomware should be a part of the safety posture. This method [21] contains skimming shapeless information for distrustful or transformed record for extensions, regarding ransomware initials and detection of 'ransom be aware' [22] content inside the repository.

In this section, we present our version checking-based technique [6] to mechanically dissect ransomware identifying feature behaviors. We use the calculus of communicating structures (CCS) of Milner to explain apps and the calculus good judgment to explicit. It is based on given steps:

Step 1: Java Bytecode-to-CCS remodel operator: Step one generates a CCS specification from the Java Bytecode of magnificence documents derived by the analyzed apps that is obtained by using defining a Java Byte code-to-CCS transform

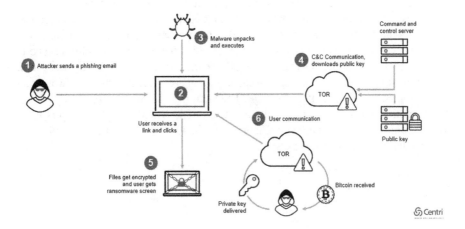

Fig. 3 Anatomy of ransomware attack [21]

operator. This function without delay applies to the Java Bytecode and translates it into CCS technique specifications. The characteristic is defined for every guidance of the Java Bytecode.

Step 2: Expressing characteristic ransomware behaviors into temporal common sense: The second one step pursuits at discovering characteristic ransomware behaviors, expressed in temporal good judgment that allows you to carry out automated ransomware dissecting.

The CCS tactics received inside the first step are used to show properties: By the use of version checking, we decide the behaviors of samples. Codes described as CCS processes are first mapped to labeled transition structures and a version checker is used. In our technique, we invoke the Concurrency workbench of the new Century (CWB-NC) as formal verification environment. When the result of the CWB-NC model checker is authentic, our methodology considers the pattern under analysis belonging to the ransomware family, fake otherwise. On this way, we will carry out an automated dissecting. We can locate all malicious behaviors in an application without manually in-section. Different properties were described to characterize the diagnosed behavior.

4 The Bitcoin System

A Bitcoin record is defined by an Elliptic Curve Cryptography key pair 2. The Bitcoin record is openly identified by its Bitcoin address, acquired from its open key utilizing a unidirectional capacity. Utilizing this open data clients can send Bitcoins to that address. A Bitcoin record is defined by an Elliptic Curve Cryptography key pair 2 [14]. The Bitcoin record is freely identified by its Bitcoin address, got from its open key utilizing a unidirectional capacity. Utilizing this open data clients can send Bitcoins to that address. The organization of Bitcoin took off without such a great amount of consideration from the exploration network and the first research papers on the theme did not show up until late 2011 in the arXiv store and later distributed gatherings and diaries. The Bitcoin framework [21] needs to disperse different sorts of data, basically exchanges and squares. Since the two information are produced in a dispersed manner, the framework transmits such data over the Internet through an appropriated distributed (P2P) arrange. The anatomy of Bitcoin explained in Fig. 4 explains the system scenario of Bitcoin attack. The Bitcoin block-chain [7] is an open grouping of timestamped exchanges that include wallet addresses, which are essentially pseudo-unknown personalities.

Fig. 4 Anatomy of Bitcoin attack [21]

4.1 Preventive Measures

There are one-of-a-kind measures that may reduce the impact of cryptoware. A quick pursuit will uncover many 'tips to avoid ransomware articles.' We will count on the important ones.

Customers

Hold your customers educated: a contamination pretty frequently begins with a human blunder. Educating your customers will be helpful in the detection of suspicious programming or connections. Be that as it can, even organized personnel is blunder willing, by no means depend upon the human issue to protect you.

On Workstation

Impair macros: Workplace statistics can include malignant macros. In order to protect the data from ransomware attack, the macros must be updated and secured. Otherwise, it may result in heavy losses. Cripple Windows Script Host (WSH): Disabling WSH [23] continues the execution of JavaScript while they are opened below windows. Nonetheless, this diploma should be painstakingly taken into consideration as it can have an effect on technology programming. There are specific styles of tool like Honeypot, HelDroid, CryptoLocker, and Sand-field.

On Mail-server
Attachments at the email portal: Block messages containing executable, yet do not postpone to rectangular attachments with document types that should not be or do not often get messaged round like.chm, .like and .js.

On the Network
Applied a middleman with net filtering: Some intermediaries permit you to channel the activity from boycotted areas. This can reduce the hazard of contamination, if the rundown is modern-day.

5   Discussion

It is widely recognized that ransomware causes damage to many people both through financial losses or destruction of files. This chance challenging to degree, but an advanced know-how of the ransomware surroundings is the key first step to identifying new potentially greater powerful intervention strategies. This could be pretty a precious useful resource. This assault is enormous without difficulty circumvented and smooth from an inflamed computer.

5.1   Future Ransomeware

After the biggest ransomware story of 2018, happened in city Atlanta, it was at its peak with WannaCry and NotPetya attacks. People are not vigilant about ransomware precautions, just after staying quiet for a while. Ransomware is not considered in a serious manner and not active too, in targeting people for a while, but cybercriminals are still targeting people. So, it clearly reflects that cyberbullies are still using ransomware attack hostile to our network for their benefits. Cybercriminals know ransomware is the easiest way to make millions and they will not leave any chance to make it. So, after all these studies, we need to beware about the security measures and speed awareness about ransomware. The organization must approach their leadership to make sure about the depth knowledge of security protocols and common tactics.

6   Conclusion

Through this paper, we have offered a categorization and an analysis of numerous key functions of ransomware. We provided the evolution of ransomware. To avoid data theft and undue extortion of ransomware, individuals and organizations need powerful network security stance. This topic is an emerging field of study in academic research. Therefore, more research effort is needed to stop the growing trend of ransomware attack. Ransomware attacks on Linux and Mac operating systems with the recent publicize, and the analysis of ransomware on these platforms is essential. Kaspersky Lab and Intel have joined forces with Interpol and the Dutch National Police to set up a Web site (www.nomoreransom.org) aimed at helping people to prevent weakening to ransomware. The Web site will host decryption keys and tools for those ransomware strains that have been fissured by security researchers. Also some detail about to safe system from attack and set some restrictions to save data from attack in future.

References

1. Salvi, M.H.U., Kerkar, R.V.: Ransomware: a cyber extortion. Asian J. Converg. Technol. (AJCT) **2**, 1–6 (2016)
2. Luo, X., Liao, Q.: Awareness education as the key to ransomware prevention. Inf. Syst. Secur. **16**(4), 195–202 (2007)
3. Alessandrini, A.: RANSOMWARE: Hostage Rescue Manual. Knownbe4
4. Chen, Q., Bridges, R.A.: Automated behavioral analysis of malware: a case study of WannaCry ransomware. In: 2017 16th IEEE International Conference on Machine Learning and Applications (ICMLA). IEEE, Cancun, Mexico (2017)
5. Al-rimy, B.A.S., Maarof, M.A., Shaid, S.Z.M.: Ransomware threat success factors, taxonomy, and countermeasures: a survey and research directions. Comput. Secur. **74**, 144–166 (2018)
6. Garg, D., et al.: A past examination and future expectations: ransomware. In: 2018 International Conference on Advances in Computing and Communication Engineering (ICACCE). IEEE, Paris (2018)
7. Huang, D.Y., et al.: Tracking ransomware end-to-end. In: 2018 IEEE Symposium on Security and Privacy (SP). IEEE, The Hyatt Regency, San Francisco, CA (2018)
8. Pathak, P.B., Nanded, Y.M.: A dangerous trend of cybercrime: ransomware growing challenge. Int. J. Adv. Res. Comput. Eng. Technol. (IJARCET) **5**(2), 371–373 (2016)
9. Liao, K., et al.: Behind closed doors: measurement and analysis of CryptoLocker ransoms in Bitcoin. In: 2016 APWG Symposium on Electronic Crime Research (eCrime). IEEE, pp. 1–13 (2016)
10. Reid, F., Harrigan, M.: An Analysis of Anonymity in the Bitcoin System. Springer, New York, NY (2013)
11. Androulaki, E., Karame, G.O., Roeschlin, M., Scherer, T., Capkun, S.: Evaluating user privacy in bitcoin. In: Financial Cryptography and Data Security. Springer, Berlin, Heidelberg, pp. 34–51 (2013)
12. Ron, D., Shamir, A.: Quantitative analysis of the full bitcoin transaction graph. In: Financial Cryptography and Data Security. Springer, Berlin, Heidelberg, pp. 6–24 (2013)
13. Meiklejohn, S., Pomarole, M., Jordan, G., Levchenko, K., McCoy, D., Voelker, G.M., Savage, S.: A fistful of bitcoins: characterizing payments among men with no names. In: Proceedings of the 2013 Conference on Internet Measurement Conference, pp. 127–140. ACM (2013)
14. Spagnuolo, M., Maggi, F., Zanero, S.: Bitiodine: extracting intelligence from the bitcoin network. In: Financial Cryptography and Data Security. Springer, Berlin, Heidelberg, pp. 457–468 (2014)
15. Moser, M., Bohme, R., Breuker, D.: An inquiry into money laundering tools in the bitcoin ecosystem. In: eCrime Researchers Summit (eCRS), 2013, pp. 1–14. IEEE (2013)
16. Kinder, J., Katzenbeisser, S., Schallhart, C., Veith, H.: Detecting malicious code by model checking. In: Detection of Intrusions and Malware, and Vulnerability Assessment, Springer, Berlin, Heidelberg (2005)

17. Ducklin, P.: "Locky" ransomware what you need to know. https://nakedsecurity.sophos.com/2016/02/17/locky-ransomware-whatyou-need-to-know/
18. Kan, M.: Paying the WannaCry ransom will probably get you nothing. Here's why. https://www.pcworld.com/article/3196880/security/paying-thewannacry-ransom-will-probably-get-you-nothing-heres-why.html
19. Hampton, N., Baig, Z.A.: Ransomware: emergence of the cyber-extortion menace. Edith Cowan University, Joondalup Campus, Perth, Western Australia (2015)
20. Surati, S.B., Prajapati, G.I.: A review on ransomware detection and prevention. IJRSI **IV**, 86–91 (2017)
21. www.centrify.com
22. Song, S., Kim, B., Lee, S.: The effective ransomware prevention technique using process monitoring on android platform. Mob. Inf. Syst. **2016**, 1–9 (2016)
23. Truta, F.: The evolution of ransomware in 2018. Bitdefender 2018 (2018)

A Survey of Cryptographic Techniques to Secure Genomic Data

Hiral Nadpara, Kavita Kushwaha, Reema Patel and Nishant Doshi

Abstract Human genome is the unique identifier of the individual, which contains highly sensitive information about every individual. The sensitivity, longevity, and non-modifiable nature of genetic data has attracted much attention from the research community. Research on genomic data has taken a tremendous speed for finding a solution for curing various fatal diseases as like Alzheimer's, cancer, etc. Due to advances in the genome sequencing technology, it is now possible to obtain accurate and highly detailed genotypes at much-reduced costs. Such a huge research work in this domain has led us to cure some very serious problems, but at the same time, it comes up with a risk of providing security to this huge amount of highly sensitive data. In this paper, we have described some of the cryptographic applications of how we can make genomic data sharing and storage more private and secure.

Keywords Genomic data · DNA · Personalized medicine · Honey encryption · Brute-force attack

1 Introduction

Every cell in our body contains a cell nucleus as shown in the diagram (see Fig. 1) that holds 23 pairs of chromosomes that make up our genetic material. A chromosome is a thread-like structure of nucleic acids and proteins found in nucleus. Every chromosome has double-helical structure called DNA (deoxyribonucleic acid) within

H. Nadpara (✉) · K. Kushwaha · R. Patel · N. Doshi
Pandit Deendayal Petroleum University, Gandhinagar, Raisan 382007, India
e-mail: hiralnadpara45@gmail.com

K. Kushwaha
e-mail: kushwahakavita207@gmail.com

R. Patel
e-mail: reema.mtech@gmail.com

N. Doshi
e-mail: doshinikki2004@gmail.com

© Springer Nature Singapore Pte Ltd. 2020
P. K. Singh et al. (eds.), *Proceedings of First International Conference on Computing, Communications, and Cyber-Security (IC4S 2019)*, Lecture Notes in Networks and Systems 121, https://doi.org/10.1007/978-981-15-3369-3_57

777

DNA Structure

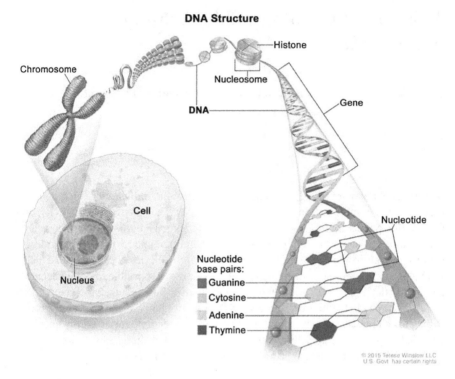

Fig. 1 Structure of genes [5]

it which is made up of nucleotides [1]. Nucleotides are mainly made of four types: adenine (A), guanine (G), cytosine(C), and thymine (T) as shown in Fig. 1. DNA carries genetic information in the form of genes [2]. All humans share the same set of genes (0.99%) and differ with very slight changes (0.1–0.2%) [3]. Those slight differences in genes result in different hair, different eyes, or skin color and may even cause the genetic disease susceptibility. A disease which causes a change in the DNA of a gene is called a mutation.

Human genome's most prevalent variant is single-nucleotide polymorphism (SNP). Based on short tandem repeats (STR), data can be obtained from the genome. As for instance, GATAGATA is an STR with the period of four which is getting repeated two times [4]. Advancement in the genomic research and reduction in the sequencing costs has boosted the use of genomic data in the domains like: (i) health care (e.g., personalized medicines); (ii) forensics (e.g., paternity test, ancestry test, etc.). In order to facilitate these kinds of biomedical research on a large scale, it often involves the sharing of genomic and clinical information gathered by different organizations. In such cases, ensuring that the sharing, management, and analysis of the data does not reveal the identity of the individuals is important. In this paper, we will overview a few cryptographic techniques and how these techniques can be used to ensure privacy-preserving genomic data sharing. The paper includes techniques like

(i) homomorphic encryption (HE)—HE varies from other encryption techniques by enabling encrypted data operations directly without a private key access. The outcome stays encrypted and can be revealed to the proprietor at a subsequent stage with their private key. (ii) Honey encryption—Honey encryption is an encryption system that provides valid, but false plaintext for every faulty key that an attacker uses to decrypt a signal. (iii) Garbled circuit—Garbled circuit is a cryptographic protocol that enables secure two-party computing in which two suspects can jointly evaluate a function on their private inputs without a trusted third party being present. Section 1 includes introduction about the genomic data. Section 2 includes the application of genomic data in various fields like forensics, health care, and research. Section 3 includes various challenges to secure genomic data and some techniques like homomorphic encryption, honey encryption, and garbled circuit to secure genomic data. Section 4 includes a comparison of the three techniques and conclusion.

2 Genomic Data in Various Fields

In this section, we are explaining the use of genomic data in various fields like forensics, research, and health care.

2.1 Use of Genomic Data in Forensics [6]

2.1.1 Paternity Test

The paternity test determines whether or not a certain male is another individual's dad. Paternity test is based on the STRs [4]. It is based on the elevated resemblance between the genomes of a father and child in comparison to two unrelated human beings. In paternity test with a single parent, let us assume that one STR $S = [\{x_{i,1}, x_{i,2}\}]_{i=1}^{n}$ is of father and another STR $S' = [\{x'_{i,1}, x'_{i,2}\}]_{i=1}^{n}$ is of the child. To determine whether S' corresponds to the child's father, the intersection of sets $\{x_{i,1}, x_{i,2}\}$ and $\{x'_{i,1}, x'_{i,2}\}$ must be non-empty for each i.

2.1.2 Ancestry Test

Ancestry testing can give information about the person's ancestors [6]. It can also provide information about the person's relatives. Various clues about the relationship between families can be obtained by examination of DNA variation. There are three kinds of ancestry testing (1) Y-chromosome testing: Since Y-chromosome is present only in men, it is used to explore questions linked to whether or not two families are linked with the same surnames. (2) Mitochondrial DNA testing: Mitochondrial DNA is in male and female. So, it can be used for testing of either gender. But it is mostly used for genealogy because it preserves data about female ancestors that can be lost from historical records due to how often they pass down surnames. (3) SNP testing: These tests evaluate large number of variations across person's entire genome. So, this test can provide an estimate of person's ethnic background. For example—Ancestry test can identify a person who may be 40% African and 60% Asian.

2.2 Use of Genomic Data in Research

Many essential information about an individual like possibilities of having a specific disease such as breast cancer, diabetes, and Alzheimer's in future can be known by the analysis of the human genome. So, various researches are going on with a huge collection of the human genomes. Genome-wide Association Study (GWAS) is one such study of the genetic variation in the different individuals [7]. It studies and identifies the variations in DNA which are associated with the disease. GWAS investigates the entire genome instead of a small pre-specified genetic region. GWAS collects the DNA of various individuals across the globe and then compares the DNA of participants having different phenotypes with the DNA variations responsible for the particular disease. GWAS also studies genetic linkage. GWAS has examined approximately 1800 diseases and traits since 2017 [8].

2.3 Use of Genomic Data in Healthcare [6]

2.3.1 Genetic Compatibility Test

Genetic compatibility test gives the information of chances of transmitting the disease from potential or existing parents to an offspring. Minor mutation in parents may not cause any problem. But if the mutation is minor and if it will transmit to the child or not can be known through this test. There are chances that the parents are having minor mutation but the child is born with major mutation. So, by doing this test we can come up with the early knowledge of the disease that is transmitted to child and can have the chance to cure it [4].

2.3.2 Personalized Medicine

The genetic makeup of an individual plays an important role in how well they react to a particular therapy [9]. Due to advancement in the genomic research and technologies, it is now possible to get information about the variation in the genomic sequencing of an individual. The mutation is sometimes responsible for the severe disease and there are chances of it getting transferred to the offspring in the future.

In medicine today, it is prevalent for doctors to use a trial-and-error approach until they discover the most efficient treatment therapy for their patient [10]. The idea behind the personalized medicine is to know the genetic makeup of an individual and treat the disease according to that makeup instead of trial-and-error treatment. These treatments can be tailored more specifically to an individual with personalized medicine and give insight into how their body will respond to the drug and whether it will work on the basis of their genome. Personalized medicine, based on one or even several genes, can also be used to forecast a person's risk for a specific disease.

3 Challenges to Secure Genomic Data [11]

Reduction in the sequencing costs of genomes has resulted in a huge collection of highly sensitive genomic data. Samples collected for the research work also form a huge genomic data. This genomic data collection comes up with various challenges like:

- How can genomic information be shared in a manner that safeguards information donor privacy?
- How to secure data from loss, theft, or misuse?
- How to conduct a privacy-conserving assessment and computing of encrypted genomic information across various users in an untrusted cloud setting?
- How to do secure outsourcing of it?

In the following section, we are discussing some of the cryptographic approaches for secure computation on genomic data.

3.1 Homomorphic Encryption

In [12], the authors proposed homomorphic-based encryption scheme for secure computation genomic data. Homomorphic encryption (HE) is a form of encryption where functions, f, can be evaluated on encrypted data $x_1, \ldots, x_n$ yielding ciphertexts that decrypt $f(x_1, \ldots, x_n)$.

We can apply this homomorphic operation directly to the plain text to encrypt it or we can encrypt the plaintext first and then apply this homomorphic operation to the ciphertext. The diagram (see Fig. 2) explains the mechanism of homomorphic operations.

As shown in Fig. 2, we are taking two sets of objects. Here, multiplication operation is performed on the sets of objects. Figure 2 demonstrates the multiplication operation. Here, the numbers 4 and 2 are used as the plaintext inputs. They are also transformed to ciphertext by multiplying by 2 but we have to split the product by half of the encrypted values in order to achieve the product of 4 and 2.

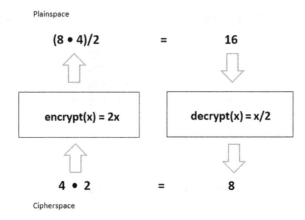

Fig. 2 Illustration of homomorphic property [12]

3.1.1 How We Can Do Secure Computation on the Encrypted Genomic Data Using Homomorphic Encryption?

As shown in the below picture (see Fig. 3), we can see how we can perform the secure computation for the disease risk of a patient through homomorphic encryption.

As shown in Fig. 3, patients will use both genomic data and clinical and environmental factors to provide their DNA sample for privacy-preserving disease risk computation. Here, CI (Certified Institution) will do the sequencing and the encryption of the DNA sampling which is assumed to be trusted party. Sequencing is done to get the genetic variation in the patient's sample. The patient will provide his/her clinical data to the Medical Unit (MU) and Storage and Processing Unit

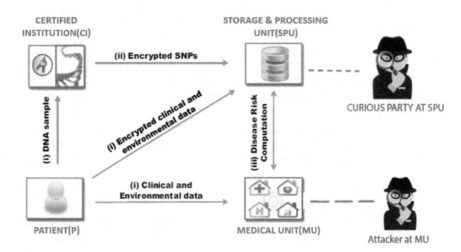

Fig. 3 System model for privacy-preserving computation of the disease risk [13]

(SPU). The patient may not want to reveal all the data to the Medical Unit (MU). But still, the computation can be done on the hidden data to check for the disease risk. CI forwards encrypted genomic data to the Storage and Processing Unit (SPU). There are chances that the employee of a hospital or any other careless person can attack the data and there are chances that there is a curious party at SPU who wants to steal the data. So, we need to keep the data at the SPU in the encrypted form. Now, the computation for the disease risk is done by comparing the data given by the patient at the Medical Unit (MU), and the encrypted SNPs are sent by the CI to the SPU. So that the computation can be done securely without revealing any data to the MU or SPU using homomorphic encryption. Each client's cryptographic keys are produced and circulated to the patients. Then, the secret key x of the patient is randomly split into x1 and x2 so that $x = x1 + x2$ is added using the homomorphic property and each share is assigned, respectively, to the SPU and MU. Similarly, the SPU's public and private keys are produced and the MU shares the public key. Finally, symmetrical keys between the parties are created to protect communication between the parties from an eavesdropper [13].

Throughout the entire process, there is no instance where the server can access the data in an unencrypted form and thus preserves the privacy of the data.

3.2 Honey Encryption

In [15], the authors proposed honey encryption to secure genomic data from brute-force attack. Honey encryption is a cryptographic technique that requires the transformation of plaintext messages into a different space with uniform element distribution. The diagram (see Fig. 4) explains the working of the honey encryption. Distribution-transforming encoder (DTE) maps the message to a seed range in a seed space as shown in Fig. 4 and then randomly selects a seed from the range and XORs it with the key to obtain the ciphertext. For decryption, the ciphertext is XORed with the key and the seed is obtained. The seed position is then used by DTE to map it back to the initial plaintext message. Even if the key is incorrect, a message from the message space is output by the decryption process and confuses

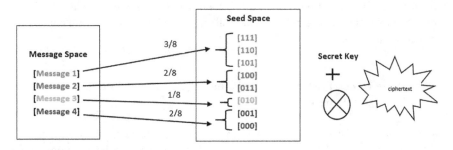

Fig. 4 Procedure of honey encryption algorithm [14]

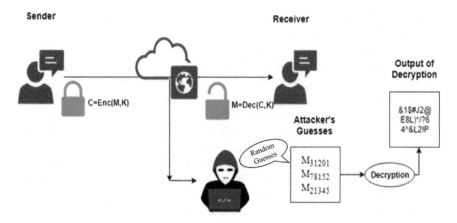

Fig. 5 Model of conventional encryption in brute-force attack [15]

the attacker. Suppose the opponent intercepts the encrypted file, fees from either the user's device (e.g. desktop, tablets, mobile devices, etc.), public cloud data storage or data transmission, and then attempts to decrypt it. The decryption algorithm, with its assumed password, produces a false file in the reaction to any wrong assumption of the user's password or symmetric shared key, K. It is therefore not distinguishable from the view of the attacker, thus increasing the difficulty of determining whether or not the attacker properly conjectured a password.

3.2.1 Brute-Force Attack

Genomic data remains stable and sensitive life-long [16]. The password-based techniques are not reliable to share the sensitive data like genes due to weak passwords. Here, as shown in the above diagram (see Fig. 5), the sender who wants to send the data to receiver encrypts his/her genomic information using the conventional password-based encryption techniques. The adversary tries to get the data by guessing the key. The adversary will get the meaningless (non-uniform distribution) data if the key guessed by him/her is wrong [15]. So, meaningless data makes the adversary to guess other combination of the keys. This is said to be brute-force attack. So, at one-point adversary will get the meaningful data and so the security of the genomic data is compromised.

3.2.2 How Honey Encryption Can Protect the Genomic Data from the Brute-Force Attack?

We have discussed earlier about how brute-force attack can compromise the security of the genomic data. The above diagram (see Fig. 6) explains how genomic data

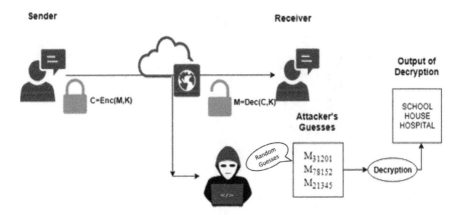

Fig. 6 Model of honey encryption in brute-force attack [15]

can be made secure from brute-force attack using honey encryption. Here, in this cryptographic technique, when the adversary tries to guess the key, he/she will get the meaningful (uniform distribution) data as like SCHOOL, HOUSE, and HOSPITAL as shown in the given diagram (see Fig. 6). So, the adversary will think that it has got the desired information [15, 16]. Even if the adversary feels that it has got the wrong information then it will keep on trying the various combinations of keys and will get a list of data. And it is difficult to find the correct or desired information from that list generated as output. In this way, honey encryption provides security against brute-force attack.

3.3 Garbled Circuit

In this section, we have discussed garbled circuit which is a cryptographically secure and effective protocol for collaborative two-party computation on genomic data. Garbled circuit is a cryptographic protocol that enables safe, semi-honest cloud computing between the two sides. Without the presence of a trusted third party, two mistrusting parties can collectively calculate a function on their personal inputs.

In the garbled circuit protocol, the function has to be described as a Boolean circuit. Suppose two sides, Alice and Bob as shown in Fig. 7, have their own private data x and y, respectively, and want a $F(x, y)$ function to be calculated safely. This method enables Alice and Bob to perform a protocol that will output the outcome $R = F(x, y)$, so that Alice will not know any data about Bob's input y, nor will Bob learn any data about the input x of Alice [17]. The working of the garbled circuit involves the following steps:

(I) The function $F(x, y)$ is described as a Boolean circuit with $\log x + \log y$ input gates. The circuit is known to both Alice and Bob.

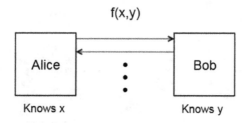

Fig. 7 Idea behind garbled circuit [8]

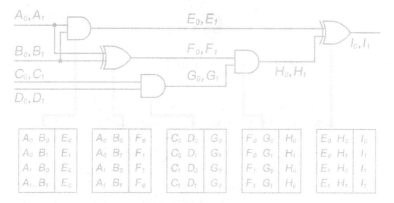

Fig. 8 Labelled Boolean circuit table [18]

A circuit compiler can be used to convert the function f into Boolean circuit that consists of AND and XOR gate as shown in the below diagram (see Fig. 8).

(II) One party (out of the two parties) will encrypt the circuit as shown in Fig. 9. That party is known as garbler. Here, Alice is called the garbler.

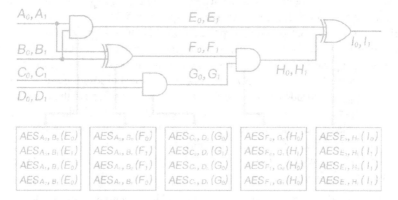

Fig. 9 Garbled (encrypted) table [18]

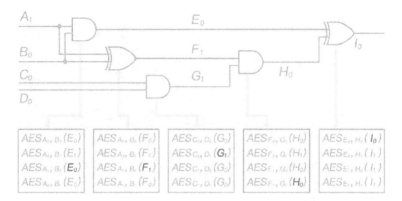

Fig. 10 Whole evaluation process [18]

(III) The garbled circuit is sent to Bob by Alice along with her encrypted input.

(IV) Bob will receive his encrypted input (encryption of y) from Alice through oblivious transfer.

An oblivious transfer (OT) protocol is a type of protocol in which a sender transfers to a receiver one of the potentially many pieces of information, but remains unaware of which piece (if any) was transferred.

(V) Bob will decrypt the circuit (NOT the inputs) and obtain the encrypted output as shown in Fig. 10. Hence, Bob is called the evaluator.

(VI) Alice and Bob communicate to learn the output.

Table 1 shows the comparison of all the above-discussed cryptographic approaches.

4 Conclusion

Genetic information is one of the most sensitive types of personal information. We have shown in this paper, how genomic information is incorporated into a broad spectrum of areas such as health care, research, forensics, etc. Cheap computing and high-throughput sequencing techniques made it simple for genomic information to be collected, stored, and processed. Large amount of genomic data is now available on the Web due to its large applications. Availability of genomic data on the Web comes up with the challenges of storing and secure computation of the genomic data on the cloud. Our survey provides some insight into some techniques for secure computation on the data stored on the cloud. Here, we have discussed few cryptographic techniques like: (a) homomorphic encryption; (b) honey encryption; and (c) garbled circuit.

Table 1 Pros and cons of cryptographic techniques

Cryptographic techniques	Pros	Cons	Future scope	Tools/algorithm
Homomorphic encryption	• No one can read or modify the data even though the cloud provider may not be secure or may even work against you • One other advantage is that the data can be decrypted less often, which is good for security	• Homomorphic encryption is very slow • Computationally expensive	• To make it faster • To make it computationally cheaper	• SIG-DB algorithm utilizes homomorphic encryption that allows genomic sequence databases to be searched with an encrypted query sequence without exposing the query sequence to the proprietary database [19]
Honey encryption	• Eliminates brute-force attack • It also gives protection from DoS attack	• Computation time—high	• Reduce the computation time	• Geno Guard is a honey encryption-based tool for safe genomic information storage [16]
Garbled circuit	• Receiver will be able to decrypt only the gate which has the output	• Simple operations for small operands also demand the generation of circuit with millions of gates. So, it becomes expensive for small operations	• Reduce the number of garbled circuits	• METIS is a cryptographic system which is based on garbled circuit that allows secure computing of encoded genomic records with the significant feature that the information owner controls the sort of features that can be calculated [20]

References

1. Savage, S.R.: Characterizing the risks and harms of linking genomic information to individuals. IEEE Secur. Priv. **15**, 14–19 (2017)
2. Bradley, T., Ding, X., Tsudik, G.: Genomic security (lest we forget). IEEE Secur. Priv. **15**, 38–46 (2017)
3. Ayday, E., Humbert, M.: Inference attacks against Kin genomic privacy. IEEE Secur. Priv. **15**, 29–37 (2017)
4. Blanton, M., Bayatbabolghani, F.: Improving the security and efficiency of private genomic computation using server aid. IEEE Secur. Priv. **15**, 20–28 (2017)
5. https://www.cancer.gov/images/cdr/live/CDR761781-750.jpg. Accessed 06 Mar 2019
6. Naveed, M., Ayday, E., Clayton, E.W., Fellay, J., Gunter, C.A., Hubaux, J.-P., Malin, B.A., Wang, X.: Privacy in the genomic era. ACM Comput. Surv. **48**(1), 6 (2015), Article 6. https://doi.org/10.1145/2767007
7. Li, J.: Genetic information privacy in the age of data-driven medicine. In: IEEE International Congress on Big Data, San Francisco, CA, pp. 299–306 (2016)
8. https://www.alexirpan.com/2016/02/11/secure-computation.html. Accessed 27 Mar 2019
9. Ayday, E., De Cristofaro, E., Hubaux, J., Tsudik, G.: Whole genome sequencing: revolutionary medicine or privacy nightmare? Computer **48**, 58–66 (2015)
10. Deznabi, I., Mobayen, M., Jafari, N., Tastan, O., Ayday, E.: An inference attack on genomic data using kinship, complex correlations, and phenotype information. IEEE/ACM Trans. Comput. Biol. Bioinform. **15**, 1333–1343 (2018)
11. Kapil, G., Agrawal, A., Khan, R.A.: Security challenges and precautionary measures: big data perspective. ICIC Express Lett. **12**, 947–954 (2019)
12. Ogburn, M., Turner, C., Dahal, P.: Homomorphic encryption. Procedia Comput. Sci. **20**, 1–562 (2013)
13. https://www.usenix.org/conference/healthtech13/workshop-program/presentation/ayday. Accessed 05 Apr 2019
14. Mok, E., Samsudin, A., Tan, S.-F.: Implementing the honey encryption for securing public cloud data storage. In: First EAI International Conference on Computer Science and Engineering (2017). https://doi.org/10.4108/eai.27-2-2017.152270
15. Omolara, A.E., Jantan, A., Abiodun, O.I.: A comprehensive review of honey encryption scheme. Indones. J. Electr. Eng. Comput. Sci. **13**, 649–656 (2019)
16. Huang, Z., Ayday, E., Fellay, J., Hubaux, J., Juels, A.: GenoGuard: protecting genomic data against brute-force attacks. In: IEEE Symposium on Security and Privacy, San Jose, CA, pp. 447–462 (2015)
17. Huang, Z., Lin, H., Fellay, J., Kutalik, Z., Hubaux, J.-P.: SQC: secure quality control for meta-analysis of genome-wide association studies. Bioinformatics **33**, 2273–2280 (2017)
18. https://medium.com/@PlatON_Network/privacy-preserving-computation-secure-multi-party-computation-ii-b9d7c32be8a1. Accessed 05 Apr 2019
19. Titus, A.J., Flower, A., Hagerty, P., Gamble, P., Lewis, C., et al.: SIG-DB: leveraging homomorphic encryption to securely interrogate privately held genomic databases. PLOS Comput. Biol. **14**(9), e1006454 (2018). https://doi.org/10.1371/journal.pcbi.1006454
20. Battke, F., Egger, C., Fech, K., Malavolta, G., Schröder, D., Thyagarajan, S.A.K., Deuber, D., Durand, C.: My genome belongs to me: controlling third party computation on genomic data. Proc. Priv. Enhanc. Technol. **2019**(1), 108–132 (2019)

Establishing Trust in the Cloud Using Machine Learning Methods

Akshay Mewada, Reetu Gujaran, Vivek Kumar Prasad, Vipul Chudasama, Asheesh Shah and Madhuri Bhavsar

Abstract In the modern era of digitization, cloud computing is playing an important key role. Tons of user's data is on the cloud and to manage it with all the odds is a mesmerizing task. To provide user's data integrity and security are the major task for any cloud service provider. In the last few decades, researchers have been motivated to provide security on a cloud with reference to the categorization and classification of security concerns. In the cloud, trust is a major issue with respect to cloud security. The facilities provided by the cloud are too attractive for customers. Cloud has distributed and non-transparent services due to that the users may lose their control over the data and they are not sure whether the cloud provider is trusted one or not. This paper mainly focuses on establishing trust in the cloud using machine learning methods. The trust parameters like availability, confidentiality, accuracy, integrity, resource management, non-repudiation, risk management, protecting communication, hardware security, reliability, and secure architecture were analyzed, and the result shows the value of trust for system can be predicted based on the proposed algorithm.

Keywords Supervised learning · Cloud computing · Security · Trust · SLA · Availability

A. Mewada (✉) · R. Gujaran · V. K. Prasad · V. Chudasama · M. Bhavsar
Computer Science Department, Institute of Technology, Nirma University, Ahmedabad, India
e-mail: akshay.mewada_jrf@nirmauni.ac.in

R. Gujaran
e-mail: 17mcei12@nirmauni.ac.in

V. K. Prasad
e-mail: vivek.prasad@nirmauni.ac.in

V. Chudasama
e-mail: vipul.chudasama@nirmauni.ac.in

M. Bhavsar
e-mail: madhuri.bhavsar@nirmauni.ac.in

A. Shah
Mewar University, Chittorgarh, Rajasthan, India
e-mail: asheesh.shah@gmail.com

© Springer Nature Singapore Pte Ltd. 2020
P. K. Singh et al. (eds.), *Proceedings of First International Conference on Computing, Communications, and Cyber-Security (IC4S 2019)*, Lecture Notes in Networks and Systems 121, https://doi.org/10.1007/978-981-15-3369-3_58

1 Introduction

In the present era of digitization, the principle motivational factor is the trust in a cloud. Cloud organizations are recognizable under private, open, and business territories. An impressive part of these organizations is to maintain security, protecting communication, secure architecture, confidentiality, and quality are continuously basic edges [1, 2]. Trust management in cloud service is a highly prioritized issue today. Trust plays an important role in the cloud for security, availability, turnaround efficiency, reliability, etc. [1]. Nowadays, machine learning (ML) approaches are very useful for detection and analysis. The accuracy of the ML methodology is pretty higher than traditional techniques [3]. Our prime motive is to apply ML methodology to achieve noticeable accuracy with less human interaction and high automation for cloud security and quality. There are several parameters studied by researchers for analyzing trust establishment in the cloud, like availability, reliability, security, scalability, identity management (IDM), data indignity, and resilience [1]. In this research study, the availability is selected as one of the trust parameters of a cloud [1]. To measure the availability of cloud resources, we have used the ML-based linear regression (LR) approach for accurate measurement. The rest of the paper is organized as follows. Section 2 indicates about background and research challenges of cloud trust. Section 3, we have discussed cloud trust parameters. In Sect. 4, we proposed the cloud trust model. In Sect. 5, data integrity and SLA verification are introduced. Experimental results have been discussed in Sect. 6.

2 Background and Research Challenges for Cloud Trust

Trust plays an important role in the cloud for security, availability, turnaround efficiency, reliability, etc. [1]. There is a possibility of a malicious insider in the cloud, which decreases the acceptance of the trust in the customer's mind. So, with the help of machine learning method, we establish the trust method which increases the acceptance [4]. When cloud downtime occurs, trust cannot be built. The reason for downtime is power loss and network connectivity issues. Due to a network connectivity problem, data on the cloud may be lost [2]. Another issue is unprotected communication, due to unprotected communication the sender may not send to the authenticated person or it may be lost. The trusted cloud depends on many parameters like confidentiality, availability, risk management, resource management, secure communication, and secure architecture, etc. [5]. These all are parameters and its value is most important for formulating the cloud equation. And also, service-level agreement (SLA) has some key elements like hardware availability, power availability, data center network availability, backbone network availability, outage notification guarantee, internal latency guarantees, etc. [6]. These elements are most important for SLA. These elements are required for better trust management. Cloud provides infrastructure service to guarantee a quality of service (QoS).

2.1 Defining Trust

Trust means to increase productivity, reduce the cost of conducting business, and provide the backup option. Trusted cloud can be defined by the following points:

- Cloud provides the private platform to the users or companies, so companies store their data on a private cloud. Therefore, it reduces the cost of infrastructure.
- Make use of service-level agreement (SLA) for infrastructure service provider guaranteed quality of service. The QoS is related to data storage, CPU memory, etc. [7].
- Cloud makes the life of people very easier because they can share data or any information at high speed [8].
- Automatically update and maintain when new things emerge in the cloud environment [5].
- Customers do not need to use a pen drive or hard disk to store or for backup [9].
- Data integrity helps customers to check that received data is coming from the cloud service provider (CSP) side or third person [10].

2.2 Cloud Security

Cloud security is the fast-growing service that provides many functionalities like secure the confidential data form the theft, so that authorization people can use the services. Hence, our data is secure with data centers [11]. The cloud security has issues like confidentiality, availability, integrity, and privacy [10]. These issues are associated with cloud infrastructure. Data integrity means it provides the actual data which is required. Security (Fig. 1) is built on confidentiality, availability, and data Integrity (CIA):

Fig. 1 Security of cloud

Data integrity

– Confidentiality: Confidentiality means the need to protect the data form unauthorized access. Only authorization people can access the data.
– Availability: Information needs to available to authorized person only.
– Data integrity: To verify that received data is exactly the same data. Any corrupted or deleted data can be timely identified and this is the major point for data recovery.

2.3 Machine Learning Methods

Machine learning (ML) is a part of artificial intelligence (AI). Artificial intelligence means to learn automatically from experience and improve the system ability. The machine learning focuses on the development of a computer program that can access data and use it to learn for themselves. Machine learning has mainly three types: Supervised learning, unsupervised learning and reinforcement learning [4]. These methods are mostly used nowadays. Machine learning methods provide an accurate solution and it is easy to implement [3].

Supervised learning (SL): System already knows what is the output, if the system gets the desired output then closed the process. In supervised learning to calculate trust in the cloud, we are using metrics: Confidentiality matrix (C), integrity matrix (I), availability matrix (A), and reliability matrix (RM). And after that find the rank of these all metrics [12]. There is a limitation when trust is calculated using this method, it uses space and time. If data is not received in a timely manner then this method is failed [4]. Trust = Avg (Rank(C), Rank(I), Rank(A), Rank (RM)).

Unsupervised Learning (UL): The system does not know what is the output. For trust purpose, Hosseini et al. used the Naive Bayes trust model. This model is composed of three factors: (1) The cloud provider which performs the service requests. (2) The cloud user which makes the service requests. (3) The trust manager which helps the cloud users in selecting the most trustworthy providers. In this model, there are two phases that happen within the trust manager, namely: Handling service requests and trust computing. We have not considered the situation where the cloud users may give unfair high or low ratings to the cloud providers [13]. So, at this point, this method cannot be implemented [4].

Reinforcement Learning (RL): The RL method depends on the reward-based and feedback-oriented. Reward means machine self-set and feedback-oriented means to take action based on the environment. It has the ability to learn and re-learn the model as shown in Fig. 2. To achieve more efficient detection in the cloud, the RL method is used [3].

Fig. 2 Elements of RL

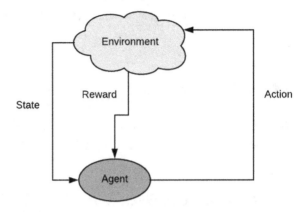

3 Cloud Trust Parameters

In this section, we are describing some parameters which are used by cloud service provider (CSP) for maintaining cloud resources to the period of time on the cloud.

3.1 Availability

Availability means system or data is accessible when required to use. In availability, there are two terms mainly: MTBR—mean time between failure and MTTR—mean time to repair. When a resource is too busy and resource is shut down then cloud resources said to be unavailable [5].

3.2 Data Integrity

Verify that received data is exactly the same. Any corrupted or deleted data can be timely identified and this is the major point for data recovery. Data integrity is an important and efficient term, and it includes security, accuracy, and data safety [10].

3.3 Resource Management

Resource management is based on the virtualization of available resources in distributed environment. Virtualization provides flexible and on-demand resources. When the task is completed, all the resources will be released. The best performance is achieved by cloud service provider's efficient resource management techniques [5].

3.4 Risk Management

There are three terms associated with this, risk assessment, risk control, and risk treatment.

3.5 Protecting Communication

The communication is based on encryption protocols like Secure Socket Layer and Transport Layer Security. With the help of these protocols, which provides the basics of confidentiality and integrity.

3.6 Hardware Security

Hardware security means a infrastructure security. Hardware security is important for monitoring network traffic and management of the system. For hardware security, it is necessary to consider the vulnerability of the existing system and provide security against it [14].

3.7 Secure Architecture

Performance management tools are required for the evaluation of secured cloud architecture.

4 Cloud Trust Model

For the business relationship, the cloud makes use of service-level agreement. The service-level agreement provides a framework for both seller and buyer and also provides protection to both. In the cloud, there are two types of service-level agreement: Infrastructure SLA and application SLA. SLA has some key elements like hardware availability, power availability, data center network availability, backbone network availability, outage notification guarantee, internal latency guarantees, etc. These elements are most important for SLA [8]. If these elements are not available when the process is going on than the trust cannot be calculated. System architecture of any framework describes the overall flow of the system in which manner it will work and flow to perform the given task. In this section of cloud system architecture, we have described the flow of cloud while providing various services to the users [15]. Figure 3 describes the flow of cloud toward trust parameters. Figure 4 describes the security of the trusted cloud. There are mainly three elements; authentication, authorization, and data integrity. Authentication means every user has its own user id and password, so when users want to communicate with CSP, this information is required for login purpose. Authorization means a person who has no authority to access the data of cloud then it cannot be accessed, only authorized person can access the data from the cloud. In a trusted cloud, the data integrity is the most effective parameter. It verifies that received data is exactly the same data that sender has provided. Any corrupted or deleted data can be timely identified and this is the major point for data recovery.

5 Data Integrity and SLA Verification

5.1 Data Integrity

The data integrity in the cloud is the most effective parameter. It verifies that received data is exactly to same data. Any corrupted or deleted data can be timely identified and this is the major point for data recovery. Also, it detects the unfair behavior and secures the user's confidential data [9]. Data integrity means a guarantee that the data can only be accessed by an authorized person [16]. Also, it provides the guarantee that data is of high quality, correct, and unmodified. Data integrity has the following properties:

Dynamic data handling: There are two types of data: static and dynamic. Static data means no change or fix data, whereas dynamic data means data maybe change, it uses some operation like deletion, insertion, and modification. Dynamic data more challenging compare to static data. Because dynamic data requires data integrity should remain unbroken even when insertion, deletion, and operation implement [7].

Data recovery: In data integrity, it is not sufficient to only determine or find corrupted data or act inappropriately. Cloud client also want a proper recovery of data. All property of data integrity identifies data corruption but some property also recovers the data. Mostly error-correcting code is used for data recovery. So data recovery used to determine or find corrupted data and also recover such data [14].

Robustness: Means identify the data corruption even the data is in the different size. Not applicable for large data set which is a limitation for data integrity, so for this limitation data integrity adopts a probabilistic approach. This approach works independently even the data corruption is too small [17].

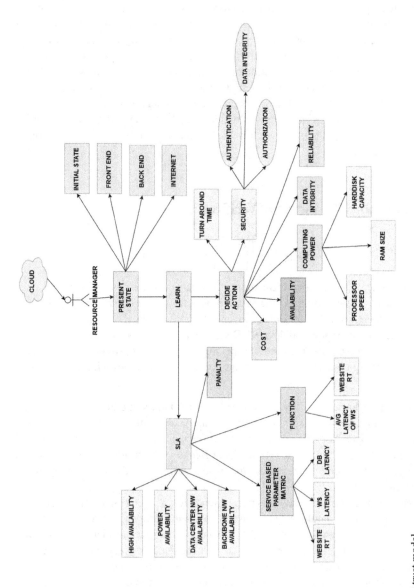

Fig. 3 Cloud trust model

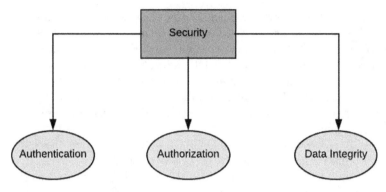

Fig. 4 Cloud security parameters

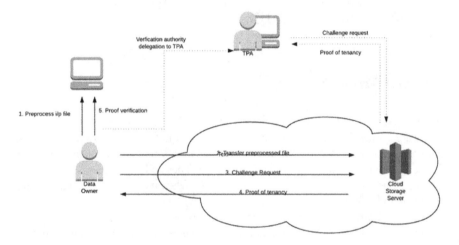

Fig. 5 Flow of data integrity

Privacy-preserving: When the analysis process is going on, the confirmation process should protect data privacy. In privacy-preserving property, third-party auditor cannot effect on confidential information of the client. So, privacy-preserving secures the user's confidential data [18].

Stateless verification: Client and server do not need to store previously results to verify future results. Because every challenge request is self-determining form all the past results. This is the essential requirement of data integrity [19].

Unrestricted challenge frequency: There is no limitation for several challenges made by the client to verify weather the data integrity is verified or not. Data integrity is not a one-time-activity. When client executes verification, at that time data integrity identifies the corruption or deletion [10] of data.

Soundness: The entrusted server should not be able to capture the challenge request. This property of data integrity ensures data reliability. If the CSP sends a challenge request with corrupted data to the client than the client never identify corruption data exactly [10]. So, the reliability of the data is more required in such cases.

Public/Private audit ability: Two approaches are there, first one for data owner analysis and second one for third-party auditor. The analysis process for both approaches is implemented without recouping the remote data. Outsourced data can verify by data owner only. Not all the time data owner may remain online for data integrity confirmation so data owner can give the responsibility to the third-party auditor. So, this supports public audit ability [6].

Fairness: To protect fair CSP, not to unfair users. If data integrity does not support the fairness that means unfair users harm the CSP reputation. So, fairness parameters are mostly used for security purpose. Also, it used for intrusion detection system to detect unfair users. So, the data integrity detects the unfair behavior and secures the user's confidential data. Data integrity means a guarantee that the data can only be accessed by an authorized person. Also, it provides the guarantee that data is of high quality, correct, and unmodified [11].

5.2 SLA Verification

Figure 6 contains information based on the SLA. SLA has some key elements like hardware availability, power availability, data center network availability, backbone network availability, outage notification guarantee, internal latency guarantees, etc. These elements are most important for SLA [7]. These all are elements that make a trusted cloud. If these elements are not available when the process running on than trust cannot be built (Fig. 5).

The service-level agreement provides a framework for both seller and buyer [5].

So seller and buyer both can pursue a profitable service business relationship.

5.3 Algorithm

This section mainly focuses on the algorithm. Mainly it covers the cloud trust parameters.

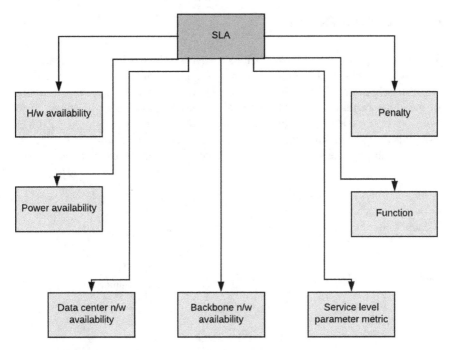

Fig. 6 SLA parameters

Algorithm 1 Algorithm for Trust

1: Inputs: CPU_Usage, Memory_Usage, RAM_Usage
2: Initialization: $CPU\_Usage \leftarrow 0, Memory\_Usage \leftarrow 0, RAM\_Usage \leftarrow 0$
3: Setup a Cloud Environment
4: $P1 \leftarrow CPU\_Usage$ & $P2 \leftarrow Memory\_Usage$ & $P3 \leftarrow RAM\_Usage$
5: $Threshold \leftarrow t1, t2$ [Dynamic Threshold based on current performance has been measured to identify the parameter t1 & t2]
6: Let P1 or P2 or P3 of new job= P.
7: **if** $P > t1$ & $P > t2$ **then**
8:     Start monitoring the process P
9: Calculate Ava, R, DI
10: Availability = MTBF/MTBF+MTTR
11:
$$R_{k(Ava)} = A_k/N_k, R_{k(D)} = D_k/C_k, R_{k(RE)} = C_k/A_k$$
12:
$$T = w_1 * Ava + w_2 * RE + w_3 * D + w_4 * TE$$
---

6   Experiment Results

This paper mainly focuses on establishing trust in the cloud using a machine learning method. Using one parameter is availability (Av) to calculate trust. Sample data set for availability as per Table 1 has been considered. The results shows that as uptime of the cloud increases, the availability of the resources increases as shown in the Figs. 7 and 9. As depicted in Fig. 8, the predicted value of availability is calculated using linear regression and mean absolute error is calculated as 0.28.

Use multiple linear regressions (supervised learning algorithm) in Jupyter (python).

– Step 1: Import the packages.
– Step 2: Check the current directory.
– Step 3: Set the dataset.csv le directory (Figs. 7, 8 and 9).

Table 1 Availability table

Monitoring (h)	Downtime (h)	Uptime (h)	Availability
24	0.79	23.2	0.96
24	1.6	22.3	0.92
24	2.4	21.6	0.9
24	3.19	20.8	0.86
24	4.0	19.9	0.82
24	4.7	19.2	0.8
24	5.59	18.4	0.76
24	6.4	17.5	0.72
24	7.2	16.8	0.7
24	7.9	16.0	0.66
24	8.8	15.1	0.62
24	9.6	14.4	0.6
24	10.3	13.6	0.56
24	11.2	12.7	0.52
24	12	12	0.5
24	12.7	11.2	0.46
24	13.6	10.3	0.42
24	14.4	9.6	0.4
24	15.1	8.8	0.36
24	16.0	7.9	0.32
24	16.8	7.2	0.3
24	17.5	6.4	0.26
24	18.4	5.5	0.22
24	19.2	4.8	0.4

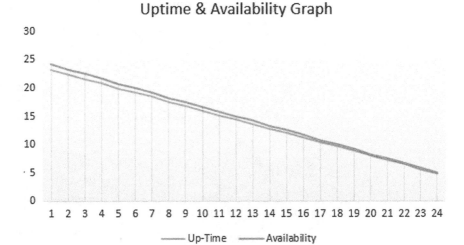

Fig. 7 Uptime and availability

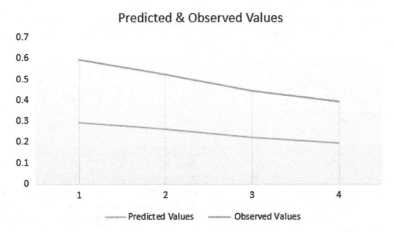

Fig. 8 Predicted and observed values

7 Conclusion and Future Work

Cloud computing can act like a utility for the industries, and its resources management is an important challenge. The results discussed here deal with different scenarios such as high availability and low availability. So, we can say that when the availability is high then the trust is achieved. This project usages the concepts of supervised learning to achieve trust. Also, this paper provides the knowledge about the various cloud parameters related to trust. As a future work, other parameters may be explored to compute trust in the cloud.

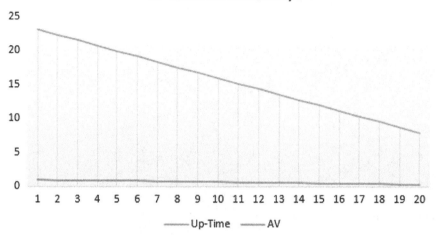

Fig. 9 Up-time and availability data for LR

Acknowledgements This research paper is a part of the project approved under DST (Department of Science and Technology), Government of India, New Delhi (Grant: DST/ICPS/CPS-Individual/2018/). We thank DST for funding the project.

References

1. Manuel, P.: A trust model of cloud computing based on Quality of Service. Ann. Oper. Res. **233**(1), 281–292 (2015)
2. Abderrahimand, W., Choukair, Z.: Trust assurance in cloud services with the cloud broker architecture for dependability. In: 2015 IEEE 17th International Conference on High Performance Computing and Communications, 2015 IEEE 7th International Symposium on Cyberspace Safety and Security, and 2015 IEEE 12th International Conference on Embedded Software and Systems, New York, NY, pp. 778–781 (2015)
3. Ling, M.H., Alvin Yau, K.-L.: Reinforcement learning-based trust and reputation model for cluster head selection in cognitive radio networks. In: The 9th International Conference for Internet Technology and Secured Transactions (ICITST-2014). IEEE (2014)
4. Wang, J.-B., et al.: A machine learning framework for resource allocation assisted by cloud computing. IEEE Netw. **32**(2), 144–151 (2018)
5. Bohlol, N., Safari, Z.: Systematic parameters vs. SLAs for security in cloud computing. In: 2015 9th International Conference on e-Commerce in Developing Countries: With focus on e-Business (ECDC), Isfahan, pp. 1–8 (2015)
6. Horvath, A.S., Agrawal, R.: Trust in cloud computing. In: SoutheastCon 2015, Fort Lauderdale, FL, pp. 1–8 (2015)
7. Ye, L., Zhang, H., Shi, J., Du, X.: Verifying cloud service level agreement. In: 2012 IEEE Global Communications Conference (GLOBECOM), Anaheim, CA, pp. 777–782 (2012)
8. Wang, H., Yu, C., Wang, L., Yu, Q.: Effective BigData-space service selection over trust and heterogeneous QoS preferences. IEEE Trans. Services Comput. **11**(4), 644–657 (2018)

9. El Makkaoui, K., Ezzati, A., Beni-Hssane, A., Motamed, C.: Data confidentiality in the world of cloud. J. Theor. Appl. Technol. **84**, 305–314 (2016)
10. Saad, M.I.M., Jalil, K.A., Manaf, M.: Achieving trust in cloud computing using secure data provenance. In: 2014 IEEE Conference on Open Systems (ICOS), Subang, pp. 84–88 (2014)
11. Saadat, S., Shahriari, H.R.: Towards a process-oriented framework for improving trust and security in migration to cloud. In: 2014 11th International ISC Conference on Information Security and Cryptology, Tehran, pp. 220–225 (2014)
12. Dey, S., Sen, S.K.: SVM—a novel trust measurement system in cloud service. In: 2018 Emerging Trends in Electronic Devices and Computational Techniques (EDCT), Kolkata, pp. 1–5 (2018)
13. Hosseini, S.B., Shojaee, A., Agheli, N.: A new method for evaluating cloud computing user behavior trust. In: 2015 7th Conference on Information and Knowledge Technology (IKT), Urmia, pp. 1–6 (2015)
14. Gonzales, D., Kaplan, J.M., Saltzman, E., Winkelman, Z., Woods, D.: Cloud-Trust—a security assessment model for Infrastructure as a Service (IaaS) clouds. IEEE Trans. Cloud Comput. **5**(3), 523–536 (2017). https://doi.org/10.1109/tcc.2015.2415794
15. Varadharajan, V., Tupakula, U.: On the design and implementation of an integrated security architecture for cloud with improved resilience. IEEE Trans. Cloud Comput. **5**(3), 375–389 (2017)
16. Neisse, R., Holling, D., Pretschner, A.: Implementing trust in cloud infrastructures. In: 2011 11th IEEE/ACM International Symposium on Cluster, Cloud and Grid Computing, Newport Beach, CA, pp. 524–533 (2011)
17. Deshpande, S., Ingle, R.: Trust assessment in cloud environment: Taxonomy and analysis. In: 2016 International Conference on Computing, Analytics and Security Trends (CAST), Pune, pp. 627–631 (2016)
18. Fugini, M., Hadjichristo, G.: Security and trust in Cloud scenarios. In: 2011 1st International Workshop on Securing Services on the Cloud (IWSSC), Milan, pp. 22–29 (2011)
19. Tripathi, M.K., Sehgal, V.K.: Establishing trust in cloud computing security with the help of inter-clouds. In: 2014 IEEE International Conference on Advanced Communications, Control and Computing Technologies, Ramanathapuram, pp. 1749–1752 (2014)

Basics for the Process and Requirements of Ethical Hackers: A Study

Anushka Garg and Vandana Dubey

Abstract Hacking is the act of breaking down into someone else PC for the cause of stealing data. Presently, hacking is majorly utilized in pessimistic practice. Through the proper utilization of ethics of hacking, we can apply it for superior motives, i.e., for the welfare of the society, it can be turned into ethical hacking. Before starting hacking, one should have proper knowledge about it—the attacks, its methods, its past experience, and then only a person can be a good ethical hacker. To be more precisely said, ethical hacking is the creativity in which we can enhance and help organizations to be secure and give a pathway for their growth.

Keywords Ethical hacking · Information security · Cyber security · Phases of hacking · Footprinting · Scanning · Exploitation · Tools for hacking

1 Introduction

Organizations confront different difficulties to monitor the framework. Cyber threats have become additional aggressive, complicated, and complex challenges. Attackers can be anyone whether it is a discontented worker, cyber terrorists, criminals, crime rings, and nation states [1–4]. The assaults can incorporate digital wrongdoing, hacktivism and reconnaissance. Hacktivism is the act of hacking, or enter into a computer system, for a politically or socially impel purpose. The person who performs an act of hacktivism is said to be a hacktivist. Reconnaissance is a primer review to pick up data; particularly, an exploratory military study of enemy territory.

Each organization is a potential target including corporates like Microsoft Inc., Sony, Fox, Lockheed Martin and financial institutes like, trade center, stock exchanges, bank headquarters, defense and civil supplies, and numerous others. The security attacks are exceedingly composed of network-based attacks and have

A. Garg (✉) · V. Dubey
Amity School of Engineering & Technology, Amity University, Lucknow Campus,
Noida, Uttar Pradesh, India
e-mail: garganushka20@gmail.com

V. Dubey
e-mail: vdubey@lko.amity.edu

© Springer Nature Singapore Pte Ltd. 2020

P. K. Singh et al. (eds.), *Proceedings of First International Conference on Computing, Communications, and Cyber-Security (IC4S 2019)*, Lecture Notes in Networks and Systems 121, https://doi.org/10.1007/978-981-15-3369-3_59

brought about. Huge measures of delicate information, for example, credit cards, medical data, passwords, and state privileged insights are being uncovered.

This paper basically discusses about the fundamental concepts and issues related to hacking along with requirements of ethical hacking. This paper is organized in a total of five sections. Sections 1 and 2 discuss introduction of the concept and fundamental issues of hacking, respectively. After that, stages of hacking and potential security threats for our systems are explained in Sects. 3 and 4, respectively. The paper is concluded in Sect. 5 at the end.

2 Hacker

The word hacker is usually used with negative connotations [4–8]. However, a hacker is someone who has the knowledge and technological expertises to understand and if necessary tamper with, software or electronic systems in general. While most hackers may have the ability to break into computer systems with malicious identity for the benefits of their parent companies or the common people. Considering the current trend of interpreting the term hackers in a negative way, many find it very important to differentiate themselves from the term. Therefore, the term ethical hacker is used by many software engineers to create a distinctive identity for them.

As stated above, hacking is a common problem nowadays to any computer system. Interestingly, the skills a person need to break into a system can also be used to detect any such vulnerability a computer system may possess. As such, an ethical hacker uses his skills to find out security flaws in electronic systems so that they can be eradicated before the system is compromised by a hacker. In simple words, an ethical hacker can test his own system to check how secure the system is.

Ethical hacking is also known as white hat hacking (as opposed to black hat hacking) or is sometimes referred to as penetration testing. While hacking is considered a criminal offense, ethical hacking is completely legal as it is done only after taking required permissions from the concern entities. Most organization hire ethical hackers to strengthen the security of their systems; however, some organizations may also allow outside attempt at breaching their security in order to establish the legitimacy of their security claims.

Types of Hackers:

As shown in Fig. 1, different types of hackers are discussed in the subsequent sections.

(i) **White Hat Hacker**: They use their skills to find weakness in a computer system. They are very much important to security testing. They are ethically bound not to harm the system and use their skills legally and are thus known as ethical hackers.

(ii) **Black Hat Hacker**: They are criminal hackers, commonly known as crackers. Whether they bring down a system to show off their skills or to steal money or information, their actions harm the affected systems and are generally considered illegal.

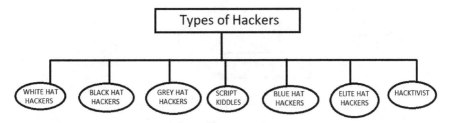

Fig. 1 Types of hackers

(iii) **Gray Hat Hacker**: Gray hat hackers are those who cannot be properly defined as either white hat or black hat hacker. While they do not cause harm by stealing money or information, they do often deface prestigious companies by bringing into notice their weaknesses or simply for fun. Even though most people know about black hat hackers or crackers for the media attention they get, it is gray hat hackers that comprise most of the hacking community.

(iv) **Script Kiddie**: The term is used for inexperience (or kid) script users who hack into a system with the help of scripts written by others. They do not possess good hacking skills, but use known and pre-existing tools without even understanding them fully.

(v) **Blue Hat Hackers**: These hackers are comparable to white hat hackers in the sense that they also use their skills to test the security flaws in a system. However, they often work as an outsider checking a system for vulnerability before it is released.

(vi) **Elite Hackers**: Apart from the broad groups mentioned above, there are elite hackers who are simply those with the highest skill set in the industry. They are often the ones who learn about the latest exploits and circulate the information.

(vii) **Hacktivist**: This group is comprised of a varied range of hackers who often act based on their ideological, religious, or political beliefs. While they sometimes bring down a system owned by people of opposing beliefs with DoS attacks, they can also act against organization by releasing sensitive information to the public.

3 Stages of Hacking

Basically, there are five stages of hacking as shown in Fig. 2 [9–11]. All these stages are discussed in following subsections in brief:

(i) **Reconnaissance**: This is the initial stage where the hacker tries to gather as much documentation as possible about the victim. It includes recognizing the target, discover out the victim's IP address range, network, DNS records, etc.

(ii) **Scanning**: It demands taking the information locate during reconnaissance and using it to inspect the network. Tools that a hacker may engage during

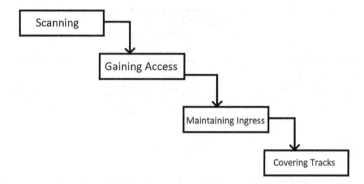

Fig. 2 Stages of hacking

the scanning stage involve dialers, port scanners, network mappers, sweepers, and vulnerability scanners. In this stage, hackers seek the informations which can assist them to commit assault such as computer names, IP addresses and utilizer's accounts.

(iii) **Gaining Access**: After scanning, the hackers draft the blueprint of the network of the victim with the assistance of data gathered during Stage 1 and Stage 2. This is the stage where the actual hacking starts. Susceptibilities uncovered during the reconnaissance and scanning stage are now utilized to gain access. The approach of relation the hacker utilizes for an advantage can be a localized area network (LAN, either wired or wireless), local key to a PC, the Internet or offline. Examples include stack-based buffer deluges, disapproval of service (DoS), and sessional hijacking. Gaining path is known in the hacker world as owning the system.

(iv) **Maintaining Ingress**: Once a hacker has gained access, they want to conduct that ingress for future exploitation and incursions. Sometimes, hackers amalgamate the system from other hackers or pact helpers by securing their absolute access with backdoors, rootkits, and Trojan. Once the hacker controls the system, they can utilize it as a base to bung additional incursions. In case, the controlled system is sometimes referred to as zombie system.

(v) **Covering Tracks**: Once hackers have been adept to gain and control access, they hide their tracks to avoid disclosure by pact helpers, to advance to utilize the owned system, to remove evidence of hacking, or to avoid constitutional action. Hackers try of abolish testimony of left off of the aggression, such as log files or intrusion detection system (IDS) alarms. Examples of action during this stage of the incursions include steganography, the utilize of tunneling protocols, and altering log files.

4 Potential Security Threats to Systems

There are various threats to the security of our valuable information available on our systems [12–15]. These threats are discussed in the following subsections in brief:

(i) **Footprinting**: Often termed as "hacker's best friend," Footprinting is a technique through which hackers can create a complete security profile of an organization. This technique primarily aims at finding info. related to Internet, intranet, remote access and extranet. Using footprinting, hackers can know as enough as they can about a system; from its remote ingress efficiency to its ports. In simple language, footprinting can be referred to as the unethical way to accumulate data regarding a specific work environment. It is also known as reconnaissance. In other words, fine art of gathering target information.

Footprinting is a streamlined process in which the particular steps are performed to get complete information about the networking environment of a system. In the first step, identifying the various domain names is performed in which the hacker is interested. Afterthat, in second step, analysis of the publicly available information of the target, like company name, domain name, business subsidiary, IP address, phone numbers, etc., is performed. Next step includes the process of identifying the capacity and size of the target, like number of potential entrances, security structure, etc., while in the final step, determining the scope of attack like range of network address, host names, exposed hosts, applications exposed on those hosts, OS and application-related patch information, structure of the applications, and the backend servers is done (Fig. 3).

Footprinting is very important because it is the first step for a hacker to begin damage to your organization [16–18]. This technique provides complete information about the system of an organization and helps the hacker determine the scope of attacks, that is, whether he is going to attack a specific server or the whole system. There are various tools and mechanism used in footprinting. For example, port scanners, like Nmap and ScanLine, to identify the live hosts on the Internet, TCP, UDP, and OS installed on the live hosts. Whereas, trace route to identify the relationship between hosts and the probable security mechanism between these and the attackers. Further, Nslookup (found in Windows System) for DNS (Domain Name System) queries and transfers zone, whereas Tracert creates network map of the target network.

Fig. 3 Methods of footprinting

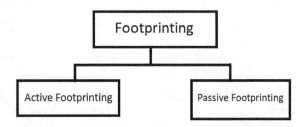

Methods of footprinting are categorized as passive and active footprinting methods. Passive footprinting is a method in which the attackers never make any direct contact with the target system. They gather information from anywhere where information is beforehand given. For example, Google search, Who is queries, DNS lookup, social networking sites. Whereas active footprinting is a method in which the attackers make any contact with the target system. They gather information from anywhere where information is beforehand given.

Some methods of active footprinting are mirroring website, e-mail tracking, Server Verification, etc. In mirroring website method, we download all available contents for offline analysis. We can copy websites as it is using website mirroring tools like—Teleport Pro, iMiser, HTTrack. After mirroring website, offline vulnerabilities are found. Due to offline mode, risk becomes zero. Whereas in e-mail tracking, the process is to examine e-mail processing path. From where the email has been sent, who has sent, ip address, etc., are known. For this, some tools are used like—e-mail tracker pro, MSGTag, PoliteMail, and Zendio. We can get every information of a fake mail, like it's location (actual), ip address, etc.

Moreover, the server verification method involves two step, initially determination of reachability of servers and then enumeration of network path from attacker to the target [19–21]. In the process of determination that servers are reachable or not, "ping" command is used, E.g., ping 192.168.1.2. It identifies connectivity with target. program that system administrators and hackers or crackers used to determine if a specific Computer is presently online and accessible is as follows: If host = 0%, Server is online and connected. Whereas, if host = 100%, Server is offline and disconnected. Moreover, to enumerate network path from attacker to Target the steps are from our system to the victim's system, the total number of routers we will cross, to find the path server verification is needed. Tools used—Tracert, Visual Traceroute, Sam Spade, TCR Trace Route, Tracert (trace route) C:\tarcert target(ip)/domain name. For example, target www.google.co.in. Furthermore, to identify connectivity with target for server verification, "ping" command may be used. Here, tools used are Ping, Tracert, Visual Traceroute, Sam Spade, TCR Trace Route. Generally, footprinting can be done through search engine, tools, and command prompt.

(ii) **Proxy**: A proxy server is essentially a middle computer that sits between the incursion and the victim. Mask your IP under a proxy server to show that you live in a different country to make yourself untraceable. Normally, when we open a website, our PC send our IP address to the website, so that it can send the webpage to your computer. But while using proxy, first the date will be sent to proxy and proxy sends it to our PC. Web server gets ip address of proxy server. If we have to hide our ip address from any website, then we can use www. anonymizer.com, www.hidemyass.com/proxy or https://hide.me/proxy. When we type a Web address into these websites, it retrieves the page we requested and displays it (Fig. 4).

There are basically three ways to attack named as direct attack/no proxy, logged attack, and using proxy chaining. Figure 5 shows direct attack in which there are suicide attackers. Whereas Fig. 6 provides logged proxy attack where only one proxy

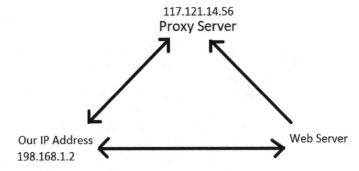

Fig. 4 Process of retrieval of requested pages through proxy server

Fig. 5 Direct attack

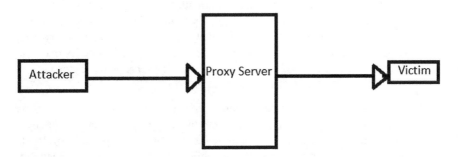

Fig. 6 Logged proxy

is used. Afterthat, Fig. 7 provides proxy chaining where different proxies will be from different countries to make victim next to impossible to find real attackers. These attackers are masters of hacking.

Attacker, keeping in mind from which country do he belongs. For example, if the attacker is from USA and diplomatic unfriendly countries. These are those countries

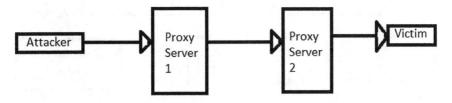

Fig. 7 Proxy chaining

with whom one country's political term is not good. For example, USA does not have any good terms with IRAN, North Korea, Cuba, Taliban, etc.

The major cause for proxy utilization is to cover the authority IP address so that an incursion can hack without any constitutional corollary or to mask the actual source of the incursion by impersonating a fake source address of the proxy. Moreover, to wirelessly ingress intranets and other websites assets that are basically off limits or to interfere every request send by an incursion and change them to a third destination, hence attackers will only be able to analyze the proxy address. Furthermore, attackers utilize chain multiple proxy server to avoid detection.

Some generally used proxy tools are Proxy Workbench, Proxifier, Proxy Switcher, SocksChain, and The Onion Routing (TOR). Moreover, Proxifiers are bypass firewall restrictions, Full IPv6 support, work through chain of proxy servers using different protocols, Secure Privacy by hiding your IP address, proxy protocols (SOCKS v4 HTTPS, SOCKS V4A HTTP, SOCKS v5, etc. Basically, Proxy Switcher is an easy way to change proxy setting on the fly, hides your IP address from the websites you visit, full support of password protected servers and automatic proxy server switching for improved anonymous surfing.

(iii) **Scanning**: Scanning is important to both attackers and those responsible for security hosts and networks. Through only footprinting, we cannot collect all the information. To collect more information, hacker use scanning. Scanning is additional information to footprinting [6]. For example, to identify operating system of victim such as XP, W, Linux, etc., services that are running on victim's pc which are OS inbuilt or to identify open ports.

Attackers like penetration tester, cyber security expert, network administration, white hat attacker penetration testers are hired by an organization or a company to secure their database. They first do scanning on that company's pc and identify the open ports or services, or we can say vulnerability. After scanning, the penetration tester closes all the vulnerability so that attackers cannot attack on it.

As shown in Fig. 8, basically there are three types of scanning named as port scanning, network scanning, and vulnerability scanning. In port scanning, attackers identify open ports and through it, he enters the system. Whereas, in network scanning, attacker identifies IP address, live host, and system architecture. Furthermore, vulnerability scanning has major importance. If attacker

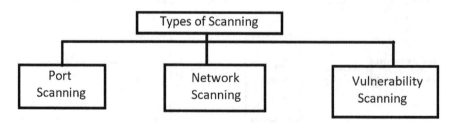

Fig. 8 Types of scanning

gets information about target system's software and its version, then he can get every/whole information of the system and then he tries to search for vulnerabilities.

(iv) **Enumeration**: Enumeration or more accurately network enumeration is a method of retrieving username or host or devices in a network. Information of a group, shares or services in a network domain can be found out in this method. This is an activity used by hackers and attackers where they create connection with the system and makes direct queries in order to get more information; This helps them to identify the points of attack for the system which is necessary for password attack. An attacker can conduct this in an intranet environment and get unauthorized access to data. Information Enumerated by Intruders are basically aimed towards retrieving information about users and their belongings to further use in other directions.

There is several information that the attacker can enumerate such as network resources, routing tables, network shares, audit and service settings, machine names, SNMP and DNS details, users and groups, applications and banners, etc [22, 23]. Some common techniques used for the purpose of enumeration are extracting user names using emails IDs, extracting information with the help of default password, extracting the user name with the help of SNMP, brute force activity directory, extracting user group from windows, extracting information with the use of DNS zone transfer.

Every network device is identified through a string of 16 ASCII characters in the TCP/IP platform. Among these, 15 characters signify the name of the device while the 16th one is kept for service or name record type. Attackers use the NetBIOS enumeration in order to obtain information on policies and password, list of computers in the domain and list of shares of the hosts in the network. Some common tools of NetBIOS enumeration are SuperScan, Hyena, Winfingerprint and NetBIOS enumerator and Netauditor (Network Security Auditor). Some common tools for SNMP enumeration are OpUtlis, Engineer's toolset, SNMP scanner, Getif, SNMP Informant, NET-SNMP, SNScan and Spiceworks. Moreover, some common tools for LDPA enumeration Softerra LDPA administrator, JXpolorer, LDPA search, LDPA account manager, Activity directory domain and LEX, the LDPA explorer, etc.

Enumeration can be stopped by taking correct measures in respect to the platform that is vulnerable. Check out the measures taking for different protocols. Enumeration pen testing is done by the attacker before an attack is made. It can identify the valid user accounts in a poorly protected resource share where active connection is established with systems and directed queries. The data collected in the reconnaissance phase is used for the method to work fine.

Pen testing starts with finding the network range which can be done with the use of tools like Whois Lookup. Once that is done, subnet mask is calculated and it goes through host discovery. Port scanning is done in the final stage to find out the open ports. Then NetBIOS, SNMP, LDPA, NTP, SMTP, and DNS enumerations are performed one by one. All the findings are documented in the end.

(v) **Hacking and Exploitation**: Exploitation could be a piece of programmed package or script which might enable hackers to require management over a system, exploiting its vulnerabilities. Hackers commonly use vulnerability scanners like Nessus, Nexpose, OpenVAS, etc., to seek out these vulnerabilities. Metasploit could be a powerful tool to find vulnerabilities in an exceedingly system. Considering the vulnerabilities, we discover misuses. John the Ripper can be used for the password cracking. [7].

Basically, there are two sorts of adventures names as remote exploits and nearly exploits. Remote exploits are the kind of endeavors where you don't approach a remote framework or system. Programmers utilize remote endeavors to access frameworks that are situated at remote spots. Whereas in nearby exploits, local adventures are for the most part utilized by a framework client approaching a neighborhood framework, however, who needs to bridge his rights.

5 Conclusion

To be a good ethical hacker, in-depth knowledge is the most important key for success. There are many types of hackers; it totally depends on us who we want to become. But to be an ethical hacker, we should follow the procedure step by step from footprinting to covering tracks without escaping any in between. We may also learn the fact that the ethical hacking is not only good but also helps us to understand the vulnerabilities in our computer that can put us in major trouble if not taken seriously. It helps us to protect from attacks that may put out integrity, confidentiality, authentication, and availability in danger. Ethical hacking ensures the prevention of attack from the black hackers and thus, securing our data and identification. This paper basically discusses all these issues and concepts related with hacking.

References

1. Kephart, J.O., Sorkin, G.B., Chess, D.M., White, S.R.: Fighting computer viruses. Sci. Am. **277**(5), 88–93 (1997)
2. Engebretson, P.: The Basics of Hacking and Penetration Testing Ethical Hacking and Penetration Testing Made Easy, 2nd edn, Elsevier (2013)
3. Harper, A., Harris, S., Ness, J., Eagle, C., Lenkey, G., Williams, T.: Gray Hat Hacking. McGraw-Hill, New York (2011)
4. Domingues, P., Frade, M.: Digitally signed and permission restricted pdf files: a case study on digital forensics. In: ARES 2018 Proceedings of the 13th International Conference on Availability, Reliability and Security, Article No. 51, 27–30 Aug 2018
5. Carnevale, D.: Basic training for anti-hackers: an intensive summer program drills students on cybersecurity skills. Chronicle Higher Educ. **52**(5), 41–41 (2005)
6. Sanders, A.D.: Utilizing simple hacking techniques to teach system security and hacker identification. J. Inf. Syst. Educ. **14**(1), 5 (2003)

7. Murugavel, U., Dr. Shanthi: Survey on ethical hacking process in network security **3**, 76–81 2014
8. Idimadakala, N.: Ethics in ethical hacking **4**, 876–881 (2013)
9. Logan, P.Y.: Crafting an undergraduate information security emphasis within information technology. J. Inf. Syst. Educ. **13**(3), 177–182 (2002)
10. Farsole, A.A., Kashikar A.G., Zunzunwala, A.: Ethical hacking. Int. J. Comput. Appl. **1**(10), 14–20 (2010)
11. Mortensen, C., Winkelmaier, R., Zheng, J.: Exploring attack vectors facilitated by miniaturized computers. In: Proceedings of the 6th International Conference on Security of Information and Networks (SIN '13), pp. 203–209 (2013)
12. Boulanger, A.: Catapults and grappling hooks: the tools and techniques of information warfare. IBM Syst. J. **37**(1), 106–114 (1998)
13. Danish, J., Muhammad, A.N.: Is ethical hacking ethical? Int. J. Eng. Sci. Technol. **3**(5), 3758–3763 (2011)
14. Smith, B., Yurcik, W., Doss, D.: Ethical hacking: the security justification redux. In: IEEE Transactions, pp. 375–379 (2002)
15. Reto, B.: Ethical hacking. In: GSEC Practical Assignment, Version 1.4b, Option 1 (2002)
16. David, H.M.: Three different shades of ethical hacking: black, white and gray. In: GSEC Practical Assignment, Version 1.4b, Option 1 (2004)
17. Cherdantseva, Y., et al.: A review of cyber security risk assessment methods for SCADA systems. Comput. Secur. **56**, 1–27 (2016)
18. Sepehri, M., Trombetta, A., Sepehri, M., Damiani, E.: An efficient cryptography-based access control using inner-product proxy re-encryption scheme. In: Proceedings of the 13th International Conference on Availability, Reliability and Security—ARES 2018, Article No. 12, Aug 27–30 2018
19. Behera, D.: Ethical hacking: a security assessment tool to uncover loopholes and vulnerabilities in network and to ensure protection to the system. Int. J. Innov. Adv. Comput. Sci. **4**, 54–61 (2015)
20. Castiglione, A., De Santis, A., Soriente, C.: Security and privacy issues in the portable document format. J. Syst. Softw. **83**(10), 1813–1822 (2010)
21. Roy, A., Karforma, S.: A Survey on digital signatures and its applications. J. Comput. Inf. Technol. **3**(1), 45–69 (2012)
22. Sepehri, M., Trombettai, A., Sepehri, M.: Secure data sharing in cloud using an efficient inner-product proxy re-encryption scheme. J. Cyber Secur. Mobil. **6**(5), 339–378 (2018)
23. Wu, R., Ahn, G.J., Hu, H.: Secure sharing of electronic health records in clouds. In: 8th International Conference on Collaborative Computing: Networking, Applications and Worksharing (CollaborateCom), pp. 711–718 (2012)

Privacy-Preserving Scheme Against Inference Attack in Social Networks

Nidhi Desai and Manik Lal Das

Abstract Social networks analytics provide enormous business values for organizational and societal growth. Massive volumes of social network data get collected in every moment and released to other parties for various business objectives. The collected data play an important role in designing policies, plans and future projection of business strategies. These social network data carry sensitive information, and therefore, the adversary can exploit users profile and social relationships to disclose their privacy. Inference attack using the mining technique poses a crucial concern for privacy leakage in social networks. In this paper, we present a privacy-preserving scheme against inference attack. The proposed scheme adds spurious data in the published dataset such that sensitive information is not predicted using mining techniques. The proposed scheme is analyzed against a strong adversarial model, where an adversary is allowed to gather background knowledge from different sources. We have experimented the proposed scheme on real-social network dataset and the experimental results show that the privacy-preserving property of the proposed scheme outperforms in comparison with other related schemes.

Keywords Social networks · Data privacy · Inference attack · Background knowledge

1 Introduction

Social network data primarily consist of user profiles and their social relationships. This data is helpful for analyzing the behavior of users that could help in designing different policies for the betterment and at the same time highlight the problems faced by the society. However, with the umpteen advantages, social network data is prone to privacy disclosure. The privacy of individual(s) is under constant threat, as the

N. Desai (✉) · M. L. Das
DA-IICT, Gandhinagar, India
e-mail: nidhi_desai@daiict.ac.in

M. L. Das
e-mail: maniklal_das@daiict.ac.in

© Springer Nature Singapore Pte Ltd. 2020
P. K. Singh et al. (eds.), *Proceedings of First International Conference on Computing, Communications, and Cyber-Security (IC4S 2019)*, Lecture Notes in Networks and Systems 121, https://doi.org/10.1007/978-981-15-3369-3_60

social network data, in many cases, is shared with the third-party users for research, advertising and designing applications. Therefore, preserving trade-off between privacy and utility in social networks in current scenario is a challenging research direction. Anonymization technique has been widely used for preserving privacy of individual's data while the data is being shared or published in digital platform. There are several instances, where the social network data got de-anonymized, and privacy got disclosed [1–4]. Due to de-anonymization threat, either many users do not publish the sensitive information or publish less accurate sensitive information. However, the adversary can still infer the unpublished sensitive information from the published dataset using the background knowledge.

There are two types of approaches used for privacy disclosure in the existing literature of social networks. The first approach is de-anonymization. It is applicable when sensitive information is published in the social network data. De-anonymization occurs when an adversary is able to identify a user in the published social network data using background knowledge [4]. Background knowledge can be considered from specific personal identification information, social relations to a more generalized voter list. The adversary using background knowledge is able to identify users in published social network data. This, in turn, will disclose the sensitive information associated with the users. The second approach is inference. It is applicable when sensitive information is not published in the social network data. Inference attack is when an adversary is able to know the sensitive information which is not published. The adversary uses different data mining techniques to infer the sensitive information. Both the above approaches are capable enough to disclose sensitive information. Even though de-anonymization would be able to identify the users, it still will not be able to disclose privacy when sensitive information is not published. As a result, the inference attack is more severe and powerful in terms of privacy disclosure in social network data.

Inference attack in social networks has garnered much attention in recent times. Several research problems have modeled the inference attack [1, 2] using different approaches. At the same time, adversary's capability has also been enhanced, as there is a large volume of data on individual's knowledge available from sources like social networking profiles, social relations and factual information and research observations. Furthermore, an adversary using mining techniques on the social network data can predict sensitive information. In particular, the adversary uses classifier to predict unpublished sensitive information, where input is available/published social dataset and output is the predicted unpublished sensitive information. Specifically, an adversary can learn rules from neighbor's dataset, which acts as a training dataset to predict and infer unpublished sensitive information. Recently, Cai et al. [1] modeled the inference attack in social networks, wherein it infers unpublished sensitive information by mining information from social circles. Their work considers profile as well as social relationship together for inference. Their basic intuition behind inference is more are common attributes in the social circle; more are the chances for similar sensitive information too. Cai et al. [1] have also proposed data sanitization method to address the inference attack. However, the method in [1] is application-dependent where the publisher needs a priori to specify privacy-dependent and utility-dependent attributes. Privacy and utility attributes are dependent on the application as for different applications the role of attributes also change, which itself is a rigid assumption, as the publisher may not have complete conviction about privacy and utility attributes before the release of data. Furthermore, the scheme in [1] removes privacy-dependent attributes and perturbs attributes common to privacy and utility prior to publishing. It is noted that removing privacy attributes negatively affects utility if the publisher is not sure about the application. Having observed the existing literature, we thought to devise a more practical and generic privacy-preserving scheme that can withstand against inference attack.

Our Contributions: In this paper, we present a privacy-preserving scheme to protect individual's privacy against inference attack. The proposed scheme adds spurious data in the published dataset such that sensitive information is not predicted using mining techniques. The proposed scheme assumes strong adversarial capability which can collect background knowledge from different sources. The proposed scheme is analyzed and the experimental results show its privacy-preserving strengths in comparison with other related schemes.

Organization of the paper: The structure of paper is as follows: Sect. 2 summarizes the related work. Section 3 presents the proposed scheme. Section 4 provides the analysis and implementation results. We conclude the paper in Sect. 5.

2 Related Work

De-anonymization in social network has been widely studied in past decade. There are existing de-anonymization [3–5] techniques that aim not to reveal identification information to adversary. De-anonymization will help in understanding the adversary's perspective in terms of privacy disclosure. This indeed will help in devising comprehensive and strong privacy models and techniques in social networks. Various anonymization techniques were proposed in the literature regarding social networks [5]. They have considered selective background knowledge. This is impractical in current scenario where the adversary has access to numerous online resources. Apart from the resources like identification information, the capabilities have also increased in terms of machine learning and data mining techniques. However, in modern resource-rich digital universe, an adversary is equipped with a large pool of background knowledge [2] by which he/she tries to infer unpublished sensitive information. Typically, background knowledge in [2] consists of social relations of individual, factual information, specific individual information and statistical information. Using the above-specified knowledge, the adversary is able to link individual and infer unpublished sensitive information using available knowledge. As a result, the inference attack in social networks has become a challenging research problem [1, 2, 6–9] in recent times. Initially, the inference attack has been studied and modeled based on community structure, user attributes and social relations [6, 7, 9]. As the computational capabilities of adversary increased, the complexity in modeling has also been increased. Nowadays, the inference attacks are studied and modeled using knowledge graph, rough set classifier and social-behavior-attribute network model. Qian et al. [2] modeled the inference attack using knowledge graph, with assumptions that the adversary has more specific information about individual whose privacy is to be disclosed. Specifically, it has focused on correlational knowledge in terms of inference. For example, a correlation like doctor earns more than 50 K can help in inferring salary. First of all, it de-anonymizes the graph based on the personal information and infers based on the correlation knowledge it possess to undermine the unpublished sensitive information of user. Gong et al. [5] modeled the inference attacks using social-behavior-attribute network model. In their work, the adversary generates a social-behavior-attribute network which incorporates social structure, user behavior and user attributes. It infers based on the behavior by collecting and analyzing different attributes present in various public domains. Zhong et al. [10] address demographic inference from location information available on social networks. It specifically targets the human mobility to infer the demographic information of user. The demographic user is specifically gender, age and many more in the direction. Nie et al. [11] predict occupation by collecting all occupation information available on various online social platforms using user attribute learning model. Jurgens [12] infers location based on social relationships on online social platforms. It takes a small amount of initial locations for prediction. Initially, the inference attacks used de-anonymization approaches for identification of users. Once, identified, users' unpublished sensitive information was inferred by different approaches like path ranking algorithm. Eventually, the location-based inference came into picture. Location inference was modeled using social relations and social data on heterogeneous platforms. Subsequently, user location obtained from different social network platforms helped in inferring demographic attribute information. Similarly, other attributes like occupation were collected from various platforms to infer where the attribute was not published. Then, gradually, the inference attack expanded its scope and capability by using different approaches like mining techniques where dependence upon background knowledge decreased. Recently, Cai et al. [1] modeled the inference attack using attribute and relation-based classifier. Initially, an attribute-based classifier generates rough set theory-based decision rules to infer unpublished attributes. Subsequently, attribute- and relation-based classifiers are applied together to infer the unpublished attributes, where relation-based classifier is based on the intuition that more common attributes the neighbors share, more are the chances of sharing the same sensitive information. The modeling approaches in [2, 5] use substantial background knowledge to infer the sensitive information as compared to [1]. Therefore, the inference using mining technique is more challenging in terms of adversarial capabilities. To overcome the inference attack using mining techniques, Cai et al. [1] proposed a data sanitization method. The data sanitization method takes as an input the social graph, privacy attributes and utility attributes and outputs a sanitized social graph. The privacy-dependent attributes are the attributes that can help in generating rules using the RST concepts [15]. Utility attributes are the attributes that are important in maintaining the usefulness of the data. It removes privacy attributes and perturbs attributes that are common in terms of privacy and utility. For categorical attributes, perturbation replaces the current attribute value with a broader attribute. Numerical attributes are perturbed by new numerical values. New numerical value is ratio of difference of the maximum and minimum value for a particular attribute to the range of the attribute set with respect to the generalization level [1]. It is noted that the data sanitization method can only be applied when privacy and utility attributes are decided a priori. This is not a flexible assumption as the publisher will not always initially select privacy and utility attributes with greater conviction. We further note that removing entire attribute set (specifically privacy-related attributes) is not a practical approach. Cai et al. [1] also suggests a method for link anonymization when a specific application is not known. However, the classification using the attributes is still possible as no nodes are removed so the rough set theory-based rules are generated. Also, the conventional methods like k—anonymity, l—diversity works when the sensitive attribute is published [1]. But, in the inference attack, the main notion is to infer the unpublished sensitive information from the given set of attribute information. Therefore, there is a need to devise a privacy-preserving scheme against inference attack.

3   The Proposed Privacy-Preserving Scheme

We propose a privacy-preserving scheme against the inference attack using mining technique. We first provide a brief context followed by the proposed algorithm.

Definition 1 *Social Network Graph* [1] A social network graph is represented as G (G_u, $C \cup D$, E) where G_u is a set of users, C is a set of quasi-identifier attributes, D is a set of sensitive attributes and E is a set of edges.

Here, E consists of two types of edges, namely relation edge E_R and attribute edge E_A. Relation edge E_R is an edge e_{ij}^R which connects two users u_i and u_j, where u_i, $u_j \in G_u$. Attribute edge E_A is an edge e_{ij}^A which connects a user u_i with its corresponding quasi-identifier attribute value C_j or sensitive attribute value D_j. The sensitive attribute consists of two types (1) D_K and (2) D_U where D_K is set of known sensitive attribute values in G and D_U is set of unknown sensitive attribute values in G.

We extend the definition described of social network graph given in [1] with respect to attribute edge. We add spurious nodes in the social network, where each spurious node should capture attributes as well as relations together. To simplify the above context, we capture the social graph into a social network data table to better understand the dynamics.

Definition 2 *Social Network Data Table* Social network data table is represented as T (T_u, QI $\cup$ SA, E_R) where T_u is a set of users, QI is a set of quasi-identifier attributes, SA is a set of sensitive attributes and E_R is a set of relation edges.

The social data table T retains the same information that is present in the social graph G. We define a mapping function $f\colon G \rightarrow T$ that maps the information/data present in graph G to the information/data in data table T. Here, function f is an invertible function. The mapping of G to T is as follows: $f(G_u) \rightarrow T_u, f(C) \rightarrow QI, f(D) \rightarrow SA$ and $f(E_R) \rightarrow E_R$ and $f(E_A) \rightarrow (T_u, QI \cup SA)$. In the same way, mapping of T to G can be achieved and data table can be recaptured to a social graph.

The proposed privacy-preserving technique adds spurious nodes to overcome the inference attack using mining techniques. The spurious nodes are added in such a way that the prediction of the unpublished sensitive information is decreased using mining techniques. Anonymized graph G' follows the same mapping function f as discussed above and thus can be converted from G' to T' (T' to G') without data loss. Spurious nodes preserve the same structure as G and spurious records as T.

3.1 Algorithm

The proposed algorithm works as follows. Firstly, the social network graph is converted to social network data table. Secondly, rules are generated using the attribute-based rough set classifier as in [1]. Thirdly, the spurious nodes are added in the social network data table and converted back to social network graph.

Algorithm 1 Privacy-Preserving Scheme

INPUT: Social Network Graph G, PDA, UDA, D_{SA}.
OUTPUT: Anonymized Social Network Graph G'.
 1: Capture Social Network Graph G to Social Network Data Table T.
 2: T' is a U x R Matrix.
 3: $T' \leftarrow T$.
 4: Rules is a N x M Matrix.
 5: $Rules[N][M] = Learn\_RST\_Rule(T(QI \cup SA), PDA, UDA)$.
 6: R is a U x (R-M) Matrix.
 7: $R[U][R - M] = Extract(T(QI \cup SA, E_R), Rules[N][M])$
 8: sp is a N x R Matrix.
 9: **for** i = 1 to N **do**
10:     $sp[i][1] \leftarrow U_{u+1}$
11:     **for** j = 2 to q **do**
12:         $sp[i][j] \leftarrow Rules[i][j]$
13:     **end for**
14:     **for** j = q+1 to M **do**
15:         $D'_{SA} \leftarrow D_{SA}/Rules[i][j]$
16:         $sp[i][j] \leftarrow Random(D'_{SA})$
17:     **end for**
18:     num is an l dimensional array.
19:     $num[l] \leftarrow Match(Rules[i], R)$
20:     $a = Random(num[l])$
21:     **for** j = M+1 to R **do**
22:         $sp[i][j] \leftarrow R[a][j - M]$
23:     **end for**
24:     Insert $sp[i]$ into T'.
25: **end for**
26: Recapture Anonymized Data Table T' to Anonymized Social Network Graph G'.

The algorithm takes as an input the social network graph G, domain of the sensitive attribute D_{SA}, the set of privacy-dependent attributes PDA and UDA is the set of utility-dependent attributes (here, PDA and UDA are optional). We have incorporated PDA and UDA from [1]. The output of the algorithm is anonymized social network graph G'.

The steps of algorithm are as follows: Line 1 captures the social graph G into a data table T. Lines 2-3 create an anonymized table T'. Lines 4-5 generate rules by using attribute-based rough set theory classifier [1]. Lines 6-7 extract the relation edges from the table T with respect to rules generated in step 5. Lines 8-25 generate the spurious records and inserts into anonymized table T'. Line 26 recaptures the table T' to anonymized graph G'.

4 Analysis and Experimental Results

4.1 Analysis

We analyze the proposed scheme against an adversary Adv_{RST} with the following assumptions. The adversary is given anonymized social graph G', which consists of incomplete social data, where some users do not publish sensitive information. The adversary Adv_{RST} has the capability of inferring unpublished sensitive attribute values by using the attribute classifier as in [1]. It also has access to the relation-based classifier [1]. The function [1] available with the adversary is as follows:

$F(\mathbf{G'}, \mathbf{Id}) \rightarrow \mathbf{D_U}$: The function takes input as social graph G' and particular information about individual (optional). The output of function is D_U, which is either the unpublished sensitive attribute value of individual "Id" or set of unpublished sensitive attribute values in general. The function F consists of two consecutive steps for predicting the sensitive attribute value, one is with attribute-based classifier and second is with attribute and relationship-based classifier collectively.

Goal: Given an anonymized social network graph G' consisting of attributes and relationship information of users, the goal of adversary is to predict the unpublished sensitive attribute value of user I in graph G' with high probability.

Privacy: Privacy is quantified with respect to inference attack using mining techniques in two ways: (i) attribute-based classifier and (ii) attribute- and relation-based classifier together, as defined below.

Definition 3 *Privacy of Social Network Graph G' Given an individual I in graph G', the probability P_A of the sensitive attribute value D_m using attribute-based classifier is equal to 1. Secondly, given an individual I, the ratio of probability of sensitive attribute value D_m using the attribute-based classifier P_A and relation-based classifier P_R should be approximately 1.*

1. $P_A(D_m^I) = 1$
2. $\alpha(P_A(D_m^I)) + \beta(P_R(D_m^I)) \approx 1$ where $\alpha + \beta = 1$

If the adversary is able to predict the sensitive attribute value D_m of individual I with the above-defined probability criteria, privacy is breached of graph G'. Otherwise, privacy is not breached.

Using the above foundations, we prove privacy of the proposed privacy-preserving technique against adversary having the capability of inference attack using mining techniques.

Theorem 1 *Anonymized Social Network Graph G' preserves privacy against adversary* Adv_{RST}.

Proof The adversary Adv_{RST} triggers the function F on graph G' as follows:

$$Adv_{RST} : F(G', I) \rightarrow D_m^I$$

The adversary Adv_{RST} is successful if it predicts sensitive attribute value D_m of individual I with probabilities as per Definition 3 otherwise unsuccessful. We elaborate function F to know adversary's success or failure in the following steps:

1. The initial step is to predict using attribute-based classifier. The first step is to generate the decision rules using quasi-identifier attributes and sensitive attributes in graph G'. However, when spurious records are added, the C positive region of D is empty as C lower approximation of D is an empty set. As a result, there is no decision rules generated. Therefore, the attribute-based classifier is unable to predict the sensitive attribute value D_m^I using the attribute-based rough set theory classifier. Privacy against inference attack using attribute-based classifier is preserved as $P_A (D_m^I) \neq 1$.
2. The next step is to predict the sensitive attribute value (D_m) of individual I using both attribute as well as relation-based classifier. As we have proved in step 1 that attribute-based classifier will not be able to predict the sensitive attribute value, there is only relation-based classifier to predict the sensitive attribute value. Also, the individual I requires at least one neighbor whose sensitive attribute value is known for prediction, which is a drawback. We have added spurious edges so the numbers of neighbors have increased which will decrease the prediction probability. As a result, privacy against inference attack using attribute and relationship-based classifier is preserved as $\alpha(P_A(D_m^I)) + \beta(P_R(D_m^I))$ not ≈ 1.

Therefore, the adversary using the attribute- and relation-based classifiers is not able to predict the sensitive attribute value of individual I with non-negligible probability. Hence, G' preserves privacy against Adv_{RST}.

4.2 Experimental Evaluation

To evaluate the proposed scheme, we have performed experiments on the Facebook Dataset [13] by taking two ego networks and naming it DATA1 and DATA2,

Table 1 Summarization of social network data sets

Information	DATA1	DATA2
Number of nodes	348	226
Number of edges	5038	6384
Number of attributes	44	28
Number of decision/sensitive attribute	1	1
Number of class labels	2	2

which consists of the attribute information as well as social relationships. The presence and absence of attributes for a particular user are depicted as 1 and 0, respectively. The username is replaced by numeric value. The attributes consist of basic information, education qualification and work-related information. In [14], gender has strong privacy concerns with respect to their location and employment information. Based on the above observation, we have considered four basic attributes, namely hometown, current location, work information and gender. The quasi-identifier attributes are current location, work information and hometown, whereas sensitive attribute is gender. Table 1 describes the structure of the datasets.

We analyze the effect of spurious records on two datasets and compare the proposed scheme and data sanitization in terms of prediction accuracy. We have not given any PDA and UDA as an input in the scheme. Instead, we have assumed that the application is not given a priori. We have implemented the proposed scheme in Python.

In Figs. 1 and 2, the effect of spurious records on two datasets, DATA1 and DATA2 are shown. The cases where the numbers of spurious records are $N/4$, $N/2$, N, 2 and 4 N have been considered. In DATA1, it is observed that the spurious records in proportion $N/4$, $N/2$, N, 2 and 4 N have higher prediction accuracy as compared to N. However, the accuracy prediction difference between N and 2 N is on lower side

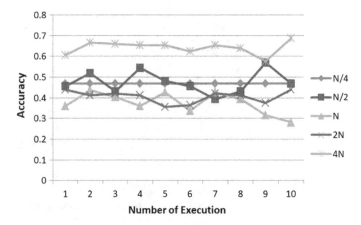

Fig. 1 Effect of spurious records on social network datasets—DATA1

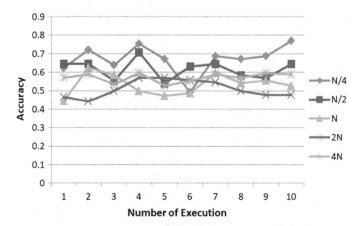

Fig. 2 Effect of spurious records on social network datasets—DATA2

as compared to the other proportions. In DATA2, it is observed similar trend in terms of $N/4$, $N/2$ and $4N$. In some executions, the spurious records in $2N$ are showing less prediction accuracy as compared to N. Therefore, N is optimal proportion of spurious records in general, but in some cases, $2N$ proportion shows good results.

We further compare the proposed scheme against the data sanitization algorithm on DATA1 and DATA2 in terms of privacy. We measure the accuracy using the RST classifier. The results are shown in Figs. 3 and 4. In 3, both the approaches are compared on DATA1. As we can see, there is a substantial decrease in the prediction accuracy in the proposed algorithm as compared to the data sanitization. Similar results are reflected in DATA2 as well.

Fig. 3 Prediction accuracy in social network datasets—DATA1

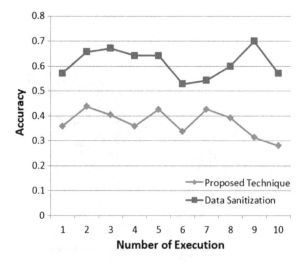

Fig. 4 Prediction accuracy in social network datasets—DATA2

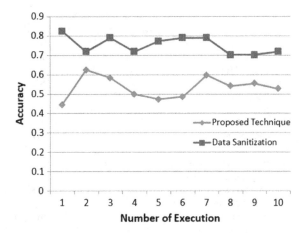

5 Conclusion

Privacy concern in modern connected world is a real challenge to deal with while supporting analytics-based business goal and services to our society. Large volumes of data get collected in every moment and released to other parties for various business goals. These social network data consist of sensitive information, so an adversary can exploit users profile and social relationships to disclose their privacy. We have proposed a privacy-preserving scheme secure against the inference attack. The proposed scheme uses additional spurious data in the published dataset such that sensitive information is not predicted using mining techniques. We have shown the privacy-preserving strengths of the proposed scheme in comparison with other related schemes.

References

1. Cai, Z., He, Z., Guan, X., Li, Y.: Collective data-sanitization for preventing sensitive information inference attacks in social networks. IEEE Trans. Dependable Secure Comput. **15**(4), 577–590 (2018)
2. Qian, J., Li, X., Zhang, C., Chen, L., Jung, T., Han, J.: Social Network de-anonymization and privacy inference with knowledge graph model. IEEE Trans. Dependable Secure Comput. (2017)
3. Li, H., Chen, Q., Zhu, H., Ma, D., Wen, H., Shen, X.S.: Privacy leakage via de-anonymization and aggregation in heterogeneous social networks. IEEE Trans. Dependable Secure Comput. (2017)
4. Narayanan, A., Shmatikov, V.: De-anonymizing social networks. In: Proceedings of IEEE Symposium on Security and Privacy, pp. 173–187 (2009)
5. Zhou, B., Pei, J., Luk, W.: A brief survey on anonymization techniques for privacy preserving publishing of social network data. ACM SIGKDD Explor. Newsl **10**(2), 12–22 (2008)
6. He, J., Chu, W.W., Liu, Z.: Inferring privacy information from social networks. In: Proceedings of Intelligence and Security Informatics, pp. 154–165 (2006)

7. Mislove, A., Viswanath, B., Gummadi, K., Druschel, P.: You are who you know: inferring user profiles in online social networks. In: Proceedings of the Third ACM International Conference on Web Search and Data Mining, pp. 251–260 (2010)

8. Gong N., Liu, B.: Attribute inference attacks in online social networks. ACM Trans. Privacy Secur. **21**(1) (2018)

9. Ryu, E., Rong, Y., Li, J., Machanavajjhala, A.: CURSO: protect yourself from curse of attribute inference. In: Proceedings of the ACM SIGMOD Workshop on Databases and Social Networks, pp. 13–18

10. Zhong, Y., Jing Yuan, N., Zhong, W., Zhang, F., Xie, X.: You are where you go: inferring demographic attributes from location check-ins. In: WSDM (2015)

11. Nie, L. Zhang, L., Wang, M., Hong, R., Farseev, A., Chua, T.: Learning user attributes via mobile social multimedia analytics. 8(3) (2017)

12. Jurgens, D.: That's what friends are for: Inferring location in online social media platforms based on social relationships. In: ICWSM, pp. 273–282 (2013)

13. https://snap.stanford.edu/data

14. Farahbakhsh, R., Han, X., Cuevas, A., Crespi, N.: Analysis of publicly disclosed information in Facebook profiles. In: Proceedings of Advances in Social Networks Analysis and Mining, pp. 699–705 (2013)

15. Pawlak, Z.: Rough set theory and its applications to data analysis. J. Cybern. Syst. **29**(7), 661–688 (1998)

Markov Model for Password Attack Prevention

**Umesh Bodkhe, Jay Chaklasiya, Pooja Shah, Sudeep Tanwar
and Maanuj Vora**

Abstract With the rapid increase in multi-user systems, the strength of passwords plays a crucial role in password authentication methods. Password strength meters help the users for the selection of secured passwords. But existing password strength meters are not enough to provide high level of security that makes the selection of strong password by users. Rule-based methods that measure the strength of passwords fall short in terms of accuracy and password frequencies differ among platforms. Use of Markov model-based strength meters improves the strength of password in more accurate way than the existing state-of-the-art methods. This paper describes how to proactively evaluate passwords with a strength meter by using Markov models. A mathematical proof of the prevention of guessable password attacks is presented. The proposed method improves the accuracy of current password protection methods significantly with a simpler, faster, and more secure implementation.

Keywords Proactive · Markov models · Accuracy · Password-based authentication

U. Bodkhe (✉) · J. Chaklasiya · P. Shah · S. Tanwar
Department of Computer Science and Engineering, Institute of Technology,
Nirma University, Gujrat, India
e-mail: umesh.bodkhe@nirmauni.ac.in

J. Chaklasiya
e-mail: 14bce014@nirmauni.ac.in

P. Shah
e-mail: Pooja.shah@nirmauni.ac.in

S. Tanwar
e-mail: sudeep.tanwar@nirmauni.ac.in

M. Vora
SaralSoft LLC, Pleasanton, USA
e-mail: maanujvora@gmail.com

© Springer Nature Singapore Pte Ltd. 2020   831
P. K. Singh et al. (eds.), *Proceedings of First International Conference on Computing,
Communications, and Cyber-Security (IC4S 2019)*, Lecture Notes in Networks
and Systems 121, https://doi.org/10.1007/978-981-15-3369-3_61

1 Introduction

Passwords are used from the beginning of digital age and even before, and it is a traditional and widely accepted method for authentication. They are easy to implement and comprehend, so they are likely to spend another century for the use of authentication. In most of the applications, the user is responsible for generating passwords as randomly generated passwords are not widely accepted and are also hard to remember. Generally, users tend to create weak and insecure passwords, as an example, many people choose passwords from a set of passwords which can be easily be guessed upon. To mitigate the problems arising from this fact, in this work, we experimented with implementation of simple rule-based systems [1] to ensure security. The said system enforces the user to choose some number of symbols from specific defined symbol sets. However, this method also has limitations as day by day processing power increases, the computer became faster to generate possible combination with more symbols, so even with rule-based enforcement, the brute force attack cannot be stopped. As a part of an advancement, some of the applications uses an evaluation mechanism for a password to pass an accepted level of security. It is known as proactive password checkers, which evaluate the strength of the password and reject the weak ones [2]. The accuracy of rule-based passwords checkers is low, and it is likely to reject strong passwords as well as accept weak ones. This adversely affects both security and usability. It is not feasible to implement a generalized password evaluation mechanism. This is because password distributions differ from use to use, application to application, person to person, and even language to language (we are likely to incorporate some native speaking words). We propose to use Markov model-based strength meter for passwords, which can evaluate strength in more accurate terms. This measures strength based on conditional probability with different gram models [1]. The base accuracy can be increased by training them on an actual password database. Fine-tuning can also be done to cater more personalized usability and security by real-time training. The most important finding is that even if the actual database is required for training, the security if database is not sacrificed. The reason being only a mathematical parameter is exposed for the working of it and not entire database. It is also computationally impossible to obtain entire database by these parameters. We would call them adaptive password strength meters [3], also known as APSMs, as they are able to react dynamically in changes to how users choose passwords.

To evaluate strength, we are assuming an attacker model with two findings. To start off, we would assume that the hacker/opponent is not interested in just user/account but is interested in guessing any password for that platform. This would mean that the attacker would not use private information [4], such as date of birth, age, relationships, so on and so forth. Next, we can contend that a sensible investigation of the secret word's quality would need to expect that the assailant needs to know the dispersion of the passwords situated inside the platform.

As we know that the database of previously occurred attacks is available, so attacker probably obtains the distribution based on available materials. So, assuming that the attacker knows the exact distribution of the passwords, then we can only conservatively estimate the knowledge that the attacker holds.

1.1 Motivation

With the extensive use of multi-user systems, possibility of password attacks increases. In such scenario, we should have strong and more robust password authentication mechanism. Hence, the strength of passwords plays a prominent role in password-authentication methods [5]. Recent authentication mechanisms and all state-of-the-art works have focused on either use of rule-based methods or methods that use password frequencies differ among platforms. Due to these frail techniques, strength of passwords is sometimes inaccurate and vulnerable to various attacks. Thus, motivated from above facts, paper proposes how to proactively evaluate passwords with a strength meter by using Markov models which improves the accuracy of password protection methods.

1.2 Research Contributions

The paper proposed the following major research contributions:

- A new approach based on Markov models to determine the complexity of user-generated passwords. It has accuracy higher than the current golden standards for password meters.
- The accuracy of the scheme has been proven by various trials extensive in nature with the results outperforming pre-existing schemes.
- Explanation of the potential position of Markov models to nullify the dictionary-based password attacks, and how the feedback came from an attack helps the network to improve its overall strength and security mechanism.

1.3 Organization of the Paper

The paper is organized as follows. Section 2 discusses the password strength evaluators. Section 3 elaborates on construction of dynamic password meters. Experimental setup, results of the experiment, and usability of the proposed model are discussed in Sect. 4. Finally, Sect. 5 concludes the paper.

1.4   Background: Markov Model

A Markov model is a sequence of states in which the probability is dependent only on the event immediately preceding it. That is only on the single previous event, not the history of events. For example, if it is known that the previous event was T, then the probability that the next event will be P is based only on T. Thus, the T-P combination forms a unity, and a probability can be assigned to it. The diagram mentioned as 1 represents a count of letters with their relations to each other. The arrows indicates that the letter is-followed-by relation. Thus, P is followed by Q, and S is followed by T and so on.

Some letters are followed by more than one letter. This indicates an OR, that is Q is followed by S or Q is followed by R. Using this diagram, one can now construct words where a word is a number of letters following each other. For example, the word PQSTRS can be formed by following the arrows.

In this context, some words can be formed called valid words, and (remaining) others are known as invalid words. An example of an invalid word would be PQSTRR because R must be followed by a S and cannot be followed by another R. Such a model is often called a finite-state model. A Markov model is an extension from the finite-state model. In a Markov model, the transitions—or arrow—represent the probability that that transition will occur. It is possible to modify the above finite-state model in such way that it becomes a Markov model [6]. This can be done by adding probabilities to the transitions.

The probability that Q will follow an P, also written as $P(P \cdot Q) = 1.0$ and the probability that a R will follow an Q also written as $P(Q \cdot R) = 0.5$ and so on. As mentioned, the transitions are probabilities, and the sum of all outgoing transitions must add up to one. There are two transitions from T: T to P and T to R. The probabilities of these are 0.9 and 0.1, respectively. These probabilities add up to 1.0. It is possible to construct a Markov model of common passwords by interpreting each letter in a password as a state or event, and then, the probability of the transitions is given by the probability that some letter follows another (Fig. 1).

1.5   Related Work

Design of Markov model for password prevention is a relatively new emerging field, with only few work of noteworthy importance. Heerden et al. [7] discussed that Duffy and Jagota developed password quality checker in 2002 to measure the quality of password. Authors in [8] attempt to crack the hashed passwords to check the quality of password. Later on, National Institute of Standards and Technology (NIST) proposed proactive password checkers using certain rules sets that exclude the weak password. But proposed rules were not strong enough [3]. Chen et al. [9] proposed hidden Markov models to Keystroke Pattern Analysis for Password Verification. Ma et al. [5] surveyed systematic study of many password models and obtained

Fig. 1 **a** Markov model [7], **b** Markov model with probabilities [7]

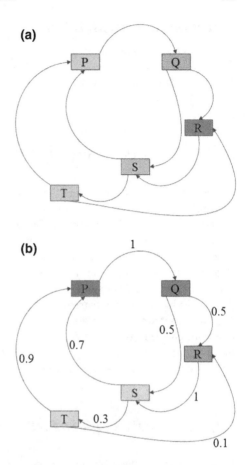

a number of findings. The authors suggested to use new methodology based on Markov models which could benefit future password research. Galbally et al. [10] also discussed probabilistic model which was able to measure the strength of password from the public datasets of passwords. Vaithyasubramanian et al. [11] analyzed the effect of Markov model-based password against brute force attack (BFS) for Web applications. Later on, Han et al. [12] proposed improved version of Markov model and probabilistic context-free grammar (PCFG)-based password guessing method [13]. Also, Khan et al. [14] discussed the attempt-based password technique which is three times more secured than existing techniques from any type of brute force attack.

2 Password Strength Evaluators

This section discusses about different password strength evaluators such as ideal passwords strength value, adaptive password strength meters along with its requirement [15].

2.1 Ideal Passwords Strength Value

In this subsection, we formulate the problem with a space of passwords and the probability P: $\Sigma \times [0, 1]$. The most ideal password checker ever formulated has been $-\log(P(x))$ with $P(x)$ is a password. The shown meter as a password strength which can be said to be ideal due to the order of passwords guessed in an optimal guessing attack is very similar to the results that can be obtained from the function. There are also more functions that can fall under the category of an ideal password checker, which are $f(x) = 1/P(x)$ and $f(x) = RP(x)$, in which $RP(x)$ is when x gets a rank that would be due to the distribution of P. For example, if the likelihoods of $p_i = P(x_i)$ are ordered with $p_i < p_j$ for the duration of $i < j$, then $RP(x_i) = i$. Almost every password checker, as well as password strength meter, can be a calculation of this function, which can often be grouped into small amounts of different buckets like "insecure/secure" or "insecure/medium/high."

2.2 Adaptive Password Strength Meters

In this subsection, the problem with ideal password strength function is taking into account a larger picture, and it will count probability on the base of the worst case and try to evaluate in comparison with all possible combinations. The problem with the scheme is some passwords are never going to be used because they are too hard to remember or some user-specific issues, they should not be counted when we are finding the probability. So, our task becomes conditional probability instead of an absolute one, which can be proved more efficient measure. The function never know the behavior of the user as it is very dynamic so to make a constant function which can measure the strength of the password cannot be addressed by a static function like ideal password strength value, we need something dynamic to solve the issue. An adaptive password strength meter (APSM) would be defined as $f(x, L)$, which would be the function f: $\Sigma \times x(\Sigma \times)k \rightarrow R$, which would get inserted with some characters, which would signify a password x over an alphabet, Σ, and a password file, L, which would contain a set number of passwords, and use them as inputs and will then output, a score based upon the strength, S. Starting off, the database of passwords, L, encapsulates a set amount of characters/passwords/strings that would be sampled using the distribution itself, and then, the process that would be assigned to the password strength meter (PSM) would be to signify the strength of the string, x, utilizing the sample P. It can be identified that f, the adaptive password strength meter, would not need apriori information of the distribution contained within P; however, the non-adaptive password strength meter necessarily needs the apriori information of the distribution of P.

2.3 Need of an Adaptive Password Strength Meter

In this subsection, one of the biggest motivations to use this scheme is that the distribution of passwords can be differed by many criteria including place, application, and even time. We have confidence in saying that those differences are being caused by the different languages being spoken as well as the different cultural backgrounds of each individual user, differences in the detected importance of the passwords, different password checkers that are in place, so on and so forth. Our task is to validate the passwords strength taking into account that we do not know the probability of distribution of the passwords in advance, but using password checkers would influence the distribution of different types of passwords, which would mean that there is not a set of rules, which is fixed, that would be able to capture the vital changes in the creation of passwords, including habits that are continually utilized by users.

3 Constructing Dynamic Password Meters

In this section, we have utilized techniques from statistical language processing, most predominately Markov models, to help implement an effective, adaptive password strength meter which would both accurately estimate certain probabilities from a data set not significant enough and which is also secured in order for the locally stored data not to be leaked. Suppose we are using the hash-based mechanism to stop leaks of passwords, then during adding entry to database just before the hashing process, we pass the password through the Markov model, by this way, our model will train itself and change the values of its parameter, and password is stored as hashed value, so it is still secure. We can use different gram models to find probabilities, and these probabilities are nothing but parameters for Markov models. The model works best when n is high in n-grams model, but it is relatively complex in terms of space and time.

3.1 Markov Models [16]

In the course of previous years, the concept of Markov models has been shown as well as utilized and has been deemed to be useful in the field of security, most predominately computer as well as password-wise. For instance, Narayanan et al. [4] have proven the value of the use of Markov models in cracking passwords. The notion that in human-generated passwords, any two consecutive letters are not independently chosen but actually follow certain patterns (for example, the two-character (gram) string "t_h" is more likely than t_q to occur, and the letter e is extremely probable to follow "t_h").

With an n-gram Markov model, it would be possible to model the odds of the next letter in the given string based upon the number of characters, denoted as n, just before it. So, for a given string $c_1, \ldots, c_m$, we can write

$$P(c_1, \ldots, c_m) = \prod_{i=1}^{m} P(c_i | c_{i-n+1}, \ldots, c_{i-1}) \tag{1}$$

The equation is limited to be able to keep a record of the n-gram counts count $(x_1, \ldots, x_n)$, than the conditional probabilities that would also be needed can be calculated using the formula listed below:

$$P(c_i | c_{i-n+1}, \ldots, c_{i-1}) = \frac{\text{count}(c_{i+n-1}, \ldots, c_{i-1}, c_i)}{\sum_{x \in \Sigma} \text{count}(c_{i-n+1}, \ldots, c_{i-1}, x)} \tag{2}$$

In addition, take into account the number of data within the passwords data set (Σ) is significant. We can discern that some of the characters are not being used as often than others, which lead to a problem in scarceness. We have to keep in mind that, generally, we would choose passwords from ten numeric digits, 56 alphabets, and less than ten special symbols. This leads to a reduction of the frequencies with the dataset which leads to having better probability approximation as the result.

3.2 Construction and Updation of n-Gram Model [17]

If theoretical n-gram database was to be established, it would be ameliorated as shown below:

(1) The algorithm continues with the same state the n-gram starts to count, count $(x_1, \ldots, x_n)$, for all $x_1, \ldots, x_n \in \Sigma$. All these counts are initialized as zero. (2) Whenever a password, $c = c_1, \ldots, c_l + n - 1$, was to be incorporated, it would be then separated into different n-grams. Furthermore, the database would be updated due to the incrementing counts that are corresponding to each of the passwords n-gram [18] by a value of 1, which would be denoted in the formula that follows: (3)

Additional noise would then be inserted into the database for invulnerability purposes which would not be indispensable.

3.3 Password Strength Evaluators [7] (PSE Algorithm)

The strength of a password $c = c_1, \ldots, c_m$, in which each character, c_i, would be chosen from the alphabet Σ, was to be estimated as shown below:

(1) For $i = 1, \ldots, m$, the conditional probabilities shown below are to be evaluated:

$$P(c_i|c_{i-n+1}, \ldots c_i)$$
$$= \frac{\text{count}(c_{i-n+1}, \ldots, c_i)}{\text{count}(c_{i-n+1}, \ldots, c_{i-1})}$$
$$= \frac{\text{count}(c_{i-n+1}, \ldots, c_i)}{\sum_{x \in \Sigma} \text{count}(c_{i-n+1}, \ldots, c_{i-1}, x)}$$

(2) Ultimately, the strength approximation of $f(c)$ for a given password c is

$$f(c) = -\log_2 \prod_{i=0}^{m} P(c_i|c_{i-n+1}, \ldots, c_{i-1})$$

3.4 Illustration

Assume the string password (with $n = 5$). The probability is computed as. Probability(maanuj) = Probability(m) $\times$ Probability(a—m) $\times$ Probability(s—ma) $\times \ldots \times$ Probability(d—maanuj) Picking one of the elements as an example: $p(o$—aanuj$)$ = count(aanu) count(aan) = 94243 344141 = 0.98. The example results in the comprehensive estimation of Probability(maanuj) = 0.0016, with the actual frequency of the certain password, located in the RockYou database, would be 0.0018. So, this goes to show that Markov models are able to closely estimate the actual probability/strength of each password. Adding on, the more complex and random passwords would be correctly estimated as stronger and harder to crack passwords than their easier and weaker counterparts.

4 Experimentation Setup, Results, and Usability

This section details the experimentation setup, results, and usability of proposed model.

4.1 Experimentation and Setup

For the experiment, any random users are chosen with a limit of common 20 distinct features. This feature might be a place, address, designation, city, country, etc. From these features, we are forming random passwords and some user-generated passwords. The experiment involves an attacker program which tries to generate random guesses based on the dictionary given to it, which has 20 features in common.

On the other hand, there are two evaluating mechanisms, and after the first batch of human-generated passwords, we are generating passwords by specific schemes like one scheme will follow the rule-based system, and another scheme will choose best-ranked password generated by a Markov model [19, 20].

For experimenting, due to the lack of required resources in terms of time and processing power, we were unable to evaluate any passwords with respect to actual human behavior mapping and also on RockYou database (RockYou is a very useful database as it was obtained by an actual leakage of passwords, which correctly maps the distribution of human behaviors.).

4.2 Results

The results of the experiment do not only cover the aspect of password strength evaluation but also provide us with the mechanism to detect a dictionary-based password guessing attack. We had tried some rule-based systems like NIST, MS, and Google for password strength evaluation and Markov model as well. As an ideal candidate database, we are using the RockYou database as it is based on natural findings, so we found that Markov models give very near evaluation outcome as obtained from RockYou database (Table 1).

While attacking the database, when we had tried on passwords generated by Markov models, they are 40,000 times secure. Based on the randomly generated user password database, it is also found that as the length of password increases, it can generate even more secure ranks. Markov models are also eligible to work on sufficiently small dataset for training, as we had only fed the network with 20 random guesses and still it is generating secure passwords, which are harder to crack by dictionary attack.

Table 1 Comparison of proposed model and other techniques in terms of strength of password

String	Optimal	Markov	NIST	Microsoft	Google
Password	9.100	9.262	2.110	1	1.000
maanuj12	11.500	11.830	22.522	2	1.022
kadsaf1n	16.150	17.080	28.532	3	1.000
Password2	22.370	21.670	27.012	3	1.000
v9or23	21.960	28.420	19.534	1	0

4.3 Usability

As we had described in results, the Markov models can be trained on publicly available dictionaries for passwords so that it can accept the guessable passwords generated by users. Here, a two-way mechanism to evaluate passwords for better strength is proposed: One Markov model will rank the user-generated passwords for initial evaluation, and then, that password is fed to an another Markov model which will measure its occurrence in attacks performed before and validate its acceptance as a password [21].

As the placement of Markov model comes before authentication, every password has to pass through the Markov model before authentication. Thus, if an attacker tries and password is accepted by model, then the attacker will be dismissed, and that system can be blocked. There are mathematical proofs available that say that even with a low probability, and the guess is still possible, the attacker can pass the Markov models. But when passing through the second-tier database authentication, if password is proven to be illegitimate, then the feedback will be used to train the Markov model so that next time, the same password will not pass. Even if said attacker gets access to parameters of the Markov model, he/she will not be able to generate actual passwords on which the model was trained. As for mathematical proof in our experiment, we had 21 * 21 matrix for a dictionary of 20 features, and we are able to generate the 11,125,676 passwords, so it is computationally unfeasible to find one of the passwords except for random guesses.

One way to stop the dictionary attack is to make a bad password directory, but due to the size and time complexity to transfer the dictionary to checker and time to traverse the dictionary, it is unfeasible. Markov model even aids here as it only requires small matrix to transfer and even lesser steps to traverse it.

5 Conclusion

This paper discusses about the importance of how to proactively evaluate passwords with a strength meter by using Markov models. Within this paper, we cataloged a new approach to determine the complexity of user-generated passwords. The assembly has been primarily based on Markov models and achieves an accuracy higher than the current golden standards for password meters. The formulation of the method has been proven theoretically, while the local data storage has been compromised, hence meeting the standard practices that have been proven optimal. The accuracy of the scheme has been proven by various trials extensive in nature with the results outperforming pre-existing schemes. In addition to that, we also tried to explain the potential position of Markov models to nullify the dictionary-based password attacks, and how the feedback came from an attack helps the network to improve its overall strength and security mechanism.

References

1. Guo, Y., Zhang, Z.: LPSE: lightweight password-strength estimation for password meters. J. Comput. Secur. **73**, 507–518 (2018)
2. Blundo, C., DArco, P., De Santis, A., Galdi, C.: Hyppocrates: a new proactive password checker. J. Syst. Softw. **71**(1–2), 163–175 (2004)
3. Castelluccia, C., Drmuth, M., Perito, D.: Adaptive password-strength meters from markov models. In: NDSS Symposium (2012)
4. Vu, K.P.L.: Improving password security and memorability to protect personal and organizational information. Int. J. Hum. Comput. Stud. **65**, 744–757 (2007)
5. Ma, J., Yang, W., Luo, M., Li, N.: A study of probabilistic password models. In: IEEE Symposium on Security and Privacy, pp. 689–704 (2014)
6. Iqbal, S., Kiah, M.L.M., Dhaghighi, B., Hussain, M., Khan, S., Khan, M.K., Choo, K.K.R.: On cloud security attacks: a taxonomy and intrusion detection and prevention as a service. J. Netw. Comput. Appl. **74**, 98–120 (2016)
7. Van Heerden, R.P., Vorster, J.S.: Using Markov Models to crack passwords. In: The 3rd International Conference on Information Warfare and Security: Peter Kiewit Institute, University of Nebraska, Omaha, USA (2008)
8. Morris, R., Thompson, K.: Password security: a case history. Commun. ACM 22(11), 594–597 (1979)
9. Chen, W., Chang, W.: Applying hidden Markov Models to keystroke pattern analysis for password verification. In: Proceedings of the IEEE International Conference on Information Reuse and Integration, pp. 467–474 (20040
10. Galbally, J., Coisel, I., Sanchez, I.: A probabilistic framework for improved password strength metrics. In: International Carnahan Conference on Security Technology (ICCST), pp. 1–6 (2014)
11. Vaithyasubramanian, S., Christy, A., Saravanan, D.: An analysis of Markov password against brute force attack for effective web applications. J. Appl. Math. Sci. **8**(117), 5823–5830 (2014)
12. Han, W., Li, Z., Yuan, L., Xu, W.: Regional patterns and vulnerability analysis of chinese web passwords. IEEE Trans. Inf. Forensics Secur. **11**(2), 258–272 (2015)
13. Xia, Z., Yi, P., Liu, Y., Jiang, B., Wang, W., Zhu, T.: GENPass: a multi-source deep learning model for password guessing. In: IEEE Transactions on Multimedia (2019)
14. Khan, S., Khan, F.: Attempt based password. In: 13th International Bhurban Conference on Applied Sciences and Technology (IBCAST), pp. 300–304 (2016)
15. Golla, M., Drmuth, M.: On the accuracy of password strength meters. In: Proceedings of the ACM SIGSAC Conference on Computer and Communications Security, pp. 1567–1582 (2018)
16. Rabiner, L., Juang, B.-H.: An introduction to hidden Markov models. In: IEEE ASSP Magazine, vol. 3, pp. 4–16 (1986)
17. Ganitkevitch, J., Van Durme, B., Callison-Burch, C.: PPDB: The paraphrase database. In: Proceedings of the 2013 Conference of the North American Chapter of the Association for Computational Linguistics: Human Language Technologies (2013)

18. Sekine, S.: A linguistic knowledge discovery tool: very large ngram database search with arbitrary wildcards. In: 22nd International Conference on Computational Linguistics: Demonstration Papers. Association for Computational Linguistics (2008)
19. Mamonov, S., Benbunan-Fich, R.: The impact of information security threat awareness on privacy-protective behaviors. J. Comput. Hum. Behav. **83**, 32–44 (2018)
20. Mahmoud, M.S., Hamdan, M.M., Baroudi, U.A.: Modeling and control of cyber-physical systems subject to cyber attacks: a survey of recent advances and challenges. J. Neurocomput. **338**, 101–115 (2019)
21. Almasizadeh, J., Azgomi, M.A.: A stochastic model of attack process for the evaluation of security metrics. J. Comput. Netw. **57**(10), 2159–2180 (2013)
22. Barkadehi, M.H., Nilashi, M., Ibrahim, O., Fardi, A.Z., Samad, S.: Authenticationsystems: a literature review and classification. J. Telematics Inform. **35**(5), 1491–1511 (2018)

An Enhanced Face and Iris Recognition-Based New Generation Security System

E. Udayakumar, C. Ramesh, K. Yogeshwaran, S. Tamilselvan and K. Srihari

Abstract Nowadays, digitalization due to advancement in technology is developing rapidly. Even though security is enhanced for maintaining the personal documents, the robbery and unauthorized usage are being increased in current scenario. In this proposed system, a new generation security system which is based on face recognition-Iris Recognition is prototyped. In the meantime, high-quality image with many details has an important role in the recognition process. Face image and eye iris are used for authentication purpose. Initially, the face image of particular person is compared with the database image. If it is authorized the interrupt will be forwarded to iris recognition. Then, the comparison output result is sent to the control unit via serial communication. If the person is unauthorized, then the message is sent to the concern. Various control circuits are developed with ARM7 microcontroller. Similarly, iris recognition of the users is carried out by comparing it with the database images. If authorized, a one-time password (OTP) is sent to the authorized mobile number, else if unauthorized the image of the person is sent as MMS to the appropriate user. Finally, if all the three steps of verification are cleared, then the user can make use of the system (say an ATM).

Keywords OTP · ARM · ATM · Global system for mobile communication · One-time password

E. Udayakumar (✉) · C. Ramesh · K. Yogeshwaran · S. Tamilselvan
Department of ECE, KIT-Kalaignarkarunanidhi Institute of Technology, Coimbatore, Tamilnadu, India
e-mail: udayakumar.sujith@gmail.com

C. Ramesh
e-mail: cramezz@gmail.com

K. Yogeshwaran
e-mail: emperoryogi.yogesh@gmail.com

S. Tamilselvan
e-mail: tamilselvanece87@gmail.com

K. Srihari
Department of CSE, SNS College of Engineering, Coimbatore, Tamilnadu, India
e-mail: harionto@gmail.com

P. K. Singh et al. (eds.), *Proceedings of First International Conference on Computing, Communications, and Cyber-Security (IC4S 2019)*, Lecture Notes in Networks and Systems 121, https://doi.org/10.1007/978-981-15-3369-3_62

1 Introduction

As a component of the security inside appropriated frameworks, different administrations and assets need assurance from unapproved use. Remote validation is the most regularly utilized strategy to decide the character of a remote customer. This past work explains methodology for validating customers by three elements, i.e., to be specific secret phrase, keen card, and biometrics. A conventional and secure system was utilized to redesign two-factor verification to three-factor validation. An inserted framework has explicit prerequisites and performed pre-characterized assignments, not at all like a universally useful PC [1].

This proposed framework is utilized for the security applications. This framework with three gadgets regularly utilized for giving verified lock, banking reason, and ATM applications. Mechanized Teller Machines (ATMs) are outstanding gadgets regularly utilized by people to complete an assortment of individual and business monetary exchanges or potentially banking capacities. ATMs have turned out to be the mainstream with the overall population for their accessibility and general ease of use [2]. ATMs are normally accessible to buyers consistently with the end goal that shoppers can do their ATM money related exchanges and additionally banking capacities whenever of the day and on anytime.

Existing ATMs are helpful and simple to use for generally buyers. Existing ATMs regularly give directions on an ATM show screen that are perused by a client to accommodate intelligent task of the ATM. Having perused the presentation screen guidelines, a client can utilize and work the ATM by means of information and data entered on a keypad. Anyway, the disadvantage in the current framework is that the client should convey their ATM card no matter what. In any case, by and large we overlook it. So just we structured a framework which causes us to utilize the ATM machine without the ATM card. In this paper, a continuous implanted face-iris acknowledgment framework for validation on cell phones is proposed [3]. The framework is executed on an implanted stage and furnished with a novel face-iris acknowledgment calculation. The proposed framework comprises of three stages of check: Face picture obtaining and iris picture correlation with database, implanted fundamental board, and OTP confirmation. Not just for the ATM application, it generally utilized for assortment of utilizations that need verified access.

2 Related Work

Picture preparing on portable advanced cells is another and energizing field with numerous difficulties because of restricted equipment and availability issues. This paper builds up a continuous face acknowledgment application model for advanced mobile phones. This presented model uses a cross-breed skin shading eigenface identification technique and intrigue point limitation for highlight coordinating. The paper is coded in JAVA programming language to satisfy android advanced mobile phones. Results are appeared and contrasted and existing open-source strategies for confirmation. The point is to keep up ongoing measures with high acknowledgment rate. Applications extend from security to individuals with inabilities adjustment [4].

The regular engineering of a PC vision framework should take into contemplations fundamental modules. Initial, a sensor that contains a computerized camera situated in a CCD or CMOS gadget. Second, a securing interface for information move, for example, explicit edge grabbers, Camera Link, IEEE 1394, USB, or the later GigE standard is fundamental. The handling unit comes as a compulsory unit with various OS alternatives. The peripherals used are screen, mouse, and console for the information Mobile is chosen by the client condition. In the whole globe, any instructive association is worried in connection to the participation of people since this affects their general exhibitions. In regular technique, the participation of understudies is taken by calling understudy names or marking on paper which is incredibly time overpowering. To take out this issue, one of the arrangements is a biometric-based participation framework that can naturally catch understudies' participation by perceiving their iris. Iris acknowledgment is viewed as one of the most solid, precise, and effective biometric ID framework because of the inward qualities of iris, for example, uniqueness, steadiness, and time invariance [5].

The point of this paper is to structure and execute an iris acknowledgment-based participation the board framework with the most recent offices at an open cost to consider the money related circumstance of the enormous figure of creating nations. The paper incorporates two sections, equipment and programming. The equipment part is capable to catch pictures and run required program while the product part is in charge of allegation, preparing, iris confining, modifying, coordinating, and putting away information; send the participation report to the predefined e-mail address. At long last, the framework identifies, computes, stores, and transmits the outcomes by playing out the MATLAB program. So as to improve the presentation of iris acknowledgment, a novel technique for iris acknowledgment dependent on square hypothesis and self-versatile element choice is proposed in this paper. Right off the bat, the standardized iris picture is decayed by convolving with multi-scale and multi-direction Gabor channels, and after that isolated into a few hinders, the square component vector which incorporates mean and fluctuation of Gabor coefficients inside each square can be acquired through measurable strategies [6].

The iris of entire iris picture was developed by conjugating the square element vector in line segment request, at long last, the two-classifier of iris picture is setup depending on the most discernable highlights, and the multi-classifiers of iris picture are built up by casting a ballot system, and the presentation is tested by CASIA iris database. The outcomes demonstrate that contrasted and the conventional iris acknowledgment techniques; the proposed strategy has improved the iris acknowledgment rate. Face affirmation from picture or video is an outstanding topic in biometrics look at. Various open places as a general rule have observation cameras for video get and these cameras have their important motivating force for security reason. It is commonly perceived that the face affirmation has accepted a huge employment in observation structure as it need not mess with the thing's cooperation.

The real-focal points of face-based distinctive verification over various biometrics are uniqueness and affirmation. As human face is a unique thing having abnormal state of irregularity in its appearance that makes face area an inconvenient issue in PC vision. In this field, the precision and speed of recognizing confirmation is a key issue. The goal of this paper is to evaluate distinctive face disclosure and affirmation systems, offer a complete response for picture-based face area and affirmation with higher precision, better response rate as a fundamental development for video perception. Game plan is proposed reliant on performed tests on various face rich databases with respect to subjects, present, emotions, race, and light [7].

As a component of the security inside dispersed frameworks, different administrations and assets need assurance from unapproved use. Remote confirmation is the most regularly utilized technique to decide the character of a remote customer. This past work examined an efficient methodology for validating customers by three variables, specifically secret key, brilliant card, and biometrics. A nonexclusive and secure system was utilized to overhaul two-factor confirmation to three-factor validation.

3   System Design

In this proposed framework, another age security machine which can be worked with the face-iris recognition framework is prototyped. By utilizing this framework ATM machine can be worked by utilizing our SIM in the cell phone [8]. The camera introduced close to the ATM machine will catch the client's picture and contrasts it and the client picture in the server utilizing MATLAB [5]. Just when the picture matches it asks the stick number and further handling begins. Generally, the procedure is ended. So by utilizing this framework need of ATM card is totally disposed. By utilizing this framework breakdown can be maintained a strategic distance vector. The exchange will be greatly verified. Figure 1 outlines the stick plans of 8051 microcontroller.

The square diagram of face-iris recognition system is shown in Fig. 2. ARM7 microcontroller is used all in all control units. The camera recognizes the image of the customer and considers the database picture in iris affirmation unit sought after by face affirmation [9]. By then advances are finished reliant on endorsed or unapproved customers. DC motor is used in the model to mean the opening and closing of the confirmed room (say an ATM).

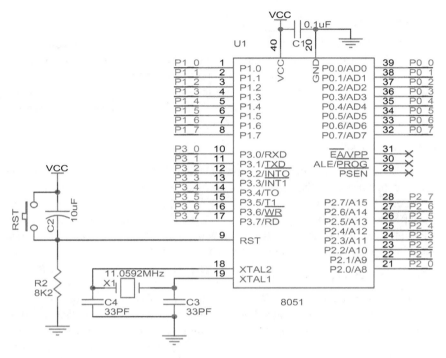

Fig. 1 Circuit diagram of 8051 microcontroller

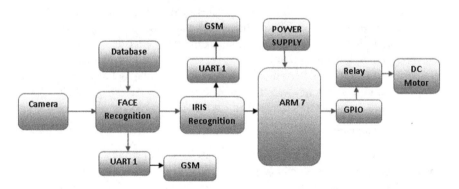

Fig. 2 Block diagram of face and iris recognition unit

(A) *GSM modem*

Worldwide framework for versatile correspondence (GSM) is an inside and out recognized standard for modernized cell correspondence. GSM is the name of a standardization social affair set to make a run of the mill phone figure judgments for a cell radio structure working at 900 MHz.

(B) *Power supply*

The AC voltage, consistently 220 V rms, is connected with a transformer that atmosphere control framework voltage down to the degree of the ideal DC yield. A diode rectifier by then gives a full-wave reviewed separated by an unmistakable capacitor channel to pass on a dc voltage. This subsequent dc voltage consistently has some swell or cooling voltage variety. A controller circuit appeared to clears the swell for more continues notice [7].

(C) *ARM7 TDMI*

The ARM7TDMI focus relies upon the von Neumann building with a 32-piece data transport that passes on the two bearings and data. Weight, store, and swap rules can get to data from memory. Data can be 8-piece, 16-piece, and 32-piece. Support for the ARM configuration today consolidates operating systems, for instance, Windows CE, Linux, palm OS and SYMBIAN OS, More than 40 steady working structures, including qnx, wind conduit's vx works and mentor outlines [10].

(D) *Fast general purpose parallel I/O (GPIO)*

Contraption sticks that are not related to a specific periphery limit are obliged by the GPIO registers. Pins may be capably masterminded as wellsprings of data or yields. Separate registers grant setting or clearing any number of yields at the same time [11].

(E) *General purpose timers/external event counters*

The timer/counter is proposed to count cycles of the periphery clock (PCLK) or a remotely given clock and on the other hand make obstructs or perform various exercises at decided clock regards, in light of four match registers. It similarly fuses four catch commitments to trap the clock regard when a data sign advances, then again making a meddle. Different pins can be picked to play out a singular catch or match work. In this plan, unused catch lines can be picked as standard clock catch inputs, or used as outside interferes [12].

(F) *Serial Communication*

Sequential correspondence is basically the transmission or social occasion of data one piece at some random minute. The present PCs generally address data in bytes format. A byte contains 8 bits. A piece is basically either a rational 1 or zero. Each character on this page is truly imparted inside as one byte. The successive port is used to change over each byte to a surge of ones and zeroes, similarly as to change over a flood of ones and zeroes to bytes. The successive port contains an electronic chip called a universal asynchronous receiver/transmitter (UART) that truly does the transformation [9].

This suggests each piece has the range of 1/9600 of a second or about 100 μs. When transmitting a character there are various characteristics other than the baud rate that must be known or that must be the course of action. These characteristics portray the entire comprehension of the data stream. The essential trademark is the length of the byte that will be transmitted. This length overall can be some place in the scope of 5–8 bits. The resulting trademark is correspondence. The balance trademark can be even, odd, engraving, space, or none. In the occasion that even balance, by then the last data bit transmitted will be a savvy 1 if the data transmitted had an even proportion of 0 bit. If odd equity, by then the last data bit transmitted will be a sound 1 if the data transmitted had an odd proportion of 0 bit. In case of MARK uniformity, by then the last transmitted data bit will reliably be a genuine 1. In case of SPACE correspondence, by then the last transmitted data bit will reliably be a real 0. If no fairness, by then there is no balance bit transmitted [13].

Half-duplex consecutive correspondence needs in any event two wires, signal ground, and the data line. Full-duplex consecutive correspondence needs at any rate three wires, signal ground, transmit data line, and get data line. The RS232 detail manages the physical and electrical qualities of successive correspondences. These sign are the carrier detect signal (CD), verified by modems to hail a productive relationship with another modem, ring indicator (RI), expressed by modems to signal the phone ringing, data set ready (DSR), pronounced by modems to show their quality, clear to send (CTS), confirmed by modems in case they can get data, data terminal ready (DTR), expressed by terminals to exhibit their quintessence, request to send (RTS), announced by terminals if they can get information [5].

(G) *Flash Magic*

Streak charm can control the area into ISP technique for some microcontroller contraptions by using the COM port handshaking sign to control the device. Usually, the handshaking sign is used to control such sticks as Reset, PSEN, and VCC. The unmistakable pins used depend upon the specific device. Exactly when this part is maintained, Flash Magic will normally place the device into ISP mode close to the beginning of an ISP action. Streak Magic will by then normally cause the device to execute code toward the piece of the deal activity [12].

4 Results and Discussion

(A) *Face Recognition Unit*

The face recognition unit is shown in Fig. 3. It has the image found by the camera along with the data that is already stored in the database. If it matches with the programmed image then the user is authorized and moves forward for the next step of verification [14].

(B) *Iris Recognition Unit*

The system setup for iris recognition is shown in Fig. 4. Iris is scanned and compared with the database images. In this paper, only a prototype of the database images is used for iris recognition [9].

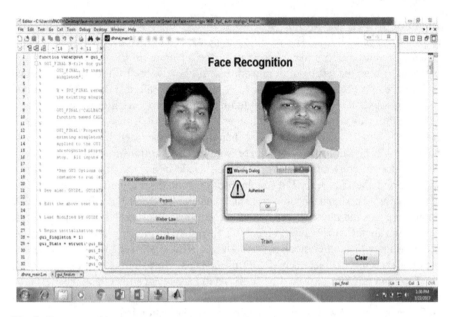

Fig. 3 Face recognition system setup

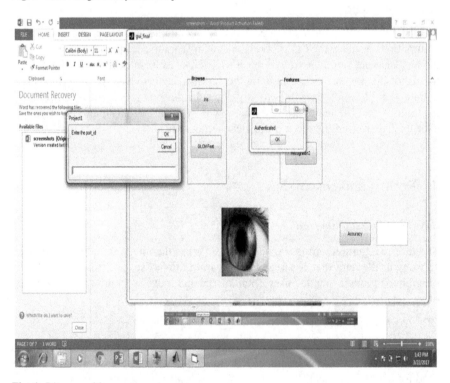

Fig. 4 Iris recognition system setup

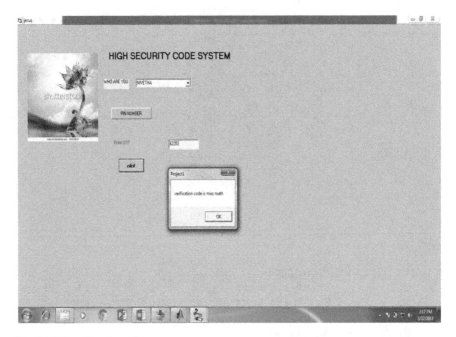

Fig. 5 OTP verification unit

(C) *OTP Verification Process*

If a person is authorized after completing face and iris recognition procedures, a third step of verification is involved as shown in Fig. 5. One-time password (OTP) is sent to the specified users as programmed. If unauthorized, the MMS of the fraudsters is sent to the authority [10].

(D) *Resulting Window after Authentication*

After a person is authenticated the window appears as shown in Fig. 6.

(E) *System setup of Recognition Unit*

The final system setup is shown in Fig. 7. Web camera is used for capturing the images to send for recognition unit. Thus, a highly secured system is developed for maintaining the personal documents, Banking applications, etc. (say an ATM).

5 Conclusion

The paper is developed with a new generation recognition unit for detecting face and iris images, the authorized people are provided with the data of the users resulting in a highly secured system. This system is used for maintaining personal documents in

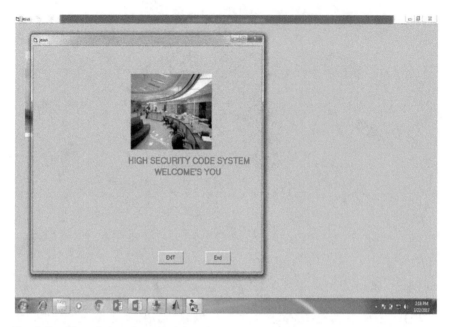

Fig. 6 Resulting window after authenticated

Fig. 7 Implemented recognition system

banking applications (say an ATM) to safeguard from fraudsters. It also has a wide range of applications in the military for preserving important weapons.

References

1. Aru, O.E., Gozie, I.: Facial verification tech in ATM transactions. Am. J. Eng. Res. 188–193 (2013)
2. Ahonen, T., Hadid, B., Pietikainen, M.: Face description for local patterns: application to face recognition. IEEE Trans. Pattern Anal. Mach. **28**, 2037–2041 (2006)
3. Corcoran, P., Cucos, A.: Techniques for secure media content in electronic by biometric sign. Trans. Electron. **51**, 545–551 (2005)
4. Jain, A.K., et al.: Biometrics: a grand challenge. In: Proceedings of the International Conference on Pattern Recognition, vol. 2, pp. 935–942 (2004)
5. Kim, Y., Yoo, J., Choi, K.: A motion & similar fake detecting for biometric face recognise sys. Trans. Consum. Electron. **57**, 756–762 (2011)
6. Srihari, et al.: Certain Inves of humans using various algorithms. Aust. J. Basic Appl. Sci. **8**, 559–564 (2014)
7. Santhi, S., et al.: SOS emergency wireless network. In: Computational Intelligence & Sustainable System. Springer, pp. 227–234 (2019)
8. Udayakumar, E., et al.: Automatic detection of diabetic retino through optic disc for morphological. Asian J. Pharm. Clin. Res. **10**, 28–31 (2017)
9. Vetrivelan, P., et al.: A neural network based automatic crop monitoring robot for agriculture. In: The IoT and the Next Revolutions Automating the World. IGI Global, pp. 203–212 (2019)
10. Kanagaraj, T., et al.: Foot pressure measure using ATMEGA 164 Microcontroller. Adv. Nat. Appl. Sci. **10**, 224–228 (2016)
11. Ojala, T., et al.: A study of text measures with classify featured distributions. Patt. Recognit. **29**, 51–59 (1996)
12. Santhi, S., et al.: Design and development of smart glucose monitoring system. Int. J. Pharma Biosci. **8**, 631–638 (2017)
13. Santhi, S., et al.: Design & development of smart glucose monitor system. Int. J. Pharma Biosci. **8**, 631–638 (2017)
14. Derman, E., Salah, A.A.: Short term based face recognition of ATM users. In: ICECCO, Turkey, pp. 111–114 (2013)

Implementing a Mobile Voting System Utilizing Blockchain Technology and Two-Factor Authentication in Nigeria

Temidayo Peter Abayomi-Zannu, Isaac Odun-Ayo, Barka Fori Tatama and Sanjay Misra

Abstract The voting system in Nigeria has always been filled with different forms of manipulation, and m-voting was suggested as the solution but has a major problem which is securely storing the casted votes. Blockchain was proposed to mitigate this problem and with the utilization of two-factor authentication to prevent illegible voters from casting their votes. The objective of this paper is to develop a mobile voting system that utilizes two-factor authentication to authenticate the voters and blockchain technology to securely store the votes. The system was then evaluated using the ISO 9241-11 usability model, and the results showed that the proposed system had a good usability rating which implies that it can be utilized in a voting procedure.

Keywords Blockchain · Mobile devices · Mobile voting · Two-factor authentication

1 Introduction

Voting is a process whereby an individual or group of individuals expresses their opinions or choices which can be either for or against a person (candidate) that could be either political (elections) and social or public [1]. Ballot/Paper voting is the most popular means of voting and is still being used in multiple countries around the world. Ballot voting is a type of voting procedure whereby the choices made by the voters must be kept secret. Voters tick out their preferred candidate on a piece of paper at

T. P. Abayomi-Zannu · I. Odun-Ayo · B. F. Tatama · S. Misra (✉)
Covenant University, Ota, Nigeria
e-mail: sanjay.misra@covenantuniversity.edu.ng

T. P. Abayomi-Zannu
e-mail: temidayo.abayomi-zannu@stu.cu.edu.ng

I. Odun-Ayo
e-mail: isaac.odun-ayo@covenantuniversity.edu.ng

B. F. Tatama
e-mail: barka.tatama@stu.cu.edu.ng

© Springer Nature Singapore Pte Ltd. 2020
P. K. Singh et al. (eds.), *Proceedings of First International Conference on Computing, Communications, and Cyber-Security (IC4S 2019)*, Lecture Notes in Networks and Systems 121, https://doi.org/10.1007/978-981-15-3369-3_63

the polling station and then drops or submit them into the ballot box provided. This process is usually the standard method still being practiced in Nigeria and is often at times inaccurate, tedious, and risky [2]. Occasionally, the last tally or vote count can be altered, and this manual procedure is susceptible to political dishonesty, political fraud, and errors [3]. Sometimes citizens do not want to take part in elections due to multiple reasons such as the requirement to wait in line to cast their votes which can be over an hour or more which is very time-consuming, inability to visit the polling station where their registration took place due to their jobs or change in their place of residence, inadequate ballots papers, political intimidation, voters can be coerced to cast their votes for candidate's that is not their choice, rigging of the votes, and brazen falsification of the results [4, 5].

This has led to the introduction of m-voting (mobile voting) which is a subset of electronic voting as a means to mitigate this problem [6]. With m-voting, voters only require a mobile device which has been the most adopted means of communication and an Internet connection to take part in the electoral procedure [7]. One of the most problematic phases in m-voting is the managing of the votes which is the process of ensuring that casted votes cannot be altered or tampered with [8]. A centralized database was utilized to store the votes but is susceptible to DDoS attacks, and since it is being managed by an administrator, the stored votes could be changed, manipulated, or tampered with by the admin or a malicious insider [9]. Blockchain technology was proposed as a means to mitigate this problem which is a distributed ledger that manages an ever-increasing list of records protected from any form of revision or tampering [10]. It is decentralized so as to avoid a single point of failure with the group working together to confirm genuine transactions [11, 12]. Blockchain can be utilized as a database that can be used to record anything of value such as marriage licenses, deed, titles of ownership, and votes and is almost impossible to reverse [13].

Since mobile devices do not have enough resources to be a miner/node on the blockchain network, another means was needed for mobile devices to be able to send their transaction or votes to the blockchain pool to be stored [14]. In order for their transactions to be sent, they have to be eligible to do so, and one of the means of verifying their eligibility is through the use of authentication [15, 16]. Authentication is the process of proving who you claim to be which usually requires a mechanism for identification that can verify one's identity prior to granting them access, and multiple means of authentication is far better than a single means [17]. Mobile voting is often seen as a tool for building trust in electoral management, increasing the overall efficiency of the electoral process, advancing democracy, and adding credibility to election results [18]. With the addition of blockchain technology and two-factor authentication to m-voting, a secured and transparent voting system can be created which mitigate the major problems associated with the ballot/paper voting process still being utilized in Nigeria.

The focus of this paper is to develop a mobile voting system that makes use of blockchain technology and two-factor authentication to provide a secured mobile voting system that can be used in Nigeria. The study is structured as follows: Sect. 2

presents the related works; Sect. 3 shows the methodology; Sect. 4 provides the results and discussion; while Sect. 5 concludes the paper.

2 Related Works

Different methodologies were being applied to the voting sector in order to not only simplify the voting process but also protect the votes while avoiding any form of tampering or manipulation. This section looks at some of the ways different authors have tried to accomplish this. It was stated that an e-voting system must be fully transparent, secured, mitigate duplicated votes, and protect the attendee's privacy [19]. They were able to achieve this by designing a secured e-voting system that made use of blockchain technology. A network security mechanism for voting systems based on blockchain technology was created by Wu and Yang [20] which utilized the distributed architecture of a blockchain to increase the security aspect in the voting procedure and only allow the addition of data while negating any form of modifications. Sakinah et al. [21] proposed an electronic voting system application that utilizes blockchain technology as a secured decentralized database that could keep a record of the votes cast by eligible voters. They created a virtual voting custom currency called Kinakoin with which a user uses to cast their votes in the form of a coin from their wallet to the wallets of the candidates. This exchange was then affirmed through a procedure of mining, and the information of the exchanges was kept in the blockchain with a special hash which served as the square's unique fingerprint. Purandare et al. [22] noted that voting is an important aspect for democratic countries, and elections decide the future of that country; therefore, the tools/devices being used for the election process should be as transparent as possible while also having a high level of security. They designed an application that can be used for the voting process which voters can use on their Android smart phones. They also made use of one-time password (OTP) as a form of authentication to verify a voter's eligibility to vote and also prevent voters from casting their votes twice. Kshetri and Voas [23] mentioned that blockchains can solve two of the major or prevalent problems in the voting procedure which are voter access and fraud. They proposed a blockchain-enabled e-voting system that gives every voter a wallet and a single coin which is used to vote. An m-voting system was designed by Gajabe [24] that enables eligible citizens to cast their votes using their mobile devices and increases voter's participation in the electoral process. The voters are first authenticated to verify their eligibility to vote, and after the verification process, they are then granted access to cast their votes which is encrypted and then sent to the central server to be stored in the database which tallies the votes and shows the results. Most authors focused on utilizing blockchain technology in terms of cryptocurrency, but our focus is on making use of blockchain technology as a database to securely store the casted votes while avoiding any form of modifications. Also, we are making use of a two-factor authentication approach as a secured means of authenticating the voters to prove their eligibility and grant them access to cast their votes, while also preventing them from casting multiple votes.

3 Methodology

This section presents the existing system, the proposed system, design analysis, voter's activity diagram, blockchain's proof-of-work algorithm, pseudocode for the two-factor authentication sequence, and implementation screenshots.

3.1 Existing System

Ballot/Paper voting is the most popular means of voting and is still being used in many countries, especially in Nigeria [2]. Each voter would have to visit the polling station where their registration took place. They would then cast their vote by placing a mark or their fingerprint dipped in ink beside their preferred candidate and then proceed to place it in the ballot box. Once the voting process is completed, officials will then count the votes manually and tally the results (there may be a need for a recount in some case). These procedures are often inaccurate, risky, and dreary, and the results are susceptible to tampering or modifications. This manual procedure leaves room for blunders, political untrustworthiness, and political extortion [25].

3.2 Proposed System

The m-voting system would make use of blockchain technology as a database to safely store the casted votes mitigating any form of tampering and two-factor authentication to verify a voter's eligibility to cast their votes, thereby making sure that only eligible voters can cast their ballot. For the two-factor authentication, voter's identification number (VIN), personal identification number (PIN), and one-time password (OTP) would be utilized.

The system architecture is based on a three-tiered layered architecture which is made up of the presentation layer, logic/business layer, and the data layer, all having their roles in the total functionality of the system. Figure 1 presents the three-tier showing the different tiers.

- Presentation Tier: The presentation tier is where the voting application runs. It provides an interface for voters to interact with the voting system using their smart phones. From this layer, the user can log in and cast their votes which will be processed at the logic tier and stored at the data tier.
- Logic Tier: The system functionalities are carried out here and it is regarded as the most important layer. It is made up of the system service, middleware layer, and application services. The application is developed using Android Studio IDE, XM, Java, Ethereum Virtual Machine (EVM), Python, C++, and SHA 256.
- Data Tier: The data tier contains all the knowledge sources required to provide the needed voter's information and also store their casted votes. It consists of two

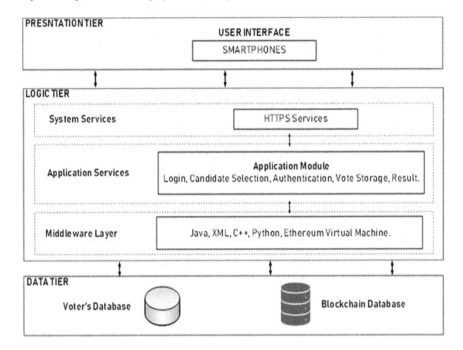

Fig. 1 m-voting architecture

(2) separate databases. The first database which is the voter's database contains every registered information about the voter which is used to identify each voter connecting to the system, and the second database which is the blockchain database is used to securely store the casted votes.

3.3 Design Analysis

In the design for this study, the most important requirements of the system were described below:

- Authentication of voters before access is granted to the application using their unique VIN and PIN.
- Enabling voters to select their preferred candidate.
- Authentication of voters before their vote is cast using the OTP.
- Storing and hashing of the casted votes in the blockchain database.
- Results can be viewed by everyone after the voting period.

```
For Blockchain{
        this.difficulty = 4;
}
For mineBlock(difficulty){
        While(this.hash.substring(0, difficulty) → Then Array(difficulty increased
        1).join("0")){
                this.nonce;
                this.hash = this.calculateHash();
        }
        }
}
newBlock.mineBlock();
```

Fig. 2 Proof-of-work (PoW) algorithm

3.4 Blockchain's Proof-of-Work Algorithm

The blockchain database/ledger would require a proof-of-work (PoW) algorithm which alludes to the computational riddle/puzzle that nodes/miners need to solve that empowers the blockchain networks to stay decentralized and secured. PoW makes use of cryptographic functions that basically ensure a specific number of computer cycles which were used to solve the puzzle which proves that you did some measures of work to solve that puzzle. The amount of work is specified by the difficulty which is proportional to the amount of work needed to solve the puzzle. Miners/nodes challenge each other to identify a nonce (also called a golden nonce) that can create a hash with a value less than or equivalent to the set network difficulty. Once such a nonce is found by a miner/node, they get the privilege to add that block to the blockchain. The nonce is a focal piece of the PoW algorithm for blockchains. In this study, we would be utilizing a difficulty of four (4) and the proof-of-work is shown in Fig. 2.

3.5 Pseudocode for the Two-Factor Authentication Sequence

The two-factor authentication pseudocode sequence used in the design of the system is shown in Fig. 3. When a voter wants to make use of the application, they would have to be authenticated at both the login and voting phases. The two-factor authentication sequence shows how the authentication processes would take place at different stages.

```
BEGIN
    Voter attempts to login and vote using the application
    IF voter's VIN and PIN is valid THEN
        Voter chooses their election choice and candidate
        IF candidate has been chosen THEN
            Requests for an OTP
            Voter enters the OTP
                IF OTP is valid THEN
                    Voter choice will be casted
END
```

Fig. 3 Pseudocode of the authentication phases in the m-voting system

3.6 Voter's Activity Diagram

See Fig. 4.

3.7 Implementation Screenshots

We are making use of five modules in the implementation process. The modules are represented as follows:

1. Login and candidate selection: The system would authenticate the voter before granting them access/login them into the application using their VIN and PIN to select their preferred candidate as seen in Fig. 5.
2. Candidate selection: The system would enable the logged in voters to select their preferred candidate as seen in Fig. 6.
3. Authentication: The system would authenticate the voter before enabling them to cast their vote using OTP as seen in Fig. 7.
4. Storing of casted votes: The system would allow the storage of casted votes to the blockchain database. The hash for each vote is generated and added to the blockchain as seen in Figs. 8 and 9.
5. Results: The system would preview the results once the voting period has ended as seen in Fig. 10.

4 Results and Discussion

To test the usability of the system, ISO 9241-11 usability model with the aid of a questionnaire was utilized. Three (3) constructs were tested which are effectiveness, efficiency, and user's satisfaction. A sample of nineteen (19) citizens was asked to

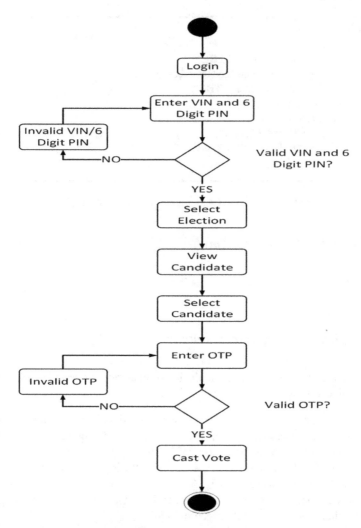

Fig. 4 An Activity Diagram showing the Activities being carried out by the Voters when utilizing the m-voting system

participate in the survey using their mobile devices. Some assumptions have to be made in order for the system to be viable, and these include the following:

1. Assumption 1: The voter's mobile device can be trusted. We accept that it is conceivable to safely run the voting application on the voter's mobile device.
2. Assumption 2: The election is correctly set up. We accept that only eligible voters can cast their votes and nothing is compromised before and during the voting procedure.
3. Assumption 3: Voters have acquired their voter's identification number (VIN) and PIN. We assume that all necessary information like age, date of birth, address, phone number, etc., have been registered at any voter's registration center in

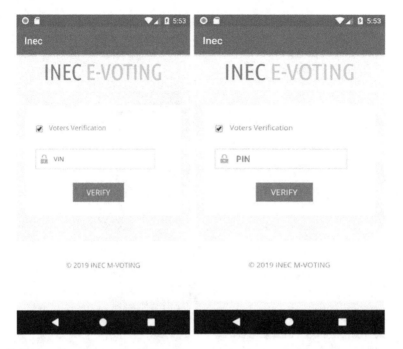

Fig. 5 Login and candidate selection interfaces

Fig. 6 Candidate selection
interface

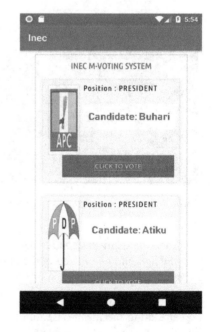

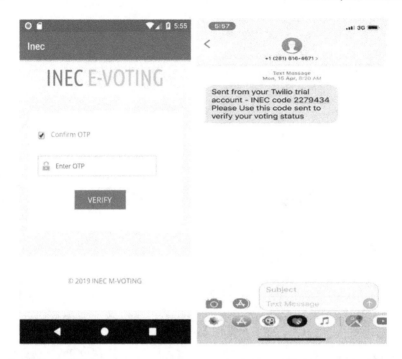

Fig. 7 OTP authentication

```
Block mined: 0000d6ba87712084989bfe77416c67a3fea7af7c01f9c150118dd13e2036de2b
Block mined: 00002fdd91c92b775ae655695c66a68db3aaa598c6887f7d5c499da9604a99ab
Block mined: 000063133999157d7d809c2d50775b163ce821a8662a871cfffae82363ceea86
Block mined: 0000c27787d9e90df7a12d57e8f4e023b673b424071da9828227a170283f608b
Block mined: 00006199592464dbc983478e2525723f671789e6ed1d53b5700b1b0bca5be451
Block mined: 0000878d821078f5006ead0ae4e0547f91acf35258370de214e571afa9025540
Block mined: 0000693cad593f5fe6cae5d0217be06353722de21300d8b05856b85c61a39241
Block mined: 0000d6351fb804ff0312bc7b5fa9c55d0f32c2c549f84b45ece6e414b661edea
Block mined: 00002584c1bb7a4be14daff0d7638db16fa6402334ed1e2002775555833b632a
Block mined: 0000008ba475e41d1ef7fcdcbecfbc484eb3da27a507484ec68064558222904f
```

Fig. 8 Hashes of the stored votes in the blockchain

order for them to acquire their voter's identification number (VIN) and PIN which would be utilized in the registration and voting process on the proposed system.

The overall score of each respondent was calculated for each usability constructed by calculating the mean based on the administered questionnaires survey scores. Strongly Agree (SA) = 5, Agree (A) = 4, Undecided (U) = 3, Disagree (D) = 2, and Strongly Disagree (SD) = 1 were utilized for the rating presented in Table 1.

```
{
    "chain": [
        {
            "index": 0,
            "timestamp": "12/05/2019",
            "vin": "200000",
            "vote": "Genesis block",
            "previousHash": "0",
            "hash": "55916185a7d6d572cf4d9bd54b386382ef5a4368fabd85801386d84fdfd99cfb",
            "nonce": 0
        },
        {
            "index": 1,
            "timestamp": "12/05/2019",
            "vin": "200055",
            "vote": "Buhari",
            "previousHash": "55916185a7d6d572cf4d9bd54b386382ef5a4368fabd85801386d84fdfd99cfb",
            "hash": "0000d6ba87712084989bfe77416c67a3fea7af7c01f9c150118dd13e2036de2b",
            "nonce": 41212
        },
        {
            "index": 2,
            "timestamp": "12/05/2019",
            "vin": "200005",
            "vote": "Atiku",
            "previousHash": "0000d6ba87712084989bfe77416c67a3fea7af7c01f9c150118dd13e2036de2b",
            "hash": "00002fdd91c92b775ae655695c66a68db3aaa598c6887f7d5c499da9604a99ab",
            "nonce": 85519
        },
        {
            "index": 3,
            "timestamp": "12/05/2019",
            "vin": "200051",
            "vote": "Buhari",
            "previousHash": "00002fdd91c92b775ae655695c66a68db3aaa598c6887f7d5c499da9604a99ab",
            "hash": "000063133999157d7d809c2d50775b163ce821a8662a871cfffae82363ceea86",
            "nonce": 54071
        },
```

Fig. 9 Contents of each block in the chain

Fig. 10 Result interface

The evaluation of the ISO 9241-11 usability model is carried out using a rating from one (1) to five (5). A system with very bad usability has 1 as its overall mean rating, bad usability has 2 as its mean rating, average usability has 3 as its mean rating, good usability has 4 as its mean rating, and excellent usability has 5 [26, 27]. From the result of the usability evaluation given in Table 1, the average effectiveness was given as 4.83, efficiency as 4.79, and user's satisfaction as 4.83. From the above analysis, it can be established that the prototype system developed has a "good usability" rating proven by the overall rating of 4.82. This system can offer multiple benefits in the electoral sector, and some of the benefits are:

- On the day of the election, eligible voters would be given access to cast their votes between a period of time and thanks to this, the government will not experience any losses since they did not declare a public holiday or the day was not called off.
- A polling station does not need to be set up due to the fact that the voters can make use of their mobile device to partake in the voting procedure which helps to reduce the high cost that is required in setting up polling stations.
- Multiple means of authentications would be utilized in order to enable the eligible voters to cast their voters and prevent illegible voters from casting theirs.
- Voters are not required to leave their homes or place of work in order to cast their votes.
- This could also provide easier accessibility of the voting procedure to multiple individuals including the elderlies and those with disabilities.
- Problems associated with the double casting of votes can be mitigated.

Table 1 Usability evaluation results collected from users

S/n	Questions	SD	D	U	A	SA	Mean
Effectiveness							
Q1	I was able to login successfully into the mobile application using my VIN and PIN	0	0	0	4	15	4.79
Q2	I was able to cast my vote successfully using the mobile application	0	0	0	5	14	4.74
Q3	I was able to use the OTP (one-time-password) to complete the voting process	0	0	0	1	18	4.95
Total							4.83
Efficiency							
Q4	I was able to use the application on my mobile device without any issue	0	0	0	7	12	4.63
Q5	I did not have to carry out too difficult steps before completing the voting process	0	0	0	2	17	4.89
Q6	The application response time was satisfactory	0	0	0	3	16	4.84
Total							4.79
User's satisfaction							
Q7	I feel comfortable using the mobile application compared to the existing voting procedure	0	0	0	0	19	5.00
Q8	There was no difficult step (s) involved in the use of the mobile application	0	0	0	1	18	4.95
Q9	I would use the application for casting my vote over the existing method	0	0	0	9	10	4.53
Total							4.83
Total mean							4.82

- The casting of votes and tallying of the votes would be a lot quicker while also providing better transparency and accuracy.
- The casted votes would be safely and securely stored in the blockchain database which would mitigate any form of manipulation or tampering like the deletion of legitimate votes or the addition of illegitimate votes.
- There will be a progressive increment in the number of youths taking part in the voting procedure.

With this voting system, a level of resiliency can be attained and is presented in the graph as shown in Fig. 11. This is measured on a scale of 1–5. 1 meaning "Worst", 2 meaning "Bad", 3 meaning "Average", 4 meaning "Good", and 5 meaning "Very Good".

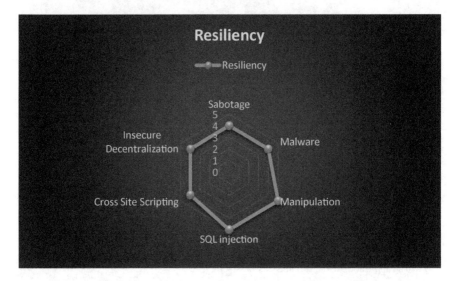

Fig. 11 Resiliency graph of the m-voting system

5 Conclusion

In this article, VIN, PIN, and OTP were utilized as the credentials needed for the authentication process. Blockchain technology was also utilized as a database to securely store the cast votes and mitigate any form of tampering in the process. The result of this study has shown a good usability rating which implies that the system is viable and can be implemented to solve the identified problems. This system is expected to securely protect the casted votes and verify voter's eligibility to cast their votes while providing an easily accessible voting system. This study has contributed to the body of knowledge by developing a mobile voting system that integrates blockchain technology and two-factor authentication to provide a secure and easy means of election participation which enables citizens to cast their votes using their mobile devices, storing, and securing the casted votes while offering better transparency. For future works, the system can be made to support multiple language types so as to better improve the usability rating and provide a better multilingual system to different language speaking types.

Acknowledgements We would like to acknowledge the sponsorship and support provided by Covenant University through the Centre for Research, Innovation, and Discovery (CUCRID).

References

1. Shuaibu, A., Mohammed, A., Ume, A.: A framework for the adoption of electronic voting system in Nigeria. Int. J. Adv. Res. Comput. Sci. Softw. Eng. **7**(3), 258–268 (2017)
2. Folarin, S., Ayo, C., Oni, A., Gberevbie, D.: Challenges and prospects of e-Elections in Nigeria. In: European Conference on e-Government, pp. 93–100. ACPI, England (2014)
3. Ayo, C.K., Ekong, U.O., Ikhu-omoregbe, N.A., Ekong, V.E.: M-voting implementation: the issues and trends. http://www.academia.edu/download/3258019/EEE4041.pdf. Last accessed 22 Sept 2019
4. Ananti, M.O., Onyekwelu, R.U.: Election and conundrum of sustainability of democracy in Nigeria. J. Policy Develop. Stud. **11**(5), 57–62 (2018)
5. Shaibu, M.E.: Nigerian election management bodies and their associated election challenges. Asian Int. J. Soc. Sci. **18**(1), 21–42 (2018)
6. Inuwa, I., Oye, N.D.: The impact of e-voting in developing countries: focus on nigeria. Am. J. Eng. Res. (AJER) **30**(2), 43–53 (2015)
7. Ekong, U., Ayo, C.: The prospects of m-voting implementation in Nigeria. In: Proceedings of the International Conference & Workshop on 3rd Generation (3G) GSM & Mobile Computing: An Emerging Growth Engine for National Development, pp. 172–179. Covenant University, Nigeria (2007)
8. Kayode, A.A., Olalekan, I.A.: A biometric e-voting framework for Nigeria. Jurnal Teknologi **77**(13), 37–40 (2015)
9. Vince, T.: Databases and Blockchains, the Difference is in Their Purpose and Design. https://hackernoon.com/databases-and-blockchains-the-difference-is-in-their-purpose-and-design-56ba6335778b. Last accessed 22 Sept 2019
10. Thomas, M.: Security and privacy via optimised blockchain. Int. J. Adv. Trends Comput. Sci. Eng. **8**(3), 415–418 (2019)
11. Curran, K.: E-voting on the blockchain. J. British Blockchain Assoc. **1**(2), 1–6 (2018)
12. Bailon, M.R.M.: International roaming services optimization using private blockchain and smart contracts. Int. J. Adv. Trends Comput. Sci. Eng. **8**(3), 544–550 (2019)
13. Fusco, F., Lunesu, M.I., Pani, F.E., Pinna, A.: Crypto-voting, a blockchain based e-voting system. In: 10th International Proceedings of the Joint Conference on Knowledge Discovery, Knowledge Engineering and Knowledge Management, pp. 223–227. SCITEPRESS, Spain (2018)
14. Shaan, R.: The Difference Between Blockchains & Distributed Ledger Technology. https://towardsdatascience.com/the-difference-between-blockchains-distributed-ledger-technology-42715a0fa92. Last accessed 20 Sept 2019
15. Uzedhe, G., Okhaifoh, J.E.: A technological framework for transparent e-voting solution in the Nigerian electoral system. Nigerian J. Technol. (NIJOTECH) **35**(3), 627–636 (2016)
16. Mpekoa, N., Greunen, D.: m-voting: understanding the complexities of its implementation. Int. J. Digital Soc. **7**(4), 1214–1221 (2016)
17. Nwabueze, E.E., Obioha, I., Onuoha, O.: Enhancing multi-factor authentication in modern computing. Sci. Res. Publish. Commun. Netw. **9**(3), 172–178 (2017)
18. Alausa, D.W.S., Ogunyinka, O.I.: The effect of e-voting and the alternative ballot box In Nigeria. Am. J. Eng. Res. (AJER) **6**(8), 46–55 (2017)
19. Yavuz, E., Koc, A.K., Cabuk, U.C., Dalkilic, G.: Towards secure e-voting using ethereum blockchain. In: 6th International Symposium on Digital Forensic and Security (ISDFS), pp. 1–7. IEEE, Turkey (2018)
20. Wu, H.-T., Yang, C.-Y.: A blockchain-based network security mechanism for voting systems. In: 1st International Cognitive Cities Conference (IC3), pp. 227–230. IEEE, Japan (2018)
21. Sakinah, B.N., Hafizhelmi, K.Z.F., Ihsan, M.Y.A., Nooritawati, M.T.: Blockchain in voting system application. Int. J. Eng. Technol. **7**(4.11), 156–162 (2018)
22. Purandare, H.V., Saini, A.R., Pereira, F.D., Mathew, B., Patil, P.S.: Application for online voting system using android device. In: International Conference on Smart City and Emerging Technology (ICSCET), pp. 1–5. IEEE, India (2018)

23. Kshetri, N., Voas, J.: Blockchain-enabled e-voting. IEEE Softw. **35**(4), 95–99 (2018)
24. Gajabe, J.: Implementation of mobility based secured e-voting system. Int. J. Res. Appl. Sci. Eng. Technol. **6**(3), 3449–3454 (2018)
25. Gentles, D., Sankaranarayanan, S.: Biometric secured mobile voting. In: 2nd Asian Himalayas International Conference on Internet (AH-ICI), pp. 1–6. IEEE, New Jersey (2011)
26. Sauro, J., Kindlund, E.: A method to standardize usability metrics into a single score. In: Proceedings of the SIGCHI Conference on Human factors in Computing Systems—CHI 2005, pp. 401–409. ACM Press, New York (2005)
27. Nielsen, J., Levy, J.: Measuring usability: preference versus performance. Commun. ACM **37**(4), 66–75 (1994)

An Improved Reversible Data Hiding Using Pixel Value Ordering and Context Pixel-Based Block Selection

M. Mahasree, N. Puviarasan and P. Aruna

Abstract In recent years, pixel value ordering (PVO) and prediction-error expansion (PEE) methods are combined to better exploit the pixel correlation in reversible data hiding (RDH) techniques. In these techniques, the cover image is split into non-overlapping blocks of equal size. The pixels of each block are sorted in ascending, according to their intensity values. The maximum and minimum pixels within each block are used to calculate prediction errors which are either expanded or shifted for embedding data. The data are embedded into smooth blocks which are identified using every two adjacent pixels of neighbor block. However, this block classification is not very effective since it considers only the neighbor block and ignores the pixels in current block. In this paper, to overcome this drawback and to improve the performance further, a novel reversible data hiding using context pixel-based block selection (CPBS) strategy is proposed. After the block division and sorting process, the complexity of each block is measured using the context pixel chosen from the current block and the connected pixels of neighbor block. Smooth blocks with low complexity are determined using a threshold value to achieve higher efficiency. When the desired capacity is achieved, the index orders of all pixels in the block are preserved and thus the reversibility is guaranteed. Experimental results verify that the proposed scheme achieves better capacity-distortion trade-off and thus outperforms with the existing PVO-based techniques.

Keywords Pixel value ordering · Prediction-error expansion · Reversible data hiding · Pixel correlation · Context pixel · Smooth blocks

M. Mahasree (✉) · P. Aruna
Department of Computer Science and Engineering, Annamalai University, Chidambaram, Tamilnadu, India
e-mail: mahasree05@gmail.com

P. Aruna
e-mail: arunapuvi@yahoo.co.in

N. Puviarasan
Department of Computer and Information Science, Annamalai University, Chidambaram, Tamilnadu, India
e-mail: npuvi2410@yahoo.in

© Springer Nature Singapore Pte Ltd. 2020
P. K. Singh et al. (eds.), *Proceedings of First International Conference on Computing, Communications, and Cyber-Security (IC4S 2019)*, Lecture Notes in Networks and Systems 121, https://doi.org/10.1007/978-981-15-3369-3_64

873

1 Introduction

In this technological world, protecting secret from intruders continues to be one of the challenges in the field of computers and telecommunication system. Data hiding provides transmission from one place to another in a secured manner. Recently, reversible data hiding (RDH) techniques have gained much attention in many sensitive fields such as remote sensing, archive management, military and medical image processing systems. According to reversibility, the data hiding schemes are classified into two types: They are (i) reversible data hiding and (ii) irreversible data hiding. Reversible data hiding (RDH) methods allow embedding data inside a cover media and afterward not only the hidden data can be extracted but the exact cover media can also be received [1–3]. In the irreversible methods, it is not possible to recover the original cover media exactly, after the hidden data is retrieved. RDH schemes are divided into three types based on the domain in which they are used. They are spatial, transform and encrypted domains. The goal of RDH is to increase the stego quality and assuring the reversibility after embedding. There are three main categories under which RDH is carried on. They are difference expansion (DE), histogram shifting (HS) and prediction-error expansion (PEE)-based RDH schemes [4]. In DE, the correlation between two neighbor pixels is used for embedding secret bits [5–7]. This method achieved good imperceptibility but low embedding rates. Thus, histogram-based RDH schemes were proposed to overcome this limitation. HS-based schemes work on the histogram of the cover media [8]. The peak and empty bins are utilized to embed secret bits. PEE is the extension of DE where the target pixel is predicted by context pixels from its neighbor and then the prediction error is used for expansion embedding [9–11]. Other than these techniques, PVO is the recently developed RDH technique where block-wise embedding is done based on largest/smallest pixel in the block. Though PVO gives better performance than other traditional methods, it lacks in optimal block selection for efficient embedding. This motivates to propose a better PVO method exploiting dependency between the current and neighbor blocks.

Based on the background and motivation given above, the proposed work made an improvement over PVO methods by introducing a novel block selection technique. The main contribution of this work is that the proposed CPBS method has higher embedding capacity when compared to other traditional PVO techniques. Also, the stego quality is maintained without any degradation.

The rest of paper is organized as follows. Section 2 briefly describes the related works. Section 3 presents the proposed reversible data hiding using PVO and CPBS method. Experimental results are provided in Sect. 4. Finally, Sect. 5 concludes this paper.

2   Literature Survey

He et al. [12] proposed a pixel-based pixel value grouping (PPVG) in which embedding is done based on the difference between maximum valued and minimum valued context pixels. Priority is given to pixels present in smooth regions. Ou et al. [13] discussed the PVO embedding in two-dimensional (2D) space. A 2D mapping is used to get back the original cover as well as secret data. He et al. [14] extended the original PVO into a general form, namely k-pass PVO in which both PVO-1 and PVO-2 embedding capacity are increased. Jana [15] proposed a reversible data hiding scheme for achieving better capacity using interpolation technique. Maniriho and Ahmad [16] presented an improvised lossless information hiding based on difference expansion method and modulus operation. This method considers both positive as well as negative differences to hide secret information with the help of a tracing table. Weng et al. [17] proposed a reversible technique for embedding secret using flexible block partition and adaptive pixel modification. Cai and Gui [18] proposed a data hiding scheme based on reference pixel and histogram shifting. Embedding is done to the histograms of each block. Jung [19] proposed a scheme that considers blocks with three pixels aligned in increasing order. The maximum and minimum prediction errors are calculated to check whether the data could be embedded or not. However, the pixels correlation is not effectively utilized. Sachnev et al. [20] method is a prediction error-based technique where the prediction errors are calculated using its four nearest pixels. Li et al. [21] proposed a novel prediction method namely pixel value ordering. In this method, two larger values and two smaller valued pixels are used for calculation. Their technique is based on the closeness of pixels within the block. Peng et al. [22] proposed a method which utilized the relative location relationships of pixels into prediction. Then, the bins with values one and zero are utilized for embedding to obtain an improved PVO. Liu and Lee [23] proposed an increased capacity-based reversible steganography based on the relative performance in row or column of the cover image. Hence, with improved pixels correlation, PVO-based RDH is proposed to increase the embedding capacity.

3   Proposed Scheme

Figure 1 depicts the block diagram for the proposed CPBS-based reversible data hiding. In PVO-based reversible data hiding techniques, block division produces equal-sized blocks and the embedding is done in block-wise manner. Smaller blocks posses highly correlated pixels and provide high embedding rates with significant degradation in image quality. Larger blocks also have many pixels with high correlation among them but they do not provide high embedding rates. Large blocks perform better only for low embedding capacity. This can be explained as follows:

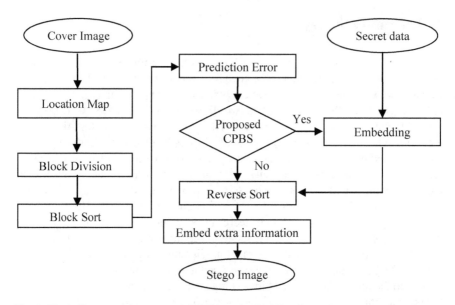

Fig. 1 Block diagram of the proposed CPBS-based embedding phase

- Embedding is done only at the maximum and minimum valued pixels of the block. Apart from reference pixels which are used for calculating the prediction errors, other pixels are not utilized well.
- Block complexity is not considered in traditional PVO techniques. In recently proposed techniques, block complexity is measured as the sum of absolute difference of every two adjacent pixels of the neighbor block. This calculation ignores the local complexity of the current block.

To address these issues, varied block sizes and a novel CPBS strategy for calculating block complexity are considered in our proposed method.

3.1 Location Map Construction

The embedding process changes the pixel value utmost by '1'. This leads to overflow and underflow conditions. The '0' valued pixels are modified to '−1' causing underflow and the '255' valued pixels are changed into '256' creating the overflow situation. This condition is seen in images where there is a lack of empty bins in image histogram. To avoid such cases, the image is checked for '0' and '255' valued pixels. The location map is created to store the locations of these pixels as follows:

- When the pixel with value '0' is encountered, change the pixel to '1' and mark this modification as binary value **1** in location map vector LM.

- When the pixel with value '1' is encountered, mark it as binary value **0** in location map indicating the unaltered pixel.
- Similarly, the pixel '255' is changed to '254' with binary **1** and the pixel '254' is kept unchanged by marking with binary **0**.

Thus, the location map generated consists of a binary sequence of length equal to the total number of pixels in set {0, 1, 254, 255}. This location map is compressed in lossless manner by arithmetic encoding technique to reduce the size. The compressed location map is denoted as CLM and its length is indicated by l_{CLM}.

3.2 Block Division

For a RXC sized image I, it is split into non-overlapped $B_1, \ldots, B_N$ blocks containing $n = \text{rxc}$ pixels in each. To analyze the performance of varied block sizes and to obtain the optimal results, six different block sizes are considered. The combinations of rxc used are 2×2, 1×3, 3×1, 2×3, 3×2 and 3×3. These block sizes are selected such that both row-wise correlation and column-wise correlation of pixels are utilized.

3.3 Context Pixel-Based Block Selection

For every block B_k where $k \in \{1, .., N\}$, their corresponding pixels are sorted in ascending order. The complexity of a block is a measure of smoothness. The smooth blocks are those containing highly correlated pixels with minimal differences in their intensity values. The rough blocks contain pixels with intensity values far apart from each other. The smooth blocks are more suitable for proposed method because the embedding is done with pixels having difference values as 1.

Figure 2 shows a block with context pixel and its neighbor pixels. The context pixel belongs to the current block. The median value of this sorted block is referred here as the context pixel, CP. The pixels of the neighbor blocks, surrounding the current block at right and bottom, are termed as neighbor pixels (P_{neighbor}). The block's complexity level CL is calculated as the sum of absolute difference between every neighbor pixels P_{neighbor} and the context pixel CP. The formula for CL is as follows:

$$CL = \sum \left| P_{\text{neighbor}} - CP \right| \tag{1}$$

When the block size is even, there are two median pixels. In such case, the maximum valued pixel between the two median values is considered as context pixel CP.

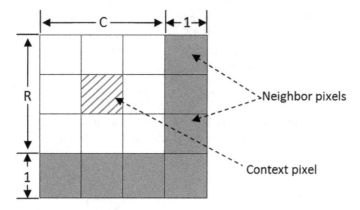

Fig. 2 3 × 3 sized block with its context pixel and neighbor pixels (shadow pixels)

3.4 Threshold Selection

The threshold value $T (0 \leq T \leq 255)$ is selected based on whether it achieves the minimum distortion for a given payload during the embedding step. Thus, it is the optimal threshold and it decides whether a block can withstand distortions while embedding. This threshold value differs for different block sizes and thus included as additional information parameter required at extraction.

- *Case 1*: When CL $\leq T$, the block is classified as smooth block. The difference between median and its neighbors is small and hence the possibility of containing consecutive pixels. Except median, all minimum and maximum pixels are either shifted to next value or expanded to carry bits.
- *Case 2*: When CL $> T$, the block is classified as rough block so unchanged.

3.5 Embedding Phase

The selected blocks in the previous stage undergo the embedding of secret data. The embedding process continues until all the desired message bits are completely embedded and we take p_{end} as the index value of the last block carrying the last bit of the secret data. The detail of the embedding algorithm is produced in Algorithm 1.

Algorithm 1: *Proposed Embedding Algorithm*

Input: Cover image, Secret message
Output: Stego image
1: $C \leftarrow imread(Cover\ image)$
2: $[r, c] \leftarrow size(C)$
3: $S \leftarrow fread(Secret\ message)$
4: $s \leftarrow binary(S)$
5: **for** $x = 1\ to\ r$ **do**   /* Location map (LM) generation*/
   for $y = 1\ to\ c$ **do**
     if $(C[x, y] == 1)\ or\ (C[x, y] == 254)$ **then**
       $LM.append(0)$; $C'[x, y] == C[x, y]$
     else if $C[x, y] == 0$ **then**
       $C'[x, y] == 1$; $LM.append(1)$
     else if $C[x, y] == 255$ **then**
       $C'[x, y] == 254$; $LM.append(1)$
     end if
   end for
   end for
6: $\{B_n\} \leftarrow block(C')$ /*Block division*/
7: **for** $i = 1\ to\ n$ **do**
   $[P_i, Index_i] = sort(B_i)$
   $P_{ref} \leftarrow median(P_i)$
   $d_{\min j} = P_{\min j} - P_{ref}$ /* $\min j$ refers to number of pixels less than median pixel*/
   $d_{\max j} = P_{\max j} - P_{ref}$ /* $\max j$ refers to number of pixels greater than median pixel*/
   $CL = \sum |P_{neighbor} - P_{ref}|$ /*Context Pixel based Block Selection*/
   if $CL \leq T$ **then**   /*Threshold $0 \leq T \leq 255$ such that required capacity is achieved */
     for $p = 1\ to\ j$ **do**
$$d'_{\min p} = \begin{cases} 0, & if\ d_{\min p} = 0 \\ d_{\min p} - s, & if\ d_{\min p} = -1 \\ d_{\min p} - 1, & if\ d_{\min p} < -1 \end{cases} /*\ s \in \{0,1\}\ */$$
$$d'_{\max p} = \begin{cases} 0, & if\ d_{\max p} = 0 \\ d_{\max p} + s, & if\ d_{\max p} = 1 \\ d_{\max p} + 1, & if\ d_{\max p} > 1 \end{cases}$$
     end for
   else
     $d'_j = d_j$
   end if
   $P'_i = P_j + d'_j$
   $B'_i = rev\_sort([P'_i, Index_i])$
   end for
8: $R \leftarrow \{B'_n\}$
9: $R' \leftarrow add\_info(R)$ /*additional parameters are embedded using steps 6 and 7 */
10: $Stego\ image \leftarrow R'$

After embedding the secret information, the additional data required for extraction is also embedded into the stego image. First, record the least significant bits of the first $30 + \log_2 N$ pixels and save as binary sequence S_{LSB}. Then replace these LSB by the additional information including

- Block size n_1 (2 bits) and n_2 (2 bits)
- Complexity level threshold T (8 bits)

- End position p_{end} with x and y values ($\log_2 N$ bits)
- Length of the location map after compression l_{CLM} (18 bits)

Finally, embed the sequence $S_{\text{LSB}} + \text{CLM}$ into the remaining blocks leaving the first $30 + \log_2 N$ pixels. This completes the embedding process and guarantees the reversibility of both cover image and secret message.

3.6 Extraction Phase

At the receiver point, the stego image is processed similar to the embedding steps in a reverse manner. Firstly, the additional information is extracted from the stego image. Then, the secret data are extracted with the help of additional information. The detailed algorithm for extraction phase is described in Algorithm 2.

3.7 An Illustration of the Proposed RDH Method Using CPBS

An example of the proposed technique is shown in Fig. 3. A 3×2 sub-block with values (173, 175, 174, 174, 176, 178) is taken from Lena image. After sorting, the sorted pixels are (173, 174, 174, 175, 176, 178). The difference values are calculated as $d_{\text{min0}} = -2, d_{\text{min1}} = -1, d_{\text{min2}} = -1, d_{\text{max0}} = 1, d_{\text{max1}} = 3$. Let the secret bits be $s = (0, 1, 0)$. According to embedding rule, the difference values with one are embedded as follows:$d_{\text{min0}} = -3, d_{\text{min1}} = -1, d_{\text{min2}} = -2, d_{\text{max0}} = 1, d_{\text{max1}} = 4$. The new embedded pixels are calculated as (172, 174, 173, 175, 176, 179). The median pixel 175 is kept unchanged so that it is used as a reference pixel at the receiver side. It is observed that due to embedding, the sorting order is changed here (173 comes after 174). This will be solved during extraction process. The pixels are put back to their original positions forming the stego pixels (172, 175, 174, 173, 176, 179).

Now, at the extraction process, the stego pixels (172, 175, 174, 173, 176, 179) are sorted (172, 173, 174, 175, 176, 179). The difference values are $d'_{\text{min0}} = -3, d'_{\text{min1}} = -2, d'_{\text{min2}} = -1, d'_{\text{max0}} = 1, d'_{\text{max1}} = 4$. According to extraction rules, the recovered pixels are (173, 174, 174, 175, 176, 178) and the secret bits are $s = (1, 0, 0)$. The second and third pixels are similar. So, check for their order of indices. The second pixel 173 is at 4th position. The third pixel 174 is at 3rd position. Hence, swap their secret bits as $s = (0, 1, 0)$. Finally, the original pixels are recovered by reverse sorting, (173, 175, 174, 174, 176, 178).

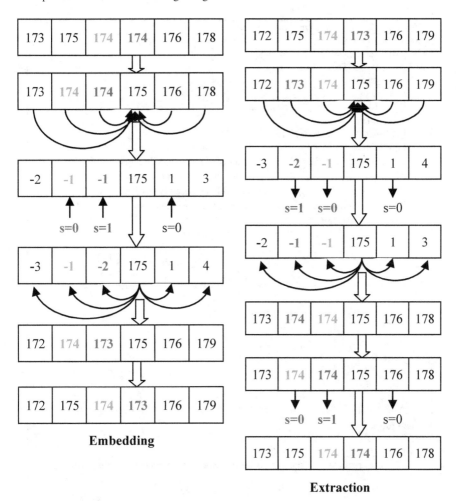

Fig. 3 Embedding and extraction phases of the proposed method

---

Algorithm 2: *Proposed Extraction Algorithm*

Input: Stego image
Output: Secret message, Cover image
1: $R' \leftarrow imread(Stego\ image)$
2: $[r, c] \leftarrow size(R')$
3: $[R, info] \leftarrow extract(R')$
4: $\{B'_n\} \leftarrow block(R)$ /*Block division*/
5: **for** $i = n\ to\ 1$ **do**
   $[P'_i, Index_i] = sort(B'_i)$
   $P_{ref} \leftarrow median(P'_i)$
   $d'_{min\ j} = P'_{min\ j} - P_{ref}$ /* *min j* is number of pixels less than median pixel*/
   $d'_{max\ j} = P'_{max\ j} - P_{ref}$ /* *max j* is number of pixels greater than median */
   $CL = \sum |P_{neighbor} - P_{ref}|$ /*Context Pixel based Block Selection*/
   if $CL \le T$ **then** /* Threshold received as one of the additional parameters */
     for $p = 1\ to\ j$ **do**

$$d_{min\ p} = \begin{cases} 0, & if\ d'_{min\ p} = 0 \\ d'_{min\ p}\ and\ s = 0, & if\ d'_{min\ p} = -1 \\ d'_{min\ p} + 1\ and\ s = 1, & if\ d'_{min\ p} = -2 \\ d'_{min\ p} + 1, & if\ d'_{min\ p} < -2 \end{cases} \quad /* s \in \{0,1\} */$$

$$d_{p\ max} = \begin{cases} 0, & if\ d'_{max\ p} = 0 \\ d'_{max\ p}\ and\ s = 0, & if\ d'_{max\ p} = 1 \\ d'_{max\ p} - 1\ and\ s = 1, & if\ d'_{max\ p} = 2 \\ d'_{max\ p} - 1, & if\ d'_{max\ p} > 2 \end{cases}$$

     end for
   else
     $d_j = d'_j$
   end if
   $P_i = P'_j + d_j$; $R_i = rev\_sort([P_i, Index_i])$
   end for
6: $C' \leftarrow \{R_n\}$
7: **while** $g \le len(LM)$ **do**   /* g=0 */
   for $x = 1\ to\ r$ **do** /* compute original pixels with extracted Location map (LM) */
     for $y = 1\ to\ c$ **do**
       if $(C'[x,y] == 1)\ or\ (C'[x,y] == 254)$ **then**
         if $LM[g] == 0$ **then**
           $C[x,y] = C'[x,y];\ g = g + 1$
         else if $LM[g] == 1$ **then**
           if $C'[x,y] == 254$
             $C[x,y] = 255;\ g = g + 1$
           else if $C'[x,y] == 1$
             $C[x,y] = 0;\ g = g + 1$
           end if
         end if
       end if
     end for
   end for
  end
8: $Cover\ image \leftarrow C$; $Secret\ message \leftarrow S$

4 Experimental Results

This section discusses the performance of the proposed reversible data hiding using CPBS method. Experiments are conducted on standard grayscale images. The implementation was done using Python programming. The cover images are of size 512×512 with bit depth 8. The results are compared with existing reversible data hiding techniques based on PVO. The quality of the stego image is evaluated using PSNR. Table 1 gives the comparison of stego quality of proposed method with some state-of-the-art methods. For various payload capacities, the proposed method yields better PSNR values. Compared with five methods, the maximum PSNR gains of Lena image by using the proposed method for the capacity of 10,000 bits is 3.06 dB for sachnev et al. method. Similarly, for the other images, the proposed method gives better gains. To show the performance of CPBS for a given image, the details of optimal parameters for Lena image from a capacity of 5000 bits to 40,000 bits are given in Table 2. It is observed that smaller block sizes give optimal results for lower embedding capacity (EC). And for higher embedding rates, larger block sizes are

Table 1 Comparison of PSNR (in dB) between proposed method and existing methods for an EC of 10,000 bits

Images	Sachnev et al. [20]	Peng et al. [22]	He et al. [14]	Liu and Lee [23]	Proposed
Lena	58.18	60.04	60.64	58.94	61.24
Baboon	54.15	53.55	54.00	52.67	54.25
Barbara	58.15	60.54	60.37	58.90	61.54
Jetplane	60.38	62.96	63.45	62.57	64.10
Peppers	55.55	58.98	59.29	58.27	59.00
Boat	56.15	58.27	58.28	56.38	59.38

Table 2 Optimal parameters of the proposed method of Lena image

Capacity (bits)	PSNR (dB)	Threshold T	Block size $n_1 \times n_2$
5000	64.73	6	3×1
10,000	61.24	8	3×1
15,000	58.85	11	3×1
20,000	57.27	13	3×1
25,000	55.93	17	3×2
30,000	54.70	21	3×1
35,000	53.67	28	3×2
40,000	52.59	42	3×2

optimal. Also, the threshold values are set with least values for lower embedding rates and this is the reason for high PSNR values of the proposed CPBS method.

Figures 4 and 5 show that the proposed method mostly gives better yield than the other methods. The significant gain in capacity for all images is due to the

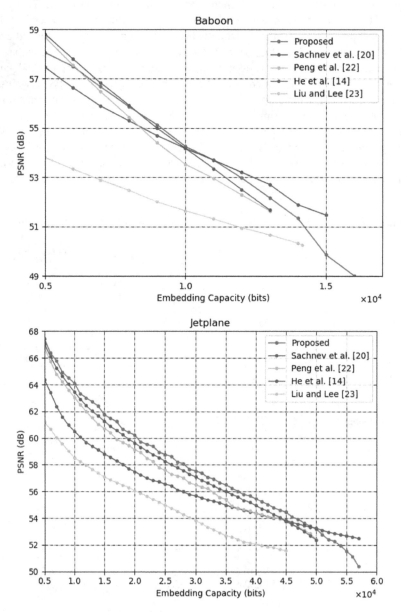

Fig. 4 Comparison of proposed CPBS and existing methods for Baboon and Jetplane images

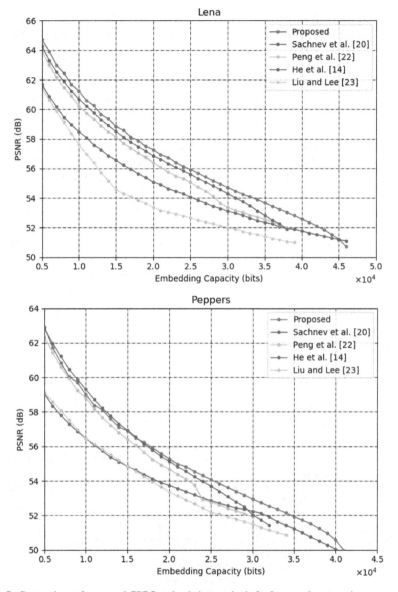

Fig. 5 Comparison of proposed CPBS and existing methods for Lena and peppers images

effective block sizes that are preferred in our proposed method. The better PSNR gains are obviously due to the novel CPBS selection technique. For Baboon image, the proposed method is not very effective for the embedding capacity above 12,000. This is due to the complex pixel arrangement of the image. For peppers image, the proposed method gives similar results with He et al. method up to a maximum of 20,000 bits and then increased performance is shown for the rest of the higher embedding rates. For Lena and Jetplane images, the superiority of the proposed method can be seen. Maximum capacity of around 60,000 bits is achieved for Jetplane image.

5 Conclusion

This paper proposed a novel reversible hiding scheme using context pixel-based block selection (CPBS) algorithm for smoothness classification. Varied block sizes such as 2×2, 1×3, 3×1, 2×3, 3×2 and 3×3 are used in the proposed method to adaptively select the block size for embedding data. By using CPBS, the proposed method makes better use of the correlation among neighbors inherent in the cover image. CPBS classifies the smooth blocks more accurately and thus provides superior performance than other existing methods in terms of embedding capacity as well as stego quality. Experimental results showed that the proposed reversible data hiding using CPBS technique outperforms the existing methods.

References

1. Shi, Y.Q.: Reversible data hiding. In: Cox I.J., Kalker T., Lee H.K. (eds.) Digital Watermarking. IWDW 2004. Lecture Notes in Computer Science, vol. 3304, pp. 1–12. Springer, Berlin, Heidelberg (2005)
2. Fridrich, J., Goljan, M., Du, R.: Lossless data embedding—new paradigm in digital watermarking. EURASIP J. Appl. Signal Process. 185–196 (2002)
3. Celik, M.U., Sharma, G., Tekalp, A.M., Saber, E.: Lossless generalized-LSB data embedding. IEEE Trans. Image Process. 14(2), 253–266 (2005)
4. Zhao, H., Wang, H., Khurram Khan, M.: Statistical analysis of several reversible data hiding algorithms. Multimed. Tools Appl. 52(2), 277–290 (2011)
5. Tian, J.: Reversible data embedding using a difference expansion. IEEE Trans. Circ. Syst. Video Technol. 13(8), 890–896 (2003)
6. Alattar, A.M.: Reversible watermark using the difference expansion of a generalized integer transform. IEEE Trans. Image Process. 13(8), 1147–1156 (2004)
7. Wang, X., Li, X., Yang, B., Guo, Z.: Efficient generalized integer transform for reversible watermarking. IEEE Signal Process. Lett. 17(6), 567–570 (2010)
8. Tai, W.L., Yeh, C.M., Chang, C.C.: Reversible data hiding based on histogram modification of pixel differences. IEEE Trans. Circ. Syst. Video Technol. 19(6), 906–910 (2009)
9. Li, X., Yang, B., Zeng, T.: Efficient reversible watermarking based on adaptive prediction-error expansion and pixel selection. IEEE Trans. Image Process. 20(12), 3524–3533 (2011)
10. Lee, C.F., Chen, H.L.: Adjustable prediction-based reversible data hiding. Digit. Signal Proc. 22(6), 941–953 (2012)
11. Hong, W., Chen, T.S., Shiu, C.W.: Reversible data hiding for high quality images using modification of prediction errors. J. Syst. Softw. 82(11), 1833–1842 (2009)
12. He, W., Cai, J., Xiong, G., Zhou, K.: Improved reversible data hiding using pixel-based pixel value grouping. Optik 157, 1–15 (2017)
13. Ou, B., Li, X., Wang, J.: High-fidelity reversible data hiding based on pixel-value-ordering and pairwise prediction-error expansion. J. Vis. Commun. Image Represent. 39, 12–23 (2016)

14. He, W., Zhou, K., Cai, J., Wang, L., Xiong, G.: Reversible data hiding using multi-pass pixel value ordering and prediction error expansion. J. Vis. Commun. Image Represent. **49**, 351–360 (2017)
15. Jana, B.: High payload reversible data hiding scheme using weighted matrix. Optik **127**(6), 3347–3358 (2016)
16. Maniriho, P., Ahmad, T.: Information hiding scheme for digital images using difference expansion and modulus function. J. King Saud Univ. Comput. Inf. Sci. **31**(3), 1–13 (2018)
17. Weng, S., Liu, Y., Pan, J.S., Cai, N.: Reversible data hiding based on flexible block-partition and adaptive block-modification strategy. J. Vis. Commun. Image Represent. **41**, 185–199 (2016)
18. Cai, S., Gui, X.: An efficient reversible data hiding scheme based on reference pixel and block selection. In: International Conference on Intelligent Information Hiding and Multimedia Signal Processing. IEEE, Beijing, China (2013)
19. Jung, K.H.: A high-capacity reversible data hiding scheme based on sorting and prediction in digital images. Multimed. Tools Appl. **76**(11), 13127–13137 (2017)
20. Sachnev, V., Kim, H.J., Nam, J., Suresh, S., Shi, Y.Q.: Reversible watermarking algorithm using sorting and prediction. IEEE Trans. Circ. Syst. Video Technol. **19**(7), 989–999 (2009)
21. Li, X., Li, J., Li, B., Yang, B.: High-fidelity reversible data hiding scheme based on pixel-value-ordering and prediction-error expansion. Sig. Process. **93**(1), 198–205 (2013)
22. Peng, F., Li, X., Yang, B.: Improved PVO-based reversible data hiding. Digit. Signal Proc. **25**, 255–265 (2014)
23. Liu, H.H., Lee, C.M.: High-capacity reversible image steganography based on pixel value ordering. EURASIP J. Image Video Process. **54**, 1–15 (2019)

Activity Recognition Using Temporal Features and Deep Bottleneck 3D-ResNeXt

Akash Panchal, Harshal Trivedi, Mihir Rajput and Dhwani Trivedi

Abstract Due to the advancements in the field of artificial intelligence, object detection and classification are becoming easier. New deep learning and machine learning-based approaches are being used for video analysis, activity recognition and activity detection, too. But, training of deep learning-based neural networks for large video datasets is computationally expensive. Also, it lacks in terms of either speed or correctness and has drastic variations due to the background, scale, etc. Thus, while deep learning has given outstanding results for image classification and detection, activity detection and recognition can still use better approaches for faster and more accurate results. In this paper, we built a model based on ResNet classifier and ResNeXt classifier, which gives improved performance in action recognition. We start with introductory topics and network architecture. Then, we discuss the experiments that we have tried, and also, we show the process of how we get to deploy this method in edge devices like Nvidia Jetson Nano and Intel UP Squared board.

Keywords 3D convolution · ResNeXt classifier · Spatio-temporal feature extraction

A. Panchal (✉) · H. Trivedi · M. Rajput
Wobot Intelligence Private Limited, Ahmedabad, India
e-mail: akashpanchal94@gmail.com

H. Trivedi
e-mail: harshal201286@gmail.com

M. Rajput
e-mail: mihirrajput9@gmail.com

D. Trivedi
Stony Brook University, 100 Nicolls Rd, Stony Brook, NY 11794, USA
e-mail: dhtr004@gmail.com

© Springer Nature Singapore Pte Ltd. 2020   889
P. K. Singh et al. (eds.), *Proceedings of First International Conference on Computing, Communications, and Cyber-Security (IC4S 2019)*, Lecture Notes in Networks and Systems 121, https://doi.org/10.1007/978-981-15-3369-3_65

1 Introduction

The purpose of activity recognition is to classify or identify a particular activity in a video. Action detection means to detect every occurrence of a given action in a video and to localize them in both space and time. Action recognition has been of very importance due to its use in video surveillance, smart home systems, patient monitoring, child monitoring, human–computer interaction, driver behaviour detection [1], etc.

Image- or video-based social media sites such as YouTube are also interested in automatically classifying millions of uploaded videos to provide semantic-aware video indexes to facilitate easy search by viewers. And due to advancements in embedded systems, video analysis techniques, cloud computing infrastructure [2–4], necessary computational becomes easier. Action recognition is only going to be needed more in coming days.

Traditional deep learning-based approaches have significantly improved the performance of video action recognition. The procedure begins with extracting the useful features and encodes them. Now the classifier identifies the action and their spatio-temporal locations from the extracted encoded features. Deep learning architectures are used for action classification, spatial (using segmentation or frame-level region proposal), temporal, spatio-temporal (using selective search-based region proposals on each frame of the trimmed video) localization or both. Its main advantage is its capability to perform end-to-end optimization.

Another option was to use 2D ConvNets. It was based on hand-crafted features and feature descriptions. Convolutional neural network (CNN) learns filters based on a stack of N input frames. Generally, in CNN, 3D data is converted into 2D and then it is used for superimposition of frames. Those frames are fed to the CNN model for action classification. But such fixed length approaches cannot learn to recognize complex video sequences, e.g. cooking or dancing due to lack of temporal structure. Using 3D CNN is one of its solutions.

From several years ago [5], 3D CNNs are useful and effective tool to recognize the actions accurately, It's also used for the low-level feature representations and the 3D filters are applied to extract low level features (see Fig. 1). The produced output

Fig. 1 C3D feature extraction [6]

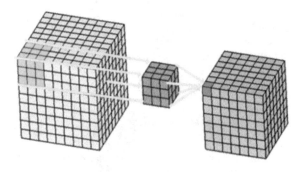

shape is a 3D volume space (i.e. cube or cuboids). The 3D CNN is not limited to 3D space, but it can be used for 2D space inputs such as images.

To overcome these limitations, T-CNN or Tube CNN (see Fig. 2) is also an essential model used for image and video classification and for localization.

The video (both trimmed and untrimmed) is divided into fixed length cuts. Using them, tube expert proposals are produced with 3D CNN. They are then connected together (using system stream and spatio-transient activity identification).

In [8], network predicts "micro-tubes" of two frames and then links them to a complete action tube (using an algorithm). Temporal encoding is used on them. In a novel framework [9, 10] by leveraging the descriptive power of 3D ConvNet, an input video is divided into equal length clips first. Then, the clips are fed into tube proposal network (TPN) and a set of tube proposals are obtained.

Next, tube proposals from each video clip are linked according to their action-ness scores and overlap between adjacent proposals to form a complete tube proposal for spatio-temporal action localization in the video. Finally, the Tube-of-Interest (ToI)

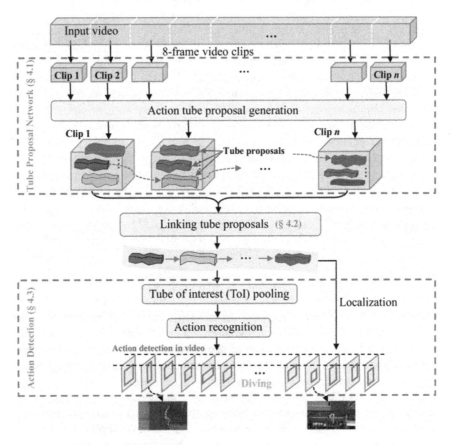

Fig. 2 Model architecture T-CNN [7]

pooling is applied to the linked action tube proposal to generate a fixed length feature vector for action label prediction.

Our work includes activity recognition using ResNet architecture in which ResNeXt is also used. Even in this architecture, of course, hyper-parameter tuning was used to gain the highest performance, speed and accuracy. For training, we used 16 frames as sample duration, while, for testing, we used 8. We measured the accuracy at different learning rates of 0.1, 0.01 and 0.001. Batch size tuning impacted the speed. We measured the speed at different batch sizes of 8 and 16.

2 Background

Many approaches have been proposed for this purpose. Deep learning architectures have been increasingly applied of late to action classification, spatial, temporal and spatio-temporal action localization. Spatial action localization has been mostly addressed using segmentation or by linking frame-level region proposal.

In [11, 12], human action recognition is performed based on the sequences of 3D skeleton data, in which 3D CNN is combined with euclidean distance matrices (EDMs). The limitation of EDM is that permutation (random shuffling) of different skeleton joints gives many representations. The advantage is that the architecture, simultaneously, in an end-to-end manner learns the optimal representation of joints and optimizes other parameters. For better performance, it needs to unify skeleton detection form RGB-D video and action recognition in single, end-to-end differentiable architecture.

Also, according to [13], performing activity recognition by most of the presently employed experimental observations is not enough with reference to the wearable sensor data, wherein the accuracy of the recognition will drop to a large extent as compared to a suitable evaluation method. It handles the necessary issues which compromise the understanding of the performance in the human activity recognition regard to the wearable sensor data. The main issue in the use of wearable sensors is the lack of some standard protocols which can be used to perform the experiments and report its result.

In [6], action micro-tube regression is used to classify and regress whole video subsets for optimal action detection. Here, 3D region proposal networks (RPNs) are used; they span two consecutive video frames [14], and they encode the temporal aspect of actions by appearance. It is an end-to-end trainable model, which has localization and classification as a single step of optimization. The advantage is that computation time can be reduced by 50%. To further boost the performance, optical flow can be integrated with this framework.

In [8, 15], 3D CNN, T-CNN refer Fig. 2 and segmentation is used for action detection and recognition. It is a unified deep network able to localize and recognize action based on 3D convolution features, which is a top-down approach. It requires large number of annotated samples to train the bounding boxes. As a solution, extend

3D CNN to encoder–decoder structure to formulate localization as segmentation, which is a bottom-up approach.

Multitask deep learning of [11, 16] is a single architecture to solve two problems: 2D/3D pose estimation from still images and action recognition from video sequences. It predicts 2D and 3D localization of body points from raw RGB frames and uses them to predict action performed in videos. Afterwards, the temporal evaluation can be used for the recognition of activities.

We also tried using the sliding window approach for action recognition. We experimented with public datasets for checking its accuracy. We tried it on the UCF dataset [17], but found that it gave only 30–40% accuracy. So, we had to find a more accurate method.

It was shown by Kay et al. that the results of 3D CNNs on their kinetics dataset were comparable to the results of the 2D CNNs which were trained in advance on ImageNet. On the other hand, the results of 3D CNNs on the HMDB51 [18] and UCF101 had an inferior outcome as compared to the 2D CNNs. The inception model [18] introduced by Carreira et al. is a very deep network having 22 layers, and it has attained the state-of-the-art performance [19]. We have introduced the ResNet architecture in this paper, which produces superior results as compared to the inception model and the 3D CNNs.

3 ResNeXt

ResNet architecture was utilized in this model, and it can prepare hundreds or thousands of layers without decreasing the exhibition quality. Because of the vanishing gradient problem, as the angle is back-propagated to prior layers, rehashed multiplication may make the value of gradient very little. Subsequently, deep systems are difficult to prepare with expanding system profundity. Along these lines, with more system profundity, the presentation would begin debasing quickly. To comprehend this, ResNet essentially stacks character mappings (layer that does not do anything) upon the present system, and the subsequent engineering can play out the equivalent (Figs. 3 and 4).

In this model, we utilized bottleneck class of ResNet. It is three-layer deep block structure. ResNet is one of the best structures in state-of-the-art image classification. Here, we investigate the accompanying models: ResNet (fundamental and bottleneck squares) [19], pre-activation ResNet [21], wide ResNet [22], ResNeXt [23] and DenseNet [24].

A fundamental ResNets blocks have two convolutional layers, and each has batch normalization and a ReLU a shortcut pass later. The top block of the model is associated with the layer just before the last ReLU through an easy route pass. ResNet-18 and 34 receive the basic blocks. Identity connections and zero padding are used for the shortcuts of the basic blocks (type A in Xie et al. [23]) to avoid increasing the number of parameters of these relatively shallow networks.

Fig. 3 ResNet (BottleNeck) [20]

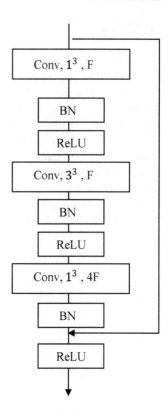

Bottleneck ResNet blocks are made of three convolutional layers. The first and third convolutional layers have part sizes of $1 \times 1 \times 1$, while those of the second are $3 \times 3 \times 3$. The alternate route goes of this square, and the main blocks are the same. ResNet-50, 101, 152 and 200 embrace the bottleneck.

The pre-actuation ResNet and bottleneck ResNet designs are comparable; however the convolution, batch normalization and ReLU orders are extraordinary. The top of the layer is associated with the layer soon after the last convolutional layer in the block structure through a shortcut route pass. In their examination, He et al. demonstrated that such pre-activation encourages enhancement in the preparation and decreases overfitting [25]. This also helps in effective steepest gradient calculation and optimizes the accuracy of the network.

ResNeXt includes the new dimensional; cardinality, which is not quite the same as deeper and more extensive. Not at all like the first bottleneck hinder, the ResNeXt block includes group of convolutions, which partitions the feature maps into little gatherings. DenseNet makes associations from early layers to later layers which are not the same as the ResNet. Also, some ResNet architectures use batch normalization layers after the activation functions to ignore the negative values and take only positive values into the count. Adding batch normalization layers after the activation function has shown that it helps to find global minimum. This link interfaces

Fig. 4 ResNet (basic) [20]

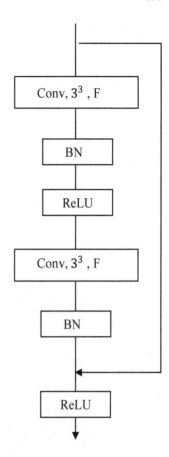

each layer densely in a feed-forward design. In their examination, Huang et al. demonstrated that not many err parameters accomplish preferred precision over ResNets [24] (Figs. 5 and 6) (Table 1).

4 Proposed Work

It is a basic, highly flexible network architecture for image classification. The system is made by repeating a structure hinder that totals a lot of transformations with a similar topology. It has a basic structure which makes a homogeneous, multi-branch design that has just a couple hyper-parameters to set like batch size (bunch of input images), dimensions (height and width of input images). On the ImageNet-1 K dataset [26], even under the confinement of looking after multifaceted nature, expanding cardinality can improve order accuracy. In any case, when the limit is expanded, expanding cardinality is more viable than going further or more extensive.

Fig. 5 DenseNet flow
architecture [20]

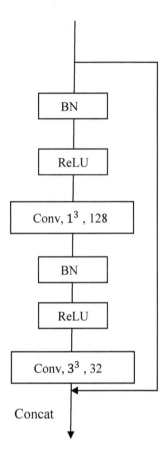

Our design was taken a stab at 4 unique plans, including 18, 32, 50 and 101 layers. We found that action recognition worked best on 101 layers. Here, the video is utilized for removing fleeting features. These features are utilized for learning and dependent on that preparation; action is identified. For training, the example length sizes of 16, 20, 32 and 52 edges were utilized. In preparing, we utilized 900 recordings of irregular sizes as positive mopping information and 800 recordings of no mopping information.

For augmentation, neither grayscale, nor RGB was used. We combined standard deviation and RGB mean for training, which had a very good overall impact. We have also tried to run the activity recognition trained model on edge device using various methods.

Fig. 6 ResNet flow
architecture [20]

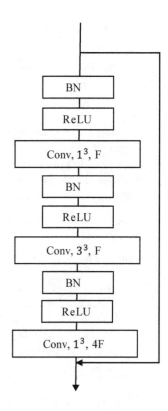

Table 1 Architecture details

Layer name	18-layer		34-layer
Conv1	$7 \times 7 \times 7$, 64, stride 1 (T), 2(XY)		
Conv2_x	$3 \times 3 \times 3$ max pool, stride 2	$\begin{bmatrix} 3 \times 3 \times 3,\ 64 \\ 3 \times 3 \times 3,\ 64 \end{bmatrix} \times 2$	$\begin{bmatrix} 3 \times 3 \times 3,\ 64 \\ 3 \times 3 \times 3,\ 64 \end{bmatrix} \times 3$
Conv3_x	$\begin{bmatrix} 3 \times 3 \times 3,\ 128 \\ 3 \times 3 \times 3,\ 128 \end{bmatrix} \times 2$		$\begin{bmatrix} 3 \times 3 \times 3,\ 128 \\ 3 \times 3 \times 3,\ 128 \end{bmatrix} \times 4$
Conv4_x	$\begin{bmatrix} 3 \times 3 \times 3,\ 256 \\ 3 \times 3 \times 3,\ 256 \end{bmatrix} \times 2$		$\begin{bmatrix} 3 \times 3 \times 3,\ 256 \\ 3 \times 3 \times 3,\ 256 \end{bmatrix} \times 6$
Conv5_x	$\begin{bmatrix} 3 \times 3 \times 3,\ 512 \\ 3 \times 3 \times 3,\ 512 \end{bmatrix} \times 2$		$\begin{bmatrix} 3 \times 3 \times 3,\ 512 \\ 3 \times 3 \times 3,\ 512 \end{bmatrix} \times 3$

Basically, the following hyper-parameters were tuned:

- Sample Duration (how many frames to consider): we used 16 frames in training and 8 frames in evalution. It decided if there is an activity or not.
- With learning rate 0.1, accuracy was 82%; with learning rate 0.01, accuracy was 86; with learning rate 0.001, accuracy was 88, whereas reduction on plateau was 90.
  Using stochastic gradient descent (SGD), we achieved the momentum of 0.9 and 0.1 weight decay.
- Batch size: By tuning batch size, interestingly, we found that it impacts the speed. Batch size 16 gave us 1 min to process 5 min of video, and batch size of 8 gave results in 40 s for that same video.

To achieve the inference on edge devices (to see Figs. 7 and 8), we have used following steps:

1. Dataset Optimization

    a. First, we reduce the size of the image data by squeezing the frames.
    b. We reduce the dataset frames resolution by half.
    c. Then, we remove extra pre-processing and augmentation techniques to fit the data into the memory of the edge device.
    d. At the inference time, we use the same approach as above.

2. Network Architecture Optimization

    a. In network architecture, we use shallow architecture like ResNeXt with 34 layers to train the model.
    b. This enables fast processing but compromises accuracy.
    c. We also remove batch normalization layers from each residual layer.

Specification:
GPU:128-Core Maxwell
CPU: Quad-Core ARM A57 @1.43GHz
Memory: 4GB
OS: Ubuntu image

Fig. 7 Nvidia Jetson Nano

Specification:
CPU: Intel Atom®X7-E3950 processor
RAM:8GB
OS: Ubuntu image (kernel 4.15)
Toolkit: Open VINO™ toolkit R5

Fig. 8 Intel UP Squared board

3. Inference Optimization

   a. We take 180×180 input image to run the forward pass. In original network, we used 224×224, 320×320 and 400×400.

   b. We decrease the step size and sample duration. This helps a lot to boost the inference speed.

   c. We also take advantage of python, FFmpeg and other libraries to optimize the code and improve the performance.

We have used the two edge devices to run the trained activity recognition model. Below are details of the edge devices' specification. (1) Nvidia Jetson Nano and (2) Intel UP Squared board.

5 Training

SGD with momentum was utilized to prepare our system. We haphazardly produce training sets from recordings in training data to perform image augmentation. To start with, uniform testing is utilized to choose temporal features of each example. Then, we generate 16 frame cuts around the chosen temporal features. In the event that the recordings are shorter than 16 frames, the recordings are looped the same number of times as essential.

We at that point haphazardly select the spatial positions from the four corners or one focus, like [9]. Likewise, the spatial sizes of each example are chosen to perform multi-scale editing [15]. The scales are chosen from $\{1, 1/2\ 1/4, 1/\sqrt{2}, 1/2\ 1/4, 1/2\}$.

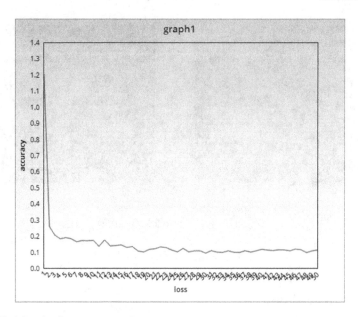

Fig. 9 Training details

The scale 1 means maximum scale (for example, the size is the length of short side of edge).

The aspect ratio of cropped picture is 1. The produced tests are on a level plane flipped with 0.50. We likewise perform mean subtraction for each example. The weight decay is 0.001 and the momentum is 0.9. We start from learning rate 0.1 and isolate it by 10 for multiple times after the validation loss immerses. In our few endeavours on UCF101 dataset, enormous learning rate and cluster size were helpful in getting great execution. The relationship between loss and accuracy for 1–50 epochs is shown in the graphs below for both training and validation data (Figs. 9 and 10)

6 Results

We found the following results while testing on UCF101 dataset test-1 split. We have attached some of the results. The 5-minute videos took 5 minutes and 35 seconds on Intel i5 8th Generation board, and the same video took 2 min 35 s on Nvidia K80 12 GB GPU. We have also compiled and executed the code on edge devices such as Nvidia Jetson Nano and Intel UP Squared board, and both the devices were able to process 10 FPS (frame per second) (Fig. 11).

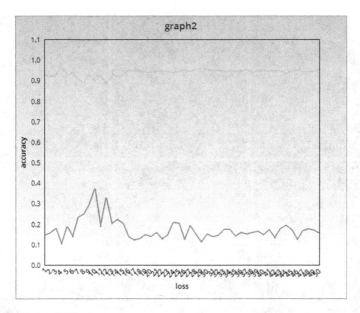

Fig. 10 Validation details

Table 2 Details of different architecture training accuracies

Architectures	Layers	Accuracy (%)
DenseNet	202	72
DenseNet	121	70
ResNeXt	101	82.3
Wide ResNet	50	77
ResNet	18	66
ResNet	34	70
ResNet	50	72
ResNet	101	73
ResNet	152	73.8
ResNet	200	74
CNN + LSTM	–	60
Two-stream I3D	27	79
C3D w/BN	50	64

7 Conclusions

In this work, we examined the various CNN architectures. Based on the analysis, we found that 3D convolution neural network is highly efficient to learn the spatio-temporal features for action recognition task. Based on the experiments made

Fig. 11 Activity results

on UCF101 datasets using ResNet-18, ResNet-50, ResNet-152 and ResNeXt-101 architectures, it can be concluded ResNeXt-101 proves to be most efficient in classifying activities at 93.6% accuracy. In addition, we can consider applying transfer learning methods like fine tuning on using pretrained network. We have used model that is trained on HMDB dataset that gives us good activity classification accuracy. For future work, we will try to improve more parameter tuning and use different architectures. We will also try to improve network accuracy using images with increased height and weight; this can help the model to learn more concrete features.

References

1. Bhatia, J., Modi, Y., Tanwar, S., Bhavsar, M.: Software defined vehicular networks: a comprehensive review. Int. J. Commun. Syst. **32**, e4005 (2019). https://doi.org/10.1002/dac.4005
2. Jaykrushna, A., Patel, P., Trivedi, H., Bhatia, J.: Linear regression assisted prediction based load balancer for cloud computing. In: 2018 IEEE PuneCon, pp. 1–3. Pune, India (2018)
3. Bhatia, J., Govani, R., Bhavsar, M.: Software defined networking: from theory to practice. In: 2018 Fifth International Conference on Parallel, Distributed and Grid Computing (PDGC), pp. 789–794. Solan Himachal Pradesh, India (2018)
4. Bhatia, J., Patel, T., Trivedi, H., Majmudar, V.: HTV dynamic load balancing algorithm for virtual machine instances in cloud. In: 2012 International Symposium on Cloud and Services Computing, pp. 15–20. Mangalore (2012)
5. Ji, S., Xu, W., Yang, M., Yu, K.: 3D convolutional neural networks for human action recognition. IEEE Trans. Pattern Anal. Mach. Intell. **35**(1), 221–231 (2012)
6. Saha, S., Singh, G., Cuzzolin, F.: Amtnet: action-micro-tube regression by end-to-end trainable deep architecture. In: Proceedings of the IEEE International Conference on Computer Vision, pp. 4414–4423 (2017)
7. Kiran, S.V., Singh, R.P.: Contextual action recognition using tube convolutional neural network (T-CNN)
8. Hou, R., Chen C., Shah, M.: Tube convolutional neural network (T-CNN) for action detection in videos. In: Proceedings of the IEEE International Conference on Computer Vision, pp. 5822–5831 (2017)
9. Das, S., Chaudhary, A., Bremond, F., Thonnat, M.: Where to focus on for human action recognition? In: 2019 IEEE Winter Conference on Applications of Computer Vision (WACV), pp. 71–80. IEEE (2019)
10. Behl, H.S., Sapienza, M., Singh, G., Saha, S., Cuzzolin, F., Torr, P.H.: Incremental tube construction for human action detection. arXiv preprint. arXiv:1704.01358 (2017)
11. Luvizon, D.C., Picard, D., Tabia, H.: 2d/3d pose estimation and action recognition using multitask deep learning. In: Proceedings of the IEEE Conference on Computer Vision and Pattern Recognition, pp. 5137–5146 (2018)
12. Luvizon, D., Tabia, H., Picard, D.: Multimodal deep neural networks for pose estimation and action recognition (2018)
13. Jordao, A., Nazare Jr, A.C., Sena, J., Schwartz, W.R.: Human activity recognition based on wearable sensor data: a standardization of the state-of-the-art. arXiv preprint. arXiv:1806.05226 (2018)
14. Wang, L., Xiong, Y., Wang, Z., Qiao, Y.: Towards good practices for very deep two-stream convnets. arXiv preprint. arXiv:1507.02159 (2015)
15. Hara, K., Kataoka, H., Satoh, Y.: Learning spatio-temporal features with 3D residual networks for action recognition. In: Proceedings of the IEEE International Conference on Computer Vision, pp. 3154–3160 (2017)
16. Li, X., Chuah, M.C.: Rehar: robust and efficient human activity recognition. In: IEEE Winter Conference on Applications of Computer Vision (WACV) pp. 362–371. IEEE (2018)
17. Soomro, K., Zamir, A.R., Shah, M.: UCF101: a dataset of 101 human actions classes from videos in the wild. arXiv preprint. arXiv:1212.0402 (2012)
18. Kuehne, H., Jhuang, H., Garrote, E., Poggio, T., Serre, T.: HMDB: a large video database for human motion recognition. In 2011 International Conference on Computer Vision, pp. 2556–2563. IEEE (2011)
19. He, K., Zhang, X., Ren, S., Sun, J.: Deep residual learning for image recognition. In: Proceedings of the IEEE Conference on Computer Vision and Pattern Recognition, pp. 770–778 (2016)
20. Hara, K., Kataoka, H., Satoh, Y.: Can spatiotemporal 3d CNNS retrace the history of 2d cnns and imagenet? In: Proceedings of the IEEE conference on Computer Vision and Pattern Recognition, pp. 6546–6555 (2018)

21. He, K., Zhang, X., Ren, S., Sun, J. Identity mappings in deep residual networks. In: European Conference on Computer Vision, pp. 630–645. Springer, Cham (2016)
22. Jahn Heymann, L,D., Haeb-Umbach, R.: Wide residual BLSTM network with discriminative speaker adaptation for robust speech recognition. In: Proceedings of the 4th International Workshop on Speech Processing in Everyday Environments (CHiME)
23. Xie, S., Girshick, R., Dollár, P., Tu, Z., He, K.: Aggregated residual transformations for deep neural networks. In: Proceedings of the IEEE Conference on Computer Vision and Pattern Recognition, pp. 1492–1500 (2017)
24. Huang, G., Liu, Z., Van Der Maaten, L., Weinberger, K.Q.: Densely connected convolutional networks. In: Proceedings of the IEEE Conference on Computer Vision and Pattern Recognition, pp. 4700–4708 (2017)
25. Tran, D., Bourdev, L., Fergus, R., Torresani, L., Paluri, M.: Learning spatiotemporal features with 3d convolutional networks. In: Proceedings of the IEEE International Conference on Computer Vision, pp. 4489–4497 (2015)
26. Deng, J., Dong, W., Socher, R., Li, L.J., Li, K., Fei-Fei, L.: Imagenet: A large-scale hierarchical image database. In 2009 IEEE Conference on Computer Vision and Pattern Recognition, pp. 248–255. IEEE (2009)
27. Carreira, J., Zisserman, A.: Quo vadis, action recognition? A new model and the kinetics dataset. In: Proceedings of the IEEE Conference on Computer Vision and Pattern Recognition, pp. 6299–6308 (2017)

Essential Learning Components from Big Data Case Studies

Pranjal Shikhar Sinha, Vineet Singh, Pallavi Asthana and Pragya

Abstract Global information generates trillions bytes of data every single day via e-mails, chats, e-commerce and media feeds. This structured and unstructured data is often referred as Big Data. The ability of big data lies in analyzing and capturing the information and quickly converting it into actionable insights. Big Data identifies business use cases with measurable outcomes to develop organization-wise big data strategy, through right tools and architecture for implementation with existing data for quick success. This paper presents the multiple case studies in some of the biggest data-generating platforms like e-governance, retail sector, healthcare sector, social networking sites, and astronomy. It also discusses important software-based tools that are employed for analyzing such vast amount of data for predicting the future performance of these platforms based on feedback and popularity.

Keywords Data mining · Predictive analysis · Hadoop · Data visualizations

1 Introduction

Data is the structure on which any association flourishes. The size, assortment and the quick difference in such data require another sort of big data analytics, just as various storage and analysis techniques. Anything extending from customer names and addresses to things available, to buy made, to workers hired and so forth has turned out to be fundamental for daily agreements. Such sheer measures of big data should

P. S. Sinha (✉) · V. Singh · P. Asthana
Amity University, Lucknow Campus, Lucknow, India
e-mail: shikhar.pranjal3@gmail.com

V. Singh
e-mail: vsingh@lko.amity.edu

P. Asthana
e-mail: pasthana@lko.amity.edu

Pragya
MVPG College, Lucknow, India
e-mail: de.pragya2011@gmail.com

© Springer Nature Singapore Pte Ltd. 2020
P. K. Singh et al. (eds.), *Proceedings of First International Conference on Computing, Communications, and Cyber-Security (IC4S 2019)*, Lecture Notes in Networks and Systems 121, https://doi.org/10.1007/978-981-15-3369-3_66

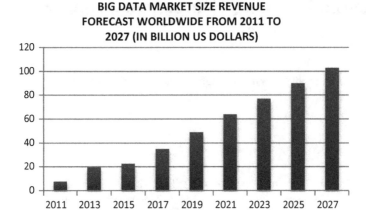

Fig. 1 Graph shows the big data market size revenue forecast worldwide [3]

be appropriately analyzed, and relating data should be separated [1]. "Big data" is a field that treats ways to deal with separate, proficiently extricate information from or by and large oversee informational indexes that are too much enormous or complex to be managed by conventional data-preparing application programming. Datasets with numerous cases offer more noteworthy measurable power, while datasets with higher complication may prompt a higher false discovery rate. Enormous information challenges consolidate getting information, information accumulating, information assessment, search, sharing, move, portrayal, addressing, invigorating, information security and information source. Big data was initially associated with three key concepts: volume, variety and velocity. Other thoughts later attributed to big data are veracity (i.e., how much noise is in the data) and value [2] (Fig. 1).

2 Literature Review

Faster and better decision making, with the speed of Hadoop and in-memory investigation, united with the ability to separate new wellsprings of information, associations can break down news rapidly—and choose decisions subject to what they have understood. New products and services, with the ability to quantify customer needs and satisfaction through examination, comes with the capacity to give customers what they need.

Big Data can be summarized as:

- Volume—It refers to the sheer size of the consistently detonating data of the registering scene. It brings up the issue of the amount of data.
- Velocity—It refers to the handling speed. It brings up the issue of at what rate the data is handled.

- Variety—It refers to the kinds of data. It brings up the issue of how unique the data organizations are.

Sources of Big Data: Machine Learning, a particular subset of AI that prepares a machine how to learn, makes it conceivable to rapidly and naturally produce models that can analyze more significant, progressively complex data and convey fast, increasingly precise outcomes—even on an exceptionally colossal scale. Furthermore, by exact structure models, an association has a superior shot of recognizing beneficial chances—or keeping away from unknown dangers. Data management, the information ought to be of high caliber and well spoken to before it will, in general, be continuously investigated. With information persistently spilling all through an affiliation, it is essential to set up repeatable systems to manufacture and keep up models for information quality. At the point, when information is reliable, associations should develop a piece of ace information, the executives' program that gets the entire undertaking in understanding. Data mining, information mining innovation allows a lot which makes you take a gander at a great deal of information to discover structures in the news, and this information can be used for further examination to help answer complex business questions. With information mining programming, you can channel through all the loud and repetitive tumult in information, pinpoint what is vast, that information can be used to assess likely outcomes, and a while later, enliven the pace of choosing instructed decisions [4, 5].

3 Case Studies

Definition of economy has evolved drastically over the years; dependent variables in monetary equations have changed with the advent of fourth industrial revolution. Online commerce has impacted in a big way, many vital constants contributing majorly for economy have been either changed or altered. Consider following case studies to build upon a predictive model with necessary components extracted from the case studies.

1. **E-Governance in India using Big Data**

- Analytics and data mining to detect tax evaders—Until now, the Indian government was shy about using machine learning and big data. With its help, it became very much more comfortable to catch the tax evaders in the year 2017. Nearly about 10 billion rupees are being found from the evaders [6–9].
- Track the flow of goods—As G.S.T. was introduced with the help of big data analysis, and with the use of G.S.T. and Railways, the flow of goods is being checked.

- Know the public mood—With the help of website *mygov.in*, the Prime Minister Office tries to know the sentiments of the public. By performing analysis, the Modi government was filtering the point to raise in the election campaign, and with the help of social media like Facebook and Twitter, they got to know the mood of the public [10, 11].
- Promising the tech-friendly future—With the help of big data, the government can track tax evaders, which proved that this government is going to give the tech-friendly future [12].

In the latest announcement, they try to make a national policy used in education, agriculture, retail, etc. Figure 2 depicts big data in Indian elections and described the platforms on which it has been used.

2. 5G Technology

Dissimilar to 4G/LTE, 5G will be something beyond a pipe, it speaks to a reason-constructed innovation, and it is structured and built to encourage associated gadgets just as computerization frameworks. From numerous points of view, 5G will be a facilitator and a quickening agent of the following modern unrest, frequently alluded to as Industry 4.0 [13]. 5G guarantees to convey high data rates (in the scope of Gbps) with ultra-low inertness (not as much as millisecond delay) for applications in industrial automation, tactile Internet, robotics, AR/VR apps, and so on.

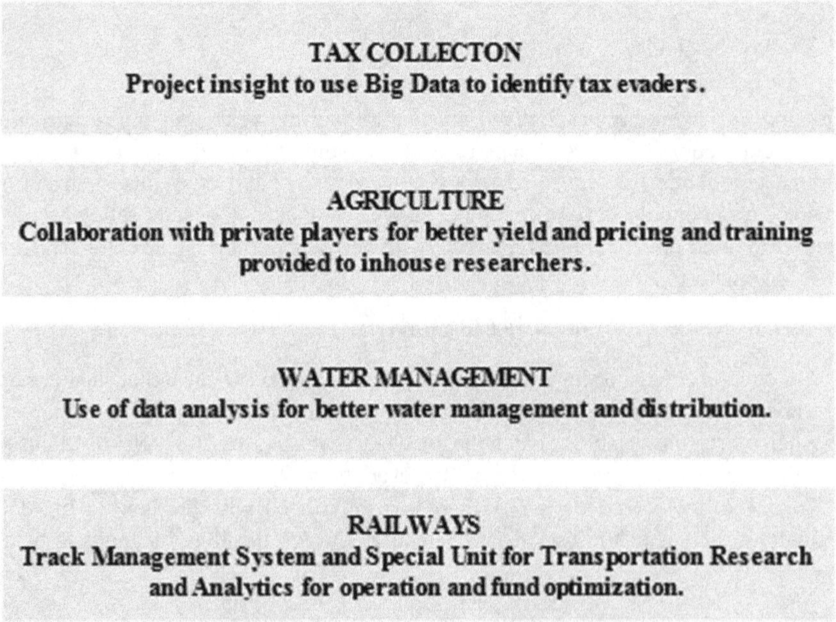

TAX COLLECTON
Project insight to use Big Data to identify tax evaders.

AGRICULTURE
Collaboration with private players for better yield and pricing and training provided to inhouse researchers.

WATER MANAGEMENT
Use of data analysis for better water management and distribution.

RAILWAYS
Track Management System and Special Unit for Transportation Research and Analytics for operation and fund optimization.

Fig. 2 How big data has been used in planning in the Indian election

Data analytics is at the sweet-spot exploiting 5G [14] arranges qualities, for example, high-transfer speed, low-latency, and mobile edge computing (MEC). 5G's capacity to help vast network crosswise over assorted gadgets (sensors/entryways/controllers), upheld by the appropriated register designs, makes the capacity to decipher the big data very still and the data-in-movement into constant bits of knowledge with unique insight.

Key Technology and Business Drivers are

- IoT over 5G (industrial IoT)—The mass measure of data being made by the IoT which can reform everything from assembling to human services to the format and working of shrewd urban communities—enabling them to work more effectively and gainfully than any time in recent memory. A fleet management company, for example, found that it had the option to lessen the expense of dealing with its armada of 180,000 trucks from 15 pennies for every mile to only 3 pennies [15].
- Data monetization—Telco's until 4G/LTE has been just utilizing information to improve administration quality and client experience. However, with the numerous conceivable outcomes of 5G system administrations joined with IoT and AI, they will investigate new business models of adaptation, for example, shrewd endeavor application administrations. For telco's, business openings lie in adapting information as well as the worth conveyed to ventures through application and system knowledge layers [16].
- Predictive maintenance—Predictive maintenance is the primary use-case of Industry 4.0. According to a statistical surveying report, predictive maintenance is an $11B showcase opportunity in the next five years. Predictive maintenance helps in predicting collapse before they happen by using AI [17].

Technical challenges and path to 5G involve following key factors:

- High-speed data-in-motion—On a very high-scale industrial IoT, smart cities and autonomous cars can pump the petabyte data within a minute. The 5G connectivity and the low-latency transmission will sum up the data throughput. Advanced cloud framework will be expected to help exceptionally fast read/write with the low-latency figure and the storage facility on the cloud.
- End-to-End security—Big data brings up various security issues, likewise with any applications today. In this way, it is essential to protect the client's security or enterprise data without any compromise. Building a robust, secure foundation from systems to applications will be critical in 5G structure and engineering [18].
- Real-time actionable insights—While low-latency is a characteristic for 5G systems, it turns into an extremely basic prerequisite for 5G to help quick data moves into the cloud, analytics at the edge and progressively, and continue the data at ultra-low-latencies to take ongoing activities in mission-critical applications, viz public safety, emergency care, and security surveillance.

3. Walmart

Walmart is one of the largest retailing company, which is based in America and was established in 1962, by Sam Walton. Until now, Walmart has 11,348 stores globally, and the annual turnover is $514.4 billion.

Walmart and Big Data—Walmart has a tremendous big data environment. The big data works in the petabytes. The gathered massive informational indexes are breaking down and mined toward the prescient examination, for streamlining the activities and business by the forecast of the habits for the clients. The retailers planned to attain maximum profit by knowing about the customers and, the factors and the consequences on the sales [19], the examinations cover millions of products and consumers. Walmart records a rise of 10% to 15% in online sales, which gives a turnover of about $1 billion. Big Data analysts very easily recognized this change in sales before and after the big data analysis.

Some techniques used by Walmart are:

- Savings catcher. The company notifies to customers, with the help of an application when its competitor reduces the price of the item. This application sends a gift voucher to its particular customer.
- Mapping. It works on Hadoop, to look after the Walmart stores most recent maps globally. This application can recognize even a small candy in the store globally [20–22].
- Track customer data. With the help of data mining technique, the sales pattern of Walmart has been observed. This pattern gives suggestions for the products to the customers, based on their previous purchase. The big data algorithm studies customer's interests, their purchase in-store and online, and also look after what is trending in the market.
- Launch new products. Walmart is utilizing Internet-based life information to discover about the slanting items with the goal that they can be acquainted with the Walmart stores over the world.

  - Social genome. It reads a large number of messages on Facebook and Tweets, watches YouTube videos and numerous blogs and analyzes the customer's requirements and the trending topics in the market, which help it to lead in the market.

Mobile big data solutions. More than 75% of customers are smart phones users. Moreover, with its help, they can have five more trips to the store online, which is about 70% more than the store visits [23]. With the help of cookies, the big data algorithm in mobile sections works more actively as per the user's previous activity, and there is no hassle created. Figure 3 depicts how Walmart used big data to become the leading retail store globally.

4. Healthcare

The healthcare field is vast, and it becomes essential to take the patient care and innovate medicines; so, the new technologies have been accepted by the industries. Big data in healthcare means gathering, dissecting, and utilizing the patient's physical and clinical data. According to research, the big data in the healthcare market is to be $34.27 billion by 2022 [24].

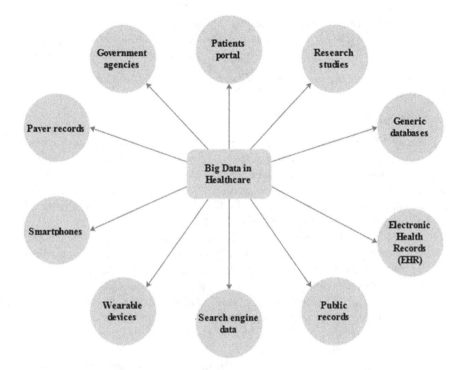

Fig. 3 Use of big data in the healthcare sector

Big Data can change the entire world of the healthcare field in the following ways:

- Tracking health. Big Data analytics and Internet of Things (IoT) can easily track anyone's statistics. Despite various wearable, there are many inventions which can detect the patient's blood pressure, heart rate, and other physiological parameters. With continuous monitoring of the body, people can be aware of their disease before the situation gets worst [25, 26].
- Cost reduction. With the help of big data, there could be a control over the staff members, which will decrease the investment rate in the staff. It will save time and money of the patients, and all the facilities will be easily available for the patients because of less crowd [27].
- Take care of high-risk patients. As all the data are digitalized, so it will be very helpful for the doctors to study the pattern of the high-risk patients, who are suffering from any chronic disease [28].
- Errors prevention. Any error due to wrong diagnosis in the medicine or wrong reporting may lead to even loss of lives sometimes. With the help of Big data techniques, medical practitioners would be able to counter check the prescriptions but making the treatment more effective and automated.

- Advancement in Healthcare Department. With the help of artificial intelligence (AI), it will be very easy to search for various data within seconds. Diagnosis of diseases and providing appropriate treatment will become very easy.

Figure 3 depicts big data in the healthcare department, so that the things do not become a hassle.

5. Facebook

It has been 11 years since Facebook started, and it is still so vigorously expanding that it has near about 1.59 billion accounts, which is about 1/5th of the world's total population. Facebook gathers a tremendous measure of information—a specific server sends out tens or hundreds of measurements that can be diagramed. This is not simply framework-level things like CPU and memory, and its additionally application-level measurements to comprehend why things are occurring.

Developing with Big Data, Facebook is one of the big data specialists and deals with the data in petabytes, including the analysis and the analytics. When people are coming closer on the platform of Facebook, it develops an algorithm to detect those relations. Facebook examines every bit of data to provide better services [29, 30].

Technologies behind Facebook's Big Data—Hadoop—Facebook holds the largest cluster of Hadoop in the world. It works beyond 4000 machines and stores data in thousands of petabytes [31, 32]. The developers can freely write map-reduce programs in any language.

Most of the files, n Hadoop are in tabular form. So, it becomes very easy for the developer to manage the subsets of SQL.

As Hadoop for Facebook is very efficient and reliable, it includes searching, video and image analysis, and data warehousing. The Facebook Messenger, developed by Facebook, is based on the Hadoop database, i.e., Apache HBase, which has an architecture that supports the overflow of messages.

- Scuba—Facebook faces numerous unstructured data on daily basis. To handle it, SCUBA was developed, which could help the Hadoop designers plunge into the enormous informational indexes and continue impromptu investigations progressively. Initially, Facebook was not able to handle such a huge amount of data single-handedly, and which cause the entire platform to crash. So, another platform for big data, Scuba allows the developer to store big data sets [33].
- Hive—In the world of unstructured data, Facebook became very popular by using Hive and the subsets of SQL, which makes Facebook run very fast and smooth without being a crashed [34, 35].
- Corona—It performs multiple tasks at the same time on a single Hadoop cluster without being crashed. Map-Reduce was designed based on pull-based scheduling model, which lags in performing the small tasks. Hadoop was limited by its slot-based resource management model, which was wasting the slots each time. Developing Corona helped in separating the clusters' job coordination.

6. Astronomy

Astronomy is the study of the physics, chemistry, and evolution of celestial objects and phenomena that originate outside the Earth's atmosphere, including supernovae explosions, gamma-ray bursts, and cosmic microwave background radiation.

Data mining innovation makes you take a gander at a great deal of information to discover structures in the news, and this information can be used for further examination to help answer complex business questions [36].

Some of the data mining software and tools which are used in the Astronomy:

- *StatCodes*—It is a Web meta Webpage that gives hypertext connects to countless factual codes helpful for space science and related fields. It is being kept up at the Center for Astrostatistics site.
- *VOStat*—It is a GUI wrapper in the R language [37]. It does not just perform different examinations, including plotting, information smoothing, spatial investigation, time arrangement examination, outline, fitting appropriation, relapse, various kinds of factual testing, and multivariate procedures; however, it likewise plots intuitive 3D visualizations [38]. The main goals of VOStat are to encourage astronomers to use statistics and spread the use of R among astronomers.
- *Weka*—It executes machine learning algorithms for different information mining undertakings, for instance, information pre-preparing, characterization, relapse, bunching, affiliation standards, and perception. Likewise, it can grow new machine learning plans [39–41].
- *AstroML*—It is a Python module produced for AI and information mining, which is based on numpy, scipy, scikit-learn, matplotlib, and astropy and is disseminated under the three-condition BSD permit. To adequately break down cosmic information, it incorporates a developing library of measurable and AI schedules in Python and a few transferred open galactic datasets and gives an enormous suite of instances of examining and envisioning galactic datasets. The objective of AstroML is to give a network storehouse to quick Python usage of basic devices and schedules utilized for measurable information investigation in space science and astronomy and to give a uniform and simple to-utilize interface to openly accessible galactic datasets [42].

4 Proposed Predictive Model

The main proposal of predictive modeling is the use of Big Data analytics with the help of various technologies in various sectors like media, government, manufacturing, healthcare department, etc., and establishing a better future by analyzing and visualizing the data and predicting the nearest best possible outcomes. Implementing the various technologies in the analysis so that there will be a bright future and better outcomes to be predicted for the better results, in the respective fields,

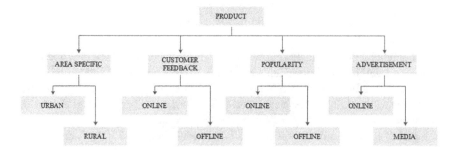

1. **Area Specific**.

   a. **Urban**. Analyzing the types of jobs which can be required for the better establishment of the cities, and planting the factories regarding it.
   b. **Rural**. Big Data can be used in the agricultural field, analyze and assure the quality and variety of grains for the particular geographical conditions; corresponding the output of every respective grain.

2. **Customer Feedback**. To improve the services based on the customer feedback. Feedback can be online or off-line and can be further analyzed for improvement in any particular area of service.
3. **Advertisement**. Based on the online activities of a customer, choice of products is streamlined and advertisement is customized according to choose of customers.
4. **Popularity**.

   a. **Online**. Using big data analytics and data visualizations can raise the popularity of the product can be analyzed using online platforms, reading the messages on Facebook, checking the tweets on Twitter, watching the videos on YouTube, and many more.
   b. **Off-line**. The feedbacks of the specific product by the neighbor or the relatives, and comparing with other similar products of the same retail price, get to know which has better assurance and is economical.

Proposed Algorithm:
Step 1: *Identify the specific area of Big Data*
Step 2: *Data analytics applied for,*

   a. *Rural*
   b. *Urban*

Step 3: *Obtain customer feedback,*

   a. *Online*
   b. *Off-line*

Step 4: *Evaluate popularity,*

   a. *Online*

b. *Off-line*

Step 5: *Advertisement methodology,*

a. *Online*
b. *Media*

Step 6: *From analytics, the outcome is predicted employing,*

a. *Data visualizations*
b. *Hadoop*
c. *Map-Reduce*
d. *A.I.e.*
e. *Machine learning.*

5 Conclusion

Big data which is generated through the online services can be further utilized for creating some meaningful results. Through the use of low-cost technologies, big data analytics can predict the future that helps in making important decisions in terms of right advertisement, finding precise market for products and in improving the services of existing platforms. It is a revolution where technology and management combine for predictive analysis and business intelligence. We have successfully presented the case studies of the most popular platforms generating the Big Data and also elaborated on various software tools that are utilized in the analysis of Big Data.

References

1. Elgendy, N., Elragal, A.: Big data analytics: a literature review paper (2014)
2. Singh, D., Reddy, C.K.: A survey on platforms for big data analytics (2014)
3. Liu, S.: 09 Aug 2019. https://www.statista.com/statistics/254266/global-big-data-market-forecast/
4. Bifet, A.: Mining big data—current status, and forecast to the future (2013)
5. Wu, X., Zhu, X., Wu, G.Q., Ding, W.: Data mining with big data (2014)
6. Banumathi, A., Pethalakshmi, A.: A novel approach for upgrading Indian education by using data mining techniques (2012)
7. Li, H., Nie, Z., Lee, W.C.: Scalable community discovery on textual data with relation. http://www.ics.uci.edu/~mlearn/MLRepository.html. University of California, Department of Information and Computer science, Irvine, CA
8. Guha, S., Rastogi, R., Shim, K.: CURE: an efficient clustering algorithm for large databases. In: Proceedings of 1998 ACM6SIGMOD International Conference Management of Data (SIGMOD'98), pages 73–84 (1998); Zhang, S., Zhu, C., Sin, J.K.O., Mok, P.K.T.: A novel ultrathin elevated channel low-temperature poly-Si TFT. IEEE Electron Device Lett. **20**, 569–571 (1999)

9. Zhang, C., Xia, S.: K-means clustering algorithm with improved Initial center. In: Second International Workshop on Knowledge Discovery and Data Mining (WKDD), pp. 7906792 (2009)
10. Sharma, P., Moh, T.S.: Prediction of Indian election using sentiment analysis on hindi twitter (2016)
11. Almatrafi, O., Parack, S., Chavan, B.: Application of location-based sentiment analysis using twitter for identifying trends towards indian general elections 2014. In: Proceedings of the 9th International Conference on Ubiquitous Information Management and Communication, Article No. 41 (2015)
12. Sawhney, D.: Technology integration in indian schools using a value-stream based framework (2016)
13. Bassi, L.: Industry 4.0: hope, hype or revolution? (2017)
14. Muralidhar Somisetty.: Big data analytics in 5G (2018)
15. Khurpade, J.M., Rao, D., Sanghavi, P.D.: A survey on IoT and 5G network (2018)
16. Gregory, I., Landryová, L., Soldán, P.: The monetization of highly automated systems in SMEs (2011)
17. Roukounaki, A., Efremidis, S., Soldatos, J., Neises, J., Walloschke, T., Kefalakis, N.: Scalable and configurable end-to-end collection and analysis of IoT security data towards end-to-end security in IoT systems (2019)
18. Singh, M., Ghutla, B., Jnr, R.L., Mohammed, A.F., Rashid, M.A.: Walmart's sales data analysis—a big data analytics perspective (2017)
19. Dean J., Ghemawat, S.: MapReduce: simplified data processing on large clusters. Assoc. Comput. Mach. (2008)
20. Riyaz, P.A., Surekha, V.: Leveraging map reduce with hadoop for weather data analytics. OSR J. Comput. Eng. (2015)
21. Sharma, M., Chauhan, V., Kishore, K.: A review: MapReduce and spark for big data analysis. In: 5th International Conference on Recent Innovations in Science. 5: Engineering and Management (2016)
22. Nivargi, V.: Big data: from batch processing to interactive analysis, [Online]. Available: Accessed Sept 2017 (2013)
23. Singh, M., Bhatia, V., Bhatia, R.: Big data analytics solution to healthcare (2017)
24. Mian, M., Teredesai, A., Hazel, D., Pokuri, S., Uppala, K.: Work in progress—In memory analysis for healthcare big data (2014)
25. Rahman, F., Slepian, M., Mitra, A.: A novel big-data processing framework for healthcare applications big-data-healthcare-in-a-box (2016)
26. Li, L., Bagheri, S., Goote, H., Hasan, A., Hazard, G.: Risk adjustment of patient expenditures: a big data analytics approach (2013)
27. Koppad, S.H., Kumar, A.: Application of big data analytics in healthcare system to predict COPD (2016)
28. Dragan, I., Zota, R.: Collecting Facebook data for big data research (2017)
29. Bronson, N., Lento, T., Wiener, J.L.: Open data challenges at Facebook (2015)
30. singh Bhathal, G., Dhiman, A.S.: Big data solution: improvised distributions framework of hadoop (2018)
31. Bhandarkar, M.: MapReduce programming with Apache hadoop (2010)
32. Verel, S., Collard, P., Clergue, M.: Scuba search: when selection meets innovation (2004)
33. Wang, K., Bian, Z., Chen, Q., Wang, R., Xu, G.: Simulating hive cluster for deployment planning, evaluation and optimization (2014)
34. Fuad, A., Erwin, A., Ipung, H.P.: Processing performance on apache pig, apache hive and MySQL cluster (2014)
35. Yan, L.X.Y.: Machine learning for astronomical big data processing (2017)
36. Zhang, Z., Barbary, K., Nothaft, F.A., Sparks, E., Zahn, O., Franklin, M.J., Patterson, D.A., Perlmutter, S.: Scientific computing meets big data technology: an astronomy use case (2015)
37. Keka, I., Çiço, B.: Data visualization as helping technique for data analysis, trend detection and correlation of variables using R programming language (2019)

38. Rubel, O., Weber, G.H., Huang, M.Y., Bethel, E.W., Biggin, M.D., Fowlkes, C.C., Hendriks, C.L., Keranen, S.V., Eisen, M.B., Knowles, D.W., Malik, J., Hagen, H., Hamann, B.: Integrating data clustering and visualization for the analysis of 3D gene expression data (2010)
39. Charalampopoulos, I., Anagnostopoulos, I.: A comparable study employing weka clustering/classification algorithms for web page classification (2011)
40. Jenitha, G., Vennila, V.: Comparing the partitional and density based clustering algorithms by using weka tool (2014)
41. Saad, S., Ishtiyaque, M., Malik, H.: Selection of most relevant input parameters using weka for artificial neural network based concrete compressive strength prediction model (2016)
42. VanderPlas, J., Connolly, A.J., Ivezić, Ž, Gray, A.: Introduction to AstroML: machine learning for astrophysics (2012)

CPSIA information can be obtained
at www.ICGtesting.com
Printed in the USA
LVHW081056100520
655303LV00003B/454